INTERNATIONAL PRAISE FOR PETER WATSON'S

The German Genius

"Watson's account of the flourishing of German science and letters between the 1770s and the early twentieth century is a tour de force.... Watson is a master of miniaturization whose enthusiasm for the subject matter never flags. It is impossible not to be impressed by his range and versatility as he bounds across the disciplines.... This intelligent book presents a breathtaking panorama." —*Sunday Times* (London)

"A journalist of heroic industry.... [Watson] is never not good company.... [He] derives the German genius from deep springs."
—*The Guardian* (London)

"Watson's story is vibrating with life. It is unputdownable. It contains a lot one didn't know. So much enlightenment and so much that moves."
—*Frankfurter Rundschau*

"Admirably accessible and flowingly legible.... Watson is absolutely right in saying that [Germany] is a very different place from what it once was, that it 'has fashioned its own democratic revolution, albeit one that—surprising as it may seem—has gone very largely underappreciated by the world outside.'"
—*Irish Times*

"[Watson] deserves gratitude for the astonishing comprehensiveness of his survey.... [He] has provided copious materials with which to correct the current... image of Germany." —*Times Literary Supplement* (London)

"Reveals several surprises.... A remarkable book on many levels. The research is first-rate and it is surprisingly accessible."
—TucsonCitizen.com

"No one who reads this book can be left in any doubt as to the reality of 'the German Genius' proclaimed by the title. [Watson] has an enviable gift of explaining lucidly and cogently ideas that are complicated or profound (or both). The Germans may well 'dive deeper but come up muddier,' as Wickham Steed famously observed, but Watson strips off the ordure with crisp common sense. . . . In particular, the chapters on science . . . are very impressive. . . . This is a big book in every sense . . . everyone interested in the sufferings and greatness of modern culture will be informed, entertained, and provoked by it."

—*Literary Review* (London)

"The outstanding quality of this book is that it places scientific discoveries at the core of cultural history, linking them with dramatic technical and industrial developments. . . . Watson's account . . . assembles such a wealth of information, based on an impressive range of sources, that *The German Genius* will be an essential work of reference for years to come."

—*The Independent* (London)

"A joy, for its ambition, its seriousness, and its moral integrity."

—*The Scotsman* (Edinburgh)

"A powerful and vivid opus. . . . Watson's story is brimming with life. You can barely put the book aside." —*Berliner Zeitung*

"There is hardly another nonfiction book out now that is as worth reading as this one." —*Hamburger Morgenpost*

"Few wasted words—a welcome resource for students of modern history, literature, and cultural studies." —*Kirkus Reviews*

"[An] engrossing, vast chronicle. . . . English now dominates the arts and sciences, but Watson writes an absorbing account of a time not so long ago when German ruled." —*Publishers Weekly* (starred review)

THE GERMAN GENIUS

ALSO BY PETER WATSON

Ideas: A History of Thought and Invention, from Fire to Freud

The Modern Mind: An Intellectual History of the 20th Century

The Medici Conspiracy: The Illicit Journey of Looted Antiquities From Italy's Tomb Raiders to the World's Greatest Museums

Sotheby's: The Inside Story

The Death of Hitler

From Manet to Manhattan: The Rise of the Modern Art Market

Wisdom and Strength: The Biography of a Renaissance Masterpiece

The Caravaggio Conspiracy

The German Genius

Europe's Third Renaissance, the Second Scientific Revolution, and the Twentieth Century

PETER WATSON

NEW YORK • LONDON • TORONTO • SYDNEY • NEW DELHI • AUCKLAND

HARPER PERENNIAL

Jay W. Baird, *To Die for Germany: Heroes in the Nazi Pantheon* (Indiana University Press, 1992). Copyright © 1990 by Jay W. Baird. Translations appearing on pages 635–637 reprinted with the permission of Indiana University Press.

Bertolt Brecht, "Hounded Out by Seven Nations" from *Bertolt Brecht: Poems 1913–1956*, edited by John Willett and Ralph Manheim. Copyright © 1976, 1979 by Methuen London Ltd. Translation appearing on page 715 reprinted with the permission of Routledge, Inc., part of The Taylor & Francis Group.

A hardcover edition of this book was published in 2010 by HarperCollins Publishers.

THE GERMAN GENIUS. Copyright © 2010 by Peter Watson. All rights reserved. Printed in the United States of America. No part of this book may be used or reproduced in any manner whatsoever without written permission except in the case of brief quotations embodied in critical articles and reviews. For information, address HarperCollins Publishers, 195 Broadway, New York, NY 10007.

HarperCollins books may be purchased for educational, business, or sales promotional use. For information, please e-mail the Special Markets Department at SPsales@harpercollins.com.

Designed by William Ruoto

FIRST HARPER PERENNIAL EDITION PUBLISHED 2011.

Library of Congress Cataloging-in-Publication Data is available upon request.

ISBN 978-0-06-076023-6 (pbk.)

16 17 18 OV/RRD 10 9 8 7 6 5

The split between Germany and the West will of necessity always be an important theme for historians.

—Hajo Holborn

The word "genius" in German has a special overtone, even a tinge of the demonic, a mysterious power and energy; a genius—whether artist or scientist—is considered to have a special vulnerability, a precariousness, a life of constant risk and often close to troubled turmoil.

—Fritz Stern

Geographically, America is for us among civilised countries the most distant; intellectually and spiritually, however, the closest and most like us.

—Adolf von Harnack

Asked in 1898 to choose a single defining event in recent history, the German Chancellor Bismarck replied, "North America speaks English."

—Nicholas Ostler

Our intellectual skyline has been altered by German thinkers even more radically than has our physical skyline by German architects.

—Allan Bloom

The Germans dive deeper—but they come up muddier.

—Wickham Steed

For those [Germans] born during and after the Second World War the cultural history of Germany before 1933 is that of a lost country, one that they never knew.

—Keith Bullivant

For countless Americans, Germany remains the ultimate metaphor of evil, the frightening reminder of the fragility of civilisation.

—Deidre Berger

The German, at odds with himself, with deep divisions in his mind, likewise in his will and therefore impotent in action, becomes powerless to direct his own life. He dreams of justice in the stars and loses his footing on earth. . . . In the end, then, only the inward road remained open for German men.

—Adolf Hitler

The Nazis are un-German.

—Victor Klemperer

Patriotism for the Frenchman is such that it warms his heart dilating it and expanding it, so he embraces in his love not only those closest to him but the whole of France, the entire country of his civilisation. The patriotism of the German on the contrary is such that his heart becomes narrower and shrinks like leather in the cold. He hates foreigners and no longer wishes to be a citizen of the world, but a mere German.

—Heinrich Heine

People in England want something to read, the French something to taste, the Germans something to think about.

—Kurt Tucholsky

The Germans are not bastardised by alien people, they have not become mongrels. They have preserved their original purity more than many other peoples and have been able to develop slowly and quietly from this purity according to the lasting laws of time; the fortunate Germans are an original people.

—Ernst Moritz Arndt

It was [Hippolyte] Taine, the Frenchman, who said that all the leading ideas of the present day were produced in Germany between 1780 and 1830.

—John Dewey

The Planet is in flames . . . Only from the Germans can come the world-historical reflection, provided that they find and preserve their German element.

—MARTIN HEIDEGGER

We see in Germany, even more than elsewhere, a division of labour between genius and tradition; nowhere are the types of the young rebel and the tireless pedant so common and so extreme.

—GEORGE SANTAYANA

I cannot think of a nation that is more torn than the Germans, you see workmen but no human beings, masters and servants, young people and sedate, but no human beings . . .

—FRIEDRICH HÖLDERLIN

Anyone who still said that they liked Caspar David Friedrich stood accused for decades of not being sufficiently critical with regard to German history.

—FLORIAN ILLIES

Nazism owes nothing to any part of the Western tradition, be it German or not, Catholic or Protestant, Christian . . .

—HANNAH ARENDT

German suffering and Jewish suffering are not equal . . . They are, however, both real.

—STEVE CRAWSHAW

In several respects, American intellectual life is today closer to the German than the British.

—HENRI PEYRE

The German language unfortunately permits a fairly trivial thought to declaim from behind a woollen curtain of apparent profundity and, conversely, a multitude of meanings to lurk behind one term.

—ERWIN PANOFKSY

Freud is better in German.

—Frank Kermode

Death is a master from Germany.

—Paul Celan

Whoever begins to question this society [Germany], eventually questions himself out of it.

—Ralf Dahrendorf

German problems are rarely German problems alone.

—Ralf Dahrendorf

It has always struck me as particularly interesting that so many of the great debunking analysts of modern culture have been German or Austrian, not English or French.

—Fritz Ringer

The Allies won [the Second World War] because our German scientists were better than their German scientists.

—Sir Ian Jacobs, military secretary to Winston Churchill

As a result of the Second World War, being German became an international stigma that had to be borne stoically and, at best, could be attenuated through good behaviour.

—Konrad Jarausch

The cultural legacy of German Jewry is German.

—Barbara John, Berlin Foreigners Commissioner

But under what suspicion one comes, if one says, the Germans are now a completely normal people, a regular society.

—Martin Walser

There is too much music in Germany.

—Romain Rolland

And the world may finally be healed by Germanism.

—Emanuel Geibel

The way Germans confront their history will be of crucial significance not just for Germany but for all of Europe as well.

—Heinrich August Winkler

Don't you guys know you are in Hollywood? Speak German.

—Otto Preminger, to a group of Hungarian émigrés

All of German literature [has] settled in America.

—Thomas Mann

We poor Germans! We are fundamentally lonely, even when we are "famous"! No one really likes us.

—Thomas Mann

So long as the Germans speak German and I speak English, a genuine dialogue between us is possible; we shall not simply be addressing our mirror images.

—W. H. Auden

Hitler was "the mirror of every German's unconscious . . . the loudspeaker which magnifies the inaudible whispers of the German soul."

—Carl Jung

I heard a Californian student in Heidelberg say, in one of his calmest moods, that he would rather decline two drinks than one German adjective.

—Mark Twain

To this day we hardly recognise that a phenomenon occurred in the eighteenth and nineteenth centuries that was as remarkable as that outburst of creativity we call the Renaissance in Italy. It was the German Renaissance—the renaissance of a culture mutilated by the Thirty Years War.

—Noel Annan

No one is a Nazi, no one ever was . . . It should be set to music.

—Martha Gellhorn

Germany is not part of the West. But Germany will never be able to do without it.

—Gregor Schöllgen

[Germany] is probably the most grown-up country in the world today.

—Mark Mardell

The memory of the Third Reich has intensified with increasing temporal distance to the Nazi past.

—Hermann Lübbe

It is characteristic of the Germans that the question "What is German?" never dies out among them.

—Friedrich Nietzsche

Defeated in two world wars, Germany appeared to have invaded vast territories of the World's mind.

—Erich Heller

Can one be a musician without being German?

—Thomas Mann

The United States and Great Britain may speak English but, more than they know, they *think* German.

—Peter Watson

1989 was the brightest moment in Europe's darkest century.

—Fritz Stern

Contents

Introduction: Blinded by the Light: Hitler, the Holocaust, and "the Past That Will Not Pass Away" 1

PART ONE

THE GREAT TURN IN GERMAN LIFE

1. Germanness Emerging 41
2. *Bildung* and the Inborn Drive toward Perfection 65

PART TWO

A THIRD RENAISSANCE, BETWEEN DOUBT AND DARWIN

3. Winckelmann, Wolf, and Lessing: the Third Greek Revival and the Origins of Modern Scholarship 91
4. The Supreme Products of the Age of Paper 111

5. New Light on the Structure of the Mind 135

6. The High Renaissance in Music: The Symphony as Philosophy 153

7. Cosmos, Cuneiform, Clausewitz 167

8. The Mother Tongue, the Inner Voice, and the Romantic Song 189

9. The Brandenburg Gate, the Iron Cross, and the German Raphaels 207

PART THREE

The Rise of the Educated Middle Class: the Engines and Engineers of Modern Prosperity

10. Humboldt's Gift: The Invention of Research and the Prussian (Protestant) Concept of Learning 225

11. The Evolution of Alienation 239

12. German Historicism: "A Unique Event in the History of Ideas" 261

13. The Heroic Age of Biology 271

14. Out from "The Wretchedness of German Backwardness" 289

15. "German Fever" in France, Britain, and the United States 311

16. Wagner's Other Ring—Feuerbach, Schopenhauer, Nietzsche 327

17. Physics Becomes King: Helmholtz, Clausius, Boltzmann, Riemann 341

18. The Rise of the Laboratory: Siemens, Hofmann, Bayer, Zeiss 355

19. Masters of Metal: Krupp, Benz, Diesel, Rathenau 369

20. The Dynamics of Disease: Virchow, Koch, Mendel, Freud 383

PART FOUR

The Miseries and Miracles of Modernity

21. The Abuses of History 401

22. The Pathologies of Nationalism 417

23. Money, the Masses, the Metropolis: The "First Coherent School of Sociology" 439

24. Dissonance and the Most-Discussed Man in Music 459

25. The Discovery of Radio, Relativity, and the Quantum 475

26. Sensibility and Sensuality in Vienna 489

27. Munich/Schwabing: Germany's "Montmartre" 503

28. Berlin Busybody 519

29. The Great War between Heroes and Traders 531

30. Prayers for a Fatherless Child: The Culture of the Defeated 547

31. Weimar: "Unprecedented Mental Alertness" 567

32. Weimar: The Golden Age of Twentieth-Century Physics, Philosophy, and History 595

33. Weimar: "A Problem in Need of a Solution" 611

PART FIVE

Songs of the Reich: Hitler and the "Spiritualization of the Struggle"

34. Nazi Aesthetics: The "Brown Shift" 629

35. Scholarship in the Third Reich: "No Such Thing as Objectivity" 649

36. The Twilight of the Theologians 673

37. The Fruits, Failures, and Infamy of German Wartime Science 689

38. Exile, and the Road into the Open 699

PART SIX

Beyond Hitler: Continuity of the German Tradition under Adverse Conditions

39. The "Fourth Reich": The Effect of German Thought on America 713

40. "His Majesty's Most Loyal Enemy Aliens" 743

41. "Divided Heaven": From Heidegger to Habermas to Ratzinger 757

42. Café Deutschland: "A Germany Not Seen Before" 789

Conclusion: German Genius: The Dazzle, Deification, and Dangers of Inwardness 817

Appendix: Thirty-five Underrated Germans 851

Notes and References 857

Index 927

Author's Note

In *The Proud Tower*, her splendid book about Europe in the run up (or run down) to World War I, Barbara Tuchman, the American historian, describes an incident in which Philip Ernst, the artist father of the surrealist Max Ernst, was painting a picture of his garden when he omitted a tree that spoiled the composition. Then, "overcome with remorse" at his offense against realism, he cut down the tree.

It is a good story. If one had to make a criticism it might be that it falls into the trap of stereotyping Germans—as sticklers for exactitude, as pedantic and literal-minded. Part of the point of the book you are holding (as with the quotations given before the Table of Contents) is to go beyond stereotypes but also to show that the stereotypes peoples have *of themselves* can be as misleading—and as dangerous—as the stereotypes their neighbors, rivals, and enemies have of them.

That is far from being the only point of the book, of course, which aims to be a history of German ideas over the past 250 years, from the death of Bach. No one can be an expert on such a long period, and in the course of my research I have been helped by a number of people whose assistance I would like to acknowledge here, some of whom have read all or parts of the typescript and offered suggestions for improvement. None of the names that follow, all of whom I thank warmly, is responsible for such errors, omissions, and solecisms that remain.

My first debt is to George (Lord) Weidenfeld, who encouraged me in this project and opened countless doors in Germany. I next thank Keith Bullivant, an old friend, now professor of Modern German Studies at the University of Florida but someone who, in 1970, with R. H. Thomas, founded the first ever Department of German Studies, at Warwick University. This is a direction now followed throughout the English-speaking world. But I also extend my gratitude to: Charles Aldington, Rosemary Ashton, Volker Berghahn, Tom Bower, Neville Conrad, Claudia Amthor-Croft, Ralf Dahrendorf, Bernd Ebert, Hans Magnus Enzensberger, Joachim Fest, Corinne Flick, Gert-Rudolf Flick, Andrew Gordon, Roland Goll, Karin Graf, Ronald Grierson, David Henn, Johannes Jacob, Joachim Kaiser, Marion Kazemi, Wolf-Hagen Krauth, Martin Kremer, Michael Krüger, Manfred Lahnstein, Jerry Living, Robert Gerald Livingston, Günther Lottes, Constance Lowenthal, Inge Märkl, Christoph Mauch, Gisela Mettele, Richard Meyer, Peter Nitze, Andrew Nurnberg, Sabine Pfannensteil-Wright, Richard Pfennig, Werner Pfennig, Elisabeth Pyroth, Darius Rahimi, Ingeborg Reichle, Rudiger Safranski, Anne-Marie Schleich, Angela Schneider, Jochen Schneider, Kirsten Schroder, Hagen Schulze, Bernd Schuster, Bernd Seerbach, Kurt-Victor Selge, Fritz Stern, Lucia Stock, Robin Straus, Hans Strupp, Michael Stürmer, Patricia Sutcliffe, Clare Unger, Fritz Unger, and David Wilkinson.

At the end of this book there are many pages of references. In addition to those, however, I would like to place on record my debt to a number of books on which I have relied especially heavily—all are classics of their kind. Alphabetically by author/editor they are: T. C. W. Blanning, *The Culture of Power and the Power of Culture: Old Regime Europe, 1660–1789* (Oxford, 2002); John Cornwell, *Hitler's Scientists: Science, War and the Devil's Pact* (Penguin, 2003); Steve Crawshaw, *Easier Fatherland: Germany and the Twenty-First Century* (Continuum, 2004); Eva Kolinsky and Wilfried van der Will, eds., *The Cambridge Companion to Modern German Culture* (Cambridge, 1998); Timothy Lenoir, *The Strategy of Life: Teleology and Mechanics in Nineteenth-Century German Biology* (Chicago, 1982); Bryan Magee, *Wagner and Philosophy* (Penguin, 2000);

Suzanne L. Marchand, *Down from Olympus: Archaeology and Philhellenism in Germany, 1750–1970* (Princeton, 1996); Peter Hanns Reill, *The German Enlightenment and the Rise of Historicism* (Berkeley, 1975); Robert J. Richards, *The Romantic Conception of Life: Science and Philosophy in the Age of Goethe* (Chicago, 2002). I also wish to thank the staff of the Goethe Institute, London, as well as the staffs at the cultural and press sections of the German Embassy in London, at the London Library, the Wiener Library, and at the German Historical Institutes in London and Washington, D.C.

A few paragraphs of this book overlap with material used in my earlier books. They are indicated at the appropriate places in the references. A handful of German words, difficult to translate, are used throughout the book. They are shown in italics at their first occurrence in each chapter, in roman thereafter.

THE GERMAN GENIUS

INTRODUCTION

Blinded by the Light: Hitler, the Holocaust, and "the Past That Will Not Pass Away"

By one of those profitable accidents of history, in 2004 two German brothers were living in London, each in a high-profile position of influence that enabled them, together, to make some very pointed observations about their temporary home. Being brothers but in very different occupations more than doubled their impact.

Thomas Matussek was the German ambassador in London. In that year he complained publicly that, almost sixty years after the end of World War II, English history teaching focused excessively on the Nazi period. He said he had found many British people who had an "obsession" with the Third Reich, "but there are very few people who actually know Germany." He said Britain's history curriculum was "unbalanced"—it had nothing to say about the successes of postwar Germany, ignored reunification, and glossed over other aspects of German history. He told the *Guardian* newspaper that he was "very much surprised when I learned that at A-level one of the three most chosen subjects was the Nazis."[1] His brother, Matthias Matussek, was at the time the London correspondent for the German weekly *Der Spiegel*, and he went further. He said it was "ridiculous" to reduce Germany—the country of Johann Wolfgang von Goethe, Friedrich von Schiller, and Ludwig van Beethoven—to the twelve years of Nazi rule, and he joked sarcastically that one of the defining characteristics of Britishness was now "resistance to Nazi Germany." His undiplomatic wording occasioned a "frost" between the brothers, but at much the same time even Germany's foreign minister Joschka Fischer accused British teachers of perpetuating a "goose stepping" image of Germany that was "three generations out of date."

Matussek was not the first. In an interview in 1999 before his departure as Germany's ambassador to Britain, Gebhardt von Moltke, Matussek's predecessor, said "one has the impression that the teaching of history in this country stops in 1945," and he too regretted the reluctance of young Britons to learn German or visit Germany.[2]

The German government does seem concerned about the country's image, at least in Britain. In July 2003 a conference was held at the Goethe (cultural and language) Institute in London to explore how Germany might be "branded" better—i.e., sold as an attractive place to travel to, study in, do business in, learn the language of—much as Quebec and Australia have been successfully branded in recent years. A survey of the *Radio Times*, a television listings magazine, carried out in the week preceding the conference, showed that no fewer than thirteen programs had been broadcast over a period of six days, "all dealing with topics related to the Second World War." A poll carried out ahead of the conference showed that while 81 percent of young Germans could name a living British celebrity, fully 60 percent of Britons could not name a living German.[3] In October 2004 the German government paid for twenty British history teachers to visit Germany—putting them up at top hotels—to discuss the issues. One of the teachers on the visit said, "Kids find the Nazi period interesting. A lot of things happen. There is plenty of violence." He thought that postwar German history was, by comparison, "a bit dry." A colleague from Newcastle thought his pupils "bigoted and uninterested. The general impression is that Germans are all Nazis who steal sun loungers. This is all a cartoon-style view. The problem is that if you ask them seriously they have no view of Germany at all."[4]

There is some evidence that the German government is right to be concerned. A survey in July 2004 found that whereas 97 percent of Germans have a basic knowledge of the English language, and 25 percent are fluent, only 22 percent of British students have any knowledge of the German language and just 1 percent are fluent. Whereas 52 percent of young Germans had been to Britain, only 37 percent of young Britons had been to Germany. A 2003 *Travel Trends* survey showed that U.K. residents made 60 million foreign visits a year but only 3 percent were to Germany, the same figure as to Belgium and half the figure for the United States, one sixth the number for France, and a seventh of those going to Spain. Over the previous four years the figure for travel to Germany was static and had fallen behind visits to the Netherlands, Italy, and Greece.[5]

Possibly, the situation was getting worse. In 1986, in opinion poll figures, 26 percent of people had seen Germany as Britain's best friend in Europe, but by 1992 that had fallen to 12 percent. When Britons were asked, in 1977, if "Nazism or something like that" could again become powerful in Germany, 23 percent said yes, 61 percent said no. By 1992 the pattern had reversed, with 53 percent voting yes and 31 percent no.[6] A *Daily Telegraph* editorial in May 2005 concluded sixty years after VE Day, "[W]e are a nation fixated with the Second World War and are becoming more so . . ."[7]

In the short run, this is unlikely to change. Another survey, this time of 2,000 private and state schools in Britain and published in November 2005, showed that "thousands" of fourteen-year-olds had given up German in favor of "easier" subjects (such as media studies) since the British government made the study of foreign languages optional in the autumn of 2004. More than half the schools in the survey said they had dropped classes in German in the preceding year. Another survey, published in 2007, showed that the number of institutions in Britain providing courses in German had fallen by 25 percent since 1998 and the number of undergraduate degrees in German awarded in London had fallen by 58 percent.[8]

Ambassador Matussek, not unnaturally, didn't like these results, but he didn't think that xenophobia accounted for the change—more likely it was ignorance. He did point out that, since Germany is Britain's biggest trading partner, it was a potentially "dangerous" development. "It's risky to ask fourteen-year-olds whether they want to drop out of languages," he warned, adding that teenagers think of Spanish as being "easy" and German as "difficult." "Most pupils think of the beaches of Spain rather than the museums and castles of Germany."

But the ambassador's concerns about "imbalance" in British education were borne out at Christmastime 2005, when the annual report of Britain's Qualification and Curriculum Authority (QCA) concluded that the teaching of history in secondary schools "continues to be dominated by Hitler . . . There has been a gradual narrowing and 'Hitlerisation' of post-14 history . . . post-14 history continues to be dominated by topics such as the Tudors and the twentieth-century dictatorships." The QCA subsequently issued guidance on teaching postwar history to provide "a more balanced understanding of twentieth-century Germany."[9]

So Ambassador Matussek was right in saying that the teaching of history in British schools is "unbalanced." Was he right to link that with a British "obsession" with Nazi Germany? Speaking of his own country, he said, "People don't take holidays there. Youth exchange is a one-way street . . . Our younger generations are slowly drifting apart and are listening less to each other. I can only speculate as to why this is. But I talk to a lot of British people and one answer that comes up repeatedly is that every country needs to go through an identity-building process. In 1940, Britain was practically confronted with an overpowering enemy and through the sheer mustering of British virtues, Britain finally managed to turn it round. That is very important in the collective psyche: to look back and think you really can do it.

"Like the conquering of the West is part of the American myth, so it is the same with Britain and the defeat of Nazism. That coincided with Britain losing her Empire, which certainly rankled with some people and led to this obsession with Germany and not always in a very funny way. We have to make a distinction between the clichéd stereotypes that are outright funny—like in *Dad's Army* or *Fawlty Towers* [TV comedy shows]—and something that goes a little deeper. The humour stops when I hear that German children are regularly beaten up and abused by British youngsters who don't know what Germany is about."

Here again the ambassador is supported by some independent research. A survey in Britain in 2004 found that when British ten- to sixteen-year-olds were asked what they associated with Germany, 78 percent said the

Second World War and 50 percent mentioned Hitler. A study at Aberdeen University showed how, especially above the age of twelve, a sample of children would react much more negatively to a photograph of a person when told it was of a German than when shown the same photograph two weeks earlier without any mention of nationality.

These reactions, says Matussek, are a particularly British problem. "This attitude isn't prevalent in other countries. A lot of our neighbours suffered much, much more than the British. But among young Russians, young Poles or young Czechs, you don't get this. Perhaps a country with nine neighbours is constantly forced to make compromises and is much more in contact than a country that lives as an island."[10]

His brother, Matthias—again—put it more strongly. "The British behave as if they had conquered Hitler's hordes single-handedly. And they continue to see us as Nazis, as if they have to refight the battles every evening [i.e., on TV]. They are enchanted by this Nazi dimension." Gisela Stuart, a German-born British Member of Parliament for the Birmingham Edgbaston constituency, said that the Matusseks were "quite right to say the British are still obsessed with the Nazi period."[11]

In 2006 John Ramsden, professor of modern history at Queen Mary University of London, published an entire book, *Don't Mention the War*, a study of the relationship between Germans and the British since 1890. He concluded that there had been several periods of friction during that time—around the turn of the twentieth century, in the run up to World War I, in the midst of that war—but that the British had thought highly of Weimar Germany and had not shown the same level of hate during World War II that they had in the earlier conflict (it was a clash more of ideologies than of peoples). Since 1945, war films and novels had kept the friction warm, however, aided by the Thatcher government when "Britain experienced . . . more open anti-German prejudice among her rulers than at any time since 1945."[12] He concluded that the defeat of Germany "seemed still to be essential to the English sense of who they are, and how they got here."[13]

That obsession shows no sign of diminishing. In July 2005, Cardinal Joseph Ratzinger of Bavaria became pope. The following day the London *Sun*, a tabloid newspaper, splashed its front page with the headline "From Hitler Youth to Papa Ratzi." Several other tabloids had a similar reaction and the *Daily Mirror*, in an article exploring the new pope's conduct in wartime, quoted an eighty-four-year-old woman from his hometown, Marktl am Inn, who said that, contrary to His Holiness's claims that he had no choice but to enroll in the Hitler Youth, "it was possible to resist." She said her own brother, a conscientious objector, had been sent to Dachau for his beliefs.[14]

In Berlin, Franz Josef Wagner, a columnist on the popular newspaper, *Bild*, was beside himself with anger. In an open letter to the British tabloids he warned them that "the devil seems to have slipped into your newsrooms . . . Anyone reading your British popular newspapers must have thought Hitler had been made pope."

All this seems to make Ambassador Matussek right on both counts—Britain *is* obsessed by the Nazis and history teaching in British schools *is* unbalanced, concentrating too much on the years 1933–45.

But this fascination with the Third Reich has done more than unbalance British education and foster an obsession with twelve years of dictatorship, helping to create an ignorance of the reality of modern Germany. It may well be the case that, as the Matusseks say, defeating Nazism is now part of Britain's self-identity. More than that, there is now a much wider sense that the Nazi period operates as an obstacle, a stumbling block, a reflecting mirror, that hinders us from looking back beyond that time, which has closed British minds to the Germany that preceded Hitler, an extraordinary country that he—a product of the Vienna gutter—on assuming office set about dismantling in a shocking and unprecedented way. Though the Russians and Poles and Czechs may not be as obsessed as the British, this blindness does apply in certain other countries as well. Wherever you look, Hitler still makes history but he also distorts it.

On February 20, 2006, in Vienna, Austria, David Irving, a British historian who has specialized in writing books about the Second World War, was sent to prison for three years, found guilty of denying the Holocaust.* Irving pled guilty to delivering two speeches in Austria in 1989, sixteen years before his trial, in which he had denied that Hitler was aware of the Holocaust and that millions of Jews had been murdered. Irving was arrested in November 2005, when he reentered the country, where it has

* In Germany it is also punishable by law to tell "the Auschwitz lie."

been a crime since 1946 to deny the Holocaust. It was by no means the first occasion Irving had crossed the legal line on this matter. He was already banned in a dozen countries from Canada to South Africa for broadcasting these views. In 2000 he was forced into bankruptcy in Britain when he unsuccessfully sued Deborah Lipstadt, an American academic who, in her book *Denying the Holocaust*, branded him one of the worst culprits. He was ordered to pay £3 million in legal costs and forced to sell his home in the fashionable Mayfair area of central London.[15]

Irving's trial came barely two months after Mahmoud Ahmadinejad, the president of Iran, called the Holocaust a "myth," claiming he did not believe that 6 million Jews had perished at the hands of the Nazis. Given the incendiary context of Middle Eastern politics, President Ahmadinejad's statement is perhaps not strictly comparable with David Irving's—we do not hold politicians to the same level of truth (unfortunately) as we do historians. But these two nearly contiguous events do underline how the Holocaust has become—and continues to be—an important focus of debate, even now, more than sixty years after it happened. If we are obsessed with Hitler, as we seem to be, can it be said we are likewise obsessed with the Holocaust?

At first sight that may seem a contentious and insensitive statement in itself. Can the murder of 6 million people—simply because they were members of a particular ethnic group—ever *not* be an important focus of debate and memory, however long after it occurred? But there is more to it than that. Of particular relevance is the fact that the Holocaust was *not* a focus of debate for many years immediately following World War II. It has become so only in recent decades, to the point where, it will be argued here, this "focus" (if it is not an obsession) is also distorting our view of the past, especially in the United States.

The Holocaust: An Obligation to Remember; the Right to Forget

In his level-headed study, *The Holocaust in American Life* (published in Britain as *The Holocaust and Collective Memory*; 2000), Peter Novick examines—as he puts it—how "the Holocaust has come to loom so large in our life." He begins with the observation that, generally speaking, historical events are most talked about shortly after their occurrence and then, about forty years afterward, they "fall down a memory hole where only historians scuttle around in the dark." This was true about events such as the Vietnam War, he says, but "with the Holocaust the rhythm has been very different: hardly talked about for the first twenty years or so after World War Two" but, from the 1970s on, "becoming ever more central in American public discourse—particularly, of course, among Jews, but also in the culture at large."[16] He records how, in recent years, "Holocaust survivor" has become an honorific title, "evoking not just sympathy but admiration, even awe." This was by no means the case in the immediate aftermath of war, where the status of Holocaust survivor was far from being honorific. Novick quotes the revealing comments by the leader of one American community in Europe, in a letter to a colleague in New York: "Those who have survived are not the fittest . . . but are largely the lowest Jewish elements, who by cunning and animal instincts have been able to escape the terrible fate of the more refined and better elements who succumbed."[17] No less a figure than David Ben-Gurion, Novick says, wanted to play down the magnitude of the tragedy because of the effect he thought it would have on Zionism—it might seem to others there would not be enough Jews to create Israel. In the United States, in 1946, 1947, and 1948, the main Jewish organizations (including the Jewish War Veterans) unanimously *vetoed* the idea for a proposed Holocaust memorial in New York City, on the grounds that such a monument would result in other Americans thinking of Jews as victims, and the monument become "a perpetual memorial to the weakness and defencelessness of the Jewish people." In the first postwar years, "much more than nowadays," the Holocaust was historicized—thought about and talked about as *just one* terrible feature of the period that had ended with the defeat of Nazi Germany. "The Holocaust had not, in the postwar years, attained transcendent status as the bearer of eternal truths or lessons that could be derived from contemplating it. Since the Holocaust was over and done with, there was no practical advantage to compensate for the pain of staring into that awful abyss." In his 1957 book, *American Judaism*, a scholarly survey of Jews in the fifties, Nathan Glazer observed that the Holocaust "had had remarkably slight effects on the inner life of American Jewry."[18]

In the immediate aftermath of World War II "everything about the contemporary presentation of the reports, testimonies, photographs, and newsreels was congruent with the wartime framing of Nazi atrocities as having been directed, *in the main*, at political opponents of the Third Reich." (Italics added.) The words "Jew" or "Jewish" did not appear in Edward R. Murrow's (horrifed, awestruck) radio broadcast about entering Buchenwald. General Dwight Eisenhower, disturbed by the camps, said he wanted "legislators and editors" to visit these locations where the Nazis had incarcerated "political prisoners"—again, no mention of Jews. Other reports spoke of "political prisoners, slave laborers and civilians of many nationalities." Jews did not go unmentioned, and some reports observed that they had been treated worse than others. "But there was nothing about the reporting on the liberation of the camps that treated Jews as more than *among* the victims of the Nazis . . . nothing, that is, that associated them with what is now designated 'the Holocaust.'" (Italics in the original.)[19]

Attitudes only began to change, Novick says, with the Eichmann trial in 1961–62, the Six-Day War in the Middle East in 1967, and, most of all, after the Yom Kippur War in October 1973, when Israel—for a brief time—looked as though she might be defeated. Novick again: "As part of this process, there emerged in American culture a distinct *thing* called 'the Holocaust'—an event in its own right, not simply a subdivision of general Nazi barbarism."[20] It was now that the word "Holocaust" entered the language as a description of all manner of horrors.

It was now, Novick says, that the Holocaust became in effect sacralized, so that it was almost above criticism. Almost, but not quite. Amos Oz, the Israeli novelist and author of *Touch the Water, Touch the Wind*, about two Holocaust survivors who fall in love, was one who asked whether, alongside the obligation to remember, there wasn't also a right to forget: "Are we . . . to sit forever mourning for our dead?" In the first year of the Intifada (1987), the distinguished Israeli philosopher Yehuda Elkana, who had been interned in Auschwitz as a child, published "A Plea for Forgetting." The Holocaust's "lesson," that "the whole world is against us," that the Jews "are the eternal victims," was, for Elkana, "the tragic and paradoxical victory of Hitler." This lesson, he thought, had contributed to Israeli brutalities on the West Bank and to the unwillingness to make peace with the Palestinians.[21] This change in feeling culminated in 1998 when, in a survey of American Jewish opinion, respondents were asked to rate the importance of various activities to their Jewish identity. This was the first year that "remembrance of the Holocaust" was included (a revealing development in itself)—and it won hands down, chosen many more times than "attending synagogue" or "observing Jewish holidays."[22]

Novick further observed that since the 1970s the Holocaust has come to be presented as not just a Jewish memory but as an American one. In a 1995 poll, to gauge Americans' knowledge of World War II, 97 percent knew what the Holocaust was, substantially more than could identify Pearl Harbor or knew that the United States had dropped two atomic bombs on Japan, and far more than the 49 percent who knew that the Soviet Union fought on the American side in the war.[23] By 2002, in a growing number of states, the teaching of the Holocaust in public schools was mandated by law.

Norman G. Finkelstein was much more acerbic than Novick. In *The Holocaust Industry*, published to some acclaim in 2000, stimulating great interest (and criticism) in Germany but relative silence in the United States, Finkelstein, whose own mother was in Majdanek concentration camp and the slave labor camps at Czestochowa and Skarszysko, accused American Jewry in particular of exploiting the Holocaust, of being "Holocaust hucksters," exaggerating the numbers who suffered *and* the numbers who survived, for their own ends, mainly to benefit Israel. He described what he called a "sordid pattern" and detailed the large salaries and fees being drawn by officials administering compensation claims, far larger than the claims themselves. Again, his theme underlines the fact that interest in the Holocaust is a *recent* phenomenon.[24]

The Historians' Dispute

Just how extreme or unique *was* the Holocaust? This is a sensitive question that the Germans themselves have had difficulty adjusting to. Whereas in America, as Novick has shown, the Holocaust has grown in salience as the years have passed, in Germany there have been some equally forceful attempts to take the debate in the opposite direction and play down its extent, significance, and singularity. Charles Maier is just one American historian who has remarked on how the German scholarly community has been polarized by this subject.

It was a division that first revealed itself in the 1980s in a phenomenon known as the *Historikerstreit*, the "historians' dispute," an acrimonious debate that was carried on among distinguished historians, such as Helmut Diwald, Ernst Nolte (a student of Heidegger), and Andreas Hillgruber, who had each produced solid, "regular" histories before. When it broke open, it comprised the following arguments:

- It was argued that Fascism was not a totalitarian system in the mold of Stalinism, but a *response* to it;
- Auschwitz was not a unique event but a copy of the Gulag; other, earlier, genocides had taken place in the twentieth century;
- More Aryans than Jews were killed in the death camps;
- Poles and Romanians were just as anti-Semitic as Germans;
- The worst excesses of the war—the invasion of Russia and the extermination of the Jews—came about because one man, Hitler, intended them to happen.

There are good answers to these arguments, not least, as Charles Maier dryly observed, "The Final Solution must not be made into a question of bookkeeping."[25] Beneath the surface, however, was there more to it? Was the Historikerstreit the symptom of a deeper malaise that, forty years after the end of the war, was at last beginning to surface?

There were those who thought that it was. The German philosopher Jürgen Habermas observed that, "In the recent past, the memories accumulate of those who for decades could not speak about their suffering and we do not really know whether one may really still believe in the redemptive power of the word." He thought that, in the historians' dispute, the "floodgates of memory" had finally been opened "and made the [German] public realise that the past was not simply fading." In 1986, in a German historical journal, Hermann Rudolph agreed that the Germans were more concerned with the war just then than they had been in the past. That concern, he said, was "apparently not wearing thin; rather the opposite . . . the question that is now thrown open is: should the Third Reich be treated historiographically so that it no longer blocks the way to our own past like some sombre and monstrous monument . . . ?"[26]

Is there something to this? In an account of the Historikerstreit, Richard Evans, professor of history at Cambridge University in the United Kingdom, has noted that, after World War II in Germany, "very little was said about Nazis. Next to nothing was taught about it in the schools. The Nazi affiliations of major figures in the economy were never mentioned. Even in politics, there was no great stigma attached to a Nazi background, so long as this did not become the embarrassing object of public debate."[27] The desire, in West Germany, for a more determined confrontation with the German past only began, Evans says, with the Eichmann trial in Israel in 1960 and the Auschwitz trials in 1964.[28] So here is a tidy parallel with the growth in interest in the Holocaust in the United States.

The importance of the Historikerstreit, in our context, is that it is yet further evidence of the obsession with Hitler and the Holocaust and of a particular pattern of forgetting or, more appropriately, not forgetting. Opinion polls in Germany showed that while 80 percent of Americans were proud to be Americans, and 50 percent of Britons were proud of being British, only 20 percent of Germans were proud of being German. Michael Stürmer, another historian, argued that only by restoring their history to themselves could Germans recover their pride again. He added that Germans were "obsessed with their guilt," and that this obsession was interfering with their ability to develop a sense of national identity, which by implication had political and cultural consequences. He resented the implication, he said, that Germany "must be viewed continually as a patient in therapy."[29] As historian Charles Maier put it, "There has been no closure in this debate, only exhaustion."[30]

This was underlined by the Jenninger affair. In November 1988, at a ceremony to commemorate the fiftieth anniversary of *Kristallnacht*, Philip Jenninger, president of the Federal German parliament (and therefore the second-highest official, after the president of the republic himself), delivered a speech in which he treated the Holocaust as an historical event and therefore not necessarily unique, and as one in which, moreover, many Germans were "bystanders"—i.e., not directly responsible. Although many people, including many Americans, thought his speech was courageous, many others were outraged and Jenninger was forced to retire.[31]

The same memory pattern repeats itself in respect to art. It was only in the mid-1990s that the world belatedly woke up to the fact that thousands of paintings—old masters and Impressionists alike—which had been looted by the Nazis from their Jewish owners, were circulating freely on the auction market, and had been doing so since shortly after 1945. Auction catalogs had for years openly printed the provenance of paintings, stipulating that they had been acquired by prominent Nazis, from Hermann Göring down to well-known dealers, but for sixty years no one had paid proper attention. It was only after two Russian art historians discovered a cache of pictures in Moscow—pictures that had been thought destroyed in Berlin—and the strengthening feeling about the Holocaust, that this scandal was fully exposed. The same was true about "dormant accounts" in Swiss banks. Here too, countless accounts belonging to Jews who had been sent to the death camps were "rediscovered" in Switzerland in the late 1990s, when almost anyone could have spotted this outrage much earlier. (One of the reasons the Swiss refused earlier claims was that claimants had no death certificates, as if the SS issued death certificates in the camps.) In March 2006, a Swiss book, *Observe and Question*, alleged that, during World War II, the Swiss authorities had turned away thousands of Jewish refugees who attempted to cross into neutral Switzerland. Swiss nationalists vowed to block distribution of the book. Here, too, this information could have been exposed much earlier.

The same argument applies to Belgium. The country's prime minister formally apologized to the Belgian Jewish community for its role in the Holocaust—but not until 2002. The conclusions of a government-sponsored report, 1,116 pages long and titled *Submissive Belgium*, were read before Parliament in Brussels in February 2007, concluding that its top civil servants had acted in a way "unworthy for a democracy." The Belgian government, exiled in London during World War II, had advised its civil servants to work with the occupying Nazis to prevent economic breakdown but in many cases, the report said, that had "deteriorated into collaboration with persecution of the Jews and their deportation to concentration camps." After the war, it went on, many cases (of reparation) were considered "too delicate to handle" by the military courts and "every responsibility of the Belgian authorities in the persecution and deportation of Jews was rejected." Again, an inordinate delay.[32]

Despite making it illegal to deny the Holocaust as early as 1946, Austria, too, has had a problem in assimilating its role in World War II—and not just because Hitler was, of course, not German but Austrian.* Forty percent of the personnel and most of the commandants of the death camps at Belzec, Sobibor, and Treblinka were Austrian, as were 80 percent of Eichmann's staff—and Eichmann himself. Despite these unwholesome statistics, the country's first postwar president, the veteran Socialist leader Dr. Karl Renner, emphasized that there was "no room" for Jewish businessmen in Austria, and he did not think that "Austria in its present mood would allow Jews once again to build up these family monopolies." In an American survey in 1947–48, nearly a quarter of all Viennese thought that the Jews had "got what they deserved" under Nazism, while 40 percent thought that the "Jewish character" was responsible for anti-Semitism. For decades Austrians presented themselves as "the first victims" of the Nazis and used this argument to rebuff Jewish claims for restitution, many of which they insisted were fraudulent. (Although the notion that Austria was the "first victim" was accepted by the Allies at the Ottawa Conference in 1943, following the *Anschluss* the SS was "swamped" with Austrian applicants.)

Perhaps the most ludicrous—and embarrassing—episode of this kind occurred during the filming of *The Sound of Music* in Salzburg in 1965, when the local authorities refused to allow swastika flags to be hung in the Residenzplatz as a backdrop. They argued that Salzburgers had never supported the Nazis—at least they did until the producers of the film said they would use instead real newsreel footage, after which the city fathers backed down.[33]

At least three prominent Austrian politicians—Hans Öllinger, Friedrich Peter, and Kurt Waldheim—were exposed (as often as not by the Nazi hunter Simon Wiesenthal, who received death threats for his pains) as erstwhile SS or Wehrmacht officers (very junior in Waldheim's case) and it was not until July 1991 that the Socialist federal chancellor

* David Irving's imprisonment in 2006, under the Holocaust-denial law, was unusual. In 2004, the latest year for which figures are available, 724 people were prosecuted for denying the Holocaust in Austria, a figure which can be interpreted in two ways. On the one hand, it shows that the Austrians were assiduous in enforcing the statute. On the other, it could be held to show that, sixty years after the end of World War Two, such a law was still badly needed.

Franz Vranitsky publicly acknowledged Austrian "co-responsibility" for what had happened in the Third Reich—rather late, one might think. The growth in popularity of the radical right Freedom Party (FPÖ) under Jörg Haider's leadership belied the fact that the country was really attempting to deal with its past. FPÖ propaganda at times verged on Holocaust denial, claiming it was in any case no different from the Soviet Gulag, while the party's attitude to immigrants resurrected the terms of biological racism so redolent of the Nazis.

All this was underlined by what happened at Mauerbach. At the end of October 1997, Christie's Auction House Vienna sold the contents of Mauerbach Monastery, an old Carthusian building in a sleepy village about thirty minutes west of the Austrian capital. Some 8,400 art objects, which had been looted from Austrian Jews, had been stored in the monastery since the 1960s. It was a dismal affair that did the Austrian authorities no credit at all. From 1945 until 1969 the government made no attempt whatsoever to trace any Holocaust survivors. At one stage the man charged with disposing of the art was the very individual who had masterminded its confiscation in the first place. On two other occasions the Austrian government passed strict laws that made it all but impossible for Jews to identify their property—and this at a time when much of this "Holocaust art" decorated Austrian embassies abroad. In one case where the claimant was eventually successful, he was charged $8,000 for years of storage—for a painting that had been *confiscated*. Only 3.2 percent of works were ever returned to their rightful owners, and it was not until the American magazine *ARTnews* exposed what was sequestered in Mauerbach that any action was taken.[34]

The Closed Archives of Vichy

Although France was one of the more liberal nations in the interwar years, opening its doors to Jewish refugees from Poland, Romania, and Germany, since the war it has fought its own set of demons relating to that difficult time. The classic, but nevertheless defensive statement about France's role in the Holocaust came from President François Mitterrand in 1992 when, with breathtaking insouciance, he declared that the collaborationist, pro-German Vichy regime that governed unoccupied France from 1940 to

1944 was illegal and "aberrational" and had "nothing to do with France today." "The French nation was not involved in that," he said, "nor was the Republic."[35]

As this implies, French collaboration during World War II has had its own memory pattern. Henry Rousso has given it a name, *The Vichy Syndrome*. Rousso found that his thesis—that the internal quarrels among the French left deeper scars than either the defeat or the German occupation—was "largely confirmed." Two of his chapters had "Obsession" in the title and in a "temperature curve" of the syndrome, a year-by-year chart of the "temperature" of the obsession with Vichy, as measured by political events, books published, films screened, and so on, he identified an "acute crisis" from 1945 to 1953, relative "calm" from 1954 to 1979, and "acute crisis" ever since (the book was published in 1991).[36] This memory pattern is not dissimilar to that for the Holocaust in America.

The actual extent—and even enthusiasm—of French collaboration was finally and fully exposed in the landmark 1981 study by Michael R. Marrus and Robert Paxton, *Vichy France and the Jews*, which established, "virtually beyond doubt," that the Vichy government went well beyond even what the Germans required of it in its persecution of the Jews. Some 75,000 Jews were deported from France during the war, the great majority seized by French police. Only 3,000 survived.

Then, in November 1991, Serge Klarsfeld, a French Nazi hunter and president of the organization Sons and Daughters of the Jewish Deportees of France, claimed to have discovered the so-called Jewish file in the basement of the French Veterans Ministry. These documents, allegedly compiled by the Paris police following the census of October 1940, were supposedly used to identify all Jews living in France. A commission of professional historians later confirmed that the real file had been destroyed in 1948, but the case raised doubts about public access to official documents relating to the Vichy regime, doubts that were sharpened in 1994, when Sonia Combe, in her book, *Archives Interdites* (Closed Archives), accused the French government archival service of restricting public access to historical documents about Vichy. She alleged that a combination of insufficient funding and a "specific effort to avoid scandal" had combined to limit access to wartime documents.[37]

None of this was eased by the four trials that took place in France in the early 1990s for "crimes against humanity." Klaus Barbie, former head of the Gestapo in Lyon, went on trial in 1992 for the arrest and deportation to Auschwitz of forty-four Jewish children. In 1994, Paul Touvier, one of the leaders of the French militia, was tried in Versailles for organizing the killing of Jewish men in Rieux-la-Pape, near Lyon. In 1998, Maurice Papon, who oversaw the deportation of 2,000 Jews from the Bordeaux region, was eventually tried and convicted. In the meantime he had enjoyed a successful career in public life. No trial received more attention than that of René Bousquet, accused of coordinating with the Gestapo to organize the infamous roundup of Jews in Paris in July 1942, when 13,000 were gathered in the Vel' d'Hiv' bicycle stadium and shipped to transit camps in France and then on to Auschwitz. Not the least controversial aspect of this case was the fact that Bousquet's role in the roundup had been reported as early as 1978, but it took the French legal system twelve years to do anything about it. Bousquet was assassinated in 1993 before his trial.

There was also the scandal that surrounded the French president, François Mitterrand himself. In a 1994 biography of the president, Pierre Péan revealed that Mitterrand had been both a civil servant in the Vichy regime *and* a leader in the French Resistance—indeed, he had held both positions at the same time for several months in 1943. Mitterrand had always denied his participation in the Vichy regime, so this was embarrassing—more than embarrassing—all around. The revelations certainly put his comments about Vichy not being the true France into a sanctimonious light. It was not until 1995 that the French state apologized for its role in the Holocaust—half a century after the events themselves and a delay longer even than that in Austria and Germany.[38]

Against this background, there was a series of cases filed in U.S. courts in the mid-1990s, targeting French companies that had profited from the plight of Jews during World War II (such as the state railway, SNCF, and a number of banks). This case followed the similar suit filed in U.S. courts against Swiss banks holding Holocaust-era assets. The French cases were thrown out, but in March 1997 the French government under Alain Juppé responded to these concerns by setting up the Mattéoli Commission, to investigate the allegations. The commission hired 120 researchers, at state expense, and produced twelve reports on Jewish experiences during Vichy.

As a result, a Foundation of Remembrance was announced in 2000, endowed with 2.4 billion francs ($342 million), the estimated total value of assets that have not yet been returned to their Jewish owners. It is the largest charitable foundation in the country.[39]

Finally, in Europe, we may mention Poland where, in the general election of 2007, the Second World War was an issue, or made an issue, by the Kaczynski twins, Lech and Jaroslaw, who, as president and prime minister, had set their country on an ultranationalist course, picking fights with both Germany and Russia and trying to use its new membership in the European Union to, as one observer put it, "mop up all the unfinished business of the Second World War." In particular, they claimed that the Polish population would have been substantially larger had not the Nazis murdered so many people and that therefore Germany "had a moral duty to give ground" in regard to reparations claimed.[40] Likewise the massacre of several hundred Jews in May 1941 at Jedwabne in Poland by their fellow citizens, even classmates, which had been the subject of a trial in 1949, was brought to the attention of the world only in 2000, with the publication of Jan Gross's book *Neighbours*. Again, the same memory pattern.

As all of these recent events confirm, Hitler, Nazism, and the Holocaust have disobeyed all the normal rules of historical forgetting and assimilation. Official apologies, reparations, and trials of former Nazis have been more in evidence *since* 1990 than before.

The Executioners' Song

No better example of the muddy waters surrounding Holocaust remembrance is provided than by Daniel Goldhagen's book, *Hitler's Willing Executioners: Ordinary Germans and the Holocaust*, published in 1996. This title, a best seller on both sides of the Atlantic, put forward what its author claimed was "an entirely new answer to the question of how it was that Germans rather than some other European people turned anti-Semitic prejudice into mass murder." The Germans did so, he argued, "not because they were forced to, nor because German traditions of obedience enabled a handful of fanatics at the top to do whatever they liked, nor because they were succumbing to peer-group pressure from their comrades-in-arms, nor because they were ambitious careerists, nor because they were acting automatically, like cogs in a machine, nor because they faced death themselves if they refused to obey the order to do so."[41] Instead, Goldhagen said the Germans killed millions of Jews because they enjoyed doing so and they enjoyed it because "their minds and emotions were eaten up by a murderous, all-consuming hatred of Jews that had been pervasive in German political culture for decades, even centuries past." He identified a "simmering hatred" of the Jews as a "cultural norm" in Germany in the nineteenth century and found that it was given social expression "as a matter of routine." He discovered nineteen publications produced in Germany between 1861 and 1895 that called for the physical extermination of the Jews and himself "reconceptualised" modern German anti-Semitism into a new framework that envisaged anti-Semitism as "deeply embedded in German cultural and political life and conversation, as well as integrated into the moral structure of society."[42]

In a postscript written for the paperback edition of his book, published in 1997, Goldhagen set out some of the reactions to the hardback. He said his book had been the subject of vitriolic attacks, by both journalists and academics but that their arguments "consisted almost wholly of denunciations and misrepresentations of the book's contents . . . The critics presented no serious argument and no evidence to support their contentions . . . They did not do so because such arguments and evidence do not exist." On the other hand, he said, the public had embraced the book, it had become a number one best seller in Austria and Germany, and he maintained that on a series of panel discussions called to discuss his thesis his critics conceded many points.[43]

It is one thing to select the best sentences from reviews when seeking to embellish the cover of a paperback edition: the function of a jacket is to sell the book. It is quite another matter, when discussing the substantive issues in a serious argument, to ignore cogent and substantial criticisms that have been leveled. There is no question that, in connection with Daniel Goldhagen and his book, his sins of omission are considerable, evincing a serious disregard of inconvenient data.

The first thing that professional historians pointed out was that Goldhagen's theories were, despite his claims, emphatically not new. A central aspect of his book was an examination of the men of Reserve Police Battalion 101, mostly older German men who had moved through occupied eastern Europe carrying out mass shootings of at least 38,000 Jews over a considerable period of time. In 1992, not so very long before Goldhagen's book was published, Christopher Browning of the University of North Carolina at Chapel Hill, in *Ordinary Men*, had studied this self-same unit, arriving at very different conclusions. Browning observed that "ordinary men" were indeed involved in the killing but went on to describe how these police reservists were shocked and surprised by their orders to kill Jews when they first received them. Their commanding officer, Major Wilhelm Trapp, was so unnerved that he allowed those who preferred not to take part to pull out of the operation; as a result, one of Trapp's own officers obtained a transfer.[44]

Goldhagen's further argument, that Germany had been deeply anti-Semitic since the Middle Ages, was also torn apart. As Richard Evans, one of Goldhagen's sternest, best-informed, and fair-minded critics, wrote: "If the German population and elite were so deeply anti-Semitic, as Goldhagen says, why did Jews actually gain civil equality by legislative enactment all over Germany in the course of the nineteenth century?" Fritz Stern described "the ascent of German Jewry" in the nineteenth century as "one of the most spectacular social leaps in European history." Before World War I, both France and Russia were more anti-Semitic than was Germany. In France, the Dreyfus affair sparked anti-Semitic riots in more than thirty towns and in Russia there were 690 documented pogroms with over 3,000 reported murders and 100,000 made homeless. In tsarist Russia, Jews were made to live in a "Pale of Settlement." In contrast, Evans records one telling vignette—that the pub-and-inn surveillance reports from Hamburg in the late 1920s revealed "virtually no" anti-Semitic feeling by rank-and-file supporters of the Social Democrats. More to the point, anti-Semitism was not an important factor in generating votes for the Nazis in the elections of 1930–33. William Allen, who carried out an in-depth study of one German town, Northeim, found that, from 1928 on, Nazi propaganda actually played *down* the anti-Semitic aspects of the party's ideology, for the very good reason that it was unpopular with the electorate. Why did Heinrich Himmler need to keep the "Final Solution" secret if ordinary Germans were as murderous as Goldhagen insists? Why did Himmler complain at one point that "every German has a Jew they wish to protect"?[45]

Goldhagen cites as compelling evidence for popular German anti-Semitism the recurrence of "ritual murder" accusations against Jews and quotes this sentence from Peter Pulzer's *The Rise of Political Anti-Semitism in Germany*: "In Germany and the Austrian Empire twelve such trials took place between 1867 and 1914." But this was not the complete sentence; Goldhagen leaves out the remainder, which reads: "eleven of which collapsed although the trials were by jury."[46] Goldhagen referred to Thomas Mann who, he said, though a long-standing opponent of the Nazis, "could nevertheless find some common ground with [them]" when he wrote: " . . . it is no great misfortune . . . that . . . the Jewish presence in the judiciary has been ended." Fritz Stern pointed out that Mann was married to Katia Pringsheim, the daughter of a prominent Jewish family, and in the very next sentence to that quoted above, which Goldhagen omitted, Mann expressed his distaste at his own thoughts, characterizing them as "secret, disquieting, intense."[47]

No less damaging to Goldhagen's scholarship was the fact that he had mistranslated some German passages—and in telling fashion. In one instance, he refers to a poem written by a member of an *Einsatzkommando* and writes that this individual "managed to work into his verse, for the enjoyment of all, a reference to the 'skull-cracking blows' . . . that they had undoubtedly delivered with relish to their Jewish victims." Although the verse was indeed extremely anti-Semitic, the phrase in quotes actually referred to "the cracking of nuts."[48] Evans concluded—and he was just one among many—that Goldhagen's book was disfigured by a "startling failure of scholarship," that it was written in the "pretentious language of dogmatism," and betrayed a "disturbing arrogance that is of a piece with the exaggerated claims for novelty."[49]

Hitler's Willing Executioners conforms exactly to what Novick is saying. Far from interest in the Holocaust declining in our (or Goldhagen's) historical consciousness, its enormity—its singularity—has now grown in salience to the point where it was caused, not by Hitler alone, or his elite entourage, or the SS, but by all Germans, including the ordinary ones, and this was so because, throughout history, Germany has always been anti-Semitic, far more so than any other country. This comes close to making the Holocaust inevitable in Germany.

It also alerts us to a phenomenon we shall have occasion to examine and criticize and where the Germans (among others) have been at fault: the writing of *meta*-history, by which I mean the attempt to understand the past via simple, all-embracing theories, the "dangerous simplifiers" as Jacob Burckhardt called them.

The "Goldhagen affair" shows how history writing can be distorted. Given the distortions and omissions he employed, one is entitled to wonder whether he could not see past or around the Holocaust to begin with, and this author at least is prompted to suspect instead that he started with his conclusions and then found the "facts" to fit his theory. Goldhagen's account is not so crude as that of the British tabloids, but it does have the same obsessive quality. As Fritz Stern saw it, "the book also reinforces and reignites earlier prejudices: latent anti-German sentiment among Americans, especially Jews; and a sense among Germans that Jews have a special stake in commemorating the Holocaust, thereby keeping Germany a prisoner of its past."[50] As the German historian K. D. Bracher has said, all modern developments in Germany are inevitably linked back to events in the Third Reich. The Germany of before that time, for most people, simply does not exist.

Dismaying as all this is, there is another perspective, put by two British observers, Ian Kershaw and Steve Crawshaw. History, particularly in the age of television, is almost as much about perception as about reality, and one of the misrepresentations about Germany in the world at large is the ignorance in other Western countries in regard to the events of 1968. The Prague Spring, the student riots in Paris and elsewhere in France in May 1968, and the student sit-ins at American universities are well remembered. Much less well remembered—hardly remembered at all, it seems—are the events in Germany in that same year. Those events are covered in more detail in Chapter 41, p. 757, of this book. Here we need only say that 1968 in Germany saw a new generation of sons and daughters (*die Achtundsechsiger*) confront their fathers and mothers about their "brown" past, their involvement with the Nazis. This was a genuine upheaval in Germany, a searing and serious attempt by those born in the wake of the war to force the nation to confront its past. Many Germans believe they began to "move on" then and are now well past the traumas. Not everyone agrees that this has happened, of course: the Bader-Meinhof violence lasted through much of the 1970s, the historians' dispute did not erupt until the 1980s, German novelists were still writing about the war in the early twenty-first century. Older Germans say the youthful rebellion was a myth, that the young were jealous of their elders—with a brown past or not—who had made such a success of the "economic miracle." But Kershaw and Crawshaw believe this helps explain the "Goldhagen phenomenon"—that his book was welcomed by the public, despite being censured by more knowledgeable critics. The book, they say, helped a fresh generation, the *grandchildren* of the Nazis, come to terms with the past. "The acceptance of any and all attacks on the old Germany provided a yardstick for modern Germans to remind themselves that they had indeed confronted the terrible past, thus helping to neutralise its demons. Goldhagen became a player, at just one remove, in Germany's own arguments with itself. The details of his arguments—untenable or otherwise—mattered less to the Germans than his readiness to be tough on Germany." In 2002 a sociological analysis of family discussions about the Third Reich was published, entitled *Grandad Wasn't a Nazi*. This revealed "the unsettling extent" to which children in Germany were inclined to "blank out" the evidence that their grandparents were complicit, "even when that evidence is acknowledged and uncontested."[51]

At the same time, their elders have become progressively more interested in the war. Wulf Kansteiner's studies of German television, especially the broadcasts of ZDF, the Zweites Deutsches Fernsehen, and the documentary films of Guido Knopp, show—among many other things—that programs about the "Final Solution" have risen from less than 100 minutes a year in 1964 to more than 1,400 minutes in 1995, with far more interest being shown after 1987. Kansteiner says there was in Germany a "memory revolution" in the 1980s and 1990s as Germans "retrieved and reinvented their history," and that there was a "repackaging of the Nazi past" around 1995, and a reorientation of Holocaust studies after the fiftieth anniversary of the liberation of Auschwitz, when the "elusive goal" of normalization had become a "tangible reality." This was essentially the same point as that made by Hermann Lübbe: "The memory of the Third Reich has intensified with increasing temporal distance to the Nazi past." Again, the crucial decade, the turning point, is the 1990s.

An explanation for the delay inside Germany in facing up to its past has been constructed by A. Dirk Moses, a historian at the University of

Sydney (though he has also worked in Freiburg), who gives a "generational account." His study, published in 2007, will be discussed in more detail in Chapter 41 but, essentially, Moses—whose references are admirably copious—says that the generation known in Germany as "Forty-fivers," people who were born in the late 1920s, who received their socialization in the Third Reich in the 1930s, and were on the edge of adulthood in 1945, had no other sociopolitical experience than National Socialism to go by, did not feel personally responsible for the atrocities (because they weren't yet old enough), *but* afterward withdrew into the "private spheres" of family life and work, their psychological rivalries with their fathers remaining unresolved; "emotionally bound" to Hitler, they threw themselves into rebuilding the country, and remained largely silent about what had gone on in Nazi times—lest that disrupt the task of reconstruction. This meant, he said, that the nation in the 1960s was largely the same one as had existed in the final years of National Socialism, that the hierarchical and authoritarian cast of mind continued and this "silent majority" "remembered the sufferings of its own rather than those of its victims." Furthermore, he said, many of the younger generation felt that the educated middle class incarnated these pathologies "in a particularly virulent way." All this accorded well, he said, with the picture painted in 1967 by two psychoanalysts, Alexander and Margarete Mitscherlich, in their book *The Inability to Mourn*, who had argued that, even at that late date, Germany was gripped by a "psychic immobilism," unable to admit its culpability in the crimes of National Socialism, because that would involve the admission of shame and guilt on such a scale that "the self-esteem needed for continued living" would be unattainable.

The "psychological" explanations are plausible. At the same time, the studies carried out at the Potsdam Institute for Military Research into "Germany and the Second World War," the ninth volume of which was published in 2008, provide two new lights on this aspect of affairs. In the first place, this meticulous project (volume nine is 1,074 pages long) removes any lingering doubt that "almost every German" in the Third Reich knew what was happening to the Jews. The evidence is now too overwhelming, from the public auctions in Hamburg, where the property of 30,000 Jewish families was sold to 100,000 successful bidders, to the prisoners in Bremen who worked in full sight of the population, especially to clear bomb damage, and were known as "zebras" on account of their striped uniforms, to the ship moored in the Rhine at Cologne, filled with Jews held ready to clear the bomb damage as soon as the air raids were over, to Düsseldorf, where the mayor wanted the captive Jews worked harder. The historians concluded that, after the war, there was a "collective silence" in Germany, protecting former Nazis who had taken part in the Third Reich's crimes, "because everyone had, before 1945, benefited from the Nazi regime in one way or another." At the same time, and as Max Hastings concluded in a review of the book, this study is a "notable tribute" to a new generation of Germans, most born long after the war, who are ready at last to compile a totally objective picture of the Third Reich, in so doing passing judgment on their parents' generation in a way few other countries have managed.[52]

Germany's "Wrong Turn"

There is one final sense in which the Holocaust exerts its influence on the writing of history and therefore on our understanding of the past. The Nazis in general, and the Holocaust in particular, were so extreme, and so unique (notwithstanding what Professors Nolte, Hillgruber, and Diwald say), that there is a tendency among some to see every episode of the past 250 years as leading up to the Holocaust, as if it were the culmination (as Goldhagen implied) of all events and ideas that occurred in modern Germany. This has had a further effect—that, because of this, because of the very nature of Nazism and the Holocaust, modern German history is inevitably seen as *political* history, the pattern and outcome of domestic and foreign policy, party-political, diplomatic, and military affairs. Here too the very existence of the Holocaust has had a narrowing and constricting influence.

The most important example of this is the work of the German historian Hans-Ulrich Wehler. In a massive four-volume investigation published between 1989 and 2003, he advanced the view that the sources of Germany's "descent into barbarism" in 1933 were to be found, not in its geographical position, at the center of Europe and threatened on all sides, as other historians had often argued, but in the "special path," or *Sonderweg*, taken by German society as it evolved to modernity between the middle of the nineteenth century and the middle of the twentieth (Leopold von Ranke, the eminent German historian, had spoken of a German Sonderweg as early as 1833).[53] In this account, Germany had taken "a wrong turning" at some stage. One view had it that the deviant path began with the fragmented Reich of the Middle Ages. In another view Martin Luther was to blame—it was his vehement rejection of Rome that was the fatal turning point. Then there was the view that the German philosophers—beginning with Immanuel Kant—had considered the concept of freedom only in a narrow, intellectualized way, as concerning the realm of ideas and demoting politics to a less important role.

More plausible, Wehler said, were certain specific political features and events of Germany's history. Fundamental were the ravages suffered by Germany in the Thirty Years' War, which devastated the infrastructure and decimated the population, which took generations to recover. Furthermore, in the seventeenth century, for example, British parliamentary elites won out over the Stuarts at a time when Prussian towns and provincial estates were in thrall to the Great Elector.[54] At a later stage, so this argument went, in 1848 the German bourgeoisie failed in its attempt to usurp political power from the aristocracy as had happened with its counterparts in, for example, England in 1640, and France in 1789. This was A. J. P. Taylor's famous historical turning point, "at which history failed to turn." Because of this, the Prussian aristocracy maintained its sociopolitical dominance. It continued to consolidate its influence through a conservative "revolution from above," in which Germany was united (under Prussian domination) from 1866 to 1871. Although industrialization provoked social changes that put still more pressure on the upper classes, the monopolization of important positions of power in the army, the civil service, and the Reich administration enabled them to keep a grip on government. These maneuvers were reinforced by a "feudalisation of the bourgeoisie," who were lured into aping the aristocracy (dueling, the scramble for titles, and, most critically, the rejection of democracy and parliamentarianism). A third aspect of this Sonderweg came in the realm of big industrial conglomerates. As a result of the "great depression" of 1873–96, these industrial behemoths sought an alliance with one another (in cartels), and with the government, which intervened more and more.[55] This strategy, says Wehler, transformed Germany from liberal competitive capitalism (as practiced in France, Britain, the United States, and elsewhere) to "organised oligopolistic capitalism."

Wehler's was an impressive and coherent thesis. It was controversial but it was so in the best sense, provoking thought but also susceptible of research. And research there was, masses of it. Historians in Germany paid Wehler the compliment of setting up a comprehensive research project on the history of the German middle classes, centered on Bielefeld.

The Sonderweg was a theory of interest as much outside Germany as within, and some of the earliest criticisms came from foreigners. This was partly because Wehler had held up Britain's political development, its path to modernity, as "normal" at a time when, inside Britain itself, controversy raged about why the country was "the sick man of Europe." So it was no real surprise when two British professors of modern history (both of whom taught in the United States), published *The Peculiarities of German History*, which was nothing other than a full-scale attack on the Sonderweg thesis. David Blackbourn and Geoffrey Eley argued that there was no general path to "modernity"; each nation had its own peculiar experience based on a particular mix of factors. The elements in the mix were the same in all countries—it was the proportions and interrelations that were different. They also pointed to the fact that German industries had produced very many modern technologies (see Chapters 17–20. pp. 341–383)—how could such industry be backward when they were such a practical, innovative success? The same was true in the academic sphere—how could the professors have been so conformist when nineteenth-century Germany gave us so many new disciplines—cell biology, sociology, non-Euclidean geometry, quantum physics, and art history among them?[56]

At first Wehler rejected these—and other—criticisms. However, by the time he published later volumes of his history, he had radically amended his theory. As one critic remarked, Wehler's *theory* was now replaced by a *list* of twelve aspects "in which the German Reich's experience was unique among that of Western European states." These had to do with the army, the legislative assemblies, the civil service, the labor movement, the power of the nobility—in other words, strictly political matters though, as an afterthought, Wehler did include as important the role of the Catholic Church and of the educated middle class, the so-called *Bildungsbürgertum*. "Thus [Wehler] abandons a central element of the *Sonderweg* thesis—

namely, the argument that society as well as politics failed to modernise. The entire thesis is now concentrated in the political sphere."[57]

Many of the issues raised by Wehler's important volumes will be referred to later, but for now the issue to bear in mind is that, however successful or otherwise his theory is judged to be, it was above all an attempt to explain the special—the peculiar—path of German history, a *political* path to modernity that led to Nazism and the extremities and catastrophes of the Holocaust. As Richard Evans, again, has remarked, this led Wehler, if not to distort his history, then to leave out a mass of important and relevant material, a not dissimilar charge to that leveled against Goldhagen, Nolte, Hillgruber, and the history curriculum in British schools.[58]

Hitler and the Holocaust are preoccupying the world to such an extent, I suggest, that we are denying ourselves important aspects elsewhere in German history. We must not forget the Holocaust—this surely does not need underlining—but at the same time we must learn to look past it. Charles Maier, an American Jewish historian, wrote that "the effort to benefit from history [keeping the Holocaust alive] has disadvantages . . . Nietzsche feared that history could interfere with life . . . Can there be too much memory?"[59] He also asked—and not rhetorically—if the Holocaust had not become an asset for the Jews, admitting that, "It is possible to make a fetish of Auschwitz."[60]

Germany's Cultural "Sonderweg"

There can be no decisive break with Germany's past, as the activities of Martin Walser demonstrate. Walser who, with Heinrich Böll and Günter Grass, is one of Germany's most distinguished postwar novelists, delivered a speech in 1998 in which he berated those who used Auschwitz as a "moral club" to continually remind Germany of its past, arguing that although he "would never leave the side of the victims," he preferred to grieve and look back in private. Many who could sympathize with this must have been distressed subsequently to read that his next novel, *Tod eines Kritikers* (*Death of a Critic*), was denounced as anti-Semitic.

Other episodes show that the Nazi past continually intrudes. The works of much younger modern novelists such as W. G. Sebald and Bernhard Schlink are about the way the war, or the memory of the war, still colors people's lives (see Chapter 42, p. 789). In 2008 Volker Weidermann, literary editor and head of features of the *Frankfurter Allgemeine Sonntagszeitung*, produced *Das Buch der verbrannten Bücher*, The Book of the Burned Books, a detailed examination of the authors whose books were burned by the Nazis at the celebrated auto-da-fé in Berlin on May 10, 1933. At almost exactly the same time, a plan to reintroduce the Iron Cross as a military award for bravery was withdrawn, the award being seen as too closely linked to the Nazis. In early 2008 also, plans to produce a definitive edition of *Mein Kampf* were discussed, as a way to prevent far-right groups from using the book for their own ends. Germany, as *Focus* magazine observed, is permanently on a tightrope walk between "the right to innocence and the duty of remembrance."[61]

This is true and, conceivably, it will remain true for the foreseeable future. Nonetheless, and although it won't please everyone, it is an argument of this book that it is high time we looked back, beyond Hitler and the Holocaust (or Shoah). There is more, much more, to modern Germany than the Third Reich, and there are important lessons to be learned from that history. From the splendors of Bach to the theology of the present pope, we are surrounded by German-born ideas.

The above argument should be tempered by the observation that, so far as Britain is concerned, there are other reasons why Germany and its achievements have been underplayed and/or underrecognized. As Nicholas Boyle has pointed out, English-speaking readers are not helped in their assessment of German literature because of a lack of contemporaneous literature of their own with which they could make comparisons: "The period of Germany's greatest cultural flowering—from about 1780 to about 1806—coincides with a relatively fallow time in their own literature and, understandably, that of France."[62] A further factor is that the turbulence of the 1790s—the aftermath of the French Revolution and the Napoleonic Wars—diverted attention from the achievements of many prominent Germans. The fact that in Germany the ancien régime passed away, its place taken by a society "as peculiarly German as it was clearly post-Revolution," a middle-class variety of Victorianism *without* industrial capitalism (until the middle of the nineteenth century at any rate), created a gulf in understanding that, it will be argued here, has never been entirely bridged, and that the excesses of the Nazis traded on and exacerbated.

Even without Hitler, even without the Holocaust, traditional German history has by and large told a one-sided story. History as it is now practiced was initially a German idea (see Chapter 12, p. 261), and all the great German historians, from Leopold von Ranke (1795–1886) to Friedrich Meinecke (1862–1954), argued that the creation and maintenance of the German nation-state was the "big story" of the "long" nineteenth century (1789–1914). Given the political changes that took place in Germany during those years, it is—to an extent—understandable why so many historians should take this view. In a more fundamental sense, however, and this needs to be said loud and clear, it was only ever half the picture. While the political narrative was unfolding, another no less dramatic, no less important, and equally impressive story was also emerging. Thomas Nipperdey, in his magisterial history of Germany, concluded that music, the universities, and science were the three great achievements that brought recognition to that country in the nineteenth century. Between the publication of Johann Joachim Winckelmann's groundbreaking *Geschichte der Kunst des Altertums* (*History of the Art of Antiquity*), in 1754, and the award of the Nobel Prize for Physics to Erwin Schrödinger in 1933, Germany went from being the poor relation among Western countries, intellectually speaking, to the dominant force—more influential in the realm of ideas than France or Britain or Italy or the Netherlands, more so even than the United States. This remarkable transformation is the subject of *The German Genius.*

Here too, however, a word of caution is necessary, because the situation is more complicated than it at first appears, certainly to non-Germans. This book is a cultural history—it examines Germany's achievements in what ordinary British, French, Italian, Dutch, or American readers understand as "culture." It is important to say at the outset that, among Germans, the concept of "culture" has traditionally been very different from what other nationalities mean by that word. In fact, there are those who argue that this very difference in the historical understanding of "culture" actually comprises Germany's real "Sonderweg." It makes sense, therefore, to consider this difference before proceeding.

This difference has been most recently and thoroughly explored by Wolf Lepenies, professor of sociology at the Freie Universität of Berlin but someone who has also spent several years at the Institute for Advanced

Study in Princeton and therefore has had, so to speak, a foot in both camps. In his book, *The Seduction of Culture in German History* (2006), Lepenies begins by quoting Norbert Elias who, in *The Germans*, published in English in 1996, wrote this: "[E]mbedded in the meaning of the German term 'culture' was a non-political and perhaps even anti-political bias symptomatic of the recurrent feeling among the German middle-class elites that politics and the affairs of the state represented the area of their humiliation and lack of freedom, while culture represented the sphere of their freedom and their pride. During the eighteenth and part of the nineteenth centuries, the anti-political bias of the middle-class concept of 'culture' was directed against the politics of autocratic princes . . . At a later stage, this anti-political bias was turned against the parliamentary politics of a democratic state."*[63] And this showed itself in a German obsession for distinguishing between "civilization" and "culture." "In German usage, *Zivilisation* means something which is indeed useful, but nevertheless only a value of the *second* rank [italics added], comprising only the outer appearance of human beings, the surface of human existence. The word through which Germans interpret themselves, which more than any other expresses their pride in their own achievement and their own being, is *Kultur*." Lepenies adds: "Whereas the French as well as the English concept of culture can also refer to politics and to economics, to technology and to sports, to moral and to social facts, the German concept of *Kultur* refers essentially to intellectual, artistic and religious facts, and has a tendency to draw a sharp dividing line between facts of this sort, on the one side, and political, economic and social facts, on the other."[64]

In the nineteenth century in particular, the sciences, by their very nature, formed a natural alliance with engineering, commerce, and industry. At the same time, and despite their enormous successes, the sciences were looked down upon by artists, philosophers, and theologians. Whereas in a country like England or America the sciences and the arts were, to a much greater extent, seen as two sides of the same coin, jointly forming the intellectual elite, this was much less true in nineteenth-century Germany.

This division, between *Kultur* and *Zivilisation*, was underlined by a

* The book was published in German in 1989, a year before Elias's death.

second opposition, that between *Geist* and *Macht*, the realm of intellectual or spiritual endeavor and the realm of power and political control.

In other words, Germany has traditionally been afflicted by what C. P. Snow, speaking about Britain in the 1950s, characterized as a "two-cultures" mentality, only much more so. The two cultures Snow identified were those of "the literary intellectuals" and of the natural scientists, between whom he claimed to find "a profound mutual suspicion and incomprehension." Literary intellectuals, said Snow, controlled the reins of power both in government and in the higher social circles, which meant that only people with, say, a knowledge of the classics, history, and English literature were felt to be educated. The division was not quite the same in Germany—where sociologists and politicians were lumped in with scientists as aspects of *Zivilisation* and opposed to *Kultur*—but it was from the same family and even more profound.

There is more to it even than that. The appeal of "culture" in Germany, Lepenies says, accompanied as it is by a "scorn" for everyday politics, has been based on a belief in the "deeply apolitical nature of the 'German soul,'" and this, he insists, nurtured Germany's claim, as a *Kulturnation*, to superiority over the merely "civilized" West from the late nineteenth century on. The resulting "strange indifference" to politics has been much more in evidence in Germany than anywhere else, he says, and involvement in culture at the expense of, and as a substitute for, politics "has remained a prevailing attitude throughout German history—from the glorious days in eighteenth- and nineteenth-century Weimar through, though now in considerably weaker form, the re-unification of the two Germanies after the fall of communism."[65] Germany's cultural achievements, the belief that it was traveling a special path, a Sonderweg, "was always a point of pride in the land of poets and thinkers. The inward realm established by German Idealism, the classic literature of Weimar, and the Classical and Romantic styles in music preceded the founding of the political nation by more than a hundred years. They gave a special dignity to the withdrawal of the individual from politics into the spheres of culture and private life. Culture was seen as a noble substitute for politics."[66] Many other observers have remarked on Germany's inwardness, that "strange indifference to politics," and some have gone further, arguing that it is this which accounts for the "nightmarish consequences" of one or other of the two world wars. The Germans took on board Thomas Hobbes but not John Locke. On this reckoning then, there *was* a special path in Germany history, but it was cultural, not political, as Wehler claimed. Karl Lamprecht remarked on this in his *German History*, published as early as 1891.[67]

Gordon Craig, the great American historian of Germany, noticed the same tendency.[68] "The alienation of the artist in Imperial Germany . . . was in large part self-willed. Towards the real world, the world of power and politics, the German artist, in contrast to the French, always had an ambivalent attitude. He was . . . repelled by a belief that to participate in politics or even to write about it was a derogation of his calling and that, for an artist, the inner rather than the external world was the real one . . . Not even the events of 1870–71 succeeded in shaking their indifference. The victory over France and the unification of the German states inspired no great work of literature or music or painting . . ." Speaking of the Naturalist writers and painters of the end of the nineteenth century, Craig adds that they "never turned their attention to the political dangers that were inherent in the imperial system. Indeed, as those dangers became more palpable . . . under William II . . . the great majority of the country's novelists and poets averted their eyes and retreated into that *Innerlichkeit* [inwardness] which was always their haven when the real world became too perplexing for them."[69]

On October 4, 1914, two months into the Great War, ninety-three German intellectuals published a manifesto, known as the Manifesto of the 93, addressed "An die Kulturwelt" (To the Civilized World) in which they defended the actions of the Reich against criticism from abroad. These individuals, among them Max Liebermann, the painter, and Wilhelm Wundt, the founder of experimental psychology, made it clear they viewed the war not as a campaign against German militarism but above all as an assault on German culture. "What was not understood abroad was that German militarism and German culture could not be separated from one another . . . The signatories of the manifesto vowed that they would fight the war as members of a cultural people (*Kulturvolk*) for whom the legacy of Goethe, Beethoven and Kant was as sacred as German soil . . . Germany's unity had been achieved not by politics but by culture." German thought, the ninety-three said, was an indispensable element of the European spirit, "precisely because it differed so much from values and ideals that were pertinent for countries like France or England. The Germans insisted on the unbridgeable difference between culture and civilisation."[70] (See p. 535 for Max Weber's view on why the Germans fought the Great War.)

Nearer our time, many Germans regarded the Weimar Republic—the attempt to establish a democratic regime in Germany for the first time—as a betrayal of German political ideals. In his "Gedanken im Krieg" (Thoughts on War; 1918), Thomas Mann wrote that the democratic spirit was "totally alien to the Germans, who were morally but not politically inclined. Interested in metaphysics, poetry and music but not in voting rights or the proper procedures of the parliamentary system, for them Kant's *Critique of Pure Reason* was a more radical act than the proclamation of the rights of man." Mann returned to the theme at the end of World War II, when he was in exile in the United States. He believed the triumph of politics in Germany—the rise of Bismarck, the role of the Kaiser, the Weimar Republic, the Nazi movement—had all (all, not just the Nazis) led to cultural impoverishment.[71] Later, Mann changed his tune and in a speech to Congress argued that "inwardness and the romantic counterrevolution had led to the disastrous separation of the speculative from the socio-political sphere that made Germans unfit for modern democracy."[72]

To a non-German this all sounds somewhat strange—dare one say it, unreal. The Western but non-German view of "culture" was aptly summed up by T. S. Eliot, in his *Notes Towards a Definition of Culture* (1948), where he famously said: "The term *culture* . . . includes all the characteristic activities and interests of a people; Derby Day, Henley Regatta, Cowes, the twelfth of August [the beginning of the shooting season], a cup final, the dog races, the pin table, the dart board, Wensleydale cheese, boiled cabbage cut into sections, beetroot in vinegar, 19th-century Gothic churches and the music of Elgar." None of this necessarily implies any particular "inwardness" on the part of participants, or any great education, come to that. It is a much less hierarchical, more ecumenical view of human affairs than the German concept of culture. What the elite of Germany meant by *Kultur* until at least the Second World War is what we, in the West, outside Germany, traditionally call "high culture": literature, theater, painting, music and opera, theology, and philosophy.[73]

But, and it is an important "but," this need not be taken as a criticism of Germany. It may well be that this different understanding of the way our intellectual activities should be organized is a crucial point, an instructive difference. At the very least, lessons are to be learned from difference. Consider, for example, these statements.

"The twentieth century should have been the German century." The words were written by the American academic Norman Cantor; he was speaking at the time about the devastating effect the Nazi regime had on Germany's leading historians, such as Percy Ernst Schramm and Ernst Kantorowicz. Next, there is this sentence, in Fritz Stern's *Einstein's German World*: "It could have been Germany's century." This time it was Raymond Aron speaking, the French philosopher talking to Stern when they were in Berlin in 1979 to visit an exhibition commemorating the centenary of the births of the physicists Albert Einstein, Otto Hahn, and Lise Meitner. What Cantor and Aron meant, in asserting that the twentieth century should/could have been Germany's, was that, left to themselves, Germany's thinkers, artists, writers, philosophers, scientists, and engineers, who were the best in the world, would have taken the freshly unified country to new and undreamed-of heights, were in fact *in the process of doing so* when 1933 came along. In January 1933, when Hitler became chancellor, Germany was—without question—the leading force in the world intellectually. It could not perhaps match the United States in sheer economic numbers—America was, even then, a far more populous entity. But in all other aspects of life, Germany led the way. Had a historian of any nationality published an intellectual history of modern Germany at the end of 1932, it would have been very largely a history of triumph. By 1933 Germans had won more Nobel Prizes than anyone else and more than the British and Americans put together. Germany's way of organizing herself intellectually was a great success.[74]

But the German genius was cut off in its prime. All the world knows *why* this happened. Much less well known is why and how the Germans achieved the pre-eminence they did. Yes, people know that Germany lost a lot of talent under the Nazis (according to one account, 60,000 writers, artists, musicians, and scientists were sent either into exile or to the death camps by 1939). But even many Germans appear to have forgotten that their country was such a dominant power intellectually until 1933.

The Holocaust and Hitler get in the way, as the work of A. Dirk Moses, referred to earlier, shows and as Keith Bullivant said explicitly: "For those born during and after the Second World War the cultural history of Germany before 1933 is that of a lost country, one that they never knew."

I don't think many people alive today grasp this fundamental point about German pre-eminence in the pre-1933 period. I exclude, of course, specialists. Among them, the situation is, if anything, reversed: the enormities of the Nazi atrocities mean that—in particular—post–World War II English-language scholarship about Germany is deep and detailed. As part of the research for this book, I visited the German Historical Institute in Washington, D.C. These institutions exist in London, Paris, Washington, and elsewhere. The Washington institute, besides its splendid library of German- and English-language books and periodicals, also has its own publishing program, which includes a massive work, *German Studies in North America: A Directory of Scholars*. This volume, 1,165 pages long, lists the projects of—roughly speaking—1,000 academics. Subjects range from German war novels to an atlas of Kansas German dialects to a study of precision in German society to a comparison of Berlin and Washington as capital cities between 1800 and 2000. There is no shortage of research interest in German topics, at least among scholars in America. But this only reinforces the central point: among the general public the ignorance of German affairs is widespread.

We are used to being told that the twentieth century was the American century, but the truth is more complex and, as this book aims to show, more interesting than that. This book's intent is to reinsert into both the non–German-speaking consciousness *and* the German-speaking consciousness the names and achievements of a people who, for historical reasons having to do with war and genocide, have been neglected—even shunned—over the past half-century.

This then is a book about the German genius, how it was born and flourished and shaped our lives more than we know, or care to acknowledge, how it was devastated by Hitler *but*—another "but" that is crucial—how it has lived on, often unrecognized, not just in the two postwar Germanies, which have never received full credit for their achievements—cultural, scientific, industrial, commercial, academic—but in how German thinking shaped modern America and Britain and *their* culture. The United States and Great Britain may speak English but, more than they know, they *think* German.

A brief note on what I mean by "German." I use it in the sense that Thomas Mann did when he spoke of "German spheres," a cultural world where he felt at home, to include Germany itself plus other German-speaking lands—Austria, parts of Switzerland, parts of Hungary, Czechoslovakia, and Poland. There was certainly, for a time, a Vienna-Budapest-Prague German-speaking and German-thinking sphere. At other times, parts of Denmark, the Netherlands, and the Baltic states came within the German sphere of influence too, when scientists or writers looked to Berlin, or Vienna or Munich or Göttingen as intellectual centers. Sigmund Freud, Edmund Husserl, and Gregor Mendel all came from Moravia, part of what is now the Czech Republic, but each spoke and thought and wrote in German and lived their lives as part of overwhelmingly German-speaking traditions. Evangelista Purkyně was also Czech and campaigned on behalf of the Czech language, yet in his science he wrote in German and contributed almost exclusively to German journals; the thrust of his intellectual work—the nature of the cell—was an intellectual area in which German scientists were preeminent. Karl Ernst von Baer was Estonian but wrote in German and held positions at the University of Göttingen; Timothy Lenoir, in his history of early nineteenth-century German biology, counts Baer as the central figure. Georg Cantor, the mathematician, was born in St. Petersburg of parents who had emigrated from Denmark, but he moved to Frankfurt when he was eleven, studied at the universities of Zurich, Berlin, and Göttingen and taught for most of his career at the University of Halle. Karl Mannheim, one of the founding fathers of classical sociology, was born in Budapest but was much influenced by Georg Simmel and wrote his most important books in Germany (and in German) at Heidelberg and Frankfurt. Hugo Wolf, who according to Harold Schonberg "carried the German art song to its highest point," was born in Windischgraz, Styria, later Slovenjgrade in Yugoslavia, now in Slovenia. I adopt the same principle as Georg Lukács, who said of the Swiss novelist Gottfried Keller, that he was just as much a German writer as Rousseau, who came from Geneva, was a French author.[75]

I do not of course mean to suggest for a minute that books could not be written titled *The French Genius* or *The British Genius* or *The American Genius*: they could. Small nations like New Zealand, Denmark, and Trinidad have their geniuses too (Ernest Rutherford, Niels Bohr, V. S. Naipaul). My point is that these contributions to the development of modern thought are well recognized. The French Enlightenment, the British Empiricist philosophers such as Thomas Hobbes, John Locke, David Hume, Adam Smith, and John Stuart Mill, and the American pragmatists, are important paths but well trodden. On the other hand, modern German cultural history is much less well known to a general readership. I hope this book goes some small way to rectifying that imbalance.

PART I

The Great Turn in German Life

I.

Germanness Emerging

One Sunday evening in the spring of 1747, as his court musicians were gathering for their regular concert, an aide handed Friedrich the Great, king of Prussia, a list of visitors who had arrived at the Potsdam town gate that day. When the king scanned the list, he suddenly cried out: "Gentlemen, old Bach is here!" Later accounts had it that there was "a kind of agitation" in the king's tone.[1]

Johann Sebastian Bach, the composer, then sixty-two, had journeyed from Leipzig, eighty miles away, to visit his son Carl, chief harpsichordist in Prussia's royal Kapelle. Ever since Carl had been in Potsdam, the Prussian king had let it be known that he would like to meet "old Bach." Carl, however, knew how different his father and the king were and had done nothing to bring them together. He was not wrong. The encounter, when it did occur, proved to be a collision between two very different worlds.[2]

Bach was an orthodox Lutheran who believed the biblical tradition that music was Hebrew. He was a widower and a family man who had twenty children between his two wives. "Frederick, a bisexual misanthrope in a childless, political marriage," says James Gaines in his description of the meeting, "was a lapsed Calvinist whose reputation for religious tolerance arose from the fact that he held all religions equally in contempt." Bach wrote and spoke German. At the king's celebrated court, everyone spoke French. Friedrich boasted he had "never read a German book."[3]

Their differences carried over into their tastes in music. Bach was the most brilliant exponent of church music, in particular the "learned counterpoint" of canon and fugue, an ancient craft that had evolved such sophistication that many musicians of the day thought of themselves as "custodians of a quasi-divine art." Friedrich considered such claims overblown. Counterpoint, to him, was old-fashioned. He dismissed music that, as he quipped, "smells of the church."[4]

Despite their differences, when the king saw "old Bach's" name on the arrivals list, he ordered that the composer be brought to the palace that very night, not even giving him a chance to change his clothes. When Bach arrived, weary after his trek, he was presented by the king with a long and complex musical motif and a request (except it wasn't really a request) that the composer make a three-part fugue of it. Despite the hour, despite his weariness, Bach rose to the task, "with almost unimaginable ingenuity," so much so that all the virtuosi in the king's orchestra were "seized with astonishment."[5] Still Friedrich wasn't done, perhaps even a little disappointed that old Bach had performed so well. He now asked the composer if he could rearrange the theme into a fugue for *six* voices. This was a hoop Bach wouldn't jump through, not there and then anyway. He insisted he would work out the arrangement on paper and send it to Friedrich later on.

In July, two months after the evening at Potsdam, the proud Bach completed the six-part fugue and dispatched it. There is no evidence that Friedrich ever had the piece played but had he done so, the king—a subtle, astute man—would have been more than a little affronted. For this composition contained what one historian described as a "devastating attack on everything that Frederick stood for."[6] In the first place, the music was deeply religious. Elsewhere it contained a subtle form of sarcasm, where the score was annotated with references to the king's rising fortunes—though in practice the music descended into melancholy.[7] Counterpoint and other forms of music that smelled of the church were interspersed throughout, all of which has allowed musicologists to conclude that, in the "Musical Offering," Bach was having the last word—defying, chiding, even satirizing the king, reminding him that "there is a law higher than any king's which is never changing and by which you and every one of us will be judged."[8]

This entire exchange—subtle, clever, but pointed—epitomized the clash between two very different worlds, a clash that, in 1747, was sharper than ever. Three years later Bach was dead. The last great achievement of his life, completed during his final months, was the Mass in B Minor, one of the great masterpieces of Western music ("titanic" in the words of the critic Harold Schonberg), and one that Bach himself would never hear. With the Mass in B Minor and Bach's death, a whole artistic, spiritual, cultural, and intellectual world was at an end. The baroque had essentially been the style of the Counter-Reformation church, and its aim, in the visual arts, as summed up by Cardinal Gabriele Paleotti in Rome, one of the great reformers of the Catholic Church, was "to set on fire the soul of her sons," to place "a sumptuous spectacle before the eyes of the faithful," and to make the church "the image of heaven on earth." Bach's aim in his music—although it was Protestant music—was much the same. Such an understanding, such an aesthetic, died with him.

But if the baroque fire was cooling, new beliefs, new tempers, new ways of thinking were taking its place. Some of these innovations were fundamental, reconfigurations in thought that were as profound and as revolutionary as anything expressed for a thousand or even two thousand years. Many of the new ideas transformed Europe in its entirety, and North America too. Several, however, were specific to Germany or applied there more than anywhere else.

Until the middle of the eighteenth century—1763—the German-speaking lands were, in the words of the Harvard historian Steven Ozment, "Europe's stomping ground." Their location, at the geographical heart of Europe, had made them a crossroads of international trade since the Middle Ages, a circumstance not entirely without its beneficial effects. In the early sixteenth century, for example, the imperial free cities of Germany—Augsburg, Ulm, Cologne, Hamburg, Bremen, Lübeck—boasted a civic culture second only to their Italian and Swiss counterparts. At that time, Nuremberg, as Tim Blanning has pointed out, was home to Albrecht Dürer, Veit Stoss, Adam Krafft, Peter Vischer, and Hans Sachs. In the seventeenth century, however, that same geographical centrality conspired to make the German lands, as Ozment's phrase implies, a battlefield for Europe's great powers—France, Russia, Sweden, the Hapsburgs of Austria-

Hungary, and Britain. The Thirty Years' War (1618–48), a bitter conflict between Catholics and Protestants, was fought largely on German territory, and was so vicious that atrocity stories became commonplace—see, for example, Philip Vincent's contemporaneous *The Lamentations of Germany*, which featured plates showing "Croats eating children," "Noses and Eares cut off to make Hatbandes," and other delicacies. At the end of that time, the Treaty of Westphalia—a peace of exhaustion as much as anything else—hammered out a new political reality, a loose confederation of states of very unequal size and importance: 7 (later 9) electors (a reference to the office, largely ceremonial, of electing the emperor and his heir apparent, the king of the Romans), 94 spiritual and temporal princes, 103 counts, 40 prelates, 51 free cities, all equally sovereign (or half-sovereign), and around 1,000 knights, all claiming authority but ruling collectively barely 200,000 subjects.[9] The main innovation among this morass was the fact that sovereign (and mainly Protestant) German states "spun away" one by one from their former historical hub, the Catholic Austrian/Hapsburg Empire. By using their new territorial rights, which gave them an independent foreign policy and armaments, Bavaria, Brandenburg-Prussia, Saxony, and Württemberg emerged from the Austrian shadow (though only Brandenburg-Prussia had a professional army worth the name).[10] In 1667, Samuel Pufendorf, a jurist and the very man who coined the phrase "the Thirty Years' War," described what had been the former empire as a political and constitutional "monstrosity."[11] Population had collapsed, Württemburg's falling, as an example, from 445,000 in 1622 to 97,000 in 1639. The German states were now so fragmented that in the 1690s barges navigating the Rhine to the Channel paid border tolls on average every six miles.[12] Trade patterns had shifted to the North Atlantic following the voyages of discovery, and the German economy withered.

This new world didn't endure. Over the next 200 years, the single most important political, cultural, and social development in central Europe was the rise of Brandenburg-Prussia, the cell from which what followed emanated. In 1700 Hapsburg Austria was 9 million strong and still the preeminent part of the "Heiliges Römisches Reich deutscher Nation" (Holy Roman Empire of the German Nation). Prussia at the time could boast a population of barely 3 million and, in terms of territory, ranked only eleventh in Europe. Yet, by the middle of the eighteenth century it boasted Europe's third-strongest army and was breathing down Austria's neck.[13] The root cause of this change owed something to the Peace of Westphalia because under it Brandenburg-Prussia had acquired the territories of East Pomerania, Magdeburg, Minden, and Silesia. Prussia's successes also owed something to a line of rulers who lived long and productive lives. But the most important development, the development that came to shape and characterize Prussia, in her own eyes as well as in the eyes of others, was a new variety of the Christian religion. Germanness, as we now understand it, emerged in the late seventeenth and early eighteenth centuries and cannot be understood without a firm grasp of Pietism.

Pietism and Prussianism

Friedrich the Great has, with good reason, attracted most attention from historians as a forceful personality who, by a combination of military prowess, personal charisma, and an intellectual/artistic cast of mind, helped catalyze the German renaissance that is the subject of the next few chapters. There is no doubt that he played an important role. Recent scholarship, however, has focused rather more on his father, Friedrich Wilhelm I. Without the achievements of this man, and the reforms he initiated, Friedrich the Great might not have had quite the glittering career that he did have.

When Friedrich came to power in 1740, aged twenty-eight, what gave the Prussian state the special character *it already had* was the unparalleled emphasis among state employees on the conscientious fulfillment of their official duties. "Whereas in other European capitals monarchs reigned over court establishments characterised by ostentatious luxury, the Prussian kings wore military uniforms and promoted an official ethic of parsimony and frugality." The Prussian bureaucracy was even then well known for its commitment to much higher standards of honesty and efficiency than its European counterparts.[14]

In the 1950s the German historian Carl Hinrichs advanced the thesis that the source of the Prussian state-service ideology can best be understood as the fruit of the Pietist movement, and he highlighted several significant connections between Pietism and the major policy initiatives introduced by Friedrich Wilhelm I. Hinrichs's central book on the subject, *Preussentum und Pietismus*, was released only after his death and comprised a series of essays rather than a fully evolved thesis. These shortcomings have recently been addressed by Richard Gawthrop, whose arguments I have adapted in what follows.[15]

Pietism first appeared around 1670, when its strong emphasis on discipline began to attract adherents who formed the view that Martin Luther's church had itself become infected with the very corruptions it had been founded to avoid. These early Pietists sought a return to Luther's "pristine simplicity" by "stressing the priesthood of all believers against the hierarchy, the inner light against doctrinal authority, the religion of the heart against the religion of the head . . . and practical acts of charity, not scholastic dispute."[16] It should be said that there were other reasons that made Pietism attractive, especially to the political authorities. Chief among them was the fact that, in emphasizing the "inner light," the Reformed churches, which had emerged after the 1648 Treaty of Westphalia, when the Papacy had lost a great deal of its worldly power, were much less of a political threat than the older, more orthodox, and more organized churches. The change to a more "internal" faith enabled the authorities to use the confessional as a way to impose a stricter moral discipline on the laity. The aims of Pietism and Friedrich Wilhelm coincided.

Pietism was deeply influenced by English Puritanism, notable for its "intrusive moralism," advocating that good works *in this life*, right here on earth, helped to determine what happened at the day of judgment. God, argued the Puritans, actually *wanted* people to perform good works here on earth—this was how He revealed himself. Friedrich Wilhelm was not formally a Pietist, but had grown up with a sensibility and work ethic not too dissimilar from that of the Pietists. Between 1713 and 1740, as a result of this complementarity of outlook, the king gave the Pietists unprecedented opportunities for realizing their ambitions, and these reinforced and helped legitimate the king's fundamental restructuring of the administrative, military, and economic life of his realm. Thomas Nipperdey thought this had another effect: Protestantism was essentially pessimistic about human nature; this made it conservative and set against modernity. Such an attitude was to have momentous consequences.[17]

The first person who had called attention to the new approach was

Philipp Jacob Spener, born in Alsace in 1635, in his *Pia Desideria*, published in 1685.[18] But it was August Hermann Francke (1663–1727) who conceived the form of Pietism that was to transform Prussia. In the 1690s Spener induced the powers that be in Berlin to appoint Pietists to two professorships on the theological faculty of the newly established University of Halle. In the eighteenth century, the University of Halle, together with the University of Göttingen, would transform ideas about learning and scholarship in Germany (and eventually across the world). Francke, born in Lübeck, had been barred from teaching in Leipzig and joined the Halle faculty as professor of Near Eastern (then called Oriental) languages and, from 1695, this gave him the position and opportunity to rethink the state's role in the light of Pietist aims.[19]

His own earlier crisis of faith, and a "born-again" conversion, convinced him that "the cultivation of the heart," prayer, Bible reading, heartfelt repentance, and daily introspection were the basic ingredients for a truly religious life, rather than intellectual sophistication and doctrinal wrangling. He insisted that piety was not to be sought in isolation: to fulfill the biblical injunction to love one's neighbor, "one should seek to improve society through practical acts of charity."[20] It was a short step to Francke's view that vocational labor must become the main sphere of activity through which Pietists could serve their fellow citizens.[21] Theologically, Francke's approach was quite daring: he justified such activism by arguing that the Creation "could be improved upon," moreover that this improvement must form the central plank in the individual's quest for salvation.[22]

Francke's vision evolved in what we might call the long wake of the Thirty Years' War. Doubt was, if not yet widespread, certainly growing.[23] This helps explain why the main feature of his educational regime was strict—very strict—supervision, designed to inculcate in the pupil a habit of asceticism and unquestioning obedience to God's authority. At the same time, Francke emphasized education in the more worldly, practical disciplines, so that pupils from Pietist schools could "produce something useful for their neighbours."[24] The institution primarily responsible for these reforms was to be the clergy, which Francke now reconceived as the "teaching estate." Networks of Halle graduates found their way to most north German cities.[25]

Intellectual Centralization and a New Collective Mentality

This was the state of educational/pedagogical affairs when Friedrich Wilhelm I became king in 1713. He had undergone his own conversion in 1708, producing in him a vision not unlike Francke's. On his accession, he lost no time in becoming the chief patron of theology graduates from Halle, as he aligned the forces of Halle Pietism with his own priorities.[26]

In order to ensure this the king needed to mobilize not just the churches and the schools but the entire state apparatus, every socializing institution in Prussia. In this way was conceived "State Pietism."[27] To help encourage it, in 1729 Friedrich Wilhelm decreed that all Lutheran pastors in his realm must have studied at the University of Halle for at least two years, a remarkable act of intellectual centralization. In 1725 Abraham Wolff and Georg Friedrich Rogall, important Halle Pietists, were made professors of theology at the University of Königsberg. The resulting influx of Pietists changed forever the character of the church in northeastern Germany.[28]

But it was in the military and in the bureaucracy that Pietist influence was most far-reaching. The military church was reorganized in 1718 and eventually more than 100 Pietist pastors were employed among the regiments.[29] Encountering ignorance on a massive scale, the pastors taught reading and writing to soldiers and their wives, at the same time introducing them to the Bible and through that to Pietist beliefs and values. The military church also educated the soldiers' children—hundreds of regimental schools were built in the 1720s. (To facilitate matters, Friedrich Wilhelm ordered chaplains not to confirm anyone who could not read.)[30] The very concept of honor (*Ehre*) was itself transformed. Honor was no longer only a reflection of distinction in purely military matters: it now became necessary for an officer to fulfill his duty to others more widely—as a quartermaster, say, as a drillmaster, even as an accountant. What mattered was how much an officer had helped his neighbors, albeit subordinates.

The same culture permeated the bureaucracy. Following the Thirty Years' War, the local princes, newly independent, required more money to maintain their courtly life on the French model, and this meant there was a demand for a relatively efficient bureaucracy to administer princely affairs efficiently.[31] The *Beamtenstand*, the "estate of bureaucrats," became established in the German lands, and in 1693 examinations were introduced for admission to the upper reaches of the judicial system. Then in 1727, the king created two professorships in cameralist studies, one at the University of Halle and the other at Frankfurt an der Oder. They were the first such professors in the history of German universities, and the lectures offered covered the technical and legal side of the Prussian state's economic, finance, and police systems. As Hans Rosenberg put it in his 1958 book *The Prussian Experience*, the three dominant elements were bureaucracy, aristocracy, and autocracy. At the same time, the king was a strong promoter of meritocracy, continually underlining the opportunities for lowly clerks to attain the highest level of tax commissar or departmental head.[32] In this milieu, the bureaucrat became an advocate of a militant ideology dedicated to raising the level of civilian society through education. By 1742, a royal commission reported that no fewer than 1,660 schools had been built or repaired. (This shouldn't be exaggerated, however. Schooling for everyone was not established until the mid-nineteenth century.)

No less important, over time the educational improvements brought about by Friedrich Wilhelm I and the Pietists created an entirely new collective mentality: in the words of Walter Dorn, the Prussians became "the most highly disciplined people of modern Europe."[33] Friedrich the Great had the good sense to keep this military-bureaucratic-educational-economic structure intact. By his death in 1786, State Pietism was the core of the culture. It would prove stable enough to survive the depredations of Napoleon—Stefan Zweig wrote approvingly of it a hundred years later.

The Rise of the University: "The Great Turn in German Life"

Together with the Beamtenstand, the Prussian universities combined to give Germany another distinction all its own: a special kind of intelligentsia that was to have long-term consequences. The eighteenth-century German universities differed from the British ones in a number of important ways. In the first place, early eighteenth-century Germany had far more universities—about fifty, as compared with, for example, just Oxford and Cambridge in England. Although many were small (Rostock, with some 500 students when it was founded in 1419, now had only seventy-four students, while Paderborn had forty-five), their number and local character meant that it was much easier in Germany for the gifted sons from poorer families to obtain higher education.[34]

At the turn of the eighteenth century, however, teaching methods were backward. The norm was the teaching of static truths, not new ideas; professors were not expected to produce new knowledge, and the arts and philosophical faculties in particular had deteriorated. In many of the Catholic universities, theology and philosophy were the only subjects offered. Moreover, they were under threat from the new *Ritterakademien* (*Ritter* means "knight"), intended for the well-born, which offered a more fashionable curriculum that stressed mathematics, modern languages, social graces, the martial arts, and a smattering of science—worldly breadth rather than scholastic depth. What scientific research there was tended to be carried out in the new royal academies of science (such academies, on the French model, were founded in Berlin in 1700, Göttingen in 1742, and Munich in 1759). The German universities were, moreover, at the disposal of the princes, theirs to command for secular (i.e., very practical) purposes; they were not self-governing communities of scholars devoted to the study of classics and mathematics as were Oxford and Cambridge.[35]

Paradoxically, however, although many people around 1700 regarded the universities in Germany as irrelevant and moribund, at the end of the seventeenth and during the first half of the eighteenth century, four new universities were opened that would transform the intellectual climate in Germany. These were Halle in Prussia (1694), Breslau in Silesia (1702), Göttingen in Hanover (1737), and Erlangen in the Frankish margravate of Bayreuth (1743). Heidelberg was also important but, founded in 1386, it was hardly new.

The University of Göttingen was to have more of an impact than any other except Halle. The leading figure in the establishment of Göttingen was Gerlach Adolf von Münchhausen.[36] Born in 1688, he studied abroad in Utrecht and subsequently took a grand tour in Italy; it was the necessity to leave Germany to acquire "polish" that struck him as unfortunate and produced in him the desire for university reform. When he became a member of the Hanover Privy Council in 1728, he began agitating for the foundation of a university there and was so successful that he was himself appointed *Kurator* of the new institution. He soon introduced several innovations that were to prove influential.[37]

In the first place, Münchhausen ensured that theology played a relatively quiet role. Göttingen became the first university to restrict the theological faculty's traditional right of censorship and, as Thomas Howard says in his study of German universities, "It is hard to overstate the historical importance of this measure." As a direct result, the confessional age ended for the universities. Götz von Selle was just one who characterized this measure as "the pivot for the great turn in German life, which moved its centre of gravity from religion to the state."[38] By this enlightened measure, Göttingen's freedom to think, write, and publish became unparalleled in Germany.

Crucially, Münchhausen changed the relative weight enjoyed by the theology and philosophy faculties. Traditionally, philosophy was a distinctly inferior discipline, for both professors and students it was an "antechamber" to the higher faculties. Münchhausen added to the weight and importance of the "philosophical" subjects—such as history, languages, and mathematics—by his insistence that these fields were more than remedial areas for poorly prepared students.[39] Eventually, the philosophical faculty at Göttingen offered, in addition to the traditional subjects of logic, metaphysics, and ethics, lectures on "empirical psychology," the law of nature, physics, politics, natural history, pure and applied mathematics (surveying, military, and civilian architecture), history, geography, art, and modern languages. On top of these "philosophical" subjects, Göttingen offered the best training in the courtly arts available at any European institution—dancing, fencing, drawing, riding, music, and conversation in foreign languages.* Observers noted the new desire among young nobles to acquire a university education and a preference for "study and scholarship," which could pave the way for "important posts." It was at Göttingen that history, philology, and antiquity ceased to be minor, subordinate fields of study and began to acquire respect as autonomous

* The stallmaster was held in such high regard that his place in academic processions came before that of the associate professors.[40]

disciplines. Alongside history, classical philology underwent a dramatic rise at Göttingen, and it and its sister discipline—*Altertumswissenschaft*, the study of antiquity—became the "German science" par excellence.[41] Johann Matthias Gesner (1691–1761) and his successor, Christian Gottlob Heyne (1729–1812), transformed the experience of the classics. They removed the emphasis on grammar, replacing it with the appreciation of the texts as examples of the creative energies of antiquity; in so doing, the purpose of the new scholarship was transformed into an evaluation of the classics for what they revealed about culture, civic life, religion. "Above all, Greek antiquity—hitherto neglected—became *the* central focus."[42] Other innovations at Göttingen included publication of the first professional journals.[43]

Göttingen also developed and refined the seminar. This was another innovation whose importance it is difficult to exaggerate. The seminar, as we shall see, led to the modern concept of research, to the modern PhD, to the academic and scientific "disciplines" or subjects, and to the modern organization of universities into "departments," divided equally between teaching and research. Originally introduced in Halle by Francke, the seminar differed in important ways from the lecture, reflecting a profound change in the concept of knowledge and learning. The crucial distinction was that between late medieval notions of knowledge, or *scientia*, and the post-Enlightenment idea of *Wissenschaft*. Scholastic-Aristotelian logic took it as read that there was/is a single, correct method of thinking, a method that, when properly employed—through syllogistic reasoning, disputation, correct definition of terms, and "the clear ordering of arguments"—could be applied to any scholarly subject.[44] Different areas of interest did not require different methods, for all could be approached and understood through right reason (*recta ratio*), apprehended through the study of logic. The main purpose of instruction in the lecture was to help the student acquire general reason.

In the seminar, however, there were fewer people, criticism was encouraged, knowledge was regarded as mutable, less fixed, and new knowledge was there to be discovered. The aim of the teachers in the seminar was not to reproduce "static knowledge" but to promote the "taste, judgement and intellect" of their charges.

Seminars evolved over time. They embodied a more intimate form of teaching, where the *exchange* of ideas and knowledge was more valued, where the students were expected to have more input. Gradually, the passive mastery of a canonically prescribed corpus of materials gave way to the active cultivation of participation, and the early seminars in Germany began to require the submission of written work beforehand as a basis for discussion and evaluation.[45] This fostered the concept of research, with a premium on originality, which—again as we shall see in more detail later—reached its apogee in the Romantic period, when original research was regarded as a form of art. In some Göttingen seminars, the practice evolved whereby the original paper had to be delivered a week in advance so that other students could prepare their responses.

In line with all this, it was at Göttingen, in the late eighteenth and early nineteenth centuries, that the term *Wissenschaft* first gained its modern meaning. In its Göttingen sense, *Wissenschaft* incorporates science, learning, knowledge, scholarship, and also implies a research-based element, an idea that knowledge is a dynamic process, discoverable *for oneself,* rather than something that is handed down.[46] The practice of written submissions for the seminars—organized along the lines of the new scientific disciplines then emerging—led to the distinction between dissertation and thesis, and to the degree of PhD. The dissertation was essentially a display of erudition (a student would be asked to locate and assemble all known fragments of this or that minor classical author), whereas a thesis was a piece of research testing or leading to a hypothesis. The PhD eventually became a recognized degree in the German civil service and from this time on, its ascendancy—which again we shall examine in more detail later—was assured.[47]

The development of the seminar and the transformation of the PhD went hand in hand with the evolution of classical and philological disciplines and biblical criticism, and so had an even bigger impact than all this implies. The neohumanism that these developments promoted helped to redefine the image of the educated man, changing it from the rather external one of the first university reform movement (at Halle, under Francke) to a more internal one, expressed in the concept *Bildung.*[48] There will be a great deal to say about Bildung in this book. Difficult to translate, in essence it refers to the inner development of the individual, a process of fulfillment through education and knowledge, in effect a secular search for perfection, representing progress and refinement both in knowledge and in moral terms, an amalgam of wisdom and self-realization.

Together, Halle and Göttingen helped to fashion a new kind of education that prepared the way for a new stratum in German society, which will require not a little attention. This stratum, too small to be a class, in Tim Blanning's words, nevertheless achieved a prominent position in Germany by means of its domination of the state bureaucracy, the church, the military, the professoriate, and the professions. The self-understanding of this new stratum, which more than any other group helped account for the revival of German culture, set it apart from the traditional, more commercial middle classes.[49] The progressive, rationalizing, meritocratic, and statist social vision that this new stratum brought to these institutions influenced the entire sweep of nineteenth-century history.[50] In the early part of the century, in the words of Thomas Howard, it even worked toward the establishment of a particular kind of state, "one often described as a culture state (*Kulturstaat*) or tutelary state (*Erziehungsstaat*), a state that numbered among its paternalistic duties the goal of inspiring and educating its people to become 'appropriate citizens' . . . who understood that their aspirations should coincide with the high and morally serious purposes of the emergent nation-state." After 1871, says Howard, "*Kulturprotestantismus*" or "*Bildungsprotestantismus*" functioned as the "civil-religious foundation" of the German empire.[51]

An important observation comes out of all this: the German intelligentsia differed sharply from its counterparts in other countries. In France, the intelligentsia became estranged from the royal regime, so much so that it eventually attacked the traditional authorities. In Russia the intelligentsia consisted almost entirely of nobles, and in Britain neither the term nor the concept existed until the twentieth century. In Germany, because a university education was needed for a government position, the intelligentsia was drawn from all social levels. Not irrelevant either was the fact that Germany at that time lacked a metropolitan capital to rival London or Paris. This left the German intelligentsia dispersed yet far more intimately involved in *practical* state administration than anywhere else. Whereas British and American sociologists have characterized "remoteness from the practical world of government and administration" as one of the identifying features of the intelligentsia, this is manifestly *not* true of Germany.[52]

The Reading Revolution, a New Public Space and Nationalism Emerging

As late as May 1775 Christian Schubart reported in his *Deutsche Chronik* (German Chronicle) an encounter with a Neapolitan lady who was, he said, "under the impression" that "Germany must be a large city." No less vividly, Joseph von Sonnenfels, one of the most distinguished figures of the Austrian Enlightenment, said this in a letter: "It is well-known how the French are accustomed to speaking and writing with unseemly contempt about German traditions, intellect, society, taste and everything else that blossoms under the German sun. Their adjectives '*tudesque*,' '*germanique*,' and '*allemande*' are for them synonyms for 'coarse,' 'ponderous' and 'uncultivated.' "[53]

It *had* been true, in the late seventeenth and the early eighteenth centuries, that most educated Germans regarded French literary and artistic culture as superior to their own, and that British political freedoms and parliamentary practices were likewise to be envied. But that was before the changes introduced by Pietism and the country's various rulers had taken hold and the universities had undergone their radical transformation. In that same period a number of economic, political, social, and intellectual changes had occurred in Europe that impacted disproportionately on German speakers and helped ensure that, before the eighteenth century was out, German culture had caught up with French and British achievements—and in some areas had outstripped them.[54]

The first was the reading revolution. This had partly to do with the gradual removal—or lightening—of censorship, harder to enforce in Germany because of its many different self-governing states, and is seen in both the anecdotal and statistical evidence. One account, written in the late eighteenth century, reads: "In no country is the love of reading more widespread than in Germany, and at no time was it more so than at present . . . The works of good and bad writers are now to be found in the apartments of princes and alongside the weaver's loom, and, so as not to appear uncultivated, the upper classes of the nation decorate their rooms with books rather than tapestries."[55]

Robert Darnton has shown that although book publishing suffered drastically after the Thirty Years' War, by 1764 the Leipzig catalog of new books had regained its prewar level of about 1,200 new titles a year; by 1770 (when Georg Wilhelm Friedrich Hegel and Friedrich Hölderlin were born) it had grown to 1,600 and by 1800 to 5,000. Reading was also encouraged by another phenomenon of the eighteenth century—the lending library, which put a limit on the time a reader had access to any particular title. By 1800 there were nine lending libraries in Leipzig, ten in Bremen, and eighteen in Frankfurt am Main. Jürgen Habermas tells us that, by the end of the eighteenth century, there were 270 reading societies in Germany and some described a new illness, *Lesesucht*, or "reading addiction."[56] Literacy rates in Prussia and Saxony in the early nineteenth century were unmatched anywhere except New England.[57]

An associated factor was the increased use of the vernacular language. "It was in the eighteenth century that the domination of the printed word by Latin was finally broken," the percentage of titles published in Latin in Germany falling from 71 percent in 1600 to 38 percent in 1700 to 4 percent in 1800.* [58] The same period also saw a marked shift in taste in Germany, with the proportion of theological titles dropping from 46 percent in 1625 to 6 percent in 1800, philosophy rising at the same time from 19 percent to 40 percent, and belles lettres up from 5 percent to 27 percent. Furthermore, the cultural decentralization of Germany, arising from its many political entities, made it—again in Tim Blanning's words—"the land of the periodical *par excellence*."[59] Whereas the number of periodicals published in France rose from 15 in 1745 to 82 in 1785, the equivalent figures in Germany were 260 and 1,225 (of course many periodicals in France had higher circulations than those in Germany, but German periodicals also had wood-cut illustrations ahead of almost everywhere else). "[I]n Austria, the chief of police had to concede in 1806 that newspapers had become a 'genuine necessity' for the educated classes, anticipating Hegel's celebrated remark that reading the daily newspaper

* In France, even pornography was published in Latin in the eighteenth century.

represented the morning prayers of modern man." The reading revolution brought with it a more critical approach to affairs, through the growth of "moral periodicals."[60]

Not only was Germany becoming emancipated from Latin, its own language was developing. In 1700 the reputation of German had never been lower. In 1679, at the height of the French Sun King's influence, Gottfried Leibniz (1646–1716) composed a pamphlet titled *Ermahnung an die Deutschen, ihren Verstand und ihre Sprache besser zu üben* (*Exhortation to the Germans to Exercise Their Reason and Their Language Better*). Contrary to his usual practice, in his scientific and philosophical writings, which were written in Latin or French, this pamphlet was in German.[61] The philosopher's exhortation was taken seriously in a series of new periodicals, in particular one published in Zurich in the 1720s by a group of friends of whom Johann Jakob Bodmer (1698–1783) and Johann Jakob Breitinger (1701–76) were the leading spirits.[62] This, known as the *Discourse der Mahlern*, was particularly concerned with the German language. Both Bodmer and Breitinger—after a few false starts—composed articles designed to make German a less ponderous language, more intimate, more pleasurable, and less like a sermon—a development that should, they claimed (and this was an important observation), have more appeal to women.

These innovations were built on by others. At Halle, Christian Thomasius became the first German professor to deliver his lectures in German rather than in Latin.[63] Christoph Gottsched (1700–1766), a native of Königsberg, who moved to Leipzig and became professor of poetry and then of philosophy, formed a German society devoted to linguistic integrity: "At all times the purity and correctness of the language shall be promoted . . . only High German shall be written, not Silesian or Meissen, Franconian or Lower Saxon, so that it can be understood right across Germany." German societies modeled on the Leipzig original were founded in several other cities.[64] Gottsched also did his best to encourage the development of the novel and the drama. In the same way, in 1751 Christian Gellert published a popular treatise on letter writing, with the intention of encouraging young people, "especially women," to cultivate a natural style of writing and of removing the "widespread misapprehension" that the German language was not supple and flexible enough "to treat of civilised matters and express tender emotions."[65] Shortly afterward, the first epistolary novels in German began to appear.

A final effect of the reading revolution was on self-consciousness. Print-as-commodity, says Benedict Anderson, generates the "wholly new" idea of simultaneity, as people throughout society realize—via their reading—that others are going through the same experience, having the same thoughts, at the same time. "We are . . . at the point where communities of the type 'horizontal-secular, transverse time' become possible." In this way public authority was consolidated, helped along by the depersonalized nature of state authority.[66] These developments were more important than they might seem at first because it was these (vernacular) print languages, says Anderson, that laid the basis for nationalistic consciousness. Anderson's conclusion is that print-capitalism operated on a variety of languages to create a new form of "imagined community," setting the stage for the modern nation, in which a "national literature" was an important ingredient.[67] In Johann Wolfgang von Goethe's *Götz von Berlichingen mit der eisernen Hand* (*Götz von Berlichingen with the Iron Hand*), a play about liberty, which describes the decline and fall of an Imperial Knight, the author himself said that the theme of the play was "Germanness emerging" (*Deutschheit emergiert*).* In the nineteenth century, says Thomas Nipperdey, all this would lead to Germany becoming "the land of schools."

However much they proliferated in the eighteenth century, novels, newspapers, periodicals, and letters had all existed in some form in the past. At the same time an entirely new cultural and intellectual medium emerged in the field of music: the public concert. By 1800 it had replaced all other forms of the art and become "the main medium for music *per se*."[69] Furthermore, because the concert took place outside the princely or ecclesiastical courts, composers were free to invent their own musical forms and compositions. "The result was the conquest of the musical world by the symphony, the symphony concert, and the concert hall. This appar-

---

* In the late nineteenth century, Kaiser Wilhelm II described himself as "the No. 1 German," but in saying this, as Anderson points out, "he implicitly conceded that he was *one among many of the same kind as himself*, that he could, in principle, be a *traitor* to his fellow-Germans, something inconceivable in, say, Frederick the Great's day."[68]

ently natural progression has led many historians to present the rise of the concert as the cultural equivalent of the French Revolution, in which the rising bourgeoisie tore down the barriers and fences which had reserved cultural goods for the feudal elite."[70] This change boosted the sale of instruments and sheet music and the opportunities for music teachers, stimulating a virtuous cycle of which Germany as a whole would be the main beneficiary.

The practice grew up in the first quarter of the eighteenth century for musicians to frequent the music rooms of inns, and it was these gatherings that eventually evolved into more formal concerts. Blanning says this took place in particular in Frankfurt am Main, Hamburg, Lübeck, and Leipzig and that what these four cities had in common was their commercial character, a factor which links the rise of concerts with the rise of the bourgeoisie. Concert halls proliferated as never before in the 1780s.

All this refers to musical *consumption*. Musical *innovation*, on the other hand, innovation in instrumental music, and in particular the symphonic form, occurred in Mannheim, Eisenstadt, Salzburg, Berlin, and Vienna, *residential* cities centered on the courts, where such public as existed consisted mainly of state employees, mostly nobles rather than "bourgeois." The high educational level of the Viennese nobles, musically speaking, made them particularly receptive to innovation. This, argues Blanning, helps account for the speed with which Franz Joseph Haydn, Wolfgang Amadeus Mozart, and Ludwig van Beethoven evolved the different forms of their art. In 1784, in the thirty-seven days between February 26 and April 3, Mozart played twenty-two benefits in Vienna.[71] The sheer *quantity* of music helped the evolution of new forms, as innovation was demanded.

The symphony—purely instrumental music—was seen as a particularly German art form at the turn of the nineteenth century. Immanuel Kant had dismissed instrumental music—music without voices, without words—as simply pleasure, a form of "wallpaper," not culture. But, as we shall see in Chapter 6, the rise of the symphony brought about a new way of *listening* to music, as people began to think of instrumental music as having great philosophical depth. A final factor was the performance of sacred texts in the vernacular, accompanied by music, a practice that had spread in the first place to Catholic Austria from Italy. In Protestant Germany it was adapted to the Lutheran tradition of the *historia*, in which biblical stories were set to music (Georg Friedrich Handel's oratorios in particular). The importance of the genre was that it made public music making *respectable*. "The oratorio was edifying, lending itself admirably to the raising of money for charity, so it overcame the old association of listening to music in public with ale-house 'musique rooms' or dance halls . . . Here is seen the beginning of a phenomenon that is very much still with us: the sacralization of art."[72]

And so, just as the German language was developing and reading and educational standards were rising, music also helped to change the image of the Germans as backward in cultural matters. The proliferation of distinguished composers could not be ignored, as the names of Johann Pachelbel (1653–1706), Georg Philipp Telemann (1681–1767), Johann Sebastian Bach (1685–1750), and Georg Friedrich Handel (1685–1759) confirm. Tim Blanning quotes a periodical published in 1741 in Brunswick, titled *Der musikalische Patriot* (The Musical Patriot), which boasted: "Must not the Italians, who previously were the tutors of the Germans, now envy Germany its estimable composers, and secretly seek to learn from them? Indeed, must not the high and mighty Parisians, who used to deride German talent as something provincial, now take lessons from Telemann in Hamburg?"[73]

The German Moses

A quite separate factor specific to Germany, which helped finalize its transformation into a great political and cultural power in Europe, was another king, Friedrich Wilhelm I's son, Friedrich the Great. The generally accepted view of Friedrich, now, is that he was a divided soul, devoted on the one hand to monarchical autocracy, yet at the same time a lifelong admirer of John Locke, who at least in theory favored liberalism for its cultural and political freedoms. In reality, this division within Friedrich did no more than reflect the evolving politics of the eighteenth century. His was a conservative administration in comparison with the political systems then in existence elsewhere in Europe and in North America, reflecting above all the German idea that freedom and equality could best be achieved under conditions of order, such order being maintained by an established authority, in the person of the monarch.

Although he was a conservative by European standards, Friedrich did bring about great change. After his accession in 1740, his many battlefield successes (achieved because of the strength and excellence of the army he inherited from his father, built up still further by him), combined with other civic reforms, completed Prussia's great transformation into Germany's foremost state and one of Europe's great powers.[74]

Friedrich's mother, Sophie Dorothea, was a Hanoverian princess whose brother was King George II of England. Her husband's Pietist, masculine world, effective though it was, was not by any means to her taste and she was anxious lest it smother her children. Friedrich's education was placed first in the hands of Huguenot soldiers, who introduced him to mathematics, economics, Prussian law, and modern history but also to fortification, tactics, and the other arts of war. His mother nevertheless insisted he be given his own library of several thousand books. As a result, even in his teens Friedrich became familiar with the leading French, English, and German writers (in more or less that order).[75]

As soon as he became king, he set up an Academy of Arts and Sciences in Berlin. Directed by a distinguished French mathematician, Pierre de Maupertuis, one aim of the academy was to attract to Berlin the best minds, who would form a learned circle around Friedrich. Day-to-day government was run from the Charlottenburg Palace on the outskirts of Berlin, while Friedrich's circle of intellectuals met at Sans Souci, a specially built retreat in an area of lakes southwest of Berlin in Potsdam. Here the king entertained and argued with great minds such as Voltaire and Jean le Rond d'Alembert, one of the editors of the multivolume French *Encyclopédie*.[76] Five copies were bought of every book that Friedrich wished to read, since he possessed identical libraries at Potsdam, Sans Souci, Charlottenburg, Berlin, and Breslau.[77]

Amazingly, to our modern way of thinking, the Prussian king and his courtiers always spoke French to one another (Voltaire wrote home that he never heard German spoken at court).[78] Friedrich shunned his native tongue as "barbaric," feeling its literary time had not yet arrived. In 1780 he went so far as to publicly criticize the German language in a pamphlet and admitted that the German books he wanted to read must be translated into French first. This was a man whose own writings included poetry, political and military tracts, philosophical treatises, and hundreds—if not thousands—of letters exchanged with leading intellectuals (645 letters with Voltaire alone, spanning forty-two years and filling three volumes).[79]

Friedrich was, nevertheless, unable to appreciate great swaths of contemporary culture. He was, for example, ignorant of Mozart and dismissed Haydn's music as "a shindig that flays the ears." He complained to Voltaire in 1775, the year after Goethe's *Die Leiden des jungen Werthers* (*The Sufferings of Young Werther*) was published, that German literature was nothing more than a "farrago of inflated phrases." He despised new forms—the *drama bourgeois*, for instance—and he equally loathed ancient German epics such as the *Nibelungenlied*.[80] In his notorious 1780 essay, *Concerning German Literature; the faults of which it can be accused; the causes of the same and the means of rectifying them*, Friedrich argued that in material terms Germany was flourishing, having recovered finally from the ravages of the Thirty Years' War, but that its culture was still suffering. What was needed now, he argued, was geniuses, but until they revealed themselves Germans must continue to rely on translations from classical and French authors. Friedrich thought that Germany's cultural level was about two-and-a-half centuries behind that of France. "I am like Moses, I see the promised land from afar, but shall not enter it myself."[81]

Despite Friedrich's pessimism, many German artists and intellectuals were convinced that it was the king's forging of Great-Power status for Prussia that had given German culture the decisive kick-start. Goethe even thought that the widespread infiltration of French culture into Prussia, through Friedrich's tastes, was "highly beneficial" for Germans, spurring them on by provoking a reaction. Many others agreed.

Then there was the fact that Friedrich, like no other king before him and few since, entered the public sphere. As Goethe was sharp enough to notice, by simply publishing a pamphlet about German literature, Friedrich gave intellectual debate a momentum that no other living person could have matched.[82] Moreover, he encouraged others to enter the public sphere in a critical spirit by having the Academy organize annual prize-essay competitions, setting such ambitious questions as: "What has been the influence of governments on culture in nations where it has flourished?" (won by Johann Gottfried von Herder), "Can it be expedient to deceive the people?" and "What has made French the universal language of Europe and does it deserve this supremacy?"

These paradoxical achievements in the literary/intellectual world were matched in the military/political sphere. Through Friedrich's many battle successes, Prussia became a major European power, a status it maintained (other than 1806–13 if we are being finicky) until World War I. His victories were followed by initiatives in other realms of government: an agency dedicated to strategic economic development, greater freedom of the press, a reduction in the number of capital crimes, and advanced codification of Prussian law. He insisted that education become compulsory for all and urged (some) religious toleration. "So far did a new middle class and civil society advance by the end of Friedrich's reign that German intellectuals could look on the revolutions in America and France as belated efforts to catch up with Prussia."[83]

Friedrich's forty-six-year rule undoubtedly helped Prussia's rise to power and, culturally and intellectually speaking, between Bach's death in 1750 and Friedrich's own in 1786, Germany without question witnessed the stirrings of its own renaissance, a rival even of the Italian Renaissance of the fourteenth to sixteenth centuries.

2 .

Bildung and the Inborn Drive toward Perfection

While these specific developments were taking hold in Germany—changes to its religion, to its language, in its universities, in its public space, in its image of itself, and in its standing as a political power—Europe itself (and North America, too) was undergoing a set of no less profound changes, perhaps the most important change in thinking since the advent of Christianity. This was the advent of religious doubt.[1]

The period between 1687, when Isaac Newton's discoveries in *Principia Mathematica* confirmed and systematized the earlier observations of Nicolaus Copernicus, Johannes Kepler, and Galileo Galilei, and 1859, when Charles Darwin published *On the Origin of Species,* comprises a unique time span in the history of Western thought, though it is not always seen as such. It was a time when a purely religious purpose to life (salvation in a future state) was called into question *while there was as yet no other model to replace it,* when Darwin's biological understanding of man had yet to appear. The fact that so much of Germany's golden age came between these two dates—1687 and 1859—was to have profound consequences, consequences that affected Germany more than anywhere else. Intellectually speaking, the country was shaped during this crucial—unique—

transitional period. In particular, and most important, this transitional period saw the development of historicism and the rise of biology.

Even by the end of the seventeenth century, fifty years before our starting point, there was no shortage of people in Europe who felt that the Christian religion had been gravely discredited. Protestants and Catholics had been killing each other in the hundreds of thousands, or millions, for holding opinions that no one could prove one way or the other. The observations of Kepler and Galileo transformed man's view of the heavens, and the flood of discoveries from the New World promoted an interest in the diversity of customs and beliefs found on the other side of the Atlantic. It was obvious to many that God favored diversity over uniformity and that Christianity and Christian concepts—like the soul and a concentration on the afterlife—were not necessarily crucial elements since so many lived without them. It was in the sixteenth and seventeenth centuries, as the invention of printing matured, that vernacular translations of the Bible brought the book before a lay audience who now discovered that many traditions were actually nowhere to be found in the scriptures. The Bible also came under more systematic criticism when it was shown that the original Old Testament had been written not in Hebrew but in Aramaic, meaning the scriptures could not have been dictated to Moses by God: the Old Testament was not "inspired."

As more and more people began to lose faith in the Bible, the calculations of the earth's age based on the scriptures lost support also. The new science of geology suggested that the earth must be a great deal older than the 6,000 or so years it said in the Old Testament, and Robert Hooke, at the Royal Society in London, observed that fossils, now recognized for what they were, showed animals that no longer existed. This too suggested that the earth was much older than the Bible said: these species had come and gone before the scriptures were written. This had implications for the significance of the Creation.

The effect of all this was to produce a world where the very nature of doubt (or the reasons for it) was itself always changing. In fact, the growth of doubt went through four distinct stages: rationalistic supernaturalism, deism, skepticism, and, finally, full-blown atheism.

Deistic thought was the most important stage. It came into existence first in England, from where it spread to both the Continent and America. The actual word "deist" was coined by the Genevois Pierre Viret (1511–71) to describe someone who believed in God but not in Jesus Christ. The anthropological discoveries in America, Africa, and elsewhere only underlined that all men had a religious sense but that on the other continents there was no awareness of Jesus. The deists were also influenced by new discoveries in the physical sciences, which suggested that God was not an arbitrary figure, as in ancient Judaism for example, but the maker of the laws that Copernicus, Galileo, Newton, and the others had uncovered. The deists in fact achieved a major transformation in the concept of God, arguably the greatest change in understanding since the development of ethical monotheism in the sixth century B.C. God lost his "divine arbitrariness" and was now regarded as a lawmaking and law-abiding deity.

The atheists were predominantly French and were known as mechanists (as the intellectual heirs of Newton they were inspired by the idea of a mechanical universe). Voltaire was just one who thought that science had shown that the universe was governed by "natural laws," which applied to all men and that countries—kingdoms, states—should be governed in the same way. Voltaire convinced himself that, through work, religious ideas would eventually be replaced by scientific ones. Man, he insisted, need no longer lead his life on the basis of atoning for his original sin; instead he should work to improve his existence here on earth, reforming the institutions of government, church, and education. "Work and projects were to take the place of ascetic resignation."

These new attitudes, grounded in the recent advances of science, together with the fact that more and more people could *read* of these discoveries, meant that the optimistic idea of progress was suddenly on everyone's mind, and this too was both a cause and symptom of changes in religious belief. Until the likes of Michel de Montaigne and Voltaire, the Christian life had been a sort of intellectual limbo: people on earth tried to be good in the manner laid down by the church but, in effect, they accepted the notion of perfection at Creation, followed by the Fall and decline ever since. The faithful expected fulfillment only in "a future state."

Pietism was of course a response to this, a *religious* response. It stressed the (moral) rewards available in *this* life. A quite different response, however, which matured as the century wore on, was the idea that if the rest of the universe was governed by (relatively) simple laws—accessible to figures like René Descartes, Isaac Newton, Gottfried Leibniz, Antoine Lavoisier, and Carl Linnaeus—then surely human nature itself should be governed by equally simple and accessible general laws.

With this went a further profound change—the reconceptualization of the soul as the mind, the mind increasingly understood by reference to consciousness, language, and its relationship with this world, in contrast to the soul, with its immortality and preeminent role in the next world. This was, in other words, the replacement of theology by biology (a word not introduced until 1802). As we shall see—and if an ugly neologism be allowed—the "biologification" of the world took place preeminently in Germany.

The individual mainly responsible for this approach, at first, was the Englishman John Locke, in his *Essay Concerning Human Understanding*, published in 1690. In this book, prepared in draft in 1671, Locke himself used the word "mind," not "soul," and referred to experience and observation, rather than some "innate" or religious (revelatory) origin, as the source of ideas. Locke further argued that motivation was based on experience—nature—which helped form the mind, rather than derived from some transcendent force operating on the soul. One unsettling effect of this was to further remove God from morality. Morality has to be taught, said Locke; it is not innate. Arguably most important of all, he said that the sense of self, the "I," was not some mystical entity relating to the soul, but an "assemblage of sensations and passions that constitutes experience." This was a key ingredient in the birth of psychology, even if that term was not much used yet.

Alongside the rise of psychology, in Locke's hands, and the (gradual) replacement of the concept of the soul with that of the mind, went a closer study of the brain. Thomas Willis had carried out numerous dissections of brains, helping to show that the ventricles (the central spaces where the cortex was folded in on itself) had no blood supply and was therefore unlikely to be the location of the soul, as some believed. Madness was increasingly being explained as a *Gemütskrankheit*, "failure of the mind," understood as housed in a bodily organ, the brain. Yet more biologification.

These changing beliefs were embodied, perhaps inevitably, in a work that took them to extremes. *L'homme machine* (*Man a Machine*) by the French surgeon Julien Offroy de La Mettrie, published in 1747, argued that thought is a property of matter "on a par with electricity," coming down on the side of determinism, materialism, and atheism, all of which were to land the author in hot water. His nonetheless influential view was that human nature and animal nature were part of the same continuum, that human nature equated with physical nature; and he insisted that there were no "immaterial substances," thus casting further doubt on the existence of the soul. Matter, he said, was animated by natural forces and had its own organizational powers. This, he said, left no room for God.

La Mettrie's book was as controversial as it was extreme, and it provoked a mighty backlash. That backlash was led from Germany.

The Rise of Historicism

In Germany there were two important areas of particular interest that would have a long-term effect on the country's intellectual life. These areas were history and biology, though aesthetics and the concept of genius also formed part of the picture.

Just as Richard Gawthrop has recently recovered a number of Pietistic writers and writings from obscurity and given them a new prominence, Peter Hanns Reill has done the same with seventeenth- and early eighteenth-century German historians. In doing this, he makes clear that the *Aufklärung*—to distinguish it from the French, English, and Scottish Enlightenments—had a number of achievements to its credit by the time of Bach's death, and certainly by the time of Friedrich the Great's.

The Aufklärung, he points out, came later than the "Western" Enlightenment (i.e., in France, England, and Scotland), "and so it could and did borrow from its neighbours." Though they borrowed from Voltaire and Hume, the *Aufklärer* (as Reill calls them), did so selectively, to address problems of specific concern in German intellectual life. Mostly, these stemmed from the impact of Leibnizian philosophy.[2] According to Leibniz, both the physical and spiritual realms were characterized by *change*. This is an unexceptional thing to say in the twenty-first century, but it was very different then: the Christian worldview implied not exactly a static state of affairs, as the Greeks had viewed their environment, but a world in limbo—Christians, even Pietists, were waiting for perfection in the *next* world. Moreover, and this is a point we shall return to time and again, the change envisioned by Leibniz was teleological: it was understood as development *toward a specific goal*, a goal that was vaguely inherent in the nature of the entity undergoing change.

Here then is the crucial point: change was accepted in late seventeenth- and early eighteenth-century Europe, and Germany in particular, but it was expected to have a *direction*, though no one knew what that direction was or what it entailed. Moreover, the discovery of that direction now lay in activities *outside* the church.

Once the principle of change had been accepted, the concept of history also changed (and so too did the understanding of politics, considered later). Until about the middle of the eighteenth century, the normal stance for German historians was similar to that of Sigmund Baumgarten, who argued that history's main purpose was to confirm man's impotence in the face of God's will—in other words to show that history underlined the veracity of Christianity.[3] In 1726, the Halle-trained historian and jurist Johann David Köhler announced that "the best chronologists date the beginning of the world on the 26th of October in the year 1657 before the Flood and 3,947 years before Christ's birth. To be sure," he continued, "the ancient Egyptians and Chaldeans, as well as the modern Chinese, make the world many thousands of years older, but Holy Scripture is more believable than all other books of heathen fables founded on the ancients' search for fame."[4]

By 1760, however, a definite shift was discernible. Instead of using history to confirm specific Christian episodes, these thinkers (known in Germany as Neologists) attempted to steer a path between orthodox, deist, and Pietist beliefs. The Neologists did not deny the importance of dogma, but nor did they accept that it was universally valid. For example, they felt able to surrender Christian chronology without rejecting the rest of Christianity, and this was an important milestone in the development of doubt.[5] The German Neologists argued that the Bible should be understood as a *collection* of books written at different times and in response to different circumstances. They took what we might regard as an anthropological approach: they accepted that God's commands were transmitted in these books, but they also conceded that the transmission was carried out by human agents who were responding to specific circumstances. The importance of the books lay in the fact that they always expressed a moral law, but the message was dressed in what Johann Salomo Semler called a "local" or "provincial" dialect. On this reckoning, it would have been inappropriate "for God to have his message transmitted in Newtonian language at a time when that language would have been totally incomprehensible." Similarly, it would be equally anachronistic for someone living in the eighteenth century to accept that the world was created in six days just because "this was the way a primitive nomadic people grasped and expressed God's majesty."[6]

Johann David Michaelis expanded this view. He argued that the way in which the ancient Israelites had transmitted their sacred knowledge was very different from that of eighteenth-century Europeans. Chronology, he insisted, was relatively unimportant to the Israelites of the Mosaic era. Instead, Moses provided his people with a selective genealogy, recording "only those events that had meaning in the memory of his people and revealed God's message." The rest was unimportant.[7] Moreover, given that the Bible was a collection of books compiled by single individuals, who lived at different times and places, it was only natural that contradictions would occur. With this bold move, the Neologists overturned the assault on Holy Scripture by asserting that the contradictions in the text actually *confirmed* its validity.

This new view enabled imaginative scholars to suggest a fresh understanding of chronology. Johann Christoph Gatterer, for example, related the age of the people in the Bible to the Fall of man. Man's life span during the biblical chronology, he observed, was divided into six levels, in the course of which life span declined from an average of 900–969 years (until the Flood), to 600 years, 450 years, 239 years (building the Tower of Babel), 120 years (the Mosaic era), to 70–80 years (since David's time). Gatterer explained the change in life spans against the background of a hypothetical natural history. According to this, the earth—created perfect by God—took some time after Adam's original sin to arrive at its present stage of imperfection. "The immediate post-Adamite air had been cleaner and healthier, the earth richer and more fertile, the fruits and vegetables bigger, better and more nourishing." That is why people lived longer and why, he said, the earth had more people in it before the Flood than at any time since.[8]

In this way, then, as Peter Hanns Reill has noted, the chronology of the Christian creation myth became "hermetically sealed off" from the remainder of historical analysis.[9] This allowed people to keep their faith, but it also allowed the development of historical understanding outside the biblical chronology. Although scholars might have disagreements about the actual course of history since biblical times, it *was* now accepted that, in the interim, there had been development, evolution (though not yet in a Darwinian sense), and that development was accessible to the diligent historian. "The rise of historicism is one of the great intellectual revolutions of the modern age."[10]

Another intellectual revolution of the seventeenth century, with important implications for historians, was the triumph of Natural Law. This came about partly through the astronomical/physical/mathematical discoveries of Copernicus, Kepler, Galileo, and Newton, and partly through the biological and anthropological discoveries in the New World, Africa, and elsewhere. Given the laws discovered by Newton and others, and the fact that "primitive" tribes in the newly discovered parts of the world lacked Christianity but still had religion and lived in civil societies, these patterns fostered the idea that there must be in human affairs fundamental regularities—laws—that existed much as gravity existed, only needing a Newton to uncover them. Natural Law was understood in this way "as the force that arranged things."[11]

The meaning of the words "nature" and "natural" were not always obvious. For classical thinkers, unaware for the most part of "primitive peoples," the natural state was life in a healthy civil society. Christians, on the other hand, always made a distinction between the state of nature (itself divided into the "pure state" of nature and the "state of fallen nature") and the state of grace. Men such as Thomas Hobbes and Hugo Grotius, however, attempted to redefine the idea of nature so as to arrive at a new explanation for the origin of things. Thomas Hobbes in particular concerned himself with the state of nature that existed *before* civil society. This too implies change, development, evolution.[12]

Reill identifies three German scholars in particular who all taught at Göttingen and all built on Hobbes and embraced Natural Law to explain the evolution of society. Johann David Michaelis (1717–91) conceived of primitive society as a collection of states so small they "resembled families."[13] Ruled by elected judges, these small states were so uncorrupted that authority could be exercised in a simple "parental" manner. The experience of such peoples was expressed in their sacred poetry. Gottfried Achenwall (1719–72) said that ever-larger states were created "by contract" between smaller ones; the aim was to ensure maximum and mutual happiness. The agreements that made up the contract formed the basic constitution which shaped the character of the state. Achenwall's colleague Johann Stephan Pütter (1725–1807) thought there were other types of social organization that came between the family and the state. He called these *Gemeinde* (a loose grouping of people) and *Volk* (a collection of families and/or *Gemeinde*), which bodies he thought lacked sovereignty. None of this is totally satisfactory but it *is* the beginning of an explanation of the formation of civil society in developmental terms. August Ludwig von Schlözer (1735–1809) summed up the new thinking: "Since the beginning of the human race, a beginning we do not know and cannot rationally reconstruct, three basic types of social organisation have developed in succession: familial [*häusliche*] organisation, civil [*bürgerliche*] groups, and state-society [*Staatsgesellschaften*]." For him, the moment of state formation was the point separating history from prehistory. Here too the scriptural account of the Creation was sealed off. The biblical account was now held to apply solely to the Israelites, which allowed the Aufklärer to argue that the basic principle governing mankind as a whole was Natural Law.[14]

And so, in this way, history gradually acquired a new function—it was to discover how society in the past had developed so that future evolution might be understood.[15]

What Knowledge Is Conveyed by Art?

If societies developed over time, what force or forces propelled that change? Natural Law might be operating at some level but the Aufklärer were attracted by the notion that perfection was not a static quality inherent in the nature of things. Instead, they understood that perfection was to be achieved by "the forces of the spirit." To them, the mind (itself a relatively new concept) was not a merely passive reflector of sensations but "possessed an inherent creative energy . . . Increasingly, they located the motor element of history in the actions of man's spirit."[16]

The "science" of the relationship between experience and creation was called "aesthetics," a word coined by Alexander Baumgarten in 1739. The link between aesthetics and history was that both disciplines, for the Aufklärer, "assumed the possibility of a leap on to a higher plane of understanding . . . Perfectibility, genius and the phenomenology of the spirit were the main elements in formulating a more comprehensive theory of historical development."[17]

Baumgarten, Christian Wolff's "most brilliant disciple," was the first to investigate the field he himself identified. What, he asked, was "the type of knowledge conveyed by art?" Baumgarten conceived the view that the senses must be capable of perfection, just as reason was. But he did not think that this perfection corresponded to the way mathematics was perfectible. A picture or a poem was for Baumgarten "a sensuous representation of an image of perfection." Perfection could be achieved through the act of creation—the perfection of a work of art lying in its unique ability "to weld diverse impressions and confused apperceptions into an individual whole that conjured up a pure image."[18] Baumgarten was joined by Johann Jakob Bodmer, who argued that poetry (and by implication other forms of art) was a form of truth equal, if not superior, to philosophy (to include what we call science), the more so because it was more closely related to history. This was an important insight because it suggested that the unique and distinctive essence of a nation is best found in its poetic and mythic traditions.[19]

For Bodmer the artist became a Promethean figure, a "wise creator," whose vision "forces his contemporaries to think and act in a new mould," someone who epitomizes his own times while attempting to change and improve them. Bodmer also introduced a teleological element: each creation of genius results in an expansion of consciousness, opening the path to the apprehension of a better—more perfect—world, enabling us to transcend the present.[20]

In fact, says Reill, by the 1760s this Leibnizian idea of perfectibility had become one of the central concepts of German aesthetics. Moses Mendelssohn (1729–86) was just one figure who was specific in his claims along these lines. In 1755 he had applied the idea of perfectibility to artistic understanding, claiming that "the healthful, the tasteful, the beautiful, the practical, all pleasures stem from the idea of perfection." He made a distinction between the perfection of man's physical nature, which he regarded as more or less complete, and the perfection of his inner nature, which had not yet been achieved: "Alone the inner man is incomplete . . . men have to work, to work tirelessly for improvement." Mendelssohn, referred to as the "Jewish Socrates," argued that there is a special faculty in the soul that functions solely in regard to beauty, enabling man to respond to beauty, to "know" it and recognize it in a way that analysis can never achieve. On this view, it was the soul that predisposed man to higher culture.

For the aestheticians of the Aufklärung, then, all artistic creation, and by extension all historical creation, is the result of the inborn drive toward perfection, referred to by the German deist philosopher Hermann Reimarus as the *notion diretrix*. Furthermore, the idea of perfectibility linked all individual creations together. Perfection was defined as "the achievement of a harmony between inner life and outer life," and that is what a masterpiece was, a harmony between spirit and nature.

This vision of the creative genius of the Aufklärer was developed still further. It was in the genius that the two realms of the individual and the general came together. More and more, the genius was considered to have the qualities of a prophet. By 1760, the Aufklärer across a wide range of disciplines were involved in trying to understand the exact nature of genius. In his 1760 book, *Versuch über das Genie* (*Truth about Genius*), Friedrich Gabriel Resewitz argued that genius was characterized by "intuitive knowledge" (*anschauende Erkenntnis*), defined as the ability to grasp the general and the individual simultaneously. Resewitz was saying, in effect, that the product of genius is itself a form of perfect knowledge. In asserting this he implied that genius "samples" divine knowledge.[21]

The evolving concept of genius had a number of ramifications. First, the new understanding implied that historical change resulted from spiritual change, but it also carried the implication that change was not automatic, for genius was notoriously unpredictable. And since by definition (to the devout at any rate) every image of perfection could only be ultimately incomplete, direction was implied but the destination could never be reached. "Art as well as history had an infinite realm of future possibility."[22]

Isaak Iselin, in *Über die Geschichte der Menschheit* (History of Mankind), published in revised form in 1768, characterized history as man's spiritual struggle to overcome nature. Conceived in this way, he was led to consider three ideal types of human behavior: man ruled by his senses, man ruled by his imagination, and man ruled by his reason, producing a threefold periodization of history: the state of savagery (senses), of cultivation (imagination), of human maturity/harmony among the three faculties (reason). In arriving at this organization, Iselin contributed to the German (as opposed, say, to the British or American) idea of freedom. For him, freedom was to be acquired through knowledge; it was an *internal* freedom that concerned him, in contrast to an outward—political—freedom.* Furthermore, the realization of the future was for him, as for other Aufklärer, possible only through a conscious act. The future didn't just happen; it was fashioned, fostered, crafted, and geniuses were to be the primary agents of this advance.[23] Here were two ideas that were to have powerful ramifications in German intellectual history.

Poetry vs. Mathematics

At the center of the historicist approach is the conviction that a fundamental difference exists between the phenomena of nature and the phenomena of history, from which it follows that the social and cultural sciences are inherently different from the natural sciences.[24] The Aufklärer also made a further distinction—between rational or abstract understanding on the one hand, and moral or "immediate" understanding on the other. Rational thought, they believed, is best suited to exploring the world external to man, while immediate understanding lends itself to the exploration of the human world. On this view, mathematics represents the ideal form of rational understanding, whereas poetry is the ideal manifestation of intuitive understanding. History, which is concerned with both the external world *and* the spiritual world, must draw from both. For the

* Whereas English distinguished between freedom and liberty, in German there is only "*Freiheit.*"

Aufklärer, the genius is not so much the great speculative philosopher but more likely a great poet. "Poetry both preceded and was superior to reflection . . . The great poet provides his people with an intuitive representation of the truths of their times at a level approaching divine understanding."[25] The historian's task, then, becomes an investigation of a people's national character according to its sacred and creative writings. For the Aufklärer, historical understanding came to be regarded as on a par with the achievements of poets and artists because it enabled people "to understand their own humanity by apprehending the humanity of others."[26]

The importance of the poetic approach was central to the Romantic movement, and the difference between the cultural and natural sciences has been an important concern in Germany right up until the present day. In the late eighteenth century, the significance of poetry was highlighted early by the short-lived but intense flourishing of the Sturm und Drang (Storm and Stress) movement. The title was taken from a play by Friedrich Maximilian Klinger, actually a rather junior member of the movement but all of them—Johann Georg Hamann, Johann Gottfried von Herder, Johann Heinrich Merck, and Johann Michael Reinhold Lenz, in addition to Klinger—were characterized by extreme youth (Lenz was nineteen in 1770, Klinger eighteen) and by, in general, temperamental instability, the defiance of accepted modes of thought and norms of behavior, restlessness, discontent, even maladjustment. Their works, essentially middle class (they were all university men), disparaged the modern state and all mercantile enterprises, and they delighted in physical exercise and nature (the wilder the better). They attacked "polite" society and followed their intuition, believing life to be both tragic and exhilarating.

It is possible to see the Sturm und Drang movement as very young and very tiresome but, as we shall see, in their more mature years, most of them went on to create great works. As we shall also see with the Nazarenes, the existence of an early group identity gave them a self-confidence they might otherwise have lacked.

The final and distinctive achievement of the historicist Aufklärer approach was the conception of a *Bildungsstaat*—a state whose main ideal was to enrich the inner life of man.[27]

The Origins of Modern Biology

This new idea of nature had another important set of ramifications which made a basic contribution to the revolution in European thought in the eighteenth century, and here too German writers helped lead the way. Gotthold Ephraim Lessing, Moses Mendelssohn, Johann Sulzer, and Thomas Abbt all criticized—and criticized bitterly—the shortcomings of the mechanistic approach and pointed instead to the biological world where, they felt, the timeless nature of Newtonian-type laws was completely inappropriate and inadequate. The study of living forms, they insisted, offered the opportunity for what they called "immediate" or "experiential" understanding. The experience of other people, animals, and plants was direct, unlike the experience of, for example, mathematics. This mode of understanding, Resewitz's *anschauende Erkenntnis*, became the major approach to knowledge in the latter half of the eighteenth century. Aestheticians abandoned the study of the eternal rules of composition to examine instead the *process* of artistic creation; jurists turned away from their attempts to discover the eternal laws of civil association, preferring instead to focus on the *development* of law within society; perhaps most important of all, natural scientists turned to the study of *growth* and development.[28] This underlines just what a great intellectual revolution historicism was in helping to create the modern age.

"The word 'biology' is a child of the nineteenth century." Until the seventeenth century, biological science as we understand it now comprised two fields: natural history and medicine. As the seventeenth century gave way to the eighteenth, natural history began to break up into zoology and botany, although many people as late as Linnaeus and Jean-Baptiste Lamarck moved freely between the two. At much the same time, anatomy, physiology, surgery, and clinical medicine also diverged. To begin with, both anatomy and botany were practiced primarily by physicians (they dissected the human body and collected medicinal herbs), and animals were studied mainly as an aspect of natural theology.[29] The underlying reality is that the so-called scientific revolution of the fifteenth, sixteenth, and seventeenth centuries actually occurred only in the physical sciences, leaving the biological sciences largely unaffected.[30]

Long before the eighteenth century, the ancient Greeks had conceived the idea that there is a purpose—a predetermined end—in nature and its processes. By the seventeenth and eighteenth centuries, this ideal had coalesced around the notion of the *scala naturae*, the Great Chain of Being, culminating in man. The manifold adaptations of organisms to their environment—everywhere apparent—fostered the idea of a "harmony" in the natural world that could only have been produced by God. The apparent goal-directed processes in the development of individuals were just too conspicuous to be discounted. Final causes must be involved, as Immanuel Kant, among others, acknowledged (see p. 82).

Overall, the concept was known as cosmic teleology—the universe is proceeding toward some particular end, predetermined by God. Until the mechanism of natural selection was identified, many biologists (Lamarck was one) argued for the existence of nonphysical (even nonmaterial) forces that drive the living world "upward toward ever greater perfection."[31] This was known as orthogenesis. Leibniz, Linnaeus, Herder, and almost all British scientists shared this view, some of them as late as the middle of the nineteenth century.

So, from the middle of the sixteenth to the middle of the nineteenth century, two schools of thought coexisted: the physical scientists believed that God, at the time of the Creation, had instituted eternal laws governing the processes of this world (essentially the deist view). Against that, devout naturalists—familiar with living nature—concluded that, so far as the diversity and myriad adaptations of living creatures are concerned, the mathematically based laws of Galileo and Newton were meaningless.[32] Germany was one of the main centers of this latter group.

Within biology (to use the modern term), a new era of observation had begun with the work of the so-called German fathers of botany—Otto Brunfels (1488–1534), Hieronymus Bock (1489–1554), and Leonhart Fuchs (1501–66). The study of medicinal plants was popular throughout the later Middle Ages and was reflected in the publication of a number of herbals. Then, as a result of the great voyages in the age of exploration and the discovery of the New World, the immense variety of plant and animal life across the globe was realized.[33] These German botanists provided a break from medieval works, which were endlessly copied myths and allegories. Instead, their descriptions were based on real plants observed in their natural habitat, with the result that their realistic drawings played much the same role in botany as those of Vesalius did in anatomy. Hieronymus Bock's descriptions—in meticulous if colloquial German—were vividly drawn from his own observations. Importantly, he also broke with the alphabetical arrangements of earlier herbals, describing instead his own method "to place together, yet to keep distinct, all plants which are related and connected, or otherwise resemble one another."[34] The German herbals are worth singling out because of the new classificatory principles they introduced. This early tradition of classification reached its climax in 1623 with the release of Caspar Bauhin's *Pinax*, in which 6,000 kinds of plants were arranged in twelve books and seventy-two sections.[35] Related plants were often put together because of their common properties, and each plant assigned to a genus and a species, though genera were not defined. In addition, there was in *Pinax* an implicit separation of the monocotyledons, and some nine or ten families of dicotyledons were brought together also. Already, reproduction was recognized as crucial.

Botanists from Conrad Gesner (1567) and Andrea Cesalpino (1583) to Linnaeus all recognized the importance of fructification for classification, but this still left great scope for argument owing to the multitude of characteristics available, all bearing on fructification.[36] Debate was and was not helped by the fact that the number of known plants increased at an astonishing rate during the sixteenth and seventeenth centuries. In 1542, Leonhart Fuchs identified some 500 species, Bauhin in 1623 referred to 6,000 species, while John Ray in 1682 listed no fewer than 18,000.[37] The need for order and classification was greater than ever, but the welter of new material was overwhelming. At much the same time, while all others around him were fixated on the concept of essentialism (according to which each species is characterized by its unchanging essence—*eidos*—and separated from all other species by a sharp discontinuity), Leibniz stressed the opposite: continuity. Ernst Mayr, the German-born Harvard historian of biology, argues that Leibniz's interest in the *scala naturae*, and the links between various life forms (as revealed in the earliest attempts at plant classification), helped prepare the ground for Linnaeus and, ultimately, for evolutionary thinking.

A key figure here was Albrecht von Haller (1707–77) who began a number of wide-ranging animal experiments, examining the operation of various internal organs. Haller found no evidence for a "soul" governing physiological functions, but his studies did convince him that bodily organs have certain properties (irritability, for example) which are absent in inanimate nature.[38] Though it may sound primitive to us, Haller's irritability concept was important because he was not a vitalist: for him organic matter was different from inorganic matter but the difference, however mysterious, was a *natural* and not a supernatural process. This helped form a climate of opinion whereby it was in Germany, toward the end of the eighteenth and the early nineteenth century, that the strongest resistance developed against the purely mechanistic understanding of the followers of Newton (though this is not to dismiss the role played in eighteenth-century biology by the Frenchman George Buffon, 1707–88, and the French-Swiss Charles Bonnet, 1720–93).[39] Three biologists in particular may be mentioned, not forgetting the important role played by Immanuel Kant in Königsberg.

Johann Friedrich Blumenbach (1752–1840) led the way. The influence of his experiments and observations was immense—roughly half the important German biologists during the early nineteenth century studied under him or were inspired by him: Alexander von Humboldt, Carl Friedrich Kielmeyer, Georg Reinhold Treviranus, Heinrich Friedrich Link, Johann Friedrich Meckel, Johannes Illiger, and Rudolph Wagner, several of whom we shall meet again. Friedrich Schelling and Kant agreed that Blumenbach was "one of the most profound biological theorists of the modern era."[40]

His foundational theories were set out in a short work, *Über den Bildungstrieb und das Zeugungsgeschäfte* (On the Formative Drive and the History of Generation). In this book, Blumenbach considered how the sperm, "by the subtle odour of its parts which are particularly adapted to causing irritation," awakens the germ "from its eternal slumber."[41] And he identified a crucial question: "Why is it that progeny always differs from its original progenitor?" while observing too that offspring often display a *blend* of parental traits. He was homing in on the idea of both genetics and evolution, except that he had parts of the theory upside down: in his view all the various peoples around the world were a *degeneration* from the Caucasian race.

Blumenbach's central idea, the one that influenced Kant and Schelling so much, was that there is a kind of "Newtonian force" in the biological realm, which is the agent for organic structure and which he called the *Bildungstrieb*.[42] He had conceived this model after several experiments with the humble polyp. What struck Blumenbach about this organism was, first, that it could regenerate amputated parts "without noticeable modification of structure"; and second, that the regenerated parts were always *smaller* than the originals. Furthermore, this seemed to be true more generally. Where humans had suffered serious flesh wounds, Blumenbach observed that the repaired area was never quite as good as new but always retained a depression. He was led to conclude "First that in all living organisms, a special inborn *Trieb* [drive, or motivating force] exists which is active throughout the entire lifespan of the organism, by means of which they receive a determinate shape originally, then maintain it, and when it is destroyed repair it where possible. Second, that all organized bodies have a *Trieb* which is to be distinguished from the general properties of the body as a whole as well as from the particular forces characteristic of that body. This *Trieb* appears to be the cause of all generation, reproduction, and nutrition. I call it the *Bildungstrieb*."[43]

Blumenbach believed that the Bildungstrieb was teleological in character and "immanent" in the material constitution of the organism. In a way, of course, the Bildungstrieb doesn't explain anything—it is merely a name for a mysterious process. But that is what appealed to Kant. For what he insisted upon was that, even if nature somehow uses mechanical means to construct organized bodies, humans can never understand that process even from a theoretical point of view. The problem for Kant was that human understanding can only construct scientific theories that use the "linear" mode of causation. In the organic realm, on the other hand, "cause and effect are so mutually interdependent that it is impossible to think of one without the other . . . This is a teleological mode of explanation, for it involves the notion of a 'final cause.' " Kant became convinced that it is impossible to produce functional organisms by mechanical means—for example, by chemical combination. He was impressed by the examples of misbirth, for him powerful evidence to suggest that something analogous to "purpose" operates in the organic realm, "for the goal of constructing a functional organism is always visible in the products of organic nature, including its unsuccessful attempts." For Kant, therefore, it was self-evident that the life sciences rested on a different set of principles from those of the physical sciences.[44]

Johann Christian Reil (1759–1813) studied in Göttingen during 1779 and 1780 and came into contact with the young Blumenbach. Timothy Lenoir, in his study of early German biologists, says Reil was possibly more original than Blumenbach. His treatise "Von der Lebenskraft," in which he introduced his own conception of the vital force within a Kantian framework, was published in 1795 in the first volume of the new professional journal *Archiv für die Physiologie*. Reil too believed that each organism shows "purposive organisation" (*zweckmässige Form*) and that this was determined by the chemical affinities between the organic materials, "just as the seed [*Kern*] of a salt crystal attracts particles according to a particular law in which the basis of its cubic shape is to be found."[45] This was, then, a sort of halfway-house theory, between Blumenbach's and Kant's. In Reil's view the germ, in the mother, "slumbers without developing, probably because its organisation has too little irritability [*Reizbarkeit*]. The father enhances the animal force of the dormant germ perhaps through the addition of the fluid of his semen to the matter of the germ."[46]

Carl Friedrich Kielmeyer (1763–1844) moved from Stuttgart to Göttingen, where he also was a pupil of Blumenbach from 1786 to 1788. His contributions helped to establish *Pflanzenchemie*, the beginnings of organic chemistry. In the course he taught on comparative zoology Kielmeyer conceived what he called the *Physik des Tierreichs*, the aim of which was to uncover the laws of organic form by comparing the anatomy of birds, amphibians, fish, insects, and worms. Kielmeyer also broke new ground when he used embryological criteria to establish affinities between animal forms. He realized that the patterns revealed in embryological development were confirmation that the system of animal organization did not require "the assumption of a special directive force *existing outside* of the individual organism, through which the life and economy of organic nature is maintained." (Italics added.) There was no need for any "supramaterial" organizing force. Kielmeyer, like Blumenbach, was unimpressed by the traditional idea of a Great Chain of Being; instead, he became convinced that species were transformed into others, albeit in a distinctive way: "Many species have apparently emerged from other species, just as the butterfly emerges from the caterpillar . . . *They were originally development stages and only later achieved the rank of independent species*; they are transformed developmental stages. Others, on the other hand, are original children of the earth. Perhaps, however, all of these primitive ancestors have died out." Kielmeyer noted that smaller organisms tended to have more offspring than larger ones, and from this concluded that there are "internal forces," specific to species, that give rise to their characteristic structure and behavior.[47]

In purely biological terms, then, these late eighteenth-century scientists and philosophers had three operating conclusions/beliefs.[48] First, it was the task of the new fields of zoology and botany to reproduce in the organic realm what physics had done in the inorganic realm—namely "to investigate the most universal phenomena of matter and the special classes of phenomena which are not further reducible to others."[49] Second, they identified (or assumed) a *Lebenskraft* or Bildungstrieb as the shaping principle of every organized body. Finally, Kant emphasized that man's reason was insufficient ever to discover these "natural purposes," or "teleological agents," in the organic realm.

The Rise of Evolutionism

This battle between mechanist thought and vitalist thought would continue throughout the eighteenth and nineteenth centuries, and even up until the first quarter of the twentieth century. But, as Ernst Mayr points out, the years between the publication of the tenth edition of Linnaeus's *Systema Naturae* in 1758 and Darwin's *On the Origin of Species* in 1859 were a time of transition. During this transition period Lamarck published his theory of transformation, in 1809, which argued for an "intrinsic tendency of organisms to strive toward perfection" together with an ability to adjust to the environment (the inheritance of acquired characteristics). It was also during this transition period that "downward classification" was phased out, to be replaced by "upward classification." In "downward classification," the organic realm was divided/organized according to its internal logic, as it appeared to this or that theoretician, using his view of what nature actually consisted of, and in the belief that species differed in their very essence, an essence that reflected their *eidos*, their special substance. In "upward classification," observations started with species, the irreducible basic building blocks, and then their similarities with other organisms were observed and codified, working upward to higher taxonomic groups.[50]

But the very idea of classification was itself evolving. For centuries the *scala naturae*, the scale of perfection, had been virtually the only conceivable way of bringing order into diversity. The idea was less popular with botanists than with zoologists, however, since hardly any trend toward perfection was observable among plants, except for a general advance from algae to the phanerograms (the subkingdom of flowering plants). And so, other approaches to classification were tried. Organisms were placed on the scale of perfection according to their *affinity* with less perfect or more perfect neighbors.[51] There was a conviction that similarity (of whatever kind) reflected an underlying causal relationship. Two kinds of similarity in particular were identified by German writers such as Friedrich Schelling and Lorenz Oken: true affinity and *analogy*. "Penguins are related to ducks by true affinity but to the aquatic mammals by analogy. Hawks show affinity with parrots and pigeons, but are analogous to the carnivores among the mammals." Such a reconceptualization seems bizarre but this approach proved crucial in the ensuing history of biology, influencing Richard Owen in his ideas of homology and analogy, which came to dominate comparative anatomy.

Without these developments in thinking about classification, the theory of evolution probably could not have developed. Yet there was still some way to go. The great problem with evolution was that it could not be observed directly, unlike the familiar phenomena of physics, such as falling stones or boiling water. Evolution, plainly, can only be inferred and only then can such evidence as fossils or stratification be adduced. [52]

To us, the time it took between the first glimmerings of evolutionism by Leibniz in his *Protogaea* (1694) and the full-blown theory of Lamarck in 1809 seems inordinately long. Like Buffon, who had flirted with evolutionism all his life, Lamarck was French, and Darwin himself, of course, was British. Yet evolutionism was far more popular in Germany than anywhere else.[53] Just how widespread it was there has been explored by several historians. Henry Potonié, Otto Heinrich Schindewolf, and Oswei Temkin are just three who have rescued the names of numerous early German evolutionists from oblivion: besides Blumenbach, Reil, and Kielmeyer, there were Friedrich Tiedemann, Reinecke, Voight, Tauscher, and Ballenstedt. Although it may come as a surprise that, with all these figures devoting their time to evolution, it should be an Englishman, Charles Darwin, who conceived the idea of natural selection, we should remember that, among the many people who set the stage for Darwin, the Viennese botanist Franz Unger stands out. Unger argued that the simpler aquatic and marine plants preceded the most complex varieties, that there must have been an original germ of all kinds of plants, that new species must have originated from already-existing ones and that all plants are united with each other "in a genetic manner." Among Unger's students was Gregor Mendel.[54]

And so, in the late eighteenth century, in Germany, doubt, deism, Pietism, and the drive toward perfection—in history, in art, in biology—all came together to create a way of looking at the world, looking within, looking back, and looking forward all at the same time. It was a transitional period, when people were groping, tentatively attempting—perhaps without being aware of it—to replace the theological concept of mankind with a biological understanding.

One man of influence who took up these ideas early was Wilhelm von Humboldt.[55] Later on, Humboldt would be instrumental in the creation of the University of Berlin, an institution so important that it needs a chapter all to itself in any cultural history of Germany. To begin with, he was a student of Blumenbach and was much taken with his concepts of Bildung and Bildungstrieb.[56] Nature, for Humboldt, consisted of specific individual centers of energy and activity, each center revealing its own character in the activity it displayed. Activity—sheer movement—was key here. In classical (Newtonian) physics, motion was always the result of some outside source. However, many thinkers, dissatisfied with the application of Newton's science as an explanation for living systems, preferred what they called "the living order of nature," where nothing stood still, where "self-generated motion" meant that every living part of nature was constantly in movement and, moreover, "this movement was not haphazard." Matter, to them, contained an immanent principle of self-movement. "Unlike mechanical concepts of force (magnetism, electricity, gravitation), these internal powers were thought to operate directionally: they had an implicit goal towards self-realisation (*Vervollkommnung*)."[57]

This revised definition of matter required a redefinition of nature. In this new view, there is in nature an inner character which speaks through it. "Nature's telos could only be intuited, never fully revealed as transparent."[58] Humboldt's gloss, which was essentially Blumenbach's view, was that matter was composed of general and individual *Kräfte* (powers or forces), each having its own nature. The most important of these immanent qualities were the general forces of Bildung, generation (*Zeugung*), and habit (*Trägheit* or *Gewohnheit*). These qualities produced the individuals of which a nation was composed, making the nation an analog of an organized body. "Reality was defined as the striving of active powers or ideas to actualize themselves, that is to acquire form."[59]

Biology apart, the root concept of Bildung, a neologism of the eighteenth century, lay in Martin Luther's use of *Bild*, meaning "image," in two seminal biblical verses:

> And God said, Let us make man in our own image, after our own likeness . . . So God created man in his own image, in the image of God created he him; male and female created he them.
>
> (Genesis 1: 26–7)

> But we all, with open face beholding as in a glass the glory of the Lord, are changed unto the same image from glory to glory, even as by the Spirit of the Lord.
>
> (2 Corinthians 3: 18).

It was, of course, the Pietists who had introduced the idea. For them it had an exclusively religious sense, but during the reigns of Friedrich Wilhelm I and Friedrich the Great, Bildung was secularized without losing the subjective ideal of personal perfection. "Even for those who rejected revealed religion and scriptural authority, Bildung offered a means of secular salvation through culture."[60] Moreover, Bildung was open to everyone in a country where the public sphere was rapidly expanding (see Chapter 1, p. 41).

Bildung "was the culture of an emerging group that did not conceive of itself as bourgeois so much as it thought of itself as cultivated, learned and, most importantly, *self-directing* . . . a man or woman of Bildung was not merely learned, but was also a person of good taste, who had an overall educated grasp of the world around him or her and was thus capable of 'self-direction' that was at odds with the prevailing pressure for conformity."[61] Bildung was in effect a secular form of Pietism: both embodied Leibniz's and Christian Wolff's notion of perfection.

Bildung, then, for someone like Humboldt, was partly a biological force, partly a spiritual necessity, partly an aspect of the natural world, like gravity. It also had religious overtones, in that it had grown out of Pietism: just as the Pietists could "improve on the Creation" and move closer to God by practically helping their neighbors in *this* world, so Bildung was an interior process whereby an individual could work on himself, or herself, to improve his or her self-consciousness, to move closer to perfection. The concept of genius—individuals whose creations offered glimpses of divine wisdom, glimpses of perfection—meant that self-cultivation, through studying the achievements of geniuses, offered the cultivated individual the prospect of achieving an approximation of divine wisdom right here on earth.

This was very much a halfway house of ideas that could only have existed in the transitional time between *Principia Mathematica* and the *Origin of Species*, between doubt and Darwin. This historical/artistic/biological view of the world, within the framework of striving for perfection, was to shape many of Germany's thinkers, not a few of whom were themselves the sons of Pietist pastors.[62]

Bildung was, in its way, the most ingenious by-product of the development of doubt.

PART II

A Third Renaissance, between Doubt and Darwin

3 ·

Winckelmann, Wolf, and Lessing: The Third Greek Revival and the Origins of Modern Scholarship

The Italian Renaissance was a German idea. The man who formulated it most clearly, Jacob Burckhardt (1818–97), author of *Die Kultur der Renaissance in Italien* (*The Civilisation of the Renaissance in Italy*; 1878), was born in Basel, Switzerland, in 1818, but studied at the University of Berlin, where he attended the seminars of the most famous historian of the time, Leopold von Ranke. Burckhardt returned to Basel in 1843 and began to lecture there at the university, as well as edit a newspaper, the *Basler Zeitung*. Growing disillusioned with journalism, he abandoned it for full-time historical research, a move that led to his first major book, *Die Zeit Konstantins des Grossen* (*The Age of Constantine the Great*; 1853), soon followed by a historical guide to the art treasures of Italy, *Der Cicerone* (*The Cicerone*; 1855). These two works were so well received that they earned him a chair—in architecture and art history—at Zurich Polytechnic when it opened in 1855. Three years later he returned to the university at Basel and remained there for the rest of his life, spurning the invitation to become Ranke's successor at Berlin. It was from Basel that, in 1860, he published his most famous book.[1]

Before Burckhardt, other writers and historians had introduced the phenomenon of the Renaissance. Petrarch (1304–74) was the first to recognize, on paper at least, the idea of the "Dark Ages," that the thousand years—more or less—before he lived had been a period of decline, and that ancient history, poetry, and philosophy were "radiant examples" of a civilization that was the highest form of life before Christ appeared. Voltaire, Saverio Bettinelli, the French historian Jules Michelet, and Georg Voigt, professor of history at Munich, in his 1859 book *Die Wiederbelebung des classischen Altertums: oder, das erste Jahrhundert des Humanismus* (The Revival of Classical Antiquity, or the First Century of Humanism), had all drawn attention to Renaissance Italy. Burckhardt's ideas did not come out of nowhere.

Nevertheless, his understanding of the Renaissance was much more coherent and complete than that of any of his predecessors.[2] It was Burckhardt who confirmed that the Italian Renaissance was far more than the rediscovery of antiquity: it had seen the development of the individual, it was then that the lineaments of modernity first appeared. Burckhardt maintained that society was now a self-conscious—and therefore a *secular*—entity as it had never been before.

As Peter Burke, the Cambridge historian of ideas, has emphasized, *The Civilisation of the Renaissance in Italy* did not lack for critics. After 150 years of increasingly specialized research, he said, "it is easy to point out exaggerations, rash generalisations and other weaknesses." But though Burckhardt's view of the Renaissance may be flawed, Burke agreed that "it is also difficult to replace." Perhaps the single most important revision of Burckhardt's argument is that of Charles Homer Haskins, professor of history at Harvard in the early decades of the twentieth century. Haskins's contention was that the essentially Platonic revival in Italy in the fourteenth and fifteenth centuries, which gave rise to the Italian Renaissance, was in fact, the *second* such classical revival in the West. The first, associated with the rediscovery not of Plato but of Aristotle, took place in the twelfth century and was marked by, for example, the new science of law and a unified legal system, which promoted the idea of *shared* knowledge that could be argued over, the wider use of Latin, the development of universities, and the growth of organized skepticism in scholarship. With the philosophy of Albertus Magnus and Thomas Aquinas, which envisaged a secular world, came a unification of thought in theology and the liberal arts, giving rise to *Summae*, encyclopedic treatises aimed at synthesizing all knowledge, changes in worship, which promoted a rise in self-expression and individuality, and, perhaps most significant of all, the rise of the experimental method, giving birth to science as we know it. Among historians, then, if not yet among the general public, there were *two* renaissances, not one, with the first rather more important than the second.

Against this background, a closer reading of Burckhardt's book reveals some interesting further observations.[3] The Renaissance in Italy, he said, was characterized by the following elements: the revival of antiquity, the rediscovery of the texts of Plato and the civilizations of ancient Greece and Rome, before the years of Christian fundamentalism, "as the source and basis of culture . . . as the object and ideal of existence." The recovery of the classics, Burckhardt said, led to the growth of textual criticism and the more advanced study of languages—there was a revival of new learning in which philology played a central role. It was in the High Renaissance (1513) that Pope Leo X reorganized Rome's university, La Sapienza. The Florentines, Burckhardt said, "made antiquarian interests one of the chief objects of their lives," accompanied by advances in the sphere of science. The treatise was revived, as was history writing, two forms of literature and inquiry that were felt as new. In philosophy, the Florentine Platonists had a massive influence on thought and on literature, aesthetics in particular. In poetry, ancient Greece and Rome were again the model, stimulating imitation but also more imaginative works by poets who were, in addition, often scholars. In natural history there were advances in botany (the first botanical gardens), and in zoology (the first collections of foreign animals). In art it was the era of "many-sided men," individuals such as Leon Battista Alberti and Leonardo da Vinci, giants who shone in many different fields.

In other sections of his book, Burckhardt said that attitudes to and beliefs about war changed in the Italian Renaissance. In a section on "War as a work of art," he argued that "War assumed the character of a product of reflection." And from Dante and Petrarch onward, there was in Italy a ferment of patriotism and nationalism. "Dante and Petrarch, in their day, proclaimed loudly a common Italy, the object of the highest efforts of all her children."

Finally, in music Burckhardt identified a characteristic of the Italian

Renaissance as "the specialisation of the orchestra, the search for new instruments and modes of sound, and, in close connection with this tendency, the formation of a class of *virtuosi,* who devoted their whole attention to particular instruments or particular branches of music." This all amounted to a celebration of humanism—the glories that humankind is capable of, without specific and continual reference to God.

There is a saying in the military that the darkness is deepest under the light and it is the opening argument of this chapter that such is the case here. That Burckhardt, in shining the light of his intellect and his historical imagination on fourteenth- and fifteenth-century Italy, cast a shadow over the culture of which he himself was a part. It will be argued here that, beginning in the middle of the eighteenth century, there was a *third* classical revival in Europe, that it resulted in a flourishing—a renaissance—of the arts and sciences, that it saw great reflection and innovation in military affairs, and that it stimulated an unparalleled philosophical revival. This promoted a surge in new aesthetic theory (already introduced in the previous chapter), including advances by poets—such as Goethe and Schiller—who were also scholars and many-sided men. It was accompanied by a great surge in patriotism and a demand for unification—this time of Germany. Other parallels may be found in music and in *Humanität,* the German form of humanism. The greatest names in musical history—from Mozart to Arnold Schoenberg—were all German. The links between *Wissenschaft, Bildung,* and *Innerlichkeit,* formulated most forcefully in the brand-new University of Berlin (founded in 1810), were to be the clearest embodiment of the German idea of humanism (all of which are discussed below).

Just as, in the Italian Renaissance, Pope Leo X reorganized La Sapienza in Rome, so in Germany a completely new idea of learning, which fundamentally shaped the modern world, was evolved. There were new forms of literature and new forms of inquiry, in which philology once again formed the core. Archaeology—the modern equivalent of antiquarianism—underwent its heroic age. This third renaissance was without question primarily German.

The Father of Classical Archaeology and the Founder of Art History

If the Aristotelian renaissance was sparked by the rediscovery of Arabic translations of his masterpieces in Toledo, Lisbon, Segovia, and Cordoba, after the *Reconquista* of the Iberian Peninsula by the Christians, and if the Platonic revival owed a great deal to scholars such as Giovanni Aurispa, who brought back from just one visit to Constantinople on the eve of the Turkish conquest no fewer than 238 Greek manuscripts, the same honor in the eighteenth century goes to Karl Weber (1767–1832) and Johann Joachim Winckelmann (1717–68). Winckelmann is the better-known figure but recent scholarship credits Weber, a military engineer in the Swiss guard, with being the man whose great efficiency and devotion to detail ensured that the excavations south of Naples—at Herculaneum, Pompeii, and Stabiae, in particular the Villa dei Papiri—were actually carried out in a workmanlike way, and thus enabled the groundwork to be completed on which Winckelmann would base his groundbreaking survey of classical art.[4]

Born in Stendal in Prussia in 1717, the son of a cobbler, Winckelmann grew up in a house with just one room, which was also his father's workshop. He pestered his parents to give him an education that was beyond their means and, in one way and another, found his way to Berlin, to study under Christian Tobias Damm, "one of the few men then alive in Germany who exalted Greek above Latin at a time when the study of the Greek language was almost entirely neglected."[5] After Berlin, Winckelmann transferred to the universities of Halle and Jena, where he studied medicine, philosophy, and mathematics, supporting himself as a tutor.[6] He would read Greek till midnight, sleep in an old coat in an armchair until four in the morning, when he would resume reading.[7] In the summer months he slept on a bench with a block of wood tied to his foot which fell down at the slightest movement and wakened him.

Winckelmann's interest in art and antiquities was nurtured after he obtained employment as a research assistant (as we would say) to Count Bünau near Dresden (which boasted more art than did any other city in Germany), but the crucial episode was his meeting with the papal nuncio in the city, who offered Winckelmann the opportunity to work in Rome—provided he convert to Catholicism.[8]

Winckelmann arrived in Rome in 1755. For him and others like him, the statues in Rome were invariably regarded as the most important masterpieces of ancient art.[9] He began in the service of Cardinal Alessandro Albani, who had a villa just outside Rome, where he was made librarian and given charge of the antiquities collection. But the fame he would soon acquire had much more to do with several visits he made to Herculaneum and Pompeii, just then attracting widespread interest.

They had been "rediscovered" in 1738 when the Spanish military engineer Rocque Joachin Alcubierre was ordered to survey a site and prepare plans for a new summer palace for King Charles VII of Bourbon at Portici, on the Italian coast south of Naples. This was not entirely accidental. Local residents in the nearby town of Resina had long obtained their water by drilling artesian wells and were fully aware that there were ruins underground—chance finds of antiquities had been occurring since Renaissance times. Alcubierre was instructed by the king to "make some grottoes and see what might be discovered."[10]

Excavations began in October 1738. In some places the volcanic lava was fifty feet thick and it was not until November that a marble *Hercules* was recovered, and it was the middle of January the following year before an inscription—of L. Annius Mammianus Rufus—revealed that a structure originally believed to be a temple was in fact a theater.[11] This reorientation was important, for a theater—unlike a temple—implied that the building was part of a city. It was Rufus's name for his theater, Theatrum Herculanense, that confirmed the city as Herculaneum. Excavations at nearby Pompeii began in April 1748.

Winckelmann visited Herculaneum and Pompeii twice, and although his excursions were not popular (the excavators were worried that he would steal their thunder), Winckelmann managed to familiarize himself with the discoveries at the Vesuvian cities, with the internal politics of the excavations, and with the contents of the more important finds at the Villa dei Papiri.[12]

It was this series of coincidences, rivalries, and sensational discoveries that provided the background to Winckelmann's publications that would prove so important in stimulating the third Greek revival. These consisted of, first, a series of *Sendschreiben*, or Open Letters, on the discoveries south of Naples; second, Winckelmann's main work, *Die Geschichte der Kunst des Altertums* (*The History of the Art of Antiquity*; 1764); and third, his *Monumenti antichi inediti* (Unpublished Relics of Antiquity; 1767). As E. M. Butler, puts it, however, "His *magnum opus* is in a class apart, for it completely revolutionized the study of art by treating it organically (Winckelmann was the first to do so) as part of the growth of the human race."[13]

The History of the Art of Antiquity is divided into two parts. The first is more conceptual, examining the phenomenon of art itself, its very "essence." Here, in a broad way, Winckelmann compares the art of different periods and peoples. The second part concentrates particularly on the tradition of Greek art, from early times to its decline with the fall of the Roman Empire, and it was Winckelmann's beautifully written description of this "trajectory" that had such impressive consequences.

For his argument, Winckelmann relied on the new statuary being excavated south of Naples though he did his best to amalgamate the discoveries there with the writings of Pliny (Pliny's *Natural History*, completed in A.D. 79, is as much an art history of the classical world as a geography). Pliny argued that most of the famous ancient Greek artists produced their masterpieces in the fifth and fourth centuries B.C. For him, Greek sculpture achieved classic perfection while Phidias was at the peak of his powers, in the mid-fifth century B.C., but Pliny also insisted, famously, that after the age of Alexander the Great "cessavit deindre ars" (art thereafter was inactive or ceased). Building on what he had seen south of Naples, Winckelmann refined Pliny's argument, discerning, he said, a "high" austere "early classical" style, associated with artists such as Phidias, and a "beautiful or graceful late classical style associated with subsequent masters such as Praxitiles and Lysippus." His identification of an evolution, from one style to the other, a refinement from the "hard stylized" forms of the archaic, to the "austere, early classic," with the graceful "late classic" leading to overelaboration, and then to decline, suggested a pleasing, organic, coherent system, the sheer symmetry of which many found irresistible.[14]

It didn't matter that Winckelmann used the evidence of statues that have subsequently come to be recognized as inferior Roman copies of earlier Greek masterpieces. What mattered is that, whereas classical scholars had previously speculated in a vague way about the rise and fall of art in antiquity, Winckelmann identified instead a sequence of clearly defined phases. More than this, Winckelmann also argued that the classical period of art in antiquity coincided with what other historians called the golden age of Greek culture, that period between the close of the Persian Wars in the early fifth century B.C. and the Macedonian invasion of Greece toward the end of the fourth century B.C. Whereas previously, ancient monuments had invariably been classified according to their iconography, or subject matter, following Winckelmann they were categorized stylistically, with reference to their period of origin. This transformed connoisseurship.

Winckelmann's other innovation was his fusion of history and aesthetics, "in which the *essence* of a tradition would be located historically at a single *privileged* moment when it supposedly achieved perfection." In linking artistic "perfection" with a particular historical period, he transformed the history of art, making it important in a sense that it had not been before. That gave it a contemporary relevance, too, suggesting that there was little prospect of any real revival in Winckelmann's own time.[15]

Especially famous was his description of the Laocoön group in the Vatican. "The universal and predominant characteristic of the Greek masterpieces is *a noble simplicity and tranquil grandeur*, both in posture and expression. Just as the depths of the sea remain for ever calm, however much the surface may rage, so does the expression of the Greek figures, however strong their passions, reveal a great and dignified soul." ' (Italics added.)*[16] This analysis had repercussions far wider than such words—however apposite—could have now. The Laocoön's importance lay in the fact that, having been specifically referred to by Pliny, when it was rediscovered in an excavation in Rome in 1506, it provided a direct link with the past. Now in the Vatican, this classic marble sculpture shows the Trojan priest Laocoön and his sons being attacked by a fierce sea serpent. Whatever our reactions today to what many people think is an overwrought monument,

* Some of this is lost in translation. The phrase sounds better in German: *edle Einfalt und stille Grösse.*

at the time Winckelmann's arguments had "the force of revelation." His arguments were considered so original, and so incisive, that he became a national figure almost overnight: "Except for Frederick the Great of Prussia, he was the most renowned German between Leibniz and Goethe." The Laocoön itself became a cult object, discussed everywhere.[17]

Among the implications of Winckelmann's argument, picked up on by Herder, Goethe, Friedrich and August Wilhelm Schlegel, and Hegel, was the notion that there is a historical divide separating ancient from modern culture, where modern culture is in fact "the *antithesis* of the integrated wholeness of ancient Greek culture, of its naïve simplicity and centredness, and of its unmediated relation to itself and nature."[18] Whether one agrees or not, Winckelmann's achievement was that he took beauty seriously, as the center of existence, not as embellishment. Above all, he suggested that if we allow the Greek ideal to influence and permeate our lives, we can hope to reproduce the conditions necessary for great art; we can attain, in other words, a form of perfection.[19]

This was all very heady. As George Santayana was to tease: "How pure the blind eyes of statues, how chaste the white folds of the marble drapes."[20] But beyond Winckelmann's inimitable style, beyond his idea of the ennobling power of beauty, it is possible to see an even deeper significance in his work. He ignored the other side of Greek life—the tragic suffering, the priapism, the orgiastic festivals of the wine god, everything Nietzsche was to call "Dionysian"—and this is surely because the stoicism that Winckelmann admired in the Greeks had qualities of Puritanism about it. Greek art, for Winckelmann, was the very opposite of baroque exuberance, of the "hedonism and licentiousness" of the rococo, which he and the emerging German middle classes associated with aristocratic decadence and the courtly French culture whose grip on Friedrich the Great and the ruling classes in Germany they resented. Winckelmann set an example to an entire generation of poets and thinkers of the golden age and helped them to accomplish something in the shadow of their King: the remaking of German culture and cultural institutions. The fact that Greece, a "powerless and almost extinct nation," should have such an influential *cultural* legacy appealed to the German *Bildungsbürger*. It had parallels with their own predicament.[21]

"The 'Greek Revival,' which Winckelmann initiated," says Henry Hatfield, "profoundly altered the course of German literature: many of its greatest writers from Lessing to our own times would have written differently without his precept and example." It is not too much to say that it affected the entire history of Western taste, as far afield as Thomas Jefferson. Not only is Winckelmann looked upon now as the founder of classical archaeology; he may be said to be one of the fathers of historicism; he had a formative influence on Herder and through him on the writing of history. Philhellenism took over as one of the defining characteristics of the *Bildungsbürgertum*, the educated middle class, influencing not only the universities but even the state bureaucracy. To Hegel, " . . . Winckelmann is to be seen as one of those who managed to open up a new organ and a whole new way of looking at things for the human spirit."[22] "By 1871," said someone else, "Graecophilia had become part of the national patrimony."[23] To Goethe, Winckelmann was like Columbus.

Winckelmann's brutal murder by stabbing in Trieste (one of the origins of Thomas Mann's *Der Tod in Venedig* [*Death in Venice*]) shocked the educated elite across Europe, adding a final dark twist to a remarkable career.[24]

"The Tyranny of Greece over Germany"

Winckelmann's reputation has lasted. He was criticized most notably during a competition organized in 1777 by the Academy of Antiquities in Kassel, which specifically examined Winckelmann's contribution to antiquarian studies, and in which Christian Gottlob Heyne argued forcefully that Winckelmann's claim that ancient art declined following its classical phase of the fifth and fourth centuries B.C. was not supported by the available evidence. But, such criticisms notwithstanding, to underline how enduring his ideas proved to be we may say first that, in 1935, in the shadow of World War II, E. M. Butler published *The Tyranny of Greece over Germany*, an examination of the influence of Winckelmann—and Greece—on Lessing, Goethe, Schiller, Hölderlin, Karl Friedrich Schinkel, Carl Gotthard Langhans, Heinrich Schliemann, Friedrich Nietzsche, and Stefan George.[25] "If I were constrained to write a history of German literature from 1700 onwards, I could only do so from this angle; for it seems to me that Winckelmann's Greece was the essential factor in the development of German poetry throughout the latter half of the eighteenth century and the whole of the nineteenth century . . . Greece has profoundly modified the whole trend of modern civilisation, imposing her thought, her standards, her literary forms, her imagery, her visions and dreams wherever she is known. But Germany is the supreme example of her triumphant spiritual tyranny. The Germans have imitated the Greeks more slavishly: they have been obsessed by them more utterly . . ." Butler did not think this obsession was entirely healthy. "Only among a people at heart tragically dissatisfied with themselves could this grim struggle with a foreign ideal have continued for so long."[26] Henry Hatfield did not agree. In *Aesthetic Paganism in German Literature* (1964), he concluded that, from *Faust* to *The Magic Mountain*, "From Winckelmann to Rilke, from Goethe to George, the majority of the greatest German writers have been 'Hellenists' to some significant degree."[27]

The Return of the "Many-Sided" Men

Winckelmann may have been the German equivalent of Petrarch, but it was Gotthold Ephraim Lessing (1729–81) who was the Marsilio Ficino of the north, the Renaissance figure who wrote on everything from philosophy to Christianity to astronomy to magic to mathematics. Generally regarded as the founder of modern German literature, Lessing too was a scholar, an antiquarian, a philosopher, a philologist, even a theologian, the first of the "many-sided men" who would characterize the third renaissance in Germany. Above all, Lessing was a symbol of the new, of the new world that existed in Germany—and to an extent throughout Europe—in the eighteenth century, which we have been exploring. This was nowhere more evident than in the fact that Lessing became the first famous German writer to *live* by his pen.

Born in 1729 in Kamenz, northeast of Dresden, he was the son of a pastor and one of twelve children, five of whom died in childhood (not an unusual casualty rate in those days).[28] Lessing had a precocious passion for books and, at the age of six, it is said, refused to be painted with a birdcage in his hand, demanding instead a stack of books. He attended the University of Leipzig in 1746. Then known as "Little Paris," Leipzig was the center of fashion and publishing, and where Johann Gottsched promoted his literary reforms in the Deutsche Gesellschaft.[29]

For the reasons we have been considering, the first generation of creative German writers with a voice of their own emerged around 1750. The most celebrated of these figures, who was slightly ahead of Lessing, was Friedrich Gottlieb Klopstock (1724–1803), who published the first three cantos of his religious epic, *Der Messias* (*The Messiah*), in 1748.[30] The publication of these cantos "of astonishingly sustained power, discipline and abundant imagery," had a profound effect upon German readers. To be understood sympathetically today, they must be read against mid-eighteenth-century theories of genius—that the products of genius are glimpses of the divine.[31] Classical in form, the cantos switch from religion to science to abstract philosophical subjects, interspersed with vivid real episodes, all sustained by brilliant language, the aim—one aim—being to show that the poet can foment as much enthusiasm and faith as the Messiah.

Klopstock was a many-sided man, too. In his treatise *Die deutsche Gelehrtenrepublik* (The German Republic of Scholars; 1774), which impressed young writers like Goethe, he broadcast his vision of a "republic of learning," which found expression, among other places, in his metaphor of the "*Hain*" or "grove," the German equivalent of the Greek *Helicon*.[32] This idea prompted a number of young writers at the University of Göttingen to form a circle called the *Hainbund*, where the natural and social sciences, literature, and the arts were discussed in equal measure.

Though Lessing was interested in—and to an extent stimulated by—Gottsched and Klopstock, it was a less well-known figure, Christlob Mylius (1722–54), a cousin of Lessing's, who introduced him to the theater. Lessing wrote a number of early plays though they were overshadowed, to begin with, by his hack (but often brilliant) journalism and the ambitious quarterly he inaugurated that dealt with the drama, as a result of which Lessing was offered a position as reviewer for the *Berlinische privilegierte Zeitung* (which later became the well-known *Vossische Zeitung*).[33] He now had a regular income and abandoned playwriting to concentrate on criticism. In doing so, he formed a firm friendship with two other many-sided men, Friedrich Nicolai (bookseller, editor, publisher, writer, philosopher, satirist) and Moses Mendelssohn (philosopher, mathematician, critic—a man who even risked criticizing the poetry of Friedrich the Great), and all three began to be talked about in Berlin.

Over the years, Lessing started, or had a hand in, no fewer than five periodicals, designed to raise the standard of German literature and rescue it from mediocrity.[34] He studied Winckelmann (disagreeing with many of his conclusions about Greek art), dipped into archaeology, and investigated what to him were the crucial differences between art and poetry.[35] In 1765, he was offered the chance to become a dramatist and consultant to a new theater company in Hamburg. This was nothing less than an attempt to create a national theater in Germany. It opened in April 1767, at which time Lessing published the first issue of his fourth periodical, *Die Hamburgische Dramaturgie* (*The Hamburg Dramaturgy*), the aim of which was to stimulate general interest in the theater. Lessing's best-known advocacy in the *Dramaturgy* was that stories about those whose circumstances are nearest to our own move us most and that the presence of kings and princes on stage, though adding grandeur, removes an element of familiarity, making identification with the characters more difficult and therefore less affecting.[36]

Neither the theater at Hamburg nor the *Dramaturgy* was as successful as Lessing hoped. This setback was compounded when, on a trip to Italy to survey its antiquities, he met and married Eva König. In January 1778 she gave birth to a child who died within twenty-four hours, the mother herself dying five days later.[37] In the midst of his despair, Lessing found himself locked into one of the great fights of his life. The previous year, he had begun to publish in *Zur Geschichte und Literatur* (*To History and Literature*), his fifth periodical, excerpts from the manuscripts of Hermann Reimarus's *Apologie oder Schutzschrift für die vernünftigen Verehrer Gottes* (*Apologia for the Reasonable Worshippers of God*). Reimarus (1694–1768) was a respected Hamburg schoolteacher who, in his manuscript, argued that Jesus was "a noble-minded but imprudent agitator," that the Resurrection was an invention of the disciples, and that, therefore, at root, Christianity is based on deceit. The problem with the manuscript was that although Lessing wanted to see it published, Reimarus only ever intended publication in a later, more tolerant time. The two men realized there would be reprisals if the book were published openly and so it was released in installments, anonymously.[38] And indeed, the *Apologia* brought about a storm of protest from orthodox Protestants. Lessing's opponents, realizing they could not defeat him intellectually, importuned the Duke of Brunswick, where Lessing was living, to censor him. Financially dependent on the duke, Lessing was forced to comply.

It was a blow, but during the fight Lessing exchanged letters with Reimarus's daughter and, in so doing, conceived a notion for reviving an earlier exercise. So came about *Nathan der Weise* (*Nathan the Wise*), his masterpiece, published in 1779.

Lessing's "masterpiece among his masterpieces" was written in blank verse almost a decade before Goethe and Schiller.[39] The plot was taken from a Boccaccio fable in which a father possessed a ring "which had the power of making the wearer, who believed in it, agreeable to God and men."[40] This father loved his three sons equally and so, unwilling to favor one above the others, commissioned two replica rings and bestowed one on each son. After his death, the sons were unable to agree on who had the genuine ring and took their dispute to a judge. This wise man's verdict was that none of the rings was genuine.

Nathan the Wise is set in Palestine during the Crusades and, to begin with, it is a play about a man, a Christian Templar, and a woman, the adopted daughter of a Jew, who fall in love not knowing that they are brother and sister.[41] The blood link is revealed soon enough to prevent them from marrying, but, at the same time, they discover that their father was Saladin's brother. Saladin the Muslim is in effect one of the three sons in the Boccaccio fable. With him, the other main protagonists are Nathan, the Jew, and the Christian Templar. Saladin asks Nathan which of the three great religions is the true one, and Nathan replies with the tale of the three rings. Throughout the twists of the plot, Nathan is portrayed as a wise soul—tolerant and understanding. Saladin, proud and noble himself, recognizes Nathan's qualities. It is the Christians who, to begin with, are contemptuous and intolerant, in particular toward the Jews. The changing fortunes of the characters eventually soften the Templar's intolerance, though the Christian Patriarch doesn't change. And this is the point of the play: Lessing shows us that there are three kinds of individual, morally speaking: those incapable of moral judgments; those who can see the right course of action, yet do nothing; and those who see what is right *and* act accordingly. The middle group suffers Lessing's unmitigated scorn.[42]

Lessing is now recognized as the dominant figure in German literary life before Goethe. His plays helped bring to an end the chronic provincialism of German literature and his criticism ended the hold that French literary models had over Germany, in particular what he called in one of his letters Gottsched's "slavish adherence" to Jean Racine, Pierre Corneille, and Voltaire. Lessing realized that Shakespeare was a much better model; he argued that *Othello*, *King Lear*, and *Hamlet* were the first modern dramas to achieve the same emotional impact of Sophocles. He argued that Dr. Faust, known to German audiences since the Middle Ages through a puppet play, would lend itself to the Shakespearean approach. This "provided a foundation on which the Weimar classicism of Goethe and Schiller was to build in the closing decade of the century" (Goethe's original version of the drama—the *Urfaust*, written in 1772–75, wasn't discovered until late in the nineteenth century, and was never performed. But *Faust, Der Tragödie erster Teil* is widely viewed now as the decisive moment of innovative change on the German stage).[43]

Not least, Lessing's meticulous investigation of the Gospels was the first dispassionate scientific examination of their origins, boosting scholarship. In the words of one critic, he was "the most admirable figure in the history of German thought and literature between Luther and Nietzsche."[44]

The Origins of Modern Scholarship

We now need to examine two others who were to convert Winckelmann's theories into definite institutional innovation.

The development of classical studies as we understand them today owes much to the work and "uncompromising vision" of Friedrich August Wolf (1759–1824). Nineteenth-century classical scholarship—the "conquest of the ancient world by scholarship," as Ulrich von Wilamowitz-Moellendorff, the eminent nineteenth-century philologist put it—properly commences with Wolf. Wolf from the beginning rejected any prospect of theological training and was determined to rid classical studies of any clerical control. He was not the first modern philologist, but his rigorous methods of source criticism shaped and promoted philology as the new queen of disciplines. His 1795 study of Homer has been described by Anthony Grafton as "the charter of classical scholarship as an independent discipline."[45]

Born in 1759, the son of a schoolteacher, Wolf could read some Greek at the age of six, and rather more Latin and French. At Göttingen, although he kept his distance from the most famous classicist there, Christian Gottlob Heyne (1729–1812), he nonetheless emulated the older man's dedication: like Heyne, he slept for just two nights a week for six months, so as to immerse himself in his beloved classical authors as quickly as possible, keeping himself awake by sitting with his feet in a bowl of cold water. He would bind up one eye with a bandage to rest it, while he used the other. His dedication was reminiscent of Winckelmann and the block of wood he attached to his foot.

Wolf's devotion paid off. In 1783, at the age of twenty-four, he was offered the position of professor of pedagogy and philosophy at the University of Halle.[46] At Halle, the original home of the seminar, he introduced his own, aimed at turning out specialist classical scholars. He succeeded so well that his seminar became the model for the new German universities of the nineteenth century. Wolf, says Suzanne Marchand, was haughtily convinced of the power of philological study "to instil self-discipline, idealism, and nobility of character," a conviction that spread throughout the civil service and professions as the nineteenth century wore on.

His best-known works, the *Prolegomena ad Homerum* (*Prolegomena to Homer*; 1795), and the *Darstellung der Altertumswissenschaft* (*Classical Scholarship: A Summary*; 1807), were not especially original but his painstaking textual interpretations, combined with some sharp common-sense thinking on Homer and his times, placed philological expertise above the philosophical. "[Wolf] was the first to show that access to the Greek mind was to proceed by means of strict attention to linguistic, grammatical and orthographical detail."[47]

In order to support his argument that Homer's poems were written down only in the mid-sixth century B.C., when Pisistratus was tyrant of Athens, Wolf used the fact that linguistics varied in the past to show how, in the earliest manuscripts, entire sections had been interpolated; and, by studying what was missing ("the argument from silence"), he deduced further conclusions—for example, classical commentators found no mention of writing in the Homeric poems. In the process, Wolf was open about his methods, admitting the difference between what he knew and what he merely conjectured, identifying which authorities he trusted and which he did not.[48]

In the *Darstellung*, he distinguished between Greeks and Romans on the one hand, and Egyptians, Israelites, and Persians on the other. He said unequivocally that only the Greeks and Romans possessed "a higher *Geistescultur* (intellectual culture)." The "Orientals," as he described the rest, had merely reached the level of "*bürgerliche Policirung oder Civilisation* [policed civility or civilisation]." He thought cultures need "security, order and leisure" so as to evolve "noble perceptions and knowledge" and this had not happened in antiquity outside Greece and Rome. Literature in particular was vital for a culture—it was the free, untrammeled product of a nation. For Wolf, therefore, Greek and Roman civilization alone constituted *Altertum* (antiquity). Egyptians, Israelites and the rest were "Barbari."[49] In his full-fledged scheme, *Altertumswissenschaft*, for Wolf, comprised no fewer than twenty-four disciplines—from grammar to epigraphy to numismatics to geography, all being needed for full access to a text.

As a scholar, Wolf's reputation was supreme. Goethe attended his lectures, and in 1796 Wolf was offered the chair at the University of Leiden in the Netherlands, then the summit in classics. He turned it down and for the next decade continued at Halle until the French occupation of the city in 1806 changed everything. It might have been a disaster, but only three years later he was offered and accepted the first professorship of Altertumswissenschaft at the new University of Berlin.

Suzanne Marchand argues that Wolf's pursuit of philological expertise "contributed to the turning inward of the university community after 1800." This was an important innovation in scholarship. "Wolfian haughty insistence on 'disinterestedness' and scholarly autonomy imbued philology and *Altertumswissenschaft* with a kind of social detachment rare among eighteenth-century scholars, many of whom had depended on the patronage of aristocrats or income from a second job outside the university. Eighteenth-century professors, too, had generally been esteemed for their lecture skills rather than for their independent research."

Winckelmann had made more of the comparison between the Greeks and the moderns than between the Greeks and the Germans. The advent of the wars with France changed all that too. Amid comprehensive defeat, the parallels between the German predicament and that of ancient Athens—politically fragmented, conquered by force of arms, yet having a superior culture (to Rome) united by a single language—became more plausible. "In the shadow of Prussia's defeat in the Battle of Jena in 1806, German philhellenism underwent a profound change; its anti-aristocratic aspects were transformed into pronational sentiments, and a new form of pedagogy, built on the notion of *Bildung*, made its peace with the state and the status quo."[50] Instead of birth or position, neohumanism—the foundational belief of the new Bildungsbürgertum—judged an individual according to his or her cultural capabilities.

One of Wolf's close friends was Wilhelm von Humboldt. Humboldt shared with Wolf a belief in the study of the ancients. For him, the study of classical texts provided a means of meeting the more impressive self-educated individuals who had existed in the past and it was also, he felt, a way to "discipline the mind." Both Wolf and Schiller, to whom Humboldt was also close, convinced him of the suitability of ancient Greece as a countervailing influence to the social fragmentation of the late eighteenth and early nineteenth centuries. In 1802, Humboldt was appointed Prussian ambassador to the Holy See, providing him with ample opportunity to live among the antiquities of Rome. This became practically relevant in 1808, when a number of Prussian schemes for reform were initiated, stimulated by Napoleon's comprehensive defeat of the Prussians at the Battle of Jena. These reforms, carried through between 1806 and 1812, were implemented under the aegis of two energetic nobles, Karl August von Hardenberg (1750–1822) and Karl vom Stein zum Altenstein (1770–1840). The most important of the reform measures were the freeing of serfs, the granting of a (limited) form of citizenship to Jews, certain economic reforms, and a rethinking of the bureaucracy—which is where Humboldt came in. A new department of the Interior Ministry—for educational and ecclesiastical affairs—was created, and Humboldt, a friend of Hardenberg and Altenstein, was made minister in charge. Until then, educational institutions in Germany (in particular primary and secondary schools) had been administered by the church, but Hardenberg and Altenstein were convinced a new relationship between the state and the schools was needed.[51] Humboldt's responsibilities included the supervision of schools, universities, the art and science academies, cultural associations, and the Royal Theater, all close to his heart. He now became patron and guardian of the educational ideal that had shaped himself.

As a fundamental measure, he centralized funding. He introduced a requirement that all prospective university students pass a new examination, the *Abitur*, the main element of which was testing translations of Greek and Latin texts. Furthermore, he made the Abitur the sole prerogative of a particular type of classical school—known as the *Gymnasium*. Only these could prepare pupils for university, a state of affairs that continued for close to a hundred years. The culmination of Humboldt's reforms was his design of the University of Berlin, founded in 1810. Consolidating the trend begun at Göttingen under Münchhausen, Humboldt promoted Berlin's philosophical faculty (containing philology, philosophy proper, and the natural sciences) over and above the more "practical" faculties of medicine, law, and theology. More than that, within the philosophical faculty he subordinated the natural sciences to the humanities, "fearing that the former would otherwise slide into mindless empiricism." By offering large salaries (again taking a leaf out of Münchhausen's book) Humboldt soon lured to Berlin a raft of brilliant young scholars across many disciplines. "Berlin quickly became known as an *Arbeitsuniversität*, an institution for industrious, mature and unsocial scholars such as Wolf and Humboldt themselves had been."[52]

Humboldt wasn't at the ministry for long—he left in June 1810. By then, however, through the university in Berlin, the Abitur, and the Gymnasium, he had made neohumanist Bildung "the cultural philosophy of the Prussian state."

Humboldt had a clear understanding of what, exactly, Bildung was. For him, the development of social morality within an individual was an all-important progression which depended on that individual's "self-transformative progress" from a natural state of ignorance and immaturity to "self-willed citizenship": a shared understanding of civic harmony and loyalty to the state, a belief that spiritual emancipation through education in the humanities was the true path to (inner) freedom and "willing citizenship." It was a vision both egalitarian and elitist at the same time and this paradox was to have far-reaching consequences.[53]

Partly under Wolf's influence, and partly under Herder's, Humboldt specified that language was to be the main focus of education. He insisted that the very shape and structure of a language revealed a nation's character. Bildung, for Humboldt, could therefore be achieved only by study of the Greeks, allied to an understanding of language. For Humboldt, Bildung—true (inner) freedom—involved three things: *Zwecklosigkeit*, *Innerlichkeit*, and *Wissenschaftlichkeit* (non-purposiveness—in the sense of non-utilitarian—inwardness, and scholarliness). All (male) students at the Gymnasium must attempt this historico-philological form of learning.[54]

He both did and did not succeed. By the time of his death in 1824, Wolf's vision had prevailed: classical philology was the recognized foundation for professionalized humanistic scholarship.[55] Later on in the nineteenth century, when German scholarship became the envy of the world, many disciplines would seem to have little to do with ancient Greece. But the *methodology* on which their successes were based went back to Humboldt, Wolf, and—ultimately—Winckelmann.

4 ·

The Supreme Products of the Age of Paper

In the fifteenth century, at the height of the Italian Renaissance, it took twenty minutes to cross Florence on foot, fighting the crowds from the Ponte Vecchio to the Piazza di San Marco. Into this area, 95,000 people were crammed.[1] By today's standards, Renaissance Florence was not a large city but even so it dwarfed Weimar, which perhaps fulfilled a similar role in the German renaissance.

As one approached the city in the eighteenth century, one saw—standing out above the town's 600 or 700 houses and their enclosing wall—the towers of a couple of churches and the ducal *Schloss* (fifteenth-century Florence had one cathedral and 110 churches).[2] There were two inns, the Erbprinz and the Elefant, three shops worth the name, and the streets were lit at night by 500 lanterns, though they were so expensive to maintain that the order to light them all was seldom given.[3] In 1786, the population was approaching 6,200, of which 2,000 were courtiers, bureaucrats, soldiers, or pensioners supported by taxation.[4] There was no trade, no tourism, and of course no factories. No wonder Madame de Staël felt Weimar to be "not a small town but a large château."[5]

Though it was small and unprepossessing physically (drainage was still very primitive), Weimar was a capital and it had a court. The original star—or "muse" (Goethe's word)—of this court was Princess Anna

Amalia of Brunswick, who had been married in 1756, while still a girl, to Ernst August Konstantin of Weimar, himself no more than eighteen. His small duchy was undistinguished, just one of 300 not dissimilar entities that stretched north to the Baltic and south to the Alps.[6] Weimar actually consisted of the combined duchies of Weimar and Eisenach, together with the former duchy of Jena and the bailiwick of Ilmenau. All four areas still maintained their separate tax systems.

When Anna Amalia was married, she was not yet seventeen and her husband was under nineteen. Before he died, two years later, she had borne him one son, Karl August, with another on the way. She became Weimar's regent during her son's minority, and it was in the nineteen years before his accession that she changed the court and made later developments possible. Anna Amalia's mother was a sister of Friedrich the Great and shared his views on the importance of art, literature, and theater.[7] On first meeting Anna Amalia, Schiller found her mind "very limited" but, in her efforts to keep up with rival courts nearby, she brought troupes of actors to Weimar, then musicians—and then "literati." There would be four men of world stature who Anna Amalia helped bring to Weimar, of whom the first was Christoph Wieland (1733–1813).

Appointed in 1772 as tutor to Karl August, then fifteen, Wieland was already one of Germany's leading authors. He was of middle-class origin (his father was a pastor) and this association, of aristocrats and the middle class, unwittingly initiated a process that was to result in the partial fusion of two culture groups that would together provide what became known as "Weimar *Klassik*."* Writers whose origins were lower down the social scale could earn a good living in either Paris or London, but in Germany social distinctions were still relatively rigid and amid all the other changes that Weimar brought about, the social change was as important as any.

Wieland's early work, however, had earned him distinction among aristocrats, who counted. He had been a senator in his native town, Biberach in Württemberg, and a professor of philosophy at Erfurt (in Thuringia, central Germany); in his novel *Der goldene Spiegel* (The Golden Mirror; 1772), he had presented a political philosophy in the tradition of

---

* *Klassik*, to distinguish from the classicism of antiquity.

Lettres persanes in that it criticized current affairs in Europe in a fanciful Oriental disguise (this was a practice that had begun in France). The novel emphasized education even for princes and especially the importance of history. But Wieland was also known for his novel *Geschichte des Agathon* (The Story of Agathon; 1766–67), a narrative of a young man who learns through personal experience that the excessively spiritual "enthusiasm" of his youth was folly. This was, in its way, a first sighting of rudimentary *Bildung*, and Wieland's importance lies in his early grasp of this concept. The loss of faith experienced by many figures of the Enlightenment seems to have been more intense, sooner, in Germany than in France or England. Following the Earl of Shaftesbury (a profound influence in Germany), Wieland understood that, in the age of doubt, a man can still live for knowledge, art, and reflection—"the enlargement of his mind"—*and* continue to fulfill traditional duties.[8]

Wieland took up his position in Weimar in September 1772 and immediately embarked on a project for a new literary monthly, *Der deutsche Merkur*, to accompany his primary responsibility, teaching Karl August. The first issue appeared in 1773 and was an immediate success. It continued publication for nigh on forty years and provided central Germany with a literary culture, in the process making Weimar, tiny as it was, the cultural capital. Wieland's views overlapped with those of Friedrich the Great, who was still alive. He too thought Germans showed "a chronic uncertainty" in matters of taste, in marked contrast to somewhere like England, which had its "classics," as we would say now. Wieland translated and published several Shakespeare plays in an attempt to show Germans what a "classic" from the post-classical world "looked like."[9]

Wieland was always convinced of the cultural importance of the theater. He pointed out that the stage had been a political institution in ancient Greece and it was, even now, in the enlightened, nonabsolute parts of Europe, a moral institution, "capable of exercising a wholesome influence on the thought and manners of an entire people." Wieland, we must remember, was writing at a time when theater had still to contend with the opposition of the church. Even if others didn't, Wieland recognized the theater as a venue where people could experience new ideas in a *shared* capacity. (This is one reason why the church—and other authorities—objected.) The theater helped establish the actuality of "imagined communities," as Benedict Anderson calls them; it helped spark the self-consciousness, and the self-confidence, of the middle classes.

As for many others, for Wieland the German nation was "not really a nation, but an aggregate of many nations, like ancient Greece."[10] He nevertheless thought that modern Germany had a character all its own, meaning he had sympathy with—and encouraged in his journal—the early Sturm und Drang poets then emerging. He also welcomed Goethe's *Götz* as "the most beautiful and interesting of monstrosities, worth a hundred of our sentimental comedies."[11] Here was the kind of new voice that, he felt, was needed.

The First Great Tragic Novel

Johann Wolfgang Goethe's arrival in "the large chateau" stemmed from a chance meeting he had with Karl August, the son of Ernst August and Anna Amalia, in Frankfurt. Having chosen his bride, Luise of Hesse-Darmstadt, as he approached his eighteenth birthday, the prince regarded himself as now free to embark on his grand tour. This began, sensibly enough, with a visit to Karlsruhe, where Luise was living, but en route Karl August stopped at Frankfurt, where he was introduced to Goethe, already well known because of *Götz* and *Werther* (see p. 115). This unlikely pair got on surprisingly well together and met for a second time months later, this time in Karslruhe. The prince had overnighted there on his way back from Paris, while Goethe was bound for Switzerland. Paris had seduced Karl August. His tastes and his ambitions had grown more sophisticated and more cosmopolitan: Goethe was invited to Weimar.[12]

The difference between Frankfurt and Weimar was larger than Goethe expected. In Frankfurt, a center of commerce, position was largely determined by financial criteria. At Weimar, in contrast, a basic distinction was made between those admissible at court (*die Hoffähige*) and the rest. To be admitted at court, a title was necessary. And so was established a pattern that was to be repeated. Goethe was raised to the nobility, becoming Johann Wolfgang von Goethe, as happened with Schiller and Herder some years later.[13]

At first, Goethe thought of himself as a visitor in Weimar, and this seems to be how others regarded him. Portraits show a man with large eyes, a slight crown on the bridge of his prominent nose, sensual lips. When he arrived in Weimar, Goethe was twenty-six to the prince's eighteen—a large gap at that age and not the only important difference between them. Karl August might be a prince but Goethe was already much more famous. The previous year, he had produced *Die Leiden des jungen Werthers* (*The Sorrows of Young Werther*), which, it is no exaggeration to say, had taken Europe by storm. Generally regarded as the first piece of "confessional" literature, this novel is of added interest because of its autobiographical element. The plot line of *Werther* is closer to Goethe's life than that between almost any other author and their works.

In his early twenties, Goethe had spent time in Wetzlar, a small town forty miles north of Frankfurt. Ostensibly looking for a law practice, but not trying too hard, he spent a good deal of his time reading and writing poetry and falling in love with a young woman, Charlotte Buff, who was already engaged to someone else. It took Goethe a while to realize he could never win Charlotte, but then he moved on to Koblenz where, as was his way, he soon fell for someone else. He kept in touch with Lotte and her fiancé and through them learned the details of the suicide of a mutual friend, Karl Wilhelm Jerusalem. Jerusalem had fallen in love with a married women who hadn't returned his feelings, whereupon the young man had borrowed some pistols (from none other than Charlotte's fiancé) and shot himself. It therefore comes as no great surprise, as Michael Hulse says, that *Die Leiden des jungen Werthers*, published in Leipzig in 1774, was received in its time (and continues to be read) as partly autobiographical, partly biographical.[14] Goethe, who took barely four weeks to write the book, later referred to it often as a "confession."

The plot is simple and, says Nicholas Boyle, could only have been written because of Goethe's religious emancipation.[15] Werther falls in love with Charlotte (Lotte), who is betrothed to another man. Although his love is requited by Lotte, it is doomed, and the unresponsiveness of the world and the couple's sufferings eat away at him, so that he can see no alternative but to shoot himself with Lotte's husband's pistols. An "editor" then "gathers" his letters and publishes them with the occasional comment.

Werther was almost immediately translated into every major European language, but the cult of the book went much wider. In Vienna there was a Werther fireworks display and in London there was Werther wallpaper. Meissen porcelain was designed, showing Werther scenes, and in Paris perfumières sold Eau de Werther. In Italy there was a Werther opera. Napoleon took the French translation with him on his Egyptian adventure in 1798 and, Hulse says, "when he met the author in 1808 he told him he had read the book seven times" (though he also added some criticisms).[16]

Not everyone shared the rapture. There were those who thought the novel risked sparking a suicide epidemic, but the fear of a wave of *Liebestode* seems to have been exaggerated. In Leipzig, nevertheless, the book was banned, as it was in Denmark. Elsewhere the book was derided, one critic suggesting sarcastically that "The smell of pancake is a more powerful reason for remaining in this world than all young Werther's supposedly lofty conclusions are for quitting it." Now that the dust has settled, *Werther* has come to be regarded as "the first great tragic novel, a work of exhilarating style and insight."[17]

Despite Goethe's fame, despite the turbulence his novel sparked across Europe, his friendship with Karl August was solid and genuine, and the writer joined in readily with those activities of the court that the younger man enjoyed—in particular, riding, shooting, and dancing. Bit by bit, however, Goethe's very presence induced a change, and he began reading from his works in progress (most of what he wrote, he read aloud to friends), in particular his unfinished *Faust* (the *Urfaust* as it is now called).[18]

Time was passing and after about a year, when it was becoming clear that Goethe's "visit" was no such thing, Karl August moved to bring his friend closer still. Goethe was persuaded to join in another popular aspect of court life: amateur theatricals.[19] In this way he became the prince's unofficial *maître des plaisirs*, and it was this appointment, informal to begin with, that shaped Goethe's immediate future and, indeed, that of Weimar, as other—more onerous and more responsible—duties followed. More than one historian has observed that Karl August's liking for Goethe owed rather more to his personal qualities than to his fame and skill as a writer. (Jürgen Habermas reminds us that Weimar was a special case, that most men of letters were little more than servants at that stage.) Goethe's elevation at court was not universally welcomed (he was considered a "half-baked Voltairian know-all" by some), but the prince was an absolute monarch in an age of absolutism—and that was that.[20]

The big change, from Goethe's point of view, came in June 1776, when he was appointed a member of Karl August's *conseil*, or Privy Council, which consisted of the duke and three advisers. Goethe was required to take an oath of allegiance, and that gave him the right to wear a distinctive laced coat.[21] Now his duties widened further. He took on the mining commission, the military commission, and was even made temporary head of the treasury. He was involved in road-building schemes and helped devise a new system of taxation. By all accounts, Goethe was a safe pair of hands, invariably aware of what was and was not practicable, and this transformed his popularity and made him respected. He was one of those responsible for the belief in Germany that "intellectuals need not live in ivory towers with their heads in the clouds."[22]

Goethe gained a lot from being in Weimar. Being forced to involve himself in mining, he found he needed to develop his interests in, and knowledge of, chemistry, botany, and mineralogy, and this moved him in the direction of science more generally. He was soon collecting plants and studying Carl Linnaeus's *Philosophia botanica*. Goethe exchanged a number of letters with Linnaeus and at the same time asked Karl August to send one of his assistants, J. C. W. Voigt, for training at the Freiberg Academy of Mining in Saxony, run by the foremost mineralogist of the day, Abraham Gottlob Werner. (Werner, who visited Goethe in late summer 1789, is considered in Chapter 7, p. 167.) Later Goethe turned to anatomy, which he studied under Professor Loder in Jena and then passed on, in his own lectures, at the academy of drawing in Weimar.[23]

At the same time that Goethe was pursuing his multifarious activities, he was also writing 1,800 letters to another Charlotte, Charlotte von Stein (the standard edition of Goethe's works and letters runs to 138 volumes). She carefully hoarded these, "well aware that they were a unique record of an exceptional man's inner life." Their relationship is alluded to in Goethe's two "Charlotte" plays, *Iphigenie auf Tauris* and *Torquato Tasso*, where she is presented as the best in German womanhood, "a German Beatrice," who aids the development of the immature poet and introduces him to "the pleasures and responsibilities of Humanity."[24]

Goethe never really lost the central interest he explored in his "Charlotte" period: his pursuit of "Bildung" (a word he used quite a bit). The inward pursuit of perfection is never again mentioned so directly as in his letters, but Goethe never lost his concern with the individual's responsibility for his own inner development.[25]

From 1781, by which time Goethe had been in Weimar for six years, he confided to Charlotte that he no longer felt able to address her as "*Sie*," and must use the more intimate "*du*." This brought about a sea change. As one critic put it, Goethe's letters now became "prose poems of happy love with few parallels in any literature." However, at the very moment their relationship should have matured, it didn't—and the results were catastrophic. So far as we know, Charlotte, in her "strange ménage," never made any move to leave her husband. When she did confess her love for Goethe, Goethe responded by leaving for Italy without even telling her that he was going, and by the time of his return from the warm south (Verona, Venice, Ferrara, Florence, Arezzo, Rome, Naples) the situation had deteriorated.[26] He had enjoyed his time ("this journey is really like a ripe apple falling from the tree"), but his views on love had been transformed (from a romantic view to a "pagan" view, it was said by some).[27] A sketch by his friend Tischbein, made in Rome, shows Goethe irritably pushing away "the second pillow"—i.e., Charlotte's husband.[28] Also, Charlotte found it difficult to come to terms with the fact that Goethe had begun another affair, with Christiane Vulpius. ("I am only interested in the real thing, eager eyes and smacking kisses.") Charlotte made a lame attempt to pillory Goethe in her play *Dido*, but it was not a success.[29] Their once-beautiful relationship had turned sour, and though they patched it up eventually, things were never the same again.

Through it all, Goethe continued to write. A mixture of the realist and the romantic, not given overmuch to abstract speculation, Goethe subscribed to the view, explored in Chapter 2, p. 65, that "God does not exercise influence on earth except through outstanding chosen men."[30] He knew he had it in him to be an outstanding man, "a great soul," as he had described the character of Iphigenia.[31] He was also deeply affected by his discovery of the Greeks (thanks to Herder) and their idea that individuals—even geniuses—may contain within them unconscious creative urges that other people will find superlative, but those works still need to be realized, to be produced, and that task involves craft, perseverance, individual effort. The idea that life is a *task*, was, of course, Pietist in origin, but the Greek influence seems to have induced Goethe to *craft* his next masterpiece, *Wilhelm Meister*, and so successfully that even a sarcastic skeptic like James Joyce had to put him on a par with Shakespeare and Dante (Joyce's trinity was "Shopkeeper," "Daunty" and "Gouty").[32]

In 1798, in a famous "Fragment" published in the *Athenäum*, the periodical of the early German Romantics, Friedrich Schlegel identified the French Revolution, Johann Gottlieb Fichte's *Wissenschaftslehre* (1794), and Goethe's *Wilhelm Meister*, as "the three greatest 'tendencies' of the age."[33] Schlegel was being deliberately provocative, but even in retrospect this takes some swallowing. Schlegel we shall encounter presently, where we can examine what, exactly, he meant by his choice. Fichte we shall also come to later, where we can explore the meaning and significance of his *Wissenschaftslehre*. But whatever Schlegel meant, when he put *Wilhelm Meister* in this exalted company, he was doing us a favor of sorts. There is no question that, as a novel, as a story, *Wilhelm Meister* was not only a masterpiece: it was also the first of a genre, a particularly *German* genre, which became known as *Bildungsroman*.

A *Bildungsroman* is typically a novel of ideas. William Bruford, professor of German at Cambridge after World War II, devoted an entire book to the German Bildungsroman, in which he shows that Goethe's model was followed by many others in Germany. Here is his definition of the form: "[W]e are shown the development of an intelligent and open-minded young man in a complex modern society without generally accepted values . . . We see him acquiring a point of view but above all a '*Weltanschauung*,' a lay religion or general philosophy of life . . . In a Bildungsroman the centre of interest is not the hero's character or adventures or accomplishments in themselves, but the visible link between his successive experiences and awareness of worthy models and his gradual achievement of a fully rounded personality and well tested philosophy of life."[34] It is a journey inward as much as forward.

In Goethe's story, Wilhelm is born into a bourgeois family and, as his adventures go by, he comes to understand the limitations of his original existence as a "carefully brought-up son" in the middle classes. He lives for a time among theater people, engrossed by the charm of their spontaneity; elsewhere, he is introduced to "the lesser talents" of being a gentleman, talents that are mainly negative—a gentleman "does not show his feelings," he understates everything, he never hurries; later, Goethe has Wilhelm wounded in an attack by armed bandits. Through it all, he meets a raft of women—older women, capricious women, women from a higher social class. He observes which men are successful with the opposite sex and what their secret is.[35] He immerses himself in the works of Shakespeare, discovering a rich world he never knew existed. Eventually, he marries a woman from a large family and—part of the point of the book—begins to achieve a measure of control over, and understanding of, his life.

Goethe had a serious aim. He had told Caroline Herder that he had lost his belief in divine powers in the summer of 1788 and the purpose of life, when there is no god, he is saying in the book, is to *become*, to become much more than one was.[36] "The ultimate meaning of our humanity is that we develop that higher human being within ourselves, which emerges if we continually strengthen our truly human powers, and subjugate the inhumane."[37] Some non-Germans have found it too much. Henry Sidgwick, the Cambridge-based late nineteenth-century philosopher, is said to have reprimanded a German visitor who observed there was no word in English equivalent to "*gelehrt*" (cultivated). "Oh yes there is, we call it a prig."

Goethe's most famous masterpiece, however, and this is true both inside and outside Germany, is *Faust*. This, "the most characteristic product of his genius," was written at intervals over sixty years, in four bursts of creative energy.[38] It was by no means a new story, being a well-known medieval legend, made into a play by Christopher Marlowe, though Goethe wasn't aware of Marlowe's work until he had written more than half of his own version.[39]

The legend may even be grounded in fact. There was a Georg Faust alive at the turn of the sixteenth century who wandered through central Europe claiming to possess recondite forms of knowledge which gave him special healing powers. After his death he gradually acquired a slight change of name and an academic title, as Dr. Johannes Faustus, a professor at Wittenberg. In his lectures, he was alleged to "conjure up at will" personages from classical Greece, and he was notorious too for allegedly playing tricks on both the pope and the emperor. There was a price to pay for this license and in his case it was said that he had agreed to "a term" of twenty-four years with the devil, whereupon his body would be "torn to pieces by demons." Faust was often featured in puppet plays, which is where Goethe may have encountered the story as a child.[40]

According to the legend, Faust becomes disillusioned with the many forms of secret knowledge he has tried out, and the devil, Mephistopheles, makes a wager with God that he can tempt Faust into his world. Mephistopheles sees to it that, in the course of his researches into magic and alchemy, he himself is conjured up by Faust, whereupon he broaches his famous proposal: Mephistopheles will introduce Faust to all the pleasures the world has to offer, the only proviso being that if, at any point, Faust should wish to sample any delight "for longer than the moment it is on offer," his life will end and he will belong to the devil. Bored, unfulfilled, Faust accepts.

In Part One the main theme is Faust's seduction and subsequent desertion of Margareta (Gretchen), a beautiful girl he meets outside a church. In Part Two—written decades later—Faust, waking from a long sleep during which he has forgotten Margareta, now becomes enamored of Helen of Troy (this is a magical world, after all).[41]

No brief summary can do justice to the attractions of *Faust*—its language, its wit, its pithy insights into human nature, not to mention Mephistopheles' brand of cynicism, "an original and highly effective, not to say sympathetic conception of the devil." *Faust* and Mephistopheles have even been compared to the Book of Job as a meditation on the nature of evil (in Job too there is a compact with God). Goethe also took a leaf out of Shakespeare's book: like the Bard's plays, *Faust* resists Christianization. God is not the petty-minded jealous God of the Israelites but a more generous—and yes, even witty—deity.

Goethe began writing the book in the early 1770s, later destroying the manuscript—or so he thought. The very existence of this early manuscript, the *Urfaust,* was unknown until 1887, when it was discovered sixty-five years after his death. Apparently it was copied by a young lady at the Weimar court, and her manuscript was never destroyed. Goethe's description of *Faust* as "fragments of a great confession" should not be forgotten. He himself wrote:

. . . Our play is rather like the life of Man:
We make a start, we make an end—
*But make a whole of it? Well, do so if you can.**

Was he saying that was the best way to treat life? Absorb its disparate parts, "Kiss the moment," as Schiller was to put it, but don't try too hard to impose too much unity.[42] For Faust, it is not the search for unity that matters, it is movement, creation, *activity*, over and above and before mere enjoyment. Mere contemplation of beauty is empty. In this Goethe is a pre-Romantic.

Nicholas Boyle, in his biography of Goethe and his age, argues that "more must be known, or at any rate there must be more to know, about Goethe than any other human being . . . Nearly 3,000 drawings by him survive, as do the villa he built, the palace he rebuilt, and the park he first laid out. He amassed very substantial private collections of mineralogical specimens, [and] incised gems . . . After he moved to Weimar the daily chronicle of his doings, now being put together for the first time in seven large volumes by Robert Steiger, is practically continuous, especially once he began to keep a regular diary in 1796. Accounts of conversations with him . . . run to some 4,000 printed pages, over 12,000 letters from him are extant, and about 20,000 letters addressed to him." As for his writing, Boyle concludes, "As the age of paper passes, so he comes to seem its supreme product."

A New Meaning for "Nation" and "Culture"

Johann Gottfried Herder (1744–1803) was five years older than Goethe and, like Winckelmann, like Heyne and like Fichte, a poor boy who had risen out of his class through sheer ability, along with a chance meeting with a Russian army surgeon who, on his way back from the Seven Years' War, was quartered in 1761–62 in the east Prussian town where Herder lived. Conceiving a liking for Herder, the surgeon proposed to take him to Königsberg, where the young man could study medicine at the university.

* *Des Menschen Leben ist ein ähnliches Gedicht: es hat wohl einen Anfang, hat ein Ende, allein ein Ganzes ist es nicht.*

In return, Herder would translate a medical treatise into Latin. Herder accepted the proposal but, once in Königsberg, he found medicine uncongenial and switched to theology.

This was a second fortuitous turn, for it brought Herder to study under Kant, and it was Kant who introduced Herder to Rousseau and Hume, who were to have such an effect on his thinking. Ordained in 1767, Herder found his way to Paris where he was received by some of the leading figures of the Enlightenment, including Denis Diderot and d'Alembert. But he was still chronically poor and so accepted a position as the tutor and traveling companion to the son of the Prince of Lübeck and Holstein. This was the third fortuitous turn. En route to his appointment, he stopped off in Hamburg, where he met Lessing and, soon afterward, in Strasbourg with the prince, he met Goethe (this was July 1770). Realizing at once that Herder was less than content with his position, Goethe prevailed upon Karl August to invite his friend to Weimar as head of the clergy in the state. Herder kept the position for the rest of his life and seems to have been content—his children used to go looking for painted eggs in Goethe's garden.[43]

Herder is not nearly as famous as Goethe, but in many ways his ideas and influence were more immediate, more direct, and more widespread.[44] "Like Max Weber over a century later, Herder was preoccupied with the problem of social relations in a world that increasingly came to resemble for him a vast machine in which men were like cogs, whose lives were governed by the inexorable operations of mechanical bureaucracies." He addressed this predicament in two books, *Ideen zur Philosophie der Geschichte der Menschheit* (*Reflections on the Philosophy of the History of Mankind*; 1784–91; four volumes), and *Auch eine Philosophie der Geschichte zur Bildung der Menschheit* (This Too a Philosophy of History for the Formation of Humanity), published in 1774. Here, his main concern was, as he put it, to discover the rules of the morals of association. What invisible hand is there that shapes spontaneous political association?

Hume and Kant aside, the main influence on Herder was Leibniz. Herder looked upon Leibniz as "the greatest man Germany ever possessed" and saw himself as part of that tradition—Leibniz, Thomasius, Lessing, and Herder. These men had conceived the idea of "becoming," when the universe was held to be an "organic" entity and in which Leibniz's understanding of history as "a continuous process of development, energised by human striving," profoundly affected Herder's historical thinking. For Lessing, moral striving, "moral becoming," as he put it, was the central concern of all education, of all cultivation. "Man could only be truly himself by consciously realising his individuality." Herder refined this eloquently. He insisted that, for him, humanity was not a state into which man was born "but rather a *task* demanding fulfilment by conscious development." This idea of Bildung as a *task* dominated the philosophy of the majority of subsequent German writers, from Goethe, as we have seen, to Humboldt and Fichte.[45] Again, one can imagine Henry Sidgwick snorting, as he turns in his grave, but there is a unity here in German thought: Bildung as a *task* comes from the recognizably Pietist lineage and looks forward to Weber's concept of the Protestant work ethic. It was also subversive, because it placed a premium on the intrinsic worth of individual judgment, and accordingly rejected (external) authority as the fundamental source in religious and moral matters.

Herder also thought that Rousseau had in effect put the cart before the horse. A "social contract" was a misnomer for him because, as he saw it, the state of society *is* man's state of nature. Man is born into a family whether he likes it or not. But man isn't just a social animal—he is also a political animal, for life in society needs order, organization.[46] This is where Herder joined forces with Bodmer, Gottsched, Wolf, and Humboldt: the underlying "sustaining force" of all this organization, he said, is *language.* As Locke had done, Herder dismissed the divine origin of language. There was, he said, no stage or epoch in prehistory when language had been invented, nor had it developed from animal sounds. We cannot think without words, he insisted, therefore language must have emerged when consciousness developed.[47] Which meant that, for Herder, language reflects the history and psychology of a distinctive social heritage, and this was by far his most influential argument—language identifies a *Volk* or nationality, and *this,* the historico-psychological entity of the common language, is for him "the most natural and organic basis for political organisation . . . Without its own language a *Volk* is an absurdity (*Unding*). For neither blood and soil, nor conquest and political fiat can engender that unique consciousness which alone sustains the existence and continuity of a social entity."[48] Language, as well as *unifying* a community, also identifies that community's consciousness of *difference* from those speaking other tongues. "A *Volk*, on this theory, is a natural division of the human race, endowed with its own language, which it must preserve as its most distinctive and sacred possession." Language was given a potency it had never had before.

It was this close association, between language and self-consciousness, that brought about such a drastic change in the most commonly accepted idea of a "nation." "A nation no longer simply meant a group of citizens united under a common political sovereign."[49] It was now regarded as a separate *natural* entity whose claim to political recognition "rested on the possession of a common language."[50]

Herder went further. For him, in any *Volk* grouping, there were two elements. There was first the bourgeoisie (*das Volk der Bürger*), and second a minority of intellectuals (*das Volk der Gelehrsamkeit*). The *Bürger* were not only the most numerous but also the most useful (he called them "the salt of the earth"), and he sharply distinguished them from what was in fact a third group, the "rabble" (*Pöbel*). What distinguished the Bürger, and what was the chief reason for their social inferiority and political impotence (until that point), he said, was their lack of education. The fact that this was so, he insisted, could not be put down to innate ability, or lack of it. "It was rather the harvest of persistent and wilful neglect." He therefore came out boldly and blamed the ruling aristocracy for this state of affairs. "That those as yet unborn should be destined to rule over others not yet born, simply by virtue of their blood, seemed to Herder the most unintelligible of propositions."[51]

Herder, therefore, did not expect the people at the top of the tree ever to *do* anything to jeopardize their position. Instead, he published his argument in the hope that he might inspire the emergence of popular leaders, "men of the people," who would spread the message of education (Bildung). It was the job of the state, he felt, to help each individual to develop and fulfill his or her propensities. "To fail to make use of man's divine and noble gifts, to allow these to rust and thus to give rise to bitterness and frustration, is not only an act of treason against humanity, but also the greatest harm which a state can inflict upon itself." This shows Herder's very modern grasp of the links between economics, politics, and education or, more particularly, Bildung. For Herder (as for Humboldt), the development of the self, the humanization of the self, will not only make people better individuals but also better—and more willing—members of the community. Reciprocity is for Herder the whole point of human association.[52]

This fit Herder's overall aim, to give culture—as well as nation—a new meaning. With his deeply historical view of human affairs, and his Leibnizian inheritance—that change was of the essence—he formed the view that whatever the "collective consciousness" of a Volk was at any particular time, *was* its culture.[53] This was wholly at odds with the prevailing Enlightenment tradition where culture was aligned with civilization and understood as a reflection of intellectual sophistication. It was thus Herder who was responsible for our modern usage of such phrases and concepts as "political culture," "peasant culture," and so on. At root, Herder suspected that culture was not simply the result of experience alone but owed something to a genetic component (though of course the modern understanding of genetics did not then exist).[54] This combination of genetics and experience helped in turn to generate the main ideas that shape history, he said, and, at any particular stage, a Zeitgeist, a "spirit of the time," can emerge. Herder is credited with coining this term.[55]

Herder's view was essentially an updated and secularized version of Francke's Pietist theology: the Creation could be improved upon. It could be developed, evolved, helped to become more than it had been, all via the process of Bildung, which elevated knowledge as the highest good, the all-important foundation for human association. A belief in the perfectibility of man was a sine qua non, he insisted, if we are to accept a role for the human will in the shaping of history.[56] For Herder, recognizing that there *is* an inborn drive to perfection is part of man's developing self-consciousness, part of his evolution, in a doubt-ridden, pre-Darwinian world.

As Aristotle—and Leibniz—had before him, Herder sought to integrate the inner development of man with his outer, social arrangements.[57] In his lifetime (he died in 1803) the advent of the industrial revolution was not yet well enough advanced to have an impact in Germany (or anywhere for that matter), but the disjunction between the inner world and the outer, social world, would come to dominate nineteenth-century thought above everything else, as the concept of "alienation," preoccupying everyone from Hegel to Karl Marx to Freud. Herder was the first to outline these lineaments in a language we recognize today.

New Forms of Nobility

Johann Christoph Friedrich Schiller (1759–1805) was born in Marbach am Neckar. His father was a military surgeon, and Friedrich was given a medical training in addition to a good general education. The medical faculty at the military academy at Marbach was excellent, and Schiller proved to be an exceptional pupil: his thesis—on physiology—was approved for publication by none other than the duke himself. (In small states dukes and princes sometimes took a personal interest in their clever charges.) Schiller's thesis was accepted only on the third attempt—but even so he broke new ground, first bursting on the world as a scientist. In his 1780 thesis, titled *On the Connection between the Animal and Spiritual Nature of Man*, Schiller not only advanced the view that the mind regulates the body, and vice versa, but he also argued that "harmony" between the two was not the "default" position, as we would say today, but rather that human physiology "is a tension-filled process," a precarious balance that needs to be nurtured and maintained. Within this view is the implication that such things as diet, physical circumstances, and personal relationships have an effect on "mental health" and not just one's relationship with God.

Medicine, important as it was for Schiller, was not his first love. At school, the Karlsschule, he was introduced to Kant, particularly his writings on aesthetics, and to Shakespeare (Schiller became known, later in life, as "the German Shakespeare"). Together with Goethe, Schiller's works comprise the main elements in the "canon" now known as "Weimar classicism."[58]

As with Herder, Schiller's early adult life was unsettled—he moved around, from Mannheim to Dresden to Leipzig. His first play, *Die Räuber* (*The Robbers*), was written in 1781, printed at his own expense, and performed the following year in Mannheim. The central character is the leader in a band of robbers, but the chief theme is that leader's rejection of the values of his father. The play is essentially about the nature of liberty—

to what extent is it inner freedom, and to what extent outer social/political freedom?—such a focus being new and daring (at least in Germany with its many absolutist states).[59]

Among all Schiller's plays, *Kabale und Liebe* (*Intrigue and Love*, 1784), has always been the most popular, surviving everywhere as *Luisa Miller* in Giuseppe Verdi's opera of that name. The play is an attack on the cruelty and oppressiveness of absolutism. Ferdinand and Luise, the two main characters, try to escape their class and the bourgeois and aristocratic conventions that imprison them. They fail. Ferdinand is prepared to risk all but not Luise, who (correctly) anticipates there will be reprisals against her father. Schiller's point is that, under absolutism, characters are unable to become autonomous.[60]

Schiller's first "overwhelming" masterpiece came next, *Don Carlos* being welcomed by a critic as "one of the world's greatest pieces of literature."[61] Here the theme of conflict between father and son is continued. Set in sixteenth-century Spain during the reign of Phillip II, Don Carlos, heir to the throne, is in love with his childhood friend, Elisabeth of Valois, to whom he was once betrothed but who is now his stepmother following her marriage to Phillip. Carlos is determined to resolve his passion for Elisabeth, and to do so enlists the aid of his friend, Rodrigo, Marquis of Posa, whose task is to engineer a meeting between the two would-be lovers. But Posa, charged with arguing the case of the oppressed people of Flanders, sees in this an opportunity for Carlos to foment a full-scale rebellion against his father's tyrannical regime, which extends not just to his political subordinates but to his family as well. More than this, Schiller's deeper design is to show that weakness is as much the basis of tyranny as is strength. Phillip may have raw political power but he is lonely, jealous, and miserable.[62]

In 1787, the year *Don Carlos* was performed, Schiller visited Weimar, hoping to meet Goethe. At the time, Goethe was in Italy, but Schiller did succeed in meeting both Herder and Wieland and spent time in the company of Duchess Anna Amalia. Two years later, on Goethe's recommendation, he was invited to become professor of history at the University of Jena.

Although Goethe was responsible for Schiller's presence in Jena, and in Weimar, we now know that, to begin with, each kept his distance. Rivalry, or respect, probably played their part though Schiller certainly thought that Goethe was too self-important. But, early in 1794, Schiller began looking for people to contribute to his periodical, *Die Horen* (The Horae, 1795–97), and Goethe was, not unnaturally, among those asked.* This got them talking, so that it was natural for them to leave together after a meeting of the Society for Natural Science in Jena. During that conversation, Schiller attacked—but politely, respectfully—Goethe's notion of a primal plant, the *Urpflanze*, from which all others are derived. Schiller followed this up with a letter in which he contrasted his own critical-analytical ("sentimental") approach to reality, with Goethe's more organic ("naïve") belief in the simplicity of nature and in natural genius (whose intuitions were implicitly above criticism). Schiller's argument owed something to his medical training, something to Kant, and something to the German intellectual traditions introduced in Chapter 2, and it won Goethe's respect: from then on they were firm friends. Although very different, their letters confirm that they subsequently each had a hand in the writing of the other's works, notably *Faust* and *Wallenstein*.[63]

Schiller, probably more than Goethe, was much affected by news of the Revolution and subsequent Reign of Terror in France (1793–94). The execution of well over a hundred members of the ancien régime sickened him, as it sickened many other Germans, but he did not follow the path of many fellow intellectuals in Germany, who turned away from the massacres to the inner life. Schiller was not given to Pietistic inwardness or, for that matter, political nihilism.[64] For him, the main threat facing the world was barbarism, which he thought had always been with us, as common in the ancient world as it was in his own time. These reflections led to one of his major theoretical works, *Über die ästhetische Erziehung des Menschen* (*On the Aesthetic Education of Man*), in which Schiller offered an alternative to the predicament he saw around him. For Schiller, education was the best—the only—way forward, but it was to be education of a special kind: it was *aesthetic* culture that produced the "healthiest" relationship between reason and emotion. For him, art and literature, images and words, offered the best hope of showing how the imagination and the

* There were three Horae, goddesses of the (three) Greek seasons; they also guarded the gates of Olympus.

understanding can work collaboratively together, one limiting the other to help us avoid extremes, which Schiller saw as the main problem underpinning barbarity. For Schiller, Bildung of the individual through knowledge of aesthetic culture had an ennobling effect on character.[65]

Though Schiller was not conspicuously Pietist, like Herder he shared Francke's view that the Creation can be improved upon. Schiller seems to have thought that our minds are divided into two: the instrument of understanding and the imagination. The purpose of imagination, creativity, is to expand understanding and self-awareness. That being so, he distinguished three epochs in the evolution of civilization—the natural state, when the individual is subject to the forces of nature; the moral state, when man has identified the rules of nature and uses those rules as the basis of living together; and the aesthetic state, when he is free of these forces. In the first epoch, raw force prevails; the "ethical" state is governed by law; and in the "aesthetic" state individuals are free to treat each other as in a play—that is, people can choose their own roles.[66] In aesthetic society, beauty "acquaints us with our full potential."

This sounds impossibly idealistic, but in *Über naïve und sentimentalische Dichtung* (*On Naïve and Sentimental Poetry*) Schiller carried his argument still further and produced a thesis that some, at least, have seen as "one of the founding documents of literary modernity." His argument here is that the naïve poet is preoccupied with nature, whereas the sentimental poet is preoccupied with art and that something is lost in the latter process.[67] For Schiller, human beings in antiquity were closer to nature, and therefore were "more human" than they are now, primarily because they were less corrupted by culture (here is one origin of the German distinction between culture and civilization). For him the Greeks were more noble than we are, precisely because they disclaim "any desire to be more than human."[68] Poetry therefore ennobles us only insofar as it keeps us close to our true nature. This aspiration to nobility, to self-improvement, to Bildung, on the part of the middle classes was for him (as it was for others in the eighteenth century, such as Edward Gibbon, Hume, and Adam Smith), "the most important element in modern history."[69]

Having made his mark as a medical scientist, a playwright, a theorist about poetry, and a philosopher of aesthetics, in 1792 Schiller turned historian and published his *Geschichte des Dreissigjährigen Krieges* (*History of the Thirty Years War*), in which he painted some unforgettable pictures of the main protagonists, most notably Gustavus Adolphus and Wallenstein.[70] Work on this book seems to have given him new ideas about the dramatic realm because, four years later, he began his three great late dramas, which were to join the canon of the Weimar classics. These were *Wallenstein*, *Maria Stuart*, and *Wilhelm Tell*.

Wallenstein, completed in 1799, when Schiller was forty, shows him gathering strength as a tragedian. Lessing, in his *Hamburgische Dramaturgie*, had argued in favor of bourgeois tragedy, on the grounds that "the concept of the state is far too abstract to appeal to our senses." Schiller, in complete contrast, took a leaf out of the Greek book, realizing that the stage is essentially an aesthetic public space, making it perhaps the only location where we can overcome the alienation between the state and the individual.

Count Albrecht Wallenstein, a (real) Bohemian Protestant who has become a Catholic, serves Emperor Ferdinand, becoming the Thirty Years' War's most famous commander. Wallenstein is as ferocious as any of those who have committed atrocities during the conflict but, in 1643, he sees a chance of concluding a peace with the (Protestant) Swedes, despite the fact that such a peace is against the will of the emperor. Wallenstein's initiative is not carried out only from the highest motives, of course—he has been as corrupted as the next man—and he himself is accused of treason and murdered on the emperor's orders, and the peace negotiations founder. But Schiller is asking here whether motives *need* to be pure for peace, suggesting that war is so corrupting that even a peace achieved for less than pure motives is still a noble aim.

Every hand is raised
Against the other. Each one has his side.
No one can judge. When will it end and who
Untie the knot that endlessly adds to
Itself.

Wallenstein seizes the moment, but the moment backfires. The experience of revolution, Schiller is saying, teaches us that any attempt to destroy an existing state based on pure reason (i.e., ignoring other political realities and the existing power structure and the emotions they engender) produces only catastrophe and chaos.[71] The plot, and indeed the character of Wallenstein, are much closer to Schiller's own time than to the Thirty Years' War, which is the setting. Wallenstein himself is closer to Napoleon than to any personages in the earlier conflict, and contemporary audiences recognized this. *Wallenstein* is important because it is an early sighting of that (predominantly German) view that reason is not the be-all and end-all of forces shaping the human condition, a tradition that was to consume Marx, Arthur Schopenhauer, and Richard Wagner, and culminate in Nietzsche, Freud, and Martin Heidegger.

Schiller has given us some of the most magnificent women of the stage. In *Intrigue and Love*, the skirmishes between Luise and Lady Milford, who occupies a much higher social standing, are a remarkable rhetorical duel. In *Maria Stuart* the rivalry between Elizabeth and Mary is, if anything, an even higher level of combat. The play dramatizes an encounter between Queen Elizabeth I of England and Mary Queen of Scots in her last days, when she was held captive in the Castle of Fotheringhay. In real life Elizabeth and Mary never met, but the play imagines the meeting and, at the outset, the two queens are "sisters"—they are not, to begin with, complete opposites. As the play develops, however, Elizabeth's behavior is increasingly governed by her senses, the physical here-and-now, whereas Mary moves to the spiritual/intellectual plane. Although both women are equally formidable, equally noble, equally isolated, Mary's "state of sublimity" creates a growing—ultimately unbridgeable—gap between the two. Despite her political pre-eminence, Elizabeth, Schiller is saying, is essentially a prisoner of her office, which prevents her from being herself. Mary, though politically emaciated and physically shackled, is still morally free. These two female monarchs, so similar in so many outward ways, have very different inner natures—and that is what counts.[72] Do Elizabeth's actions against Mary stem entirely from the exigencies of the political situation, or is it more personal and, if so, in what way? Can Elizabeth ever know? Can *we* ever know? Is such self-knowledge possible?

For many people, the characters and predicaments of *Don Carlos*, *Wallenstein*, and *Maria Stuart* are even more overwhelming than those of Goethe's *Faust* or *Werther*. Giuseppe Verdi drew on Schiller for four operas

(*Luisa Miller*, *Joan of Arc*, *I Masnadieri*, and *Don Carlos*), Beethoven used "An die Freude" (Ode to Joy) in his choral symphony, and Schiller's poems inspired Johannes Brahms, Franz Liszt, Felix Mendelssohn, Franz Schubert, Robert Schumann, Richard Strauss, and Pyotr Ilich Tchaikovsky. More Schiller has been set to music than has Shakespeare.

5 ·

New Light on the Structure of the Mind

At one stage in his life, the philosopher Immanuel Kant formed a firm friendship with a certain Joseph Green, an English merchant in Königsberg (the city was a port, with many foreigners). According to Reinhold Jachmann, another friend of the philosopher and one of his earliest biographers, Kant would go to Green's house almost every afternoon. There, he would "find Green asleep in an easy chair, sit down beside him and, lost in meditation fall asleep himself. Then Bank Director Ruffmann usually came in and followed suit, till finally, at a certain time, Motherby [Green's partner] entered the room and woke the sleeping company, who then engaged in the most interesting conversation till seven o'clock. They used to part so punctually at seven that people living in the street were in the habit of saying it could not be seven o'clock yet, because Professor Kant had not gone past."[1]

This story, like so many others about Kant, has been dismissed as fanciful nonsense by his modern biographers. Which means that we now need to doubt all the other colorful details credited to him over the years, such as whether he really was an adept at both billiards and cards but gave up the latter because no one he knew could keep up with him. Did he really move house because the crowing of his neighbor's cock disturbed him—only to occupy a mansion too near the prison, where the singing of the prisoners' choir was likewise distracting? And no doubt he didn't always match his waistcoats to the colors of the flowers in season, as some "observers" have said. No matter. Kant was still an original genius in all manner of ways. Ernst Cassirer said that the fundamental "spiritual forces" in Prussia in the eighteenth century were Winckelmann, Herder, and Kant. Paintings and portrait busts (which, presumably, we *can* trust) inevitably depict him as though his features are about to break out in a smile. He was the first great philosopher who was a university professor and who has had a great impact—on philosophy and on academic life.

Kant (1724–1804) is, for many people, the most important philosopher since Plato and Aristotle. One reason for this—an argument that underlies the first half of the present book—is that he was living in a crucial era, when the old certainties attaching to the Christian faith were being washed away and before Darwin published *On the Origin of Species by Natural Selection* in 1859, which gave us a new, and in this case a biological understanding of ourselves, bringing with it a measure of intellectual agreement that simply did not exist at the time Kant was alive. Theology, as we have seen, was no longer the queen of the sciences.

This context helps explain the emergence of so-called German Idealism in the late eighteenth century. Insofar as these things can be understood at all, Idealism probably emerged in Germany rather than anywhere else because it was—or had been—the most fiercely Protestant country, with a virile tradition of looking inward to search for the truth, a strong, uncompromising semi-mystical form of self-examination.[2] In Königsberg, there was in addition particular awareness of the ideas of the English and Scottish Enlightenments. This had a lot to do with the British navy's need for a certain kind of timber for its ships' masts—flexible and sturdy at the same time. Baltic timber, the trade centered on Königsberg, was just right. This made for a strong British presence in the port and, as is often the case, ideas followed commerce.

We have seen that among the new sciences taught in the reorganized philosophical faculty at Göttingen was what we would now call "empirical psychology," though that term did not exist then. The shift to psychology in Germany—in all Europe—was to culminate in Kant, but three other Germans led the way: Christian Thomasius, Christian Wolff, and Moses Mendelssohn.[3]

Thomasius, one of the founders of the University of Halle, who daringly lectured in German, not Latin, famously argued that Nature, the source of law, exists independently of God's will, and that ethics stem from a "special physics"—the empirical science of (human) nature.[4] He devised what he called a "calculus of the passions" as a result of which rational judgments about conduct are (should be) made possible. He went so far as to assign numerical grades to the various passions on a scale from five to sixty. The precision of this system seems absurd now but its importance lies in the fact that Thomasius conceived human nature as a psychological entity, not a theological one.[5]

Christian Wolff, the son of a tanner, is sometimes called the *prelector* or teacher of Germany. Notoriously ordered out of Halle in 1723, because he argued, unwisely, that "reason does not allow itself to be ordered about," he was much taken with mathematics because it comprised *connected* knowledge, connected logically. He tried to apply a similar reasoning to psychology; he thought the soul's nature could be understood empirically, scientifically, so he too was replacing theological with psychological understanding.

Moses Mendelssohn was born in Dessau in 1729 and in 1743 went to Berlin, where he met Lessing, who published his first philosophical tract, the *Philosophische Gespräche*, in which he argued that genius creates what nature cannot, in the process bringing about new perfections.[6] "A beautiful object enhances the perfection of our bodily state," and this perfection impacts on the soul. For Mendelssohn, too, individual psychology replaces universal theology.[7]

These were important innovations, radical for their time and, with hindsight, all of a piece. Set beside Kant, however, they are simply confused.

The Limits to Reason

The sheer intellectual difficulty of the task, to discover what man is and should become, in the absence of a traditional creator or a clear biological understanding—the historical novelty of the predicament—is hard for us to grasp 200 years later. But this difficulty is very evident in the work of Kant, Fichte, Schelling, and Hegel, for example. Many aspects of their thought are hard to grasp, and this is only partly to do with the fact that they were, admittedly, hardly the most elegant of writers. What they were seeking to uncover and describe was difficult; they tried to isolate phenomena that they themselves only glimpsed in moments of lucidity. Nonetheless, "The period of German Idealism constitutes a cultural phenomenon whose stature and influence has been frequently compared to nothing less than the golden age of Athens." This is Karl Americks, the well-known Kantian scholar, writing in the *Cambridge Companion to German Idealism.*[8] Americks is referring to the overall transformation in thinking achieved by the Idealist philosophers lasting from the 1770s into the 1840s rather than to any particular style. "The texts of German Idealism continue to be an enormous influence on other fields such as religious studies, literary theory, politics, art, and the general methodology of the humanities."[9]

Idealism was developed in Königsberg, Berlin, Weimar, and Jena. Only Berlin was a city of any size—130,000 or so then. Both Herder and Fichte studied under Kant, later moving on to live near Goethe, who was sympathetic to Kant's approach. Karl Leonhard Reinhold proved to be an excellent popularizer of Kant in the nearby university town of Jena, and he was followed by Fichte, Schelling, and, eventually, Hegel. They developed their own varieties of Idealism and at the same time forged alliances with the literary giants of the era—Schiller, Hölderlin, Novalis (Friedrich von Hardenberg), and Friedrich Schlegel. They were further augmented by the arrival of a new generation of talented individualists: Friedrich Heinrich Jacobi, Friedrich Schleiermacher, Ludwig Tieck, Jean Paul Richter, August Wilhelm Schlegel, Friedrich Schlegel, Dorothea (Veit) Schlegel, Caroline (Böhmer) Schlegel, and Wilhelm and Alexander von Humboldt, "a relentlessly creative group."[10] Most of them moved on eventually to settle in Berlin when the new university was established there (see Chapter 10, p. 225). Following Napoleon's stunning victory at Jena in 1806, German Idealism contributed to Prussia's recovery and in particular to the rise of nationalism and conservatism within Germany.[11]

"German Idealism deserves the attention it has received. It fills an obvious gap generated by traditional expectations of philosophy and problems caused by the rise of the unquestioned authority of modern science." Idealism had the highest aims, seeking a synoptic understanding of all our most basic predicaments in a unified and autonomous approach. For the Idealists, philosophy should not be a series of ad hoc solutions to abstract technical puzzles. Ultimately, Idealism saw "culture" and "nation" as "higher" moral communities, stretching beyond individualism, the wholesome reflection of Christian duty.[12] It went beyond religion and incorporated politics.

At its simplest, Idealism argues that the bodily organs that allow humans to understand the structure of nature must be phenomena that are "built in" to nature to begin with. It follows from this that there must be limits to reason and therefore limits to what we know and to what we *can* know. Idealism echoes clearly the Platonic notion of "ideas," that "there is another level or realm of reality that exists beyond the common sense level in which we normally 'experience' life. For the idealists the world exists not quite in the manner that we assume it does . . . there is a set of features or entities that have a higher, more 'ideal' nature."[13] Kant called this realm the "noumenal" realm to distinguish it from the "phenomenal" realm, the realm of phenomena as we perceive them.

Kant's early works had more to do with science than philosophy.[14] Following the Lisbon earthquake of 1755, he produced a theory of earthquakes; he also conceived a theory of the heavens which predated Pierre-Simon Laplace's nebular hypothesis—that the solar system was formed by a cloud of gas condensing under gravity. But it is as a philosopher that Kant is chiefly known, and in his philosophy he identified—and then sought to clarify—what were for him the three most important questions facing mankind. First, he addressed the problem of *Truth*: How do we know the world and is it a true representation? Second, *Goodness*: What principles should govern human conduct? Third, *Beauty*: Are there laws of aesthetics, conditions which nature and art must satisfy in order to be beautiful?[15]

Kant addressed the first question in what is generally regarded as his most important book, *Kritik der reinen Vernunft* (*Critique of Pure Reason*), published in 1781 in Riga. It came after ten years of rumination and reflection—years that, as more than one critic has observed, did not improve his writing style. Kant rarely seems to have thought it necessary to give illustrations of his abstract points, never imagining that it would make his arguments easier to follow. His starting point was what for him was the crucial difference between two kinds of judgment. When someone says: "It is warm in this room," what he or she really means is: "It seems to me warm in this room—others might not find it so." On the other hand, the mathematical proposition that the sum of the angles in a triangle are equal to two right angles of 180 degrees is correct irrespective of the person making the measurement. It is true, as Kant put it, without reference to experience: it is universally true and "from the first" (a priori).[16]

How does this difference arise? Kant's answer was that the shapes of geometry are "ideal constructions" of our mind. Geometry is in effect a creation of the human mind, insofar as no one has ever seen a "pure" triangle, say, without any other attributes. Such a phenomenon does not—could not—exist. The figures and triangles that we see about us are only imperfect representations. This, for Kant, was very important for it showed that cognition of the world, how we know the world, "need not necessarily be the product of experience, of the mere functioning of our senses." Experience is the raw material but those experiences only become fully intelligible through the "productive activity" of the mind. Thought creates *concepts*.

Kant is saying that we do not have in our heads, as it were, an image of the world "out there"; instead we have an idea of how it *appears* to us, according to the laws of our intellectual make-up, which are present a priori, and which—invariably, inevitably, and necessarily—shape experiences a posteriori. Because of this we can never know anything "in itself."[17]

Kant identified several a priori aspects of our minds, of which the two most important were space and time. He was saying that we are born with an intuition of space and time, we understand them without experiencing them, in advance of any real sensation. Space and time, he argued, are not properties belonging to objects, but are merely subjective ideas *we* impose on them. Kant thought his case was proved by our idea that space is infinite, "which no one can experience or demonstrate." Though we can imagine space with nothing in it, we cannot imagine the absence of space itself.[18] The same is true of time. As with space, we can imagine not much happening over a certain period but we cannot imagine the absence of time itself. Time as we understand it—as with space—has no beginning and no end, it is infinite. It cannot stem from experience.

Kant's underlying point was that our minds are "living, actively operative organisms," not passively receiving information from without, through the senses and summed through experience; instead our minds shape our perceptions according to their own laws. He didn't stop at space and time but identified twelve categories or laws of thought, which shape how we understand the world. Among them he included "unity," "multiplicity," "causality," and "possibility." "Things in themselves possess neither unity nor multiplicity . . . we ourselves, through the operation of our understanding, combine certain impressions *a priori* into a unity or multiplicity (trunk, branches, twigs, leaves, into the concept tree)." Kant did not say that there is *no* connection between the outer and inner world. Scientific experiments, for example, proved that there is a close connection. Insofar as we are able to manipulate phenomena in ways that others can replicate, "There must exist common ground between the sensuous world and the understanding."[19]

This approach raised an intriguing set of questions in the world between doubt and Darwin. For example, where did it leave the question of God? Was what Kant was saying evidence for a metaphysical world that exists beyond reality, beyond our senses and our understanding? Many people cannot imagine a world without God, just as they cannot imagine a world without space, so did that make God an a priori intuition as real as space or time? Kant thought that the intuition to recognize the connection between external phenomena led to the idea of the *universe*, the absolute whole. This idea of the "whole" carried with it the further idea of an ultimate *cause* of the whole. By the same token, the fact that the inner structures—or laws—of our minds form a whole, a connected, interlocking, understandable whole, produces an equivalent idea, holding everything together—this is the concept of the soul. From there it was no great jump to say that the inner and outer worlds, soul and universe, point to an ultimate common basis, embracing both. The entity that "holds and unites" everything we give the name of God.[20]

It is not quite as simple as that. The universe might be an "absolute necessity," given the structure of our minds, but we cannot forget that the universe is not an object that we can experience in its entirety, but merely an *inference*—and this produces its own problems. For example, the very concept of a universe implies that it has a boundary. If that is so, what is there beyond this boundary? How then can the universe be infinite? The universe, in other words, is a "contradictory and hence impossible idea." The same argument applies to time and the ideas of "before" and "after."

Time without end is simply inconceivable; so is an end to time. "Space and time are simply forms of our thought."[21]

In analogous fashion, for Kant the existence of God can never be proved rationally. God is a notion, *our* notion, like space and time, and that is all. "*God is not a being outside me, but merely a thought within me.*" He was careful not to deny the existence of God—instead he denied our cognition of Him (for which the king reprimanded Kant). God, he argued, can be conceived only through the moral order in the world. Kant thought that humans "are compelled" to believe in God (and immortality), not because any science or insight leads them in that direction but because their minds are built that way.

Evolution toward Morality

The *Critique of Pure Reason* is Kant's most basic work. In the *Kritik der praktischen Vernunft* (*Critique of Practical Reason*; 1788) he spread himself to examine the "faculty of desire," morality. He started by conceding that morality may be judged in two ways. On the one hand, an action may be viewed as good if its *consequences* are good. On the other, an action will be good if it stems from good *motives*. Complications arise because it is easier to observe the consequences of an action than its motives. More complex still, good intentions may produce disaster, while evil intentions may have beneficial side effects.[22]

Kant's first step was to eliminate religion and, very largely, psychology from the picture. Goodness, ethical behavior, does not deserve the name if, in performing some action, we expect to benefit personally or in a religious sense, for then such action is selfish and not, in and of itself, good (though good may result).[23] This led Kant to his assertion that "*There is nothing in the world which can be unreservedly regarded as good, except a good will.*" But, he immediately asks, how is good will to be recognized? His answer is: *duty.* What he means by this is: follow your conscience, his famous concept of the "*categorical* [absolutely valid] *imperative*" (command). The categorical imperative or "inner command" is the voice of the conscience. "Conscience is the awareness of an inner seat of judgement within man," says Kant. Our internal ethics do not stem from experience, but are inherent in reason, a priori, and have two elements. A person must determine his or her conduct from within him- or herself. And the ethical basis for that is very similar to the biblical injunction: do not give way to "the weather in your soul"; do as you would be done by; act as you would wish others to act in equivalent situations.[24]

Again, this was more radical then than it sounds now. Goodness was embodied in justice and "Justice is the limitation of the liberty of each in the interest of the liberty of all, in so far as this can be achieved by a general system of laws."[25] Greater justice follows from greater self-knowledge and this is where, for Kant, education comes in. For him, the most important difference between man and animals is that man possesses the ability "to set himself aims and goals and to cultivate the raw potentialities of his nature . . . Behind education is concealed the great Arcanum of the perfection of human nature."[26]

Kant thought that a central concern of man—maybe the main one—was an "evolution toward morality," toward the moral character which is guided by good principles.[27] Accordingly, the main elements of education for him were instruction towards *obedience*, *veracity*, and *sociality*. Absolute obedience must be imposed at first, gradually supplanted by voluntary obedience arising from an individual's personal reflection. Obedience was important to Kant because, he argued, those who have not learned to obey others will be unable to obey themselves, their own convictions.[28] *Veracity* was, he said, essential for the unity of the personality; people can be whole only if they lack inner contradictions. *Sociality*, friendliness, the third element, ought not to be overlooked either: "Only the joyful heart is capable of delighting in good."

Fine Art as the Product of Genius

The third of Kant's great critiques was the *Kritik der Urteilskraft* (*Critique of Judgment*; 1790). In this, says Ernst Cassirer, "Kant touched the nerve of the entire spiritual and intellectual culture of his time more than with any other of his works . . ."[29] Kant's starting point is the concept of purposiveness. Against the background of the Enlightenment, the scientific revolution, and all the other developments considered in Chapter 2, p. 65, Kant focused on the logic—or lack of it—in the relationship between parts and the whole. Which came first? Does that question make sense?

An organism like an animal exists as a whole but consists of parts. The whole cannot survive without the parts and the parts cannot survive without the whole. What does it mean to be a part? Different species of animals, or plants, "belong" to higher taxa. What does this mean? Do these groups (genera, say, or families, though these categories did not have their modern meanings then) exist in any real sense outside our heads or is there some a priori process within us that determines how we understand parts and wholes and their relation?[30]

Before Kant, it was assumed that the Great Chain of Being was a true reflection of God's purpose for nature. For Kant, even the purpose of nature, indeed the very concept of purpose itself, is built into *our* nature, so we can never know whether "purpose" exists outside ourselves. Our instinctive notion of purpose will determine the way we understand nature; nature's laws do not exist "out there"; *we* impose those laws on nature.

This led Kant to his reflections on art. If the order of nature, as shown by its particular laws, reflects no more than our constitutional ability to impose a unity *on* nature, this comes about because the *attainment* of such unity is always coupled with a feeling of pleasure and "the feeling of pleasure also is determined by a ground which is *a priori* and valid for all men."[31]

The phrase "valid for all men" is crucial. Art, for Kant, was a realm of "pure" forms, each complete in itself. "The work of art . . . has its own basis and has its goal purely within itself, and yet at the same time in it we are presented with a new whole, a new image of reality." Science concerns itself with superordination and subordination in a causal capacity, leading from premise to conclusion. In aesthetics we grasp the whole immediately and its parts and their relation to the whole is immediate, not causal; we surrender ourselves to pure contemplation. The aesthetic consciousness "grasps in this very fleeting passivity a factor of purely timeless meaning." For Kant, the aim of art is to evoke "disinterested pleasure." The fact that many people find the same things beautiful, that beauty is "valid" for all men, aroused in Kant the notion of "subjective universality." The fact that everyone attributes a similar pleasure to a work of art is, for him, a vital aspect of the experience. It was evidence—important evidence—of a *universal voice* not mediated by concepts.[32] Ideas in art are a more immediate kind of experience than other experiences.

The importance of this distinction led Kant to consider geniuses, building on Lessing. "The creation of genius receives no rule from outside, but it is the rule itself. In it is shrouded an inner lawfulness and purposiveness. Genius is the talent (natural gift) which gives rule[s] to art . . . fine art is only possible as a product of genius." The existence of genius differentiated artistic productivity from scientific productivity. Kant argued that "there can be no genius in the sciences." For him, the decisive difference lies in the fact that any scientific insight, as soon as it has been identified, possesses no form over and above the insight itself. The personality of the scientist doesn't matter. In art, however, "the *form* of the product is integral to the insight conveyed."[33]

Kant's theory of genius became a rallying point for the Romantic movement and its view that the aesthetic imagination is the "begetter of the world and reality."[34] We shall come to the Romantic movement in Chapter 8, p. 189, but what distinguishes Kant's own view, in purely philosophical terms, is that it went against the concept of "reason," as it had been evolved by the Enlightenment. Kant had, he felt, identified a "deeper" concept, the "spontaneity of consciousness," which was reflected in art, which went beyond reason but was just as real. This new "determinant" of consciousness was, for Kant, an important—perhaps the most important—ingredient of freedom. "Only artistic insight discloses a new path to us. In art, in the free play of the powers of the mind, nature appears to us as if it were a work of freedom, as if it were shaped in accordance with an indwelling finality . . ."[35]

This difference between art and sciences, between the geniuses and scientists, was for Kant a glimpse into the purpose of life. The very idea of purpose comes from within, and the unity that we are driven to impose on art by our inner nature, and the universal subjectivity that exists, allows us to *inflict* purpose. In doing this, we enlarge ourselves and are able to share that enlargement with others. For Kant, this is what freedom meant, an inner enlargement, a profoundly influential idea in the German-speaking lands.

Kant's range and ambition were shown in his project, in 1795, to explore the—to us—ambitious notion of perpetual peace. This side of the cataclysms of the twentieth century, such an idea verges on the preposterous, but it was not so very different when Kant made his attempt. Europe still had its share of absolute states and the blood spilled in the French Revolution and its aftermath was still wet. Kant had recently (in 1793 and 1794) evolved his idea about an ethical commonwealth, a moral community, an invisible church, by means of which the highest good, "the autonomous will of men," would be achieved. In his plan for perpetual peace he set down various conditions—standing armies shall be governed by conditions of universal hospitality—many, if not all, of which sound to us (as no doubt to his colleagues and neighbors) as impossibly idealistic. But there was one that, as it turned out, was not impossibly idealistic and, in time, was at least half realized. This was his proposal that "The Civil Constitution of Every State Should Be Republican." This, he thought, was the original basis of every form of civil constitution and it was, in its time, radical. But it lives on now, not just in the spread of democracies, or republics, but in the notion (which feels modern but dates back to Kant) that democracies are reluctant to declare war on each other.

The Rise of Jena

Jena was and was not like Weimar. It had always been a small town, like countless others, populated mainly by artisans, and with a second-rate university: nothing exceptional. Then, all at once, as the eighteenth century drew to a close, it suddenly blossomed as the center of a new revolution in German intellectual life.[36]

Goethe himself was partly responsible. By his position, his character, by his very presence, he made Weimar and Jena rise in profile, and the university at Jena became the very model of a reformed, even what has been called a "Kantian" university. This was still an age when many universities were considered irrelevant and unruly nuisances, but Jena adopted the more successful and more modern Halle/Göttingen model—the union of teaching and research where students were brought into contact with leading minds working on the latest ideas. Also following the Göttingen model, the philosophical faculty, rather than the theological faculty, was the main focus of activity. A new periodical, the *Allgemeine Literatur-Zeitung*, was founded there and it soon became the most widely read intellectual journal in Germany.[37] According to Terry Pinkard, in his study of the legacy of Idealism, the public that subscribed to journals like the

Allgemeine Literatur-Zeitung read and discussed Kant "with the same intensity as novels and more popular literature."

One of the first post-Kantians there was Friedrich Heinrich Jacobi (1743–1819), who, in 1785, turned Kant's critical approach on the master himself. He developed the view that, fundamentally, reason "takes its first principles" from the heart and not from the head and that, contrary to what Kant said, "*all* knowledge must rest on some kind of faith." What he meant was that, underlying all thought, whatever it is, there must be something, a first principle, that cannot itself be proved by reference to something else, that exhibits what he called "immediate certainty."[38] For example, he said that we have immediate certainty of our own bodies. Therefore, Jacobi said, if this is so, why should we not trust the "immediate certainty" we have of God? He became convinced that Idealism was a form of nihilism, a term he coined.[39]

In 1786, and again in 1790, Karl Leonhard Reinhold released a series of letters, later brought together as a book, *Briefe über die kantische Philosophie* (*Letters on the Kantian Philosophy*). These letters supported Kant's viewpoint and did so with such panache that Reinhold was briefly—but only briefly—regarded as an even brighter star in the philosophical firmament than Kant himself. A Jesuit novitiate who had converted to Protestantism, he was appointed a professor at Jena in 1787, and one of his self-appointed tasks there was to systematize Kantian thought into a formal science.[40] This is perhaps the origin of the tendency in German thinking at that time to attempt to construct elaborate interlocking systems, exploring as much as possible from first principles, and in an internally consistent way, an approach that would culminate in Fichte, Hegel, and Marx. Reinhold added to what had gone before by asserting that consciousness had the quality of immediate certainty, and this moved it center stage as the entity to be explained, over and above Kant's emphasis on experience and intuition.[41]

God Replaced by the Self

There was another side to all this. As we have seen, Königsberg had good links with Britain, and many shared the views of Scotland's down-to-earth school of common sense. To such people, the entire paraphernalia of "transcendental Idealism" seemed far-fetched, and there was in Germany no shortage of critics and skeptics.[42] More fruitfully, perhaps, there was a whole generation of people who, while not accepting all that Kant had to say, found enough in his philosophy to try to take it further. Among these, Fichte, Schelling, and Hegel were the most interesting. Schelling will be considered in the section on Romanticism, Hegel in the chapter on alienation. Fichte is another matter.

Bertrand Russell thought that Fichte's system "seems almost to involve a kind of insanity."[43] Certainly, Fichte represents above all the example of the speculative philosopher trying to build a whole system on the basis of one central idea or construct. He is also an important stage in the emergence of what we now call psychology.[44]

Johann Gottlieb Fichte (1762–1814) was the poor son of a ribbon weaver in Saxony, and, like Herder, was given an unexpected chance for an education when, as an eight-year-old, he showed total recall of that day's sermon in church, a feat witnessed by a local noble who was so impressed that he decided to give the boy a proper schooling.[45] This was not a complete success but did help in that Fichte eventually made his way to Königsberg to meet Kant. At first the master was not overly impressed and so, in order to improve his standing, Fichte composed a short piece, "Ein Versuch einer Kritik aller Offenbarung" (An Attempt at a Critique of All Revelation). Kant liked what Fichte had written and helped to get it published. However, the publisher—deliberately or otherwise—left Fichte's name off the finished product and, since the text showed such a command of Kantian theory, everyone assumed the author was the master himself. After the truth came out, Fichte's fame was assured and when Reinhold was offered a better-paying job at Kiel in 1794, Fichte was seen as his natural successor, a meteoric rise from nowhere. He was thirty-two.

In Jena he threw himself into the fray, taking on a work that had itself created a commotion. This was the *Aenesidemus*, by G. E. L. Schulze, professor of philosophy at Helmstedt. Schulze's argument, refuting Reinhold, and therefore Kant, was that we cannot know with certainty anything of things-in-themselves. Instead, he insisted, all we can be certain of is our own mental states. Fichte argued against this but in doing so he constructed, or tried to construct, a whole system of thought with interlocking parts.[46] The subsequent book, *Die Grundlage der gesamten Wis-*

senschaftslehre (*The Foundations of the Whole Doctrine of Science*), was that which Schelling included with Goethe and the French Revolution as one of the three great "tendencies" of the age—see p. 119.* [47]

Fichte's key insight, which he thought deepened Kantianism, was that the distinction we make between subject and object is itself *subjectively* established.[48] Fichte accepted Jacobi's idea of immediate certainty, Reinhold's immediate certainty of consciousness, and Kant's subjective universality, but added what was for him the most important element, the immediate certainty of *self*-consciousness. Self-consciousness, he insisted, is a basic ingredient of consciousness and together they are the irreducible elements by which we grasp reality. Moreover, a basic ingredient of self-consciousness and consciousness is the "not-self." "The self is not a static entity—it develops through time as its awareness of itself grows and changes, through encounters with the 'not-self'" (i.e., other selves and objects "out there"). Reason is in effect a by-product of consciousness and self-consciousness—we infer the world around us, its interconnections and dependencies.[49]

Now at one level, to us in the twenty-first century, these arguments of Fichte's seem like a very confused and overelaborate way of stating the obvious, even repeating much of what Locke had said far earlier, and that is the way some people have construed him, with many others viewing Hegel as by far the more important post-Kantian. But, thinking our way back into late eighteenth-century forms of understanding the world, Fichte's theories, as set out in his lectures and books, were significant in two ways that are not immediately obvious to us. First, his was the ultimate "psychologizing" of human nature (to return to the anachronistic term). His emphasis on the self, the "I," and the "not-self," without any reference to religion in general or Christianity in particular, was an important stage in the revision of our understanding, from the theological to the psychological, which would lead in time to the Freudian and post-Freudian world. At the same time, his understanding of the centrality of the self, and its understanding of—and interactions with—the "not-self," had important implications for ideas about freedom. In Germany, and in Kantianism in

* *Wissenschaft* in German means both science and scholarship.

particular, as was mentioned above, freedom had been seen as an "inner" phenomenon, a psychological freedom to be achieved by learning, by education, by a journey inward. Fichte realized that (a quite different idea of) freedom depended on the relation between the self and the "not-self," that the self could be free only to the extent that its freedom did not impinge on, or curtail, the freedom of other selves. In a land of small absolute states, this was far more controversial—revolutionary even—than it seems to us now.

Likewise, Fichte's theories threw a fresh light on the state and its responsibilities. "The state functions as the 'objective' viewpoint that precipitates out of the various subjective viewpoints of the citizenry as they each keep score on each other."[50] This begins to sound like Jeremy Bentham's "felicific calculus," with the state's virtue judged by the contentedness of the greatest number. Insofar as one self is the equal of another, this also took on the color of a democratic, even republican, viewpoint.

Fichte was a charismatic teacher whose lectures often overflowed, with students standing on ladders at the windows to hear him. But his career at Jena came to a sudden end when he reacted to criticism in a high-handed way and his threatened resignation was accepted.[51] He transferred to Berlin where he taught privately for a while, before being chosen as the first philosophy professor at the newly formed University of Berlin in 1810 (see Chapter 10, p. 225).

It is worth pointing out that the *Wissenschaftslehre* went through sixteen different editions.[52] This had something to do with his charisma but also owed something to the fact that a new form of understanding of man was being born. That new understanding was a psychological approach to mankind. Locke and Francke played their part in this new understanding, and Pietism, too. But Kant had introduced one other change that should not go unnoticed. Though Idealism is sometimes referred to as a form of speculative philosophy, that isn't wholly fair. Kant had introduced a rigorous new way of observing, of observing *ourselves*. This sometimes got out of hand—it may well have got out of hand with Fichte—but this observation of ourselves, the concentration on subjective universality, consciousness and self-consciousness, was the real beginning of modern psychology, one of the beginnings certainly. The problem with this new approach was that it emerged before Darwinian understanding had been evolved. This had major consequences for psychology, which for many has always been seen more as a form of philosophy rather than a form of biology. It is one reason why the unconscious, and with it the therapeutic approach to life, was at root a German idea.

6.

The High Renaissance in Music: The Symphony as Philosophy

In Germany, as elsewhere in Europe, vocal music was more popular than instrumental music well into the sixteenth century. (Martin Luther had a lusty singing voice.)[1] But then arose, in Italy, the first great European organ school. As artists like Dürer visited Venice to learn from the masters there, so many Germans now visited *La Serenissima* to study the organ and return with new techniques of polyphonic writing.[2] This development would lead in time to Heinrich Schütz (1585–1672), one of many musicians who traveled to Venice to study under Giovanni Gabrieli, and to Johann Jacob Froberger (*c.* 1617–67), Johann Pachelbel (1653–1706), and Dietrich Buxtehude (1637–1707). Like Georg Philipp Telemann (1681–1767), Buxtehude was in his lifetime far more famous than Bach (who traveled sixty miles on foot to hear him). But their reputations have not lasted quite as Bach's has, and German music's first culmination is in fact found in the works of the Leipzig master and one other composer with whose name he is inseparably linked: Georg Friedrich Handel.

Born in the same year, in towns barely eighty miles apart, they never met but they shared the highest peaks of technical perfection, which their predecessors had been struggling toward but had not yet achieved. Beyond that, they could not have been more different. Handel was a worldly figure, cosmopolitan, at ease with success, whereas Bach was above all devout and a "thoroughgoing provincial."[3]

Handel has been described as "the greatest assimilator of pre-existing material in the history of music." What this means is that he was a magpie, who borrowed or stole—mainly from Italian composers—theme after theme, even whole movements, reworking them as his own. In his oratorio *Israel in Egypt*, no fewer than sixteen of the thirty-nine tunes rely (in some cases heavily) on themes devised by other composers. Handel invariably adds his own brand of grace and polished simplicity.[4]

For many people, professional musicians in particular, Johann Sebastian Bach (1685–1750) is the greatest composer the world has seen. In contrast to Handel, he never left Germany, remaining for many years as the cantor at St. Thomas's Church in Leipzig. Hardly any of his music was published in his lifetime, his reputation being less that of a composer and more that of an organist and improviser at the keyboard.[5] Practically, Bach played an important part in the development of the organ and had his instruments modified to meet his wishes. He was fortunate, too, in living in the great age of baroque organ builders, men such as Arp Schnitger (1648–1718) and the well-known Silberman family. Andreas Silberman, who begat the tradition, designed and built the Strasbourg Cathedral organ (1714–16), while his brother Gottfried did the same for the cathedral at Freiberg in Saxony in 1714. Gottfried brought the pianoforte to Germany after its invention in Florence by Bartolomeo Cristofori.

Bach's ability to juggle themes, to state thesis and antithesis, to explore a melody in diverse directions, returning almost unnoticed to the main thread, is a form of musical weaving unparalleled in human achievement, not just in its technical intricacy, which went—goes—beyond anything anyone else has ever been able to do, but because at the same time it retains and maintains emotional richness and satisfaction. Nor should we overlook his formal innovations: under his guidance, the harpsichord was transformed from an accompaniment to a virtuoso solo instrument.[6]

Though Johann Sebastian was a genius by any standard, in the mid- and late eighteenth century the name of Bach that most people knew was Carl Philipp Emanuel (1714–88), one of a number of talented musician-sons, who included Wilhelm Friedman (1710–84), the eldest, a composer of church cantatas and keyboard concertos, and Johann Christian

(1735–82), the "London Bach," who spent several years by the Thames and penned a score of Italian operas and concertos.

Everyone knows a little bit about the Bachs but at much the same time there was in Germany another group of composers known as the Mannheim school, a talented ensemble who helped to make up the orchestra of Elector Karl Theodor, Count Palatine and Duke of Bavaria (1724–99) at Mannheim. It was there that full orchestral scores were first developed, with parts written out and individually exploited. This innovation is regarded as the birth of the modern era as far as orchestral music is concerned.

The Origins of Grand Opera

Until the middle of the eighteenth century, in Germany as in England and in France, Italian opera was dominant. Libretti were invariably in Italian, and singers, whether or not they were Italian themselves (and they usually were, however bad), were imported to sing the principal roles, complete with "stereotypical Italian gestures."[7] Toward the end of the eighteenth century, this began to change. This period saw the spread of the comic opera, known as the *Singspiel*, one feature of which was spoken dialogue in the vernacular (i.e., German in Germany), so that the interpolated songs were also in the vernacular. This practice reached near perfection in Mozart's 1782 opera *Die Entführung aus dem Serail* (*The Abduction from the Seraglio*), but the new form actually owed more to the ideas of Christoph Willibald Gluck (1714–98).

Gluck, almost single-handedly, took on the Italian idea of opera, identifying a new approach and composing a set of powerful works which embodied his new vision. He introduced his new practice in *Orfeo ed Euridice*, in Vienna in 1762, but it was with *Alceste*, another "Italian" opera on a classical subject, that he wrote his celebrated preface, setting out his new philosophy.[8] He argued that the singers should confine their vocal displays so as to highlight and develop the course of the dramatic action rather than launch into virtuosi fireworks for the sake of it; he also argued that the overture should be "a proper emotional preparation" for the drama, "not just a set of tunes presented while the audience was finding their seats"; above all, he insisted that the music should serve the needs of the text to intensify the dramatic effect. This is another of those ideas that seem unexceptional to us today, but in its time it was very controversial and Gluck's view prevailed only because his operas achieved such dramatic intensity that it became obvious that what he said was right. Harold Schonberg commented, "One can reasonably claim that the tradition of grand opera in the modern theatre begins with Gluck."

Four Giants

In Italy, the period of the "High Renaissance" refers to those thirty years, 1497–1527, when three artists—Raphael, Michelangelo, and Leonardo da Vinci—were all supremely active. In the German renaissance the equivalent period was the last twenty-five years or so of the eighteenth century, when four magnificent musical giants—geniuses—emerged whose pre-eminence lies beyond dispute and who set the stage for the great century of German music which followed. Indeed, we may regard the following hundred years as the greatest century of all time in the history of composing.[9]

Born into a poor family in Lower Austria, Franz Joseph Haydn (1732–1800) was a choirboy at St. Stephen's Cathedral in Vienna until he was seventeen and a music teacher until his mid-twenties. The turning point in his life came in 1761, when he entered the service of Prince Paul Anton von Esterházy. Haydn continued as kapellmeister at Esterháza for the next three decades, until 1790, and they were without question golden years. The family was the most enlightened of patrons and, under them, Haydn produced his brilliant series of symphonic and chamber music triumphs, which gained him an international reputation. This reputation took him to London in the 1790s, where he wrote twelve of his finest symphonies. Despite being more in the public eye now, he rejoined the Esterházys and, during his later years in Vienna, produced his great string quartets, op. 76 and op. 77, and the two oratorios *Die Schöpfung* (*The Creation*) and *Die Jahreszeiten* (*The Seasons*).[10] Haydn's output includes more than a hundred symphonies, some fifty concertos, eighty-four string quartets, forty-two pianoforte sonatas, a variety of masses, operas, and other pieces for various solo instruments. His brilliance has a familiar, unaffected quality. This may in part reflect his wide use of folk melodies—especially those from

Croatia—which kept his music simple, direct and accessible. He was clear about his talent. "I was cut off from the world; and since there was no one to confuse or trouble me, I was forced to become original."[11]

Despite his brilliance as an orchestral composer, Haydn himself had hope of writing a great opera. It was never to be realized, though there are, in *The Creation*, moments where this possibility is hinted at. It was instead Mozart who shone in that vein. Haydn and Mozart met several times in the last decade of the latter's life (Mozart died in 1791). Comparing their output, it is clear each influenced the other, so much so that, as Malcolm Pasley says, "run-of-the-mill Haydn is barely distinguishable from run-of-the-mill Mozart." At their greatest, on the other hand, the two composers are unmistakable.

Wolfgang Amadeus Mozart's short life (1756–91) was very different from that of Haydn. He was born into a cultured musical family in which his father, Leopold (whose works are still occasionally heard today), was a musician at the court of the archbishop of Salzburg. Wolfgang was an infant prodigy, "a prodigy among prodigies": he could play the pianoforte at three and could invent short compositions at five. At seven he added the violin to the piano, accompanying his sister, Maria Anna, known as Nannerl, and together they were taken by their father on a series of sensational tours of European cities (he was eight when they started and eleven when they finished). Mozart wrote his first opera, *La finta semplice*, when he was twelve. He was appointed court musician at Salzburg and later at Vienna and, like Haydn, lived much of his career under a patron, many of his works being composed for particular court functions (this is true, for instance, of the three great string quartets, K. 575, K. 589, and K. 590). Mozart was nothing if not practical, and other works were written for specific outstanding performers, as with the clarinet concerto, written for the virtuoso Anton Stadler, the leading performer of the time on what was then still a new instrument.[12]

For musicologists, it is Mozart's development of the solo part that comprises his most distinctive contribution. Traditionally, in the early eighteenth century, in a concerto the musical argument was passed back and forth between the soloist and the orchestra, a principle originally derived from the concerto grosso, in which a group of soloists is pitched against the orchestra. Mozart evolved the independence of the solo instrument by increasing the virtuosity required to play his beautiful themes. It was also under his guidance that the concerto acquired three movements, which became standard. The first movement was generally an allegro, followed by a second, slow movement, terminating with a rondo. This too became a standard pattern throughout the nineteenth century.

Despite the popularity of his classical concertos, it is Mozart's piano concertos (twenty-five of them) and his last three great symphonies—no. 39 in E-flat major (K. 543), no. 40 in G minor (K. 550), and no. 41 in C major (the *Jupiter*, K. 551)—that are generally regarded as the most beautiful music the world has to offer. For many people, however, even they pale alongside his achievements in opera. "In opera he had—many would say has—no equal."[13]

It may have been Gluck who formally and originally spelled out what the new form of opera was trying to achieve, but it was Mozart who fulfilled the new ideal better—incomparably better—than anyone else.[14] In Mozart, musical characterization is so vivid and reflects and amplifies the text so much that it produces a psychological depth to his characters that is simply nonexistent in earlier composers, a depth from which the drama emerges "directly and urgently," so that the music itself is the means by which motivation is conveyed. Arguably, it is in the greatest of his German operas, *Die Zauberflöte* (*The Magic Flute*), composed and first produced in the last year of his life, that Mozart achieves his most sublime work. If Haydn could rival him in instrumental virtuosity and Beethoven, as we shall see, in emotional depth, in the coloratura arias of the Queen of the Night there is something that has no counterpart anywhere.[15]

Haydn, Mozart, and Ludwig van Beethoven (1770–1827) are often pooled together as the "first Vienna school" and, to be sure, like Haydn and Mozart, Beethoven died in Vienna. But there is no real sense in which these three formed a Viennese school with common artistic aims or even common methods.

Beethoven, born in Bonn, came from a family of musicians, as Mozart did. His father and grandfather had been professional players at the court of the Elector of Cologne. At the age of twenty-two Beethoven studied with Haydn in Vienna. After that, he gave lessons to the children of aristocratic families, but he always wanted to be a composer and gradually acquired fame in that direction.[16]

Beethoven had an unhappy private life and that, perhaps, made his music different from that of Bach or Mozart. The mastery, mystery, and perfection of Bach and Mozart are like polished gems in their cool, classical beauty. "Theirs is the music of the gods, but Beethoven—his is the music of *man*, of his suffering, his impatience, his exhilaration, confronting the world and yet affirming it at the same time . . . The progress of [Beethoven's] music is a passage to human greatness . . . a musical achievement which stands unchallenged as a monument to the mind of man."[17]

Beethoven's output may be understood as falling into three periods. Before 1800 his works show the influence of Haydn. In 1800 the Piano Concerto no. 1, op. 15 and the First Symphony op. 21 appeared and this marks the emergence of the Beethoven with which the average concert-goer is most familiar. In the First Symphony Beethoven's innovation in replacing the minuet of the third movement with a more lively scherzo, completed by a final allegro, vastly heightened the tension of the work, setting out Beethoven's distinctive voice: tension-filled movement and restless passion. It is this restlessness that characterizes the middle period of Beethoven's life: the first eight symphonies, the five piano concertos, the violin concerto, the opera *Fidelio*, the piano sonatas, including the *Appassionata* and the *Waldstein*. Beethoven was above all an instrumental composer. He once said, "I always hear my music on instruments, never on voices."[18]

As magnificent as all this was, it was in the final period of his life that Beethoven produced his greatest music, by common consent some of the greatest music in the world. "All music leads up to Beethoven and all music leads away from him," says Mumford Jones. This is the period of the Ninth (Choral) Symphony, the *Missa Solemnis*, the last three piano sonatas and the *Diabelli Variations* for the piano, and the last five string quartets.[19] After a life of turbulence, movement, and conflict, ending in his tragic deafness, Beethoven's late music shows a serenity and a redemption that the composer never attained elsewhere in his life.

Just as Haydn had excelled in the symphonic form, Mozart in opera, and Beethoven in instrumental music, so Schubert—the last of the four great Viennese masters—excelled in song.

Franz Schubert (1797–1828), who was born in Vienna and lived all his life there, had an even shorter life than Mozart, dying at thirty-one.

Yet he managed to be one of the most prolific of composers.[20] He never had the benefit of aristocratic patrons and lived entirely as a freelancer (as we would say). In the precocity of his musical talents he rivaled Mozart and began composing at an early age. But he did not attract attention like Mozart did and it was not until a group of his friends published twenty of his songs at their own expense in 1821 (when he was twenty-four) that he began to be noticed. By that stage he had composed seven of his nine symphonies, the string trio, the *Trout* Quintet, and several operas and masses. Many of his works were never performed in his lifetime, and many others were published only posthumously.[21]

But it is for his *Lieder* (songs) that Schubert is chiefly known. Here too he was prolific—six hundred songs, including seventy-one settings of poems by Goethe and forty-two by Schiller. Despite his choice of Goethe and Schiller, Schubert was not particularly astute in his choice of verse; still, his music showed an unparalleled ability to use a tune to go beyond a mere "setting" and devise instead its musical equivalent. In doing so he elevated the piano accompaniment to a level it had never reached before, "a level for which the term accompaniment is no longer adequate." [22]

The final new elements in music making (as opposed to *listening*, considered in the next section) were introduced by Carl Maria von Weber (1786–1826). Weber had a diseased hip and walked with a limp but he was a virtuoso of the guitar and an excellent singer, until he damaged his voice by accidentally drinking a glass of nitric acid. He was summoned to Dresden to take control of the opera house there and made the conductor (himself) the single most dominant force, setting a fashion that continues to this day. He too worked hard to counter the contemporary craze for Italian opera, based mainly on the works of Rossini. It was thanks to Weber that a German operatic tradition emerged that was to culminate in Wagner.[23] Weber's own opera *Der Freischütz*, first performed in 1821, with Heinrich Heine in the audience, opened up a new world. In *Der Freischütz* the orchestra now became far more than a background factor for the voices. The strings and wind sections, for example, were co-opted to express their own individuality, to add mood and color. This advance allowed more scope for the conductor to shape the operatic experience. Opera had, more or less, achieved the form we now know.

Music as Philosophy

The standard "backbone" of classical music consists today of Bach, Handel, Haydn, Mozart, Beethoven, Schubert, and Brahms—all German. This backbone emerged first, in Germany, at the turn of the nineteenth century and continued throughout the 1800s, with only Hector Berlioz, Frédéric Chopin, Tchaikovsky, and Verdi among the great composers who were not German.

But musical *production* is only half the picture. Just as the efflorescence of painting in the Italian Renaissance is now understood against the commercial and religious tendencies of the time, so in Germany in the late eighteenth and early nineteenth centuries, musical listening, musical consumption, musical understanding, were affected and influenced by the prevailing Idealistic philosophy. This is a long way from how we conceive the musical experience today.

A completely new understanding of the arts—and in particular music—emerged in Germany at the end of the eighteenth century. Listening, as Mark Evans Bonds has shown, took on a new seriousness, in particular in relation to instrumental music.[24]

At that time, the symphony was comparatively new. It emerged only in the 1720s, a development of the opera overture, itself often called a "symphony" at the time and a practice that did not die out until the 1790s.[25] Until 1800 or thereabouts, the symphony was much less important than the opera, and even Kant, notoriously, in the *Critique of Judgment*, dismissed instrumental music as "more pleasure than culture" ("*mehr Genuss als Kultur*"), on a par, he said, with wallpaper. He was impressed by the ability of music to move listeners but since instrumental music contained no ideas (because it used no words), he thought its effects must be transitory "which would, in time, dull the spirit."[26] His views were widely shared, but around the turn of the century the status of the symphony underwent a profound change.

One reason was the gradual shift from private to public performances, considered earlier. The wider audiences had wider tastes and consisted of the newly emerging middle classes eager to educate and improve themselves. Equally important—if not more so in the long run—were the changing attitudes toward the nature of art, in particular the relationship that came to be perceived between music and philosophy. This transformed the act of listening.[27]

This new aesthetic, which began to value instrumental music, was directly derived from Idealism, in which it was argued that the benefits to be obtained from art needed much more than "idle reception" but, rather, *activity.* Any given artwork, any product of genius, reflected a higher ideal realm that listeners had to work on, to play their part in. The rapture of music, the extent to which we are "carried away" during a performance, the "forgetting of the self," became for many the first stage in the journey toward this other, higher realm. Beethoven himself believed that art could be a bridge between the earthly and the divine.[28]

This view was expanded during the 1790s. Schelling led the way in arguing that art and philosophy addressed the same basic issue, the link between the world of phenomena and the world of ideas. For him, sound was the "innermost" of the five senses; its very incorporeality meant that its "essence" was more ideal than the other senses. August Schlegel shared this view.[29]

The link between Idealism and music without voices thus becomes clear, and this approach reached its apogee in what Mark Bonds calls "the most important piece of musical criticism ever written."[30] This was a review, published in the *Allgemeine musikalische Zeitung* in 1809, by E. T. A. Hoffmann (1776–1822), of Beethoven's Fifth Symphony. In this review, Hoffmann identified music as occupying "a separate realm beyond the phenomenal," thereby endowing music with the capacity to provide "a glimpse of the infinite." Instrumental music, he said, "discloses to man an unknown realm, a world that has nothing in common with the external sensuous world that surrounds him, a world in which he leaves behind him all feelings that can be expressed through concepts, in order to surrender himself to that which cannot be expressed" in words, "a potential catalyst of revelation accessible to those who actively engaged the work by [their] creative imagination."[31] It followed that "the onus of intelligibility" now moved from composer to listener. Bonds again: "This new framework of listening was in effect philosophical, based on the premise that the listener must strive to understand and internalise the thought of the composer, follow the argument of the music and comprehend it as a whole."

Hoffmann's review of Beethoven's Fifth Symphony was the first time such a philosophical approach had been applied to a specific composition. He claimed to have identified a "teleological progression from the childlike innocence of Haydn to the superhuman Mozart to the divine Beethoven . . . Listening to Beethoven we become aware, dimly, of a higher form of reality not otherwise perceptible to us . . . Art is no longer a vehicle of entertainment, but a vehicle of truth . . . The arts in general begin where philosophy ends."[32]

The very notion of explaining a work of instrumental music in depth was itself new. It grew out of the wider conception of Bildung but the link between Bildung and listening also had to do with the change in the understanding of listening itself. The symphony, for example, was associated with Kant's notion of the sublime, a form of art defined by reference to its vastness of scope and its "oceanic" capacity to overwhelm the senses.[33] Many philosophers and artists argued that contemplation of the infinite through the sublime offered insights that the merely beautiful could not provide. "The massed forces" of the symphony supported this idea.[34]

When Hoffmann described music as "unfolding" from Haydn through Mozart to Beethoven he was also espousing a form of historicism, even a form of Hegelianism, accepting in effect the existence of a "world spirit," evolving toward ever-higher states of human consciousness. Beethoven's symphonies represented a culmination in music, a "moment of historical timelessness," in which the composer had achieved *Besonnenheit*. This word, difficult to translate, attempts to describe a quality in which the artist has not so much *created* something, something almost divine, as that it has always been there, waiting to be *un*covered, or realized.[35]

The Symphony as Sociology

A separate element in the (wordless) symphony, particularly in the turbulent aftermath of the French Revolution, was its *communal* character, seen as contrasting markedly with the concerto. The symphony was communal and serious, whereas the concerto was showy and empty. It was this which, for a time, made the symphony particularly German. According to this view, culture arises from the relation between the individual and the state and Bildung; this process, whereby individuals come to find their creative role in a harmonious state, was seen as paralleled in the symphony. It was for this reason that singing in choruses was understood (by Goethe among others) as an appropriate training for citizenship.[36] Social harmony, like the orchestra, could exist only among a group of individuals who had worked on themselves to achieve a minimum level of personal self-realization.[37]

This was important because concepts of Germany changed decisively during Beethoven's lifetime. When his Ninth Symphony was premiered in Vienna in 1824, Germany was still an abstraction, but the idea of a pan-German state was no longer implausible and it was during the early nineteenth century that music was first recognized as having a role in the establishment of German national identity.[38] Friedrich Rochlitz, editor of the newly established *Allgemeine musikalische Zeitung*, wrote an editorial in 1799 expressing his hope/expectation that music would be used in the "education [*Bildung*] of the nation." "Without being accused of national pride," said another writer in 1805, "the German can declare that he deserves first place among all the nations in the realm of musical composition."[39] Music was both a producer and a product of nationalism, underlined by the growth of music festivals. These were more important than they might otherwise seem because in Germany at the time the rights of assembly were severely curtailed and so festivals, spread over two or three days and devoted to the symphony and the oratorio, attracted hordes of "music lovers" who, while devoted to an aesthetic, were also drawn to a microcosm of what an imagined Germany might be—a state in miniature but also a cultural rather than a territorial power. Here too the symphony was seen as a parallel to an organic community, the ideal structure of society.[40]

The symphony was the German genre par excellence for one final reason. Besides being "serious," with a sound philosophical basis, it also comprised a counterweight to opera, long dominated by the Italians and the French. This attitude/belief was to have a long and important legacy. Wagner put into the mouth of one of his characters the idea that, in writing the *Eroica* Symphony, Beethoven, who was "no general," nevertheless explored "the territory within which he could accomplish the same thing that Bonaparte had achieved in the fields of Italy." For Wagner, Beethoven's symphonies represented a stage in the progressive synthesis of the arts.

Unable to deny Beethoven's achievements with the symphony, Wagner neatly trumped him, arguing that the master himself had announced the culmination of the genre with his Ninth Symphony. By incorporating words into what was traditionally wordless, Beethoven, Wagner insisted, had implicitly conceded that instrumental music had run its course. It fell to himself, Wagner said, to take up where Beethoven had left off.[41]

7 ·

Cosmos, Cuneiform, Clausewitz

Abraham Gottlob Werner (1749–1817) was by all accounts an extremely eccentric man. At the School of Mining in Freiberg in Saxony, where he taught, he had a fire in the lecture room, "no matter what time of year." He invariably "wore fur over his bowels," fussed endlessly over the placements for his many dinner parties and the arrangement of the books in his library, and above all was "crazy about his stones." According to one of his pupils, he had amassed a collection of 100,000 rocks, each one composed of different minerals. On one occasion, when the specimen tray was being passed around his class, someone jostled it and nearly spilled the contents on to the floor. "At which point . . . Werner turned pale and could not speak . . . it was seven or eight minutes before [he] could command his voice."[1]

This singular soul was the founder of modern geology. At the end of the eighteenth century the main concern in geology (not that the term was used as we use it now) was not with basic science but with reconciling the biblical account of earth's origins with the record in the rocks.[2] Germany was at the forefront of this because of its mining history.[3] Silver provided the backbone of the money supply in Europe at that time, a period when a subsistence economy was giving way to a money economy. The explosive growth of silver mining in the German states—most of all in Saxony—stimulated the foundation, in the late fifteenth and early sixteenth centuries, of entire towns such as Freiberg, Saint Joachimstahl, and Chemnitz. The discovery of silver in the New World caused a slump in the mid-sixteenth century, but the other abundant mineral resources in Germany—which included kaolin (the raw material for the growing porcelain industry, stimulated by the introduction of high-quality Chinese porcelain into Europe in the sixteenth century)—fostered a healthy demand for mineralogists. Freiberg was the busiest region and played a leading role in the development of mineralogy and geology. Besides silver, the introduction of high-quality Chinese porcelain into Europe produced a race to find the secret of its manufacture, a search that was a boon for mineralogists. The French installed the first works at St. Cloud at the end of the seventeenth century, but were outgunned when the Germans set up enterprises at Vienna, Höchst, and Nymphenberg, not to mention Berlin and Meissen. It was soon understood that kaolin, China clay, was the crucial ingredient, and so began the search for deposits of this new precious substance. By 1710, the Meissen works, founded by Friedrich August I of Saxony, was manufacturing porcelain, helped by the discoveries of J. F. Böttger (1682–1719), the first director, who showed that certain fluxes (alabaster, marble, or feldspar) made kaolin fusible. This discovery remained a closely guarded state secret, despite no fewer than 30,000 experiments being mounted.[4] In this way, mining and chemistry became intimately related and helped determine the pre-eminence of German mineralogy.

The German universities, which had a bias in favor of the humanities, were not regarded as the best places to encourage very technical matters, and as the eighteenth century wore on, the princes began to realize that technical institutes were called for. The mining academy at Freiberg was established in 1765 and Werner was appointed ten years later.

Today, Werner is best known for his advocacy of the "Neptunist" version of the earth's history, contrasted with the "Vulcanist" or "Plutonist" account, rival versions that were intimately bound up with religious beliefs. In the Neptunist account, the surface of the earth was formed by rocks deposited out of a giant primeval ocean, which had originally covered the earth. There were serious problems with this theory. It did not even begin to explain why some types of rock that, according to Werner, were more recent than other types, were often found situated *below* them.

Still more problematic was the sheer totality of water that would have been needed to hold all the land of the earth in solution. It would have to have been a flood many miles deep, and in turn provoked an even bigger question: what had happened to all that water when it receded?

The chief rival to Werner, though nowhere near as influential to begin with, was the Scotsman James Hutton, and his theory of Vulcanism, named for the god of fire.[5] Hutton looked around him and concluded that weathering and erosion are even today laying down a fine silt of sandstone, limestone, clay, and pebbles on the bed of the ocean near river estuaries. He then asked what could have transformed these silts into the solid rock that is everywhere about us: his answer was that it could only have been heat. Where did this heat come from? Hutton believed it came from inside the earth and was expressed by volcanic action.

There was no question but that Hutton's Vulcanism fitted the facts better than Werner's Neptunism. Many critics resisted it, however, because Vulcanism implied vast tracts of geological time, "inconceivable ages that went far beyond what anyone had envisaged before."

Recent scholarship has credited Werner with a second and more important idea, one that has fundamentally shaped modern geology and, unlike Neptunism, stood the test of time. This is the linking of rock stratification and elapsed time. The most influential view to begin with was that advocated by Peter Simon Pallas (1741–1811), who identified primary, secondary, and tertiary sequences. On this account, all mountains were constructed in the same way. There were crystalline rocks that formed the center—the core—and went all the way up to their peaks. On their flanks were sedimentary rocks (limestones, marls, and shales), and finally, on the outside, lower down, looser deposits containing organic remains. These ideas were built on in Germany by J. C. Fuchsel, who identified specific stratigraphic formations, each layer having a characteristic fossil content. From this grew the idea of "formation suites," layers in predictable sequences that were similar from location to location.[6]

According to modern scholars such as Alex Ospovat at Oklahoma State University, it was Werner, living in Germany amid the rise of historicism and of evolutionism, who grasped that the essential difference between rocks was not mineralogy or chemistry but the "mode and time of formation," that rock *formation* was the basic process in geology.[7] Werner understood that there were only twenty to thirty of these "universal formations" and that they therefore reduced the chaos of mineralogy to "very distinct and determinable" proportions. It was now that fossil content rose in importance as a more specific indicator of age and sequence. Like others, Werner recognized that fossils became more varied and complex in later (higher) geological levels.[8] From 1799 he identified paleontology as a discipline of the future and offered a course in it.

This more sophisticated understanding of the meaning of stratigraphy was Werner's main lasting contribution. Rachel Laudan of the University of Hawaii has now identified what she terms the "Wernerian radiation," in which she says that Werner gave rise to a "movement" in geology, a "coherent lineage," one strand of which accepted, developed, and modified his idea of rock "formations," building on his ideas about fossils as the clearest way to understanding the past. The second branch of the Wernerian radiation was the causal school. This branch retained its interest in mineralogy but as an indication of causal processes, according an increasing role to heat in the earth's economy and in the process amalgamating Werner's and Hutton's theories.[9]

Laudan traces the Wernerian radiation, beginning in Freiberg but then, via the people who studied there, spreading to Britain, Ireland, Scandinavia, France, the United States, and Mexico. She traces Wernerian textbooks and societies to France, Scotland, and Cornwall, Wernerian academic journals, Wernerian students as teachers in the École des Mines in France, the mining school in Mexico, and Wernerian courses at Oxford and in Edinburgh. There was also a Wernerian radiation away from geology. Goethe, for example, subscribed to Wernerian theory till the end of his life, as did many of the Romantics, some of whom—Novalis among them—even attended his courses.[10]

The First Mathematician of Europe

Everyone knows the name of Isaac Newton. There are few prizes in the modern world for coming second in anything, but Carl Friedrich Gauss, according to John Theodore Merz in his *History of European Thought in the Nineteenth Century*, was, with Newton, one of the two greatest mathematicians of modern times, though that other German-speaker, the Swiss

Leonhard Euler, runs them close. Laplace called Gauss the first mathematician of Europe. Many think modern mathematics begins not with Newton but with Gauss. He in turn was much influenced by Kant, whose arguments implied that mathematics was an aspect of the *imagination* and, therefore, a form of freedom.

Carl Friedrich Gauss (1777–1855) was born into a laborer's family in Brunswick and he was as precocious as Mozart. He made simple calculations before he could talk; at the age of three he was correcting his father's arithmetic; and when he was nineteen he identified the formula that underlay the geometric construction of a 17-sided shape.[11] The Greeks had shown how, with a compass and straight edge, a perfect pentagon could be constructed, but no one between the Greeks and 1796 had been able to show how to use these simple tools to construct other "regular polygons" with a prime number of sides. Gauss was so excited by his discovery that, there and then, he decided to become a professional mathematician, and for eighteen years he would compile a mathematical diary. His family kept this diary in their possession for a century, until 1898, and it comprises one of the most important documents in the history of mathematics. Among other things, it confirms that Gauss proved—but often failed to publish—many results that other mathematicians did not discover until much later.

Gauss was, perhaps, much more than anyone else, the embodiment of the mathematical *imagination*. Understanding the behavior of numbers is as much an aesthetic matter as a utilitarian one. Number patterns don't *have* to be useful. The rest of us don't always see the point of *why* prime numbers are so fascinating or *why* it is so important to understand their behavior. Partly because of this, mathematicians are perhaps destined to inhabit their own private, solitary worlds, and that was certainly true in Gauss's case. He rarely collaborated, and worked alone for most of his life. His relationships with his wife and sons were less than ideal and he dissuaded his boys from a career in mathematics, it was said, so that there was no risk of the Gauss name being associated with inferior work. His wife died soon after bearing their third child, who also died, so Gauss spent much of his personal life stultified by despair and loneliness. Although he did not have an easy life, his entry in the *Dictionary of Scientific Biography* makes it clear that his ideas were influential in thirteen separate areas.[12]

Gauss became famous for his method of least squares, which enabled him to predict the moving orbits of planets; for his ideas about the pattern of prime numbers (divisible only by 1 and themselves), which revealed a hidden order that had totally escaped everyone else and uncovered their relationship to logarithms; for his invention of "clock arithmetic," which would eventually prove important for the security of the Internet; and for his invention of imaginary numbers, which would also transform understanding and link up, much later, with quantum physics.[13] But it is his conception of non-Euclidean geometry, commutative algebra, and the electric telegraph that really shows the extent to which his imagination was ahead of his time.

According to his mathematical diary, Gauss was still quite young when he began to consider the possibility that the ancient Greeks—Euclid in particular—had got it wrong with some of their fundamental axioms in geometry. In particular, he had begun to have doubts about parallel lines. Euclid had set out the classical paradigm and identified the classical solution: if you draw a straight line and then a point *off* that line, there can be only *one* line that is parallel to the first line and runs through the point.[14] When he was only sixteen, Gauss began to consider—daringly—whether there might be other geometries at variance with the Euclidean. He didn't publish anything for years, fearing ridicule, because—if he were right—other things followed, such as the fact that the angles of a triangle would not always add up to 180 degrees. Gauss couldn't get these subversive thoughts out of his mind: he even climbed to the summit of three hilltops to shine beams between them, to see if the angles added up to 180 degrees. This suggests that Gauss had some idea that light might bend in space, anticipating Einstein by nearly a century. It had occurred to Gauss that three-dimensional space might be curved in the way that the two-dimensional surface of the earth was. This thinking developed out of his observation that lines of longitude, along which the shortest path between two points on the surface of the earth is measured, all meet at the poles. They appeared parallel but were not. No one had considered that three-dimensional space might also bend.

Gauss was to be proved right, as Einstein was proved right, with Arthur Eddington's confirmation of the bending of light in 1919, but once again Gauss never published his ideas and the friends this troubled man shared his thoughts with were pledged to secrecy.[15]

Noncommutative algebra is the mathematical description of noncommutative geometry, which emerged in the nineteenth century in relation to physics and chemistry. At its simplest it refers to the possibility that, in mathematics, *xy*, strange as it may seem, is not always equal to *yx*. We shall meet this phenomenon again in the case of isomers in chemistry and with the benzene ring, where "rightness" and "leftness" determine chemical properties. This, plus the second law of thermodynamics, considered in Chapter 17, p. 341, which says that time is a fundamental aspect of space, shows that a purely mechanical (i.e., Newtonian) understanding of the universe has to be incomplete. Gauss's noncommutative algebra was an early attempt to come to grips with this problem. Once more he was well ahead of his time.

Though the bulk of Gauss's career was spent in the highly abstract world of numbers, it was bracketed by two very practical discoveries. The first—his calculation of the orbits of moving objects—has already been referred to. The second came when he was in his fifties and already had many abstract, imaginative discoveries to his name. Among the non-mathematical phenomena he was interested in (although of course he was interested in the mathematical aspects), was terrestrial magnetism, in particular the way it varied across the earth, and in the existence of magnetic storms.[16] In 1831, stimulated by Michael Faraday's discovery of induced current, Gauss collaborated (for once) with the brilliant experimental physicist Wilhelm Weber, one of the (liberal) "Göttingen Seven," to investigate a number of electrical phenomena. They made several discoveries in static, thermal, and frictional electricity but kept their powder dry for a time since their main interest was terrestrial magnetism. This prompted the idea that a magnetometer might also serve as a galvanometer and that, in turn, it might be used to induce a current that could send a message. Weber managed to connect the astronomical observatory at Göttingen with the physics laboratory a mile away by means of a double wire "that broke 'uncountable' times as he strung it over houses and two towers."[17] The first words, and then whole sentences, were transmitted in 1833 and the results of this first "operating electric telegraph" was mentioned (briefly) by Gauss in a notice in the *Göttingische gelehrte Anzeigen*, for August 9, 1834. Gauss grasped the military and economic significance of the invention and tried unsuccessfully to persuade the government to take an interest. It was not until Carl August von Steinheil, professor of mathematics and physics in Munich, in 1837, and Samuel Morse in the United States, in 1838, developed more user-friendly techniques that the electric telegraph caught on. Being ahead of his time was an occupational hazard for Gauss.

Nonetheless, his contemporaries referred to him as *princeps*, and he is now generally elevated to the level of Archimedes and Newton, inspiring a later generation—August Ferdinand Möbius, Peter Gustav Lejeune Dirichlet, Bernhard Riemann, Richard Dedekind, Georg Cantor, and others. As Marcus du Sautoy has said, the collaboration between Gauss and Weber on the telegraph, and Gauss's innovations with the clock calculator and its role in computer security, make them "the grandfathers of e-business and the Internet." Their collaboration is immortalized in a statue of the two of them in the city of Göttingen.[18]

The Advent of Humane Medicine

In marked contrast, statues of Samuel Christian Friedrich Hahnemann have been erected in Washington, Paris, Leipzig, Dessau, and Köthen. In North America at the turn of the (twentieth) century, other memorials existed in the form of twenty-two homeopathic colleges, while homeopathic remedies were used by one in five doctors. By 1945, homeopathic universities existed in the United States, Hungary, and India. In the twenty-first century, there is a homeopathic college in Canada, as well as a National Homeopathic Center just outside Washington, D.C., a Homeopathic Society in India, an Oxford College of Classical Homeopathy, a professional journal *Homeopathy*, edited from Luton in England and published by Elsevier in the Netherlands, one of the world's leading publishers of scientific journals. Dentists use homeopathy, it is employed in childbirth, on pets, and on farm animals.[19]

At the same time, there exist an organization and a Web site called "Homeowatch," dedicated to exposing the "quackery" of homeopathy, to showing that the science on which it is based is wrong-headed, even fraudulent, and that the many products produced in its name are medically worthless. As Martin Gumpert puts it in his biography of Samuel Hahnemann, "is homeopathy merely an excrescence of science, or is its core genuine and useful, even if we may not like its covering?"[20] Hahnemann's name elicits hatred and scorn among most regular medical practitioners but at the same time he refuses to go away. At one stage, the British royal family appeared in thrall to homeopathic medicine.

Hahnemann (1755–1843) was the son of a painter in the Meissen porcelain factory, and like many contemporaries he was precocious: at an early age he could speak Latin and Greek, classify old coins, and catalog books. But medicine was his main love, and he graduated from the University of Erlangen in 1779. His first destination was Hettstedt, a mining town that lacked a doctor, and there he came across a mysterious copper sickness that often proved fatal. It was in studying this disease that he first began to have doubts about the traditional blood-letting technique that all doctors of the time used. The basic approach was to make patients excrete, so as to void their bodies of whatever poisonous substances they had accumulated. They were induced to sweat, prescribed laxatives, forced to gargle, vomit, or salivate. The most extreme was blood-letting.[21]

When Hahnemann moved on again, this time to Gommern, near Magdeburg, he had another fraught encounter when a patient of his, a cabinetmaker, broke down suddenly (as we would say now), and Hahnemann accompanied him to the mental hospital (again, as we would say). There, the cabinetmaker was strapped to a chair attached to a mechanism that enabled it to be rotated rapidly sixty times a minute. This "treatment" used the centrifugal force of the rotating chair to send blood rushing to the patient's brain, producing dizziness, vomiting, evacuation from the bowels and kidneys, "while blood even oozed from the skin around the eyes." Such brutality reduced even the wildest inmates to catatonics. These experiences eventually led to Hahnemann's book *Freund der Gesundheit* (*The Friend of Health;* 1792), written after he had settled in Leipzig, where he put forward the idea of a public hygiene policy, becoming one of the first people to do so. And it was in Leipzig that he translated *Treatise on the Materia Medica*, by William Cullen, a professor of medicine in Edinburgh.[22] In the course of translating this work, Hahnemann had the idea for which the world now knows him.

Cullen had written the following sentence when he was discussing the properties of cinchona bark (the source of quinine), which he said was a "febrifuge," a substance that drives away fever: "In this case the bark works by means of its fortifying effect on the stomach." Hahnemann was brought up short. He knew that cinchona had never fortified *his* stomach. On the contrary, quinine made him very sick. Accordingly, he now decided on his own trial. "By way of experiment, I took four drams of good cinchona twice a day. My feet, my fingertips, at first became cold." There was no sign at all of his stomach being "fortified." "I grew languid and drowsy; then my heart began to palpitate, and my pulse became hard and small; intolerable anxiety, trembling (but without cold rigour), prostration through all my limbs. Then pulsation in my head, flushing of my cheeks, and, in short, all those symptoms which are ordinarily characteristic of intermittent fever, one after another made their appearance." It was a little while later that he noted the observation that would change everything: "Substances which excite a kind of fever extinguish the types of intermittent fever."

Fever cures fever. That was Hahnemann's new doctrine. As Gumpert says, "We must remember that this was before the germ or cell theories of disease, and that Hahnemann's new ideas were an alternative to the brutal current method of the evacuation of 'pernicious juices.' "[23] In 1796 Hahnemann offered a paper to the newly founded *Journal of Practical Medicine* titled, "Essay on a New Principle for Ascertaining the Curative Power of Drugs, with a Few Glances at Those Hitherto Employed." His central idea was now clearly set out: "*In order to cure diseases, we must search for medicines that can excite a similar disease in the human body.*" "*Similia similibus!*"—this is the essence of homeopathy.[24]

He formulated his views in full in *Die Organon der rationellen Heilkunde* (translated as *The Organon of Homeopathic Medicine*; 1810) and *Theory of Chronic Diseases* (1828–39), where he argued for the use of minute quantities of remedies that, in larger doses, produce effects similar to those of the disease being treated.[25] His wilder views are revealed in his further belief that small doses of medication could be induced to have powerful effects by vigorous shaking (called *succussion*). He referred to this increase in potency as *dynamization*, which, for Hahnemann, released an "energy" that he regarded as "immaterial and spiritual." Eventually, he thought patients need not swallow "dynamized" medicines at all; it was enough to sniff them.

Most doctors dismiss homeopathy now on the grounds that any active ingredients, such as they are, are diluted often by 10,000 times, reducing them to well below the levels at which any pharmaceutical capacity could exert an effect.

Hahnemann continued practicing homeopathy over a long life, dying in Paris (he had married a French patient in 1843, when he was eighty-eight). He was visited by patients from all over the world, and a Homeopathic Medical College opened in Philadelphia in 1848. By 1900 America had 111 homeopathic hospitals, those 22 homeopathic medical schools mentioned earlier, and 1,000 homeopathic pharmacies. Thereafter the fashion for homeopathic cures declined, only to resurface in the 1960s. It is now very popular in India, Latin America, and Europe, and Great Britain has five homeopathic hospitals, and homeopathic cures are covered—amid great controversy—on the National Health Service.

The Scientific Discovery of the New World

"[Alexander von] Humboldt has done more good for America than all her conquerors," said Simón Bolívar, the Venezuelan-born general credited with leading the liberation of Venezuela, Colombia, Ecuador, Peru, Panama, and Bolivia. "[He] is the true discoverer of South America." Ralph Waldo Emerson described Humboldt as "one of the wonders of the world, like Aristotle, like Julius Caesar . . . who appear from time to time as if to show us the possibilities of the human mind." A recent biography of Humboldt says bluntly that it is "quite possible that no other European had so great an impact on the intellectual culture of nineteenth-century America."[26] In his day, Humboldt was as famous as Napoleon. He was friends with Goethe (who shared his interests in plants and mining), Schiller, and Gauss, and his brother Wilhelm founded the University of Berlin. The paleontologist Stephen Jay Gould described him as "the world's most famous and influential intellectual." He has also been described as "one of the greatest but least remembered figures in scientific history" and that is true too.[27]

Born in Berlin in 1769, he and his brother were tutored privately (their father was technically an aristocrat but had only recently been ennobled). All his life Alexander was a restless man. He was good at drawing and his self-portraits show a handsome face, though he wore his hair so as to conceal marks sustained from a childhood bout of smallpox. His brother found him "self-centered" and a "busybody," which he feared others would construe as vanity.

Humboldt enrolled as a law student at Göttingen when he was twenty; the son-in-law of one of his professors was Georg Forster who, as a teenager, had accompanied his father on James Cook's second voyage around the world.[28] The younger Forster was already known for his highly acclaimed account of that adventure, and he and Alexander teamed up to make their own journeys across Europe, answering the latter's restlessness but stimulating it too.

Humboldt's most important teacher in his early years was someone he encountered after he left Göttingen and attended the Freiberg School of Mines: Abraham Werner. After studying with Werner, Humboldt joined the Prussian mining service where he had a distinguished career. Using his own fortune to good effect, he invented—among other things—a safety lamp and a rescue device for miners threatened with a reduced air supply belowground. (He tested these mechanisms on himself in potentially dangerous experiments.) He was an out-and-out empiricist—facts, numbers, measurement, these, not philosophical speculation, were for him the building blocks of science.

But it was Humboldt's wanderlust that was to distinguish him from everyone else, and after a number of travels within Europe itself—looking at active volcanoes—he set out on the first of his two "great journeys."

The first, and the more momentous, was to South America. On October 20, 1798, he left Paris with the French botanist Aimé Bonpland, who would be his traveling companion for the next six years. (Humboldt studied under Laplace and was fluent in French—most of his writings were in that language.)[29] They went first to Marseille and then on to Madrid, where Humboldt was introduced to the Spanish king and managed to persuade him to allow a scientific expedition to South America. This was remarkable, first because there had only ever been six scientific expeditions to Spain's New World colonies (she was almost exclusively interested in the gold and silver to be obtained there) and second, because Humboldt was a Protestant. But the royal passports Bonpland and he obtained in March 1799 gave them total freedom of movement in the colonies.[30] They left from La Coruña and after breaking through the British blockade landed on July 16, 1799, in what would become Venezuela. Now began what has been called "the scientific discovery of the New World."

The two men faced great hardships and considerable danger as they traveled—on foot, by packhorse, in native canoes, and oceangoing ships—across Venezuela, Cuba, Colombia (where Bonpland caught malaria and they were delayed for two months), Peru, Ecuador, and Mexico.[31] In the process, they "recorded, sketched, described, measured, compared and gathered" some 60,000 plant specimens, 6,300 of them unknown in Europe. Humboldt, however, was not just interested in geography, geology, and botany: he studied ancient Indian monuments, population figures, social arrangements, economic conditions. He was appalled by the slavery he witnessed and thereafter campaigned against it. He navigated the Orinoco and Magdalena rivers (which run respectively west–east through Venezuela toward Trinidad, and south–north through Colombia to the Caribbean), and confirmed the bifurcation of the Casiquiare River, showing that it really did connect the Orinoco and the Amazon as had been rumored.[32] Humboldt also set a new mountaineering record, climbing to 19,000 feet on Mount Chimborazo (Urcorazo, "snow mountain" in Quichua) in Ecuador in June 1802. He failed to reach the summit, but his record stood for nearly thirty years.

They traveled with forty-two instruments—thermometers, barometers, quadrants, microscopes, rain gauges, eudiometers (for measuring oxygen in the air)—each with its own velvet-lined box. They nearly lost all these more than once as they tried to negotiate the many fearsome rapids along the Orinoco. It was on the Orinoco and its countless tributaries that Humboldt discovered a form of rubber, and where he found that the "natives" could distinguish a river by the taste of its water.

In 1804 he returned to Europe via the United States, visiting Philadelphia and Washington, D.C., where he met President Thomas Jefferson in the White House and at Monticello and was elected a member of the American Philosophical Society.[33] On his return to Europe, Humboldt brought with him quinine, curare (the nerve poison), and *dapicho*, a substance similar to rubber. He was the first to stress the glories of the Inca and Aztec civilizations. In Paris he met Simón Bolívar, with whom he was to correspond until Bolívar's death in 1830. Both could see the need for scientists to help in the development of Bolivia, and Humboldt did all he could to help.

The journal he composed in the course of and after his travels was eventually published—in thirty-four volumes over twenty-five years. Not the least of its attractions were some 1,200 copperplates showing South American flora, fauna, and topography. He also wrote many specialist formal scientific treatises, in which he developed climatology as a science, established the specialities of plant geography and orography (the science of mountains), and initiated ideas that we still use today, such as mean temperature and the isotherm.

In 1829 he was invited to explore Siberia as a guest of the Russian government. His 9,000-mile itinerary took in Kazan, the northern Urals, western Siberia as far as the Altai Mountains, and the edge of Chinese Tungusic territory. He successfully predicted the existence of diamonds in the Urals.

In 1845, when Humboldt was seventy-six, he published the first volume of *Kosmos*, the second appearing two years later.[34] This turned out to be a triumph, a popular scientific book in the best sense. "The entire material world from the galaxies to the geography of various mosses is presented 'in pleasing language.'"[35] Four volumes in all were published, and the singular nature of the book may be shown from the fact that, although it was a massive popular success, it contained more than 9,000 references.

In his autobiography, Charles Darwin wrote: "During my last year at Cambridge, I read with care and profound interest Humboldt's *Personal Narrative*. This work and Sir J. Herschel's *Introduction to the Study of Natural Philosophy* stirred up in me a burning zeal to add even the most humble contribution to the noble structure of the natural sciences." On the centenary of Humboldt's birth in 1869 the *New York Times* devoted the *whole* of its front page to Humboldt (there were no pictures and no advertisements).[36]

Humboldt also helped advance the careers of many young scientists, but perhaps his most enduring monument is that more places around the world have been named after Humboldt than anyone else—thirty-five in all: one city in Mexico, one in Canada, ten in the United States, three counties in the United States, nine bodies of water (including the Humboldt Current in the Pacific Ocean), seven mountains and glaciers (including Humboldt Mountains in China and New Zealand), four parks or forests (including the Humboldt National Park in Cuba). There is also the Mare Humboldtianum, on the moon.[37]

A Break in the Great Wall of Languages

"Only after 1771 does the world become truly round; half the intellectual map is no longer blank." These are the words of Raymond Schwab, the French scholar, in his book *The Oriental Renaissance*; what he meant was that the decipherment of the "Great Wall of Asian languages"—Sanskrit, Hindi, the hieroglyphics, and cuneiform scripts—was, in his words, "one of the great events of the mind." C. W. Ceram agreed. The decipherment of the cuneiform script, according to him, "was one of the human mind's most masterly accomplishments," a work of true genius.[38] It came amid the golden age of translation, when many of the scripts of the ancient Near East, and India itself, were giving up their secrets. The overall impact of the age of translation on European thinking, and German thinking in particular, is discussed in Chapter 8, page 189.

Aside from its intrinsic importance, the decipherment of cuneiform has attracted interest for two colorful reasons. One, the original effort was made as the result of a wager; and two, the working out of the decipherment is starkly clear and simple: the sheer cleverness is there for all to see. The man who made the bet and did the deciphering, Georg Friedrich Grotefend, was born on June 9, 1775, at Münden in Hanover. He studied philology at Göttingen, where he became friendly with Christian Gottlob Heyne (see Chapter 1, p. 41). On Heyne's recommendation, in 1797 Grotefend became an assistant master at the *Gymnasium* in Göttingen, later promoted to vice principal of the Frankfurt am Main grammar school.

His early interest was in Latin, but in his late twenties he became fascinated by cuneiform scripts that had been discovered in the seventeenth century but were not yet understood. The idea for the wager occurred to Grotefend while he was drinking at an inn with colleagues. Never believing he could pull it off, they accepted his proposal immediately, the more so as the only scripts available to him were some poor copies of the inscriptions discovered in the ruins of Persepolis. He wasn't deterred and tackled head-on a problem that the best scholars of the day had found insurmountable.[39]

He probably would not have made the breakthrough he did without having had the traditional education in Germany which, as we have seen, stressed the classics and philology. Grotefend noted that on some of the Persepolitan tablets there were three different scripts, written side by side in three separate columns.[40] Knowing a certain amount of ancient Persian history, through his study of the Greek writers, he was aware that Cyrus had laid waste to the Babylonians around 540 B.C., and this had allowed the rise of the first great Persian kingdom. Grotefend therefore inferred that at least one of the scripts on the tablet would be written in the language of the conqueror. He judged that that would be the middle column since in antiquity the most important script was always put there.

That was his starting point. He next noticed a complete absence of curved lines in cuneiform, provoking the thought that the characters had not actually been "written," as such, but instead impressed in wet clay. We now know that cuneiform writing (from the Latin *cuneus*, meaning wedge-shaped) had originally been pictographic but had become progressively more stylized—for ease and rapidity of writing—and later Persian was almost an alphabetic system, reduced to about thirty-six characters from the original six hundred. With all this as background, Grotefend observed that most of the points of the wedges ran either downward or to the right. Furthermore, the angles formed where two wedges met always opened to the right. The implication was clear: cuneiform was written horizontally, not vertically, and it read from left to right, not the other way round.[41]

The actual act of decipherment began when he further observed that one group of signs, and another single sign, recurred frequently throughout the text. He inferred that this group of signs was the word "king" and the single sign—a simple wedge slanting upward from left to right—was a device that separated one word from another, in effect an ancient space bar. His next inference—and perhaps his greatest—was to assume that particular mannerisms could be found in the inscriptions, mannerisms that would have remained unchanged over generations. Here he drew a parallel with the practice, in his own time, of using the phrase "Rest in peace." This, he pointed out, had been carved unchanged on gravestones for centuries. Probably, something similar would have been used on the Old Persian texts. From his knowledge of such ancient texts as he had encountered through his Greek and Latin studies, he thought that, in any inscription, such phrases as "great king," "king of kings," "son of . . ." would be found. These dynastic formulas should be repeated in all three columns of the tablet. The actual phrase was familiar from languages already known:

> *X, Great King, King of Kings, King of A and B, Son of Y, Great King, King of Kings.*

If such a syntax did exist, Grotefend inferred that the first word must be a king's name. Following that would be the word divider, and then two words, one of which ought to be "king," which should be easily identified because it would be often repeated.[42] Looking down the inscriptions, Grotefend observed that there were just two versions of the *same* cuneiform groups at the beginnings of the columns—the same words were used each time but in a different order. If he was right about the word for "king," then what was written was as follows:

X (king), son of Z

Y (king), son of X (king)

Again assuming he was correct, this inscription referred to a dynastic succession *in which father and son were kings but not the grandfather.* He therefore set himself to find a royal succession that fit such a picture. Looking down the king lists for Persian kings, known from other studies, he quickly concluded that it could not be Cyrus and Cambyses, because the two names in cuneiform did not begin with the same letter, and the inscriptions could not refer to Cyrus and Artaxerxes, because these names were too dissimilar in length. That left Darius and Xerxes, each with the same number of letters, each beginning with a different initial. "They fitted so easily," said Grotefend, "that I had no doubt about making the right choice."[43] This success was underlined by the fact that Xerxes's father was Darius, and both of them were kings, but Darius's father, Hystaspes, was not a king. This is exactly what the inscription said.

Publication of Grotefend's discoveries was delayed for a while because he was felt to be too young to have made such an important breakthrough and because he was "only" a schoolteacher rather than a full-fledged university academic. In fact, another thirty years were to pass before anyone added anything of substance to Grotefend's discoveries, when the Frenchman Émile Burnouf and the Norwegian-German Christian Lassen made further inroads into the decipherment of cuneiform.

The Transformation of War

Vom Kriege (*On War*), a long and not altogether cohesive work by an otherwise unknown Prussian general of the Napoleonic era, Carl Philipp Gottlieb von Clausewitz, has achieved a supreme position in Western thinking about warfare. Dismissed by some as a narrow-minded pedant, an out-and-out militarist obsessed with war "as an instrument of policy," he has also been attacked for treating war "as a rational act."[44] On the other hand, for Bernard Brodie, the American strategist of the nuclear era, *On War* is "not simply the greatest book on war but the one truly great book on that subject yet written." It has been compared with Adam Smith's *Wealth of Nations* and Darwin's *Origin of Species* in the force of its impact.[45]

The book has undoubtedly become a classic, but it didn't happen immediately. *On War* was a product of its time, and it is true to say that the book owes almost as much to Napoleon as to its German author. We should not forget that, at the time the book's main ideas were being conceived, Prussia's survival as an independent nation was under threat. Napoleon had disturbed the European balance of power fundamentally, and the French ideas of revolution and individual rights threatened the ancient regimes everywhere.[46] The emperor's advances had sparked vituperative essays by such writers as Friedrich von Gentz but these "squibs" could do little to counter the new reality: that, so far as war was concerned, Napoleon had enlisted the aid of the general population, inventing a new mass army. Prussia and its forces, Clausewitz realized, must reform in order to combat these new circumstances. Like Wilhelm von Humboldt (see Chapter 10, page 225), Clausewitz realized that major reforms were needed, that the age of absolutism was well and truly over, and that the reforms needed to be related to the people.

His early career was amazing from a modern standpoint. Clausewitz came from a military family and was a soldier at the age of twelve, remaining one until his death in 1831. He saw combat before his thirteenth birthday and had fought in five campaigns against France by the age of thirty-five.[47]

His capacity for reflection did not go unnoticed, and he was taken under the wing of Gerhard von Scharnhorst (1755–1813), co-opted into the exclusive Military Society that Scharnhorst founded, of which Prince August, nephew of Friedrich Wilhelm III, was another member and where the arts of war were discussed. Clausewitz fought at the Battle of Jena in 1806 where he did well but was taken prisoner.[48] When he was released, he clamored for reform and began publishing articles advocating change, at first anonymously.

After those five campaigns, and despite military service with the tsar, because the Prussian king had sided with the hated French against the Russians, Clausewitz was eventually reinstated, promoted to colonel, and nominated as superintendent of the War College in Berlin.[49] Now he started work on a lengthy study of war that he had first conceived in 1816. By 1827, says Hugh Smith, in his study of Clausewitz, a draft of the first six books of *On War* was in existence—about 1,000 pages of manuscript.[50] The book was never completed. Clausewitz succumbed to cholera after putting down an insurrection in Poland in 1830. It fell to Clausewitz's widow, Marie, to complete the mammoth task of preparing her husband's manuscripts for publication.

A New, More Brutal Kind of Army

On War should be read against several significant changes in warfare that took place during Clausewitz's lifetime.[51] In his first engagement in 1793, for example, when he was just twelve, eighteenth-century strategy was still being used as armies maneuvered for limited objectives and preferred tactical skirmishes over full-fledged battles. As was often the case then, neither side emerged victorious. By the time of his second campaign, however, the battles of Jena and Auerstedt in 1806, Napoleon had changed all the rules of engagement.[52]

Battles of the Napoleonic era had a higher ratio of casualties than those of the eighteenth century because the nature of armies had changed. In the eighteenth century, armies were, in effect, a royal possession, their officers drawn from the aristocracy with a personal allegiance to the monarch. Most wars, therefore, did not involve the general population as combatants hardly at all. Troops were professional soldiers, mercenaries, and foreigners. Desertion was high.

Then came the French Revolution and Napoleon. After 1789, for France, war became "the business of the people—a people of thirty millions, all of whom considered themselves to be citizens." Because Frenchmen now identified with the nation, they allowed themselves to be called to arms in far greater numbers. "Before 1789 an army in the field rarely exceeded 50,000 men. Within a decade or so conscription and militia systems were able to raise forces of over 100,000, and in 1812 France could assemble 600,000 men for its Russian adventure."[53] With such a supply of troops, major battles could be risked more often. "Between 1790 and 1820 Europe saw 713 battles, an average of twenty-three a year compared with eight or nine a year over the previous three centuries."[54]

In line with this, and following the humiliating peace at the end of the Seven Years' War, the French army began to change its structure. Traditionally, the battalion or regiment, roughly 1,000 men, was the basic unit, but now a distinctly larger formation, the division, was conceived. This consisted of 10,000–12,000 men under independent command, comprising infantry, cavalry, and artillery, with engineering, medical, and communications support. Napoleon also put divisions together to create "corps" of up to 30,000 soldiers and then put corps together to create armies. The importance of sheer size was shown in the fact that a corps of 20,000–30,000 men, it was said, "could not be eliminated in an afternoon, being able to resist long enough for relief to arrive."[55] Sheer size meant that commanders could more easily pursue an opponent and force him to fight. Traveling was safer because the larger numbers could be spread over different roads and were therefore harder to attack. Napoleon also discovered that by pursuing defeated troops, giving his cavalry their head, he could significantly "magnify the scale of the original victory."[56] Despite these manifold advances, it was only on the eve of war, in 1806, that Prussia established permanent divisions.

And so, militarily, when Clausewitz came to maturity, war was becoming far more brutal. This profoundly shaped *On War* and made many of the theorists who immediately preceded him—Adam Heinrich Dietrich von Bülow, Georg Heinrich von Berenhorst, Antoine Henri de Jomini, and even Scharnhorst—look dated, though it was Scharnhorst who had first called for the introduction of divisions into the Germany army.[57] All agreed that chance was an important factor in war, and so was morale, but beyond that there was little agreement. If one writer influenced Clausewitz more than anyone else it was probably Niccolò Machiavelli, whose view about the unchanging nature of the human condition, and politics—constant conflict—he shared.

On War is a big book that at times is "inconsistent, obscure and opaque"; yet it has stayed influential "because it illuminates by simplifying complex issues and dramatising the human element of war . . . [Clausewitz] is passionately involved in the subject and at the same time detached and objective."[58] The key elements are probably that there are two types of war—one to overthrow the enemy, the other to secure limited objectives; and that war needs to be understood not as an independent variable but as a function of policy. If Clausewitz has one message it is that "only major engagements involving all forces lead to major success . . . *On War* never fully escapes the spirit of all-out war that values the great, decisive battle."[59] This argument—which may seem obvious and commonsensical—was newer then than it seems now, because in the eighteenth century it had always been exceptional for one battle to settle an entire war. It was as if the lesson of Napoleon had been learned most by the people he had humiliated. (This too was one of Clausewitz's arguments—that the defeated feel defeat more keenly than victors enjoy their victory, an observation that was to echo down the nineteenth century, and all the way up to 1939.)

Clausewitz also introduced the concept of "centers of gravity." This was his way of confirming that strategies "require some link between military activity and political objectives." He identifies four centers of gravity: territory, the capital of a country, its armed forces, and its alliances.[60] Among these, the pre-eminent center of gravity is a nation's army. That must be destroyed for decisive victory.

In a sense, Clausewitz's achievement was to clear the air. He realized that, with the change from engagements between battalions and regiments to engagements between divisions, corps, or armies, the whole concept of war became more terrible, and that with the growth of conscription and mass armies, commanders had to face up to the new realities.

Since the publication of *On War* between 1832 and 1834, Clausewitz has become a key figure in understanding war.[61] At the outset he was criticized, not least by de Jomini, whose reputation outshone Clausewitz's for much of the nineteenth century. But Clausewitz's book gradually won support from those who realized what he was getting at. Friedrich Engels recommended Clausewitz to Marx around the middle of the nineteenth century, and Marx familiarized himself with the main ideas of *On War* without reading the book itself. But the initial print run of 1,500 copies was still not fully sold by then, and it is fair to say that Clausewitz had fallen into "respectful oblivion." Despite this, the publishers, Dümmler of Berlin, put out a second edition in 1853 and its reception was better. In Prussia the book was taken up in earnest in the 1860s by leading generals, Helmuth von Moltke in particular. He was impressed by what he took to be Clausewitz's advocacy of "Napoleonic" war, with its emphasis on size, morale, patriotism, and leadership. Moltke's own victories against Austria in 1866 and France in 1870–71 helped to shape the view that war was "a practical, proper and glorious instrument of national policy."[62] Moltke embroidered Clausewitz's idea, even arguing that soldiers should take over from politicians in wartime.

A French translation of *On War* appeared in 1849–50 and an English version in 1873. Military colleges on both sides of the Atlantic began to adopt *On War* as a major text. After its failure to contain the irregular forces in the Boer War in South Africa (1899–1902), the British army began to take an interest in Clausewitz. The main message the British found was the need for popular militarism. And this view spread. At the beginning of the twentieth century, all the nations of continental Europe began building powerful armies and fleets, and Clausewitz was regarded as one of those responsible for these developments. Writing the introduction to an English edition of *On War* in 1908, Colonel F. N. Maude said: "It is to the spread of Clausewitz's ideas that the present state of more or less readiness for war of all European armies is due."[63]

8.

The Mother Tongue, the Inner Voice, and the Romantic Song

In France in the late 1680s King Louis XIV added six young Jesuits—all scientists as well as prelates—to a mission he was sending to Siam. The men were put ashore in the south of India, the first of the French (as opposed to Portuguese) "Indian Missions" that were to gain fame for their *Lettres édifiantes et curieuses*, which gave detailed accounts of their experiences. Abbé Jean-Paul Bignon, the French king's librarian, requested that the missionaries be alert for "Indic" manuscripts, which he was keen to obtain to form the backbone of an Oriental library. In 1733, in the *Lettres édifiantes*, the Jesuits announced their response: the discovery of the first "big game" of the hunt, a complete Veda, long thought to have been lost. (It was in fact a complete Rig-Veda in Sanskrit.) Subsequently, a whole raft of Hindu manuscripts was brought to Europe in the eighteenth century, and that movement, together with the deciphering, at much the same time, of the Egyptian hieroglyphics and cuneiform, prompted Edgar Quinet, the French anticlerical historian, to describe it in 1841 as an event "more or less comparable" with the arrival of the ancient Greek and Latin manuscripts, many in Arabic translation, that had transformed European life in the eleventh and twelfth centuries. Raymond Schwab, in *The Oriental Renaissance*, argued that the discovery of the Sanskrit language and its literature was "one of the great events of the mind."

The Oriental Renaissance

The so-called Oriental renaissance properly began with the arrival in Calcutta of William Jones, a British poet, linguist (he spoke thirteen languages), and judge, and with the establishment of the Asiatic Society of Bengal in January 1784. This was established by a group of talented English civil servants who were employed by the East India Company and who, besides their official duties helping to administer the subcontinent, also pursued broader interests, which included language studies, the recovery and translation of the Indian classics, astronomy, and the natural sciences.

Jones was president of the Asiatic Society of Bengal and it was in his Third Anniversary Discourse that he announced his great discovery which was to transform scholarship, namely the relationship of Sanskrit to Greek and Latin. In "On the Hindus," his address delivered on February 2, 1786, he said: "The Sanskrit language, whatever be its antiquity, is of a wonderful structure; more perfect than the Greek, more copious than the Latin, and more exquisitely refined than either, yet bearing to both of them a stronger affinity, both in the roots of the verbs and in the forms of the grammar, than could possibly have been produced by accident; so strong, indeed, that no philologer could examine them all three, without believing them to have sprung from some common source, which, perhaps, no longer exists."

In linking Sanskrit to Greek and Latin, and in arguing that the Eastern tongue was, if anything, older than and superior to the Western languages, Jones was striking a blow against the very foundations of Western culture and the assumption that it was more advanced than cultures elsewhere. The history of the East was at last on a par with that of the West, no longer subordinate to it, no longer necessarily *a part* of that history.

Although it was an Englishman who had discovered the all-important link between Sanskrit and Greek and Latin, and although the French carried out some of the earliest translations of the Hindu scriptures and classics, the Oriental renaissance found its fullest expression in Germany.[1]

The dimensions of this renaissance need underlining. In 1832 Wilhelm Schlegel said that his own century had produced more knowledge of India than "the twenty-one centuries since Alexander the Great." (Schlegel was, like Jones, a linguistic prodigy. He knew Arabic and Hebrew by the time he was fifteen and, at seventeen, when he was still a pupil of Herder's, he lectured on mythology.) The German translations of the Bhagavad Gita and Gita Govinda, published in the first decade of the nineteenth century, had a tremendous influence on Friedrich Schleiermacher, Schelling, the Schlegel brothers, Schiller, Novalis, Goethe, and, eventually, Arthur Schopenhauer.

The poetry of the Bhagavad Gita, its wisdom, and its complexity and richness brought about a major change in attitudes toward the culture of India and the East. In *Über die Sprache und Weisheit der Indier*, Friedrich Schlegel discussed the metaphysical traditions of India on an equal footing with Greek and Latin ideas. This was more important than we may understand now, because, against a background of deism and doubt, such an approach allowed that the Indians—the inhabitants of the far-off East—had as thorough a knowledge and belief in the true God as did Europeans. This was quite at variance with the teachings of the church. The sheer richness of Sanskrit also went against the Enlightenment belief that languages had begun in poverty and grown more elaborate. This helped to launch—in Germany first, and then elsewhere, as we have seen—the great age of philology. Many religious souls at the time remained convinced that the earliest (and most perfect) language had to be Hebrew, or something like it, because it was the language of the chosen people. Franz Bopp (1791–1867), who had studied Sanskrit manuscripts in Paris and London, turned his back on these preconceptions and showed how complex Sanskrit was even thousands of years ago, throwing doubt on the very idea that Hebrew was the original tongue. Friedrich Schelling took the ideas of Jones one step further. In his 1799 lecture, *Philosophie der Mythologie*, he proposed that, just as there must have been a "mother tongue," so there must have been one mythology in the world shared by all peoples.

One final, fundamental way in which the discovery deeply affected people was in the notion of "becoming." If religions were at different stages of development, and yet were all linked in some mysterious way—only glimpsed at so far—did this mean that God, instead of just *being*, could himself be said to be *becoming*, undergoing a process of *Bildung*?

God came to be seen, not in an anthropomorphic sense, but as an abstract metaphysical entity.

An Alternative to Classicism and the Original Language of Eden

The Oriental renaissance played a vital role in the origins of the Romantic movement. The strongest link was between Indic studies and the German form of Romanticism. Indic studies proved popular in Germany for broadly nationalistic reasons. It seemed to German scholars of the time that the Aryan/Indian/Persian tradition linked with the original barbarian invasions of the Roman Empire from the East and, together with the myths of the Scandinavians, provided an alternative (more northerly) tradition to the Greek and Latin Mediterranean classicism that had dominated European life and thought for the previous 2,500 years. Furthermore, the discovery of similarities between Buddhism and Christianity, together with the Hindu ideas of a world soul, seemed to the Germans to indicate a primitive form of revelation, the original form, out of which Judaism and Christianity might have grown, but which meant that God's real purpose was hidden somewhere in the Eastern religions. Such a view implies that there was a single God for all mankind, that there was a world mythology, the understanding of which would be fundamental. In Herder's terms, this ancestral mythology was "the childhood dreams of our species."

A further factor that influenced Romanticism was that the original Indian scriptures were written in poetry. The idea became popular, therefore, that poetry was "the mother tongue," that verse was the original way in which wisdom was transmitted from God to mankind ("Man is an animal that sings"). Poetry, it was thought, was the original language of Eden.[2]

The range of poets, writers, and philosophers who came under the influence of these views spanned the Atlantic but it was especially strong in Germany. Goethe learned Persian and wrote in the preface to the *West-östlicher Divan*: "Here I want to penetrate to the first origin of human races, when they still received celestial mandates from god in terrestrial languages." Heinrich Heine studied Sanskrit under Wilhelm Schlegel at Bonn and under Bopp in Berlin. As he wrote: "Our lyrics are aimed at singing the Orient." Both Wilhelm Schlegel and Ferdinand Eckstein, another German Orientalist, believed that the Indic, Persian, and Hellenic epics rested on the same fables that formed the basis of the *Nibelungenlied*, the great medieval German epic of revenge, which Richard Wagner was to rely on for *Der Ring des Nibelungen*. For Schleiermacher, as for the entire circle around Novalis, the source of all religion "can be found," according to Ricarda Huch, "in the unconscious or in the Orient, from whence all religions came."[3]

A Change in the Meaning of Individuality

In the history of Western political thought, says Isaiah Berlin, the Oxford historian of ideas, "there have occurred three major turning-points." The first of these took place in the short interval at the end of the fourth century B.C. between the death of Aristotle (384–322) and the rise of Stoicism, when the philosophical schools of Athens "ceased to conceive of individuals as intelligible only in the context of social life . . . and suddenly spoke of men purely in terms of inner experience and individual salvation." A second turning point was inaugurated by Niccolò Machiavelli (1469–1527) and involved his recognition that political values "not merely are different from, but may in principle be incompatible with, Christian ethics."[4] This produced a utilitarian view of religion, discrediting any theological justification for any set of political arrangements.

The third great turning point—which Berlin argues is the greatest yet—was conceived toward the end of the eighteenth century, with Germany in the vanguard. "At its simplest the idea of romanticism saw the destruction of the notion of truth and validity in ethics and politics, not merely objective or absolute truth, but subjective and relative truth also—truth and validity as such." This, says Berlin, has produced incalculable effects. In the past, it had always been taken as read that moral and political questions, such as "What is the best way of life for men?" "What is freedom?" were in principle answerable in exactly the same way as questions like "What is water composed of?" and "When did Julius Caesar die?" It was assumed that the answers were discoverable, because, says Berlin, despite the various religious differences that have existed over time, one fundamental idea united men, though it had three aspects. "The first is that there is such an entity as a human nature, natural or supernatural, which can be understood by the relevant experts; the second is that to have a specific nature is to pursue certain specific goals imposed on it or built into it by God or an impersonal nature of things . . .; the third is that these goals, and the corresponding interests and values (which it is the business of theology or philosophy or science to discover and formulate), cannot possibly conflict with one another—indeed they must form a harmonious whole."

It was this idea that gave rise to the notion of natural law and the search for harmony. The rival contention of the Romantics, stemming from Kant, was to cast doubt on the very idea that values, the answers to questions of action and choice, could be discovered *at all.* This is an important moment in the history of the European consciousness.[5] The Romantics argued that some of these questions simply had no answer. No less originally, they argued that there was no guarantee that values could not, in principle, conflict with one another. Finally, the Romantics produced a new set of values, a new way of looking at values, that was radically different from the old way.

Kant's great contribution, as we have seen, was to grasp that it is the mind that shapes knowledge, that there *is* such a process as intuition, which is instinctive, and that the phenomenon in the world that we can be most certain of is the difference between "I" and "not-I."[6] On this account, he said, reason "as a light that illuminates nature's secrets" is inadequate and misplaced as an explanation. Instead, Kant said, the process of birth is a better metaphor, implying that human reason *creates* knowledge. To find out what I should do in a given situation, I must listen to "an inner voice." According to the sciences, reason was essentially logical and applied across nature equally. But the inner voice does not conform to this scenario. Its commands are not necessarily factual statements at all and, moreover, are not necessarily true or false. The purpose of the inner voice, often enough, is to set someone a goal or a value, and these have nothing to do with science, but are created by the individual. It was a basic shift in the very meaning of individuality and totally new.

In the first instance (and for the first time), it was realized that morality was a creative process but, in the second place, and no less important, it laid a new emphasis on creation, and elevated the artist alongside the scientist. It is the artist who creates, who expresses himself, who creates values. The artist does not discover, calculate, deduce, as the scientist (or philosopher) does. The artist invents his goal and then realizes his own path toward that goal. "Where, asked Herzen, is the song before the composer has conceived it?"* Creation in this sense is the only fully autonomous activity of man and for that reason takes pre-eminence. At a stroke, art was transformed and enlarged, no longer mere imitation, or representation, but *expression*, a far more important, more significant and ambitious activity. "A man is most truly himself when he creates. That, and not the capacity for reasoning, is the divine spark within me; that is the sense in which I am made in God's image."[7]

We are still living with the consequences of this revolution. The rival ways of looking at the world—the cool, detached light of disinterested scientific reason, and the red-blooded, passionate creations of the artist—constitute the modern incoherence. Both appear equally true, equally valid, at times, but are fundamentally incompatible. As Berlin has put it, we shift uneasily from foot to foot as we recognize this incompatibility.

The dichotomy was shown first and most clearly in Germany.[8] The turn of the nineteenth century saw Napoleon's great series of victories, over Austria, Prussia, and several smaller German states, and these failures created a desire for renewal in the German lands. In response, many German-speakers turned inward to intellectual and aesthetic ideas as a way to unite and inspire their people. "Romanticism is rooted in torment and unhappiness and, at the end of the eighteenth century, the German-speaking countries were the most tormented in Europe."[9]

The route from Kant to the Romantics was not a straight line, but it was clear. For Herder, it was the "expressive power" of human nature that had produced some very different cultures across the world. Fichte portrayed the self as "activity, effort, self-direction. It wills, alters, carves up the world both in thought and in action, in accordance with its own concepts and categories." Kant conceived this as an unconscious, intuitive process but for Fichte it was instead "a conscious creative activity . . . I do not accept anything because I must," Fichte insists, "I believe it because I will." There are two worlds, and man belongs to both. There is the mate-

* Alexander Herzen (1812–70), writer and "father of Russian socialism."

rial world, "out there," governed by cause and effect, and there is the inner spiritual world, "Where I am wholly my own creation." "Contemplative knowledge," the ideal of the Middle Ages, is the wrong model, says Fichte. "What matters is action . . . Knowledge is not to be looked upon passively but is to be used, used to help us create, for creation is freedom."[10]

This was a provocative idea, says Berlin, because through Fichte it became applied to nations, and nations could only *become* nations by creating, acting, *doing*. Nationalism, active nationalism, therefore became the natural stance. "So Fichte ends as a rabid German patriot and nationalist, who thought that Germany had not been corrupted as the Latin nations had."[11] Fichte beefed up this view in his famous address to the German nation, written after Napoleon had conquered Prussia. The speeches themselves had little impact but later, when they were read, they contributed to a huge upsurge of nationalist feeling "and went on being read by Germans throughout the nineteenth century, and became their bible after 1918."

"All those who have within them the creative quickening of life, or else, assuming that such a gift has been withheld from them, at least await the moment when they are caught up in the magnificent torrent of flowing and original life, or perhaps have some confused presentiment of such freedom, and have towards this phenomenon not hatred, nor fear, but a feeling of love, these are part of primal humanity. These may be considered as true people, these constitute the *Urvolk*, the primal people—I mean the Germans. All those, on the other hand, who have resigned themselves to represent only the derivative, the second-hand product . . . and shall pay the price of their belief. They are but an annexe to life . . . They are excluded from the *Urvolk* . . . The nation which bears the name 'Germans' to this day has not ceased to give evidence of a creative and original activity in the most diverse fields."[12]

The Rise of the Unconscious

Friedrich Schelling (1775–1854), who succeeded Hegel in his professorial chair, had a more organic, less aggressive view of spiritual self-development than Fichte. For Schelling the world consisted of phenomena which varied in their degree of self-consciousness, from total unconsciousness, gradually coming to full consciousness of themselves. At its most fundamental, there are the brute rocks that form the earth, which represent the "will" in a condition of total unconsciousness.[13] Gradually life infuses them, producing the first biological species. Plants and animals follow, self-consciousness growing, leading toward the realization of some kind of purpose. Nature represents progressive stages of the will and is striving toward something "but is not aware what it strives for."[14] Man, as well as striving, becomes *aware* of what he is striving for. This is an event important for the whole universe, which, in Schelling's account, is brought in this way to a higher consciousness of itself. This was God for Schelling, a self-developing consciousness, a progressive phenomenon evolving.[15]

It had a deep effect in Germany because, according to this mode of understanding, the function of the artist now involved an ability to dive deep into the unconscious forces "which move within him," and bring them to consciousness, however difficult the struggle. For Schelling, in order for art to have value, it must tap into "the pulsations of a not wholly conscious life." Otherwise art is a mere "photograph," a piece of knowledge that, like science, is no more than careful observation. These two doctrines, Fichte's understanding of the will and Schelling's of the unconscious, formed the essential backbone of the aesthetics of the Romantic movement. The truth of art, says Thomas Nipperdey, became the great question of the nineteenth century. [16]

Friedrich Schlegel had yet another view.[17] For him, there were three elements that shaped Romanticism. He agreed that Fichte's theory of knowledge was one, but he added, for good measure, the French Revolution and Goethe's famous novel *Wilhelm Meister*. The French Revolution exerted its effect on the Germans because, as a consequence of the Napoleonic Wars, there was sparked, in Prussia in particular, "a vast burst of wounded national feeling." The events of the Reign of Terror during the Revolution were crucial, those events switching back and forth as they did in such unpredictable ways as to suggest to the Romantics that not enough was known about human behavior and that what *was* known was only the tip of a vast hidden iceberg, some unknown and uncontrollable and even undiscoverable impersonal force, the strength of which could not be deflected.[18] Schlegel's third great influence, Goethe's *Wilhelm Meister*, was admired because it showed how, "by the free exercise of his noble and unrestrained will," a man can work on himself to improve himself, increase his self-consciousness.* [19]

A Change in the Meaning of Work

The effects were momentous. For one thing, the Romantic revolution reinforced the Protestant understanding of work. Instead of being regarded as an ugly necessity, it was transformed into "the sacred task of man," because only by work—an expression of the unfolding of the will—could man bring his distinctive, creative personality to bear upon "the dead stuff" of nature. Man now moved ever further from the monastic ideal of the Middle Ages, in that his real essence was understood not as contemplation but *activity*. What mattered now was the individual's search for his freedom, in particular "the creative end which fulfils his individual purpose." What matters for the artist now is "motive, integrity, sincerity . . . purity of heart, spontaneity." *Intention*, not wisdom or success, is what counts. The traditional model—the sage, the man who knows, who achieves "happiness or virtue or wisdom, by means of understanding"—is replaced by the tragic hero "who seeks to realise himself at whatever cost, against whatever odds." Worldly success is immaterial.[20]

This reversal of values cannot be overstated. Since man's values are not discovered but created, there is no way they can ever be described or systematized, "for they are not facts, not entities of the world." They are simply outside the realm of science, ethics, or politics. Harmony cannot be guaranteed, even within one individual whose own values may shift over time. For the Romantics, martyrs, tragic heroes who fought for their beliefs against overwhelming odds, became the ideal. The artist or hero as outsider was born in this way.

The Second Self

It is an idea and ideal that leads to a form of literature, painting, and (most vividly) music that we instantly recognize—the martyred hero, the outcast genius, the suffering wild man, rebelling against a tame and phi-

* Goethe himself didn't agree. He later wrote, "Romanticism is disease, classicism is health."

listine society. As Arnold Hauser rightly says, there is no aspect of modern art that does not owe something important to Romanticism. "The whole exuberance, anarchy and violence of modern art . . . its unrestrained, unsparing exhibitionism, is derived from it."[21]

Associated with all this at the time was the notion of the "second self," the belief that inside every Romantic figure, in the dark recesses of the soul, was a completely different person, and that once access to this second self had been found, an alternative—and deeper—reality would be uncovered. This is, in effect, the discovery of the unconscious, interpreted here to mean a secret, ecstatic something, which is above all mysterious, nocturnal, ghostlike, and often macabre. (Goethe once described Romanticism as "hospital-poetry" and Novalis pictured life as "a disease of the mind.") The second self, the unconscious, was seen as a way to spiritual enlargement. The early nineteenth century was the point at which the very concept of the avant-garde could arise, with the artist viewed as someone who was ahead of his time and apart from the bourgeoisie. The concept of genius played up the instinctive spark in new talent at the expense of painfully acquired learning over a lifetime of effort.

The Marriage of Poetry and Biology: Romantic Science

The Romantic mentality that coalesced in Germany in the late eighteenth and early nineteenth centuries emerged in the first instance through the close friendship and intense passions that welded together a group that became famous as the *die Frühromantiker* or "early Romantics." They were poets and painters, philosophers and historians, theologians and scientists, and they were young.[22] As with other young revolutionaries, both before and since, they came to disdain conventional thought. Their movement was widely held to be one of resistance to or rebellion against the Enlightenment, asserting the primacy of the "poetry of the heart" above the prosaic nature of the modern world. In particular they scorned the Terror in Paris in 1793, and the optimism of that same Enlightenment, so eloquently expressed by Kant in "Zum ewigen Frieden" (Of Eternal Peace; 1784), which they believed had proved so illusory.

The "intellectual architect" of the movement, in the words of Robert

J. Richard, was Friedrich Schlegel (1772–1829). A poet, a literary critic, and historian, he provided the initial meaning of *romantisch*. The French word *roman* (novel) had entered the German language toward the end of the seventeenth century, when, as *romanhaft*, it carried the meaning of an action-packed adventure. In the late 1790s, however, Schlegel argued that literature that is *romantisch* is characterized by a "continual striving after the perfect realisation of beauty" and is always trying to attain a higher state for man, even though one can never be sure what that higher state is.[23]

The close and closed nature of the early Romantics is epitomized by Schlegel and his brother, Wilhelm. Friedrich fell in love with Caroline Böhmer, the daughter of the Orientalist Johann D. Michaelis. She was a "fiery and omni-talented woman" who seems to have been the lover of just about everyone in the Romantic circle and had three husbands, including Friedrich's brother Wilhelm, and Friedrich Schelling. (Another aspect of the closeness was the fact that Schelling shared a room with Friedrich Hölderlin (1770–1843) and Georg Wilhelm Friedrich Hegel (1770–1831).)[24]

Friedrich Schleiermacher (1768–1834), another in the group, was a theologian who, while most of the others were trying to align the arts and sciences, was concerned to establish the role that poets and artists played in religion. "They convey the heavenly and the eternal as an object of pleasure and unity . . . They strive to awaken the slumbering kernel of a better humanity, to inflame a love for higher things, to transform a common life into a higher one . . . They are the higher priesthood who transmit the most inner spiritual secrets, and speak from the kingdom of God."[25]

Novalis (1772–1801) stressed the dark side of human nature, the superiority of night to day, of death to life, and Heinrich von Kleist (1777–1811) likewise highlighted the fragility of human existence, of doubt and despair, and that, as he tried to show in *Prinz Friedrich von Homburg* (*Prince Frederick of Homburg*), man is not the master of his own destiny. These were motifs taken up later by Joseph von Eichendorff (1788–1857), E. T. A. Hoffmann, Richard Wagner, and Thomas Mann who, in his "Rede vor Arbeitern in Wien" (Speeches to the Workers in Vienna) in 1932, spelled out for his audience the importance of German Romanticism as the ultimate German art form. (In fact, as we shall see, it alternated regularly with social realist works.)

Several among the Romantics had an interest in science, and here the paradigmatic discipline was biology. The main idea, stemming from Kant but shared by Schelling and Goethe, was that living nature was organized into fundamental types, "archetypes" (*archetypi*, *Urtypen*, *Haupttypen*, *Urbilden*, etc). On this understanding there were four basic animal structures: *radiata* (starfish and medusa), *articulata* (insects and crabs), *mollusca* (clams and octopuses), and *vertebrata* (fish and human beings).[26] It was Kant's view that the archetypal structure of organisms reflected the very ideal they embodied. For him there was, in effect, a divine mind that imagined these archetypal ideas.

The *Naturphilosophen* who came after Kant believed that there were special causal factors that accounted for the "instantiation" of archetypes and their progressive variations. These causal factors were understood as special applications of the physical powers that had emerged in the eighteenth century: animal electricity, for example. They were given names such as *Lebenskraft* and *Bildungstrieb*. For them, matter and *Geist* (understood as both spirit and/or mind) were conceived as two aspects of the same underlying *Urstoff*. The natural world had an underlying unity, which remained to be discovered. This approach gave rise to several theories about the higher-order patterns in nature. Beginning with the Kantian notion of *ideal reality*, they explained variations in organisms as a result of the gradual development—the evolution—which "instantiated" progressive variations of the ideal forms. This was not Darwinian evolution, but rather *dynamische Evolution*, as Schelling termed it.[27]

The Naturphilosophen also accepted that nature was teleologically ordered. From Herder, Goethe, and Schelling on, they opposed the mechanical ideal as developed by Descartes and Newton. Instead, they believed that nature was steadily transformed from a simpler, less organized, earlier state to a higher, more developed, later state. They also accepted Kant's argument about the similarity between teleological judgment and aesthetic judgment, which he set out in the *Kritik der Urteilskraft* (*Critique of Judgment*). The most important effect of this was that the Romantics equated these two kinds of judgment, meaning that "the basic structures of nature might thus be apprehended and represented by the artist's sketch and the poet's metaphor, as well as by the scientist's experiment and the naturalist's observation."[28] Romantic biologists believed that the aesthetic comprehension of an entire organism came first, before science analyzed its respective parts.

Friedrich Schelling was the great advocate of the marriage of biology and poetry, though Friedrich Schlegel agreed.[29] In the treatise *Von der Weltseele* (*On the World Soul*; 1798), Schelling explored the latest scientific research, concluding that "teleological structures characterised all living creatures." He thought that nature was infinitely productive (he had no idea of genetics) and that it took the forms it did as a result of being continuously inhibited or limited by opposing forces. These forces—magnetism, electricity, chemical processes—brought about changes in the powers of organization (sensibility, irritability, and *Bildungstrieb*). "Schelling conceived the infinite productivity of nature as an unending evolution (*unendliche Evolution*), with its products as momentary resting places, a slowing of the evolutionary process but not a cessation of it."[30] In its way, this was a pre-Darwinian notion of adaptation.

In general terms, then, we may say that one significant achievement of Romantic science in Germany was a pre-Darwinian idea of evolution. A second was a pre-Freudian notion of insanity. Johann Christian Reil was one of the most famous medical theorists of his time. Born in the far north, the grandson of a Lutheran pastor, he worked on several studies of mental illness before his seminal work, *Rhapsodien über die Anwendung der psychischen Curmethode auf Geisteszerrüttungen* (Rhapsodies on the Application of Psychiatric Methods of Cure to the Mentally Disturbed).[31] This became perhaps the most influential work in shaping German psychiatry before Freud. Reil's view was that insanity arose from the "fragmentation of the self, from an incomplete or misformed personality and from the inability of the self to construct a coherent world of the non-ego—all of which resulted from a malfunction of self-consciousness, that fundamentally creative activity of the mind." A marked element in Reil's system was that civilization had its dark side. Self-consciousness, he maintained, "synthesises the mental man, with his different qualities, into the unity of a person." This is clearly very modern.[32]

Reil also had an evolutionary view. He believed that new species would go on occurring and that higher, more-developed forms would evolve in the wake of less-developed species. The force behind this process, he claimed (as did Kielmeyer), was the same as that which drove the development of the fetus. He believed that, over time, more fully evolved individuals would emerge that would come to epitomize more completely the potential of the species. It was not as if species were pushed from behind toward a predetermined future; instead they developed closer and closer to "the realisation of the ideal of absolute organism."[33] This was a form of biological Idealism.

Goethe's Urphenomena

Goethe had always had an interest in science; this is evident from the fact that, after he matriculated at the University of Strasbourg in 1770, he chose a liberal arts course which included political science, history, anatomy, surgery, and chemistry.[34] But his interest didn't begin to mature until he returned from his extended visit to Italy in 1786–88. For the next two decades, more or less, he spent a great deal of time studying the history of science and investigating two interests, plant morphology and color theory, even as he wrote some of his best poetry. When he died, on March 22, 1832, besides his letters and other writings and his 5,000 books, he left a museum of scientific instruments and cabinets of flora, fauna, and countless minerals that Werner would have coveted—50,000 artifacts in all. The Leopoldina edition of Goethe's scientific writings was published from 1947 on.[35]

Goethe's contributions to science fell into five main categories—geology, anatomy, botany, optics, and the nature of the experiment. Karl Fink underlines Goethe's very modern view of science: he was aware that scientific "facts" are, as often as not, interpretations that owe as much to the scientist himself as to what is "out there." Goethe was never in thrall to the experiment as others were, seeing it as less about "proof" and more about the way science "presented" itself. In his science Goethe also inhabited that world between doubt and Darwin. He thought that the nature of reality could best be glimpsed "at the borders" of objects, that *change* was where nature revealed herself. Famously, he thought that there were *Urphänomene*, primordial forms in nature that gave rise to other, later forms. Granite, for example, was for him the archetypal rock, "the basis of all geological formation." He thought that basalt (now known to be volcanic rock) was a "transitional" form of granite, in the same way that the whale was (for him) a transitional animal between fish and mammal and the polyp a transition between animal and plant forms.[36]

It was Goethe's view that crystallized granite was "the first individualisation of nature," the first step away from the *Urstoff*, and that "second-level" transitions produced the simpler organic forms, such as corals and ferns. As Karl Fink has put it, this is "becoming" as applied to nature.

Goethe became almost obsessed by the intermaxillary bone because he thought it might reveal the transition from one species of skull to another. He had familiarized himself with the theories of Carl Linnaeus (1707–78) suggesting that facial bones were the distinguishing characteristic of zoological types, and it was Goethe's argument that the intermaxillary bone appears between the two main bones of the upper jaw, containing the four incisor teeth. He identified the bone to his own satisfaction in such domestic and wild animals as the walrus, lion, oxen, and the apes, but, most important of all, he considered it the "distinguishing mark" between apes and humans, "playing no part in the facial structure of the former but important in the latter." This was another of those ideas that was more radical then than it appears now—on this system of understanding, animals were seen as on the same continuum as man, a pre-Darwinian notion that conflicted with biblical dogma.[37] Goethe believed it was self-evident that fish, amphibians, birds, and mammals were all derived from one "primeval form" ("*nach einem Urbilde*"). There had, he said, been a series of "successive changes" that had produced the variety we see about us, and he thought there were two crucial differences between organic matter and inorganic matter, namely the "indifference" of the latter and the "purpose" and organization of the former, plus the fact that organic matter has "borders" and consists of individuals, botanical or zoological. Here too, Goethe is groping to understand his world in a pre-Darwinian, nonbiblical fashion.

Goethe's researches on optics and color theory were also predicated on another set of borders, the juncture where darkness and lightness meet.[38] This produced in him the notion that there are three forms of color—one originating in the physiology of the eye, a second originating outside the eye (observed through optical mediums), and a third located in the substance observed. More than this, though, all forms of color—physiological, physical, and chemical—are for Goethe derived from a primal phenomenon, the polarity of light and darkness. For him, this polarity was equivalent to the attraction and repulsion in magnetism, plus and minus charges in electricity, even major and minor keys in music. It was analogy run wild, a perfect example of Romantic science and out of date even when it was published.[39]

Finally, there was Goethe's understanding of the scientific method, the experimental approach. He accepted the fundamental point that "nature has no system," that she "emerges from an unknown centre" and evolves "to an unrecognisable border." Abstractions conceived by the mind can therefore be misleading: "We can't force nature in this way; all we can do is try to 'overhear' her secrets."[40]

Goethe recognized that language may not—ever—exactly match nature and so, in the process, may "freeze" understanding in unnatural ways. "Through words we neither express completely the objects nor ourselves." Poetic language was for him the deepest link between language and nature, whereas the experiment was a *demonstration* of nature, "both more and less vivid than language."[41] "The mark of the modern scientist," he added, "is possession of sufficient reflective skills to distinguish between himself, his language and the object of his investigation . . . He must avoid turning perceptions into concepts and concepts into words and then operating with these words as if they were objects."[42] The debt to Kant is clear and very modern.

In some ways this was Goethe's greatest achievement: the search for the serial relationships in nature, emphasizing border experiences, the junctures where "the real joints of nature" are located, is most likely to reveal the process of change, development, organizing principles. This is also why it needed individuals who were both poet and scientist, who could combine "imagination, observation and thought in the act of language."

9 ·

The Brandenburg Gate, the Iron Cross, and the German Raphaels

We shall be describing a curious phenomenon in this chapter, a whole raft of artists who have fallen very much out of favor in the twentieth and twenty-first centuries but who were, in their own time, very fashionable indeed. In fact, they were the most famous painters, sculptors, and architects of their day. This change of fortune is nowhere more apparent than with Anton Raphael Mengs (1728–79).[1]

In the first published biography of him, Giovanni Lodovico Bianconi's *Elogio storico del Cavaliere Antonio Raffaelle Mengs* (Milan, 1780), Mengs was held to be "the most notable painter of his century and of comparable stature and importance to Raphael and Apelles in the total history of art."[2] The greatest praise was that bestowed on Mengs by Winckelmann himself, who dedicated his *History of Antiquity* to the painter, and said in the text that he was "the single modern painter who had most closely approached the taste and perfection of the ancients in his art." Mengs's studio in Rome was a meeting place and sanctuary "to which all connoisseurs and aspiring young artists of classicising taste and bent" would naturally gravitate.

Mengs's father, Ismael, was court painter in Dresden and named his son after Correggio (Anton) and Raphael. The boy was raised with a strict formal training in art, beginning at age six when his father put him to drawing simple straight lines, and from there allowed him to progress to circles "and other pure geometric forms."[3] In 1741, at the age of thirteen, Mengs was taken to Rome, where he was forced to concentrate on Raphael, but only after he had "mastered" Michelangelo's sculptures. (He had to tell his father at the end of every day what he thought he had learned.) After three years in Rome, the Mengs family returned to Dresden where, famously, Anton Raphael was "discovered" as a child prodigy (he was fifteen), and made court painter in 1745, aged barely sixteen. He proved popular with Friedrich August II, who acquired seventeen of his works. Dresden is the only place in Germany where Mengs's work can be seen.

Despite these early successes, Ismael decided that a second tour in Italy was desirable for his prodigious son, mere portraits being less worthy than history painting. This time the family—granted leave by the Dresden court—traveled via Venice, where they studied the Titians, Bologna, for the Carracci, and Parma, for the Correggios. After his second spell in Rome, Mengs returned again to Dresden and was promoted from court painter to first painter to Friedrich August II. Rather than satisfy Mengs, this rapid promotion seems only to have stimulated his ambition, and in 1752 he left for Rome a third time. He stayed nine years and never saw Dresden or his royal patron again.

The pre-eminence of Rome was partly due to Winckelmann, but not entirely. The French Academy in Rome had been established as long ago as 1666 for the reception and further training of the best young painters, sculptors, and architects, who usually spent a few years along the Tiber before returning to France (the usual term was six years, but one painter spent nearly two decades there).

Mengs formed important friendships with Monsignor (and after 1756, Cardinal) Albericho Archinto, who had persuaded Winckelmann to convert to Catholicism, and Cardinal Alessandro Albani, a nephew of Pope Clement XI. These contacts led to Mengs's first really important commission in Rome, for the ceiling of S. Eusebio in 1757.[4] This was one of the most ancient churches there, dating from the fifth century, and when his picture was unveiled it was universally welcomed as "*eine Schöpfung der Zauberkunst*," a magical creation.[5]

Mengs and Winckelmann were natural companions in Rome and sought each other out, even planning a joint treatise on the taste of the

Greeks. In fact, Mengs was as important for his taste as for his painting and he became more and more interested in the art of antiquity. On a visit to Naples in 1758–59 he had begun a collection of "Etruscan" vases, which would eventually comprise 300 pieces when he donated it to the Vatican Library. He also had another collection of plaster casts, modeled after famous antique statues. This he gave to the king of Spain in the hope that they would help improve what he regarded as "the lamentable state of public taste in Spain." A second collection was acquired, after his death, by the court in Dresden "where it influenced the Dresden and Meissen porcelain produced in the last quarter of the eighteenth century."[6]

Mengs's first painting to treat an ancient historical subject has disappeared (a lot of his work has been lost) and the earliest classical history painting of his that does survive is *The Judgment of Paris*. Paris, seated, looks out on the three nude goddesses, as described in Ovid and as shown in an existing painting by Raphael. Mengs's evolution toward neoclassicism is shown still more clearly in a small tondo painting, *Joseph in Prison*, now in Madrid, where the flat stones and smooth ashlar masonry look forward to Jacques-Louis David.[7] Mengs had by now discovered in Rome the paintings of Nicolas Poussin.

Such compositions made Mengs the obvious artist for patrons making the grand tour, but to establish himself as a serious exponent of classicism he needed a project on a grander scale. This arrived when he was given the commission to decorate the Villa Albani, which Thomas Pelzel describes as the most significant commission of Mengs's career. The contents of the cardinal's magnificent new villa, near the Porta Salaria, were so important that it was an obligatory stop for any informed visitor to Rome.[8] The ceiling took Mengs about nine months. In the center of the composition is *Parnassus*, with the laurel-crowned Apollo holding a laurel branch and a lyre. They are surrounded by the nine muses and their mother, Mnemosyne, a reference to Cardinal Albani as patron and protector of the arts. The muses are known to be likenesses of Albani's favorites among the more ravishing women of Rome. In this flattering view, Villa Albani is established as the center of the neoclassical world.[9]

In this composition, Pelzel says, Mengs was convinced he was going further than Raphael. Mengs believed that Raphael did not have "that knowledge of true beauty which the Greeks possessed."[10] Mengs, on the other hand, in his own mind, had the advantage of the new discoveries at Herculaneum. In the Albani ceiling it was therefore Mengs's aim to improve upon the style of Raphael "in the light of his own superior knowledge of Greek art." When the ceiling was unveiled in 1761, Winckelmann said he could "recall nothing in the works of Raphael to place beside it." It was in connection with this ceiling that Winckelmann called Mengs "the German Raphael."[11]

Mengs's fame was now spreading, and in 1772 he was elected president of the Accademia di S. Luca, soon after receiving a major commission from Pope Clement XIV.[12] Mengs selected an *Allegory of History*. He remained interested in antiquity to the end of his life, even when he was so ill that he was forced to paint from bed. It was then that he was awarded the biggest honor of his career, the commission for a large altarpiece for St. Peter's. This, the *Giving of the Keys*, was never carried further than the initial cartoon, for Mengs died in June 1779.

Winckelmann thought Mengs was the one modern painter who had "most clearly approached the taste of the ancients" and this too was the verdict of later authorities who studied the neoclassical movement. In Winckelmann's letters there are copious references to life in Mengs's convivial house in Rome, and several of Mengs's pupils served in academic posts all over Germany, including Heinrich Wilhelm Tischbein, who described him as "the most accomplished German painter since Dürer."[13] According to one account, as many as 500 German artists passed through Rome at this time.

But Mengs's most enduring influence was outside his own country, in the genesis of French neoclassicism. According to the French historian Jean Locquin, any Frenchman of classical or archaeological inclination who found himself in Rome in the late eighteenth century would have sought inspiration "de la bouche du Maître [Mengs], qui répond si parfaitement aux aspirations de l'époque."[14] Joseph-Marie Vien, Jean-Baptiste Greuze, and, above all, David were influenced by the general milieu radiating from Mengs and Winckelmann. Mengs was by no means the only influence on David, of course—there was a general return to Poussin in the middle of the eighteenth century, but "Mengs was at the height of his reputation when David was in Rome, where he is said to have attended drawing classes 'nel museo del cavaliere Mengs.'"[15]

The term "neoclassical" did not come into use until the 1880s, by which time this form of art was already well out of favor.[16] But in the late eighteenth century it was regarded as the "true" or "correct" form of painting and looked upon as a *Risorgimento.* The aim was a return to first principles—to antiquity. This was to be achieved by visits to Rome, by the study of Raphael and Poussin, by reading Winckelmann and the Greek and Roman classics themselves. There was an idea that there is a classicism common to all the arts, but there was no precise program and for that reason artists as varied as Jean-Antoine Houdon, Hubert Robert, Greuze, George Stubbs, Joshua Reynolds, and Francisco Goya derived inspiration from the same sources, all engaged in an exercise to see how nature could be "purified and ennobled."[17] As Winckelmann had said, line took precedence over color, and restraint was valued over passion. Between 1780 and 1795 the greatest masterpieces of neoclassicism were produced, culminating in that "icy star," David.

Neoclassicism is, therefore, particularly interesting for this brief moment of stylistic unity, as true of architecture as of painting. As Wend von Kalnein has put it, "For the best part of a century architecture spoke the same language from Rome to Copenhagen, from Paris to St Petersburg."[18] Columns and porticos became the main features of public buildings everywhere, from banks to theaters and churches to town halls.

Creating the Berlin Skyline

German neoclassicism arrived late, despite the roles played by Mengs and Winckelmann in its genesis. It began about a generation after France and Britain and was at its height from 1800 on.[19] Though Berlin and Munich led the way, Karlsruhe, Hanover, Brunswick, and Weimar all boasted neoclassical buildings. Friedrich Wilhelm II, who succeeded Friedrich the Great in 1786, brought to Berlin the architects Friedrich Erdmannsdorff (1736–1800), Carl Gotthard Langhans (1732–1808), David Gilly (1748–1808), and Johann Gottfried Schadow (1764–1850), and from then on neoclassical buildings began to dominate the Prussian capital.

Gotthard Langhans's Brandenburg Gate (1789–93) paved the way. Langhans was the chief court architect though his knowledge of Greek architecture was derived not from experience, but rather from book learning. This shows, for the Brandenburg Gate has many features that are not Greek. Even so, it was widely understood as an expression of the new style and modeled on the Propylaeum on the Acropolis in Athens. It would be much mutilated down the years, notably by Napoleon who took Schadow's bronze statue of Eirene to Paris in 1806 (it was later returned).

Langhans was followed by the Gillys, father and son. David Gilly, from Pomerania, assumed the directorship of public buildings in Berlin and in 1793 founded a school of architecture there, later turned into an academy. It was this academy that trained the younger generation of architects—Heinrich Gentz, Karl Friedrich Schinkel, and Leo von Klenze.[20]

The principal genius of early German neoclassicism was David Gilly's son, Friedrich (1772–1800). He died tragically young in 1800 at the age of twenty-eight, but in 1796 he came closer to the Greek ideal than anyone else had done with his design in a public competition for a monument to Friedrich the Great, to be erected in the Leipziger Platz in Berlin.[21] Friedrich's drawings show a solitary Doric temple on a raised platform, approached by means of a triumphal arch with a Doric colonnade. This design contrasts with the Brandenburg Gate, but nonetheless "set the standard" for German neoclassicism.

Gilly's influence was strongest on Karl Friedrich Schinkel, "thanks to whom Prussian Classicism became of European importance."[22] Schinkel, "the last great architect," as Adolf Loos described him, was honored almost everywhere architects received honors and knew personally many of the great luminaries of his time: Clemens Brentano, Fichte, the Humboldt brothers, Friedrich Carl von Savigny, and Gustav Friedrich Waagen, the art historian. He is much admired by modern architects such as Philip Johnson, James Stirling, and I. M. Pei.

The son, grandson, and great-grandson of Lutheran pastors, Schinkel was born in 1781 in Neuruppin, a town famous for its Gymnasium, about twenty miles northwest of Berlin. When he was six, his father was killed in a fire that laid waste most of his hometown, and in 1794 his widowed mother took the rest of the family to Berlin. An exhibition of drawings by the young Friedrich Gilly so fascinated Schinkel that, at the age of sixteen, he decided to be an architect. He began his studies with David Gilly, Friedrich's father, in March 1798, while the younger Gilly was abroad. When Friedrich returned, they formed a close friendship so that, by 1799,

Schinkel was living in the Gilly household. That was the year when a separate *Bauakademie* was officially opened on the first floor of Henrich Gentz's new Berlin Mint, Schinkel becoming one of ninety-five students. Carl Gotthard Langhans was on the teaching staff.[23]

Berlin in 1794, when Schinkel arrived there, had a population of 156,000, compared with 332,000 at his death in 1841. It was built on marshy ground and crisscrossed with dikes, canals, and ramshackle wooden bridges, hardly a sophisticated capital of a rising state. There were, however, a few monumental buildings: the old Stadtschloss on an island in the river Spree, had been renovated by Berlin's first great architect, Andreas Schlüter (1659–1714). North of that was the Lustgarten, or pleasure garden, dominated by Johann Boumann's Lutheran cathedral (1747–50). There was a library and a Palladian opera house but little more because Friedrich the Great had preferred Potsdam. At the beginning of the nineteenth century, the two most prominent of the very few modern buildings in Berlin were the Brandenburg Gate and the new Mint.[24]

Because of the uncertainties produced by the Napoleonic Wars, Schinkel spent the early years of his career as a stage designer and painter of romantic landscapes. But a panorama Schinkel produced in 1809 caught the attention of someone close to the royal family, and on the strength of that he was commissioned to redecorate the bedroom of Queen Luise at Charlottenburg Palace. When the queen died later that year, he submitted plans for a mausoleum. That commission eventually went to Gentz, but Schinkel had more luck with Berlin's most important war memorial, a Gothic cross he designed for the Tempelhofer Berg (subsequently known as the Kreuzberg). Cast iron was used for this war memorial, one of the first times the material was employed. Iron was on everyone's mind just then because, at the beginning of the War of Liberation in 1813, Schinkel had collaborated with Friedrich Wilhelm III on the design of the Iron Cross, which would become Prussia's most honorific military medal. In this case iron was used not so much because it reflected developing industry (Berlin had excellent foundries), but as a substitute for precious metals and symbolic of a sacrifice made for the fatherland. The crown had appealed to wealthy families to contribute jewelry to help pay for the wars, and they were given receipts in the form of iron jewelry. This often bore a small cross with the head of the king and inscriptions such as "Gold gab ich für Eisen 1813" (I gave gold for iron). Between 1813 and 1815 it is estimated that over 11,000 pieces of iron jewelry were produced, including 5,000 iron crosses.[25]

By the time the Kreuzberg monument was executed (1821), the Napoleonic Wars were long gone, and prosperity was returning to Prussia. Schinkel became fully engaged in a range of improvements in and around Berlin and was given ever greater administrative responsibility for architectural projects.[26]

Though he turned into one of the finest neoclassical architects—if not *the* finest—Schinkel was not interested only in Greek antiquity, and when he first visited Italy, in 1803–05, he paid just as much attention to Italian medieval buildings.[27] His genius was deep enough that he could express himself in a number of styles. His commissions included the Neue Wache of the Palace Guard (1816), the Schauspielhaus or Royal Theater (1819–21, the previous one having burned down), and the Altes Museum (1824–28). In each of these masterpieces, Schinkel was so at ease with classicism that he was able to use its principles to lay the foundations of a new style. For example, behind the Ionic colonnade of the Altes Museum—"the finest that neoclassicism has to show anywhere, and far superior to the British Museum or to Chalgrin's Bourse in Paris"—the lines are simple but rational, and the façade and the main building perfectly complement one another. In his later development, which took him away from the Greek and toward the Italian Renaissance and British industrial buildings, he showed himself as too good to be confined to any one inherited idea.

In 1824 Schinkel made another visit to Italy, this time accompanied by the art historian Gustav Waagen. His aim was to inspect the display of art collections there; two years later he went to England to inspect the new British Museum. In London Schinkel was more impressed by the works of art themselves than by the buildings which housed them, and he was less impressed by English architecture pure and simple than he was by the engineering structures—the tunnels and bridges (iron once more)—erected by Isambard Kingdom Brunel and Thomas Telford. This was a time when churches were giving way to museums, theaters, and even factories as the focus of architecture.[28] On his return, Schinkel introduced iron staircases into many of his new buildings.[29]

Late in life, Schinkel envisaged a "Higher Architecture," a less utilitarian form of building, though this was never realized. One might call it an ideal, almost Kantian, form of architecture. After his death in 1841, he fell from favor but was rediscovered by the generation of Loos, Peter Behrens, and the young Ludwig Mies van der Rohe, after about a century.[30]

The passion for Greek architecture in Berlin and the other cities of Germany (for example, Munich) achieved remarkable proportions, so much so that in 1835 Leo von Klenze journeyed to Greece, where he succeeded not only in having a law passed to protect the Acropolis and other Classical sites, but he also designed an entire district of Athens and a palace for Otto I, son of Ludwig I of Bavaria, who became king of Greece in 1832. The neoclassicists had finally returned whence they came. [31]

The First Artistic Secession

The world of the Romantics was a small world, and the same point may be made about the neoclassical world, even about Prussia/Germany herself, so far as the arts were concerned. For the most part, the luminaries all knew each other, and painted or sculpted or translated each other. Georg Friedrich Kersting, Mengs, and Tischbein painted Goethe; Heinrich Keller, Martin Klauer, and Traugott Major sculpted him. Joseph Anton Koch and Gottlieb Schick painted Humboldt, and Klauer, Christian Rauch, and Christian Friedrich Tieck sculpted him. Henry Fuseli (Johann Heinrich Füssli) translated Winckelmann and painted Bodmer, Joseph Koch painted August Wilhelm Schlegel, Mengs painted Winckelmann. Schadow sculpted Klopstock and Gilly; Tieck sculpted Lessing, Karl Wickmann sculpted Hegel, and Albert Wolff sculpted Friedrich Schadow. It was a world that was conscious of itself as a talented age, much as the Italian Renaissance had been.[32]

After Mengs and Winckelmann, the most notable German presence in Rome was a group of painters variously known as the Brotherhood of St. Luke, the Düreristen, and the Nazarenes. They began as a small knot of like-minded souls at the Academy in Vienna, where Mengs's pupil, Friedrich Heinrich Füger, was director. Füger was a good director—David himself was an admirer—but Johann Friedrich Overbeck, Franz Pforr, and Johann David Passavant grew irritated with the routine of the Vienna Academy. What they had in common, over and above a dislike of routine and Vienna itself, which wasn't religious enough for them, was a preference for earlier Italian old masters—Perugino, Raphael, and Michelangelo—rather than later painters such as Correggio, Titian, and the Bolognese school then so much in vogue, and especially for the so-called Italian primitives of the late Middle Ages and early Renaissance. This, the Gothic Revival, ran parallel with Romanticism.[33]

Their views were reinforced by the publication in 1797 of an anonymous short booklet, *Herzensergiessungen eines kunstliebenden Klosterbruders* (*Effusions of an Art-Loving Monk*), which, in Keith Andrews's words, "made an impact in inverse proportion to its modest size."[34] Written by Wilhelm Heinrich Wackenroder, appearing on the eve of his death at the age of twenty-five and put together by his friend, the poet Ludwig Tieck, it was not really art history as much as a series of art stories, vivid incidents taken from the lives of the great painters, together with intimate details of how early German artists used to live, based chiefly on Joachim Sandrart's biography of Dürer. In full Romantic mode, Wackenroder and Tieck explained art as divine inspiration.

It was against this background that four apprentice painters at the Vienna Academy joined the triangle of Oberbeck, Pforr, and Passavant. They were Ludwig Vogel and Johann Konrad Hottinger from Switzerland, Joseph Wintergerst from Swabia, and Joseph Sutter from Austria itself. These seven painters now met regularly to critique each other's work and were soon united as a band opposed to the policies of the Academy. Taking a lead from Wackenroder's book, they called themselves a "brotherhood," the other name being relatively easy to decide on: the Evangelist St. Luke, the patron saint of painters. In line with their religious, monkish aims, Fra Angelico, the painter-monk, was their ideal. "The artist," said Overbeck in a letter, "must transport us through Nature to a higher idealised world . . ."[35]

Because of their continual conflicts with the Vienna Academy, they planned a move south, to "Raphael's town" and when their own Academy was forced to close in May 1809 because of the French occupation, and was then allowed to reopen only as a much smaller entity, the rebels—not included among the smaller chosen few—used the opportunity to decamp. It was May 1810 when the first artistic secession of modern times came into being.

They set up shop in Rome, in the monastery of S. Isidoro, an Irish Franciscan church and college founded in the sixteenth century. Each painter had his own cell in which to live and work. At night they ate together, then they read and drew in the refectory. Drawing—like prayer—became a ritual. In the Vatican they could dwell on the frescoes by Pinturicchio and Raphael to their heart's content. "Wackenroder's art-loving monk had become a reality."

Overbeck assumed the leadership, but more than anything it was the friendship between him and Pforr that "laid the foundation for the rebirth of German art."[36] They painted for each other and in a very similar style, as a comparison of Pforr's *Friendship* and Overbeck's *Italia and Germania* will show. Before the collaboration could go very much further, however, in July 1812 Pforr died of consumption at the age of twenty-four. While Pforr had been sinking, some among the group had begun meeting at the house of Abbate Pietro Ostini, a professor of theology at the Collegium Romanum, and after Pforr's death the monastic isolation of the "Fratelli di S. Isidoro" was ended and the building itself given up. The members of the Brotherhood of St. Luke changed their name to "Düreristen," but because of their emphasis on Catholicism, and because of their monastic mode of life, not to mention the flowing cloaks and hair that they affected, they were given the nickname "the Nazarenes." Like many such satirical soubriquets in the history of art, it stuck.[37]

Despite these mixed fortunes, the Brotherhood's works were becoming known farther north, and other young painters began to head over the Alps, among them Rudolf and Wilhelm Schadow, sons of Johann Gottfried Schadow, the famous Berlin sculptor, and Johann and Philipp Veit, sons of Dorothea Veit and stepsons of Friedrich Schlegel. But the most serious new talent was Peter von Cornelius.[38]

Headstrong, single-minded, and much influenced by Goethe's *Faust*, Cornelius had produced a series of illustrations for Goethe's drama. Goethe liked the drawings well enough but encouraged Cornelius to study his Italian contemporaries. In Italy, Cornelius joined Overbeck and took Pforr's place. In fact, his arrival marked a new direction for the Nazarenes. Less overawed by Raphael, Cornelius persuaded the brotherhood to work less for their own narrow circle and instinctively grasped that if there really were to be a national artistic regeneration, a new monumental art was needed, an art that would occupy churches, monasteries, and important public buildings. He convinced himself and the others that fresco painting had that monumental quality lacking in easel paintings and so persuaded the Prussian Consul General in Rome, Salomon Bartholdy, who occupied the Palazzo Zuccari (now the Biblioteca Hertziana), to allow four of the Nazarenes a commission. As a theme they chose the Old Testament story of Joseph in Egypt.[39]

The frescoes were a great success (they were removed to Berlin in 1887). All of the artists—Overbeck, Cornelius, Philipp Veit, and Friedrich Schadow—were at their lyrical best, their figures strong and rhythmic, with an intense sense of atmosphere. "It was a collective break with what had gone before—away from Mengs, the Baroque, the Neoclassical; the vividness and striking purity and harmony of the colours was a revelation."[40] Fellow artists of all nationalities flocked to Rome to see the new frescoes. Antonio Canova and Bertel Thorvaldsen were unstinting in their praise, and the Nazarenes became the firm center of attention in the Rome art world. Among those 500 German artists mentioned earlier who either visited or actually lived in Rome full-time, one, Baron Friedrich von Rumohr (1785–1842), became the first art historian in the modern sense. He made it his business to discover how the ideas of the Nazarenes had evolved and in doing so was among the first to explore the archives and to make a first-hand examination of the early masters, "in the flesh," so to speak (this was the age of engravings, remember, before photographs). Rumohr was largely responsible for the history of art becoming an academic discipline, and his *Italienische Forschungen* (1827–32), setting out systematically the results of his inquiries, was an early sighting of the word "*Forschung*"—research.[41]

In the early 1820s the brotherhood began to fragment.[42] Cornelius, Overbeck, and Julius Schnorr von Carolsfeld were enticed to Munich by King Ludwig I. The new king turned into one of the greatest anachronisms of history, convincing himself that he could initiate a national artistic renaissance through commissions that strove to re-create previous glories—he built Greek temples, Byzantine and Romanesque churches, Gothic houses. The techniques of antiquity, such as mosaic and the encaustic method of painting, were revived.

To begin with, Cornelius enjoyed himself. His first commission was for the decoration of the Glyptothek, the museum that was to house the antique sculptures Ludwig had collected.[43] However, Cornelius had a post-Kantian, overintellectual view that paintings should consist of vast pictorial schemes, in which each part had to be understood individually before the purpose of the whole could be grasped. This (exhausting) stress on didactic was even more in evidence in his next great commission. In the new Ludwigskirche, in the center of Munich, Cornelius conceived yet another grandiose epic of the Christian religion—an ambitious scheme designed to fill the entire building and based entirely on the Bible. Even the king was put off and he reduced the commission, confining Cornelius to the apse and the choir.

That was not all. The king had by then come under the influence of his architects, Leo von Klenze and Friedrich von Gärtner, and they hated what Cornelius was trying to do (they thought Cornelius's paintings were designed to outshine the buildings they were in). As a result, Cornelius and the prince argued and, in 1840, the painter offered his services to a new patron, King Friedrich Wilhelm IV of Prussia, who immediately invited the artist to Berlin. There, Cornelius set about the crowning task of his career, which was to occupy him for the remaining twenty-five years of his life. The king had set his heart on rebuilding Berlin's great cathedral, and Cornelius was put in charge. The size of the project suited his ambitions, but even before the foundations of the cathedral had been laid, the political turbulence of 1848 overtook the entire scheme. Cornelius continued doggedly to produce design after design, huge cartoons for a scheme that he must have known would never happen, a fresco cycle in which the subject was the divine grace in face of man's sin, culminating in redemption. When Lady Eastlake, wife of the director of Britain's National Gallery and a great traveler and writer, saw the cartoons in Cornelius's studio, she was horrified by "the acres" of space they took up and concluded Cornelius was not the "great gun" of German art but "a mere popgun." Despite this, many of his fellow artists abroad—Jean-Auguste-Dominique Ingres, François-Pascal-Simon Gérard, and Eugène Delacroix, for example—admired him, for his intentions, at least. Delacroix praised "his courage even to commit big mistakes if the energy of expression demanded it."[44]

Ludwig behaved in much the same way toward Schnorr after he arrived in Munich in 1827 to join Cornelius. His first commission was to be a fresco cycle based on the *Odyssey* but Ludwig soon tired of that, preferring instead the *Nibelungenlied* for the newly built Royal Palace in Munich, modeled on the Palazzo Pitti. These frescoes took almost forty years before they were finished, mainly because Schnorr could not get enthusiastic about a nonreligious commission. But other schemes took almost as long, and when Cornelius abandoned Munich for Berlin, Schnorr became the target of the same critics who had attacked his friend. And so, when Schnorr visited Dresden in 1841 and was offered the directorship of the academy there, Ludwig made no attempt to keep him.

Schnorr produced one very successful work in Dresden, a series of 240 wood blocks—a *Picture Bible.* "If the Nazarenes left a testament, a justification of all they had wanted to accomplish, it could not have been more aptly demonstrated than in these Bible pictures . . . yet they were not a communal work but the work of one man who was not even of their immediate circle."[45] The Nazarenes never quite made it. Possibly, they were simply too theoretical.

A New Vocabulary for Painting

Many of the ideas and themes raised in the first section of this book come together in the work of the painter Caspar David Friedrich, who arrived in Dresden in 1798. His symbolism, his nationalism, his concern with the sublime, his Romanticism, his inner battle over the Christian faith . . . all these are reflected in his very distinctive form of art. "He painted mysteries and has remained something of a mystery himself."[46]

Born in the Baltic harbor town of Greifswald, in Pomerania, in 1774, Friedrich was the son of a candle maker and soap boiler.[47] After studying at the Copenhagen Academy he visited several conspicuous beauty spots in Germany, choosing eventually to settle in Dresden, and remaining there until his death. His highly distinctive style may be partly explained in personal terms: his mother died when he was seven and the brother he was closest to drowned while the two boys were ice-skating. Caspar suffered a lifelong sense of guilt.

At Copenhagen his teachers were exponents of Danish neoclassicism. Their emphasis on drawing from nature, combined with Friedrich's early love of travel, seems to have bred in him a fascination with landscape. Notably, his landscapes are populated by small isolated figures and megaliths or "heroic ruins." In time he evolved his own iconological vocabulary. "He painted Nordic images with an apocalyptic dimension, his landscapes rarely depict daylight or sunlight, rather they show dawn, dusk, fog or mist."[48] His contemporaries assumed this was his portrayal of the German "mood" as a result of the French invasion—politically weak but intellectually strong. Either way, Friedrich became convinced that the contemplation of nature leads us to a deeper appreciation of the way things are. His clear technique, mysterious scenes, and lighting effects (in this he was a forerunner of Salvador Dalí, in particular) ensured that his reputation grew quickly. He secured patrons and prizes equally and formed friendships with the main figures of German Romanticism.

One of his most typical—and most controversial—paintings was *Cross on the Mountains*, 1808, also known as the Tetschen Altarpiece. The crucified Christ is shown in profile, at the top of a mountain. Christ is alone, surrounded only by nature. In terms of size, the Cross is an insignificant element in the composition, which is dominated by the rays of the setting sun, symbolizing the old pre-Christian world, as Friedrich admitted.[49] By the same token, the mountain represented immovable faith, with the many fir trees being an allegory of hope. This was the first time anyone had produced a landscape intended as an altarpiece, and not everyone liked it. But Friedrich produced several other paintings in which crosses dominate a landscape, and whether occupied by Christian symbols or not, his landscapes are all spiritual entities first and foremost, "rife with mystical atmosphere." His friendships with Romantic writers had convinced him, he said, that "art must have its source in man's inner being; yet, it must be dependent on a moral or religious value."[50]

His other famous painting is *The Wanderer above the Sea of Fog*, showing the back of a man standing on top of a mountain, looking down on other mountaintops and clouds. Mysterious, but still technically faultless, this impressed Schinkel so much, it is said, that he gave up painting and turned to architecture.

Contemporary political events also lent themselves to Friedrich's style. Thanks to the Napoleonic Wars he developed a fierce loathing of France and an intense passion for his own country. His support for the various German liberation movements was expressed in scenes showing French soldiers lost among inhospitable German mountains. In general, though, his aim was to depict "the experience of divinity in a secular world" and this is what he tried to show in his melancholy renderings of the ruins of Gothic churches or dramatic forest landscapes. In his work, humans are more often than not helpless against the forces of overwhelming nature—Kant's idea of the sublime.

Friedrich's fame peaked in 1820 when the Russian tsarevitch, Alexandra Feodorovna (born Princess Charlotte of Prussia), bought several of his pictures. In the wake of the Prussian restoration, his political attitudes brought about increasing official attacks on his art and, like Cornelius, he became an anachronism. He died in 1837, forgotten by all but a few.

His emotional style of painting was rediscovered in the early twentieth century when the German Expressionists, Max Ernst, and other surrealists saw him as a similar visionary. Many painters in America were influenced by him, including the artists of the Hudson River school, the Rocky Mountain school, and the New England Luminists. Together with other Romantic painters—like J. M. W. Turner or John Constable—he helped to make landscape painting a major genre in Western art.

PART III

The Rise of the Educated Middle Class: The Engines and Engineers of Modern Prosperity

10.

Humboldt's Gift: The Invention of Research and the Prussian (Protestant) Concept of Learning

"The decades between 1790 and 1840 constitute the critical, formative period in the evolution of modern scholarship. By 1840 the natural and physical sciences, history and linguistics had forged the disciplinary divisions and had generated the central problems which would dominate academic learning into the twentieth century." This is R. Steven Turner, in his 1972 Princeton PhD thesis, "The Prussian Universities and the Research Imperative, 1806–1848." He continues: "Scholars of most European nations contributed to this heroic age of organised learning, but German scholars played the pre-eminent role."[1]

An ideological change took place in the first half of the nineteenth century, so that by 1850 German universities had almost entirely been converted into research institutes, "geared to the expansion of learning in many esoteric fields."[2] This "research imperative," as Turner calls it, involved four innovations: (1) Publication of new results based upon original research became an accepted responsibility of a professor and the sine qua non for even a minor university appointment; (2) the universities began to build the infrastructure—libraries, seminars, and laboratories—that would support research; (3) teaching was redirected and attempted to initiate students into the methods of research; (4) the Prussian professoriate embraced a university ideology that glorified original research. It was in the German universities of the early nineteenth century that the "institutionalization of discovery" was integrated with teaching for the first time.[3] After 1860 this ideology extended to England and the United States.

In some ways, as we have seen, the universities were the last place where this should have happened. Other institutions, such as the academies of science, had been quicker to respond to new intellectual trends (in Britain, for example). Eighteenth-century universities, as was discussed in Chapter 1, p. 41, existed to "preserve and transmit" learning. The professoriate in the nineteenth century, however, felt that a creative function must be added to their teaching obligations. This new approach was outlined in a series of treatises by Fichte, Schelling, and Schleiermacher, all published in the course of the reforms carried through by Wilhelm von Humboldt. As the historian Friedrich Paulsen wrote: "[W]hoever wishes to enter upon a scholarly career, upon him is the demand to be placed that he not merely have learned the knowledge at hand, but rather that he also be capable of producing knowledge out of his own independent activity . . ."[4]

The modern professor is a member of two communities: the institution where he teaches and the fellow scholars in his discipline. The first disciplinary community, says Turner, may be traced to Professors Johann Friedrich Pfaff at Helmstedt and Carl Friedrich Hindenburg at Leipzig, who founded Germany's first specialized mathematics journal, the *Archiv der reinen und angewandten Mathematik*. In chemistry Karl Hufbauer identified Lorenz Crell at Helmstedt and his *Chemisches Journal* as the center of the newly emerging chemists' community. In these fields, as among the classical philologists identified in Chapter 1, p. 41, an inner circle was emerging.[5]

The significant point about these self-conscious communities was that they began to acquire *authority*. In the eighteenth century such authority had been limited because the state had a monopoly over hiring and firing and often simply did not consult either the faculty or the discipline.[6] (Göttingen was an exception, which accounts in part for its pre-eminence.) Attempts were made to encourage professors to publish, but not works of original scholarship: textbooks were what counted, not specialized monographs.

That our modern concept of research had still to emerge is evident from the language of academics who, before 1790, spoke of "discoveries" (*Entdeckungen*) and "emendations" (*Verbesserungen*) in the sciences without ever using the word "research" (*Forschung*). Discoveries arose, it was assumed, from sheer force of intellect, from minds which fastened on a previously unrecognized relationship or that could order a mass of learning and so extract a higher generalization. It was, in other words, the prerogative of genius. On top of that, it was understood that some areas of the sciences were, essentially, static. J. D. Michaelis was just one who did not expect new truths to emerge in certain sciences: philosophy, law, theology, and much of history.[7]

Probably, nothing would have happened without Napoleon and the crushing defeats he imposed on Prussia. (Thomas Nipperdey opens his magisterial history of nineteenth-century Germany with the words, "In the beginning there was Napoleon.") Reformers were swept into power. The country's collapse, they believed, had stemmed from the "rotten core" of the Frederician garrison-state with its emphasis on "mechanical obedience and iron discipline." A moral renewal was needed and that included education.

This was carried through on three fronts: organizational, administrative, and ideological.[8] Old, weaker institutions were abolished, others amalgamated, and, most exciting of all, new universities founded at Berlin and Bonn.

The reforms started in Königsberg, where King Friedrich Wilhelm III was impressed by the patriotism of the university faculty when his court moved there during Napoleon's invasion. Friedrich Wilhelm turned down a delegation of Halle professors who wanted him to transfer the University of Halle in its entirety to Berlin, but he did agree to found an entirely new university in the city, and this proved crucial. Fichte, Schleiermacher, and F. A. Wolf had all migrated to Berlin, and the king's decision provoked a great explosion of theorizing about universities. The critical move took place, however, when Hardenberg, one of the reform-minded ministers, brought Wilhelm von Humboldt back from a diplomatic sinecure in Rome to head up the newly created Department of Religious and Educational Affairs. Humboldt, having himself written on philology, was closer to Wolf than to anyone else, and they set out to recruit individual scholars themselves. The university opened its doors for the winter semester of 1810 with Fichte as the first (elected) rector. This began what Nipperdey calls "the religion of education" in nineteenth-century Germany.[9]

Humboldt succeeded in attracting a number of eminent scholars—the jurist Friedrich Carl von Savigny, the anatomist Karl Asmund Rudolphi, plus Schleiermacher, Wolf himself, of course, J. C. Reil, and J. G. Bernstein. He also poached a raft of scientists from the Berlin Academy. In the early days, the faculty was strongest in philology and law, and here Berlin soon outshone Göttingen. In the sciences, however, it was not until the late 1820s that the new institute really began to shine. By then a second new university had been created at Bonn, and Humboldt had been succeeded by Kaspar Friedrich von Schuckmann.[10]

No less important than the organizational and institutional reforms of the university were the theoretical innovations, the spiritual and philosophical rejuvenation, as Turner puts it. "Their common tenets may be grouped under one name, *Wissenschaftsideologie*. This new concept went on to unprecedented success during the years following the founding of Berlin University. It became the official ideology of the German universities during the nineteenth century, endowed with an awesome, almost religious status, an ideology that has defined the 'idea' of the German university, with its emphasis on the unity of research and teaching."[11]

Besides Humboldt, five other individuals carved out the fundamentals of this Wissenschaftsideologie. Fichte was the best known, writing *On the Vocation of the Scholar*, given as a series of lectures and published as pamphlets at Jena and Berlin between 1794 and 1807. Schelling, Henrik Steffens, Schleiermacher, and Wolf all added their thoughts to his. Through them two intellectual traditions came together in Wissenschaftsideologie.[12] First, the new tradition owed a great deal to Idealist philosophy, by then centered at Jena. Following Lorenz Oken and Schelling himself, Steffens became the chief advocate of *Naturphilosophie*, regarded as the scientific branch of Idealism (covered in Chapter 8, p. 189, on Romanticism). But Wissenschaftsideologie was rooted, secondly, in the tradition of academic neohumanism associated with the University of Göttingen. Wolf and Humboldt had both studied under Heyne, while Schleiermacher had become known as a philologist through his editions of Plato. "The glories of Greece and Rome, they argued, could best enhance the moral and aesthetic sensibilities of German students. Hence their study should precede all later, professional study, not only in their gymnasiums but also in the university. Broader and deeper immersion in the classics, the neohumanists believed, would go far to eliminate both the crass utilitarianism of the eighteenth century universities and the corruption of student life."[13]

There was also the fact that these advocates of Wissenschaftsideologie believed in an important difference between school and university. At school the pupil gained information; at university he learned judgment and independence. Schleiermacher in particular thought the universities—between the schools and the academies—"were suited to 'the German genius' . . . the university is thus concerned with the initiation of a process . . . nothing less than a whole new intellectual life process, to awake the idea of learning (Wissenschaft) in youth . . . so that it becomes second nature for them to consider everything from the point of view of learning."[14] Schleiermacher and others believed that universities were more than just higher schools. Thus was born the concept of *Brotstudium*, "bread studies." Bread studies provided the student with enough information for a job, but not to advance learning.

All this, a new understanding of what it meant to be educated, was being created in Wissenschaftsideologie. The Kantian and post-Kantian philosophical systems identified two modes of Being, the Real and the Ideal. According to Schelling, Wissenschaft "is knowledge of the absolute unity existing between the Ideal and the Real. *Wissenschaft* is the philosophical insight that there *is* unity between the Real and the Ideal. Wissenschaft is innate in all men but it is a growing thing, evolving and dynamic and so central to this was the concept of *Bildung*, also drawn from idealist philosophy—the process of becoming in an educative sense. Under this system, discovery—research—was a moral act as much as anything."[15]

It wasn't far from this to Fichte's argument that the scholar, the professor, is the natural leader and teacher of mankind. "The scholar should be *morally* the *best* man of his age; he should exhibit in himself the highest grade of moral culture then possible."[16] There *were* doubters, who thought that neohumanism was atheistic and subversive but, curiously enough, in a Romantic age, scholars personified the Romantic individual. By 1817 Berlin had replaced Jena as the focal point of the new university ideology.

But the new universities coincided with a patriotic revival in Prussia and in Germany generally, and the contribution of the university students in the war of liberation also played a part in changing attitudes, making the universities more popular than they had ever been.[17]

Between 1818, when a measure of political stability returned to Prussia (as to the rest of Europe), and 1848, the year of revolution, the pursuit of scientific and scholarly research became the defining characteristic of the German university. "*Wissenschaftsideologie* glorified discovery and creativity within the universities; and . . . It assumed that one obtains academic knowledge through the rigorous application of well-defined methods of investigation which moreover means that the tools of discovery can be made available to large numbers of students."[18] This was the new Prussian concept of learning.

It is important to say what it was not as well as what it was. In Germany the revolution in scholarship began with the Kantian critiques but owed just as much to classical philology, history, and the discovery of the Indo-European languages, covered in Chapter 8, p. 189. Intellectually, this was all as innovative as the discoveries in the natural sciences that were occurring simultaneously in France through Lavoisier, Laplace, and Georges Cuvier. In Germany, though, the sciences did not play this role. Creativity in science did not begin there until after about 1830.[19]

Speculative philosophy apart, classical philology epitomized the German *Wissenschaft* from the time of Heyne and Wolf on—it was its new techniques and standards of rigor that were later transferred into law, history, and other branches of scholarship. Moreover, the fierce intellectual rigor that Heyne and Wolf brought to classical philology, and the accomplishments of this approach, stimulated new specialities. Germanic culture itself was one. Romantics such as Friedrich von der Hagen, Achim von Arnim, and Clemens Brentano had rescued from oblivion large amounts of otherwise forgotten old Teutonic literature. The brothers Grimm—Jacob and Wilhelm—produced jointly *Kinder- und Hausmärchen*, a work on which their fame was based, while Jacob alone published his no less famous *Deutsche Grammatik* of 1819–37. This, together with their etymological dictionary, the *Deutsches Wörterbuch*, known as "der Grimm," and of which they completed only four of the eventual thirty-plus volumes, helped establish a rigorous basis for the advance of Germanic philology.[20]

Likewise, in historiography, the critical tradition that began in classical philology was espoused by Barthold Niebuhr. Born in Copenhagen, Niebuhr was briefly a civil servant before taking up a position at the new University of Berlin, where he gave a famous series of lectures in 1811–12 which he then turned into a book, *Römische Geschichte* (*History of Rome*; 1811–32), a no-less-impressive multivolume work. It was here that Niebuhr employed critical analysis to the sources of Roman history so as to identify a sound narrative among the myths and oral traditions that had come down from antiquity. Niebuhr's account was overtaken and improved by the writings of Theodor Mommsen later on, but even so his *Rome* was a sensation, widely regarded as history of a new kind, and proved to be a model for Leopold von Ranke.[21] The disciplines of philology and history were, importantly, centered in the philosophical faculty, and this contributed further to the rise in importance of that faculty, a changeover in priorities that, as we saw, began in the eighteenth century at Halle and Göttingen.

"The Most Exact and Exalted of German Sciences"

Their common romantic roots gave the different branches of the new scholarship in Germany a striking unity of theme and outlook.[22] Over and above that, however, there was also a transformation in critical method.

Together with "*Wissenschaft*" and "*Bildung*," the term "*Kritik*" was emerging as a basic category of the academic approach. It had first been encountered, of course, as a more or less technical term in Kantian philosophy in the 1780s, where it exemplified a turning away from the existing content of knowledge toward a critical assessment of the *sources* of knowledge and the *validity* of existing learning. By the time the University of Berlin was fully established in the 1820s, scholars still used "*Kritik*" in this sense (the method is called *quellenkritisch*). The term implied a constant, skeptical evaluation of sources. It implied too that the critical scrutiny of evidence should precede the more constructive aspects of scholarship. More technically, it referred to the "scrupulous precision" with which sources—archives, manuscripts—were to be treated.[23]

The "recension" of a text epitomized the new approach. In this process, scholars compared different versions of a source, each of which had to be accurately dated, and all errors eliminated. F. A. Wolf did this most famously in *Prolegomena ad Homerum* in 1795, concluding (see Chapter 3, p. 91) that, in effect, there was no such person as Homer. In fact, many of Wolf's specific arguments were exploded by his own students—and fairly quickly at that. But, in a sense, that was not the point. The book—at a stroke—demonstrated the sheer power of the critical method to unearth real historical knowledge. It aroused passionate debate, debate in which Wolf played his part, but during the next two decades his methods were extended into new areas—for example, the German epics, and the biblical texts. "The *Prolegomena* established philology as the most exact and exalted of German sciences."[24]

The new Prussian learning was a highly self-conscious entity and, says Nipperdey, somewhat solitary. Turner says that a feeling of intense excitement and accelerating intellectual progress permeates the letters and papers of scholars during the early nineteenth century. "Boeckh wrote repeatedly of the 'new learning,' Wolf of *Altertumswissenschaft* as 'a new science at its birth' and Leopold von Ranke reported with awe the 'still unknown history of Europe' lying before him in the Vienna archives."[25]

In line with these changes, the "disciplinary community" began to emerge and with it the associated furniture—libraries and manuscript collections, prestigious journals and their editorships, reviews and critiques, which now became very important, as part of a scholar's output, not least because they helped maintain rigorous methods and standards.

Not everyone had the time or inclination for such an approach and so, before too long, and gradually, philologists began to write for each other in their journals. Thus was born the first instance of a professional literature. It was a development that did not go unnoticed by the public; philologists, for example, became known for their egotism and sheer arrogance. Some, like Karl Konrad Lachmann, were notorious for their acid reviews.[26]

That arrogance apart, however, the critical method had helped to produce a new attitude toward scholarly creativity and the process of discovery. There developed a dissatisfaction with mere erudition, so valued in the eighteenth century: there was now a growing emphasis on *originality* as the criterion of the value of a scholarly enterprise. One effect of this was to undermine the eighteenth-century belief that discovery "was available only to geniuses," and instead allowed that a greater number of individuals, "with lesser gifts," could achieve something worthwhile. This encouraged a prevailing sense of movement, an expectation of infinite advance, and marked a major transition from the eighteenth- to the nineteenth-century understanding of learning.[27] Out of all this came an idealizing of creativity and an ideology of original research.

"Knowledge is itself a branch of human culture," insisted Fichte. Humboldt agreed. The purpose of the universities is "to cultivate learning in the deepest and broadest sense of the word," not for some practical or utilitarian end, but for its own sake as "preparatory material of spiritual and moral education (*Bildung*)."[28]

The fragmentation of scholarship into disparate, disconnected specialities also began in earnest in the 1830s. Scholarship was now seen as a process of *accumulation*, stone-by-stone.[29] "Comprehension of the entire edifice remained an ideal, but only at one metaphysical remove." This is still more or less the attitude we have today.

The Growth of the Science Seminar

The other major change was that, during the *Vormärz* (the period between 1840 and the March revolution of 1848), the philosophical faculties of the Prussian universities—which had been the poor relations in the early eighteenth century—consolidated their advance and blossomed into a position of leadership. Between 1800 and 1854, again according to Turner, the number of students enrolled in philosophy grew from 2.4 percent to 21.3 percent, and the teaching staff showed a similar rise. Philosophy, philology, and history—"the three disciplines around which not only German scholarship but all of German intellectual life were being rapidly renewed"—all found their home in the lower faculty (i.e., they were not part of the theological, legal, or medical "higher" faculties).[30]

The philosophical faculty also owed its rise in pre-eminence to the fact that it prepared teachers for the new *Gymnasien* brought in by Humboldt's reforms. Beforehand, most teachers had been trained in the theological faculty because the church had run the schools. Humboldt's initiatives removed the schools from church control and provided an examination, the *Abiturexamen*, which a student had to pass to enter a university. In line with neohumanistic principles, the *Abitur* stressed Greek, Latin, and mathematics. These reforms produced a class of professional teachers in Prussia that spawned a rapid increase in the number of Gymnasien across the country between 1818 (91) and 1862 (144).[31]

In tandem, more and more students began to study the natural sciences—again pushing up numbers in the philosophical faculty. Science subjects expanded rapidly, especially after 1840, though to begin with the graduates mainly went into teaching because, even by 1860, Prussia did not possess sufficient industrial plants to absorb more than a fraction of the students. These changes occurred at the expense of applied subjects like *Landwissenschaft* and cameralism.[32]

By now the seminar was well established. It will be remembered from Chapter 1, p.41, that the essence of the seminar was that it was smaller and more intimate than the lecture; there was no place for rhetoric, and it was understood to be an advanced course of study, for those really committed to their subject. Normally, it worked in the following way. Every two weeks a research paper was presented by one member of the seminar and subjected to general criticism. The best papers would be published at the expense of the ministry and 500 thaler allotted for prizes. Admission to a seminar, therefore, promised substantial rewards and other disciplines, like history and theology, copied this model. New seminars spread across Germany and across the disciplines and were understood as being well tailored to transmitting the new critical methods. Gradually they became an elite "inner track" for the brightest students.

As was mentioned earlier, the 1830s were a critical decade in scholarship in that separate disciplines began to acquire their own journals and other specific infrastructure. It was in the 1830s that the sciences began to assimilate the concepts of philological and historical scholarship.[33] Until then, the sciences had taken no part in Germany's academic revolution and such science as was taught was elementary: chemistry stressed "recipes," the life sciences were devoted mainly to classification. Frankly utilitarian, they epitomized the materialistic bread study which Wissenschaftsideologie saw as the main obstacle to spiritual and intellectual rejuvenation and to Bildung. The obsession with classification was, for neohumanists, particularly deadening.[34]

In the end, however, the sciences benefited from the attacks of the neohumanists. Precisely because these attacks were made in terms of *Wissenschaftsideologie*—that Wissenschaft was an unlimited, organically unfolding "cultural good"—younger scientists began to counterargue that the sciences, no less than the humanities, trained the intellect (Geist) and led to the refinement of the individual (Bildung). One important side effect of this argument was that "pure" science was seen as superior to applied science, which was dismissed as mere "bread study."[35]

The post-1830 scientists became convinced that research not only added to the sum of learning but also that it helped the moral development of the individual doing the research. As a result, the number of frankly technological courses in science and mathematics dropped markedly, not least in Berlin and Halle. Technological education was relegated to other institutions, notably the forerunners of what were to become *technische Hochschulen*. Instead, the universities now began to teach science in a "purer" style, and chemistry and the life sciences became an integral part of the philosophical faculty. This was a major change in attitude that found institutional expression at the University of Bonn, where the first Seminarum für die gesammten Naturwissenschaften was established in 1825. "Bonn's modest seminar can be properly considered Prussia's first step toward the network of large research institutes which by 1880 had made the German organisation of science world famous."[36]

That expansion began after 1830, and the following two decades witnessed the all-important flourishing of German science—with the work of Johannes Müller, Eilhardt Mitscherlich, Peter Gustav Lejeune Dirichlet, Evangelista Purkyně (or Purkinje), Franz Neumann, Julius Plücker, and Carl Gustav Jacob Jacobi. It was during this time too that Hermann von Helmholtz, Emil Heinrich Du Bois-Reymond, Rudolf Clausius, and Ernst Brücke first attended Prussian universities. In its way, it was the beginning of a second scientific revolution, when Prussian science attained a European pre-eminence that German classical and historical studies had already enjoyed for some time.[37]

A crucial role was played by C. G. J. Jacobi, whose own career epitomized wider changes. Educated first at the Gymnasium at Potsdam, Jacobi attended the University of Berlin in 1821, where he studied philology in August Boeckh's seminar. He subsequently abandoned philology and turned instead to mathematical physics. Doing well, he was given a professorship at Königsberg, where he lectured about his research—on elliptical functions—and then, with Franz Neumann, founded in 1835 the Königsberg mathematics-physics seminar, modeled on Boeckh's. In this seminar he insisted on original work from his students and every paper submitted received a stipend of twenty thaler, increased to thirty if the paper was published. The seminar also funded the cost of instruments used by the students. In this way, Jacobi's Königsberg seminar became the focus of German mathematical physics and was widely imitated—at Halle (1839), Göttingen (1850), Berlin (1864), and elsewhere. Such science seminars as this one formed a logical transition, pointing forward to the great laboratories of the 1870s.[38]

The Idea of the University and the Kulturstaat

But there was an extra level to all this: government acceptance of the new ideology. During the 1830s, or thereabouts, professors were increasingly appointed because of their reputations among their peers, rather than as teachers. There was, in effect, says Turner, a new social contract between university intellectuals and the Prussian state: this was the theory of the *Kulturstaat*. The theory of the *Kulturstaat* maintained that society exists for the evolution of *Kultur*. "Culture attains its most highly conscious manifestation in the universities, where it is developed and preserved. The state, therefore, must serve and support its universities and depositories of culture and guarantee the academic freedom which makes the preservation and development of culture possible. A nation's universities serve as national symbols of its intellectual greatness. As long as the state does this, the universities owe the state support, respect and service . . . On this basis vast sums of money went into the universities and the theory of the *Kulturstaat* furnished the basis for the remarkable political symbiosis between Prussian intellectuals and the state which, despite many strains, lasted throughout the nineteenth century."[39]

Germany was fortunate that its state bureaucrats mainly saw eye-to-eye with its leading scholars. In 1817, Karl von Hardenberg had appointed Karl vom Stein zum Altenstein as director of a newly created Kulturministerium. A Romantic botanist, a fervent follower of Fichte and Humboldt, Altenstein kept the universities free of religious or political interference and allowed the new scholarship and new attitudes to flourish. He was followed by Johannes Schulze, who had been a student at Halle, where he attended Wolf's seminar, and had himself undertaken a new edition of Winckelmann's work, at Goethe's suggestion. Given carte blanche with Germany's Gymnasien, he embraced Wissenschaftsideologie as the ideal basis of education. It was Schulze who made Greek compulsory for all Gymnasium students, and Schulze who determined that only Gymnasien could send their graduates to the university.[40]

The figures collected by R. Steven Turner, Charles McClelland, and William Clark (all Americans, as it happens), underline this picture. In 1805 Prussia allowed 100,000 thaler for its universities, a sum that rose to 580,000 by 1853. The teaching staff grew by 157 percent over the same period. In fact, between 1820 and 1840 the number of professors increased by a bigger proportion than did the number of students (187 percent and 50 percent in philosophy, 113 percent and 22 percent in medicine). At Berlin, financial support for scientific institutions was increased—from 15.5 percent in 1820 to 34 percent in 1850. "The intellectual cannot be too highly valued. It is the basis of all that on which the strength of the state can eternally rest."[41]

The ministry frequently imposed its intellectual viewpoint. The most well-known example is Hegelianism, which before 1830 became a virtual state-philosophy in Prussia, where the ministry ensured its dominance by granting Hegel's students a near monopoly over chairs of philosophy. Altenstein, Eichhorn, and Schulze all saw to it that no one could become a professor until he had published a "solid" book.[42] The dual nature of the professoriate—teaching and research—had become a fact of academic life. Younger scholars, bloodied in the emerging "research imperative," saw specialized research as the only road to a professorship. These new values were the basis for the great new institutes and laboratories that proved to be a jewel of the Bismarckian era.[43]

11.

The Evolution of Alienation

Intellectual tastes and fashions are curious entities. In 1808, Beethoven's Fifth and Sixth Symphonies were premiered at exactly the same moment as Caspar David Friedrich's *The Cross on the Mountain*. In 1839, as Peter von Cornelius was completing his vast fresco of the *Creation, Redemption, and the Last Judgment* in the Ludwigskirche in Munich, Felix Mendelssohn was conducting the first performance of Franz Schubert's Symphony in C Major (The Great Symphony). German music remains as popular as ever, but German painting of the period 1750–1850 (and in fact somewhat later), has sunk into—if not oblivion—marked neglect. A related paradox concerns German speculative philosophy. It too was regarded at the time as a bright star in the firmament—the names of Schelling, Feuerbach, and, above all, Hegel were on everybody's, not just German, lips in the early decades of the nineteenth century. In the cold light of the twenty-first century, however, these names—and the very act of speculative philosophy itself—seem very distant.

Speculative philosophy had a special status just then because, once again, Europe was in that intellectual time frame between doubt and Darwin. Religion, Christianity in particular, was in retreat and with it the concept of revelation. Philosophy naturally filled this intellectual gap but was speculative in the sense that such thoughts and insights of the philosophers, wherever they came from, had to convince others by force of reason and internal consistency rather than by ecclesiastical authority. At the same time (and this is the paradox) this speculative philosophy, especially that of Hegel, gave rise to one of the most powerfully influential—perhaps *the* most influential—philosophy the world has ever seen: Marxism. If Marxism is fading now as a political force, it is still seen in many quarters as a useful analytic form of understanding. And in the concept of "alienation" we find one of the most powerful ideas shaping the phenomenon of modernism. It has affected painting, the novel, theater, and film, not to mention psychology.

We begin, not with Hegel or Marx but with Friedrich Schelling. A member of that small closed circle of German Romantics, he included among his most intimate friends such figures as Novalis, Ludwig Tieck, and the Schlegels. In his own thought he was chiefly concerned with what he saw as "deep and pervasive affinities" between man and nature.[1] Nature, for him and others like him, was the product of an active "organizing principle" or "world soul" (*Weltseele*). This principle "externalized" itself as objective phenomena, which together reflected a unity that was ultimately to be understood in teleological terms; in other words, and put simply, there is a continual process of creation, and its various levels are related to one another in a purposive manner.[2] These stages of creation could be investigated by the different branches of science—physics or biology—but they could not be truly understood in isolation; some overarching system was needed for a complete perspective. For Schelling there were three obvious and all-important levels. There was the inorganic level, governed by the laws of mechanics; there was the organic level, governed by the laws of biology; and there was the level of consciousness, only present among humans. The development of consciousness was looked upon by Schelling as the "culmination and goal" of the entire process.[3] In saying this, he claimed to discern a cyclical movement in history. Spirit objectified itself in the natural world as phenomena, but it "returned to itself" as mind. Investigating the nature of "mind" now became the primary task of philosophical reflection; the spirit's apprehension of itself was the "final task" of mankind.

In framing this approach, Schelling attached a profound importance to artistic creativity. The cumulative effects of creativity would lead to an ever-greater appreciation of what Schelling variously called "absolute identity," "pure identity," and "absolute reason." There is a sense in which, to our ears, this sounds almost absurd but it was, again, a concept of an ultimate reality that had no grounding in religion and, as yet, showed no real biological understanding. With hindsight, it is possible to credit Schelling with some notion of "emergent evolution," but his thoughts were really a halfway house and a dead-end.

Or not quite. Schelling is perhaps most pertinent as a forerunner of Hegel, who incorporated some of his notions, in particular his idea of the "absolute spirit," or mind.

Georg Wilhelm Friedrich Hegel was born in Stuttgart in 1770. From 1788 to 1793 he was a theology student at Tübingen, where his fellow students included both Schelling and the future great Romantic poet Friedrich Hölderlin. After graduation Hegel first worked as a tutor to a number of families in Bern and Frankfurt and seems to have looked forward to a career as a reforming educator. But, possibly as a result of his contacts with Schelling and Hölderlin, he fell under the influence of Kant and then Fichte. He moved to the University of Jena to work more closely with Schelling, with whom he edited the *Kritische Journal der Philosophie* (Critical Journal of Philosophy) and wrote about both Schelling's and Fichte's philosophy, exploring their differences. He gradually diverged from Schelling, his views becoming clear in his *Phänomenologie des Geistes* (*Phenomenology of the Spirit*), published in 1806.[4] Schelling interpreted certain passages in this book as an attack and their friendship ended. The occupation of Jena by Napoleon's troops closed the university, and Hegel was forced to leave. He worked first as a journalist in Bamberg, and then as headmaster and teacher of philosophy at a *Gymnasium* in Nuremberg, where he married and started a family. In 1816 he was appointed to the chair of philosophy at Heidelberg and then, in 1818, to the same chair at Berlin.

From his earliest days, Hegel was influenced, as were so many in the Germany of his time, by the differences between modern societies and Classical Greece, contrasting the divisions and antagonisms of modern societies with the apparent harmony of the ancients. He was distressed by the "anarchic individualism" of contemporary European life, feeling that the great majority of people had lost a sense of common purpose and dignity and were unable any longer to identify with institutions and customs that had traditionally fulfilled their aspirations. Religion—

Christianity—which at one time might have provided a remedy, had also failed. There was estrangement wherever one turned.

This estrangement was not yet called alienation but it already played a vital role in Hegel's thinking. It was such estrangement that led him to propose his vast synoptic vision where every aspect of the world, every discipline of knowledge, was allotted its position and an explanation and rationale provided. Within this vast system, two oppositional ideas reappeared, but reclothed and in more imposing form.[5]

For Hegel, as for Schelling, the development of the world, of phenomena, was to be understood as the evolution of the spirit, described as "the process of its own becoming." Hegel did not think, as Schelling appeared to, that there was any such thing as "pure undifferentiated identity," or spirit, which was in some way "logically prior" to the reality of phenomena; instead, for Hegel spirit could only exist in the multitude of ways in which it revealed itself—there was no "other world" in his scheme. He did write as if there were an "inner" reality that, as it were, existed "behind" the world of ordinary experience and expressed the truth of things "without husk," but this was a logical relationship revealed in pure thought rather than a separate existence. Explanation of these logical relations was for him the proper subject matter of philosophy.[6]

What he tried to show, what he thought was the most vital question, was that subjective consciousness and objective existence must always involve an "irreducible duality." It was the task of philosophy to go beyond this outlook, to the point where "our familiar conceptual schemes dissolved." Like Schelling, only more so, Hegel conceived spirit as an entity involving development toward an end "in which it might be said to achieve confirmation of its own being."[7] In everyday terms, the history of the world was to be understood as a teleological process, whereby spirit first revealed itself in some form that expressed its eventual possibilities. This was a process that occurred at two distinct levels. In one, spirit manifested itself "unconsciously" in producing natural phenomena, not just individual objects but societies and civilizations, entities that represented concrete expressions of different stages of evolution toward a progressively fuller consciousness and self-understanding. At the second level, Hegel emphasized that development could be understood historically, the successive patterns in life and culture being the successive embodiments of spirit. Hegel's fundamental idea was that, as distinct (social) forms and institutions evolved, together with advances in thought that provided fresh modes of interpreting experience, "spirit gradually moved toward an ever deeper comprehension of its own nature." The end point was what he called "absolute knowledge," a state of philosophical understanding "in which spirit finally came to recognise that the entire world, in all its varied manifestations, was the product and articulation of itself." "Absolute knowledge" marked a form of understanding in which spirit, as a result of philosophical reflection and understanding, "returned to itself." In this way, the external objective world, and the internal subjective world would be unified, and the original condition of self-estrangement, or self-alienation, was overcome.[8]

There are fairly obvious parallels here with the traditional Christian dogmas of the Fall, atonement, and redemption. Hegel himself conceded as much and even admitted a correspondence between the idea of God and his notion of absolute spirit. He was, if you like, struggling with a post-Christian/pre-Darwinian understanding. But in fairness, Hegel's absolute spirit was not seen as a transcendent personality, over and above the universe or independent of it.[9] This was a very important sense in which Hegel's philosophical system was *not* religious.

The details of his system were set out in two principal works: in the early (and fairly abstract) *Phenomenology of the Spirit*; and the more explicitly historical *Vorlesungen über die Philosophie der Geschichte* (*Lectures on the Philosophy of History*), published posthumously.[10] The crucial aspect of man's condition, according to Hegel, is a repeated oscillation between man's "self-created social world" and his evolving attitudes toward that environment. There is a continual "dialectic" between the creative and critical periods of the process. As social and political circumstances evolve, spirit reveals itself in the deeper understanding of the essence of man, which Hegel identifies as the development of freedom. History manifests itself in successive civilizations which reveal an increasing self-consciousness and freedom as men acquire "a fuller grasp of their own needs and a profounder recognition of the relations in which they stand to one another."[11] "Master-slave" societies give way to individualistic ones, which are in turn displaced by the notion of a social order where "mutual respect between persons supersedes antagonism and distrust." True freedom would be achieved insofar as an individual's inner potentialities are realized in a world he himself had created and in which he found himself "at home."[12]

Hegel's philosophy was comprehensive and intended to be (Nipperdey called it "the tyranny of abstraction"). It was, nonetheless, in social and historical theory that he was to exert most influence, even if that influence resulted from a reaction *against* him by radical writers who turned some of his ideas upside down.

The Hegelian Aftermath

To begin with, Hegel was regarded as a reassuring force. His account of history implied that what had evolved was for the best, confirming the institutions of society rather than seeking to change them in favor of something better.

Hegel died in 1831. Beethoven and Schubert had died not long before, and Goethe was to die the following year. The world was changing. In parallel, there emerged in the late 1830s and early 1840s a group of German intellectuals who came to be known as the "Young Hegelians" and put forward the far more aggressive view that the real meaning of Hegel's doctrines had either been overlooked or misconstrued, and that the implications were far more radical than most people wanted to believe.

It is difficult for us now to think ourselves back into the mind-set of those times, but the truth is that, in the 1820s, Hegel's philosophy had become supreme in Germany. It was strongly supported by the minister of culture, Karl Altenstein, and centered around the Berliner-Kritische Association, founded in 1827 as a home for the Hegelian periodical *Jahrbücher für wissenschaftliche Kritik*. In 1832, the year after his death, an association was formed in Berlin of Hegel's most intimate friends and pupils, which continued as the intellectual backbone of the school, to propagate Hegel's teaching and prepare an authorized edition of his works. So powerful was his system that many believed Hegel's philosophy was the ultimate one, the culmination of all philosophical thought, and that there was little else to do except work out the details Hegel hadn't had time to pursue. Inevitably, however, differences among the Young Hegelians began to show themselves.[13]

In practice, several of these young radicals were the first to raise ideas that Marx was to consolidate into his own all-embracing theory. In 1835, for example, David Strauss (1808–74) published *Das Leben Jesu* (*The Life of Jesus*). Educated to begin with at Tübingen under the iconoclastic Old Testament scholar F. C. Bauer, Strauss transferred to Berlin to attend Hegel's lectures shortly before his death.[14] Hegel seems never to have been very interested in the historicity of the Gospels, but Strauss thought they were the essence of Christianity, myths that reflected "the profound desires of the people."[15] He understood the Gospels as "imaginations of facts" thrown up by the collective consciousness of a society at a specific (Hegelian) stage of development. This implied that the very idea of revelation and incarnation was but a phase on the way to something higher, better, freer. The effect of Strauss's book was sensational—one review described Strauss as the "Ischariot of the day"—but in the context of Marx it had two consequences.[16] One, it helped confirm his loss of faith. Two, in the heavily censored society of the times, where political discussion was fraught with danger, biblical criticism allowed the development of philosophical/sociological thinking in relative safety.

The other Young Hegelians whose more specific ideas Marx would adapt included August von Cieszkowski, who was the first to argue that "it was not enough to discover the laws of past history—men must use this knowledge to change the world"; Lorenz von Stein, who first determined that industrialization meant a reduction in salaries for the proletariat who, for that reason, "can never possess private property"; and Arnold Ruge, who stressed that man was *defined* by his social relations and that it was through *work* that he expressed himself. Many of these views had emerged through the discussions in the so-called Doktor Club in Berlin, from 1837 on, where an orthodoxy had also developed that Hegel's real thought had been concealed while he was alive and that his philosophy, at one time felt to be reassuring, actually contained "revolutionary tendencies." These figures were, however, all overshadowed by three others of greater significance: Ludwig Feuerbach, Moses Hess, and Friedrich Engels.

The Importance of Ludwig Feuerbach

Ludwig Feuerbach's widely read and much-acclaimed *Das Wesen des Christentums* (*The Essence of Christianity*) was published in 1841 and carried on the transformation in the critical study of Christianity already begun by

Strauss and others.[17] However, Feuerbach was not content merely to update and polish Hegel: in the new German tradition, he subjected Hegel to a thorough critical analysis. Feuerbach thought that Hegel had made a crucial mistake, that existence comes *before* thinking. Thought, Feuerbach said, was naturally dependent on a "sensuously apprehended natural world of objects and events." Humans are part of that world and only by reference to it are meaning and content generated. Philosophy, therefore, cannot begin at the opposite end, as it were, and use pure conceptions as a starting point.[18]

In *The Essence of Christianity*, Feuerbach tried to show this process operating in religion. Religion, he argued (and this was well before Freud), "implied the projection by man of his own essential properties and powers into a transcendent sphere in such a way that they appeared before him in the shape of a divine being standing over and above himself." "The divine thing," he went on, "is nothing else than the human being, or, rather, human nature purified, freed from the limits of individual man, made objective—i.e. contemplated and revered as another, a distinct being." In worshipping God, man is worshipping himself. But this was not necessarily a bad thing or a dead end. Historically considered, Feuerbach says, worship has helped man to a greater understanding of himself and what he can hope *to become*. The negative side, he thought, lay in the fact that the idealized conception of the divine inevitably led men to diminish their own status, to a self-impoverishment of the earthly realm.[19] This gap between the possible and the actual was a description of our alienation, and destiny was to be understood "not in terms of the absolute's return to itself through self-knowledge, but in terms of man's return to himself through the recognition and realisation of his own powers and possibilities." This was Kant Hegelianized.

The young Karl Marx (1818–83) was greatly influenced by Feuerbach (in fact, for a time Feuerbach was more important than Marx). In particular, there was Feuerbach's idea that anthropology and physiology were the most fundamental of sciences. This contributed to Marx's central idea—that the "humanisation of nature" and the "naturalisation of man" are what philosophy should seek to achieve. Feuerbach produced in Marx the conception of man "as a being whose very essence is modified by his contact with nature and his fellow men in society."[20]

And this is how Marx, following Feuerbach, came to regard Hegel's conception of alienation as central. But while Feuerbach had concentrated on the idea of alienation as central to the religious experience, for Marx, alienation—man's self-estrangement—was intimately linked to his concrete social situation.

Another precursor of Marx was Moses Hess, also a Young Hegelian. His *Heilige Geschichte* was the first expression of coherent socialist thought in Germany. His aim was to explore how mankind "can regain union with God now that the original harmony has been lost," and it too was an attempt to reconfigure Hegel.[21] In a subsequent book, *Die europäische Triarchie,* Hess argued that the abolition of private property was essential to any new social order, that "spiritual alienation" could only be removed once the "servile classes" were relieved of economic exploitation. He, like others, believed that revolution would come first (or next) in England because the divisions between wealth and poverty were greatest there. Hess and Marx encountered each other in Bonn in 1841, with Hess subsequently describing Marx as "Rousseau, Voltaire, Holbach, Lessing, Heine and Hegel rolled into one."[22] "Money," insisted Hess, "is the worth of men expressed in figures, the hallmark of our slavery." For Marx, "Money is the jealous God of Israel beside which no other God may exist."[23] After Marx had moved from Bonn to Cologne in 1842, he and Hess attended the lectures of Bruno Bauer together. Hess shared with Marx the view that Germany was "a more theoretical nation" than any others, and that that too was a form of alienation. For a short time, as David McLellan has observed, Hess was setting the pace.

"Perhaps the Most Significant Intellectual Collaboration of All Time"

Marx, says Bruce Mazlish, was one of the "Essenes" of early socialism.[24] This is meant to imply a certain religious and ascetic quality but in fact Marx defies easy generalizations. At times he saw himself as a scientist, invoking the name of Darwin as an analogy to his own role in discovering laws, not of "natural technology" but "of human technology." In the late 1830s, at the very end of the Romantic period, Marx himself wrote poetry and forged a friendship with Heinrich Heine, Ferdinand Freiligrath, and Georg Herwegh, poets who will be discussed in Chapter 14, p. 289. But as Mazlish also points out, the spread of Marxism is analogous to the expansion of Christianity and Islam.[25] "Some argue that Marx is heir of the tradition of the great Jewish prophets, thundering forth at mankind . . . But Marx received that tradition in its Lutheran form, as a result of being raised a believing Christian. Marx, needless to say, did not remain a believing Christian, any more than Luther was a forerunner of communism . . . What they do share . . . is a rhetorical structure, namely the characteristic articulation of the apocalyptic tradition that moves step by step . . . from the original condition of domination and oppression to the culmination of perfect community."[26] Although he became a militant atheist, "a scoffer at the 'union with Christ,'" the function of religion, its place in our psychology, remained of central importance to Marx, and this is why he found Hegel and Feuerbach so attractive. Hegel didn't say as much explicitly, but Marx felt that he—and all of mankind—had been taken in by religion. And he thought he had advanced on Hegel when he said, in "The Eighteenth Brumaire of Louis Bonaparte" (1852), that "Men make their own history; but . . . not . . . under conditions of their own choosing."

Influenced by his father, a successful lawyer, Marx originally studied jurisprudence.[27] When he graduated from the Gymnasium in Trier, his school reports show him to have been "well grounded" in knowledge of the Christian faith, to have an aptitude for ancient languages, less for French and physics. (In fact, Marx was eventually able to write fluently in French and English as well as German.)

Inside a year or so at Bonn University he had changed course, to philosophy and, at his father's insistence, moved to Berlin. His letters to his father show that he was much influenced by Hegel and that he saw his own life in dialectical terms. He had his inner struggles but thought of them as "logical" for a man in his historical and social position. Immersed in these struggles, he came to know some of Hegel's other disciples and joined the Doktor Club composed of Young Hegelians.[28] It was there that he met Bruno Bauer and his brand of radicalism.

This radicalism was worn uneasily at times. The woman Marx married, Jenny von Westphalen, represented his first success in life. By any standards, she was a catch. In one of his letters, Marx wrote: "I am asked daily [this was 1862] on all sides about the former 'most beautiful girl in Trier' and 'Queen of the ball.' It is damned pleasant for a man when his wife lives on in the imagination of a whole city as a delightful princess."[29]

Marx insisted that on her calling card Jenny use the words "*née* von Westphalen."

The marriage endured—and she was a great practical help to him—but it went through a bad patch around 1850, about the time their first child died and when, to escape, he decamped to the British Museum, in the evenings seeking consolation with another woman, Helene Demuth, the Marxes' servant. The following year, she gave birth to an illegitimate son, Marx's role being kept secret at the time and only made public by chance much later. Engels accepted paternity of the boy and Marx never acknowledged him. (Engels told all this to Marx's daughter, Eleanor, on his deathbed.)[30]

Marx's affiliation with Bruno Bauer and other leftist Hegelians ended his chances of becoming a university teacher. However, he proved himself an able journalist with the *Rheinische Zeitung* which, under his direction, doubled its circulation. In the 1840s, industrialization was beginning to appear in Germany, and social and economic issues loomed larger and larger and were growing more complex, as they had done in England during the previous century and as Engels and Marx both realized. Socialist and communist solutions (much the same at the time) were on everyone's lips in advanced circles. But Marx had not yet embraced these theories. In 1842, in a famous article he wrote on the theft of wood, he defended—mainly in legal terms—the traditional rights of the peasants to the gathering of dead wood against the growing need of industry. Private property was yet to become his central concern.[31]

Gradually—and not so gradually at times—Marx grew irritated with the interference by the government censor in regard to the *Rheinische Zeitung*, and in March 1843, he resigned just as the paper was being closed down. His career as a full-time journalist had lasted barely a year, and he now began his life of exile as a professional revolutionary.

He went first to Paris. He thought he could continue as a journalist, agreeing with Arnold Ruge to serve as a coeditor of a new periodical. Titled the *Deutsch-Französische Jahrbücher* (the German-French Annuals), it was intended to be an international outlet. However, the *Jahrbücher* appeared only once, in a double number but without any French authors, so it was scarcely annual and scarcely international. That one issue, nevertheless, contained three seminal articles, two by Marx, "On the Jewish Question" and the introduction to "Contributions to the Critique of Hegel's *Philosophy of Right*," together with Engels's "Outlines of a Critique of Political Economy." Marx had been impressed by Engels when the latter came through Paris, when they spent ten uninterrupted days in each other's company, "thus laying the basis for the most successful and significant intellectual collaboration of perhaps all times."[32] This was no thanks to the French who, despite their own revolutionary credentials, expelled Marx and Ruge in January 1845 and closed down their periodical.

Undeterred, Marx and Engels founded the German Correspondence Committee, intended to keep communists in different countries in touch with each other (a forerunner of future Communist Internationals). The following year they organized a German Workers' Society and helped with the League of the Just, a radical secret society. Marx was growing increasingly proactive. In addition to the activities described above, he was helping to arm the workers in Brussels for revolution, making use of his father's legacy. He was found out and expelled and moved back to Paris, then to Cologne, where he founded yet another new periodical devoted to the revolution.

The revolution that had been expected (in some quarters) throughout "the hungry forties" finally erupted in 1848 in a number of cities but soon petered out. As the failures mounted, and the conservatives regained the upper hand, Marx was arrested and tried in Cologne for subversion. A brilliant speech won over the jury and he was acquitted. In May 1849 he was in trouble again and was expelled from Prussia, his new periodical being closed down. He tried Paris one more time, was expelled again and, in the summer of 1849, "acknowledging the failure of the 'deed,'" crossed the Channel to London. He remained there till he died, never losing his hunger for revolution.[33]

Marx's New Psychology

Marx thought of himself as a democrat but the Swiss historian Jacob Burckhardt condemned him as one of the "*terribles simplificateurs*" of history. To an extent, Burckhardt was right. Marx gave almost no regard in his writing to the protection of individual rights, assuming that in communist society there would be no need for such a device. "He does not move in the tradition of a John Locke, a James Madison, or a John Stuart

Mill, with their concern for a system of checks and balances on the human proclivity to power; their definition of 'liberty' and Marx's is far apart."[34]

Although he was a man without a country, Marx was a very German writer. Much influenced by Hegel and his examination of human self-alienation imposed by religion, Marx chose criticism—the German scholarly method—to examine alienation in *this* life. His main focus was Germany and the hoped-for revolution in that country. Germany, he insisted, though backward practically, was "ahead in thought." The failed revolutions of 1848 played their part in his thinking: one could not expect a revolution from the bourgeoisie—they were simply not up to playing their historical role. A new actor—or hero—must be found. This new actor was to be the proletariat. "A class must be formed," Marx declared, "which has *radical* chains, a class in civil society which is not a class of civil society . . ." He was only too well aware that the proletariat was "only beginning to form itself in Germany." "Only if the bourgeoisie would play its assigned supporting role, be a proper villain, would the new heroic class develop . . . For one class to be the liberating class *par excellence*, it is necessary that another class should be openly the oppressing class."[35] These sentiments are crucial to understanding Marx.

It was while he was in Paris in 1844 that Marx finally came to grips with English classical economic theory and began to use a material basis for his critique of Hegel, though we should never forget that it was Engels who exposed the dismal actuality of the factory system in Manchester, also in 1844, in the *Condition of the Working Class in England* (published 1845). While Engels concentrated on the grime of Manchester, Marx spent his time analyzing Adam Smith. Marx's key section is where he sets out to show that the increased wealth of society means inevitable poverty and degradation for the individual. (Smith himself wasn't deaf to this threat, though on balance he thought that the benefits far outweighed the drawbacks.) Marx, however, thought that the pursuit of self-interest by employers would always win out and would distort the market.[36]

Marx was always a philosopher as much as an economist. His basic contention was that the worker becomes "all the poorer the more wealth he produces." Marx insists that the worker is poorer "even if better paid," because of an increase in alienation. The worker has become impoverished *as a human being*. And so he developed the concept of alienation, arguing that it originated in labor and had four defining aspects: (1) labor is no longer the worker's own under capitalism, it is an alien entity, dominating him; (2) the very act of production alienates the worker from his own nature—he becomes less than a man; (3) the needs of the market—and of the factory—estrange men from other men; and (4) from his surrounding culture. Marx believed these forces of alienation were producing a new psychology.[37]

In 1845 Karl and Jenny were living in Brussels, Jenny pregnant with Laura. Engels moved next door and the two men made a six-week visit to England, mostly in Manchester, observing and reading. Back in Brussels they embarked on *Die deutsche Ideologie* (*The German Ideology*), a book that never found a publisher in their lifetime and was abandoned in 1846 (it finally appeared in 1932). Though disappointed, Marx later felt that the book had served its purpose—helping the two of them achieve a measure of self-clarification. "He was too modest," says Bruce Mazlish, "the theses of *The German Ideology* were to gnaw away at the foundations of capitalism."[38]

Marx's first achievement was to write as if he had discovered a new science, one that revealed a new stage in mankind's development, a new level of Hegelian self-consciousness. To "make history," Marx argued, men must live, and that meant they must satisfy their needs. In the industrial stage, a certain mode of cooperation is required, a certain set of social arrangements, and this mode of cooperation has consequences. He gives credit to the French and English for first grasping that history is the history of industry and exchange, making economic history central. He dismissed political history, there was no social contract as such, à la Rousseau: only economic relations "tie man to man." "Such a view marks a profound revolution in political science."[39]

Marx also argued that this financial division of labor underlies "the emergence" of the state. The state offers what is in effect an illusory communal life.[40] Families and classes exist, offering some identity, but "it follows from this that all struggles within the State, the struggle between democracy, aristocracy, and monarchy, the struggle for the franchise, etc., etc., are merely the illusory forms in which the real struggles of the different classes are fought out among one another." Political life is but "a veil" for the "real struggles" based on the division of labor and private property, and this is a further cause of estrangement. This leads Marx to a famous passage addressing the ruling *ideas* in a society: "The ideas of the ruling class are in every epoch the ruling ideas: i.e., the class which is the ruling *material* force of society, is at the same time its ruling *intellectual* force." Because of this, the alteration of men "on a mass scale" can be achieved only by an act, a *revolution*. "Only in the activity of revolution itself [does] man make himself into a new man, cleansed and purified." The division of labor, private property, and intellectual self-understanding of the state are put together "in one arching synthesis."[41]

The German Ideology was followed by the *Manifest der kommunistischen Partei* (*The Communist Manifesto*), in 1848, which was even more aggressive in predicting the coming revolution. The League of the Just had been formed in 1836 by German radical workers living in Paris. It was a small secret society dedicated to revolution in Germany. After an unsuccessful uprising in 1839, most of its members left Paris for London where, in 1847, the society's name was changed to the Communist League. At their annual congress in 1847, the league—riven by factions—commissioned Marx and Engels, recent adherents, to draft a manifesto.

Engels did most of the work on the draft but then Marx realized this was the perfect vehicle to make known their views to a wider world. In imposing his vision on Engels's draft, he made the *Manifesto* a classic "confession of faith."[42]

He began, famously: "A spectre is haunting Europe—the spectre of Communism." At the time, says Bruce Mazlish, there were perhaps between twenty and one hundred communists in London, yet Marx treats them as though they are "the only alternative to the status quo." After this stirring propaganda, Marx provides a grand sweep of history, informing us, with the certainty of a scientist, how the bourgeoisie rose in opposition and, at the expense of the feudal aristocracy, transformed the technology of production and its system of financing, expanded and transformed the market, and created a different civilization, based on international trade and exchange. In doing so, skills came under threat, and so the bourgeoisie "created the conditions of its own doom." The requirements of the bourgeoisie called into existence the modern working class—the proletarians. "The bourgeoisie has simplified the class struggle, which Marx claims to be 'the history of all existing society,' into a final Manichean struggle of only two classes: the haves and the have-nots, the capitalists and the proletarians."[43] The conflict is almost biblical in its simplicity.

Given the horrors that have gone on in—or been attributed to—Marx's name, it only seems fair to point out that, aside from the abolition of property in land and the right of inheritance, Marx's list of practical measures seems hardly radical today: a progressive or graduated income tax; centralization of credit in the hands of the state; nationalization of communication and transport; the combination of agriculture with manufacturing industries; free education of all children. These measures were perhaps overlooked because of the stirring language, the sheer firepower of the drama he claimed to see unfolding all about him, and for his conclusion: "WORKING MEN OF ALL COUNTRIES, UNITE! . . . Let the ruling classes tremble at a Communist revolution. The proletarians have nothing to lose but their chains. They have a world to win."[44]

And then, there was *Das Kapital*. There is no question, says Bruce Mazlish, that this book was Marx's greatest achievement, or that it is a great work. "The question is: in what sense is it a great work?"[45]

Its central ideas are: (1) a labor theory of value, (2) a theory of surplus value, (3) a theory of capital accumulation and its consequences, and (4) a law of "increased misery." The idea of a labor theory of value was initially conceived not by Marx but by Adam Smith—in fact, it was commonplace in the early nineteenth century and made Marx seem very up-to-date. However, no sooner had *Das Kapital* been published than another revolution overtook economics: this was the so-called marginal utility theory, a mathematical approach that undermined the labor theory of value and is one reason modern economists pay so little attention to Marx's economic theories.

Marx held that all value was created by the laborer, who wasn't allowed to keep the full value of his work but had part of it—the larger part—expropriated from him, subjecting him to a life of misery and degradation.[46] The central problem, he said, was profit. "If the capitalist gets back the capital he started with, and the laborer gets the rightful value of his work put into the commodity, how can 'profit' be extracted from the productive process?" Only by paying the worker much less than what he is worth can the capitalist secure his profit. For Marx this counted as exploitation and "Workers who never read *Capital* nevertheless could now trust that there was a scientific underpinning to their feeling of being exploited."[47]

There were many shortcomings in Marx's arguments. For example, if surplus value arose because capitalists undervalued labor, how was it that industries using a lot of machinery, and very little labor, were frequently *more* profitable than labor-intensive industries? Marx never came up with a satisfactory answer.

He kept some of his most vivid prose for his account—and critique of—the accumulation of capital. Where does it originate? For him, it arose not from the capitalist's hard work and savings, but from "brutal confiscation, slavery and rapine . . . Capital comes into the world covered in blood from head to foot . . . accomplished with merciless Vandalism, and under the stimulus of passions the most infamous, the most sordid, the pettiest, the most meanly odious." There was no way out. Capital, he insisted, will become more concentrated, in fewer hands, added to which there will also be a long-term decline in profit as competition grows. The end result, one end result anyway, is that the law of increasing misery shows itself. Marx called it a "law," but in fact in most capitalist countries the conditions of most workers have improved.[48]

But is *Das Kapital* intended to be read as a dry textbook? Not really. "It is a passionate drama, an epic poem, in which we descend into capitalism's innermost circles, go through its purgatory fires, in order to emerge at the end with a glimpse of its downfall and a promise of future salvation. It is Marx's imagery . . . which grips us."[49]

The other flaws in Marx's theories are now well known. The most important is his assumption that all political power is in the hands of the capitalist, to the exclusion of the worker, that "bourgeois democracy is a sham." Yet the very parliamentary inquiries into working conditions that Marx himself used to damn the capitalists actually produced important improvements (slow in coming, it is true) and in 1867, the year *Das Kapital* was published, urban workers were given the vote in Great Britain. Steadily, if slowly, in the bourgeois European democracies, a "welfare state" came into being. We are now at a sufficient distance from *Das Kapital* to be able to generalize that, when workers have had a clear political voice, they have never voted to overturn the capitalist industrialist system, rather for a greater share of its "surplus." Even this is to misread and misunderstand the purpose of *Das Kapital*, which was, as Engels saw, the Workers' Bible, part of a campaign to kindle revolution.

Despite these shortcomings, Marx's key insight was to grasp that the developing productive forces of a society create new social relations, binding economics and sociology together.[50] Moreover, with his Hegelian background, he offered an evolutionary perspective.

On a personal level, Marx was a great fighter, incorrigibly struggling for a better world. Marxism aside, Marx the man stressed—as Francke, Herder, and Hegel had stressed before him—that a society's ethos and values are *created* by its members. This is the other German ideology and one that we still espouse, despite everything that has happened.

"The Most Learned Man in Europe"

Marx's daughter, Eleanor, summed up Engels's character in 1890, when he was seventy: "Next to his youthful freshness and kindness, nothing is so remarkable about him as his many-sidedness. Nothing remains foreign to him," she wrote, "natural history, chemistry, botany, physics, philology . . . political economy and last not least, military tactics."[51] Theodor Cuno—founder of the Milanese section of the First International, and later a member of the American Knights of Labor, whose life had been saved by Engels when Cuno had nearly drowned swimming in the sea for the first time in 1872—also remarked of his savior that "His brain was a treasury of learned knowledge." Marx too "was proud of Engels." In fact, Marx regarded Engels as "the most learned man in Europe."[52]

Friedrich Engels was born in 1820 in the Rhenish town of Barmen, now part of Wuppertal. His father was a staunch Pietist, but Engels himself was more impressed by the incipient industrialization of the area, struck by the fact that, along the Wupper, a "vigorous, hearty life of the people," with traditional folk songs, had been lost, unlike many other places in Germany. He left the Gymnasium before graduating, to work in his father's office. He was sent to other company offices, all the while enjoying himself—riding, skating, and fencing; he joined a choral group and even tried composing.[53] In his reading he was influenced by Schleiermacher, Fichte, and David Strauss's *Life of Jesus*, which provoked his loss of faith. Then he encountered Hegel, which struck him like a religious conversion. He formed ties with the Young Hegelian circle of which Marx formed a part and in 1842 released "Schelling und die Offenbarung" (Schelling and Revelation), a pamphlet that attracted attention as far afield as Russia. Following that, he began writing more regularly for the newspapers, though his father suggested he spend some time in the company's Manchester office to flesh out his commercial acumen.[54]

In England, where, says Tristram Hunt, he found the "zest" for life much less than on the Continent, Engels nevertheless met his future common-law wife, Mary Burns, apparently a domestic servant, who introduced him to proletarian circles in Manchester, contacts that formed the background to his book *Die Lage der arbeitenden Klasse in England* (*The Condition of the Working Class in England*).[55] Returning to Barmen in 1844, Engels traveled via Paris, where he met Marx. Reaching home, he wrote *The Condition of the Working Class*, which David McLellan has called "a pioneering work in the relatively modern fields of urban geography and sociology." We now know it to be a one-sided picture of the English working classes, exaggerating their prosperity before industrialization and propagandizing the impact of the machine. Nonetheless, the text was vivid. "Perhaps no other book but Elizabeth Gaskell's *Mary Barton* provides so graphic a description of the real evils the English working class suffered in this period."[56]

By the time of the revolutions of 1848, Engels and Marx had collaborated on *Die heilige Familie* (*The Holy Family*), *Die deutsche Ideologie* (*The German Ideology*) and the *Communist Manifesto*. In the revolution, while Marx went to Paris, Engels fought as a "line" soldier in "the last stand" of the democratic revolutionaries in Baden against Prussian soldiers, who had an easy victory. Engels actually took part in four battles, "discovering that he was more courageous than he had dared hope." But then both Engels and the Marx family went into exile in England, the former taking up employment again for Ermen and Engels in Manchester, enabling him not only to support himself but to supplement Marx's meager income. Later, as he earned more, Engels's support of Marx "became quite substantial." Apart from his collaboration with Marx, Engels—the former soldier—wrote on military affairs.

At this point, and despite his experiences on the barricades, Engels was by no means a revolutionary. He rode to hounds, joined the Albert Club, named for Queen Victoria's German consort, the membership of which was half English and half German, and, from 1860, was awarded a share of the profits of Erman and Engels, "adding to the irony that Marx's principal source of income, at least at this time, was capitalistic."[57]

In 1870 Engels moved back to London and rented a house within walking distance of Karl and Jenny. He now found time to write his own books, based on his wide reading. These included *Der Ursprung der Familie, des Privateigentums und des Staats* (*The Origin of the Family, Private Property and the State*; using Lewis Morgan's well-known *Ancient Society*, which argued that production was the key to progress from savagery to civilization).[58] The last of his own important works, *Ludwig Feuerbach und der Ausgang der klassischen deutschen Philosophie* (*Ludwig Feuerbach and the End of Classical German Philosophy*), was published in 1886. This clarified his own and Marx's relationship to Hegel and Feuerbach; he reiterated Hegel's argument that truth develops over time "without ever arriving at an absolute conclusion" and Feuerbach's idea that outside nature (which includes human beings), there is nothing, that philosophy and religion are "simply the reflections of humans' own natures."

Engels's erudition is perhaps insufficiently appreciated now. He had a wider range of interests than Marx, spoke and wrote English and French as well as he did German and as well as Marx, and was eventually able to speak Greek, Latin, and some Italian, Spanish, and Portuguese.* He wore his learning lightly but it stood him in good stead, in that he predicted that Wilhelm II, grandson of Wilhelm I, would commit many blunders and prove catastrophic for Germany. He was not blind to the fact that the English working classes had enjoyed increased prosperity, although unequally, as a result of the British Empire, and he thought that explained why there had been so little socialism in England since Owenism. He thought that with the decline of empire and the abolition of the English monopoly, together with the rise of American commercial success, the English working class would lose its privileged position and socialism would reemerge. As with so much else, he was right.[59]

Engels was hardly less interesting than Marx, and there has always been a debate as to who was first with the main ideas they jointly evolved. With Engels surviving Marx by more than a decade, and editing the second and third volume of *Das Kapital* after Marx died, it is perhaps no surprise that

* He is even credited with knowing certain Irish dialects.

J. D. Hunley, in a recent critique, makes a powerful case for saying that there was in fact very little difference between Marx and Engels in their materialistic understanding of history, economics, and politics, that the "Principles of Communism," which Engels wrote, is hardly different from the *Communist Manifesto*, which they both produced, though the latter is slightly more radical.[60] Engels, Hunley says, may have been marginally less enthusiastic for the revolutionary cause than Marx, though this could have had more to do with the fact that Engels lived longer, surviving to see the gathering strength of the Social Democrats in Germany. In the preface to the English translation of the first volume of *Das Kapital*, both Engels and Marx took the view that in England revolutionary changes might occur by peaceful and legal means, and both agreed that this was preferable. But Engels, no less than Marx, was convinced that in some countries force would be necessary.[61]

Both Engels and Marx retained a form of Hegelianism to the end, believing that history was the result of impersonal forces, yet shaped by men. Engels went on record as saying specifically that it was "laughable" to explain *everything* in history in terms of economic factors. "History, to an extent, rests on the unconscious of all concerned. . . ."[62]

Marx's comment on Darwin's *Origin of Species* was perhaps revealing: "This is the book that contains the natural historical foundation for our view." Note the use of the word "our." But there is little evidence to suggest that Engels's contributions to at least the first volume of *Das Kapital* were more financial and critical than substantive. Volumes two and three were different matters, because of the messy nature of Marx's manuscripts. We don't know what additions Engels did make but historians are agreed there is no evidence of any intent to deceive. Tristram Hunt, in his 2009 biography of Engels, says that "Marx's bulldog" tried to enfold him in a "scientific turn," that the bulldog was "mezmerized" by the scientific advances of the nineteenth century and sought to position their socialism within this context. Maybe so, but their collaboration was always one of mutual respect, a decisive factor in making that collaboration "the most significant intellectual partnership of all time."[63]

12.

German Historicism: "A Unique Event in the History of Ideas"

Thanks to Johann Herder, history became the basis of all culture. Development and evolution became central to all understanding. This is Herder, in *Ideen zur Philosophie der Geschichte der Menschheit* (*Ideas for the Philosophy of History of Humanity*). "The purpose of our existence . . . is to develop this incipient element of humanity fully within us . . . Our ability to reason is to be developed . . . our finer senses are to be cultivated . . . the task incumbent upon each one is to develop his own unique personality to the fullest." Fichte, Schelling, and Hegel all argued in favor of the basic uniqueness of individuals and nations in history. For them, as for Wilhelm von Humboldt, the purpose of man's life becomes emphatically not "happiness," but the fulfillment of his potentialities.[1]

The most important element in the change from an Enlightenment to a historicist outlook was the chain of political catastrophes and recoveries acting on the German intellect between 1792 and 1815. To begin with, the educated middle class in Germany had by and large welcomed the French Revolution. But a profound unease settled in after the Terror, leading to widespread doubts about the doctrine of natural law. This was intensified by the Napoleonic occupation, reinforcing nationalistic feeling and identifying Enlightenment values with the detested French culture.

The reforms stimulated by these events changed German attitudes toward history in three ways.[2]

One, the Enlightenment belief in universal political values was shattered. German opinion now took the view that all values were of historical and national origin and that foreign institutions and ideas could not be transplanted unchanged onto German soil. History, not abstract (French) rationality, was the key. Two, the concept of the nation was transformed. Herder had been reasonably cosmopolitan, seeing a richness to life in the way nations differed. By the time of Fichte's *Reden an die deutsche Nation* (*Addresses to the German Nation*) in 1806, however, Germans were presented as a unique and original nation that, unlike the French, had not lost touch with its original genius. The French were now regarded as a "superficial nation" who, as Humboldt put it, lacked "the striving for the divine." Three, the role of the state was also transformed. Herder dismissed states as artificial entities, "detrimental" to human contentment. But after him, the state was more and more seen in power-political terms. In an 1807 essay on Machiavelli, Fichte argued that in regard to the dealings between states, "there is neither law nor right except the right of the stronger." Fichte, and Ranke after him, developed the view that "might is right."* [3]

The new attitude was nowhere more in evidence than with Wilhelm von Humboldt, who argued that there is indeed a purpose to world history: "The ends of life cannot be abstract, we must leave creativity to lead where it will . . . There is no higher purpose, no super-pattern."[4] Again, this is not an exceptional thing to say now, but in the world between doubt and Darwin it was radical and, to many, dangerous.

For Humboldt, as we have seen, the highest ethical good is to be found in *Bildung*, the development of the individuality and uniqueness of each man or woman.[5] It is a view with profound consequences. On this account, political, cultural, and historical understanding is quite different from physical nature. "Lifeless" nature may be understood by means of abstraction, and the mathematical regularity of its behavior, whereas real living forces can be known only through the energies they express, reflect-

* This corresponds with Mephistopheles' statement in *Faust II*: "For everyone knows that might is right. Not 'how' but 'what' is all one asks!" "Man hat Gewalt, so hat man Recht / Man fragt ums Was, und nicht ums Wie."

ing their inner nature. Without doubt *some* uniformities do exist in man's nature—"Without them, no statistics would be possible." But the existence of free creativity makes historical prediction impossible. Research becomes important precisely because it is itself creative. And since history is nothing but a mass of individual wills, history must be an "exact, impartial, critical examination of events."[6]

As a result, says George Iggers, Germany's historians shared a particular concept of history throughout modern times. "With much more justification than in France, Britain or the United States, we may speak of one main tradition of German historiography." This centered on the character of political power, the conflict of the great powers, and a marked emphasis on diplomatic documents, with a consequent neglect of social and economic history and of sociological methods and statistics.

Friedrich Meinecke, Ernst Troeltsch, and other German historians have recognized that historicism broke free from the 2,000-year domination of the theory of natural law, with its understanding of the universe as consisting of "timeless, absolutely valid truths which correspond to the rational order dominant throughout the universe."[7] This was replaced by a conception of the fullness and diversity of man's historical experience. "This recognition, Meinecke believes, constituted Germany's greatest contribution to Western thought since the Reformation and 'the highest stage in the understanding of things human attained by man.'" Moreover, according to Iggers's interpretation, Troeltsch and Meinecke maintained that (non-German) European thought remained committed to natural law throughout the nineteenth and into the twentieth century. This difference, he says, helped lay the basis for the "deep divergence" in cultural and political development observed between Germany and "Western Europe" after the French Revolution. Another Sonderweg.[8]

"An Epoch in the History of European Intelligence"

The German historians also stimulated change in more practical ways. Consider first something we take for granted now: access to archives, and the freedom to publish whatever is discovered there.[9] G. P. Gooch reminds us this was by no means always true and that the first scholar of heroic consequence in modern historiography, the man who improved its standing to the status of an independent discipline and who inspired many later historians, was Barthold Niebuhr (1776–1831). The "rather mawkish" son of the noted Danish traveler and explorer Carsten Niebuhr, he was introduced to the great classics of other civilizations by his father. He pursued law and philosophy at Kiel but he also studied history and knew as early as nineteen what he wanted to be: "If my name is to live, it will be as an historian and publicist, as a classicist and philologist."[10] He spent time in Denmark and Berlin, in public administration, but in 1810 he was offered a professorship at the University of Berlin, and it was there that he began his mammoth work on Rome. In the middle of the fighting he published two volumes of a book that we now recognize as inaugurating the systematic study of Roman history. Niebuhr always claimed that his time in public administration had provided him with an understanding of history that "no previous historian" had experienced and gave him a perspective that, he said, showed that history is more an account "of institutions than of events, of classes than of individuals, of customs than of lawgivers." This was a crucial shift in emphasis but not his only one, his other achievement being to identify the sources of early Roman history and assess their credibility. He had thoroughly assimilated Wolf's methods and results in the *Prolegomena ad Homerum*, which convinced him that the history of early Rome could be found in a critical examination of its literature. Goethe was impressed and so was Thomas Babington Macaulay in Britain, who declared Niebuhr's book(s) on Rome "an epoch in the history of European intelligence."[11]

Law as an Achievement of Civilization

In the German context, an important aspect of this developing historical consciousness occurred in the realm of law. Two Berlin professors were crucial in showing how laws were not "God given," as many people thought, but had evolved. Karl Friedrich Eichhorn studied law, political science (*Staatswissenschaft*), and history at Göttingen. His original intention was to be a practicing lawyer, but after being offered a chair at Frankfurt an der Oder, he turned to research and writing. The first volume of his *Deutsche Staats- und Rechtsgeschichte* (*History of German Law and*

Institutions) appeared in 1808 when he was twenty-seven, and it earned him an invitation to Berlin.

Eichhorn's aim was to show that state and public law was "the product of all the factors that influence the life of a nation." He described the links between legal ideas and institutions, showing how both had evolved. In doing so, he helped generate a spirit of nationality, but in Berlin he became identified with the view that law—like art and philosophy—is one of the defining achievements of a great civilization.

Friedrich Carl von Savigny was a lifelong friend of Eichhorn's. Also educated at Göttingen, he published a work on certain aspects of the Roman Law of Possession in 1803 and the following year set out on a prolonged tour through the libraries of Europe. These travels provided him with a unique experience and self-confidence, so much so that when, in the wake of Napoleon's victories, there was a call for a French-style German-wide code of law, Savigny effectively opposed the idea. Now a professor at Berlin, he forcefully argued instead that law had to grow, by custom and usage, that any code "imposed" on a people would necessarily be arbitrary and do more harm than good. This view was underlined in his *Geschichte des römischen Rechts im Mittelalter* (*History of Roman Law in the Middle Ages*), the first volume of which was published in 1815, where he traced the survival of Roman law in the institutions and local customs of the towns, showing how that law had survived and even proliferated under the German "barbarians." His point was that Roman law was older than German law, that it had grown through use, interpreted by experienced jurists.

In the same decade that saw the earliest works of Niebuhr, Savigny, and Eichhorn, Jacob Grimm founded the science of Teutonic origins.[12] Born in Hesse in 1785, Grimm studied law at the University of Marburg, where Savigny's lectures awakened in him an interest in history; it was in Savigny's library that Grimm first encountered early German literature, which he took to be an "uncultivated field." When Savigny made his tour of the libraries of Paris, Grimm accompanied him and began to collect his own material. This gave him the idea of collecting German sagas and fairy tales.

The first volume of *Kinder- und Hausmärchen*, written with the help of his brother Wilhelm, appeared in 1812 and made them famous. "More than any other part of the Romantic output, the *Märchen* became part of the life of the German nation."[13] The Grimms believed that the earliest history of all peoples was the folk sagas and that history had neglected them because they contained no "facts." Jacob was one of those who believed that, to the contrary, sagas contain more historical substance than anyone thought. He likened medieval literature to medieval cathedrals, the "anonymous expression of the soul of a people." In his *Deutsche Mythologie*, where he added oral history to written stories, he described a world of swan maidens, pixies, kobolds, elves, dwarfs, and giants, all retreating as Christianity spread across Europe.[14]

The most enduring aspect of this scholarly nationalism occurred when several German historians got together to establish a proper German history, for which they determined that a complete record of the archives and sources was needed. In 1819, led by Heinrich Friedrich Karl vom und zum Stein and several other professors at Berlin, the Gesellschaft für Deutschlands ältere Geschichtskunde (Society for the Study of Early German History) was founded at Frankfurt and a journal, *Monumenta Germaniae Historica*, established. When Georg Heinrich Pertz, the editor and director, retired half a century later, no fewer than twenty-five "stately" folios had been published. Again, this achievement was so seminal that we take it for granted now.[15] Around the *Monumenta* other works of German history reflected and encouraged the emerging nationalism: Heinrich Luden's *Geschichte des teutschen Volkes* (*History of the German People*; twelve volumes, 1825–37), Johannes Voigt's *Geschichte Preussens* (*History of Prussia*; nine volumes, 1827–39), Johann Friedrich Böhmer's *Fontes Rerum Germanicarum* (1843). Then came Ranke.

"The Greatest Historical Writer of Modern Times"

Leopold von Ranke, according to G. P. Gooch in his book on German historians, "was beyond comparison the greatest historical writer of modern times, not only because he founded the scientific study of materials and possessed in an unrivalled degree a judicial temper, but because his powers of work and length of life enabled him to produce a larger number of first-rate works than any other writer. It was he who made German [historical] scholarship supreme in Europe; and no one has ever approximated so closely to the ideal historian."[16]

Ranke studied theology and philology at Leipzig, where he read the Old Testament in Hebrew. He was not, as should now be clear, the first to use critical methods but, says Gooch, he, more than anyone, popularized them "and showed what could be done with them." He made his mark with his first book, *Geschichte der romanischen und germanischen Völker von 1494 bis 1514* (*Histories of the Romanic and Germanic Peoples from 1494 to 1514*) in 1824, when he was twenty-nine. But it was really his appendix, "Zur Kritik neuerer Geschichtschreiber" (In Criticism of Modern Historians), that became famous and, if anything, was regarded as more important than the main text.[17] In this appendix, Ranke applied Niebuhr's critical principles to modern sources.

Ranke also became famous for his "discovery" of archives. This, as we can now see, is something of an exaggeration. Several other people were using the archives at the same time as, or even before, Ranke, but here too it was his *use* of the materials that caught the eye. In Berlin he encountered several volumes of the reports of the Venetian ambassadors from the second quarter of the sixteenth century—the very zenith of Venetian power. His crucial insight was to see that these reports told a very different story from that revealed in the memoirs of the day, written by the leading players themselves or by observers of one kind or another, all of whom had their own axes to grind and therefore gave but a partial view of affairs.

These "objective" ambassadorial reports had a profound influence on Ranke and determined the kind of history he wrote. Sources were crucial but, for Ranke, his most important task was the "great, comprehensive narrative." He agreed with Humboldt that the historian "proceeds like a poet, who after having grasped the material has to create it anew, drawing on his own powers."[18] This is still the modern approach.

Ottomans and the Spanish Monarchy of the Sixteenth and Seventeenth Centuries was the first to use this approach and the first volume of the series *Fürsten und Völker von Süd-Europa* (Princes and Peoples of Southern Europe). Ranke's first aim was to make the behavior of the main players explicable in terms of contemporary diplomacy, trade, finance, and administration. Again, we take this approach so much for granted now that we forget it started with Ranke. Other books reinforced this approach. Because he was a Protestant, the Vatican archives were off limits but he discovered in the archives of some of the great papal families (notably the Barberini) sufficient material for his magisterial *Die römischen Päpste* (*History of the Popes*; volume one, 1834). These archives enabled him to treat the Papacy like any other institution in the development of Europe, but the fulcrum of the book is the Counter-Reformation, of which Ranke was the first authoritative interpreter. The attempt to revive spiritual life and the foundation of the great orders are brilliantly evoked. He had a further coup with *Deutsche Geschichte im Zeitalter der Reformation* (*German History in the Time of the Reformation*, 1839–47). After his work on the popes, Ranke felt he should write the history of Protestantism to put alongside his account of Catholicism and here too he discovered a mass of correspondence, this time in connection with Charles V in Brussels.[19]

Ranke's most typical project was *Neun Bücher preussischer Geschichte* (*Nine Books of Prussian History*), appearing in 1847–48. A study of Friedrich the Great based on the archives in Berlin, this book was essentially an examination of the rise of a great power, depicting Frederick as the foundational figure in the Prussian administrative machine. Ranke employed the same method as he did with his history of the popes, allowing no bias to taint his judgment and, in order to obtain a better grasp of the king's motives, he betrayed no hostility toward Austria. This is Ranke as "the father of value-free history," doing his best, as he famously put it, to tell history "as it really was" (*wie es eigentlich gewesen*), avoiding imposing modern ways of thinking on historical personages. This approach he employed with his subsequent works on Machiavelli, the history of France, and the history of England. He wrote "as a European" (his wife was English), his conviction being that European history was essentially about the rise and rivalry of the great powers, what would come to be called *Realpolitik*.

The sheer number of masterpieces set Ranke apart from other historians. (Perhaps no one has ever had as much historical knowledge.) Though his books deal with the great tendencies of whatever age he was writing about, he also acknowledged the importance of the individual actor: "General tendencies do not alone decide; great personalities are always necessary to make them effective."[20] He assumed "a divine order of things" that "cannot be proved but felt." This order manifests itself in the "sequence of periods." So Ranke's first achievement was to divorce the study of the past from the passions of the present. Before Ranke, historians assumed memoirs and chronicles to be the best authorities. After him, all scholars accepted that nothing less than the papers and correspondence of the actors in immediate contact with the events they describe was a sine qua non.[21]

In criticism of Ranke, George Iggers says that, although he was interested in power, he never considered the role of evil in history. Convinced that states were "ideas of God," ends in themselves, his perspective was that of governments. Arguably, this led Ranke and his followers to underestimate economic and emerging sociological factors. A potent side effect was that his approach helped foster a growing nationalism.[22] Historians were better scholars after Ranke, but in Germany in particular they were also more politically involved and active.

The German Idea of Freedom

Ranke's approach, we can see with the benefit of hindsight, was not without its consequences. George Iggers says that three ideas shaped German history but also the wider picture.

The state as an end in itself and the concept of the Machtstaat.

The German conception of the state always had an aristocratic and bureaucratic bias plus an appreciation of the cultured, propertied middle class as the backbone of the country. German historians maintained a much sharper distinction between government and governed than was true in France or in Britain. The state was seen as an "individual," an end in itself.

The rejection of normative thinking.

For Ranke, the main task of the state is to secure "the highest measure of independence and strength" among other powers, so that the (German) state will be able to fully develop its innate tendencies. All domestic affairs are subordinated to this end, from which it follows that "The state cannot sin when it follows its own higher interests." Sheer power becomes one and the same as morality.

Anti-Begrifflichkeit, the rejection of conceptualized thinking.

Generalizations and overarching theories in history and the cultural sciences are of limited value. History, "the area of willed human actions," requires understanding, but this is not accomplished by abstract reasoning, rather "by direct confrontation" with the subject and acknowledgment of its individuality. It follows that all historical understanding requires an element of intuition. The irrational aspects of life need to be taken into account.[23]

These notions meant that German historians moved in a world of their own, remaining largely unaffected by the great transformations of the period 1848–1914, in particular the great social and economic changes brought in by industrialization. History for them remained primarily the interplay of the great powers, and the primary solution for domestic social and economic problems was an expansive foreign policy, the main means of which was a strengthening of the nation.

It is in this sense that, as Iggers has said, German historicism is a unique event in the history of ideas. Besides its substantive scholarly achievements, its effect on politics and on Germany's self-understanding was remarkable. The increasing achievements of the natural sciences throughout the nineteenth century did not affect this. Only the disasters of the twentieth century brought change.[24]

The other important consequence of the Ranke mind-set was how it affected notions of freedom. Freedom, the historians insisted, can only be achieved within and through the state, and this was closely tied to the political and social outlook of a particular class, the academic *Bildungsbürgertum*. Historicism thus provided a theoretical basis for the traditional political and social structure of nineteenth-century Prussia and Germany. This represents a major cultural divergence of Germany from "the West." For German historians, the reformed Prussian monarchy represented a "high point" in the history of freedom: it was a society where the individual was fully free, *but at the same time integrated into a social whole*. This "German idea of freedom" was a core belief, at least among Humboldt's Bildung-loving Bildungsbürgertum and flatteringly contrasted with "the atomistic ideas of 1789."[25]

13.

The Heroic Age of Biology

On the evening of March 11, 1890, hundreds of men in white tie and tails gathered for a gala dinner at the Berlin City Hall. The chandeliered room was lined with palm trees, and everyone of note was there, including the cream of Germany's hostesses, seated at separate tables in an arcade. It was, according to one present, "A festival of magnificence perhaps unparalleled in the history of science" and, to mark the occasion, ten speeches were delivered.[1]

Each speech honored one man, the last to speak. The dinner, known as the Benzolfest, was held to mark the twenty-fifth anniversary of a discovery he had made that brought to a head one of the great adventures of the mind in the nineteenth century and that had occurred only after half a century of painstaking inquiry. The man was August Kekulé and his discovery was the benzene ring and the belated realization that there *was* such a thing as a molecule, the smallest particle of a chemical compound that can exist, that it has a structure—a shape and a size, with specific properties depending on that structure—and that this was the basic building block of organic chemistry, the chemistry of life, the chemistry that governs biology.[2]

Organic chemistry had been invented—or discovered—seventy years earlier. It was one of three breakthroughs that made the middle years of the nineteenth century a heroic age for biology. The second was the development of fertilizers, which transformed agriculture at a time when many people all across Europe were leaving the land to work in the new metropolises, so that the demand for food had reached unprecedented levels. The third was the identification of the cell, the realization that it was the basic building block of both plants *and* animals, and that its differentiation made up the various organs of living things. Between them, these developments transformed medicine, concepts of illness and wellness, industry (dyes, fertilizers, cosmetics, drugs) and—insofar as the new discoveries explained life processes and linked inert matter conceptually to living organisms—played a major philosophical and religious role in refining our understanding of ourselves at a time when traditional beliefs were under severe threat.

The vital role played by carbon in the science of natural products—organic chemistry—was known from a fairly early date. What puzzled people was why one element out of the dozens already known should account for the amazing diversity of natural substances. This strange state of affairs helps account for the fact that so many scientists of the early nineteenth century believed that chemistry was not enough to explain the diversity and that some kind of "vital force" must be operating.[3]

The term "organic chemistry" had crept into use around 1777, though understanding was rudimentary, and early textbooks did little more than list the various substances regarded as organic: gum, saliva, urine, albumen, gelatin, and blood, which many regarded as an "impossibly complex" substance.[4]

Systematic sense was first put into this field by the Frenchman Antoine Lavoisier, who showed that several natural products—alcohol, sugars, and acetic acid (from vinegar)—contain only three elements: carbon, hydrogen, and oxygen. The two men who built on this and therefore came to personify the emergence of organic chemistry were Justus von Liebig and Friedrich Wöhler. From about 1824, for roughly three decades, von Liebig and Wöhler investigated almost every area of the new science, publishing hundreds of research papers and teaching thousands of students (8,000 in Wöhler's case). Wöhler, three years older than von Liebig, was quiet and modest, slender, and always looked much younger than he was (when he met Michael Faraday on a visit to Britain, Faraday thought he was talking to Wöhler's son).[5] Von Liebig, on the other hand, was an irascible, all-too-fallible man who, as John Buckingham says in his history of early biology, had a career disfigured by more than its share of failures, mistakes, and acrimonious squabbles (at one stage his British publishers refused to print a book of his because of the libels in the text). Nevertheless, von Liebig's achievements opened a new era in the discovery of the organic molecule.

He was born in May 1803 in Darmstadt and studied chemistry under Wilhelm Kastner at Bonn and Joseph-Louis Gay-Lussac in Paris, where analytic methods were then much more rigorous than in Germany. His breakthrough came when, on the recommendation of Alexander von Humboldt, Ludwig I of Hesse appointed von Liebig extraordinary professor at the University of Giessen. Almost immediately, he and two colleagues set up their own teaching laboratory. The twenty places were soon filled, and the founding of a new chemical laboratory at Giessen signaled the beginning of the eastward migration of chemistry across the Rhine. Using new equipment that von Liebig designed himself, which allowed much quicker and more accurate analysis, he and his students analyzed many of the more mysterious natural substances, including quinine, morphine, and strychnine, finding out in the process that their molecules contained relatively large numbers of atoms but not in simple ratios.

They also discovered the important phenomenon of isomerism, which came about when von Liebig was in Giessen studying the salts of fulminic acid, and Wöhler was in Stockholm, collaborating with Jöns Jakob von Berzelius (1779–1848), the famous Swedish chemist, and examining another acid, cyanic acid.[6] Though completely different in their properties (cyanic acid was not at all explosive, as the fulminates were), Wöhler obtained exactly the same results for his analysis of silver cyanate as von Liebig did for his silver fulminate. How could that be? The two men met in Frankfurt to compare results, and, to everyone's surprise, decided that they were *both* right. This meant that two different substances could have the *same* elemental composition. In this specific case, cyanic acid and fulminic acid each contained carbon, nitrogen, oxygen, and hydrogen in exactly the same proportions. It was Berzelius who coined the term "isomerism" to describe the phenomenon, more and more examples of which would be uncovered in the coming years.[7]

It was a phenomenon slow in being grasped partly because, throughout the eighteenth and nineteenth centuries, organic chemistry, not to mention physiology, were muddied by the concept of "vital force"—the belief that living organisms could not be explained by physical laws alone, that there must be some "special influence" at work. This view was reinforced by the sheer extent and diversity of organic substances which, it was thought, only a deity could have envisaged. As more analyses were completed, and more substances found to be made of carbon, nitrogen, and water only, the mystery deepened.[8]

It was in this intellectual and religious climate that Wöhler performed the experiment for which he will always be remembered. By treating silver cyanate with ammonium chloride he was hoping to derive the ammonium salt of his cyanic acid. However, after he had filtered off the (insoluble) silver chloride and evaporated the residual solution, he found he had "colourless, clear crystals in the form of slender four-sided dull-pointed prisms." To his astonishment, they resembled nothing so much as urea. "This similarity . . . induced me to carry out comparative experiments with completely pure urea isolated from urine, from which it was plainly apparent that [urea and] this crystalline substance, of cyanate of ammonia, if one can so call it, are completely identical compounds." In fact, the two compounds were not identical—they were isomers, but even so Wöhler's was an iconic experiment: he had manufactured a substance, urea, hitherto the product solely of animals, out of inorganic materials *and without any intervention of vital force.* "Von Liebig and his successors regarded [this] experiment as [the] beginning of a truly scientific organic chemistry."[9]

The vital force did not vanish overnight but it did now come under sustained attack, not least from von Liebig, who carried out a raft of experiments on food consumption and heat production in animals. He showed beyond all doubt that the energy that characterizes a living organism is a product of the combustion of food in the tissues, with no need for other mysterious sources such as "electricity" or "nervous energy" or "vital force."[10]

Benzene: A New Era in Chemistry

Physiological chemistry was one line of biological research. Another, unlikely as it may seem, grew out of gas lighting. By 1816, twenty-six miles of metal piping had been laid in London to carry illuminating gas to factories and for street lighting.* Early gas supplies were made, not from coal tar, as they were later, but from the more obviously organic whale or cod oil, which produced a high residue of liquid waste that would condense out, either at

* Ordinary homes would not be fitted with gas lighting until later in the century.

the works or in the pipes themselves. This waste, the so-called gas oil, was produced in sufficient quantities for the owners of the gasworks in London to send a sample to Michael Faraday in 1825 to see what it was and what it might be used for. He experimented for about ten days and found it to be a hydrocarbon—a substance containing carbon and hydrogen only.[11]

At first, Faraday called the waste "bicarburet of hydrogen," later known as benzene. At that stage, there was not the slightest inkling that benzene is the stable backbone underlying a vast series of substances that would come to be known as "the aromatic compounds."* [12]

Seven years after Faraday's identification of benzene in 1832, von Liebig and Wöhler began their second important collaboration, this time working on the aromatic compounds. In the first phase they isolated from the oil of bitter almonds a substance they named hydrobenzoyl (today called benzaldehyde), which they found to contain carbon, hydrogen, and oxygen only. But it was their next move that was to prove crucial: they performed a series of transformations—and found that treating benzaldehyde with chlorine gave benzoyl chloride, a substance that could be further transformed into benzoyl iodide by potassium iodide and so forth. This was the first demonstration of a series of systematic chemical transformations that could be carried out with related organic substances. What von Liebig and Wöhler were the first to realize was that throughout this series of transformations a sizable *backbone* of the molecular structure, which they calculated as $C^{14}H^{10}O^{2}$ (as then written), remained *unchanged*. This backbone, which they labeled "benzoyl," they called a "compound radical." The idea of a "radical," in this context, meant a collection of elements "mimicking the behaviour of a single element." Lavoisier had considered the idea, but in a much simpler form, and as it related only to inorganic examples. A new era in chemistry was opening up. This was confirmed when the Berlin chemist Eilhardt Mitscherlich performed a transformation that von Liebig and Wöhler had missed. In 1834, by heating benzoic acid with lime, he obtained "nothing other than Faraday's bicarburet of hydrogen"—benzene itself. This substance, benzene, C_6H_6, in time became revered as the truly irreducible nucleus or "radical" of the aromatic compounds.[13]

* The name is a corruption of the Arabic, *Lubān Jāwi*, or "incense of java," later corrupted by Portuguese merchants to *benjawi*, *Benjamin*, *benzoin*. Used in church incenses, it was isolated as early as 1557.

As more discoveries about the properties of the many organic compounds accumulated, it became ever clearer how exceptional benzene was. All other substances with a low ratio of hydrogen to carbon were unstable. Benzene had the same ratio of the two elements (1:1) as the highly explosive gas acetylene, but the benzene nucleus, as von Liebig and Wöhler repeatedly demonstrated, "could pass unchanged through a whole series of substitution reactions that could lead back again, given the right manipulations, to benzene itself," behavior that set it "quite apart" from inorganic compounds. "The eventual solution to the nature of benzene would be one of the great attainments of the human mind."[14]

Berzelius and other older chemists never really came to grips with organic substitution reactions, and the next moves were made mainly by French, Alsatian, and German chemists who gravitated to von Liebig's Giessen laboratory. In this environment the structures and the properties of the aromatic compounds were gradually isolated, until the fundamental reality was grasped: organic chemistry is very largely the chemistry of "functional groups"—a term not yet invented—attached to a relatively inert hydrocarbon skeleton. Charles Frédéric Gerhardt, a Swiss brought up in Alsace, who studied under von Liebig at Giessen, was the first to understand how structure and function were related.[15] For example:

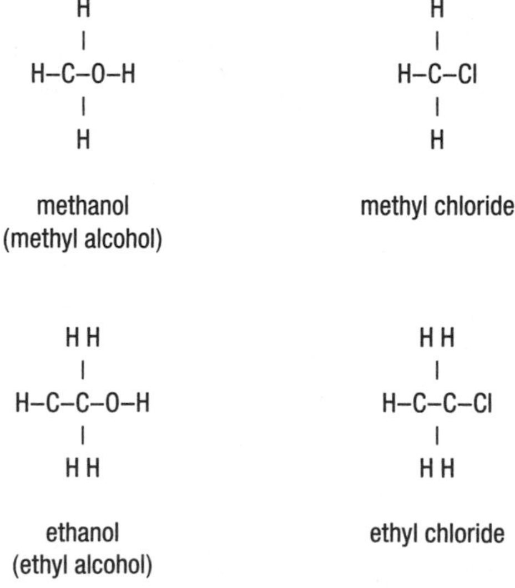

Brilliant as this insight was, this picture concealed a more complex—but more fundamental—truth. What was behind this concept was something not discovered until the 1860s: valency.[16] Valency, in everyday language, is "the combining power of one atom for another"—in a way, the number of "hooks" an atom has available to join it to its neighbors. By the 1850s, water was identified as H_2O (hydrogen monovalent, oxygen bivalent), but the behavior of carbon was still perplexing, since methane was CH_4, ethane was C_2H_6, ethylene was C_2H_4, and acetylene C_2H_2. Is the valency of carbon, 4, 3, 2, or 1? The answer, eventually, was found to be four, and what accounts for the difficulty nineteenth-century chemists faced is that carbon atoms form chains *and rings with each other.*

```
  H H            H  H
  \/              \/
H–C–C–H          C = C
  ||              /\
  H H             H H
```

ethane, C_2H_6 ethylene, C_2H_4

Once this phenomenon had been discovered, organic chemistry gave up more of its structural secrets. Here are some modern structural formulae, where "R" equals "radical," the simplest of which is "methyl":

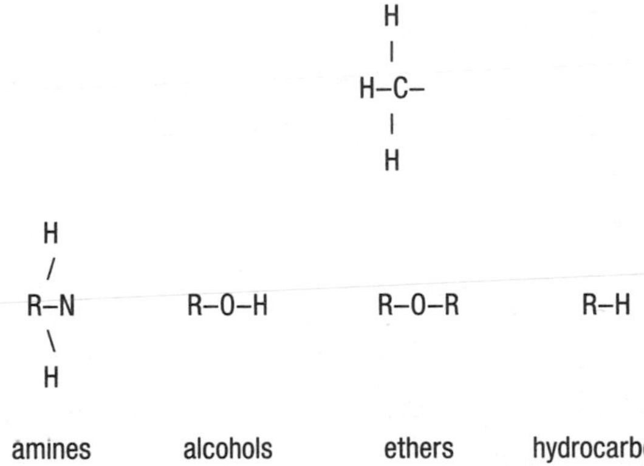

The man who did more than anyone else to explain the operating principles of organic chemistry was August Kekulé. However, the circumstances of his various "discoveries" were controversial, and even after all this time they still divide historians of science.

He was born on September 7, 1829, in Darmstadt, where von Liebig had been born a generation earlier; Kekulé sounds and looks as if it is French, but in fact the family was originally Bohemian nobility.[17] August studied architecture at Giessen, but fell under von Liebig's spell and switched to chemistry. Later he argued that his architectural training (such as it was) had helped him to think in pictures—and this played a vital role when he came to identify the structure of carbon compounds.

In 1854 he visited London and there, one summer evening, he made the first of his controversial claims. He said he had an important dream. These dreams aroused suspicion because, by their very nature, they could not be corroborated, and other scientists suspected Kekulé invented them to establish his own priority in his various claims to have identified the structure of organic substances. To give some idea of the controversy aroused, Archibald Scott Couper had written *his* first paper on organic bonds in 1858. Kekulé, however, said he had his dream about the same phenomenon in 1854 *but didn't say so* until 1890.

Organic chemistry may have had a difficult birth, but once the nature of the benzene ring was understood, the relatives of benzene—naphthalene, toluene, phenol (carbolic acid), cresols—soon became available on a vast scale, extracted from coal tar, producing a vast range of wealth-generating products: aniline dyes, trinitrotoluene, carbolic soap, creosote, naphthalene mothballs—the list is impressively long. The dyestuff industry led the way but "aromatic chemistry," a term coined by Kekulé, proliferated over the following decades, producing endless industrial chemicals, but also powerful drugs like aspirin in 1899 and Paul Ehrlich's pioneering antisyphilitic drug Salvarsan in 1909 (see Chapters 18 and 20, pp. 355, 383).[18]

Benzene was at the center of this activity. Its formula, eventually understood as C_6H_6, was so stable that it could be transformed into many derivatives by substitution reactions without itself decomposing. Kekulé said that the structure of benzene came to him in yet another dream, this time in the winter of 1861–62, in Ghent. He said he dreamed of a snake that had seized hold of its own tail, leading Kekulé to publish his theory of the ring structure in 1865. (Arthur Koestler remarked dryly that this was "probably the most important dream in history since Joseph's seven fat and seven lean cows.")

As John Buckingham has observed, "The benzene structure that emerged from the 1860s is a thing of considerable beauty and intellectual satisfaction . . . Like the DNA structure of nearly a century later, it *had* to be right."[19] The ring is the key, meaning there are no reactive loose ends. Every carbon atom has two valencies that are used to bond it to its neighbors, while a third "hook" attaches it to a hydrogen atom. This leaves one over for a bond with something else. A complete understanding of benzene's valency was not possible before the rise of quantum theory in the 1930s (see Chapter 32, p. 595), but in the mid-nineteenth century chemists did begin to suspect that three-dimensional geometry might play a role in chemical reactions. This realization would help give rise to particle physics at the end of the nineteenth century, confirming Thomas Nipperdey's point that the revolution in the natural sciences in the nineteenth century had a more far-reaching impact than the revolution brought about by Kepler, Galileo, and Newton.

This new theoretical understanding had extremely practical consequences, accounting for the heroic developments in commercial chemistry after the 1860s, which helped Germany become a world economic—and then military—power. In 1862, in a letter to von Liebig, Wöhler worried at the large array of chemists being produced by German universities and queried their fate.[20] Only three years later, when Hermann Kolbe was appointed professor of chemistry at the University of Leipzig, he asked for—and was granted permission to build—a laboratory for 132 students. Von Liebig laughed at the folly of it all, yet when it opened in 1868, it was immediately swamped by the demand.[21]

The Age of Fertilizers

Von Liebig, as we have noted, was a combustible character. It should, therefore, not be too much of a surprise to learn that, around 1840, he underwent a sudden scientific change of life and abandoned the theoretical aspects of organic chemistry for the much more practical interests of agriculture.[22] The change was, nonetheless, provoked by his interest in carbon. In an analysis of strawberries and fruits, he had found that, in a given area of land, whether it was cultivated fields or "wild forest," the same total quantity of carbon was produced each year in the composition of whatever plants grew in it. This was the starting point for what became a bitter argument as to whether such carbon derived from the atmosphere or from the humus in the soil. The dispute arose because von Liebig had long been interested in the source of nitrogen. He had found ammonia in the body of every plant he investigated and this persuaded him that it must come from ammonia dissolved in rainwater, which, he found, always contained certain amounts of that substance. The more he looked at plants, the more uniformities he found. Such uniformities, von Liebig thought, could not be accidental, and he concluded, controversially, that the nutrients in the soil and air were *inorganic* not organic.[23]

He put this together into what has been called "the most comprehensive picture of the problem of plant nutrition that had ever been presented." Von Liebig's *Organic Chemistry in Its Applications to Agriculture and Physiology*, London, 1840, beginning with the role of carbon in plant nutrition, refuted the then widely held view that humus—decayed plant matter—formed the main nutrient substance for plants. Von Liebig's second argument was that the source of carbon assimilated into plants is the atmosphere. The function of plants, he argued, was to separate the carbon and oxygen of carbonic acid, "releasing the oxygen and assimilating the carbon into compounds such as sugar, starch and gum."[24]

This argument for a chemical understanding of the internal processes of vegetables (further evidence to discredit the "vital force") was, however, not what made von Liebig's book the sensation it was. What drew attention was his view that certain nutrient materials—external sources—were essential for plant growth, for these were conclusions that impinged directly on agricultural practices. For von Liebig, the very idea of fertilizer meant adding to the soil what nutritional elements were not supplied naturally from the atmosphere. Fertilizers, he said, should comprise not humus but bases such as lime, potash, and magnesia, plus phosphoric acid, the best source of which was pulverized animal bones.

Von Liebig's book provoked intense interest among agriculturalists, particularly in Britain and America. At Rothamsted experimental farm in England, von Liebig's fertilizers were tested on wheat and found to have no noticeable effect on production, whereas ammonium salts added to the soil brought about great improvements in the harvests year after year. These results blew a hole in von Liebig's entire "mineral theory," or so it was thought, and his ideas were dismissed. Von Liebig didn't give way, however, though it would be another decade before he solved the problem. He had been too worried that soluble salts would be leached away by rainwater—in fact, the topsoil absorbed them. And, despite the early doubts, von Liebig's book changed completely attitudes to scientific agriculture. Before 1840, the conventional wisdom was that both plants and animals needed organic—previously living—material in order to survive. Following von Liebig, it began to be accepted that the nutrient substances of plants were inorganic. This utterly transformed one basic belief about agriculture, namely that the production of food had fixed limits. It was now accepted, in contrast, that no such limits applied.

The Discovery of the Cell

At much the same time as the discovery of the benzene ring and the understanding of the nature of fertilizer, German biologists were also at work on the discovery of the cell. This, the idea that all forms of life are composed of "independent, but cooperative" units that we now call cells ranks as one of the seminal discoveries in biology.[25] The first person to observe cells was Robert Hooke (1635–1703), curator of experiments at the Royal Society in London, whose *Micrographia* appeared in 1665. In later centuries, many others, benefiting from the ever-improving microscopes, observed "globules" or "vesicles," of different sizes and shapes, in both animal and vegetable tissue. We know from a letter that the Dutchman Anton van Leeuwenhoek of Delft wrote to Robert Hooke in March 1682 that he had already observed a darker body inside cells that would come to be called the nucleus.[26] By the end of the eighteenth century, most botanists accepted that *plants* were composed largely of cells, with Kaspar Friedrich Wolff (1733–94) one of the first to advocate that the fundamental subunit of all tissues—animal and vegetable—was a vesicle or globule and which, like others before him, he sometimes called a cell. However, no one had ever suggested—in print anyway—that plant cells and animal cells were homologous, and no one knew how cells divided or how new cells were formed. In 1805 Lorenz Oken (1779–1851) put forward the view that all living forms, plants as well as animals, were composed of "infusoria," these being simple organisms like bacteria or protozoa, in other words, the simplest and most primitive forms of life then known.[27]

But the first man to advance thought to its modern understanding was Jan Evangelista Purkyně. Strictly speaking, Purkyně was Czech, not German. However, since their defeat in the battle of the White Mountain in 1620, the inhabitants of Bohemia had been "inundated" by waves of Germanization, with Czech speakers gradually reduced to menial positions. The University of Prague, founded by Charles IV in 1348 and originally open to Czechs, Germans, and Poles, was, by the time Mozart made his celebrated journey to that city in 1787, a German-speaking institution.[28]

Purkyně (or Purkinje, as the name is spelled in the German literature) was educated as a choirboy in Mikulov (Nikolsburg) in Moravia. He first obtained employment as a teacher but left his order and took a medical and philosophy degree at the University of Prague, graduating in 1819. Later he accepted a chair of physiology and pathology at the University of Breslau (Wrocław), then politically and culturally a German city. The University of Breslau had been founded in 1811, a year after the University of Berlin, and there was intense rivalry between the two institutions. As part of this rivalry, Purkyně was given the first institute of physiology in Germany.

From his earliest years, he entertained the notion that there were fundamental parallels between animal and plant cells. The 1830s saw more progress, with several experiments clarifying the structure of such animal tissues as skin and bone, these papers referring to "granules," "*Körnchen*," "*Körperchen*," and "*Zellen*"; the idea that there was "homology" between *some* plant cells and *some* animal cells, says Henry Harris, was gaining strength.[29] Then there was the fact that Franz Bauer, an Austrian who was a superb botanical artist, highlighted the nucleus in his drawings. These had been made as early as 1802 but were not released until the 1830s, when he made it plain he regarded the nucleus as a regular feature of cells.[30] The nucleus was actually so named by Robert Brown, custodian of the botanical collections at the British Museum (and the man who identified "Brownian motion"), but his suggestion was made the most of in Germany, the word being used as an alternative to "*Kern*" (kernel). The nucleolus, within the nucleus, was first observed by Rudolph Wagner in 1835, though to begin with he called it a "Fleck," and then "the germinative spot" ("*macula germinative*").[31]

Purkyně's advances were not due simply to improved microscopy; he used the new dyes to perfect new staining techniques. He and his colleagues alluded several times in print to the similarity between animal and plant cells and in a lecture he gave to the Society of German Naturalists and Doctors, meeting in Prague in September 1837, Purkyně made a tour d'horizon of the animal tissues in which "*Körnchen*"—with central nuclei—had been observed: salivary glands, pancreas, the wax glands of the ear, kidneys, and testes. "The animal organism can be almost entirely reduced to three principal elementary components: fluids, cells and fibres . . . The basic cellular tissue is again clearly analogous to that of plants which, as is well known, is almost entirely composed of granules or cells."[32] His other contribution was to use the word "protoplasma" to describe the basic "ground substance" of cells.

In November 1832, Karl Asmund Rudolphi, professor of anatomy and physiology at the University of Berlin, died. The vacant chair was occupied the following year by a man who was to become one of the more famous nineteenth-century biologists, Johannes Müller.[33] In 1835 Müller published a monograph on the comparative anatomy of the *myxinidae* (hagfish), in which he described the similarity between cells in the notochord (the neural channel in the spine) and plant cells. This was a crucial observation, all the more so as Theodor Schwann became Müller's assistant. Schwann would capitalize on Müller's insight but only after his momentous meeting with the botanist Matthias Jakob Schleiden.

Schleiden's career had followed a familiar pattern. He first took up legal studies, obtaining a doctorate at the University of Heidelberg in 1827.[34] He didn't enjoy legal work, however, and changed professions, beginning a degree in natural science at Göttingen in 1833 and subsequently transferring to Berlin. Schleiden was invited to work in Müller's laboratory and it was there that he met Theodor Schwann.

Though a late convert to botany, Schleiden was always very keen on the microscope and played an important role in its introduction in biological research. (He is thought to have had a hand in the establishment of the Zeiss optical works in Jena.)[35] In 1838 Schleiden released "Beiträge zur Phytogenesis" in *Müller's Archiv*, a journal that the Berlin professor had started and that had become one of the most respected periodicals of the time. This article, immediately translated into English and French, was the first airing of the cell theory, which, according to tradition, was conceived in a conversation between Schleiden and Schwann on the subject of phytogenesis. Schleiden was impressed by Robert Brown's identification of the cell nucleus (1832) and used that as his starting point. The nucleus was then called the cytoblast, and according to Schleiden, "as soon as the cytoblast reaches its final size, a fine, transparent vesicle forms around it: this is the new cell." Schleiden described this cell as "the foundation of the vegetable world." While this paper clearly announced "the advent of plant cytology," Schleiden did so by asserting that cells are "*crystallised* inside an amorphous primary substance," which was quite wrong (italics added). Nevertheless, his botany textbook, published in 1842, the *Grundzüge der wissenschaftlichen Botanik*, gave over a large section to plant cytology, and in so doing transformed the teaching of botany, attracting many people to what they felt was a new science.[36] Schleiden himself never fully appreciated the true significance or role of the nucleus, or cytoblast, but his fellow biologists in Müller's laboratory more than made up for this shortcoming.

His friend and colleague Theodor Schwann was a biologist for fifty years yet devoted only five of those years (1834–39) to the subject for which he is best known. Schwann's most famous monograph was published the very same year, 1838, in which Schleiden released his "Beiträge" article. Schwann began by outlining the structure and growth of the cells of the notochord and of cartilage. He did so, Schwann said, because their architecture "most closely resembles" that of plants and because cell formation from the "Cytoblastem" is clearly demonstrated. The second section bore a title that reflected his argument and tone: *On cells as the foundation of all tissue in the animal body*. Purkyně and others had, of course, described cells in many tissues and had *speculated* that they might be fundamental entities, but Schwann was the first to categorically assert that cells were basic.[37]

After that, his book surveyed what was by then a lot of histological evidence to support the thesis: he discussed cells attached to each other (the epithelium, nails, feathers, and the crystalline lens); cells where the walls are amalgamated with the intercellular substance (cartilage, bone, and teeth); and cells giving rise to fibers (connective and tendinous tissue). Not everything he had to say was accurate but his main purpose was more polemical, an attempt to establish a *general principle* underlying all cell development, in plants as well as in animals. As he said in the foreword: "The aim of the present treatise is to establish the intimate connection between the two kingdoms of the organic world by demonstrating the identity of the laws governing the development of the elementary subunits of animals and plants. The main outcome of the investigation is that a common principle underlies the development of all the individual elementary subunits of all organisms, much as the same laws govern the formation of crystals despite their differences in shape." The reference to crystals was of course much the same as what Schleiden had said and was, again, wrong. This error—an important one—was compounded, in the eyes of many, because Schwann hardly referred in his own work to others who had made contributions in the same field. Purkyně reviewed Schwann's book and disputed Schwann's claim for priority.[38]

Whoever was first, Schwann's book provoked more research. One profitable line of enquiry was followed by Franz Unger (1800–70), professor of plant anatomy and physiology at Vienna (who numbered Gregor Mendel among his pupils).[39] Unger collaborated with Professor Andreas von Ettingshausen (1796–1878), a physicist in Vienna, who had an excellent Plössl miscroscope, and they observed cell behavior that would eventually lead to the recognition of the importance of cell division. This was also a preoccupation of Carl Nägeli (1817–91) at the University of Zurich.[40] Nägeli thought that his early studies of cell division, published in 1844 and 1846, showed two types of cell formation: free cell formation and the division of pre-existing cells. Two years later he changed his view in an important way, now making a simple distinction between reproductive tissues, where free cell formation was the rule, and vegetative tissue, where cell division was the norm.

In 1845 Nägeli turned to the study of vegetable growth, his investigations culminating in the late 1850s when he traced the lineage of cells to a single apical cell. It was Nägeli who showed the regular way in which the original cell cuts off daughter cells—in either one, two, or three rows—

which gave him laws that could be represented mathematically. In his later studies, Nägeli conceived the important distinction between formative tissue (*Bildungsgewebe*) and structural tissue (*Dauergewebe*) which is no longer actively multiplying. He observed that in the stems and roots of plants there was a certain type of cell that remained unaffected by differentiation and whose origin could be traced back all the way to the original "foundation cell" or zygote. This distinction between formative and structural tissue was an early sighting of the idea of heredity.[41]

To the end of his life, Nägeli continued to believe in the spontaneous generation of cells. And so, when Gregor Mendel sent Nägeli his *Versuche über Pflanzen-Hybriden* in 1866, Nägeli took Mendel's work seriously enough to repeat it. Unfortunately for him, he used *Hieracium*, a plant that reproduced asexually, and he therefore thought that Mendel's hybrid ratios and demonstrations of complete reversion, though of mathematical interest, were irrelevant to genuine species. Although Nägeli failed to recognize Mendel's genius, his pupil Karl Correns was less blind. Correns was one of the three rediscoverers of Mendel's laws.[42]

Observations on the formation of the embryo within a fertilized hen's egg were made more than 2,000 years ago, before Aristotle. But the link between cells and embryos was not really made until 1827, when Karl Ernst von Baer reported his observations about the mammalian ovum in a (Latin) letter written from Leipzig. For obvious reasons, study of the development of a single mammalian egg was virtually impossible, let alone a single human egg. The first description of segmentation in the egg was not made until 1824, by the Frenchmen Jean-Louis Prévost and Jean Baptiste Dumas. They recognized that the furrows they observed deepening on the surface of the developing egg were the first signs of its division, and that this process was repeated until the structure came to look "like a raspberry."[43] Difficult as it is for us to understand now, it never crossed their minds that what they were describing was cell multiplication.

Only in 1834 did von Baer publish his more detailed description of segmentation. He had just moved from a chair at Königsberg to St. Petersburg, where he made the important inference that what he had observed removed any idea of "preformation" of the embryo (that it existed as a fully formed miniature in the unfertilized egg). Von Baer's paper was regarded as a sensation among German scientists and from then on the biological significance of furrow formation, and subsequent segmentation, was accepted.[44] This advance was soon followed by that of the Englishman Martin Barry (1802–55) and Carl Bergmann, then an assistant to Rudolf Wagner in Göttingen, and himself later a professor of anatomy at Rostock. Using experiments with frogs and newts, Barry and Bergmann confirmed that the furrows that divided the egg gave rise to the cells that went on to form the embryo. The insight of Harald Bagge, at Erlangen, was no less important: he observed that the nuclei in the embryonic cells divided *before* the cells divided. The observation of this continuity of the nucleus, together with the demonstration that the egg was itself a cell and that it "begat" daughter cells by binary fission, marked a decisive step in the growth of what later became the science of genetics.[45] The investigation of this set of phenomena culminated in 1855, when Robert Remak published his great work on the embryology of vertebrates, in which he discovered and named the three layers of the embryo: ectoderm, mesoderm, and endoderm and, no less important, observed that cell division always begins with the nucleus. It was just four years before Darwin was to publish the *Origin of Species*.

14.

Out from "The Wretchedness of German Backwardness"

The Treaty of Westphalia in 1648 had created one set of European states. The Congress of Vienna, called in 1815 to decide the shape of Europe in the wake of Napoleon's fall, created another. The main aim of the Congress was to prevent there ever again being a revolution in Europe, and to that end the assembled diplomats and politicians set about re-creating much the same landscape as had existed immediately after 1648. But this carefully balanced European system depended on central Europe remaining fragmented and powerless. Many of the Europeans at the Vienna Congress were disturbed by the so-called Germanophiles, determined to unify Germany and turn her into a nation-state. As the French foreign minister, Charles-Maurice de Talleyrand-Périgord, wrote to Louis XVIII: "They are attempting to overturn an order that offends their pride and to replace all the governments of the country by a single authority . . . The unity of the German fatherland is their slogan . . . they are ardent to the point of fanaticism . . . Who can calculate the consequences, if the masses in Germany were to combine into a single whole and turn aggressive?"[1]

The principle of nationality was acknowledged, as Hagen Schulze has pointed out, only where it was linked to the legitimate rule of a monarch: in Great Britain, France, Spain, Portugal, the Netherlands, and Sweden—northern and western Europe. The German-speaking lands and Italy were left out. This helps explain why nationalism, *cultural* nationalism, began as a German idea. The political fragmentation of the region was actually the logical outcome of the European order—look at the map to see why. "From the Baltic to the Tyrrhenian Sea, it was Central Europe that kept the great powers apart, kept them at a distance and prevented head-on collisions." No one wanted an undue concentration of power in central Europe, for if anyone should take control, they could easily become "mistress of the entire continent."

The period from 1815, the Treaty of Vienna, to 1848, the year of revolutions, provides a neat time frame politically, but it is meaningful intellectually and culturally, too, certainly so far as Germany is concerned. During this period, and in fact beyond it, across the various bourgeois revolutions of 1848–49, in Berlin, Dresden, Prague, and Vienna, all of which failed, literature fragmented in two. There were those authors who simply ignored the social changes that were occurring in Germany (albeit later there than in Britain or France), who turned their backs on urban and bourgeois life and located their stories in the countryside, or in villages and small towns, withdrawing into a timeless—almost feudal—world, people like Heinrich von Kleist, Franz Grillparzer, Adalbert Stifter, and Joseph von Eichendorff, the latter in *Aus dem Leben eines Taugenichts* (*Diary of a Ne'er-do-well*). There was also a raft of writers who responded to the new circumstances—Heinrich Heine and Georg Büchner in particular, and the "agitatory" poets Ferdinand Freiligrath and Georg Herwegh. But this division, and the "cultural lag" in Germany, the fact that industrialization and urbanization occurred later there than elsewhere, and the fact that these writers lived in the shadow of Goethe and Schiller, meant that though their genius was well (if belatedly) recognized, nevertheless they are simply not international household names as are Victor Hugo and Honoré de Balzac, or Edgar Allan Poe and Ralph Waldo Emerson, or William Makepeace Thackeray, Lord Byron, Percy Bysshe Shelley, and Charles Dickens, all of whom were contemporaries. These German authors are classic examples of a culture that needs to be made more familiar to us all.[2]

THE SUPERIORITY OF POETRY

Friedrich Hölderlin, born in Württemberg, studied theology at Tübingen, where he formed a triumvirate of friends and roommates with Georg Hegel and Friedrich Schelling. They were a big influence on each other, and many scholars believe it must have been Hölderlin who brought Hegel's attention to the ideas of Heraclitus, whose theory about the "union of opposites" finds such an echo in Hegel's concept of dialectics. Hölderlin's stature as one of Germany's greatest poets has been acknowledged since the beginning of the twentieth century but he has only recently been recognized as a philosopher. This perhaps reflects his belief that poetry provided the best access to the truth (another view predominant in the age between doubt and Darwin).[3]

Hölderlin's life was compromised by the fact that he fell hopelessly in love with Susette, the wife in the family where he was working as a tutor, and by the fact that, early on, he showed signs of what was then termed "hypochondria." He made Susette the heroine of his novel *Hyperion*, which, in the form of letters, tells the story of one man's "eccentric path" in life. The novel reflects Hölderlin's view that too much self-consciousness (à la Hegel) is potentially dangerous, that an individual's exploration of life risks his losing the original unity with nature into which he is born and which it is the purpose of poetry to describe. Hölderlin thought that Kant's noumenal world was ultimately (as Kant himself had insisted) unknowable but that poetry could, from time to time, capture glimpses of it, and this was another of its primary functions. In *Hyperion*, the central idea is that beauty cannot be so much created as *uncovered*. It is always there, in the world, and it is the poet's task to reveal it. This view would be echoed by Heidegger and Hans-Georg Gadamer.[4]

In early 1802 Hölderlin again found employment, again as a tutor, this time to the children of the Hamburg consul in Bordeaux. This necessitated his traveling there on foot. The time for observation and reflection that this provided gave rise to one of his greatest poems, "Andenken" (Remembrance):

The northeast blows,
My favourite among winds,
Since it promises fiery spirit
And a good voyage to mariners. . .

I remember well
how the crowns of the elm trees
lean over the mill,
and a fig tree grows in the courtyard.

Hölderlin returned to Germany a few months later but was now showing frank signs of mental disorder, which got worse when he heard that Susette had died. He was, fortunately, saved when in 1807 a Tübingen carpenter and literary enthusiast, Ernst Zimmer, who had admired *Hyperion*, took him in, and gave him a room overlooking the Neckar valley. Zimmer cared for Hölderlin until his death in 1843.

His poetry was admired enough in his lifetime for Hölderlin's friends to club together and publish the work. After his death, however, he sank into oblivion, partly because of his madness and partly because he was dismissed as a "melancholy imitator" of Schiller. He was rediscovered only in the early twentieth century, by the circle around Stefan George but also by the philosopher Martin Heidegger, and Hölderlin's work is now regarded as one of the high points of German literature.[5] In his late madness, he would write poems of childlike beauty which he would sign with fantastic names such as "Scardanelli":

where shall I
When it is winter, find the flowers,
And the sunshine
And shadows of earth?
The walls persist

He influenced a raft of mainly German writers from Rainer Maria Rilke and Hermann Hesse to Theodor Adorno, and his works were set to music by Johannes Brahms, Richard Strauss, Max Reger, Paul Hindemith, and Benjamin Britten.

"Neither Animals nor Gods"

If Hölderlin was a "melancholy Schiller," Heinrich von Kleist supplanted him as the model for all dramatists. Born in Frankfurt an der Oder, he was a restless wanderer who lived in Paris, Switzerland, and Prague before finally settling in Berlin in 1810 as editor of the *Berliner Abendblätter*. There he had a short, tragic love affair with Henriette Vogel, an unstable Bohemian would-be artist, who persuaded him to join her in a bizarre suicide pact. He shot her first, then turned the gun on himself, on the shore of the Kleiner Wannsee near Potsdam. He was just thirty-four.[6]

Despite this, he has come to be regarded as the most important north German dramatist of the Romantic movement. His best work is probably the play *Prinz Friedrich von Homburg*, followed closely by the novella *Michael Kohlhaas*, set in Luther's day, in which the main character is a horse dealer.[7] Kleist's plays are above all psychological dramas, the denouement often less important than the exactitude of the language, which explores the psychology so explicitly that the audience cannot avoid the pain or embarrassment or shame which is the playwright's subject. Kleist is more popular than ever these days, looked upon by scholars as a postmodern author, though others choose to see him as a precursor of Henrik Ibsen—even, in some quarters, as a proto-Nazi because of his "rampant" nationalism. A good example here is *Die Hermannsschlacht* (*The Battle of Teutoburg Forest*), where the interests of the individual protagonist are subordinated to the service of the *Volk*. Kleist is even better known for *Der zerbrochene Krug* (*The Broken Jug*), a comedy in which a judge "gradually and inadvertently" reveals that he has committed the crime under investigation (though when *Der zerbrochene Krug* was staged in Weimar, directed by Goethe, it was a disaster). Kleist was very modern, tackling such subjects as race relations in the colonial era. But his dramas are chiefly known now for their depiction of unfulfillable longing, the barbarity of the Junkers, in particular "the wretchedness of German backwardness." He is also seen as a precursor of Richard Wagner.

Like Kleist, Franz Grillparzer (1791–1872) varied between "inner stories" and political dramas, and this produced its own problems when he released two historical tragedies that presented German-speaking monarchs in a less than favorable light, concentrating on the dilemmas that can face a prince when his duty conflicts with self-interest—both plays fell foul of the censor.[8] Born in Vienna, the son of a lawyer, he became famous after publication of his tragedy *Die Ahnfrau* (*The Ancestress*, 1817), which features brother-sister incest and parricide. This was followed by *Sappho* (1818), in which he tells the story of a poet's renunciation of earthly happiness in pursuit of her higher mission.[9]

Grillparzer suffered setbacks in his personal life too, for at much the same time he met Katharina Fröhlich, with whom he fell in love. She entirely reciprocated his feelings, and they became engaged, but Grillparzer's complicated psychology meant he could never bring himself to marry, a predicament that plunged him into despair. This so obsessed him that he poured his feelings into a diary, later composing an impressive cycle of poems, *Tristia ex Ponto* (1835), and two of his greatest dramas, *Des Meeres und der Liebe Wellen* (*The Waves of Sea and Love*; 1831) and *Der Traum, ein Leben* (*The Dream, a Life*; 1834). In *The Dream, a Life*, Grillparzer is at his best.[10] The hunter Rustan is no longer content with a quiet life with his wife and daughter and is incited to the *vita activa* by the black slave Zanga.[11] However, his dream (which takes up most of the play) is so terrifying that in the morning Rustan wants only to go back and find happiness in the quiet life.

For in greatness there is danger
And renown's an empty game
Bringing with it idle shadows,
Far too high the price of fame.

Toward the end of his life, honors were heaped on Grillparzer, and his eightieth birthday was declared a national holiday (his own comment was: "Far too late"). At his death three completed tragedies were found among his papers and one of them, *Die Jüdin von Toledo*, an adaptation from the Spanish, is now accepted as a German classic. After his death he sank into obscurity and not until the centenary of 1891 did the German-speaking world acknowledge his genius.[12] The "inner" stories have worn better than the political dramas.

The Laws of Gentleness and the Avoidance of Locomotives and Factories

Adalbert Stifter (1805–68), as well as being a writer and a poet, was an accomplished painter (his works sold well enough) and a pedagogue. He studied law at the University of Vienna, but had an unhappy family life, being prevented by his parents from marrying the woman he loved, then contracting an unhappy union with another woman who was unable to conceive. Suffering from cirrhosis of the liver and in deep depression, he slashed his throat with a razor.

Stifter wrote many long stories and short novels, the greatest of which was *Der Nachsommer* (*Indian Summer*; 1857), now counted a seminal Bildungsroman in the German canon. It describes the self-cultivation of Heinrich Drendorf, a German merchant's son and shows how he gradually acquired all the necessary characteristics to look upon himself—and to behave—with dignity. Stifter himself lived through the violence and chaos of the 1848 failed revolution in Vienna and subsequently went to live a much quieter life in Linz. His book describes how Drendorf pursues his *private* fulfillment through a range of humanistic endeavors—science, art, history, pedagogy, but all at a distance from contemporary issues, avoiding, as Stifter puts it (and as he did in his own life), "locomotives and factories."[13] Although a merchant's son, Drendorf is conspicuously uninterested in the practical world of trade and commerce. When he is out for a walk on a mountainside, a storm is brewing and he seeks shelter at the estate of an old man. Once on the estate, the Rosenhaus, Drendorf cannot help but notice that the old man, Freiherr von Risach (an important figure politically, though we are never told why), orders his life punctiliously around art, antiques, and gardening (books are always replaced on their shelves immediately after use), and that he is just one of several characters who enjoy this idyllic existence, *controlling* their passions rather than giving way to them. Nietzsche found Stifter's *Nachsommer* a "distant celestial world," with a "milky-way brightness" and, along with Gottfried Keller's *Green Henry* (see p. 296), one of the two best German books of the nineteenth century.[14]

Stifter's point is that pleasure is derived from "the laws of gentleness," that "evolution anticipated revolution," that most of the lasting beneficial changes in the world are slow to emerge and silent in their effects—this is nature's way. Friedrich Hebbel, another writer, offered the crown of Poland to anyone who could finish reading *Der Nachsommer*, but W. H. Auden, Marianne Moore, and W. G. Sebald all stressed their debt to him, while Thomas Mann said he was "one of the most extraordinary, the most enigmatic, the most secretly daring and the most strangely gripping narrators in world literature."

The Alternative Bourgeois

Like Stifter, Gottfried Keller (1819–90) enjoyed painting. He studied in Munich for two years before deciding he would never be good enough and then turned to writing. His book *Der grüne Heinrich* (*Green Henry*) is regarded by some critics as the greatest Swiss novel, and was recently admitted into *The Western Canon* by Harold Bloom, the American critic. It was this book that, together with *Indian Summer*, Nietzsche described as one of the two greatest German-language novels of the nineteenth century.

Born in Zurich, Keller was the son of a lathe operator who died when Gottfried was just five. He attended a variety of schools, including an Industrieschule until he was fifteen, when he was expelled for a misdemeanor. Forced to seek employment, he apprenticed himself to Peter Steiger and Rudolf Meyer, landscape painters in Munich. After two years, however, he abandoned art, returned to Zurich, and took up writing. He studied at Heidelberg, where he attended the lectures of Ludwig Feuerbach. Feuerbach was a great influence on Keller, who was much concerned with what Daniel Bell would call, more than a hundred years later, the cultural contradictions of capitalism, in particular how the individual could live a fulfilled life in a society where capitalism encouraged so much individualism.

Keller was in a way a transitional figure between those writers considered above, who primarily turned their backs on the new urban bourgeois world, and those considered next, who recognized that it should be the chief focus of their concern. Keller preferred legal change to revolution (he wasn't *quite* as gentle as Stifter), but he couldn't quite embrace that change in his work, and certainly not after 1848. He was one of those who developed the nineteenth-century novella, a particularly German form of short narrative, brief and highly symbolic, summarizing life in society by focusing on an "extraordinary, individual event."

Green Henry is customarily identified as a Bildungsroman, following Goethe's model, though it is also reminiscent of Balzac's "Le chef-d'oeuvre inconnu."[15] There are strong autobiographical elements in the story. The protagonist, Heinrich Lee, is called green because all his youthful outfits are made from his father's green uniforms, available because his father had died at an early age. Heinrich is expelled from school and studies painting in Munich. The other element in Heinrich's life is his love for two women, Anna, who represents "heavenly love," and Judith, a widow, who answers his "more earthly needs." The plot resolves itself when Heinrich realizes he can never achieve more than modest success as an artist but is then overtaken by the death of his mother. This forces on him the realization that, in a very real sense, he was responsible for her death because of what she sacrificed for him and that he, in his self-obsession, had impoverished her. He dies of shame.

Keller came to dislike this story—or the ending—and rewrote it years later. In the revised version, Heinrich doesn't die, but lives on in a dispiriting bureaucratic sinecure. This seems to have struck a chord. The first version had not really caught on, but the revised version received wide acclaim.[16]

This new ending was partly autobiographical too, because in 1855 Keller returned to Zurich, later becoming cantonal secretary. From this vantage point Keller was particularly aware of the growing division between capitalism and artistic individualism, the division that Marx labeled alienation, which Keller found equally abominable. He addressed this in a series of novellas titled *Die Leute von Seldwyla* (*The People of Seldwyla*; 1856 and 1873–74), a distinctly odd but not necessarily disagreeable place. Here the people are no less daring and enterprising than anywhere else but, as they gain experience of the world, they change. They become "whimsical philistines" who withdraw into the security of their own city: they refuse to see work as "a process of upward mobility," they reject speed, derive pleasure from the trivial side of life, rather than what everyone else regards as "important." They are, in effect, exploring alternative values to those of the bourgeoisie.

Coal-Smoke and Song Birds

Heinrich Heine (1797–1856) is now recognized as one of Germany's greatest writers. Yet, like Grillparzer, like Hölderlin, recognition was delayed for decades. One factor special to him was his Jewishness: anti-Semites in early nineteenth-century Germany denied that he could be "both Jewish and German." The fact that he lived in France from 1831 to his death and was enthusiastic about French culture didn't help either (in his memoirs he speaks of two passions: the love of beautiful women and the French Revolution).[17] The Nazis tried to erase his memory completely.

When Heine's first collection of verse appeared in 1821, Weimar Classicism was long gone and Romanticism was fading too. He himself wrote: "The thousand-year empire of Romanticism is at an end, and I myself was its last and fabulous king, who abdicated the throne." For him, Romanticism was "a desperate inward retreat from an unsatisfactory external world." He studied under Hegel in Berlin when the philosopher was at the height of his fame and influence. And Heine agreed with him, seeing a Hegelian progression in the arts, which had begun with the most "material" art forms (for example, the Egyptian pyramids), then progressed via Greek sculpture and Renaissance painting to the least material—poetry and music. "Our present age," he felt, "will go down in the annals of art as the age of music."

He divided his own time between prose (journalism—Georg Lukács described him as a revolutionary journalist of significance—travel writing, criticism) and poetry. His early verses were collected into the *Buch der Lieder* (*Book of Songs*), which Franz Schubert, Robert Schumann, and Felix Mendelssohn helped to make famous. In his prose of that early time, particularly *Briefe aus Berlin* (*Letters from Berlin*) and *Die Harzreise* (*The Harz Journey*)—travel notes of a sort—he showed himself as sensitive to the very different Germanies that then existed, in particular the unchanging life of the countryside as compared with the industrializing, fast-paced, always-different world of the city. After the excesses of the Napoleonic Wars, he anticipated that the great issue of the day would be emancipation, of races as much as of social classes and other oppressed peoples. "Our age is warmed by the idea of human equality . . ." In his journalism he tried hard to prepare the ideological ground for a German revolution.[18]

Though the fragmentation of Germany, as reinforced by the Treaty of Vienna, did not appeal to him, Heine was by no means a nationalist, one reason being the nationalists' espousal of a "Christian German" identity, which had no place for Jews, of which he was one. In *Die Romantische Schule* (*The Romantic School*), written with a French audience in mind, he drew attention to the cosmopolitanism of the great eighteenth-century German writers and explored what had been lost. Romantic poetry, he said frankly, was incompatible with modern life: "The railway engine shakes and jolts our minds, so that we cannot produce a song; coal-smoke is driving away the song-birds . . ."[19]

The 1840s were a complicated time politically. Food shortages in several European countries during the "hungry forties" stimulated radical activity. In Germany, the accession of Friedrich Wilhelm IV to the Prussian throne in 1840 aroused hopes of liberalization after his father's long (forty-three-year) reactionary reign. There was a flood of political verse, produced mainly by the so-called *Tendenzdichter* (committed poets) such as Ferdinand Freiligrath (later a political exile in London), and the philologist August Heinrich Hoffmann von Fallersleben, who was dismissed from his professorship on account of his political verse. This verse included "Das Lied der Deutschen" (Song of the Germans), "Deutschland, Deutschland über alles," which is not always understood as a liberal work appealing for a German national state with free institutions.[20] Heine thought these "committed" poets were banal; the proper poet-genius, for him, could belong to no party and toe no line:

Aimless is my song, Yes, aimless,
As is love, as life is aimless
As Creator and creation.

Heine wasn't apolitical; he despised capitalism and expected more "heroics" from the bourgeoisie than were there, but for him the true poet searches for those deeper and more fundamental forces that go beyond the aims of the radicals. Like Jacob Grimm he thought that folktales—the deeper poetry—contained glimpses of the ancient Germanic religion.[21] Christianity had drawn people away from earthly (and earthy) realities, as revealed in folktales, toward a more ethereal, disembodied spiritual realm. By recovering and reworking the original folktales, he believed he could revive the lost excitement (though he believed "God's opium" was passing).

The radical turmoil that existed in "the hungry forties" produced a foretaste of revolution in Germany in 1844, when the Silesian weavers mounted an insurrection. Their traditional cottage industry was simply unable to cope with the industrialized textile manufactures of Britain. Reduced to starvation, they were a pitiful group and their uprising, quickly put down, inspired Heine's famous bitter proletarian poem, "Die schlesischen Weber" (The Silesian Weavers), which would resonate throughout Germany down the century:

The shuttle flies, the loom creaks loud,
Night and day we weave your shroud—
Old Germany, at your shroud we sit,
We're weaving a threefold curse on it,
We're weaving, we're weaving![22]

The idea, stemming from Heine's deep concern about the anachronism that was Germany, was partly based on a song composed amid another uprising, that of the Lyons silk weavers in 1831: "We shall weave the old world's shroud."

Heine was famously ambivalent about his Jewishness. Like many German Jews before and since, he was first and above all else, a *German*. In 1824 he described himself to a friend as "one of the most German beasts in existence . . . my breast is an archive of German feeling."[23] At the same time, one of the features of his unfinished Jewish novel *Rabbi von Bacharach* is his loving description of kosher food.

This did not stop Heine from converting to Protestantism in 1825. It was a curious business, secret at the time, famous afterward. It was not a case of "instant bleaching," the term applied to Jewish converts who were given a baptismal gift of ten ducats by the Prussian state if they named the king as their godfather. Heine had always been uncomfortable with his Jewishness and used to describe himself only as "of Jewish ancestry." Crescence-Eugénie Mirat ("Mathilde"), who met Heine in 1834, began living with him two years later, and married him in 1840, had no idea he was Jewish. He always anticipated that German Jews would achieve full civil equality, explaining the anti-Semitism of his day as economic, not religious.[24] He thought of his own time as tolerant.[25]

In 1848, the year of revolution, Heine's health dramatically deteriorated and for the last eight years of his life he was bedridden, forced to lie on mattresses laid on the floor, which, he complained, formed his "mattress grave."[26] His speech could be labored but his mind was still agile and illness brought him back to God. He observed mordantly that the healthy and the sick need different religions and that Christianity "was an excellent religion for the sick."[27] Ritchie Robertson has described the late sick-bed poems as unlike anything else in the canon save, perhaps, the late poems of Yeats, savage and playful. In one memorable verse Heine bequeaths his ailments one-by-one to his enemies. In another he challenges God head-on:

Drop those holy parables and
Pietist hypotheses:
Answer us these damning questions—
No evasions, if you please.

From his mattress grave he tells us that poetry can be no help in this desolate world—he agreed with Hegel that the world "had entered the age of prose."[28]

Modernity and Murder

Many people—especially in Germany—believe that, had he lived a normal lifespan, Georg Büchner (1813–37) would have become the equal of Goethe or Schiller. His best known work, *Woyzeck*, is certainly an arresting masterpiece. Born in Goddelau, near Darmstadt, Büchner was the son of a doctor and the brother of the philosopher Ludwig Büchner. He studied medicine at Strasbourg, published his dissertation, on aspects of the nervous system, and moved to Giessen, the up-and-coming center for scientific research. But Büchner had always been interested in politics and, appalled by the conditions in Hesse, helped to form a secret society dedicated to revolution. He longed for the poor to attain self-consciousness, realizing that, in his day, the proletarians were not yet a "class." In a letter written from Giessen, he observed: "The political conditions could drive me crazy." He was forced into exile when one of his pamphlets was judged too incendiary, first in Strasbourg, then Zurich. He became professor of anatomy at the University of Zurich but died almost immediately of typhus at the age of twenty-three.

Büchner produced his first play—*Dantons Tod* (*Danton's Death*), about the French Revolution—in 1835, followed by *Lenz*, a novella based on the life of Jakob Michael Reinhold Lenz, a poet of the Sturm und Drang period. His second play, *Leonce und Lena*, was about the nobility but then came *Woyzeck*, unfinished, published posthumously and the first literary work in German whose main characters came not from the aristocracy or the bourgeoisie, but from the working class. (The title of the play was chosen by subsequent editors.) Büchner left four drafts, which among them allow for a proper reconstruction. The drafts were probably begun in 1836, but a performance was not arranged until 1913. The work is perhaps best known through Alban Berg's opera, *Woyzeck*, which premiered in 1925.

Woyzeck, based on true events, tells the story of a common soldier driven mad—and to suicide—by unyielding military discipline and strict hierarchical societies where, as he says at one point, "Man's an abyss; you get dizzy when you look down." Büchner had followed a lengthy debate in a medical journal regarding a convicted murderer, J. C. Woyzecjk, who had killed his lover in a fit of jealousy in Leipzig in 1821 and was subsequently beheaded in public. A soldier and barber, Woyzecjk had fallen on hard times, sliding into unemployment, beginning to hallucinate, and then showing signs of paranoia. Despite all this, the King's Counselor, who had examined Woyzecjk twice, found him "depraved but not insane." According to the counselor's moral standard, derived from Kant, Woyzecjk had deviated from society's norms and had to be punished as a deterrent to others. In the course of the play, Woyzecjk kills his lover and then himself.[29]

The play is a savage indictment of the social conditions then existing in Germany, the new forms of poverty caused by industrialization, the "atomization" that drives all individuals against each other in a society which ostensibly values individuality, and the fundamental ignorance of most people about the psychological pressures that can exist in simply getting through the day. Guilt is to be found neither in the murderer nor in his victim—nor in his tormentors, who are themselves tormented. In a letter to his parents, written in 1834, Büchner said: " . . . it lies in no one's power to avoid becoming a fool or a criminal." The jagged nature of the scenes, the way they do and do not follow each other, the use of working-class dialogue and accents, was all new on stage and meant that the play would eventually have an enormous impact—on Expressionism, for example, and on many modern and even postmodern authors.

Büchner was appalled and defeated by the fatalism of the poor. One of the poor characters in *Woyzeck* says: "I think that if we went to heaven, we'd have to help make the thunder."

The End of the Goethean Age

From 1829 onward, Heine had on several occasions addressed the significance for him of the forthcoming "end of the cultural age" that had "begun at Goethe's cradle and will end at his [Heine's] coffin." Despite his very real admiration for Goethe, he bemoaned the "quietism" that he felt characterized the bulk of the writing of the period, particularly since the turn of the century. Great periods of art in the past, he claimed, were never divorced from the great issues of the day, pointing to Phidias and Michelangelo as two great artists whose work exemplified this premise. In fact, Goethe's death in 1832 came to mark a watershed in German literature of the nineteenth century, as we have seen. A clinging to the values of the Goethean legacy—his inwardness and a turning away from the world of industrial change—characterized the writing of Grillparzer, Stifter, and Keller (all, significantly, living outside Germany proper), as well as the novellas of the Biedermeier period around the middle of the century.

But there were also a number of authors who looked up to Heine, despite the fact that he spent most of his mature years in exile in Paris. The writers of Junges Deutschland (Young Germany)—Christian Dietrich Grabbe, Karl Gutzkow, Heinrich Laube, Theodor Mundt, Rudolf Wienberg, and Ludwig Börne are the major names—were most active in the 1830s, addressing various social issues in their novels and dramas. All these writers have in common that they were writing in an age marked by strict literary censorship that was designed to quell any form of public dissent. For this reason the ascent to the Prussian throne of Friedrich Wilhelm IV in 1840 was seen as a key turning point, as he announced his intention to liberalize the constraints on writers. Although the monarch soon found himself forced to backtrack, the brief period of liberalization was a catalyst to a veritable torrent of political—at times intensely nationalistic, at times more or less Marxist—writing that marked the *Vormärz*. This period saw the emergence of the hugely popular sociocritical novels of Ernst Willkomm, who was influenced by Eugène Sue and Charles Dickens, as well as a host of other novels based on *The Pickwick Papers*. Georg Weerth was important too—he, like Friedrich Engels, drew on his personal knowledge of the conditions of the working class in England in his writing.

But the most notable writing of the time, undoubtedly influenced by the important peripheral presence of Heine (and despite his scorn for the genre), was the realm of political poetry, already alluded to, with Georg Herwegh, Ferdinand Freiligrath, and Hoffmann von Fallersleben producing sociocritical verse that was often distributed via the broadsheets and not infrequently written so as to be sung to popular tunes, especially in the workingmen's clubs that sprang up at the time. The failure of the revolution of 1848 brought all this to a rapid end, with most writers fleeing abroad—to London or the United States, which saw a huge and influential influx of German refugees (see Chapter 15, p. 311).

The Biedermeier Phenomenon

The immediate consequence of the post-1815 world was that, in an effort to avoid a repeat of the French Revolution, the reestablished monarchies of Europe kept a much firmer political grip on their subjects. There was not just strong censorship, as we have seen, but a widespread use of secret intelligence agencies to root out subversion. The restrictions in Austria were as bad as anywhere, with Prince Klemens von Metternich's actions paralleling Napoleon's—lodges, clubs, and societies were all closed down and "inconvenient" members imprisoned. This produced a medium-term reaction in that it forced people out of the public coffee houses and meeting halls and into the secluded world of their private homes. Raymond

Erickson, in *Schubert's Vienna*, tells us: "The world outside was politically dangerous, so private life, home, and social contacts were restricted to a circle of true and reliable friends."[30]

This is the background to what eventually became known as Biedermeier culture, a decisive shift from—even a reaction against—high Romanticism (a "lull" before the storm of modernism in Thomas Nipperdey's opinion). In the Romantic movement the focus had been on an individual's own experience. The Biedermeier changed that to a focus on relationships. The private world of friendship took on a significance that had hitherto been neglected and this more intimate atmosphere was reflected in the arts of the time. Biedermeier culture lasted longer—well beyond 1848—in literature than in the other arts, but was even seen in architecture, where houses became drawn back from the street. In literature it can be seen in the quiet, intimate poetry of Annette von Droste-Hülshoff, Adelbert von Chamisso, Eduard Mörike, and Wilhelm Müller, the last two being set to music by Hugo Wolf and Franz Schubert. The growing urbanization and industrialization led to a new kind of audience: the early Lieder of Schubert could be performed at the piano without a substantial musical training, signifying a far more private existence than had occurred before. All this helped ensure that one of the main forms of Biedermeier culture was furniture—objects that decorated the private home.[31]

Biedermeier furniture is less aggressive than the Empire style; it has simpler, less ambitious lines, is made of cheaper, locally available woods, like cherry or walnut, rather than than the more expensive imported mahogany. It is "reliable," "common-sensical," even—according to one authority—"boring." The very word "Biedermeier" is itself mocking. In 1848, the painter-poet Josef Victor von Scheffel published a number of sarcastic poems in the Viennese satirical magazine *Fliegende Blätter* (*Flying Leaves*), among them "Biedermann's Evening Socialising" and "Bummelmeier's Complaint." These names were combined (satirically again) by Ludwig Eichrodt into the pseudonym Gottlieb Biedermeier. "*Bieder*" is a German word meaning "common, everyday, plain," "boring but in an upright way," and Meier, or Meyer, is a common German last name, like Smith.

This surfeit of satire is a little unfair to the chief Biedermeier furniture designer, Josef Dannhauser, whose designs could be quite flamboyant. At its height, his factory in Vienna employed 350 workers, designing and manufacturing not just furniture but sculpture and devices for interior decoration. After the factory closed in 1838, several of his workmen traveled around Europe as far as Stockholm, St. Petersburg, and Budapest, where their skills were in demand, spreading Biedermeier ideas.[32]

As all this shows, Biedermeier culture was essentially a middle-class phenomenon, and a particularly *German* middle class at that. Unlike in France, the aristocracy and the administrative/middle classes in Germany rarely if ever mixed. The new furniture designs therefore enabled the newly enriched bourgeoisie to make their mark. It was all quite different from the worlds of Büchner, Keller, and Heine.

The industrial revolution, delayed in Germany but beginning to take off in the 1830s, and gathering pace in the 1840s, made this a period of technical innovation as the steamship, the steam railway, sewing machines, gas lighting, and the mass production of more and more objects became common. This had the perhaps predictable effect that the "Biedermeier person" came to take pride in his or her artistic taste, identifying craft works as superior to machine-made items. Tableware and glassware became adorned with detailed miniature paintings, and highly decorated porcelain flourished, as did handmade fashions. In painting, the laborious copying of nature became popular and in portraiture realism prevailed, with intimately observed psychological detail. Pictures of family life were popular, the bourgeois parlor a refuge from commercial and industrial reality.

This mustn't be overdone. The Viennese did get out. Theater became chiefly known for its spectacle (it became customary, for example, to announce ahead of time in the newspapers how many shots would be fired in the battle scenes). The Biedermeier period also saw the rise in popularity of Schubert's songs (considered earlier), when his friends began to organize the famous "Schubertiaden," sponsored evenings in which nothing but his music was played.[33]

Schubert wrote several symphonies and it was his last, the Ninth, the "Great" in C Major, that was famously rediscovered by one of the other great composers of the time, Robert Schumann. He had heard of its existence and ten years after Schubert's death visited Franz's brother, Ferdinand, who showed him great swaths of manuscripts, among which Schumann recognized an entire symphony, and was allowed to take it away. Just over a year later, Mendelssohn conducted the world premiere in Leipzig. When he heard it, Schumann said, "This Symphony has created a greater effect among us than any other since Beethoven . . ."

Schumann (1810–56) was himself the most complete Romantic. Surrounded by insanity and suicide in his family, he worried all his life that he too would succumb in one way or another. The son of a bookseller and publisher, he grew up suffused in the works of great writers—Goethe, Shakespeare, Byron, and Novalis—all of whom exerted a great influence on him (he burst into tears when he read Byron's "Manfred," which he later set to music). Schumann tried to write poetry himself and emulated Byron in other ways too, embarking on numerous love affairs. In the early 1850s he suffered a week of hallucinations in which he thought that the angels were dictating music to him while he was threatened by wild animals. He threw himself off a bridge but failed to kill himself and, at his own request, was placed in an asylum in 1854. His best-known work, and perhaps the best-loved, is *Carnaval*, in which he paints pictures of his friends, his wife, Clara Vieck, Chopin, Niccolò Paganini, and Mendelssohn. (*Carnaval* was a great influence on Brahms.)[34] However, Schumann's music was intensely disliked during his lifetime, and he found it necessary to earn a living as a critic. He was a good one—one of his first reviews introduced Chopin to the German public ("Hats off, gentlemen. A genius!"), and one of the last introduced Brahms. He could have been a great pianist but, in attempting to improve his fingering technique, Schumann stretched his hands so much that he permanently ruined one of his fingers.

By the time of his death in 1856, after several difficult years, Schumann's music was at last beginning to earn an international reputation and is mainly remembered for two things. The *Fantasy in C Major*, his greatest work for solo piano, is now recognized as "one of the trinity of pieces upon which all romantic piano music rests" (the others are Chopin's Sonata in B Flat Minor, and Liszt's Sonata in B Minor).[35] Schumann's second achievement occurred with his move from piano music to song. Some of his songs, such as *Dichterliebe*, now rank with Schubert's *Die Winterreise* because, in a very real sense, he took up where Schubert left off, expanding the role—formal, technical, and emotional—of the piano, adding preludes and postludes, for example. He composed 250 songs and expanded the repertoire of voices, producing a series of very melodic vocal duets.

The Invention of the Modern Musical Repertoire

Schumann himself revered Mendelssohn, for Felix Mendelssohn was possibly the most widely accomplished musician after Mozart. A fine pianist, he was also the greatest conductor of his day and the greatest organist. He was an excellent violinist and was well read in poetry and philosophy. Born in Hamburg in 1809, he came from a wealthy Jewish banking family and was the grandson of the philosopher Moses Mendelssohn. A fervent German patriot, he believed that his fellow countrymen were supreme in all the arts. Indeed, if there is such a thing, Mendelssohn was *over*cultured. As a boy he was made to get up at 5:00 A.M. to work on his music, his history, his Greek and Latin, his science, and his comparative literature.[36]

Like so many of the other Romantic musicians, he was a child prodigy, though he was doubly fortunate in that his parents could afford to hire their own orchestra and he could have them play his own compositions, where he would conduct. He went to Paris and met Liszt, Chopin, and Berlioz. For his first work he took Shakespeare as his inspiration: *A Midsummer Night's Dream*, a fairyland that was perfect Romantic material (though Mendelssohn never had much in the way of internal demons). After Paris, he went to Leipzig as musical director and quickly made it the musical capital of Germany. One of the first conductors to use the baton, he employed it to turn the Leipzig orchestra into the foremost instrument of musical performance of the day—precise, sparing, with a predilection for speed.[37] He increased the size of the orchestra and revised the repertoire. In fact, Mendelssohn was probably the first conductor to adopt the dictatorial manner that seems so popular today, as well as being the main organizer of the basic repertoire that we now hear, with Mozart and Beethoven as the backbone, Haydn, Bach (whose *St. Matthew Passion* Mendelssohn rescued from a hundred years' slumber), and Handel not far behind, and with Gioachino Rossini, Liszt, Chopin, Schubert, and Schumann also included.[38] It was Mendelssohn who conceived the shape of most concerts as we hear them: an overture, a large-scale work, such as a symphony, followed by a concerto. (Until Mendelssohn, most symphonies were considered too long to hear at one go: interspersed between movements there would be shorter, less demanding pieces.)

Mendelssohn's own music was very popular in the middle of the nineteenth century, but his reputation today is divided. There are those who feel he was the nineteenth-century equivalent of Mozart, others that he never quite lived up to his promise.

Germanistik AND THE CENTRAL DRAMA OF MODERNITY

Underneath and around Biedermeier culture, another concept was developing in nineteenth-century Germany. This was the idea of *Volkskultur*, allied to mass culture. It developed out of the ideas and activities of Herder and the Grimm brothers, described earlier, which, as the nineteenth century wore on, extended to *Volkskunst, Volksmusik, Volksliteratur, Volkstheater, Volksdichtung* (folk or national art, music, literature, theater, poetry), *Volkstum* (folkdom), and *Volkskunde* (popular culture, of which a big strand was occupied by the *Volksbuch*, or popular narrative).[39]

Behind these was the idea that there *was* such a thing as a collective genius in Germany, which gave the nation an organic unity, a *Volksgeist*, or national spirit. This enabled Germans to feel that their culture and history represented a proud alternative to the classical Latin culture of France, Italy, Spain, and the Holy Roman Empire, which had dominated European thought for centuries. On this vision, high culture and Volkskultur were seen as different sides of the same coin, different expressions of a common root, an essentially uncorrupted collective genius. This emerging speciality came to be known as *Germanistik*, German Studies. Writers like Ernst Moritz Arndt (1769–1831) and Friedrich Ludwig Jahn (1778–1852) were united—and increasingly strident—in their belief that there was a genius in the Volk "whose voice must be preserved and articulated by transposing oral traditions into a written record of folk tales, folk plays, fairy tales and folk songs."

In the years before 1848, the term "Volk" became gradually interchangeable with the term "*Masse*." The meaning of this term, "mass" in English, was subtly different in the nineteenth century from its meaning today—it meant a class of people without political representation, meaning that "mass culture" was likewise denied the standing it was entitled to. "The battle for the mind of the masses as the central drama of modernity had begun and it developed in Germany with particular ferocity."[40]

The central problem—culturally speaking—was how to create a cohesive identity of the masses in the new growing industrial conurbations, which were living entities never seen before. Germany lacked unifying symbols more than did other European nations, which made its industrial areas more disparate, more disaggregated than anywhere else. The result was a divide into "a majority which saw culture as serving the greater glory of the nation and a minority which cherished critical independence from authority, both secular and ecclesiastical."[41]

In time, this division would come to matter. Volkskultur would acquire a mystical quality, which sustained and enriched the masses. Whereas for Herder and the Grimm brothers, culture had defined and unified Germany and helped explain her to herself, as the nineteenth century progressed, and as "Volk" and "Masse" more and more came to mean the same thing, the alternative quality of Volkskultur, the alternative to the classical Latin culture, acquired increasingly triumphal overtones. In the latter half of the century, as we shall see, Germany's industrial achievements—involving the masses—inherited this understanding and this attitude. Germans thought of themselves (as they were) as leading the way in both high culture and mass culture, which was an expression of Volkskultur. That form of self-understanding first emerged in the Biedermeier period.

15.

"German Fever" in France, Britain, and the United States

"The Tempest in Petticoats"

The raft of changes in Germany that have been the subject of the opening chapters did not go unnoticed, or unremarked, in the world outside. The first—and in many ways still the most remarkable—observer of Germany of that time was Germaine de Staël, a French-speaking Swiss writer who spent most of her time in Paris, becoming the most famous woman of her day. Rich and independent, Madame de Staël suffered the misfortune of being unattractive—an unpardonable sin in Paris. All this, plus her uncompromising intellectual brilliance and her determination to be involved actively in the affairs of her day, meant that she clashed repeatedly with Napoleon.[1] She was also a Protestant and therefore always something of an outsider. Her book *De l'Allemagne* (On Germany; published only with much difficulty in 1810), is nonetheless an impressive tour d'horizon of German culture, which introduced the new literature and Romantic philosophy to the attention of the French (and then to the rest of Europe).[2]

During the Revolution, she had been enthusiastic about its aims, if not all of its methods. She left Paris during the Terror but returned and, during the 1790s, became famous for her salon and her so-called duel with Napoleon. In her several novels, she championed the role of women such as herself, which the first consul/emperor objected to, and she was directed to live forty leagues from Paris. This was what provoked her visit to Germany.

Before she left Paris, she took German lessons with Wilhelm von Humboldt, then the Prussian ambassador in France (it was he who had convinced her of the renaissance in German culture). She traveled quickly to reach Weimar. Weimar itself, for all its achievements, "trembled" at the news of her imminent arrival. It was, we should never forget, a society "which both copied and despised French culture," and her arrival could not help but be a major event.

To the surprise of all concerned, the grand duke and his wife hit it off with their exotic visitor, this "Tempest in Petticoats." They were alternately charmed and intrigued by her outlandish turbans and revealing gowns, the "whiff of Parisian *chic*" that such clothes brought with them. Goethe had been a bit standoffish at first. He had been friendly enough at a distance, helping to arrange the German translations of Madame de Staël's books. But now, he said, if they were to meet, she must come to him—at Jena. The duke, however, was so enjoying her company that he instructed Goethe to return to Weimar. At first, Madame de Staël wasn't sure the author was worth the trouble. "Goethe ruins my ideal image of *Werther*; he is a fat man without distinction to look at, who likes to think he is a man of the world but only half succeeds." She never rid herself of the criticism that, despite its eminence, Weimar was provincial, that neither Wieland, nor Schiller, nor Goethe ever read a newspaper.

Nonetheless, Weimar grew on her and, as her German improved, her reading widened. "Goethe, Schiller and Wieland have more ingenuity, more depth in literature and philosophy than anyone I have ever met," she wrote to her cousin. "Their conversation is all ideas . . . Schiller and Goethe are attempting all kinds of innovations in the theatre."[3]

Compared with Weimar, Berlin was a disappointment. She was admitted to the court and introduced to all the aristocracy, but she did not feel settled there, perceiving that it was neither suited to her literary interests nor given over to social life, in which it was far inferior to Paris. Only after several weeks did she meet someone who, as she told her father in a letter, "had more knowledge of literature than almost anyone she had ever met." August Wilhelm Schlegel "spoke French like a Frenchman and English like an Englishman and while he was only thirty-six, he had read everything in the world."[4] Among the others she met was Fichte, to whom she boldly declared that his philosophy was "beyond her" and challenged him to explain it to her "in a quarter of an hour." Fichte gallantly made the attempt, but she interrupted him after just ten minutes, conceding that she did now grasp what he was driving at—and illustrated her new understanding with an analogy, a travel story in which someone achieved an improbable feat through the exercise of the will. Fichte was furious at what he saw as a trivialization of his self-important views.

De l'Allemagne had a difficult birth. In an attempt to rehabilitate herself with Napoleon, de Staël sent him a proof. Reading it, the emperor chose to believe that it was "anti-French" and instructed General Savary, the new minister of police, to seize the book and expel its author. Ten thousand copies were pulped, though one was smuggled to Vienna, enabling it to be published eventually in 1813, when it won widespread acclaim.

De l'Allemagne, like *Corinne* and de Staël's other books, subverted all that Napoleon stood for. Besides the book's detailed discussion of German poetry, prose, and drama, and Kantian and other philosophies, the central concern was with freedom—inner freedom as well as political freedom. It showed that people who were subjugated politically could not be subjugated intellectually and implied that Kant was the starting point of resistance to oppressors. It was in this book that Madame de Staël coined the word "Romanticism" to describe the new form of poetry she found in Germany: a poetry that celebrated the individual human spirit.[5] Her discovery and her translations of the German "greats" had an immediate impact on her French and other European contemporaries (such as the British), who until then were largely ignorant of German culture. The French at that time dismissed German culture as vulgar, but she argued instead that, even if that were true, original thinking (which is what the Germans had) counted for more than good taste. Her hope was that *De l'Allemagne* would serve to rekindle French literature, which was in her view moribund under Napoleon's censorship.

She was not blind to Germany's faults, finding there a general uncongenial atmosphere of "stoves, beer and tobacco," and the aristocracy dull. She was aware that people in general were xenophobic, that there was "more imagination than wit" and a surprising contrast between their intellectual daring and submissiveness to authority.[6]

"Horae Germanicae"

One of the fellow travelers that Madame de Staël met in Germany was the Englishman Henry Crabb Robinson. Trained as a lawyer but also a fellow of the Society of Antiquaries, Crabb Robinson belonged to the Norwich circle of "intellectuals and dissenters" that included William Wordsworth, Samuel Taylor Coleridge, and Robert Southey among its fellow members. He traveled to Germany in 1802–03 to study philosophy, Kant in particular, and sent back articles on German subjects to the *Monthly Register*.[7] But Crabb Robinson was by no means the only Briton to start taking an interest in Germany. William Taylor of Norwich, also a member of the circle around Southey and Wordsworth, described himself as the "first Anglo-Germanist." He became known for his translation of Lessing's *Nathan der Weise* and wrote a number of essays on German authors—Herder and Lessing in particular—between 1790 and 1820. In Britain, as it turned out, more influential than Madame de Staël's book were John Black's translations of Schlegel's writings on dramatic art and literature, in which he adapted Kant and Schiller's aesthetics to a critique of Shakespeare. Both Wordsworth and William Hazlitt found Schlegel's "Shakespearean insights" instructive. *Blackwood's Magazine*, begun in 1817, contained a regular section, "Horae Germanicae" from 1819 on, "in which new German works were the subject of knowledgeable attention."

None of the above names, however, had anything like the impact of four writers who between them did succeed in making the contemporary developments in Germany much more widely known across the Channel and, to some extent, in America: Coleridge, Thomas Carlyle, George Henry Lewes, and George Eliot.[8]

It is not too much to say, Rosemary Ashton tells us in her study of the impact of German thought on nineteenth-century Britain, that it was Coleridge alone who, in the period between 1800 and 1820, induced his fellow Victorians—people like Eliot, Lewes, John Stuart Mill, Thomas Arnold, Richard Holt Hutton, and the philosopher James Hutchison Stirling—to come to terms with the new German developments and ideas. Though France, particularly the ideas of Henri de Saint-Simon and Auguste Comte, had a profound impact on political notions in Victorian

Britain, it was Germany—its philosophy, history, and aesthetics—which, she says, had the most enduring effects on English thinking.

Overall, Coleridge's English contemporaries were puzzled by this seeming obsession with Germany and teased him about it, but later generations took a very different view. In 1866, Walter Pater praised Coleridge for helping to identify the philosophical and literary movement in Germany as an "irresistible . . . metaphysical synthesis."[9]

Attracted particularly by Schiller's *Die Räuber*, Coleridge took up German and in 1798 crossed the Channel for a visit to Germany, where he discovered Kant. He confided to Crabb Robinson in 1812 that there was "more to Kant than any other philosopher." He was taken particularly with Kant's third critique, which considered aesthetics as a science. This led Coleridge to the ideas of the Romantics, especially the Schlegels and Schelling. Coleridge's chief impact, therefore, was on the reception of German philosophy, rather than literature, in Britain.[10]

Carlyle had much more influence regarding German literature. Known everywhere, not always flatteringly, as the *Vox Germanica* of London, Carlyle exhibited a Germanic thoroughness and a Germanic interest in history: he spent fourteen years writing a biography of Friedrich the Great. (Hitler had the German translation read to him during his last days in the bunker.) To begin with, like Coleridge, he was drawn to German philosophy as a counter to British skepticism and materialism, and it was his enthusiastic endorsement of Kant and Fichte in a series of articles he published in the *Edinburgh Review* that excited his generation. These essays were soon republished in American magazines, where Ralph Waldo Emerson, Margaret Fuller, and others absorbed the new "German philosophy," which took root in New England as Transcendentalism.[11]

Carlyle's role can be identified much more directly because so many people spoke or wrote about the effect on them of his "very German" novel, *Sartor Resartus* (1833–34). "Hardly a young person survived the 1830s without being struck by *Sartor* and by it inspired to read—even if only in Carlyle's English translation—*Wilhelm Meister*." Carlyle was tireless in his attempt to convince his readers of the value of reading German literature, especially Goethe, and was successful to the extent that he could write in 1838 that "readers of German have increased a hundredfold." Thanks to him, G. H. Lewes took up German and traveled to Germany in 1838, aware that an understanding of German literature was virtually obligatory for a budding author and critic. Once in Germany he discovered a systematic aesthetic in Hegel. Lewes returned to Britain full of the philosopher, but he also drew attention to Goethe's inquiries into botany and optics, which had gone unnoticed in Britain until then. (Lewes wrote the first complete *Life* of Goethe in any language.)[12]

Lewes traveled to Germany more than once with George Eliot, who was no less fascinated, although her focus was the higher criticism of the biblical texts. This form of study was every bit as controversial in Britain as it was in Germany and accounted for much of the prejudice north of the Channel against Germans, the names that were particularly reviled in Britain being Strauss and Feuerbach. George Eliot, however, was a "freethinker" and this made her open to ideas from abroad and gave her the courage to translate Strauss's *Das Leben Jesu* in 1846. She too thought that Britain couldn't afford to ignore German developments.

Not that she was blind to Germany's shortcomings any more than was de Staël—she found the "*Gelehrten*," the scholars of Weimar, "naively pompous." And yet, in 1865 she published an article with the title "A Word for the Germans," in which she conceded that Germans could at times be weighed down by a laborious, cumbrous writing style, but insisted: "If he is an experimenter, he will be thorough in his experiments; if he is a scholar he will be thorough in his researches. Accordingly no one in this day really studies any subject without having recourse to German books."[13] No one, she concluded, could call himself an expert on anything until he had read what the Germans had to say.[14]

The pre-eminence of Germany had been the case for some time in education. Francke's work, for example, attracted interest very early on, and charity schools were founded in Britain on the same principles as those at Halle.[15] A description of Halle and its system was published by Dr. Josiah Woodward, as *Pietas Hallensis* in 1705, and widely read in America too. Anthony Boehm, a graduate of Halle, opened a Francke-inspired school in Britain as early as 1701.

As interest in German thought grew, an increasing number of boys were sent to Germany for language training, for which the demand was so great as to encourage L. H. Pfeil, father of Goethe's secretary, to found a school dedicated to the purpose. In 1800 a dedicated periodical was founded for those interested in Germany. Called *The German Museum or Monthly Repository of the Literature of Germany, the North and the Continent in General*, it lasted for barely three years, but then some of its features were taken up by *Blackwood's Magazine*.[16]

More important in the long run was Thomas Campbell's idea that a University of London should be founded along the lines of the Universities of Berlin and Bonn, rather than on the Oxford or Cambridge model. When it was founded, the University of London had a chair of German right from the start, the professor being none other than Schleiermacher's brother-in-law. German philologists occupied both the chair of Oriental languages and that of Hebrew. Isaac Lyon Goldsmid, who became one of the University of London's most substantial benefactors, and subsequently Britain's first Jewish baronet, traveled to Bonn and Berlin to clarify his ideas about what a university should be.[17] Before the Royal Commission on Oxford and Cambridge produced its report in 1850, Oxford had already made an attempt to become more "German" by introducing lay professors and a more practical examination system.

Thomas Arnold, the famous headmaster of Rugby School, was the first man north of the Channel to acknowledge the value of philology. He recognized the advances being made in Prussia, going so far as to teach "his" boys German rather than French. After *The German Museum* had come and gone, a second periodical, *The Philological Museum*, started in 1831. This did better, becoming virtually a "parish magazine" for the Germanists.[18] Hardly less influential than Arnold was his great-nephew, Adolphus William Ward. He studied at Leipzig, caught what he called the "German fever" and, after he had gained the professorship of History and English Language and Literature at Owens College, Manchester (the future University of Manchester) in 1866, set about transforming it to a research-oriented German-style university.

The most famous British academic of that time was John Emerich Edward Dalberg (later Lord) Acton (1834–1902). He had a German mother, and partly for this reason spent eight years studying under the historian Johann Döllinger at Munich. Acton became known for his survey in the *English Historical Review* of "The German Schools of History." No less inspired by Germany was Florence Nightingale, who was much taken by the Institute of Protestant Deaconesses at Kaiserworth near Düsseldorf, which trained teachers and nurses. It was her visit to the institute in 1850 that convinced her that nursing could be a profession, much more than menial employment. She returned the following year for training herself.[19]

The interest in German—and especially Prussian—education was growing in England ("Look at Germany" was a frequent mantra). In 1861, Mark Pattison, an Oxford don who had been the London *Times* correspondent in Berlin, was appointed to a commission that was asked to report on German schools. In the published document, he argued that the real bedrock of the German success story with its schools—which had now been in existence for half a century—was compulsory attendance; this was, he concluded, "a precious tradition." Arnold himself, when he gave evidence to the Taunton Commission on Endowed Schools, also recommended German (and French) practices, his arguments proving so compelling that his report was later published separately as *Higher Schools and Universities in Germany* (1882). He advocated a much greater concentration on science, again as was true in Prussia.[20] Philology was still central, in the German way, and Max Müller, a German polymath whose knowledge of Sanskrit was such that the East India Company commissioned an edition of the Rig Veda, settled in Oxford. Despite all this, in 1860 Germany still had six times as many students per capita of the population than did Britain.[21]

Britain's German Reagent

Following the visit of Justus von Liebig, the biologist, to Britain, at the invitation of the British Association (see Chapter 13, p. 271), two "outstations" of Giessen were founded: these were the Rothamsted Experimental Station in 1843 and the Royal College of Chemistry in 1845. Von Liebig's advice was sought in regard to the presidency of the Royal College, and he recommended August Wilhelm von Hofmann. Prince Albert, Queen Victoria's German consort, met Hofmann at Brühl on the Rhine and subsequently interceded with the king of Prussia to allow the chemist a leave of absence from the University of Bonn for two years. Hofmann stayed in Britain for more than a decade, and the various aspects of his multifarious career are considered later.[22]

"The most active German reagent in Britain from 1840 to 1859 was Prince Albert." History has been kind to Albert. His great contribution, according to Hermione Hobhouse, was "to free the British monarchy from the Party allegiances which had hitherto been accepted, to pave the way for a constitutional model in which there was a place for Her Majesty's Opposition as much as for Her Majesty's Government."[23] His most tangible monuments were the royal palaces which, we are apt to forget, were built or remodeled in his lifetime: Buckingham Palace, Balmoral Castle, Osborne on the Isle of Wight, and the farm buildings at Windsor Castle.

Born in 1819 at the Schloss Rosenau, near Coburg, Francis Albert Augustus Charles Emmanuel, Prince of Saxe-Coburg-Gotha, was the second son of Duke Ernest I, and grew up in a world with intimate links to British royalty. Queen Victoria was Albert's cousin, her own mother, the Duchess of Kent, being Princess Victoria of Saxe-Coburg.

Albert was an intelligent and above all *interested* consort, who did much to stimulate proliferation of the arts and sciences in Britain. He became known for his visits to the studios of living artists, persuaded the queen to be more practically involved in philanthropic matters, and was himself president of the British Association at the association's meeting in Aberdeen in 1859, and chairman of the international conference on statistics (very close to his heart, for he had been tutored by M. Quetelet, one of the French founders of the subject) in 1860. It was the prince who suggested in 1855 that the entry to the diplomatic service should be by competitive examination, rather than by patronage, as was traditional.[24]

He was the most important figure in the British collecting world in the 1840s and 1850s, his taste enriching the National Gallery as well as the Royal Collections. Consulting Ludwig Gruner, an art expert from Dresden, among the works he acquired were Duccio's *Crucifixion*, Fra Angelico's *St. Peter Martyr*, and Lucas Cranach's *Apollo and Diana* and *Madonna and Child*. Albert set in train several major studies of his favorite painter, Raphael, the intention being to create a corpus of material on the artist. Eventually some 1,500 photographs, prints, and engravings were collected and deposited in the British Museum for the use of scholars. He was helped by two German art historians.[25]

As a clever man who could see the changes taking place around him, he recognized the need for education and industry to work together. He joined the Society of Arts (founded in 1754 "for the encouragement of Arts, Manufacturers and Commerce") very soon after his arrival in Britain and became its president in 1843. It was officials of this society who, in 1844, revived the idea of an annual exhibition of manufacturers, which would eventually lead to the Great Exhibition of 1851. This was, perhaps, Albert's greatest contribution, his role in making the Great Exhibition happen. He was chairman of the Royal Commission on the exhibition and intimately involved in all the detailed planning.

The displays of the German-speaking states at the exhibition easily outstripped those of the United States in magnitude and rivaled those of France.[26] The best work from the Prussian government's iron and zinc foundries was shown, together with Saxony's Meissen porcelain, musical instruments, and clocks, and telegraphs produced by Siemens and Halske, revealing the advanced state of communications in the country. There were textiles—dyed many colors—lenses, machines for creating newspaper type, plus sculptures from the Berlin and Munich schools. It was an early view of Germany's looming industrial power.

The Great Exhibition was a notable success, not least financially, realizing a surplus of £180,000, an enormous sum. Initially, Prince Albert wanted the profits from the exhibition (which the government promised to match) to be used to found a number of schools of science and industry at South Kensington, where he also wanted all the scientific societies to be grouped, together with the Institution of Civil Engineers, to form a Napoleonic-style technical-national university. It didn't work out but "Albertopolis" was realized to the extent that "South Kensington," "that un-English complex of museums, scientific institutions, colleges of music and art, part university, part polytechnic," advanced in fits and starts, and today is the intellectual and artistic heart of London. The Albert Memorial stands at the edge of this, overseeing the prince's great creation.[27]

While he was in Britain, the prince retained his interest in the country of his birth. He had been infected with enthusiasm for German unity while he was a student at Bonn.[28] He used his London experience to try to influence the Prussian king in the direction of a constitutional monarchy and parliamentary government along British lines, a model he genuinely approved. His advice may have been responsible for Friedrich Wilhelm's decision in 1847 to issue a patent establishing a united Diet for the whole of Prussia. Albert fell gradually out of sympathy with the authoritarian trend that was such an important element of the Bismarckian approach. However, he did form a firm friendship with the prince of Prussia, the later King and Emperor Wilhelm I. When Prince Wilhelm was forced to flee from Berlin after the Berlin uprising in March 1848 (he was responsible for the shooting of demonstrators and became known as *Kartätschenprinz*, the Prince of Cartridge Shot), Albert made use of the prince's presence in London to begin a sustained attempt to win him over to constitutionalism; Albert argued that, after 1848, a confederation was no longer adequate and that a single state was necessary.[29] He was also critical of an excessive Prussian influence over the future of Germany.

The political differences between Prince Albert and his son, Edward, the Prince of Wales, should not be overlooked. The latter reacted against what he saw as his father's "overestimation" of Germany and his marriage to Alexandra, the daughter of the future king of Denmark, ensured that he was firmly in the anti-German camp after the Schleswig-Holstein war of 1864 (by which time Albert was dead). Following this, Edward's sympathies lay with France. A parallel development occurred with Wilhelm II of Prussia, who took against his father and his English mother. In the new climate, dynasties could no longer be links between nations.[30]

These are murky waters, with ambiguous messages. Albert's very tangible influences on the arts and sciences in Britain, on its educational structure and its constitutional monarchy, are his true legacy, and quite substantial enough. Partly as a result of Albert's influence, many German businesses opened up in Britain (150 in Lancashire) and numerous German clubs established along Oxford Street in London.

The PhD Crosses the Atlantic

Germany's relationship with the United States has been very different and in many ways more intimate. It was a German, the cosmographer Martin Waldseemüller (1470–1522), who was the first to suggest, in 1507, that the name "America" be used to designate the New World.[31] Similarities between the beliefs of the English Quaker William Penn and the German Pietists had a major effect on America. The English government was in debt to Admiral Penn, father of William, on account of his military successes and because he had paid his men out of his own pocket. The amount owed was a tidy sum: £16,000. Instead of cash, William accepted instead the grant of a large area of land north of what would become Washington, which was named Pennsylvania. Then, in 1677, when Penn was in Germany to meet with Pietists, they negotiated to buy 15,000 acres, subsequently extended to 25,000 acres.[32] This land would become Germantown.

At various times, attempts were made to turn Missouri, Texas, and Wisconsin into completely German states. Such plans never succeeded, but as a result those states always had a larger than average number of Germans. In 1835 it was thought necessary to establish a society called "Germania," the aim of which was to sustain German customs, speech, and traditions against what were felt to be destructive influences. Wisconsin in particular had an attraction for Germans. The climate and soil were similar to that of northern Germany, land was cheap, and people could vote after only one year of residence. The state had a commissioner for immigration, resident in New York, an arrangement so successful that at one time two-thirds of the immigrants to Wisconsin were German, and the Wisconsin Bureau of Immigration became known throughout Europe. The Wisconsin Central Railroad sent an agent to Switzerland, where he recruited some 5,000 immigrants, mainly German speakers, promising them land along the railroad they would help to construct.[33] German immigrants into America were particularly numerous after 1848, in the wake of the European revolutions; this meant the bulk of the new people were radicals who were far more apt to side with the northern cause in the Civil War.[34]

"The earliest instance of intellectual exchange of any consequence that we know about between Germany and New England was the correspondence between Cotton Mather and August Hermann Francke. In 1709 the Boston theologian sent a collection of 160 books and tracts on Pietism to Halle, and several sums of money to support Francke's philanthropic work. Francke's reply was a Latin letter of sixty-nine pages, describing the work of the Halle institutions."[35] The sons of both men continued to correspond, during which time "orphan-homes" on Halle lines were opened in America.

Franz Daniel Pastorius was the first German teacher that we know about and he worked in the English Quaker School in Philadelphia. A friend of William Penn, he became the founder of Germantown, establishing the first German school there in 1702. He introduced two innovations which had profound consequences: his school was coeducational, and it had a night school for anyone whose work precluded their attendance during the day.

Benjamin Franklin is the first American on record to attend a German university—in his case, Göttingen in 1766 (he was made a member of the *Göttingen Gelehrte Anzeigen*). George Ticknor and Edward Everett were, however, the first two regular American students at Göttingen (1815–17). Ticknor, by all accounts, was heavily influenced by Madame de Staël's book: "All the north of Germany is filled with the most learned universities in Europe. In no country, not even in England, have the people so many means of instructing themselves and of bringing their faculties to perfection."[36] Ticknor and Everett were the first in a movement that would grow in strength in the nineteenth century and would shape American education fundamentally. Throughout the century two batches of Americans flooded into German universities: Göttingen, Berlin, and Halle until 1850, roughly speaking, including Ralph Waldo Emerson and Henry Wadsworth Longfellow; later, Leipzig, Bonn, and Heidelberg were more popular. On his visits there Everett never stopped buying books, which in the end formed the core of the German library at Harvard. This, says Albert Faust, was the beginning of the mass migration of German book collections to America (the "*Bücherwanderung*").[37]

Carl Diehl estimates that between 9,000 and 10,000 Americans studied in Germany from 1815 to 1914, not least nineteen future college and university presidents. His figures show a slightly different picture, that four universities attracted the bulk of the students—Göttingen, Berlin, Halle, and Leipzig—while Heidelberg became popular later on. Most entered the philosophical faculties to study the humanities or the natural and social sciences, with a rapid decline in the theological faculties after 1850. The American influx was led by just two American institutions in the early years—Harvard and Yale, with 55 percent of the American students having been students at one or the other institution.[38]

Then, as the nineteenth century progressed, as an interest in history and science developed in the United States, as its own literature began to emerge, and as more graduates returned from Germany—some with PhDs—Germany's universities grew even more in stature in the eyes of Americans, in particular their approach of linking teaching and research.* This later generation—from the end of the 1840s—was the first to import the ideal of German scholarship, advanced academic study as a recognized professional vocation, and it was these returning students who created the modern form of scholarship in the humanities in America. Diehl identifies such men as Francis Child and George Lane, who were to form the backbone of the German-trained faculty at Harvard, Basil Gildersleeve, the first philologist at Johns Hopkins, and William Dwight Whitney, the eminent Yale Sanskrit philologist, all of whom studied in Germany. To them may be added "a dazzling array of future college and university presidents, many of whom would be instrumental in creating the modern university in America." Charles Eliot's curriculum reform at Harvard in the 1870s and his promotion of graduate studies have generally been taken as the first indication of the emergence of the modern university system in the United States, but Diehl points out that, by that time, "there were at least nine professors of humanities out of the total Harvard faculty of twenty-three who had received advanced training in Germany."[39] They had been influential in choosing Eliot in the first place, and he himself studied chemistry in Germany. By 1870 Yale also had a half-dozen German-trained professors in the humanities, including both the outgoing president, Thomas Dwight Woolsey, and the new one, Noah Porter. In fact, says Diehl, study in Germany had become a kind of graduate school for the graduates of American universities. "By 1850 many American universities made it publicly known that they would favor applicants with German training."[40]

Around 750,000 German immigrants, known as the "Forty-Eighters," entered the United States between the mid-1840s and the mid-1850s. Among Germans, having a "Forty-Eighter" among your ancestors is almost as notable as having an ancestor on the *Mayflower* is for English-speaking Americans.[41] In 1854, 215,000 Germans immigrated to America, a record beaten only in 1882, when 250,000 crossed the Atlantic.

---

* It should also be remembered that the development of iron steamers made transatlantic travel much quicker, cheaper, and safer.

In the arts and humanities, German-Americans had their share of painters—Emmanuel Leutze (*Washington Crossing the Delaware*) and Albert Bierstadt (*In the Sierras*)—and authors—Friedrich List (*Outline of a New System of Political Economy*) and Owen Wister (*The Virginian*). Among German-American philanthropist-businessmen were John Jacob Astor (born near Heidelberg in 1763) and Francis Martin Drexel (born in the Austrian Tyrol in 1792), who spent some years as a painter in South America before traveling north and founding a bank in Philadelphia in 1837 (the New York house, Drexel, Morgan, and Company was founded in 1850). John D. Rockefeller was descended from Johann Peter Rockefeller, who came from Germany and settled among the earliest New Jersey Germans.[42]

Overall, alongside the universities, the German influence on American life was felt most strongly in the early nineteenth century in music and journalism. The large German Protestant churches promoted both vocal and instrumental music. In New England the Handel and Haydn Society had by far the most famous choir. Founded in 1815 by Gottlieb Graupner, owner of a music store, who also organized and conducted the first prominent orchestra in America, which was rivaled only by a band of musicians from Hamburg in Philadelphia. About the middle of the nineteenth century, New York took over as the center of American music making, and this had something to do with the arrival of the German Orchestra, upward of a score of young musicians, many of them "Forty-Eighters."[43] Choral societies were founded soon after in Buffalo, Pittsburgh, Cleveland, Louisville, Cincinnati, and Charleston (the Teutonenbund). These competed at annual singing festivals, some of which grew into Musikvereins. Beginning in Milwaukee, they were not just choirs but also commissioned new operas and oratorios.

The first German-American periodical was founded in 1739 by Christopher Sauer; this was the *Der Hochdeutsch-Pennsylvaniesche Geschicht-Schreiber, oder Sammlung Wichtiger Nachrichten aus dem Natur und Kirchen-Reich*. Published in Germantown, its title was (thankfully) shortened to *Germantown Zeitung* as it changed progressively from a semiannual to a quarterly to a monthly and then, from 1775, to a weekly. By the end of the eighteenth century, there were five German newspapers in Pennsylvania, one of them published half in German and half in English.

It was the early years of the nineteenth century that saw the great boom in circulation and influence: *Die New Yorker Staats-Zeitung* was founded in 1834, *Der Anzeiger des Westens* in St. Louis in 1835, and the *Cincinnati Volksblatt* in 1836.[44]

In 1813 Caspar Wistar, a glass manufacturer, took over from Dr. Benjamin Rush as president of the Society for the Abolition of Slavery and, two years after that, followed Thomas Jefferson as president of the American Philosophical Society.[45] As a result, Hegel, Schleiermacher, and the Young Hegelians received growing attention.

It should not be overlooked that this was a period of strong cultural exchange. There was also a great deal of interest *about* Britain and America inside Germany. It was at this time, for example, that Rudolf von Gneist wrote his four-volume work on local government in the United Kingdom, a work used north of the Channel. It wasn't globalization as we know it in the twenty-first century, but not everyone was as nationalistic in the nineteenth century as is sometimes made to appear.

16.

Wagner's Other Ring—Feuerbach, Schopenhauer, Nietzsche

Thomas Mann said that Richard Wagner's acquaintance with Arthur Schopenhauer was the greatest event of the composer's life. It occurred in the autumn of 1854 when he read—and was overwhelmed by *Die Welt als Wille und Vorstellung* (*The World as Will and Representation*), two volumes of well over a thousand pages, yet which he read four times in one year. Few—if any—great composers studied philosophy as seriously as Wagner did. According to Bryan Magee, himself a philosopher, neither *Tristan und Isolde* nor *Parsifal* would have taken the form they did without Wagner's absorption of Schopenhauer's ideas, and the same argument applies to entire sections of *Der Ring des Nibelungen*.[1]

The reason Wagner was so different from his fellow composers can be put down to politics, in particular his disillusionment following the revolutions of 1848, which caused him to turn inward, away from activity, thus opening himself up to other influences, of which philosophy proved decisive. A passionate—and active—left-wing revolutionary when he was young, Wagner has often been depicted as lurching to the right in middle age. It is truer to see him as someone who fell out of love with politics itself, as someone no longer convinced that the most pressing human problems are amenable to political solutions.[2]

Born in 1813, the same year as Giuseppe Verdi, Wagner died at the age of sixty-nine, in 1883, the same year as Marx. He knew very early on what he wanted to do, which was to compose operas, and he started while he was still in his teens.

At the time, three forms of opera were popular: the German Romantic opera of Weber; the Italian Romantic Realism of Vincenzo Bellini, Gioachino Rossini, and Gaetano Donizetti; and French opera, epitomized by the spectacles of Giacomo Meyerbeer and Fromental Halévy. Wagner tried his hand at all three, deciding that the best way forward lay with the German form, the genre of his three best-known early operas—*Der fliegende Holländer* (*The Flying Dutchman*; 1841), *Tannhäuser* (1845) and *Lohengrin* (1848). Joachim Köhler says *The Flying Dutchman* was Wagner's "French Revolution" though no one noticed it at the time.

There then followed, not a crisis exactly, but a period of reflection. By this stage, Wagner had been married twice, at first to a beautiful actress who had no idea what kind of genius her husband was and wanted only for him to be conventionally successful. His second marriage, to Cosima, an illegitimate daughter of Franz Liszt, was considerably happier. Much less beautiful, she devoted her life to him.[3]

His early politics had led him into friendship with Mikhail Bakunin, the Russian anarchist, and to Wagner's presence on the barricades in the Dresden uprising of 1849. In 1843, following *Rienzi* and *The Flying Dutchman*, both of which were well received, he accepted the position of Kapellmeister in Dresden, where Bakunin was living. Wagner was just twenty-nine and in his autobiography says he found the anarchist "a truly likeable and sensitive person." Bakunin of course knew Marx, though he could be anti-Semitic.[4] Wagner and Bakunin formed part of the leadership of the uprising in Dresden. As a result of the failed revolution, Wagner was forced into exile in Switzerland, a wanted man in Germany.

During his first years in Switzerland, he composed scarcely any music but produced instead a number of prose writings that made his name more widely known. Two of them are still much read: *Das Kunstwerk der Zukunft* (*The Work of Art of the Future*; 1849) and *Oper und Drama* (1850–51). These are both important works of theory; and with his theory of the "complete embrace" of art in place, he set about trying out his ideas, which were to lead to music very different from anything composed before.[5]

In the first place, he produced the libretti of the four operas that comprise *Der Ring des Nibelungen*: *Das Rheingold, Die Walküre, Siegfried,* and *Götterdämmerung* (usually translated as "The Twilight of the Gods"). At the same time, he wrote the music for the first two. Then came a long break. After Act II of *Siegfried* he gave up on *The Ring* and, for twelve years, never went back to it. He wasn't idle: he composed *Tristan und Isolde* and *Die Meistersinger von Nürnberg.* Only then did he go back to *The Ring,* finishing *Siegfried* and writing the music for *Götterdämmerung.* Only one opera followed the *Ring—Parsifal,* which premiered in 1882, the year before his death.

His personal situation was rather more dramatic than this sounds. When he composed *Rheingold, Valkyrie,* and *Tristan,* Wagner was already in his fifties, but he had little prospect of the works ever being performed. Moreover, he was in debt in Vienna and, to avoid imprisonment, he was forced to flee and so was "on the run" for the second time.[6] At this point, the Wagner story, if it doesn't have a fairy-tale ending, has at least a fairy-tale middle. Ludwig II, the king of Bavaria, was an eighteen-year-old passionate soul who felt as strongly about Wagner's music as Wagner did himself. Out of the blue he offered the composer funds to stage his operas and, using the same funds, Wagner built his own opera house and launched the Bayreuth Festivals, which continue to this day.

Before Schopenhauer, the thinker who had the greatest influence on Wagner was Ludwig Feuerbach (see Chapter 11, p. 239). In his autobiography, Wagner says he "discovered" Feuerbach while living in Dresden and that he was "the sole adequate philosopher of the modern age." In particular, and as the poet Georg Herwegh was the first to note, he was influenced by Feuerbach's *Das Wesen des Christentums* (*The Essence of Christianity*), which, it will be remembered, argued that nothing exists except man and nature, that therefore "higher beings" are merely a reflection of our own anxieties and ambitions (Wagner dedicated one of his own books, *The Work of Art of the Future,* to Feuerbach).[7] The aspect of Feuerbach's argument that appealed to Wagner was his idea that the reason religious belief has been almost universal is that it "meets basic human needs" and is not really interested in, say, biology or physics.[8] Religion has to be looked at not for what it reveals about heaven or fundamental aspects of reality, but for what it reveals *about ourselves.*[9] These ideas were incorporated into the libretto of *The Ring*, where many of the characters are "gods at an early stage of the world's development."[10] They are, in a Feuerbachian sense, projections of universal human characteristics and desires and not to be understood as inhabiting a transcendental world.

Though this is fine as far as it goes, it omits several other elements in Wagner's idea of musical drama that add levels of complexity. The first of these was nationalism, which we need to remind ourselves was then a left-of-center cause, drawing its power as a reaction to those political conservatives who wished to preserve the separateness of the smaller ancien régimes, each with its own ruling elite and, more often than not, archaic feudal institutions. In line with this, music had its own nationalist elements. Wagner, in particular, thought that, after Mozart and Beethoven, it was absurd that Germans should still place such a high premium on French opera. After Bach and Haydn, the German tradition was now the greater one. *Die Meistersinger* was his answer.[11]

Another complicating factor was what has been called Wagner's "metaphysical turn." According to his autobiography, the turning point was the right-wing antiparliamentary coup in Paris in 1851 when Louis Napoleon seized power. Wagner concluded from this that the world he wished to see exist would never come about by political action and that the human condition is essentially unchangeable. He turned away from politics, and became more inward- than outward-looking.[12] One final factor in Wagner's psychological makeup was his view of ancient Greece. When ancient Greek civilization disintegrated, he said, the essentially humanistic Greek gods, and the most important subject matter of all—myth—was no longer available to art.[13]

The "Great Event"

In many aspects of his thought, Schopenhauer had grasped most of the insights Wagner had arrived at, albeit by a different route. The minute he encountered Schopenhauer, Wagner realized not only how far ahead the other man was, but also that the philosopher's German prose was itself "a work of art." In the last half of 1854, Wagner was at work on the music for the beginning of *The Valkyrie* when he chanced upon Schopenhauer's *The World as Will and Representation*. Wagner, who was not in the best of health when he encountered Schopenhauer (he had a boil on his leg), never let go of this book and dipped into it often.[14]

The World as Will and Representation had been published as long ago as 1818 but had hardly set the Rhine on fire. In April 1853, however, in the radical *Westminster Review*, the assistant editor of which was the Germanophile George Eliot, an article titled "Iconoclasm in German Philosophy" appeared over the name of John Oxenford. Oxenford gave a commendably clear summary of Schopenhauer's philosophy, so much so that the article was swiftly translated and published in the *Vossische Zeitung*, meaning the translation was seen by a far wider public than the original. Schopenhauer caught on and suddenly, in his midsixties, he became famous after a lifetime of being sidelined.[15]

It was this sudden burst of interest in Schopenhauer that brought him to Wagner's attention, with the result that, at Christmas 1854, Wagner sent the philosopher a copy of the libretto of *The Ring*, inscribed "With reverence and gratitude." Unfortunately, Schopenhauer took offense because Wagner did not enclose a letter with the libretto, and neither then nor subsequently did he have any dealings with Wagner. Bryan Magee again: "There is something almost unbearably poignant about the fact that Schopenhauer went to his grave not knowing that one of the greatest works of art of all time had already come into being under the influence of his philosophy."[16]

Music as Metaphysics

Schopenhauer believed Kant was the most consequential philosopher of all time, certainly since the Greeks, and saw himself as carrying on the Kantian tradition. The single idea he found most seductive was Kant's notion that "total reality is comprised of a part which can be experienced by us and a part which can not," the division of "the phenomenal" from "the noumenal" (see Chapter 5, p. 135). Building on this, Schopenhauer's philosophy consisted of four intertwined entities. He thought, first, that Kant had it wrong when he said that, outside the empirical world, there are *things* in the plural. For one thing to be different from another, he said, they had to occupy space and time; but space and time are aspects of experience and must exist only in the empirical world. Even something like numbers, abstractions that "seem to exist" beyond space and time, can only be entertained in our minds because of our understanding of succession, which is itself unintelligible without the notions of space and time.[17] Schopenhauer concluded from this that, outside space and time, "everything must be one and undifferentiated." In other words, total reality consists of two aspects—phenomena, a world of many material objects, located specifically in space and time, and the "noumenal realm," which is a "single, undifferentiated something—spaceless, timeless, non-material, beyond all reach of causality." This realm is inaccessible to experience or knowledge.

Schopenhauer said further that he thought the two realms were different aspects of the same reality understood in different ways. For him, the noumenon is the *inner significance* of what we apprehend in the phenomenal world. Although Schopenhauer wasn't at all religious (he was a declared atheist, one of the first people to publicly admit this), he said he was making the kind of distinction a Christian makes in his understanding of the soul, as something significant hidden inside us. Down deep, said Schopenhauer, we are, all of us, the same something, but a something we can never fully apprehend. For Schopenhauer, there is an ultimate oneness of humanity, a realm we all share. Most important, we are compassionate because we realize that if one person injures another, that person in some way injures himself or herself. For modern tastes this is more than a little mystical. It also belies Kant's argument that ethics are rational.[18]

The second aspect of Schopenhauer's basic system—much easier to understand—is that he thought human life was bound to be tragic. Life, he said, is made up of endless "hoping," "striving," "yearning"—we are always, from our earliest days, reaching out for something. This endless yearning is inherently unfulfillable, for as soon as we get what we want, we want something else. This is our predicament.[19]

It is a predicament that is made all the worse by the third element in his thought, that we are, most of the time, selfish, cruel, aggressive, and heartless in our dealings with each other. If he was right, he said, if the noumenal and phenomenal worlds are the same reality but apprehended in different ways, this must mean that the noumenal realm itself is amoral and terrible. This was his famous—notorious—pessimism. Schopenhauer had a problem with what to call this terrible, blind, purposeless noumenal world and though he eventually came up with the word "will," he was never entirely happy with it. He chose that word, and the phrase "the will to live" because it seemed to him to be the "ultimate impulse" within us.[20] For Schopenhauer, we have to recognize the various manifestations of this will to exist, and to overcome them if we are to achieve contentment away from the world.

He thought that one of the reasons religions take the form they do is that most people cannot stomach profound metaphysical and moral truths when stated baldly—they have to be sugarcoated in parables, myths, and legends. Schopenhauer thought that religions embody the profoundest truths there are, and he also thought they had a great deal in common with creative art.[21] This led him to his fourth main argument, that the most accessible way for us to see into the heart of things—if only momentarily—is through sex and art, particularly the art of music.

Art apart, Schopenhauer's focus on sex was surprising, but to him it obviously had a wide-ranging effect on human behavior. He said: "If I am asked where the *most intimate knowledge* of that inner essence of the world, of that thing in itself which I have called the *will to live*, is to be found . . . I must point to *ecstasy in the act of copulation* . . . That is the true essence and core of all things, the aim and purpose of all existence." (He also added, "What is all the fuss about?")[22]

Art was similar: whatever a work of art is, once we are absorbed in it, we forget ourselves. At the same time, for Schopenhauer, each of the arts is representational—except music. Therefore, music is the expression of "something that cannot be represented at all, namely the noumenon." It is a metaphysical voice: "The composer reveals the innermost nature of the world, and expresses the profoundest wisdom, in a language that his reasoning faculty does not understand." It takes us away from the struggle for life.[23]

The Starting Point of Modern Music

The seriousness with which Wagner treated Schopenhauer and Kant helps inform us about his music. Disillusioned about politics—about the political *process* rather than any particular set of political views—Wagner was drawn to Schopenhauer's argument that art could be a refuge from the world, as the only way to encounter, however briefly or unsatisfactorily, the noumenal world. He was intent on creating—or uncovering—something that existed outside space and time, and the redemption of mankind, bringing it back into the fold, removing alienation, was for him the culmination of experience.[24]

Wagner was at work on the music for *The Valkyrie* when he encountered Schopenhauer; he had completed the libretti but not the music for *Siegfried* and *Götterdämmerung*. It follows that only *Tristan and Isolde, The Mastersingers*, and *Parsifal* were created after he had imbibed the philosopher.[25] In these three above all, we see how Schopenhauer influenced the musician. Wagner himself said that *Parsifal* was his "crowning achievement," after which he intended to cease writing opera and turn to symphonies, but in fact Wagner's sound-world began to change with *Siegfried*. Composed two years after he first encountered Schopenhauer, the music of *Siegfried* is already very different from what went before, *The Rhinegold* and *The Valkyrie*. The main difference, as again Bryan Magee has noted, is the relationship of the orchestra to the characters. In the earlier operas the music rises and falls—always *accompanies*—the words; in *Siegfried*, for the first time, the spectator cannot always hear the words, the sheer *weight* of orchestral sound, the massive wall of music, compels attention.

Schopenhauer's belief that music held a special place in the arts led him to make a number of specific comments about music, about acoustics as the ground for metaphysics, and to include a technical device in harmonics known as "suspension."[26] This reference seems to have found immediate resonance with Wagner, so much so that he decided to compose a whole opera based on the way suspension operates.* The idea was that "the music would move all the way through from discord to discord in such a manner that the ear was on tenterhooks throughout for a resolution that did not come." This was, in effect, pure musical Schopenhauer in that "the unassuaged longing, craving, yearning, that is our life, that is indeed us," would only be resolved in the final chord, which, in dramatic terms, would also be the end of the protagonist's life. In Bayreuth he even lowered the orchestra pit to help this effect.[27]

* Though the suspension chord (*Vorhaltsakkord*) could first be heard in the Call of the Sirens, *Ruf der Sirenen*, in *Tannhaüser*.

This is what makes *Tristan* a revolutionary composition. Consisting of almost nothing but discords, it sounds different from most of what has gone before and has, since its first night, been regarded as the starting point of "modern music," breaking all the rules. Tonality was the supreme aim—and achievement—of traditional music, and as a consequence it was composed in keys. *Tristan* was so different that the opera was not performed for five years after the score was published (for one thing, the bizarre succession of notes was impossible for the singers to sing or even remember).[28]

Exhilarating as this was, Wagner now argued plainly what was beginning to be obvious in his operas, that there is no equality between music and words. In opera, the experience is primarily a musical one, music is "the invisible world of feeling . . . As we construct the phenomenal world by application of the laws of time and space which exist *a priori* in our brain, so this conscious presentation of the Idea of the world in the drama would be conditioned by the inner laws of music, which assert themselves in the dramatist unconsciously, much as we draw on the laws of causality in our perceptions of the phenomenal world."[29] Music, as Bryan Magee has observed, is thus elevated here to a level of philosophical importance it never had before or since.[30]

Nietzsche contra Wagner

Just as Wagner looked up to Schopenhauer, so Friedrich Nietzsche looked up to Wagner. Nietzsche was twenty-four and still a student when the two men met in November 1868. Wagner was at the height of his fame. *The Mastersingers* had been premiered that year and received with greater enthusiasm than any of his previous works.

Nietzsche, like Wagner, came from Saxony in east Germany, and belonged to a family of Lutheran pastors.[31] Scholarships sent him to the universities of Bonn and Leipzig, where his subject was classics, not philosophy. He was such a brilliant student that, at the age of twenty-four, when still an undergraduate, he was offered an associate professorship in classical philology at the University of Basel, becoming a full professor a year later. So advanced was he that the University of Leipzig awarded him his degree without his having to submit a thesis or wait for the examination. He moved to Basel immediately.

These highly unusual accolades had made his reputation at the time he met Wagner. The two men remained good friends from 1868 until 1876, after which the younger man relinquished his academic post in Basel to devote himself full time to philosophy (he had tried to switch within the university but had been turned down). Following the break from Wagner, which we shall come to, Nietzsche developed his idiosyncratically itinerant lifestyle and, over a period of about twelve years, when he was between thirty-two and forty-four, he "poured out" the writings for which he is now famous.[32] His friendship was formed too late to have any influence on Wagner but he did influence other composers—Gustav Mahler, Frederick Delius, Arnold Schoenberg, and Richard Strauss, whose tone-poem *Also Sprach Zarathustra* is based on Nietzsche's best-known book. Like Schopenhauer, Nietzsche had a great interest in music—his great pleasures centering on Schumann, Schopenhauer, and solitary walks. He paralleled Wagner in that it was his discovery of Schopenhauer that proved the intellectual turning point in his life.

After they met, the friendship ripened, and Nietzsche became a frequent visitor to Tribschen, Wagner's house. Nietzsche spent Christmas there, helped with the printing of Wagner's autobiography, read the proofs, and was the copyist on the urtext of *Siegfried*. When Wagner went for walks, Nietzsche was allowed to play his piano.[33] Nietzsche's first book, *Die Geburt der Tragödie aus der Geist der Musik* (*The Birth of Tragedy out of the Spirit of Music*; 1872), is dedicated to Wagner, and Nietzsche goes so far as to say that his book is a "crystallisation" of his conversations with the composer.[34]

Nietzsche's argument is that we have—or had then—essentially misunderstood the ancient Greeks. A close reading of Greek tragedy, in particular the works of Aeschylus and Sophocles, shows that their concern is the "oceanic and irrational" feelings that swirl through human affairs—passion, eroticism, aggression, and intoxication, experiences he called "Dionysian." But this passion was rendered through mythic stories that channeled the imagination in a particular way, the "Apollonian," suitable to the linear—and therefore essentially rational—form of plays in the theater. While the plays themselves are one thing, the Dionysian side to life was destroyed by the development among the Greeks of "critical and self-critical intelligence," a relentless drive that culminated in Socrates, the "supreme critical intellect." Intellectual understanding, critical self-consciousness, became the prevailing methodology, arousing, as Nietzsche said, "terror and misconceptions."[35] Even morality, according to Nietzsche, was, in Socrates' world, a function of knowledge, the whole of human existence accessible to the "conceptualising intelligence."[36] This approach to experience culminated in Euripedes' tragedies, which made a mockery of what had been the essence of Aeschylus and Sophocles, namely their exploration of the irrational, of what, in Nietzsche's day, was already being called the unconscious. Nietzsche felt that Euripides was shallow, that his works forfeited the ability to move people and that this was how Greek art had declined and decayed. In Nietzsche's view, Wagner's music—with its emphasis on compassion as the basis for morality, and on the irrational—marked a return to Aeschylus and Sophocles, restoring drama to its former completeness as an art form.[37] Wagner loved the book and said Nietzsche was in a closer relationship with him than anyone except his wife.

This situation did not last. Around 1874, as Nietzsche was leaving his twenties, he seems to have felt an urge to strike out on his own. Or was it that Wagner, although used to associating with geniuses—Heine, Schumann, and Mendelssohn—refused to acknowledge that quality in Nietzsche? Wagner's wife, Cosima, observed that Nietzsche was "undoubtedly the most gifted of our young friends, but the most displeasing . . . It is as if he were resisting the overpowering effect of Wagner's personality."[38]

There was no decisive act of breakage and, at the Bayreuth Festival in 1883, Wagner confided to Nietzsche's sister: "Tell your brother than I am quite alone since he went away and left me." Nor did Nietzsche turn his back on Wagner right away. Two of his own books were devoted to the composer, *The Wagner Case* and *Nietzsche contra Wagner* while the title of a third, *Götzendämmerung* (*Twilight of the Idols*) was a clear echo of *Götterdämmerung.*[39]

A complicating factor was the philosopher's bad health. Nietzsche experienced chronic problems with his eyes, ferocious migraines, and terrible stomach upsets that caused him to vomit "in embarrassing circumstances." Since he would succumb to tertiary syphilis at the early age of forty-four, these symptoms are usually assumed to be the first signs of onset. Another symptom may well have been his idiosyncratic lifestyle: he lived alone and moved from rooming house to small hotel in Switzerland,

Italy, or France, devoting six to ten hours a day to walking in the open air, returning to his room only to write, eat, or sleep. The books that he composed in this way are now among the classics of philosophy, with a style all their own. Nietzsche was not concerned to give an overarching argument in the manner of Fichte or Hegel. He saw it as his task to offer startling, short, pithy insights in prose that is regarded as among the best German there is ("incandescent").[40]

"Weeds, Rubble and Vermin"

Nietzsche made the sharpest of breaks with Schopenhauer. There was no "other realm" for Nietzsche—he came to believe that this world that we experience is the only reality there is. (This is worth bearing in mind in all that follows: in modern German history the division between irrationalism and rationalism is sometimes too sharply drawn.) "The apparent world is the real world," said Nietzsche, and it was occupied by "weeds, rubble and vermin."[41]

His best known affirmation is that "God is dead," from which two important things follow. One, there is no transcendental realm; and two, our morals and values cannot stem from the other realm. Morals must have their basis in this world since there is nowhere "else" they can come from. Socially and historically, human beings create their own moralities and values, and the "ideal" is nothing but "a figment of the human imagination." He agreed with Schopenhauer that living things are basically selfish and will always seize what they want and then defend their possessions to the death. This is what he called "life-assertion," for him the most fundamental instinct of all, meaning that "the war of all against all" is the natural order. Civilization emerged from this war of all against all, in which, over the millennia, "the strong eliminated the weak, the healthy the sick, the clever the stupid, the competent the incompetent." Everything basic developed out of this struggle, but then came a crucial change. Two or three thousand years ago, in various parts of the world, there arose a generation of people who invented morality. "They taught that the powerful should not take what they wanted but should voluntarily submit to law."[42]

Nietzsche picked out in particular Socrates and Jesus. Between them, he felt, they had reversed the process that had distinguished humans from other animals and made culture possible. Their influential—but to him perverted—doctrine succeeded because it suited the interests of the majority, the masses, the "ungifted." In doing this, Socrates and Jesus, he thought, did their best to inhibit the natural processes of the onward march of civilization and were jointly responsible for decadence and decline.[43]

From such a starting point, Nietzsche embarked on a wholesale critique of contemporary culture. For him it was self-evident that mankind's institutions, our arts and sciences, our philosophy and politics—as they have grown since Socrates and Jesus—have developed on the basis of false values. It was now our task, he said, beginning with himself, to reconstruct our world. "It will be the biggest break, the biggest watershed in the whole of history so that, throughout the future, all time will be reckoned as being either before this event or after it."

His purpose was to render human beings wholly spontaneous, to make them as free and unself-conscious as animals and to realize that they were discordant beings, not harmonious ones.[44] The unrepressed and uninhibited life would be for him superior to what had gone before and, to reflect this brave new world, he minted a new name for those who lived in this way—*Übermenschen* (supermen). The *Übermensch* would be free to take full advantage of the fact that there is no soul, no God, no transcendental realm, *no world other than this.* "There are no rewards other than the joy in being. The meaning of life *is* life."[45] Life assertion became the task of superman, the supremely valuable activity. The will to live, to assert one's presence in the world, to sweep aside all obstacles—Nietzsche called this "the will to power" and we can see clearly how this turns Schopenhauer on his head and comprises the radical mutation of the notion of Bildung. Since the noumenal realm doesn't exist, our "oneness" with it cannot exist either, and therefore our compassion that arises from it, and forms the foundation of morality, cannot exist either. Morality stems from self-interest, and there is absolutely no place for compassion.[46]

In working against Schopenhauer, Nietzsche also—naturally—drew away from Wagner. This shift surfaced spectacularly in his attack on *Parsifal*, which he denounced as "An apostasy and reversion to sickly Christian and obscurantist ideas . . ."

In 1889, Nietzsche collapsed in the street in Turin, incurably insane, and was nursed for the next eleven years by his sister Elisabeth, who was to doctor his manuscripts and create her own controversy. Until that point he had never really settled in his own mind his attitude to his former friend. Just before his mental collapse, he had been quietly playing Wagner on the piano, but he was full of bitterness.

Only nine days after the composer's death in 1883, Nietzsche confided in a letter to a friend, "Wagner was by far the *fullest* human being I have known." However, he went on, "Something like a deadly offence came between us; and something terrible could have happened if he had lived longer."[47]

Details about this "deadly offence" emerged only in 1956, when correspondence first came to light between Wagner and a doctor who had examined Nietzsche.[48] It related to a consultation Nietzsche had in Switzerland in 1877. The doctor, a passionate Wagnerian, examined Nietzsche and found his health poor—indeed Nietzsche was at risk of going blind. This was when Nietzsche and Wagner were still friends and so, following the examination, Nietzsche wrote to Wagner, reporting the diagnosis, but also enclosing an essay on *The Ring*, which the doctor had written and given to Nietzsche, on the understanding that it would be passed on. Wagner replied to the doctor, thanking him for the essay, but also raising the matter of Nietzsche's health, apparently referring to the belief, common at the time, that blindness was caused by masturbation. The doctor, in his reply to Wagner, behaved extremely unprofessionally, confiding that, during his examination, Nietzsche told him he had visited prostitutes in Italy "on medical advice." (This was sometimes recommended then as treatment for chronic masturbation.)

Even at this distance, the set of events is shocking; how much worse it must have been then. It is now known that the details of this exchange circulated during the Bayreuth Festival of 1882, coming to Nietzsche's own notice later that same year. He confessed in a letter that an "*abysmal* treachery" had got back to him. More than one observer has concluded that this episode helped to unbalance Nietzsche.[49]

It is a story that diminishes two great men.

17.

Physics Becomes King: Helmholtz, Clausius, Boltzmann, Riemann

For a whole year, beginning in February 1840, Julius Robert von Mayer (1814–78) served as a ship's physician on board a Dutch merchantman to the East Indies.[1] The son of an apothecary in Heilbronn, Württemberg, Mayer graduated in medicine from the University of Tübingen in 1838 and enlisted as a ship's doctor with the Dutch East India Company.[2] It was during a stopover in Djakarta in the summer of 1840 that he made his most famous observation. In the manner of the day, he let the blood of several European sailors who had recently arrived in Java. He was surprised at how red their blood was and inferred that it was more than usually vivid owing to the high temperatures in Indonesia, which meant the sailors' bodies required a lower rate of metabolic activity to maintain body heat. Their bodies had extracted less oxygen from their arterial blood, making the returning venous blood redder than it would otherwise have been.

Mayer was struck by this observation because it seemed to him to be self-evident support for Justus von Liebig's theory that animal heat is produced by combustion—oxidation—of the chemicals in the food taken in by the body. In effect, he was observing that chemical "force" (as the term was then used), which is latent in food, was being converted into (body) heat. Since the only "force" that enters animals is their food—their fuel—and the only form of force they display is activity and heat, these two forces must always—by definition—be in balance.

Mayer originally tried to publish his work in the prestigious *Annalen der Physik und Chemie*, edited by Johann Christian Poggendorff, but was rebuffed.[3] His first published work, "Bemerkungen über die Kräfte der unbelebten Natur" ("Remarks on the Forces of Inanimate Nature"), was therefore published in the *Annalen der Chemie und Pharmacie* in 1842, and it was here that he argued for a relationship between motion and heat, that "motion and heat are only different manifestations of one and the same force [which must] be able to be converted and transformed into one another." Mayer's ideas did not have much impact at the time, though presumably the editor of the *Annalen der Chemie und Pharmacie*, none other than Justus von Liebig, thought them worth printing.[4]

Julius Mayer's story is tidy. The historian of science Thomas Kuhn has pointed out, however, that between 1842 and 1854, no fewer than twelve scientists had arrived at some version of the idea that became known as the "conservation of energy." The word "energy" was new at mid-century, but by 1900 all of physics would revolve around the concept.[5] Kuhn points out that, of these twelve pioneers, five came from Germany, one from Alsace, and one from Denmark, areas of German influence. He put this preponderance of Germans down to the fact that "many of the discoverers of energy conservation were deeply predisposed to see a single indestructible force at the root of all natural phenomena." He suggested that this root idea could be found in the literature of *Naturphilosophie*. "Schelling, for example, [and in particular] maintained that magnetic, electrical, chemical and finally even organic phenomena would be interwoven into one great association."[6] Von Liebig studied for two years with Schelling.

The Advent of Physics as We Know It

So far as physics was concerned, the first half of the nineteenth century saw some crucial changes in approach and even in vocabulary, changes that reflected the evolving nature of physics. In the late eighteenth century, for example, the term "physics" had referred to the natural sciences in general. In the early nineteenth century, the same word came to mean the study of mechanics, electricity, and optics, generally employing a mathematical and/or experimental methodology.[7] By mid-century, "there emerged a distinctive science of physics that took quantification and the search for mathematical laws as its universal aims." In 1824, for example, the curator of the University of Heidelberg proposed that a "mathematical seminarum" be established there, to be modeled on the increasingly successful seminars in philology that, as we have seen, were being credited with improving German classical education. Other universities followed.[8] Moritz Stern, extraordinary professor of mathematics at Göttingen, called for much the same thing, while the Berlin Physical Society was founded in 1845.[9]

Research, as already noted, began to acquire greater prestige, and physics was no different from other disciplines. The *Annalen der Physik* increasingly devoted its pages to the research of German scientists and less to the translations of papers from foreign journals. Founded in 1790, the *Annalen* was itself a symptom of the changes taking place: by the 1840s it was the most important German journal of physics, though many new journals proliferated in that decade, just as many new medical instruments began to be introduced. Johann Christian Poggendorff, the editor of the *Annalen* since 1824, could make or break scientific careers.[10] Poggendorff edited the *Annalen* until he died in 1877, when its pre-eminence was so assured that the Berlin Physical Society took it over, with Gustav Wiedemann as editor, and Hermann von Helmholtz as adviser on theoretical matters. In physics, a clear division of labor was already emerging between the experimenter and the theoretician. By the 1860s or 1870s, research in physics, including theoretical physics, was regarded as an end in itself, not just as an adjunct to teaching: beginning in the 1870s, professorships of theoretical physics were established at a number of German universities. Mary Jo Nye has tracked these institutions, in particular the Physikalisch-Technische Reichsanstalt (PTR) at Berlin-Charlottenburg and calculates that 800 physicists and chemists from Britain and North America earned doctorates in Germany in the nineteenth century and that thirty-nine important British scientists came under German influence.[11] By the same token, new laboratories were being built all over Europe, differing from their predecessors in kind as well as number, no longer merely teaching aids but spaces for research in their own right. "Experiment increasingly looked like the key to unlocking nature's secrets."[12]

It was in the laboratory that more and more experimenters were looking for ways to turn one kind of force into another. To Romantic natural philosophers in particular, as Kuhn has said, these "proliferating instances" of the apparent conversion of one kind of force into another seemed to confirm the underlying unity of nature—they were mutually convertible because they were different manifestations of the same underlying power. At the same time, pragmatists saw economic possibilities in these transformations. The new technology of photography used light to produce chemical reactions. The voltaic pile seemed to turn chemical forces into electricity, a major concern in industrializing and urbanizing societies. Above all, there was the steam engine, a machine for producing mechanical force from heat.[13]

The Discovery/Invention of Energy

In the eighteenth century, heat and electricity had been explained by supposing there were "imponderable fluids and ethers" which interacted with the atoms of ordinary matter. In 1812 the Academy in Paris announced it would offer its Grand Prix des Mathématiques to whoever could show how heat moved through matter.[14] Joseph Fourier's mathematical theory of heat, published in 1822, brought heat and mathematics together, while James Prescott Joule's experiments in 1843 established the equivalence of heat and mechanical work. Two years later, Julius Mayer published his observations about body heat and blood color.

With hindsight, everything can be seen as pointing toward the theory of the conservation of energy, but it still required someone to formulate these ideas clearly; that occurred in the seminal memoir of 1847 by Hermann von Helmholtz (1821–94). In *On the Conservation of Force*, he provided the requisite mathematical formulation, linking heat, light, electricity, and magnetism by treating these phenomena as different manifestations of "energy."[15]

The son of a Prussian Gymnasium teacher, Helmholtz studied medicine at the University of Berlin, funded by a Prussian army scholarship. In return, he served as a medical officer before becoming associate professor of physiology at the University of Königsberg in 1849.[16] Helmholtz's 1847 essay was privately published as a pamphlet. Like Mayer, he had sent his paper to Poggendorff at the *Annalen der Physik* but was rebuffed. Helmholtz's previous physiological publications had all been designed to show how the heat of animal bodies and their muscular activity could be traced to the oxidation of food—that the human engine was little different from the steam engine. He did not think there were forces entirely peculiar to living things but instead that organic life was the result of forces that were "modifications" of those operating in the inorganic realm.[17] In the purely mechanical universe envisaged by Helmholtz there was an obvious connection between human and machine work.[18]

While Mayer and Helmholtz, being doctors, came to the science of work through physiology, Helmholtz's fellow Prussian Rudolf Clausius approached the phenomenon, as did his British and French contemporaries, via the ubiquitous steam engine. Unlike Mayer and Helmholtz, Clausius did succeed in having his first important paper, "On the Moving Force of Heat, and the Laws Regarding the Nature of Heat That Are Deducible Therefrom," accepted by the *Annalen*; it appeared in 1850.[19] Clausius (1822–88) was born in Köslin in Pomerania and was yet another son of a pastor. From the Gymnasium he went to the University of Berlin where he was at first attracted to history (studying under Ranke), but then switched to mathematics and physics. In 1846, two years after graduating at Berlin, Clausius entered Boeckh's seminar at Halle and worked on explaining the blue color of the sky. The importance of his 1850 paper was immediately recognized and, on the strength of it, he was invited to become a professor at the Royal Artillery and Engineering School in Berlin, later transferring to the chair of mathematical physics at Zurich.[20]

In his famous paper Clausius argued that the production of work resulted not only from a change in the *distribution* of heat, as Sadi Carnot—the French physicist and military engineer—had argued, but also from the *consumption* of heat, and that heat could be produced by the "expenditure" of work. "It is quite possible," he wrote, "that in the production of work . . . a certain portion of heat may be consumed, and a further portion transmitted from a warm body to a cold one: and both portions may stand in a certain definite relation to the quantity of work produced." In doing this, he stated two fundamental principles, which would become known as the first and second laws of thermodynamics.[21]

The first law may be illustrated by the way it was later taught to Max

Planck, the man who, at the turn of the twentieth century, would build on Clausius's work. Imagine a worker lifting a heavy stone onto the roof of a house. The stone will remain in position long after it has been left there, storing energy until at some point in the future it falls back to earth. Energy, says the first law, can be neither created nor destroyed. Clausius, however, pointed out in his second law that the first law does not give the total picture. In the example given, energy is expended by the worker as he lifts the stone into place and is dissipated in the effort as heat, which among other things causes the worker to sweat. This dissipation, which Clausius was to term "entropy," was of fundamental importance, he said, because although it did not disappear from the universe, this energy could never be recovered in its original form. Clausius therefore concluded that the world (and the universe) must always tend toward increasing disorder, must always add to its entropy.

Clausius never stopped refining his theories of heat, becoming in the process interested in the kinetic theory of gases, in particular the notion that the large-scale properties of gases were a function of the small-scale movements of the particles, or molecules, that comprised the gas. Heat, he came to think, was a function of the motion of such particles—hot gases were made up of fast-moving particles, colder gases of slower particles.[22] Work was understood as "the alteration in some way or another of the arrangement of the constituent molecules of a body." This idea that heat was a form of motion was not new. The American Benjamin Thompson had observed that heat was produced when a cannon barrel was bored, and in Britain Humphry Davy had likewise noted that ice could be melted by friction. What attracted Clausius's interest was the exact form of motion that comprised heat. Was it the vibration of the internal particles, was it their "translational" motion as they moved from one position to another, or was it because they rotated on their own axes?[23]

Clausius's second seminal paper, "On the Kind of Motion That We Call Heat," was published in the *Annalen* in 1857, where he argued that the heat of a gas must be made up of all three types of movement and that therefore its total heat ought to be proportional to the sum of these motions. He assumed that the volume occupied by the particles themselves was vanishingly small and that all the particles moved with the same average velocity, which he calculated as being hundreds, if not thousands of meters per second. This brought about the objection from several others that his assumptions and calculations could not be right, since otherwise gases would diffuse far more quickly than they were known to do; he therefore abandoned that approach, introducing instead the concept of the "mean free path," the average distance that a particle could travel in a straight line before colliding with another one.[24]

Others were attracted by Clausius's efforts, in particular James Clerk Maxwell in Britain, who published "Illustrations of the Dynamical Theory of Gases" in the *Philosophical Magazine* in 1860, making use of Clausius's idea of the mean free path. However, where Clausius had assumed that every particle in a gas traveled at the same average velocity, Maxwell relied on the new science of statistics to calculate a *random distribution* of particle velocities, arguing that the collisions between particles would result in a distribution of velocities about a mean rather than an equalization. (Just what these particles *were* was never settled, not then, though Maxwell was convinced their very existence "was proof of the existence of a divine manufacturer.")[25]

The statistical—probabilistic—element introduced into physics in this way was a very controversial and yet fundamental advance. In his 1850 paper Clausius had drawn attention in the second law of thermodynamics to the "directionality" of the heat flow—heat tends to pass from a hotter to a colder body. He had not at first bothered with the implications of the irreversibility or otherwise of processes, but in 1854 he argued that the transformation of heat into work and the transformation of heat from a higher temperature into heat of a lower temperature were in effect equivalent and that in some circumstances they could be counteracted—reversed—by the conversion of work into heat, where heat would flow from a colder to a warmer body. This, for Clausius, only emphasized the difference between reversible (man-made) and irreversible (natural) processes: a decayed house never puts itself back together, a broken bottle never spontaneously reassembles.

It was only later, in 1865, that Clausius proposed the term "entropy" (from the Greek word for "transformation") for the irreversible processes whereby the tendency for heat to pass from warmer to colder bodies was also described as an instance of the increase in entropy. In doing this Clausius now emphasized the *directionality* of physical processes, and he described the two laws of thermodynamics as follows: "The energy of the universe is constant" and "The entropy of the universe tends to a maximum." Time, in some mysterious way, had become a property of matter.[26]

For some people, the second law had a much greater significance than even Clausius thought. Another Briton, William Thomson, Lord Kelvin, thought that the irreversibility that was such a feature of the second law—the dissipation of energy—also implied a "progressivist cosmogony," one that moreover underlined the biblical view about the transitory character of the universe. Thomson drew the implication from the second law that the universe, known by then to be cooling, would "in a finite time" run down and become uninhabitable. Helmholtz had also noticed this implication of the second law, but it was only in 1867 that Clausius himself, who had by then moved back to Germany from Zurich, acknowledged the "heat death" of the universe.[27]

The Appearance of "Strangeness" in Physics

The statistical notions aired by Clausius and Maxwell attracted the attention of the Austrian physicist Ludwig Boltzmann.[28] Boltzmann (1844–1906) was born in Vienna during the night between Shrove Tuesday and Ash Wednesday, a coincidence which, he half-jokingly complained, helped to explain his frequent and rapid mood swings, which tossed him between unalloyed happiness and deep depression. The son of a tax official, Boltzmann was appointed professor of mathematical physics at the University of Graz in 1869 at the age of only twenty-five. Later he worked with Robert Bunsen at Heidelberg and with Helmholtz in Berlin. In 1873 he joined the University of Vienna as professor of mathematics and remained there until 1902 when he committed suicide during one of his depressions.

Boltzmann's main achievement lay in two famous papers, describing in mathematical terms the velocities, spatial distribution, and collision probabilities of molecules in a gas, all of which determined its temperature. The mathematics were statistical, showing that—whatever the initial state of a gas—Maxwell's velocity distribution law would describe its equilibrium state. Boltzmann also produced a statistical description of entropy.[29]

What is important about the work of Mayer, Helmholtz, and in particular Clausius and Boltzmann is that, whether one can follow the mathematics or not, they brought probability into physics.[30] How can that be? Matter definitely exists, transformations (as when water freezes) obey invariant laws. What can probability have to do with it? This was the first appearance of "strangeness" in physics, heralding the increasingly bizarre twentieth-century quantum world. These early physicists also made "particles" (atoms, molecules, or something else that was not yet clearly understood) integral to the behavior of substances.

The understanding of thermodynamics was the high point of nineteenth-century physics and of the marriage between physics and mathematics. It signaled an end to the strictly mechanical Newtonian view of nature, and it would prove decisive in leading to a spectacular new form of energy, nuclear power. This all stemmed, ultimately, from the concept of the conservation of energy.

The Golden Age of Mathematics

In his history of mathematics, Carl Boyer says that the nineteenth century, more than any other preceding period, deserves to be known as the golden age of mathematics. "The additions to the subject during those one hundred years far outweigh the total combined productivity of all preceding ages." The introduction of such concepts as non-Euclidean geometries, *n*-dimensional spaces, non-commutative algebras, infinite processes, and non-quantitative structures "all contributed to a radical transformation which changed the appearance as well as the definition of mathematics."[31] While the French remained strong, and several countries supported mathematics linked to practical activities, such as surveying and navigation, research in pure mathematics—mathematics for the sake of it—was the exception rather than the rule, practiced more than anywhere else in Germany.[32]

The strength of mathematics in Germany owed something to the fact that, as in physics, the subject had an important new journal. Until the nineteenth century, the best mathematical periodicals had come from the École Polytechnique in Paris, but in 1826 August Leopold Crelle (1780–1855) launched his *Journal für die reine und angewandte Mathematik*

(Journal for Pure and Applied Mathematics), though it was often known more simply as "Crelle's Journal."[33]

Above all, the golden age—initiated by Gauss—was continued by Bernhard Riemann and Felix Klein. Riemann, frail and shy, was yet another son of a pastor. Born in 1826, he took his doctorate at Göttingen, then spent several semesters in Berlin to study under C. G. J. Jacobi and Peter Dirichlet before returning to Göttingen for a training in physics from Wilhelm Weber. (His subsequent career was split between mathematics and physics.)[34]

In 1854 he was called upon to give an inaugural lecture before the faculty at Göttingen. "The result in Riemann's case was the most celebrated probationary lecture in the history of mathematics."[35] In his lecture, titled, "On the Hypotheses which Lie at the Foundation of Geometry," Riemann urged a totally new view of geometry as the study of "any number of dimensions in any kind of space." This became known as Riemann geometry. In this paper he envisaged what he called manifolds, surfaces (now known as Riemann surfaces), which are forms of space that are non-Euclidean, where the laws of Euclid no longer apply. The idea of curved space is the best known, because the easiest to understand: a "plane" is in fact the surface of a sphere, and a "straight line" is the great circle of a sphere. Riemann's results in this area of thinking were so significant that Bertrand Russell described him as "logically the immediate predecessor of Einstein."[36] Without Riemann's geometry, general relativity could not have been formulated.

When Peter Dirichlet, another great mid-century German mathematician, died in 1859, Riemann was appointed to the chair that Carl Gauss had once occupied. In that chair he followed up Gauss's interest in number theory. In Chapter 7, p. 167, mathematicians' fascination with prime numbers was introduced, and it was noted how Gauss had uncovered the link between primes and logarithms. His invention of imaginary numbers was also discussed (see p. 172). It will be remembered that the link between primes and logarithms allowed Gauss to give a good but still *approximate* prediction of how many primes there are up to any figure, *N*. It was Riemann's achievement to establish and clarify a *definitive* prediction of the number of primes—and by using Gauss's other invention, imaginary numbers.

Riemann worked with something known as the zeta function. This, in one form or another, had been of interest to mathematicians since Pythagoras, in ancient Greece, who had pointed out the link between mathematics and music. Pythagoras found that if he filled an urn with water and banged it with a hammer, it would produce a certain note. If he then removed half the water and banged the urn again, the note had gone up an octave. As he removed water equivalent to a proper integer (½, ⅓, ¼), the notes produced sounded in harmony to his ear, whereas if any intermediate amount were removed, the sound was discordant. Pythagoras came to believe that numbers lay at the root of the order of the universe, giving rise to his famous phrase "the music of the spheres."

For other mathematicians, however, this led to an investigation of the behavior of reciprocal numbers (the reciprocal of 2 is ½ and the reciprocal of 3 is ⅓). This investigation eventually led to what mathematicians call the zeta function, zeta being represented by the Greek symbol, ζ. The zeta function is represented as follows:

$$\text{zeta } (\zeta) \text{ for } x = \frac{1}{1^x} + \frac{1}{2^x} + \frac{1}{3^x} + \ldots \frac{1}{n^x} + \ldots$$

This function turns up some interesting results, the most celebrated being the discovery in the eighteenth century by the Swiss mathematician Leonhard Euler that when zeta is 2 the sequence becomes:

$$\frac{1}{1^2} + \frac{1}{2^2} + \frac{1}{3^2} + \frac{1}{4^2} + \ldots = 1 + \frac{1}{4} + \frac{1}{9} + \frac{1}{16} + \ldots$$

and that this may eventually be written as:

$$1 + \frac{1}{4} + \frac{1}{9} + \frac{1}{16} + \ldots = \frac{1\pi^2}{6}$$

This discovery took the mathematics world by storm, for the number, $^1/_6\pi^2$, written as a decimal, produces an indefinite progression, like π itself. (This is number theory, remember, the sheer *behavior* of numbers being fascinating for mathematicians, whether that behavior has any use or not.)[37]

When Riemann was elected to the Academy in Berlin in November

1859, he wrote a ten-page paper to mark the event, as was (again) customary. This proved every bit as radical as his inaugural lecture. One of the things he did in this paper was to feed Gauss's other invention—imaginary numbers—into the zeta function, obtaining an entirely unexpected pattern, the most notable feature of which was (when the results of the equations were plotted on a graph) a series of waves and which, he found, could be used to correct Gauss's calculations regarding primes, to give an *exact*, error-free prediction of the number of primes in any sequence. And so the apparent randomness of the primes had been shown to have an order. Not a simple order, it is true, but an order nonetheless. Order—however complex—is a form of beauty for mathematicians.

Felix Christian Klein was born in Düsseldorf on April 25, 1849, and delighted in pointing out that his birth date was a collection of primes squared: 5^2, 2^2, 7^2. He made his most important contribution in group theory, another new field. The son of a government official in the Rhine province, he was appointed to a professorship, in his case at Erlangen, at an even younger age than Boltzmann, when he was twenty-three. He moved to Munich's Technische Hochschule in 1875, where he taught, among others, Max Planck (he also married Anne Hegel, granddaughter of the philosopher). In 1886 his health deteriorated, and he accepted a quieter life as professor of mathematics at Göttingen. There he consolidated Göttingen as the world's leading mathematics research center.[38]

To explain what Klein was driving at in group theory, imagine two visual experiments. First, imagine a rectangular sheet of paper, its sides measuring *A* and *B* inches. Rotate the sheet through forty-five degrees and then photograph it. The photograph will not show a rectangle and the sides will not be *A* and *B* inches long, yet the paper will not have changed. What are the mathematics of this foreshortening, the relationship between the original and the photograph? Second, consider an aerial photograph of a particular country—for example, Italy—taken from a satellite fifty miles up in space. Next, view the same country from, say, five miles up. The outline of Italy is the same but many details are now visible that weren't before—estuaries, small bays, tiny off-shore islands. Again, what transformation has taken place, what has changed, and what has stayed the same, and how can that change/staying-the-same be represented mathematically? This last example was not available in Klein's day because aerial photography didn't exist, but the problem, mathematically, is now known as "fractals" and shows how far ahead of his time Klein was. It presaged chaos theory.

Under Klein's leadership Göttingen became a mecca to which students from many lands, especially America, flocked.[39] The French had led the way at the turn of the nineteenth century, at the École Polytechnique, embracing the work of Joseph-Louis Lagrange, Gaspard Monge, and Jean-Victor Poncelet. The research and inspiration of Gauss, Riemann, and Klein ensured that that leadership passed to Germany. It held that lead—at least in theoretical terms—until the advent of Hitler.[40]

18.

The Rise of the Laboratory: Siemens, Hofmann, Bayer, Zeiss

No one better illustrates the changes taking place in Germany in the nineteenth century than Werner Siemens. As just one indicator, he became Werner *von* Siemens in 1888. Born in 1816, the fourth of fourteen children of a tenant farmer in Lenthe, near Hanover, Werner had to leave his Gymnasium in 1834 without taking his examinations because of the family's precarious financial situation, so that he could join the Prussian army and gain some engineering training in that way. He had the foresight, while at school, to drop Greek and take extra lessons in mathematics and land surveying.

He said later that the three years he spent at the Berlin Artillery and Engineering School were the happiest of his life. Among his teachers was Martin Ohm, brother of the physicist Georg Ohm.[1] While at the school Werner started to produce the first of the inventions at which he was to prove so adept. The earliest concerned gilding and plating silver, a process he sold to a German silver manufacturer.

He became interested in the theory of the conservation of energy (he was familiar with the work of both Mayer and Helmholtz) and this fanned his interest in engines (he published some of his early ideas in Poggendorff's *Annalen*)—all of which meant he was one of the first to appreciate the great importance of telegraphy.[2] His time in the army had taught him, among other things, the need for rapid, reliable communication, and so in 1847 he produced a pointer telegraph, which was notable for its reliability, this dependability laying the foundation for the Siemens & Halske Telegraph Construction Company, which he founded jointly with Johann Georg Halske, himself a mechanic, in Berlin that same year.[3]

Once he had a reliable telegraph, Siemens saw its many possibilities. He laid the first long subterranean wire, from Berlin to Grossbeeren, almost twenty miles to the southwest. He recognized that the invention of gutta-percha in Britain would enable the lines to be insulated, which meant that telegraph wires could be spread across the world, even in America following the Civil War. An underground line from Berlin to Frankfurt am Main came next, Frankfurt being where the German National Assembly was meeting. The line was buried to keep it safer at times of political trouble.[4]

In 1851 Siemens announced what would become his greatest invention—the dynamo-electrical machine.[5] He clearly foresaw the exceptional growth of power engineering, with Siemens & Halske repeatedly introducing new applications for electric current: in 1879 the first electric railway was presented at the Berlin Trade Fair, and the first electric streetlights were installed in Berlin's Kaisergalerie; in 1880 the first electric elevator was built in Mannheim; in 1881 the world's first electric streetcar went into service in Berlin-Lichterfelde; in 1886 the first electric trolley bus made its appearance; in 1887 the Berlin Mauerstrasse power station opened; in 1891 the first electric drills were made; and in 1892 the electricity meter, indicating the widespread acceptance of electrical machines, was installed. The name Siemens became synonymous with *Elektrotechnik*, a word coined by Siemens himself.[6]

In 1879 he helped found the Elektrotechnischer Verein (Engineering Society), one of whose aims was the introduction of faculties for electrical engineering at the technische Hochschule. By then he had been elected a member of the Berlin Academy of Sciences, in 1874, a rare—and perhaps unique—honor for someone who did not have a PhD.

The Color Revolution

In 1862, Queen Victoria attended the London International Exhibition in South Kensington wearing a vivid mauve gown. This choice, says

Diarmuid Jeffreys, was more significant than might appear because one of the main exhibits at the show was a massive pillar of purple dye. "Sitting next to the pile was its inventor/discoverer, William Perkin."[7]

Perkin, like Siemens, had always been interested in engineering and in chemistry. He had been a student at the new Royal College of Chemistry, established as a consequence of the growing awareness in Britain that its science was lagging behind that of its continental competitors, Germany in particular. Among the benefactors of the college was Prince Albert, the prince consort, who, as noted in an earlier chapter, had persuaded the celebrated German scientist, August Wilhelm von Hofmann (then only twenty-eight) to be the first professor at the Royal College. Perkin started as one of his students, but by 1856 had been appointed Hofmann's personal laboratory assistant.[8]

Hofmann began this collaboration by suggesting that Perkin try to synthesize quinine. Virtually every professional chemist had been trying to do this for years, as a synthetic cure for malaria (vital in an age of colonial expansion). Like everyone else, Perkin failed but then he toyed with a substance called allyl toluidine and, "by one of those flukes of science," the aniline Perkin used contained impurities. Entirely unexpectedly, he found that the black sludge left behind, when washed with water, turned a vivid purple.

Throughout history until that point, people had little choice in the colors available for their clothes. Derived from animal, vegetable, and mineral substances, the "earth colors"—reds, browns and yellows—were by far the most common, and the cheapest.[9] As a result, the rarer colors were much sought after, blue and purple in particular. On top of that, under the industrial revolution, millions of yards of cotton fabric were being produced by the new machine-driven textile mills of Lancashire and elsewhere, and opportunities for cheaper and more interesting colors were opening up.[10] Perkin therefore took out a patent on his purple dye and set up a factory in London. He called his new color *mauveine*, or mauve, and it quickly became popular, helped by Empress Eugénie, the wife of Napoleon III, who wore it because she thought it matched her eyes. By the time he was thirty-five, Perkin was rich.

The Germans, through Hofmann, had had a hand in Perkin's education. Now they saw their chance to take a dividend. With abundant coal in the Ruhr and more chemists than anywhere else, coal-dye companies sprang up all over Germany. A plethora of new synthetic dyes was rapidly discovered and in no time German dye companies led the world.[11] Coal-tar dyes expanded so quickly that, within a few decades of the end of the century, they had virtually eliminated natural colors from the market. Once the colors became standardized (not easy with natural products), a stability was introduced to the market that hadn't been there before.

The new color industry also owed its life to the simultaneous development of two other industrial/scientific innovations. One was the large-scale manufacture of illuminating gas, a by-product of which was tar. Second was the rise of systematic organic chemistry in the laboratory (see Chapter 13, p. 271). The starting point occurred in 1843 when Justus von Liebig instructed one of his assistants to analyze some light coal oil sent to him by a former student, Ernest Sell.[12] The assistant chosen by von Liebig to analyze Sell's oil was Hofmann, who had just secured his doctorate at Giessen. Hofmann's analysis revealed that coal-tar oil contained aniline and benzene, two substances that would themselves go on to become important industrially and commercially. Hofmann himself was initially more interested in teaching and research, and his interest in dyes grew only slowly. Because he was more interested in theory, it was the composition of dyes that mattered to him, and so his investigation of fuchsin, produced by the French and named after the fuchsia flower because its color was similar, was more systematic than anyone else's. Giving the substance the scientific name of rosaniline, he was soon able to demonstrate its structural relation to aniline yellow, aniline blue, and imperial purple, all of which had recently been discovered.[13] Because of these results, it now became possible to manipulate systematically the basic rosaniline structure, adding new functional groups that altered the shade produced. Hofmann himself manufactured triethyl and trimethyl rosaniline, two spectacular dyes marketed under the trade name "Hofmann's violet."[14]

According to John Beer in his celebrated study of the German dye industry, these five dyes—mauve, fuchsin, aniline blue, yellow, and imperial purple—"were the most important coal tar colors that the young aniline dye industry produced . . . It was only five years since the industry had been founded and yet already twenty-nine dye manufacturing companies in western Europe were doing well enough to risk their reputation in international competition." But Beer also shows that, over the next decade, while the German industry went from strength to strength, both the French and the British industries faded. "The French industry failed to prosper owing to a lack of trained technicians and the excessively theoretical approach of the École Polytechnique, whereas the British industry declined after 1873, partly because of the backward state of organic chemistry (which Hofmann had tried to rectify), an unwillingness by English capitalists to back research, and because the profession of chemist or engineer carried little prestige in intellectual circles or society at large."[15]

In marked contrast, the German and Swiss dye industries prospered by copying French and British processes—from Bessemer steel to waterproof paper. Scores of Germans learned their trade in Britain before returning to Germany. One important effect of this was to increase Germany's cloth output as much as fivefold in woolens between 1842 and 1864, and fourfold in cotton between 1836 and 1861.[16]

Two other factors contributed to the German and Swiss successes. These were the creation of polytechnic institutes and of factory research laboratories, which together supported the industry's ever-increasing need for trained scientists and engineers.[17]

The polytechnic institute (*technische Hochschule*) was modeled after the École Polytechnique that Napoleon founded in Paris for the training of mechanical, civil, and military engineers. It took the German *Hochschulen* quite some time to catch on and catch up, but during the 1860s and 1870s a concerted drive was begun to achieve for them full equality with the traditional universities. They were helped by the expansion of engineering owing to the advances in the understanding of electricity, magnetism, and the conservation of energy, the new forms of transport (railways and shipping in particular), and the other advances in higher mathematics, physics, and chemistry covered in previous chapters. Gradually, the matriculation standards were increased until they were on a par with the universities—so much so that, by 1900, the "Diplom Ingenieur" was the equal of a doctorate, and generally preferred by industry.[18] The polytechnics were subsequently allowed to confer the degree/title of "Doctor," removing the stigma hitherto attached to "engineer."

The creation of the factory laboratory was an event "whose historical significance . . . lies in the changes it brought about in the techniques of scientific research—changes that accelerated man's control over nature to such an extent that every major institution has since been affected." Not only that but the cooperation of several specialists produced faster results than did individual inquiry "and so arose the research team, directed by a research director . . . Places could thus be found for impractical but gifted theorisers, for purely 'gadget-minded' but skilful experimenters and for those who were poor observers but could make links between newly discovered and old facts." The German dye industry won its ascendancy "by wrenching thousands of little facts from nature by massed assault."[19]

Perhaps the great achievement of the laboratory was the way it transformed the coal-tar dye industry into the pharmaceuticals industry.[20] Pharmaceuticals came into their own during the 1880s and 1890s, partly because it was now that anesthetics began to be generally used, chloroform and ether becoming profitable substances for the dye companies to manufacture. And partly because, with the conception of the germ theory of disease (see Chapter 20, p. 383), there arose a need for antiseptics. These were almost all phenols, which the dye companies had for years been using as dyestuff components.[21]

Antipyretics and analgesics were discovered much as mauve was—by accident in the search for something else. Dr. Ludwig Knorr at Erlangen was yet another of those looking for a quinine substitute when he found that the pyrazolone compound he had just manufactured had pain-killing and fever-lowering properties. Höchst, originally a dye company near Frankfurt, bought the rights to this drug in 1883, and it was quickly followed by similarly acting substances, of which the most notable were "Antifebrine" (1885), pure acetanilide, Phenacetin or p-ethoxyacetanilide (1888), dimethylaminoantipyrine, sold as "Pyramidon" by Höchst (1893), and aspirin (1898). Sedatives appeared in the 1890s—"Sulfonal" and "Trional" (Bayer) and "Hypnal" and "Valyl" (Höchst). The work of Koch and Pasteur on immunology (also Chapter 20), led Höchst into the large-scale production of serums and vaccines to treat such dreaded diseases as diphtheria, typhus, cholera, and tetanus.

Following the lead at Höchst, and Bayer, another dye and drug company, at Elberfeld, Westphalia, interest in pharmaceuticals snowballed, with various firms employing bacteriologists, veterinarians, and other specialists.[22] The new field of insecticides saw the building of laboratory-greenhouses where botanists and entomologists tested the killing power of pesticides. Photographic film, paper, and developing chemicals comprised another new branch of laboratory specialization. But the other two momentous processes that were discovered/invented at that time were nitrogen fixation, carried out at Ludwigshafen, and artificial rubber, developed at Bayer.[23]

Nitrogen fixation was the original achievement, in 1902, of two Norwegians, Kristian Birkeland and Sam Eyde, who demonstrated that the oxides of nitrogen could be produced simply by heating air to a very high temperature by means of an electric arc. This was commercially feasible in Norway because hydroelectric power was so abundant and so cheap, but these conditions applied in few other locations. In the search for a more economical method of fixing nitrogen, Fritz Haber at Badische Anilin und Soda-Fabrik, near Mannheim, Germany's lagest dye factory, found it in 1909 by synthesizing ammonia out of nitrogen and hydrogen under high pressure and temperature. Carl Bosch refined Haber's process so that an ammonia plant was operating at Oppau near Ludwigshafen by 1913, establishing the firm's pre-eminence in the fertilizer and munitions industries.

A final sense in which the chemical/dye industry was important was its trade organization. After the unification of Germany in 1871, the dye manufacturers established an organization with an exceedingly long name, the Association for the Protection of the Interests of the German Chemical Industry—Registered Association. In existence since 1876, most people know it as the *Verein* and it was the Verein that was instrumental in the creation of the cartel and the chemical company IG Farben.[24]

Cartels proliferated in the 1880s until, by 1905, the Ministry of the Interior counted 385 in Germany, 46 of them in the chemical industry. By 1908, the Bayer company was a member of twenty-five cartel agreements.

The emergence of the cartel was a response to changed working conditions having to do with science. As commercial competition increased, profits declined to the point where capital and long-term investment and research could not be justified. As a result, the cartel fixed prices and market share. One of the first such controls occurred in 1881, fixing the price of alizarin (otherwise madder red) and allocating to each producer a certain fraction of the market. (Between 1869 and the date of the cartel, the price of alizarin had dropped from 270 marks per kilogram to 17.50.) This cartel didn't last, partly because, as John Beer says, it was impossible for former rivals to bury the hatchet overnight, but also because the Swiss dye companies did not form part of the arrangement. Later cartels did work, mainly because, instead of being defensive, they discovered it was more effective to pool their patents and share profits in predetermined ratios. Former rivals now became more cooperative as they shared a bigger cake.[25]

The German word for cartel is *Interessengemeinschaft*, or IG. The cartel of the color industry, IG Farben, would create an infamous furor after World War I.[26]

From Dyes to Drugs

Just as Werner Siemens was the best example of the link between theoretical and applied science in Germany in the nineteenth century, so Friedrich Bayer and Johann Friedrich Weskott were the best examples of the important move from dyes into pharmaceuticals. Bayer, from a family of silk weavers in Barmen, was born in 1825, an only son surrounded by five sisters. Weskott's family had moved to Barmen because the Wupper River provided excellent water supplies for their bleaching business. Both men were ambitious and in 1863 agreed to mount a joint venture, Friedrich Bayer & Company.[27]

The business was a success, but it was only when the two founders died in the early 1880s and the reins were taken over by Carl Rumpff, Bayer's son-in-law, that the firm's direction began to change. Rumpff took Bayer public and with the capital raised he recruited a number of young chemistry graduates, one of whom was Carl Duisberg. Duisberg was charged with finding new areas where the company could expand.[28]

It so happened that in the mid-to-late 1880s a new substance had appeared on the market in Germany, called Antifebrine, and this would open up for Duisberg a whole new world.

In 1886, two Strasbourg doctors, Arnold Cahn and Paul Hepp, had a patient who suffered from intestinal worms and they sent off an order to a local pharmacy for naphthalene, the standard treatment. At the pharmacy, however, there was a mix-up and without knowing it, the two doctors were sent a different substance entirely, a preparation called acetanilide. This, an acetylation of aniline, was yet another by-product of coal tar, well known in the dye industry, but very definitely *not* a medicine and in fact never given to human beings before. Only when Cahn and Hepp noticed that this "medicine" was having no effect on the patient's worms did they begin to ask questions. And what they observed, among other things, was that their patient's temperature had fallen noticeably.[29]

Paul Hepp's brother, it so happened, was a chemist at a company called Kalle. By chance, Kalle manufactured acetanilide for the coal-dye industry and Cahn and Hepp approached them to see if the company would be interested in marketing acetanilide as an antipyretic. The Kalle directors liked the idea of an antipyretic, but they had a problem because the formula for acetanilide was well known: if their drug were successful, all their commercial rivals could join in the scramble for profits. That is when someone at Kalle had the bright idea to produce a simple, easy-to-remember *name* for the drug. Until then the drugs sold by pharmacists were invariably known by their complicated chemical names, even though most general practitioners were ignorant of the chemistry involved. The point about Antifebrine, as the Kalle drug was called, was that it was much much easier to remember than acetanilide, exactly the same substance. The clever part lay in the fact that, under German law, a doctor's prescription had to be followed exactly: if the prescription specified Antifebrine, Antifebrine it had to be.

Watching this, Duisberg reasoned that if it could be done once, it could be done again. He cast his eye over a substance called para-nitrophenol, a waste product of the dye industry, that was similar to acetanilide; Bayer had 30,000 kilos of it going begging. Could this be exploited? He asked one of his men, Oskar Hinsberg, to look into it—and within a matter of weeks Hinsberg isolated a substance called acetophenatedine that, if anything, was an even more powerful antipyretic than acetanilide, with fewer side effects. Duisberg called it Phenacetin and, says Diarmuid Jeffreys, the origins of today's global pharmaceutical business "can be traced back to that moment."[30]

Other successful drugs followed, so that when Rumpff died in 1890 and Duisberg took over, his first big decision was to create a separate pharmaceuticals division with a dedicated laboratory. Duisberg's other clever move was to organize Bayer's pharmaceuticals laboratory into two sections—the pharmaceuticals group, tasked with inventing new drugs, and the pharmacology group, which tested the drugs. This was a sensible form of quality control, much copied.[31] It was in this environment that there was produced the most successful drug the world has known.

The Most Successful Drug the World Has Known

Conceptually, the drug was the work of three men—Heinrich Dreser, Arthur Eichengrün, and Felix Hoffmann, though it was Hoffmann who carried out the crucial experiments.

In the course of his literature searches, as he later told the story, Hoffmann came across several references to the synthesis of acetylsalicylic acid, or ASA, which, it was claimed, reduced the unpleasant gastric side effects of salicylic acid, the traditional remedy for rheumatic fever and arthritis. Hoffmann then began to repeat some of these experiments, varying the substances. Thus it was that, according to a note in his laboratory journal, on August 10, 1897, he stumbled on a way to make ASA that removed virtually all its gastric side effects.[32] As was now traditional, the pharmacology department tested the substance and Eichengrün found it effective. But Dreser objected, insisting that salicylic acid "enfeebles the heart" and ASA was rejected.

Matters were complicated by the fact that, in the same fortnight in which he discovered ASA, Hoffmann discovered another substance that Dreser believed had much greater potential: heroin. Diacetylmorphine, the full chemical name of heroin, was not in itself new. It was discovered in 1874 by the Englishman C. R. Alder Wright, who had been investigating opium derivatives at St. Mary's Hospital in London. Dreser had come across Alder Wright's written report in the literature and, since morphine had traditionally been used as a painkiller and in the treatment of respiratory diseases like tuberculosis, and because another opium derivative, codeine, was also used as cough treatment, he charged Hoffmann with making more experiments. "Two weeks after he had formulated ASA, Hoffmann successfully synthesised diacetylmorphine, in the process earning the curious distinction of 'discovering' in the same fortnight one of the most useful substances known to medicine and one of the most deadly."[33]

Dreser began testing the new substance on everything from frogs to rabbits and then on himself and other human volunteers. These volunteers found that the drug made them feel so "heroic" that the substance's brand name "suggested itself." Following further clinical trials, Dreser told the Congress of German Naturalists and Physicians in 1898 that it was "ten times more effective as a cough remedy than codeine but with only one tenth of its toxic effects." He added, for good measure, that it was "a completely non-habit-forming and safe family drug [and] would solve the problem of morphine addiction . . ."[34] Dreser even had plans to promote the drug as a remedy for baby colic and influenza.

Meanwhile Eichengrün went behind Dreser's back. First, he tried ASA on himself. Discovering that it had no apparent effect on his heart, he sent off batches of the drug to Bayer's representative in Berlin, who had good contacts with the general practitioners there, and arranged discreet trials. Within weeks, the doctors were returning "glowing assessments," far better than anyone at Bayer dared hope: ASA had few unpleasant side effects, and on top of that it was discovered to be an analgesic. The drug was put into production.[35]

Which meant there was the need for a name. Because salicylic acid could be derived from the meadowsweet plant, an abbreviation of the plant's Latin genus, *Spiraea*, was suggested. Someone else suggested that the letter "a" should be added at the front, to acknowledge acetylation. Many drugs at the time ended with "in," simply because it was easy to say. Which is how "aspirin" came to be.[36]

The Rise of the Microscope

The rise of the laboratory would not have been possible without a parallel rise in that most useful of laboratory instruments—the microscope. Developments in optics took place in the nineteenth century in France, Holland, Britain, and the United States—but three of the leading figures were German: Carl Zeiss, Ernst Abbe, and Ernst Leitz.

Carl Friedrich Zeiss was born in 1816 in Weimar and studied mathematics, physics, optics, and mineralogy at the University of Jena before going on to work under Professor Matthias Schleiden—the codiscoverer of the importance of the cell—at the Physiological Institute there. Zeiss opened his own shop in 1846 and did well, expanding steadily so that, in the first twenty years, the company produced 1,000 instruments. In the same year, recognizing that he needed to be more scientifically systematic if his business were to prosper, Zeiss engaged Ernst Abbe (1840–1905), at the time a young lecturer in physics and mathematics at Jena. The two men formed a partnership in which Abbe became director of research at the Zeiss Optical Works. It was Abbe who worked out the mathematical/physical basis of what became computational optics, which would lead to many new devices. The first of these, introduced in 1869, was an illumination device, providing lighting for the objects under examination. Three years later, in 1872, Abbe (who was a great social reformer in the workplace) formulated his wave theory of microscopic imaging, the "Abbe Sine Condition." This made possible a whole range of microscopic objectives all based on mathematical theory.[37]

Zeiss and Abbe were supplemented by Otto Schott (1851–1935). Brought up in Westphalia, Schott is now regarded as the father of modern glass science. His understanding of glass chemistry led to the introduction of more than one hundred new types of glass. The most important made possible the first "Apochromat" lens, in 1886. Apochromatic lenses have better color correction than achromatic lenses, making them particularly useful in astronomy.

Zeiss also led the way in the production of binocular telescopes and prism binoculars, each of which gave improved depth perception.[38] The invention of the motor car (see Chapter 19, p. 369) was to open up a new area, with the need for ever more elaborate headlights.

Just as famous was the work developed by Ernst Leitz. Born in the Black Forest in 1843, Leitz was six when the company that would eventually bear his name was founded by Carl Kellner, a twenty-three-year-old physicist, in Wetzlar. Kellner began making optics for microscopes and telescopes, in particular an orthoscopic eyepiece he himself invented (an orthoscopic eyepiece corrects for distortion and gives a very flat image). Rudolf Virchow and Justus von Liebig (see Chapter 20, p. 383) were customers of Kellner's.

After Kellner died at the age of twenty-nine, a colleague, Charles Behltle, took over the company, and it was Behltle who recruited Leitz as a partner ten years later. From 1869 on, when Behltle died, Leitz became the sole proprietor of the company, which by then had developed the microscope side of the business much more than the telescope side. By 1889 it too was expanding into binoculars and also into still and cine projectors. Just before the First World War two men—Oscar Barnack and Max Berek—joined the company and moved it more in the direction of cameras. It was Berek who computed the dimensions of the first camera lens to bear the name of the company that was to become so famous in the twentieth century—Leica.[39]

The microscope, more than anything else, is the symbol of the laboratory. Its rise also symbolized a change that took place in science as the nineteenth century turned into the twentieth. In the middle of the nineteenth century, chemistry and engineering had led the way—in the creation of electrical machines, dyes, and pharmaceuticals. These scientific industries continued to advance, but the microscope also enabled progress to be achieved in the biological sciences. In particular, it made practical the investigation of those microorganisms that caused disease.[40]

19.

Masters of Metal: Krupp, Benz, Diesel, Rathenau

Alfred Krupp, the "cannon king," was born in 1812. The firm that bore his name was only a year older than he. His father, Friedrich, was a not altogether successful businessman—when he died (Alfred was fourteen), the firm was close to bankruptcy, and Alfred had to be removed from school on financial grounds. He complained all his life that he got his education "at the anvil."[1] The son did not take after the father, however. Friedrich was a romantic and weak man, whereas Alfred was resolute, not at all sentimental. Those qualities were needed—it took two decades, half a working lifetime, for him to turn the firm around.[2]

Even then it was achieved partly by accident. After Napoleon's defeat, Prussia was awarded large parts of the Rhineland to compensate for the loss of her Polish territories. At the time, the agricultural land to the east was much more valuable, but with the growth of industry in the nineteenth century, and the burgeoning needed for coal, the situation was reversed. This reversal caused Prussia to look west and forced her into a closer association with the states separating her from France, to protect her interests. These economic factors had political consequences, aiding the eventual formation of the Reich itself and these events in turn affected Krupp. The ever-closer economic ties among the various German states, reflected in the customs union or *Zollverein* of 1834, made interstate travel, and interstate business, much easier.[3]

Krupp took advantage of these changed circumstances, traveling to the principal centers of Germany, securing orders for a variety of metal products, from coin dyes to cutlery. But the events of 1848 also played into his hands. He ordered his workers to have nothing to do with the revolutions across the country, and when these revolutions eventually collapsed, Prussia's position within Germany was strengthened, since for the most part it was Prussian soldiers whom the German princes had called on to put down the disorders. "By and large the story of how Prussia came to swallow Germany between 1848 and 1871 is also the story of how Alfred Krupp came to dominate German industry."[4]

The firm moved into armaments in 1843, albeit on a modest scale. His brother Hermann had alerted Alfred to the fact that iron musket barrels—then the staple of nineteenth-century armies—were less than satisfactory, and he therefore began to manufacture "the first mild-steel musket-barrel ever produced." This was the beginning of Germany's steel age, but even so Krupp had to overcome the prejudices of the generals who had grown accustomed to bronze or iron guns.[5] It was not until 1859 that he at last received a viable order—for 312 cannon—from the Prussian government.

Although Krupp is known to history for his guns, Peter Batty says that his real genius was probably his understanding of railways. Railways had started in Germany in 1835, in Bavaria, and by 1850 there were already 6,000 miles of track. That figure was to snowball ten times within the next half century, and Krupp was one of the first to see the opportunities for steel.[6] He secured his first railway contract, for 500 steel springs and axles, in 1849. This sparked a series of experiments which eventually matured as the weldless steel railway tire, a brilliant innovation that almost certainly made Krupp more money than all his guns put together. An inherent weakness in the early railway tire was the point at which the outer wheel, which sat on the rail, was welded to the rim, which sank down below the rail on the inside to keep the locomotive and the carriages on the tracks. As trains got faster and heavier, this welding weakness became ever-more important. To overcome this, Krupp simply adapted the technique the firm had been using in their fork and spoon machines—namely rolling *seamless* steel tires on to the wheel rims while they were still hot. Instead of two pieces welded together, wheels were now one piece of metal with no join. Alfred had the part of the factory where the tires were assembled built so far from everywhere else that it was known to his employees as "Siberia." The weldless wheel propelled Krupp to the front rank of industrialists.

Cheap Steel and the First Arms Race

In the field of guns, Krupp began to realize that his future lay more with the military—the generals—than it did with the politicians.[7] Peter Batty observes that "few bureaucrats in Berlin could stand Alfred—his high-handedness and overbearing manner had made him intensely disliked among his fellow industrialists too." Krupp began to court the officers surrounding the crown prince and "Thus began the spinning of that web of close threads between Essen and Berlin that characterised the Krupp saga." It paid off. In October 1861, Wilhelm, then the prince regent, visited the Krupp works to see for himself "Fritz, the biggest hammer in the world." A few months later, on becoming king of Prussia, Wilhelm made Alfred a privy councillor, and shortly afterward awarded him the Order of the Red Eagle with Oak Leaves, an honor usually reserved for victorious Prussian generals. All this coincided with the growth of the Krupp gun business; Alfred was now selling cannon "by the score" to Belgium, Holland, Spain, Egypt, Turkey, Sweden, Switzerland, Argentina, Austria, Russia, even Britain. The first great arms race was beginning and in the process the German press dubbed Krupp "the cannon king," a title he relished more than any other. It was an era symbolized by Bismarck's observation in 1862, that "The solution of the great problems of these days is not to be found in speeches and revolutions—but in blood and iron."[8]

Otto von Bismarck, Prussia's prime minister from 1862 on and Germany's chancellor until 1890, was three years younger than Krupp and in many ways the "Iron Chancellor" and the "Cannon King" were very similar. Both were tyrannical, misanthropic, and incapable of intimacy, and both sought refuge in things—Bismarck in dogs and trees, Alfred in horses and guns. It was once said of Bismarck: "I have never known a man who experienced so little joy," and it could equally have been said of

Alfred. "No two Prussians have been so responsible for the image, in the Anglo-Saxon mind at least, of the Prussian as someone aggressive, belligerent and destructive."[9]

Bismarck, we should never forget, was a Junker—that class of "militaristic, predatory land-owners" who had won their great estates in the east by force and who, therefore, believed in force. To preserve this class, which was the chancellor's lasting aim, he had to preserve Prussia, and that meant diminishing both Austria and France, destroying German liberalism, and replacing a primarily cultural German nationalism with a Prussian political nationalism. In doing these things, he came to be the "best hated man in Europe," with Krupp a close second.[10]

Bismarck first visited Essen in October 1864, en route from Paris to Berlin. Discovering that he and Krupp shared a passion for horses and big trees, he took Krupp into his confidence and during their rambles apparently disclosed some of his plans for Prussia. Bismarck well understood how Krupp's guns could play a part in these plans and Krupp sensed big profits, not least from the chancellor's intention to expand the navy.

The first major battle in which Krupp guns took part (on both sides, it should be said), was at Königgrätz on July 3, 1866, between Prussia and Austria. Though less than perfect, another 700 Krupp guns were ordered inside four months. And though this Austro-Prussian war of 1866 was one of the shortest of wars, it had far-reaching consequences in that, as a result, Prussia grabbed the Duchies of Schleswig and Holstein and annexed the states that had not sided with her before Königgrätz—Hanover, Hesse, Nassau, and Frankfurt. With Austria soundly defeated, Bismarck's strategy to make Prussia a world power was set in motion.[11]

Two years later Prussia announced it was intending to start a navy. At first the admirals intended to buy British guns, but Krupp, backed up by the king, prevailed in his argument that German guns should be used, and in September 1868 the new navy ordered forty-one heavy guns for their three new ironclads—from Krupp. This in itself shaped German naval policy for decades.[12]

In 1870, as cleverly as he had baited Austria four years earlier, Bismarck followed the same maneuver with France, luring Napoleon III to declare war on Prussia to restore to France those parts of the country that Napoleon I had lost. On the day Prussia mobilized, Krupp offered a consignment of guns to the armed forces as his contribution to the war. The gift was declined, but the army increased its orders for Krupp armaments to the point where Prussia for the first time was buying more guns from Krupp than from anyone else.[13]

Equipped with old-fashioned bronze muzzle-loaders, Napoleon III's troops were nowhere near equal to the Prussians. "The new Krupp steel breech-loaders and the new Krupp steel heavy mortars pulverised the forts of Metz and Sedan in no time at all and blasted a hole through the outskirts of Paris itself." This encounter was revealing about both Krupp and Bismarck. Most Prussian generals were opposed to the shelling of Paris, which had just been rebuilt by Georges-Eugène Haussmann. (The beautiful city that we know as Paris was brand new then.) However, both Krupp and Bismarck were very much in favor of attacking the capital, Krupp so much that he offered the army his 2,000-pounder. He also began to devise a giant siege gun capable of bombarding 1,000-pound shells from great distances right into the heart of Paris. They could not be built in time, but eventually they became the World War I weapons that "horrified the world." As a result, Krupp joined Bismarck and the Kaiser in the trinity of the most hated men in all France. His name "came to signify purely and simply a particular implement of destruction. From being hated by just one nation in 1871, Krupps over the next seventy-four years were to become an object of loathing on an international scale such as perhaps no other industrial organisation has ever attracted."[14]

For Krupp, Prussia's victory in the Franco-Prussian War, whatever the underlying reason for it, was wonderful publicity. Orders came flooding in. He turned down an honor, saying he preferred to be the first among industrialists than the last among knights.[15]

What emerged from the victory of 1871 was the Prussian Empire. Wilhelm, the Prussian king, now became the German emperor; Bismarck, the Prussian prime minister, became the German chancellor. Prussia took over all of Germany except for Austria, and part of France too: Alsace and Lorraine, rich in coal and iron ore. Though there were still other kingdoms (Bavaria, Saxony), Prussia was now the most powerful state on the Continent. Moreover, she had a convenient, just-defeated neighbor. Both Bismarck and Krupp were to trade on the anxiety that France would always hanker after a war of revenge.

The Prussian victors imposed on France an indemnity of five milliards of francs.* This payment, and the speed with which it was paid off (just thirty months), produced an extraordinary boom in Germany—the so-called *Gründerzeit.* The new Imperial German Government spent these francs on two things—on armaments and on repaying their debts to individual Germans who had helped fund hostilities with war loans. These individuals suddenly found themselves awash in great swaths of capital, which they promptly reinvested. Some twenty new companies had been registered each year during the two decades before the war, but in 1871 alone there were over 200 such registrations and in 1872 more than 500.[16] The boom benefited Krupp as it did every other manufacturer in Germany. As many new iron works, blast furnaces, and machine-manufacturing factories were built during the three years after 1871 as had come into being during the previous seventy.

However, Krupp, like many others in the Gründerzeit, had overreached himself, buying in just one year, 1872, more than 300 iron-ore mines and collieries and two entire ironworks; he also had commissioned four transport ships to bring to Germany the new iron-ore deposits acquired in Spain. And so, when the stock market crashed in 1873, hundreds of businesses went bankrupt, and Krupp was short by half a million pounds, more than £50 million at today's levels. The banks moved in and their representative, Karl Meyer, took over day-to-day running of the firm. The company paid off the last mark of its debt fifteen years later, the year Krupp died.[17]

Krupp's personal lifestyle was not, however, curtailed in any significant way (Meyer was an old friend). Notably, his gunnery tests continued to be great social occasions. This was still the era of the great railway expansion in America, and huge numbers of steel rails were bought from Krupp by the American railroad companies. Nevertheless, Krupp's last years were bleak. Since being sidelined by the banks, he had turned grumpy and spent his days lost in his great monstrosity of a house, the Villa Hügel, "where he hired a pianist to play to him during meals, but where no one would play dominoes or skat because he was such a bad loser."[18] When he died of a heart attack on July 14, 1887, at the age of seventy-five, only his valet was at hand.

* About £28 billion at 2010 prices, though making historical comparisons of this kind is notoriously unreliable owing to the much greater liquidity available now than in the past.

The year before he died, his first grandchild, Bertha, had been born. This was the Bertha after whom the huge gun that devastated the Belgian forts in 1914 was named. Krupp's notoriety did not die with him.

None of the other great steel giants of Germany—August Borsig (1804–54, locomotives), Hugo Stinnes (1870–1924, mining, shipping, newspapers), or August Thyssen (1842–1926, mining, steel), shared Krupp's notoriety, though their wealth more than equaled his. Thyssen and Krupp merged in 1999.

The Age of the Automobile

As early as the 1860s, in Switzerland, France, and Britain, several "horseless" vehicles were produced, though none of them went anywhere, so to speak. Only in 1885 did Karl Benz, in Mannheim, construct a machine that would lead to the automobile age.

The son of an engine-driver and the grandson of a blacksmith from the Black Forest, Karl Benz had engineering in his blood.[19] Born in 1844, by the time he was thirty he had his own small workshop where he built gas engines. His company started as the Mannheim Gas Engine Company but was changed to Benz and Company in 1883. A year later he constructed his first internal combustion engine, using a slide valve and an electric ignition. There were many possible uses for such an engine, and he had a number of flat disagreements with his partner, Emil Bühler, who would not permit any money to be spent on a "horseless carriage." Benz therefore started out again with a new partner, Max Rose. He too was skeptical about horseless carriages, but he did earmark a small tranche of capital "for experiments." It was this capital that allowed Benz to construct the vehicle from which the automobilism of the world has sprung. Benz knew that the weight of the engine was crucial to success and that it had to be a good deal lighter than any gas engine produced thus far. Until that point, his stationary engines had produced about 120 rpm (revolutions per minute), and he knew he had to more than double that. His other crucial early decision was to have four cylinders and not two, because road vehicles, he felt, would need to keep changing their speed. He situated his engine on its side, the flywheel running horizontally, so that gyroscopic action, when turning corners, would not interfere with the engine's running. His instinct was to place the engine at the rear, over the two back wheels, using the front ones for steering, as happened with tricycles, then in common use. The power would be connected to the wheels by chains. The fuel used—benzene—was vaporized by a surface carburetor, patented on January 29, 1886. The coolant was water.

According to St. John Nixon in his history of automobiles, it is "beyond doubt that the vehicle was ready for trial during the spring of 1885. It was driven by Benz around a cinder track which adjoined his workshop. His wife and children were present when this event took place."[20] The vehicle was probably first tried on public roads in October 1885. That, at least, was confirmed by an old employee in 1933. By the end of the year, Benz had clocked all of 1,000 meters, at a speed of around 12 kph. However, the vehicle suffered mechanical or electrical trouble each time it was taken out.

Benz's immediate aim was to drive his vehicle twice around Mannheim without stopping. He was forced to do it after dark; otherwise his contraption attracted huge crowds, and he was worried that the police might forbid him access to the public highways. Night after night he put someone in the passenger seat, started the engine, and traveled farther and farther before the inevitable breakdown. Then, in a journey that St. John Nixon insists ranks with George Stephenson's, he finally made the double circuit nonstop. It made news, the *Neue Badische Landeszeitung* reporting the events in its issue of June 4, 1886. This part of the story, however, does not have a happy ending. To begin with, Benz's innovations were successful and, by 1900, he was building more than 600 automobiles a year. But he failed to develop what he had given to the world, and the improvements he made to his cars were little more than tinkerings so that others like Gottlieb Daimler overtook him.[21]

Born in Schorndorf in 1834, Daimler was apprenticed to a gunmaker before becoming an engineer.[22] In 1872, at the age of thirty-eight, he was made technical director of Otto and Langen, gas-engine manufacturers of Deutz. He worked there for just short of a decade and in that time he helped develop the internal combustion engine. In 1882, however, he fell out with his fellow engineers over the direction of research and bought himself a property at Cannstatt, where he could continue in the direction he wanted to go. His old colleague, Wilhelm Maybach, was with him.

Daimler was convinced that the internal combustion engine had a spectacular future but only if two problems were solved. One, the engines constructed until that point turned over much too slowly. And two, if this were to be overcome, a different system of ignition was needed. At that stage the most commonly used ignition employed a slide valve which, for a moment, retreated and exposed the explosive mixture in the cylinder to a flame. Daimler's instinct told him that any valve system would never be able to close quickly enough at high speeds to allow for the full effects of the explosion to be conserved. In 1879, Leo Funk patented a system in which an external burner kept a hollow tube at white heat, the mixture being forced into the tube by the ascending piston. Daimler realized this was the way forward.[23]

The patent Daimler was awarded (No. 28022) on December 16, 1883, was for the first fast-running engine. Curiously, however, Daimler did not at first intend his engine to be used for anything other than stationary work. But when he and Maybach saw that this engine could run at 900 rpm, their thoughts turned toward a motorcycle. This was patented in August 1885—it had two speeds, was cooled by a fan, and had iron tires. The engine, of half a horsepower, was situated behind the seat. To start the engine, a burner was lit that heated the ignition tube and the engine was cranked in the usual manner. The power from the engine was conveyed to the rear wheel by a belt. During the winter of 1885–86, Daimler's motorcycle was tested on a frozen lake in Cannstatt, a ski being used in place of the front wheel. He too undertook road trials after dark, in his case so that teething troubles could be ironed out in private. In November 1885 his eldest son, Paul, drove from their house to Untertürkheim, three kilometers away, and made it home.

The first Daimler car took to the road in the autumn of 1886, between Esslingen and Cannstatt. In the archives of Daimler-Benz A.G., there is an account of these early-morning trials, written by Wilhelm Maybach and Paul Daimler.[24] They say the vehicle ran "quite well" and that speeds of 18 kph were attained. Daimler put his engine in boats and even designed a railcar driven by a Daimler engine. In 1889 they moved decisively ahead with a vehicle with a tubular frame through which the coolant (water) circulated. Engines remained at the rear of cars until 1896 when they were placed under the hood (Daimler being much taken with the designs of the Frenchman Émile Levassor).

It was not all plain sailing, but Daimler prospered more than Benz. Paul Daimler, Maybach, and Emil Jellinek, a rich Austrian who was consul-general at Nice (where many early car trials and rallies were held), collaborated on a model that would put all rivals in the shade; it was more stylish and had many technical innovations, not the least of which was the relative silence of its engine.[25] It was unveiled in final form in Nice in 1901 when, because there was such anti-German feeling in France in the wake of the Franco-Prussian war, it was named after Jellinek's daughter, Mercedes.

Though Benz, Daimler, and Mercedes are the best-known names in automobile history, Rudolf Diesel is almost as important. Born in Paris in 1858 to parents who were Bavarian immigrants, he was educated at Munich Polytechnic. There he heard a lecture by Professor Carl von Linde on thermodynamics, in which Linde explained that the steam engines then so popular used only around 10 percent of their fuel to perform useful work, a shortcoming that stayed in Diesel's mind.[26] When he graduated from Augsberg Technical School, he was the youngest person ever to achieve that honor *and* he achieved the highest-ever marks. He impressed Linde so much that the professor got him a job at a factory in Switzerland selling the ice machines that Linde had helped develop.

Obsessed with engines of one sort or another, Diesel soon invented a machine to make clear ice.[27] The Swiss company was not interested, but French brewers were and he found a ready market for his machine back in Paris. His real breakthrough came in 1893, when he was thirty-five, at which point he took out a patent for a "Combustion Power Engine," the engine we know today as the diesel engine.[28] The difference between Diesel's engine and the internal combustion engine is simple but profound. In the gasoline engine an air-fuel mixture is drawn into the cylinder, where it is ignited by a spark plug. In Diesel's engine only air is drawn into the cylinder. With no fuel present it can be compressed about twice as much, driving the temperature much higher. At the right moment, fuel is injected into the cylinder, where it ignites spontaneously.

It is a simpler system but, in the early days, the fact that the engine operated at very high temperatures and pressures meant that they were too much for the materials then available, making his engines unreliable. In 1897, however, the first Diesel engine factory was built at Augsberg and he prospered. Unfortunately, sloppy management of his money meant that almost all of it slipped away. In 1913 he was invited to London for the opening of a new Diesel factory. He took the channel steamer at Antwerp but disappeared during the night; his body was found in the North Sea about ten days later.

In some areas of the world, Diesel engines now have a more than 50 percent market penetration and are much preferred for such outlets as submarines, mines, and in oil fields.

Just as Daimler cars were the fruit of a father-and-son collaboration, so AEG, Germany's other great engineering company, alongside Siemens, was developed by the father-and-son team of Emil and Walther Rathenau. In fact, Rathenau and Siemens, who at one stage were partners, form brackets to this section, emphasizing how intertwined German science, business, and politics were between the middle of the century and the First World War.[29]

Emil, born in Berlin in 1838, into a wealthy Jewish family, had bought himself a successful machine factory in the north of the city two years before Walther's birth. Thus he had made one fortune by the time of the Paris Exhibition of 1881 when he saw Edison's electric lightbulb. He snapped up Edison's patents and two years later founded the Deutsche Edison-Gesellschaft (German Edison Company, or DEG). This seemed a clever move since he did so in collaboration with his greatest potential customer, Siemens. In fact, in its early years DEG was beset with technical and legal problems (over patents, mainly) and because of this the company was eventually transformed into the Allgemeine Elektrizitätgesellschaft (German General Electric Company, or AEG). The links with Siemens were dissolved and only in 1894 was Emil Rathenau able to begin to turn his firm into the biggest electrotechnical giant in Germany.[30]

The Rathenau family was Jewish only in name. Mathilde Rathenau, the daughter of a Frankfurt banker, took care to give her children (two boys and a girl) a good education in music, painting, poetry, and the classics of literature. For her, business wasn't everything.[31] Walther never formally accepted Christianity, but he did acknowledge the divinity of Jesus and, in a magistrate's court in Berlin in 1895, disassociated himself from his former "Mosaic belief." All his life he was sensitive to the second-class status of Jews in Germany, yet at the same time he advocated assimilation and hoped for equality. Like many other Jews, he regarded himself as a German: everything else was of lesser significance.

Having a PhD (from Strasbourg), Rathenau kept up with scientific developments—the behavior of metals, electrolysis, hydroelectric power—but he was always more interested in industrial organization, business *strategy*, and its links to politics, rather than the day-to-day running of companies like AEG. This made him ideal board material, and Rathenau's real significance is that he formed part of that generation of industrialists—Krupp, Stinnes, and Thyssen were others—who began to rival the army officers, diplomats, and professors at the top of the status ladder, though the rising prestige of the industrialists was opposed by a fierce anticapitalist and anti-industrial feeling in some quarters, who saw industrial power as the main cause of human misery. Although many realized that the *Industriestaat* was replacing the *Agrarstaat*, industrialists still found it difficult to progress politically and Rathenau in particular found this frustrating. Yet Germany did change fundamentally from the 1890s onward, when industry replaced agriculture, forestry, and fishing as the mainstay of the country's GDP and when more workers were employed in industry than in agriculture, and more people lived in cities than in small towns and villages.[32]

Unlike his critics, Rathenau was convinced that industrialization and capitalism were the only secure foundations for a powerful modern state, and he also felt that the German Empire had the long-term edge over Britain, because it also had a strong agricultural sector.[33] He was similarly convinced of Britain's industrial decline, for which he blamed the trade unions, the poor level of training for engineers, and weak management. He did not believe, however, that the industrial state was an end in itself. "He saw industrial domination as a transitory phase to achieve a greater 'spiritualised' period in human history."[34] He was led in this way to a relatively crude social Darwinism allied with some of Gobineau's racial beliefs. Rathenau felt that, eventually, the northern European middle classes—people like himself—would come to dominate the world.[35] Educated businessmen were the new aristocracy, who would know where to lead their fellow citizens to the higher, post-material spiritual level. Continuous industrialization, he was convinced, must be accompanied "by ethical achievement." He was therefore in favor of heavy taxes for the rich both in life and in death—he wanted to see an "uncompromising" inheritance tax and he went so far as to advocate the "abolition of luxury." "Distribution of property," he wrote, "is not a private affair, any more than is the right to consume."[36] And he argued that "richness should be replaced by prosperity which in turn is based on creativity and responsibility for one's work or one's own society." Workers should have a say in management. However, as Hartmut Pogge von Strandmann has pointed out, "there is no evidence to suggest that Rathenau pursued a markedly different line towards his own AEG employees from that of other industrialists towards theirs." He thought that better working conditions would increase productivity.

His importance lay in the clarity with which he saw—and described—what was happening in Germany, how the dynamics of modern prosperity were shaped by science and industry and how the country's traditional elite was failing to adapt. If there was a whiff of sanctimony about his stance, that too was revealing. He was better at identifying problems than at finding solutions.[37]

20.

The Dynamics of Disease: Virchow, Koch, Mendel, Freud

Rudolf Virchow (1821–1902) was the most successful German physician of the nineteenth century. Besides his clinical and theoretical achievements, his work on the social aspects of medicine mean that he had an impact much wider than the purely medical field. His long career epitomized the rise of German medicine after 1840, an ascendancy that transformed a discipline that was still largely clinical and prescientific.[1]

Born in a small market town in Pomerania, Virchow was educated privately in the classical languages, but he preferred the natural sciences. Because of his abilities, he received in 1839 a military fellowship to study medicine at the Friedrich-Wilhelms Institut in Berlin. This institution was specifically designed to provide an education for those who would not normally be able to afford one, in return for which they joined the army medical service for a specified time. Virchow studied under Johannes Müller and Johann L. Schönlein, who introduced him to the experimental laboratory and modern diagnostics and epidemiological studies, all relatively new.

Virchow graduated in 1843, his first field job being medical house officer at the Charité Hospital in Berlin, where he made microscopic investigations of vascular inflammation and the problems of thrombosis and embolism.

Always outspoken, in 1845 he delivered two speeches before an influential audience at the Friedrich-Wilhelms Institut in which he dispensed with all transcendental influences in medicine and argued that progress would only come from three main directions: clinical observations, "including the examination of the patient with the aid of physico-chemical methods"; animal experimentation "to test specific aetiologies and study certain drug effects"; and pathological anatomy, especially at the microscopic level. "Life," he insisted, "was merely the sum of physical and chemical actions and essentially the expression of cell activity."[2] While still in his twenties, in 1847 he was appointed an instructor at the University of Berlin under Müller.

But he was never just a medical man. In 1848 a typhus epidemic swept through the Prussian province of Upper Silesia, and Virchow was one of the team of physicians sent by the government to visit the afflicted region and survey the damage. While there, Virchow came face-to-face with the destitute Polish minority, struggling in appalling circumstances. And so, instead of returning to Prussia with a set of strictly medical guidelines, Virchow's report recommended political change, plus sweeping educational and economic reforms. It was hardly what the government had bargained for.

His political beliefs led him to take part in the uprisings of 1848 in Berlin, where he fought on the barricades, afterward becoming a member of the Berlin Democratic Congress and editor of a weekly entitled *Die medizinische Reform*. This was heady, but in 1849 he was suspended from his academic positions. He quit Berlin and took up the recently created chair in pathological anatomy at the University of Würzburg, the first of its kind in Germany. While there he was for a time distanced from political activity, and it was then that he achieved his greatest scientific contributions, establishing in particular his concept of "cellular pathology." In 1856 Virchow returned to Berlin as professor of pathological anatomy and director of the newly created Pathological Institute.

The Foundation of Bioethics

Now that he was back in Berlin, however, Virchow's old political instincts began to revive. He became a member of the Berlin City Council, where he concerned himself with public health and was instrumental in improving both the sewage system and the water supply of the city. Emboldened by these successes, he was in 1861 elected a member of the Prussian lower house, representing the liberal German Progressive Party, which he helped to found. Most notably, the Progressives opposed Bismarck's policy of rearmament and forced unification, a resistance that provoked Bismarck so much that he challenged Virchow to a duel. Virchow had the sense not to rise to the bait, and in the Franco-Prussian War of 1870–71 proved himself no mean nationalist, helping to organize hospital facilities and hospital trains for the wounded.

Virchow had a very modern view of epidemiology, believing that some diseases are "artificial," stressing their sociological side, arguing that political and socioeconomic factors were significant etiological elements. He even argued that epidemics could arise in response to social upheaval and could only be eliminated or alleviated through social change. No less controversial (then) was his argument that it is "the constitutional right of every citizen to be healthy." Society, he insisted, had the responsibility "to provide the necessary sanitary conditions for the unhampered development of its members."[3] This, a kind of medical *Bildung*, is now regarded as the foundation of bioethics.

He made some mistakes. He was skeptical about bacteriology. Germs, he was convinced, could not be the sole etiological agent in an infectious illness, environmental and sociological factors being clearly responsible in his view for the typhus and cholera epidemics of 1847–49.

Toward the end of his life, from about 1870 on, he turned to another science: anthropology. Co-founder of the German Anthropology Society (in 1869) he made several studies of skull shapes and carried out a nationwide racial survey of schoolchildren. From this, he concluded that there was no "pure" German race, a highly controversial result.

Anthropology led to archaeology and in 1870 he began his own excavations in Pomerania. In 1879 he traveled with Heinrich Schliemann to Hissarlik, where Troy was being excavated (see Chapter 21, p. 401), and he subsequently helped to attract antiquities to Berlin, for which the city became known.[4]

His eightieth birthday in 1901 was celebrated as far afield as St. Petersburg and Tokyo. In Berlin there was a torchlight parade. His taste for public argument, and his dogmatism, had some unfortunate side effects, most notably his opposition to Ignaz Semmelweiss's insight that hand-washing by doctors between patients would prevent puerperal fever. But Germany had progressed in less than half a century from speculative and philosophical healing to become the world center of modern scientific medicine, and Virchow was probably the most important figure in that transformation.

New Knowledge about Infection

As important as Virchow, and perhaps more so, was Robert Koch (1843–1910), the man who devised so many of the basic principles and techniques of modern bacteriology.[5] It was Koch who isolated the causes of anthrax, tuberculosis, and cholera, and in his many travels he also influenced authorities in several countries to introduce public health legislation based on new knowledge about the microbial origin of infection.[6]

Robert was one of thirteen children (two of whom died in infancy). He grew up with an intimate knowledge of animal and plant life and the new art of photography. By the time he was ready for the local primary school, he had taught himself to read and write. At Göttingen he first thought of studying philology (as he also considered immigrating to America) but enrolled in natural sciences, soon transferring to medicine.

No bacteriology was yet taught at Göttingen but Jacob Henle, the anatomist, did consider the possibility that contagious agents could include living organisms.[7] After graduation in 1866 Koch attended Rudolf Virchow's course on pathology at the Charité Hospital in Berlin, and his career in some ways mirrored Virchow's. He volunteered for service as a field hospital physician in the Franco-Prussian War and became interested in archaeology and anthropology. But his successes surely owed a great deal—more so even than in Virchow's case—to the scale of his microscopic investigations. He installed a laboratory in his own home, where he had an excellent microscope by Edmund Hartnack of Potsdam, plus a number of microphotographic devices, and a darkroom. He began by studying anthrax.

It had been known for some time that anthrax was caused by rod-like microorganisms observed in the blood of infected sheep.[8] Koch's first contribution was to invent techniques for culturing them in samples of cattle blood, which enabled him to study the microorganisms under his microscope. He traced their life cycle, identified spore formation and germination. More important, he found that although the bacilli were relatively short-lived, the spores remained infective for years. He proved that anthrax developed in mice only when the inoculum contained viable rods or spores of *Bacillus anthracis*, publishing his results in 1877, together with a technical paper that detailed his method of fixing thin films of bacterial culture on glass slides, enabling them to be stained with aniline dyes. This made possible the study of their structure by microphotography. Medicine was thus the direct beneficiary of the recent developments in three separate areas—dyestuffs, microscope technology, and photography.

Koch's next move was to equip his microscope with Ernst Abbe's new condenser and oil-immersion system (manufactured by Carl Zeiss), which enabled him to detect organisms significantly smaller than *B. anthracis*.[9] As a result (using mice and rabbits), he identified six transmissible infections that were pathologically and bacteriologically distinctive. He deduced that human diseases would derive from similarly pathogenic bacteria.

On the strength of this, in 1880 Koch was made government adviser (*Regierungsrat*) in the Kaiserliches Reichsgesundheitsamt (Imperial Department of Health) in Berlin. He shared a small laboratory with his assistants, Friedrich Loeffler and Georg Gaffky, both army doctors. They were charged with developing methods to isolate and cultivate pathogenic bacteria and to establish scientific principles that would improve hygiene and public health. (Johanna Bleker has shown that it wasn't until the 1850s and 1860s that German doctors thought of hospitals as places of effective science.)

Koch played a part in the development of the use of strictly sterile techniques, isolating new disinfectant substances, comparing their destructive action on different bacterial species.[10] He found that carbolic acid was inferior to mercuric chloride, bringing about the "dethronement" of Lister's "carbolic spray," and he found that live steam was much better than hot air in sterilization. This revolutionized operating room practices.[11]

In 1881 he turned his attention to tuberculosis. Inside six months, "working alone and without a hint to colleagues," he confirmed that the disease was transmissible (which not everyone accepted) and isolated from a number of tuberculous specimens of human and animal origin a bacillus with specific staining properties. He then induced TB by inoculating several species of animals with pure cultures of this bacterium. His lecture, to the Physiological Society in Berlin on March 24, 1882, was described by Paul Ehrlich as the "greatest scientific event."[12] The demonstration of the tubercle bacillus in the sputum was soon accepted as of crucial diagnostic significance.

In the same year, there was an outbreak of cholera in the Nile Delta. The French government, alerted by Louis Pasteur to the possibility that the epidemic could reach Europe, and told that the cause of cholera "was probably microbial," sent a four-man scientific mission to Alexandria. Koch arrived just over a week later, leading an official German commission. Within days he had observed colonies of tiny rods in walls of the small intestine in ten bodies that had died of cholera. He found the same again in about twenty cholera patients. Though promising, this organism failed to induce cholera when fed to or injected into monkeys and other animals. However, Koch's observations in Egypt were confirmed in Bengal, where his commission traveled next, and where cholera was endemic. In the spring of 1884 he identified village ponds, used for drinking water and all other domestic purposes, as the sources of why cholera was endemic in Bengal. He had, he said, observed cholera bacilli in one such pond.[13]

Although Koch and his work caught the eye (and continue to do so), the bacilli of swine erysipelas, glanders (an infectious disease of horses), and diphtheria were isolated by Loeffler, and the typhoid bacillus by Gaffky.[14] Advances were being made at such a rate that additional institutes of public health were established in Prussia, and in 1885 Koch was appointed to the new chair of hygiene at the University of Berlin. There was a hiccup when Koch announced he had developed a substance which prevented the growth of the tubercle bacilli, a substance which, it was subsequently found, didn't always work and sometimes had toxic side effects.[15] It emerged in this way that dosage was all-important. This hiccup strained relations between Virchow and Koch but, over Virchow's objections, an Institute for Infectious Diseases went ahead as planned in Berlin. The circle around Koch was by now more impressive than that around Virchow and included Paul Ehrlich and August von Wasserman.[16] As a result of Koch's work, a communicable diseases control law was passed in 1900, the year in which his institute moved to larger quarters, adjoining the Rudolf Virchow Hospital, making it the most famous medical complex in the world.

Koch achieved a level of fame for a doctor that has probably never been equaled, not even now. Toward the end of his life, he was in demand all over the world—South Africa, where he investigated rinderpest, Bombay (plague; he identified rats as the source but overlooked fleas as the vector), St. Petersburg (typhus), and Dar-es-Salaam (malaria and blackwater fever). He eventually isolated four types of malaria.[17]

In 1905 he received the ultimate accolade, the Nobel Prize for Physiology or Medicine. He died of angina on April 9, 1910. "Addicted" to chess and a great admirer of Goethe, Robert Koch probably benefited mankind—and the poor as well as the better off—more than anyone else to that point, and maybe since.

The Discovery of Antibiotics and the Human Immune Response

Despite the stirring achievements of Virchow and Koch, which would take time to work through their effects, at the beginning of the twentieth century people's health was still dominated by a "savage trinity" of diseases that disfigured the developed world: tuberculosis, alcoholism, and syphilis. TB lent itself to drama and fiction. It afflicted the young as well as the old, the well-off as well as the poor, and it was for the most part a slow, lingering death: as consumption it features in *La Bohème*, *La Traviata*, *Der Tod in Venedig* (*Death in Venice*), and *Der Zauberberg* (*The Magic Mountain*). Anton Chekhov, Katherine Mansfield, and Franz Kafka all died of the disease.[18]

The fear and moral disapproval surrounding syphilis a century ago mingled so much that despite the extent of the problem it was scarcely talked about. Despite this, in Brussels in 1899, Dr. Alfred Fournier established the medical speciality of syphilology, using epidemiological and statistical techniques to underline the fact that the disease affected not just the "demi-monde" but all levels of society, that women caught it earlier than men, and that it was "overwhelming" among girls whose poor background forced them into prostitution. This paved the way for clinical research, and in March 1905 Fritz Schaudinn, a zoologist from Roseningen in East Prussia, noticed under the microscope "a very small spirochaete, mobile and very difficult to study" in a blood sample taken from a syphilitic.[19] A week later Schaudinn and Eric Achille Hoffmann, a bacteriologist originally from Pomerania, and a professor at Halle and Bonn, observed the same spirochaete in samples taken from different parts of the body of a patient who only later developed roseolae, the purple patches that disfigure the skin of syphilitics. Difficult as it was to study, because it was so small, the spirochaete was clearly the syphilis microbe, and it was labeled *Treponema* (it resembled a twisted thread) *pallidum* (a reference to its pale color). The discoveries owed much to the invention of the ultramicroscope in 1906 by the German chemist Richard Zsigmondy at the Schott Glass Manufacturing Company, which provided specialized glass for Zeiss (see Chapter 18, p. 355). These advances meant the spirochaete was now easier to experiment on than Schaudinn had predicted, and before the year was out a diagnostic test had been identified by August Wasserman. It followed that syphilis could now be identified early, which helped prevent its spread. A cure was still needed.

The man who found it was Paul Ehrlich (1854–1915). Born in Strehlen, Upper Silesia, he had an intimate experience of infectious diseases: while studying tuberculosis as a young doctor, he had contracted the disease and been forced to convalesce in Egypt. His crucial observation was that, as one bacillus after another was discovered, associated with different diseases, the cells that had been infected also varied in their response to staining techniques. Clearly, the biochemistry of these cells was affected according to the bacillus that had been introduced. This deduction gave Ehrlich the idea of the antitoxin—what he called the "magic bullet"—a special substance secreted by the body to *counteract* invasions.

By 1907 Ehrlich had produced no fewer than 606 different substances or "magic bullets" designed to counteract a variety of diseases. Most of them worked no magic at all, but "Preparation 606" was found to be effective by a Japanese assistant, Dr. Sachahiro Hata, from Tokyo. Ehrlich called this magic bullet Salvarsan, which had the chemical name of asphenamine. He had in effect discovered the principle of both antibiotics and the human immune response. He went on to identify what antitoxins he could, to manufacture them, and to employ them in patients via the principle of inoculation. Besides syphilis he continued to work on tuberculosis and diphtheria, and in 1908 he was awarded the Nobel Prize for his work on immunity.[20]

Three Footnotes: The Discovery and Rediscovery of the Gene

Even after all this time, the coincidence in the rediscovery of the work of the botanist-monk Gregor Mendel makes for moving reading. Between October 1899 and March 1900, three other botanists—two Germans (Carl Correns and Erich Tschermak) and the Dutchman Hugo de Vries—published papers about plant biology, each of which (in a footnote) referred to Mendel's priority in discovering the principles of what we now call genetics. Thanks to this coincidence, and their scrupulousness in acknowledging his achievement, Mendel—once forgotten—is now a household name.[21]

Johann Mendel was born in 1822 in Heinzendorf in what was then Austria and is now Hynčice in the Czech Republic. His father was a farmer who fought in the Napoleonic Wars, and his mother was from a family of gardeners, meaning their whole life was dominated by plants—arable land, orchards, forest. In 1843, he entered the Augustinian monastery in Brno (Brünn), where he adopted the name Gregor. Mendel had no real Christian vocation, but the environment freed him economically and gave him peace of mind to pursue his studies. The abbot of the monastery was much concerned with the improvement of agriculture and had established an experimental monastery garden, where the director, Matthew Klácel, was interested in variation, heredity, and evolution in plants. He favored the Hegelian philosophy of gradual development, an approach that contradicted Christian orthodoxy and led to his dismissal and immigration to America, after which Mendel took over.[22]

Mendel was too sensitive for pastoral work (he was frequently disturbed by the suffering he saw among the poor). Instead, he was dispatched to the University of Vienna to expand his intellectual horizons.[23] In Vienna he was taught experimental physics by Christian Doppler (identifier of the

"Doppler effect") and by Andreas von Ettingshausen, the statistician. This proved important for Mendel's ideas about plant breeding. He also studied with Franz Unger, known for his views on evolution and lectures stressing sexual generation as the basis of the great variety in cultured plants. Unger argued that new plant forms evolved by the combination of certain elements within the cell, though he was unclear as to what exactly these were.[24]

Back in Brno, Mendel was elected abbot in 1868, and he too used his time to promote farming. In 1877 he helped introduce weather forecasts for farmers in Moravia, the first in central Europe.[25]

More to the point, he began experimenting with peas. The results for which we remember him were the fruit of ten years of "tedious experiments" in plant growing and crossing, seed gathering, careful labeling, sorting, and counting. Almost 30,000 plants were involved. As the *Dictionary of Scientific Biography* notes, "It is hardly conceivable that it could have been accomplished without a precise plan and a preconceived idea of the results to be expected." In other words, his experiments were designed to test a specific hypothesis.

From 1856 to 1863 Mendel cultivated seven pairs of characteristics, suspecting that heredity "is particulate," contrary to the ideas of "blending inheritance" to which many others subscribed. He observed that, with seven pairs of characteristics, in the first generation all hybrids are alike—and the parental characteristics (e.g., round seed shape) are unchanged. This characteristic he called "dominant." The other characteristic (e.g., angular shape), which only appears in the next generation, he called "recessive." What he called "elements" determine each paired character and pass in the germ cells of the hybrids, *without influencing each other.* In hybrid progeny both parental forms appear again and this, he realized, could be represented mathematically/statistically with A denoting dominant round seed shape, and a denoting the recessive angular shape. Were they to meet at random, he said, the resulting combination would be:

$$\tfrac{1}{4}AA + \tfrac{1}{4}Aa + \tfrac{1}{4}aA + \tfrac{1}{4}aa$$

After 1900, this was known as Mendel's law (or principle) of segregation and can be simplified mathematically as:

$$A + 2Aa + a$$

He also observed that with seven alternative characteristics, 128 associations were found—in other words, 2^7. He therefore concluded that the "behaviour of each of different traits in a hybrid association is independent of all other differences in the two parental plants." This principle was later called Mendel's law of independent assortment.[26]

Mendel's employment of large populations of plants was new, and it was this that enabled him to extract "laws" from otherwise random behavior—statistics had come of age in biology.[27] He attempted to sum up the significance of his work in *Versuche über Pflanzenhybriden* (1866). This memoir, his magnum opus and one of the most important papers in the history of biology, was the foundation of genetic studies. It was never truly appreciated because he had difficulty following up his pea work, his experiments with bees failing because of the complex problems involved in the controled mating of queen bees. He did show that hybrids of *Mattiola*, *Zea*, and *Mirabilis* "behave exactly like those of *Pisum*" but colleagues like Nägeli, to whom he wrote a series of letters, remained doubtful.[28]

Mendel had read *On the Origin of Species.* A copy of the German translation, with Mendel's marginalia, is preserved in the Mendelianum in Brno. These marginalia show his readiness to accept the theory of natural selection. Darwin, however, never seems to have grasped that hybridization provided an explanation as to the causes of variation. As a result, Mendel died a lonely unrecognized genius.

The Invention of the Unconscious

Just as some form of "evolution" was in the minds of many biologists and philosophers in Germany (and elsewhere) from the late eighteenth century on, so too the idea of the unconscious was a long time germinating. "Spirit release" rituals were common in Asia Minor as early as 1000 B.C.[29]

Among the general background factors giving rise to the unconscious, Romanticism was intimately involved, says Henri Ellenberger in his magisterial *The Discovery of the Unconscious.* This was because Romantic philosophy embraced the notion of *Urphänomene,* "primordial phenomena" and the metamorphoses deriving from them. Among the *Urphänomene* were the *Urpflanze*, the primordial plant, the *Allsinn*, the universal sense, and the unconscious. Johann Christian August Heinroth (1773–1843), described by Ellenberger as a "romantic doctor," argued that conscience originated in another primordial phenomenon, the *Über-Uns* (over-us).

A number of philosophers anticipated Freudian concepts. In *The World as Will and Representation*, Schopenhauer conceived the will as a "blind, driving force." Man, he said, was an irrational being guided by internal forces, "which are unknown to him and of which he is scarcely aware." Eduard von Hartmann argued there were three layers of the unconscious: (1) the absolute unconscious, "which constitutes the substance of the universe and is the source of the other forms"; (2) the physiological unconscious, which is part of man's evolutionary development; and (3) the psychological unconscious, which governs our conscious mental life.

Many of Freud's thoughts about the unconscious were anticipated by Nietzsche, who had a concept of the unconscious as a "cunning, covert, instinctual" entity, often scarred by trauma, camouflaged in a surreal way but leading to pathology. Ernest Jones, Freud's first (and official) biographer, drew attention to a Polish psychologist, Luise von Karpinska, who originally spotted the resemblance between some of Freud's fundamental ideas and those of Johann Friedrich Herbart (who wrote seventy years earlier). Herbart pictured the mind as dualistic, in constant conflict between conscious and unconscious processes. An idea is described as being *verdrängt* [repressed] "when it is unable to reach consciousness because of some opposing idea." Gustav Fechner (1801–87), an experimental psychologist (and yet another son of a pastor) built on Herbart, specifically likening the mind to an iceberg "nine-tenths under water."

Pierre Janet, a doctor in La Havre, France, claimed to have refined a technique of hypnosis, under which patients sometimes developed a dual personality. One side was created to please the physician while the second, which would occur spontaneously, was best explained as a "return to childhood." (Patients would refer to themselves by their childhood nicknames.) When Janet moved to Paris, he developed his technique known as "Psychological Analysis," a repeated use of hypnosis and automatic writing, during the course of which, he noticed, the crises that were induced were followed by the patient's mind becoming clearer. However, the crises became progressively more severe, and the ideas that emerged showed that they were reaching back in time, earlier and earlier in the patient's life.

The nineteenth century was also facing up to the issue of child sexuality. Physicians had traditionally considered it a rare abnormality but, in 1846, Father P. J. C. Debreyne, a moral theologian who was also a physician, published a tract in which he insisted on the high frequency of infantile masturbation, of sexual play between young children, and of the seduction of very young children by wet nurses and servants. Most famously, Jules Michelet, in *Our Sons* (1869), warned parents about the reality of child sexuality and in particular what today would be called the Oedipus complex.

The idea of "two minds" fascinated the nineteenth century, and there emerged the concept of the "double ego" or "dipsychism." The dipsychism theory was developed by the University of Berlin philosopher of aesthetics, Max Dessoir (1867–1947) in *Das Doppel-Ich* (*The Double Ego*), published to great acclaim in 1890, in which he divided the mind into the *Oberbewusstsein* and the *Unterbewusstsein*, "upper consciousness" and "under consciousness," the latter, he said, being revealed occasionally in dreams.

Freud's own views were first set out in *Studien über Hysterie* (*Studies in Hysteria*), published in 1895 with Josef Breuer, and then more fully in his work titled *Die Traumdeutung* (*The Interpretation of Dreams*), published in the last weeks of 1899.* Freud, a Jewish doctor from Freiberg in Moravia, was already forty-four.

It is in *The Interpretation of Dreams* that the four fundamental building blocks of Freud's theory about human nature first come together: the unconscious, repression, infantile sexuality (leading to the Oedipus complex), and the tripartite division of the mind into ego, the sense of self; superego, broadly speaking the conscience; and id, the primal biological expression of the unconscious. Freud saw himself in the biological tradition initiated by Darwin. After qualifying as a doctor, Freud obtained a scholarship to study under Jean-Martin Charcot, who ran an asylum in Paris for women afflicted with incurable nervous disorders. In his research,

* The book was technically released in November 1899, in Leipzig as well as Vienna, but it bore the date 1900 and was first reviewed in early January 1900.

Charcot had shown that, under hypnosis, hysterical symptoms could be induced. Freud returned to Vienna from Paris after several months and began a collaboration with another brilliant Viennese doctor, Josef Breuer (1842–1925). Breuer, also Jewish, had made two major discoveries, on the role of the vagus nerve in regulating breathing, and on the semicircular canals of the inner ear which, he found, controlled the body's equilibrium. But Breuer's importance for Freud, and for psychoanalysis, was his discovery in 1881 of the so-called talking cure.

For two years, beginning in December 1880, Breuer treated for hysteria a Vienna-born Jewish girl, Bertha Pappenheim (1859–1936), described for casebook purposes as "Anna O." She had a variety of symptoms, including hallucinations, speech disturbances, a phantom pregnancy, and intermittent paralyses. In the course of her illness(es) she experienced two different states of consciousness and extended bouts of somnambulism. Breuer found that in this latter state she would, with encouragement, describe certain events, following which her symptoms improved temporarily. However, her condition deteriorated badly after her father died—there were more severe hallucinations and anxiety states. Again, however, Breuer found that "Anna" could obtain relief from these symptoms if he could persuade her to talk about her hallucinations during her autohypnoses. This was a process she herself called her "talking cure" or "chimney sweeping" (*Kaminfegen*). Breuer's next advance was made accidentally: "Anna" started to talk about the onset of a particular symptom (difficulty in swallowing), after which the symptom disappeared. Building on this, Breuer eventually discovered that if he could persuade his patient to recall in reverse chronological order each occurrence of a specific symptom, until she reached the first occasion, most of them disappeared in the same way. By June 1882, Miss Pappenheim was able to conclude her treatment, "totally cured."[30]

The case of Anna O. impressed Freud. For a time he himself tried electrotherapy, massage, hydrotherapy, and hypnosis with hysterical patients but abandoned this approach, replacing it with "free association"—a technique whereby he allowed his patients to talk about whatever came into their minds. This technique led to his discovery that, given the right circumstances, many people could recall events that had occurred in their early lives and which they had completely forgotten. Freud concluded that though forgotten, these early events could still shape the way people behaved. Thus was born his concept of the unconscious and with it the notion of repression. Freud also realized that many of these early memories that were revealed—with difficulty—under free association, were sexual in nature. When he further found that many of the "recalled" events had *in fact* never taken place, he refined his notion of the Oedipus complex. In other words, the sexual traumas and aberrations reported by patients showed what people *secretly wanted* to happen, and confirmed that human infants went through a very early period of sexual awareness. During this period, he said, a son was drawn to the mother and saw himself as a rival to the father (the Oedipus complex) and vice versa with a daughter (the Electra complex). By extension, Freud said, this broad motivation lasted throughout a person's life, helping to determine character.[31]

A later development occurred with the death of Freud's father, Jakob, in October 1896. Although father and son had not been very intimate for a number of years, Freud found to his surprise that he was unaccountably moved by his father's death, and that many long-buried recollections spontaneously resurfaced. His dreams also changed. He recognized in them an unconscious hostility directed toward his father that he had hitherto repressed. This led him to conceive of dreams as "the royal road to the unconscious." Freud's central idea in *The Interpretation of Dreams* was that in sleep the ego is like "a sentry asleep at its post."[32] The normal vigilance by which the urges of the id are repressed is less efficient and dreams are therefore a disguised way for the id to show itself.

Freud has recently come under sustained criticism and revision and is now much discredited.[33] At the time he lived, however, in the late nineteenth century and in the early years of the twentieth, the unconscious was taken very seriously indeed and played a seminal role underpinning a transformation that was to have a profound effect on thought, in particular in the arts, in the late nineteenth and early twentieth century. This was the phenomenon known as modernism.

PART IV

The Miseries and Miracles of Modernity

21.

The Abuses of History

If this book were a theatrical production, at this point the lighting would change and the stage would become much darker. "The Germans have been superbly rational in their laboratories and industrial organisations. Their vision of politics and society, however, was blurred by clouds of evil fantasy." This is Fritz Stern in the introduction to his 1972 collection of essays, *The Failure of Illiberalism*.[1] At exactly the time that Helmholtz, Clausius, Siemens, Virchow, Koch, Benz, and Mendel were making their great innovations, a very different kind of intellectual activity was gathering pace, so different in tone, style, direction, and substance that several observers have remarked that, in the run-up to World War I, there were not one but two Germanies. It is now time to examine that other Germany.

Since this is a book more about the culture of Germany than its politics per se, we shall concentrate on the areas where this "other Germany" emerged. It emerged among the country's historians, took in a constellation of views that embraced aggressive nationalism, militarism, Darwinism, the Aryan myth, and anti-Catholicism, and culminated in a variety of sociological theories by both more and less reputable sociologists. These ideas produced a sharpening of Germany's self-image at the end of the nineteenth century as she distanced herself intellectually, culturally, and even morally from her neighbors and the rivals immediately surrounding her.

The Rise of High Culture and "Inwardness"

Although the focus of what follows is mainly intellectual history, politics—quite naturally—cannot be ignored entirely, in particular the part played by two men who between them epitomized and shaped, were both symptom and cause of, this other Germany: Otto von Bismarck and Kaiser Wilhelm II.

To recap briefly, we have seen how, in 1848, Germany's attempt at bourgeois revolution failed. Some parliamentary practices were established in the 1860s, but in general the aim of the German middle class for political and social equality and emancipation was unsuccessful. Germany failed to make the sociopolitical advances that Great Britain, the Netherlands, France, and North America had achieved, in some cases generations before. German liberalism, or would-be liberalism, was based on middle-class demands for free trade and a constitutional framework to protect Germany's economic and social space in society. When this attempt at constitutional evolution failed, to be followed in 1871 by the establishment of the Reich, led by Prussia, an unusual set of circumstances came into being. In a real sense, and as Gordon Craig has pointed out, the people of Germany played no part in the creation of the Reich. "The new state was a 'gift' to the nation on which the recipient had not been consulted."[2] Its constitution had not been earned; it was a contract among the princes of the existing German states, who retained their crowns until 1918. To our modern way of thinking, this had some extraordinary consequences. One result was that the Reich had a parliament without power, political parties without access to governmental responsibility, and elections whose outcome did not determine the composition of the government. This was quite unlike—and much more backward than—anything that existed among Germany's competitors in the West. Matters of state remained in the hands of the landed aristocracy, although Germany had become an industrial power. As more and more people joined in Germany's industrial, scientific, and intellectual successes, the more it was run by a small coterie of traditional figures—landed aristocrats and military leaders, at the head of which was the emperor himself. This dislocation was fundamental to "Germanness" in the run-up to the First World War.

It was one of the greatest anachronisms of history and had two effects that concern us. One, the middle class, excluded politically and yet eager to achieve some measure of equality, fell back on education and *Kultur* as key areas where success could be achieved—equality with the aristocracy, and superiority in comparison with foreigners in a competitive, nationalistic world. "High culture" was thus always more important in imperial Germany than elsewhere and this is one reason why—as we shall see in due course—it flourished so well in the 1871–1933 period. But this gave culture a certain tone: freedom, equality, and personal distinctiveness tended to be located in the "inner sanctum" of the individual, whereas society was portrayed as an "arbitrary, external and frequently hostile world."[3] The second effect, which overlapped with the first, was a retreat into nationalism, but a class-based nationalism that turned against the newly created industrial working class (and the stirrings of socialism), Jews, and non-German minorities. "Nationalism was seen as moral progress, with utopian possibilities."[4] Against the background of a developing mass society, the educated middle class looked to culture as a stable set of values that uplifted their lives, set them apart from the "rabble" (Freud's word) and, in particular, enhanced their nationalist orientation. The "*Volk*," a semimystical, nostalgic ideal of how ordinary Germans had once been—a contented, talented, apolitical, "pure" people—became a popular stereotype within Germany.

These factors combined to produce in German culture a concept that is almost untranslatable into English but is probably the defining factor in understanding much German thought as the nineteenth century turned into the twentieth. The word in German is *Innerlichkeit.* Insofar as it can be translated, it means a tendency to withdraw from, or be indifferent to, politics, and to look inward, inside the individual. Innerlichkeit meant that artists deliberately avoided power and politics, guided by a belief that to participate, or even to write about it was, again in Gordon Craig's words, "a derogation of their calling" and that, for the artist, the inner rather than the external world was the real one. Not even the events of 1870–71 succeeded in shaking this indifference. "The victory over France and the unification of Germany inspired no great work of literature or music or painting." In comparison with the literature of other European countries, the Germans never turned their attention to the political dangers that were inherent in the imperial system. "Indeed," writes Craig, "as those dangers became more palpable, with the beginnings under Wilhelm II of a frenetic imperialism, accompanied by an aggressive armaments programme, the great majority of the country's novelists and poets averted their eyes and retreated into that *Innerlichkeit* which was always their haven when the real world became too perplexing for them." There were no German equivalents of Émile Zola, George Bernard Shaw, Joseph Conrad, André Gide, Maxim Gorky, or even Henry James (except, just maybe, in his later work, Gerhart Hauptmann, 1862–1946, though he was the only one).

The Rise of Nationalistic History

We may open this aspect of the story with a return to Germany's historians. They were not especially inward—quite the contrary, in fact—but they benefited from the nationalism that inwardness helped to produce in Germany in the last half of the nineteenth century. First, consider Theodor Mommsen (1817–1903). Born in Garding in Schleswig, yet another son of a pastor, he enrolled at the University of Kiel in Holstein since he could not afford the more prestigious German universities. He won scholarships to visit France and Italy to study classical Roman inscriptions and made so much of them that in 1857 he was appointed a research professor at the Berlin Academy of Sciences. He helped create the German Archaeological Institute and in 1861 became professor of Roman History at the University of Berlin. He published more than 1,500 works and pioneered the study of epigraphy, being responsible for the *Corpus inscriptionum Latinarium* in sixteen volumes, of which he wrote five himself. He rose at 5:00 A.M. every day and was frequently seen reading as he walked. He had sixteen children. and two of his great-grandsons, Hans and Wolfgang, both became prominent German historians. In 1902, aged eighty-five, he won the Nobel Prize for Literature, one of few nonfiction writers to receive that accolade.

Mommsen was also a politician, a delegate to the Prussian Landstag from 1863 to 1866 and from 1873 to 1879, and a delegate to the Reichstag from 1881 to 1884, at first for the Deutsche Fortschrittspartei (German Progress Party), then for the Nationalliberale Partei (National Liberal Party). He had violent disagreements with Bismarck and with his fellow historian Heinrich von Treitschke, but was at the same time a fervent nationalist. Mommsen was a paradoxical figure—at least to us, today—because he embodied that world where nationalism was not yet the right-wing cause it became.

His most famous work was his *Römische Geschichte* (*History of Rome*).[5] Appearing in three volumes between 1854 and 1856, it was in fact unfinished, though it still made him what many regard as the greatest classicist of the 1800s. In the middle of the century Mommsen's *History of Rome* was ranked with Goethe's *Faust* and Schopenhauer's *The World as Will and Idea* as the most influential of works. His theme had a contemporary relevance because he argued that Julius Caesar was a genius and that rule by him was—and would have been, had he not been assassinated—far more just and fair and "democratic" than rule by the corrupt and self-serving senate.[6] As a fervent nationalist, Mommsen argued that "Caesarism," rule by a strong but fair-minded genius, was a less corrupt, more just guarantee of democracy than any other system.[7]

The book also contains an early sighting of what would come to be called *Völkerpsychologie*, a new science in which the "psychology of races" is used for the glorification of the country.[8] In particular, Mommsen haughtily argues in his book that the Germans are more talented than the Greeks *or* Romans. "The Greeks and the Germans alone possess a fountain of song that wells up spontaneously: from the golden vase of the Muses only a few drops have fallen on the green soil of Italy."[9] Mommsen's political position is difficult for us to understand today—a liberal who was a monarchist, a rigorous scholar whose nationalism bordered on racism. He was, for instance, rabidly anti-French. He welcomed the war of 1870 "as a war of deliverance which at last would extricate his people from that stupid imitation of the French."[10]

Heinrich von Sybel was born two days later than Mommsen, on December 2, 1817. He grew up in Düsseldorf, where his father, a lawyer, was a senior civil servant and was raised to the nobility in 1831. Their home was the site of many artistic gatherings, with Felix Mendelssohn among the regular guests.[11] At the University of Berlin, Heinrich was taught by Ranke and Savigny, in many people's eyes becoming their most distinguished pupil. As a Privatdozent at the University of Bonn, he soon made an impact, notably with the *Geschichte des ersten Kreuzzuges* (*History of the*

First Crusade) and *Die Entstehung des deutschen Königtums* (*The Origin of Kingship in Germany*), the first of which made his name by acting as a corrective to the idealized picture the Romantics had of the Middle Ages. This brought him a professorship in 1844, when he was twenty-seven, and at the same time he became a prominent opponent of the Ultramontane Party. This came about when the Holy Shroud of Turin was exhibited at Trier and attracted thousands of pilgrims. Sybel considered the shroud a fake and helped publish an inquiry into its authenticity. From then on, he was as interested in politics as in history and in 1846, when he was appointed professor at the small University of Marburg, he found time to take a seat in the Hessian Landstag. Then, in 1850 he sat in the Erfurt parliament as a member of the so-called Gotha Party, whose aim was the regeneration of Germany through the leadership of Prussia. He thought the House of Hapsburg was moved by "the Jesuit spirit" and that Austria "had nothing German in her."[12]

Sybel was, therefore, just as politically active as Mommsen and, like Mommsen, in addition to his political activities, he produced three great works, for which he is still remembered. The first was his *Geschichte der Revolutionszeit 1789–95* (*History of the French Revolution*). Sybel was much influenced by Edmund Burke and his *Reflections on the War in France*, but Sybel's contribution was to bring to bear the German brand of higher criticism on the records of the revolution. In the process, he showed, for instance, that many of the letters attributed to Marie Antoinette could not have been written by her—which aroused great interest in France itself and contributed to a new, and less romantic, vision of the Revolution, as put about particularly by left-wing French historians. While Sybel's scholarship was impeccable, and he was granted access to many archives in Paris and elsewhere in France, his conclusions suited his prejudices. He was a great believer in the idea that great men make history, that "the masses do nothing," and that therefore the real lesson of the French Revolution was the emergence of Napoleon.[13]

In 1856, on the recommendation of Ranke, Sybel became professor at Munich. There he established a Historical Seminar and the second of his great achievements, the *Historische Zeitschrift*, the original model for almost all historical journals that now exist and that itself is still going strong today. But Munich, as the capital of Bavaria, a Catholic state, was never going to be comfortable for Sybel. In the political turmoil that followed the war of 1859, he lost the support of the king and two years later transferred to Bonn.[14]

There he immediately became embroiled in politics, being elected a member of the Prussian Lower House and taking part in the attack on Bismarck. At that stage, when the press in Germany was not as independent as it was in France or Britain, professors who were politicians were a recognized phenomenon, even though Bismarck used to mock them.[15] For a time Sybel dropped out of parliament owing to eye problems but in 1867 he was back, gaining a seat as a National Liberal in the Constituent Assembly, from where he opposed the introduction of universal suffrage. This formed part of an important reconciliation with Bismarck, strengthened later when Sybel returned to the Prussian Parliament in 1874 to support the government in its fight with the Clericals and, later still, its opposition to the Socialists.[16]

Partly thanks to this, in 1875 Bismarck made Sybel director of the Prussian archives, opening up great opportunities. One of these was the correspondence of Friedrich the Great, of which Sybel was one of the editors. But by far the most important work, still impressive and useful today, was *Die Begründung des deutschen Reichs* (*The Founding of the German Empire*). As director of the Prussian archives, Sybel was allowed access to hitherto secret Prussian state papers, enabling him to give very full accounts of many self-contained episodes and events—the wars with Austria, Schleswig-Holstein, and Sadowa (Königgrätz). At the same time, his very closeness to some of the events, and his personal acquaintance with the authors or participants in those events, inevitably limited what he could say and how he said it. The history is essentially an account of Prussia's rise to pre-eminence, an explanation of why this was inevitable and why it was right: Prussia was a young, vigorous nation, Austria tired and old. The hero is Bismarck, the villains are the Austrians, the French, and the Danes (in Schleswig-Holstein).

After the fall of Bismarck in 1890, Sybel was no longer allowed access to the secret papers, so his later volumes (dealing with the years 1866–70) are less important. Which is perhaps just as well. At every turn *The Founding of the German Empire* is one step on from Mommsen's *History of Rome* in the catalog of tendentious history.

But even the tendentiousness of Sybel pales alongside that of Heinrich von Treitschke (1834–96), "a gifted phrase-maker of what are to us alarming phrases." Born in Dresden, he was the son of an officer in the Saxon army who rose to become military governor of Dresden, was raised to the nobility, and became a friend of the Saxon king. After a bout of measles and glandular fever, Heinrich's hearing was impaired sufficiently for careers in the bureaucracy and the army to be closed to him, and he had a distinctive "half-strangled" voice, not dissimilar to that of those born deaf. He therefore turned to an academic career, studying at the universities of Leipzig, Bonn, and Göttingen, where he was a student of Friedrich Christoph Dahlmann.[17] Dahlmann was an ardent German patriot, a fervent apostle of the Prussian ideal, a liberal who believed in a strong state. Treitschke imbibed these views and would build on them.

In 1863 he was appointed professor at Freiburg im Breisgau, on the southwestern edge of the Black Forest but three years later, at the outbreak of the Austro-Prussian War, he supported Prussia so strongly that he transferred to Berlin, became a Prussian subject, and was made editor of the *Preussische Jahrbücher*, where his articles—intemperate in tone—called for the annexation by force of the kingdoms of Hanover and Saxony. This hardly found favor with his father, who still lived in Leipzig and was still close to the king of Saxony. It did the son's career no harm, however: after appointments at Kiel and Heidelberg, he was appointed professor at Berlin in 1874.

By then he had been a member of the Reichstag for three years, a position he used to great advantage: between then and his death, he became one of the best-known figures in Berlin, a position further enhanced on Sybel's death, when Treitschke became editor of the *Historische Zeitschrift*. There he continued his ever-louder campaign in support of the Hohenzollerns and in the late 1870s, he grew increasingly anti-Semitic. This led him into conflict with Theodor Mommsen.[18]

In one sense, Treitschke was a descendant of Ranke, in that, for him, history was mainly about politics. He went so far as to belittle sciences such as electricity and archaeology, instead seeing the growth of the political power of Prussia as the great issue of the day, an approach that led to his greatest achievement as a historian, the five-volume *Deutsche Geschichte im 19. Jahrhundert* (*History of Germany in the Nineteenth Century*). The first volume appeared in 1879 but, at his death, sixteen years later, he had reached only 1847. His book was in the great tradition of multi-volume historical studies and, at one stage, was to be found in every middle-class home in Germany. His students included some very famous historians, among them Hans Delbrück, W. E. B. DuBois, Otto Hintze, Max Lenz (Max Planck's cousin), Friedrich Meinecke, Friedrich von Bernhardi, and the sociologist Georg Simmel. Between them, Treitschke and Bernhardi did much to turn German opinion against Great Britain. "Among the English," he said, "the love of money has killed all feeling of honour and all distinction between right and wrong."[19]

Treitschke was a great coiner of epigrams. "We have no German fatherland. The Hohenzollern alone can give us one." His most famous (and contentious) was "The Jews are our misfortune," but he also attacked Catholics ("a gang of priests"), and maintained that "No culture survived without servants." More important, however, is his position as the man who underwent the greatest change, from a liberal—opposed to the many petty restrictions of the semi-absolutism of the time in which he grew up—to a conservative in Bismarck's gradually unifying Germany, a man who argued that everything from law to economics must be understood as aspects of politics. He came close to being a demagogue, often regarded abroad as "an official mouthpiece of German policy," a policy that, he argued, included war "as the noblest activity of man."[20] Germany, he felt, should be free to express herself, to claim her rightful place in the sun. Like Sybel, he was granted access to the Prussian archives, though he was able to distance himself from what he found there far less than anyone else. It was Treitschke, among others, who Lord Acton had in mind when he wrote: "They brought history into touch with the nation's life, and gave it an influence it had never possessed out of France: and won for themselves the making of opinions mightier than laws."[21] Treitschke is regarded as one of those who helped to produce the belligerent Germany of the pre–World War I years, but he himself was produced by those years, too. He was convinced that the "purest German virtues" were the accomplishment of the king and nobility, working through the administration and the army.[22] This is the theme of his *History*.

A Treitschke monument was erected in the forecourt of the University of Berlin on October 9, 1909, three weeks before a Mommsen monument went up, both of them flanking the Helmholtz monument, in place since 1899. Treitschke's standing fell in the 1920s but soared in the 1930s. His statue was renovated in 1935–36 in the Nazi period and transferred to the Seitenhof, but in 1951 it was dismantled and melted down.

Mommsen's monument is still in its original location, and a statue of Max Planck (see below) stands between those of Helmholtz and Mommsen.*

More Fathers of Archaeology

Mommsen, Sybel, and Treitschke, though they had their differences, fall into the same pattern, as politically involved historians. Later chapters will show how their ideas and approach reverberated throughout German society and on the world stage. But though they, and others such as Johann Gustav Droysen and Dahlmann, together formed a clearly identifiable tendency, it wasn't the only one. Other German historians of the period emphasized cultural history. We met Jacob Burckhardt (an important influence on Nietzsche, whom he knew in Basel) in Chapter 3, p. 91. Ernst Curtius, Heinrich Schliemann, and Wilhelm Dörpfeld among them established the late nineteenth/early twentieth century pre-eminence of German classical archaeology.

In the middle of the nineteenth century, Greece, a newly independent kingdom, had first a German king, Otto, then a Danish one, George I, though of course Denmark had long been part of the German cultural umbrella. Many Germans lived in Athens and one of them, Ludwig Ross, was charged with restoring the Parthenon.

More important in the long run were the discoveries made by Ernst Curtius at Olympia.[23] This city meant a great deal to all Greeks—ancient chronology was related to the Olympic festivals that, it was known, had been held every four years since 776 B.C. There had been several attempts to mount an excavation, but nothing came to pass until 1874. While Curtius was in Athens as an envoy of Kaiser Wilhelm I, he helped to establish the new German Archaeological Institute, which held biweekly meetings

*As this book goes to press, there is an ongoing (unpleasant) debate in a district of Berlin about renaming Treitschke Street.

(the first on December 9, Winckelmann's birthday).[24] And it was Curtius, professor of archaeology at Berlin, who, together with the German ambassador, persuaded the Greek foreign minister and Ross's Greek successor as keeper of antiquities to sign the Olympia Agreement. This became a prototype for all such agreements from then on and stipulated that the Germans would pay all expenses, including those of the Greek police nominated to oversee the excavations, that the Germans should have the choice of where to dig, provided landowners were compensated, and that all finds should remain in Greece, though the Greek government could, at its discretion, give away any duplicates that it saw fit. Germany had the right to make copies and casts, and all publications were to be simultaneously released in Greek and German.[25]

In just over two months the temple of Zeus was exposed, followed by the Winged Victory, 42 heroic bodies, and more than 400 inscriptions. For many, however, it was the discovery of the Hermes that attracted most acclaim. This, according to Pausanias, was once taken for a work by Praxiteles (*fl.* 364 B.C.), one of the most famous sculptors in ancient Greece. The excavations at Olympia were the first to be carried out on modern scientific principles, and they made the Germans the leaders in showing the world what the archaic style (highlighted long ago by Winckelmann) was really like. Everything from the sanctuaries of Hera and the hill of Cronos to the temple of Zeus, with its rows of statue bases dedicated to famous victors, even the workshop of Phidias (now a ruined church), was beautifully laid out. It was so well done that Pierre Frédy, Baron de Coubertin, was moved to establish the modern Olympic Games in 1896.

In the same year that Curtius began work at Olympia, another German was at work elsewhere in Greece, embarked on a project that, if anything, caught the eye even more than Olympia. Curtius, the professional archaeologist, was continually upstaged by a man who, he thought, often did more harm than good. This other man was Heinrich Schliemann and the place he was excavating was, as he thought, Troy.

Schliemann had a colorful life, although—it is almost certainly true to say—not quite as colorful as he would like us to believe.[26] For Schliemann, while undoubtedly romantic, was a proven liar, and maybe a liar on an epic scale. Born in 1822 in Neu-Bukow in Mecklenburg-Schwerin, he was yet another son of a pastor. He had a whole career, a whole life, before he took up archaeology—as a grocer's apprentice, a cabin boy, an agent in an import/export firm in St. Petersburg, where he prospered and learned Russian and Greek. He started a bank in Sacramento, California, making a fortune in six months by buying and reselling gold dust, returning to Russia, where he met his first wife, Ekaterina. She thought Heinrich was richer than he was, and when she discovered her mistake, she withheld conjugal rights. This had the desired effect, and he cornered the market in indigo, to such effect that Ekaterina bore him three children. In the Crimean War (1854–56), Schliemann further cornered the market in saltpeter, brimstone, and lead, all needed in ammunition, and made a further fortune from the Russian government. Only after all this, in the late 1860s, did Schliemann turn to archaeology.

Schliemann sank himself into the Greek world, going so far as to divorce Ekaterina (in America, after a series of subterfuges) and advertise for a Greek wife in the Athenian press—he chose her from a clutch of photographs submitted.

By the nineteenth century, the actual site of Troy was far from certain (in the eighteenth century people could fall out for a lifetime arguing where it was). Nowadays, there are scholars who doubt that it ever existed or that the Trojan War ever took place, and who think therefore that Homer's classics are works of fiction. Nonetheless, over the centuries three sites have competed for the honor. Until Christian times, few doubted that Troy was identical with the "Village of the Ilians" at the hill of Hissarlik, near the Simois River. The geographer Strabo, on the other hand, opted for Callicolone, farther south and farther inland, which had two springs close together, which Homer describes and are not found at Hissarlik. Later travelers favored Alexandria Troas, an impressive set of ruins on the coast, but much farther south still.

Schliemann was not a scientific historian, nor an archaeologist, something that would rankle with Curtius. But Schliemann was the first to use excavation to test a hypothesis, and this use of the "experimental method," according to many, makes him the true father of archaeology.[27]

His excavation of Hissarlik started in 1869. Part of the land where the great mound was located was owned by the controversial American consul in the area, Frank Calvert, and he and Schliemann formed what would become an uneasy partnership.[28] Over the years, Schliemann (and others) discovered several layers on the site, as many as seven—or even eight—cities, one on top of the other. Two problems dogged him. One was the massive trench he dug to enable him to inspect all levels quickly, and that, in view of what happened—or didn't happen—later, may well have destroyed the evidence that could have decided matters one way or the other. This trench showed shards of material quite high up in the sequence that were obviously much older than the period of the Trojan War. The second problem was his discovery, in May 1873, of the so-called treasure of Priam. This was found at a level quite inconsistent with the Trojan War, and rumors quickly spread that he had bought the pieces on the black market and put them together to make a synthetic hoard. The entries in his notebooks didn't tally either.

The discovery of the treasure of course made for a romantic story, and with it Schliemann found the fame he undoubtedly sought. However, the controversy surrounding the treasure has never gone away. It was looted in Berlin at the end of the Second World War and only recently put on display in Moscow.

Nor was his reputation enhanced by his later digs at Mycenae. Here his achievement is more solid because little was known of the Mycenaean civilization at that point, and Schliemann uncovered a series of shaft graves and some wonderful gold ornaments "that are now the glory of the National Archaeological Museum in Athens."[29] Here too, however, the famous gold mask found in the fifth shaft grave does, for many people, bear a mustache that is particularly nineteenth-century in character.

Schliemann dug elsewhere—at Tiryns, Orchomenos, and Crete—before returning in 1882 to Hissarlik. By now he had the sense to enlist a proper professional archaeologist as his assistant/partner, and it was this man, Wilhelm Dörpfeld ("Schliemann's greatest find"), who had worked with Curtius at Olympia, who may well have found the real Troy.[30] In the spring of 1893, two years after Schliemann's death (on December 26, in Naples), Dörpfeld opened up the southern side of Hissarlik and immediately struck walls far more magnificent than Schliemann had ever found, with a pronounced "batter" or angle, as mentioned in Homer (when Patroclus tries to scale the wall), together with an angular watchtower and two important gates. Inside were large, noble houses, from the layout of which he was able to deduce that they were arranged in concentric circles.

No less important, Dörpfeld found everywhere the remains of Mycenaean pottery, exactly the same as Schliemann had found at Mycenae. Finally, he found evidence of a great fire that had ended the site, designated as Troy VI.

Not everyone accepts that Troy VI is Homer's city—the American Carl Blegen later argued for Troy VIIa—and maybe it never will be settled. It is also unlikely we shall ever again have an archaeologist as colorful and controversial as Schliemann.

Schliemann arranged with Bismarck for his discoveries to be displayed in the Ethnological Museum in Berlin (five halls all marked "SCHLIEMANN") and this was soon built on. The growing national consciousness of the late nineteenth century would eventually be expressed (and not just in Germany) in museums as much as in military affairs. The Berlin Museum, on an island in the Spree and designed by Schinkel, had been approved in 1823 and featured a massive Ionic colonnade of eighteen columns. Only genuine sculptures were admitted (there were to be no casts), and antiquities of all periods were gradually assembled, designed to appeal to the public, the artist, and the scholar—in that order.[31] But it wasn't for another fifty years that the Berlin Museum accepted its first work of world importance: the altar of Zeus from Pergamum.

Pergamum was, of course, one of the great cities of the Hellenistic world, rising in importance after the death of Alexander the Great; it was the site of one of antiquity's most famous libraries—2,000 volumes in all. There were many temples in Pergamum, but the glory was the altar of Zeus, on the highest hill. It was famous in antiquity for its brilliant sculptures and friezes, 120 meters in length. In the third century, Ampelius described it to his friend Macrinus as one of the Wonders of the World: "a marble altar forty feet high with very large sculptures; and it portrays the Gigantomachy" (the battle of the gods and the giants, before humans were formed).

By 1871, when Carl Humann, an engineer and architect from Essen who, like Schliemann, had a fascination with archaeology, first began to excavate at Pergamum, the city was run down, many of its ancient stone blocks plundered by Turkish builders. Humann's brother, Franz, had acquired various rail and trade concessions from the sultan, and Carl worked with him. They secured permission to dig and were visited by

Ernst Curtius as it soon became clear that spectacular discoveries were being made. Most notable were a series of slabs of bluish marble, which were sent on to Berlin. There, the new director of the sculpture gallery, Alexander Conze, came across a reference in Ampelius that caused him to realize that Humann had excavated—was *still* excavating—the Gigantomachy itself. In all, thirty-nine slabs were unearthed, ten free-standing statues, and many inscriptions. The next year another twenty-five slabs, including that of Zeus with his giant opponent, were uncovered, along with thirty-seven statues.

Because Turkey was then very poor, the sultan sold his interest to the Germans, and the altar of Pergamum became, with Priam's treasure, the first great archaeological discovery to go on display in Berlin, in 1880. The Royal Museum in Berlin, under Conze, now gained an "overwhelming lead" in the display of treasures from the Greek lands.[32]

By then, the French, British, and Americans had all established their own archaeological institutes in Greece and Italy. It was not only Mommsen, Sybel, and Treitschke who felt the call of nationalism.

22.

The Pathologies of Nationalism

Militarism as we know it, modern militarism, probably emerged in the eighteenth century when the rise of absolutism was accompanied by the growth of large-scale military organizations to support princely power.[1] Even then, though, because most of the soldiers had to travel on foot, the size of armies was limited. That, as we have seen, changed with Napoleon's introduction of the citizen army, which made fighting units much bigger and meant far more people had the experience of being *in* the military. In Prussia, after his victories, Napoleon confined the army to 42,000 men. The king responded by dismissing all the soldiers after twelve months and inducting another 42,000, a process he repeated the following year. In turn, the industrial revolution, the development of techniques of mass production, together with developments in metal technology resulting from the steam engine, meant that weapons were more plentiful as well as more terrible. The development of the railways added the dimension of greater mobility to war, marking another radical transformation. The rise of imperialism and colonial expansion later in the nineteenth century contributed a further twist, not least in the way the campaigns and victories in far-off lands helped glorify military values. These factors affected all nations.

That said, there were a number of elements special to Germany that caused people to identify "Prussian militarism" as something set apart.

There was, for example, what has been called a "military revolution" in the 1860s.[2] This was a unique short-service system, insisted upon by King Wilhelm I against Liberal opposition. It ensured three years' obligatory service in the regular army and another four in the reserve, meaning Prussia had a far larger front-line army relative to its size than any other European power.

A second factor was the Prussian General Staff. This, according to Paul Kennedy, "rose from obscurity in the early 1860s to be 'the brains of the army' under the elder [Helmuth von] Moltke's genius." Before this date there had been no general staff beyond the quartermaster, other staff officers being recruited for specific campaigns, often not far in advance. Moltke now chose the best of the products of the War Academy and taught them to prepare for possible future conflicts, updating their plans as a result of the study of history, war games, and maneuvers. "A special department was created to supervise the Prussian railway system and make sure that troops and supplies could be speeded to their destinations."[3] Above all, the officers were taught Clausewitz's doctrine of the decisive battle, and to be prepared to bring large bodies of men to converge on the crucial location, using their own initiative if communications were disrupted. This combination of factors gave the Prussians decisive—and relatively swift—victories in 1866 and 1870 where, in the latter case, within a fortnight of the declaration of war, three armies (300,000+ men) were sent to the Saarland and Alsace.

It was this system that, under Bismarck's leadership, would come to dominate Europe, what Nicholas Stargardt calls "the new militarism," shifting the European center of gravity to Berlin.[4] It was underpinned by the fact that, from the 1860s on, when the new Prussian military system took hold, the trend to industrialization intensified, a matter that cannot be ignored. How quickly the balance shifted among the Great Powers is shown most vividly in the statistics:

TOTAL POPULATION OF THE GREAT WESTERN POWERS (MILLIONS)

	1890	1913	% CHANGE
Russia	116.8	175.1	149.9

United States	62.6	97.3	155.4
Britain	37.4	44.4	118.7
France	38.3	39.7	103.6
Germany	**49.2**	**66.9**	**135.9**
Austria-Hungary	42.6	52.1	122.3

Per Capita Levels of Industrialization

	1880	1913	% change
Russia	10	20	200
United States	38	126	332
Britain*	87	115	132
France	28	59	211
Germany	**25**	**85**	**340**
Austria-Hungary	15	32	213

Relative Share of World Manufacturing Output

	1880	1913	% change
Russia	7.6	8.2	8.0
United States	14.7	32.0	18
Britain	22.9	13.6	–41
France	7.8	6.1	–28
Germany	**8.5**	**14.8**	**74**
Austria-Hungary	4.4	4.4	0.0[5]

These figures are not, of course, in themselves evidence for militarism, but taken together they do underline Germany's material progress, both a symptom and a cause of greater nationalistic feeling, and they are in any case paralleled by changes in military capabilities:

* Britain at 1900 = 100

Military and Naval Personnel of the Powers			
	1880	1914	% CHANGE
Russia	791,000	1,352,000	171
United States	34,000	164,000	482
Britain	367,000	532,000	145
France	543,000	910,000	168
Germany	**426,000**	**891,000**	**209**
Austria-Hungary	216,000	444,000	206

Warship Tonnage			
	1880	1914	% CHANGE
Russia	200,000	679,000	340
United States	169,000	985,000	583
Britain	650,000	2,714,000	418
France	271,000	900,000	332
Germany	**88,000**	**1,305,000**	**1,483**
Austria-Hungary	60,000	372,000	620

Germany was at the very heart of Europe, geographically speaking, and the very speed of its transformation was an issue in itself. "This alone was to make 'the German question' the epicentre of so much of world politics for more than half a century after 1890." The quality of Germany's military personnel was dramatically underlined by one study that showed the number of illiterate recruits in Italy was 330 out of 1,000, 220/1,000 in Austria-Hungary, 68/1,000 in France, but only 1/1,000 in Germany.[6]

The Alldeutscher Verband (Pan-German League) and the Deutscher Flottenverein (German Navy League) were only too happy to reinforce this tendency. The Pan-German League was formed after Kaiser Wilhelm II had ceded Zanzibar to Britain in exchange for the island of Helgoland.[7] This strategic withdrawal did not go down well with certain sectors of the German public, especially as it followed quickly the dismissal of Bismarck. A young collaborator of Krupp's, Alfred Hugenberg, founded the Pan-German League, and its expansionist policy soon received the support of thousands—if not tens of thousands—of Germans, some of whom, like Ernst Haeckel, Max Weber, and Gustav Stresemann, were distinguished in other fields, though others, like the British-born Houston Stewart Chamberlain, were little more than rabid racists. In 1908 Heinrich Class became the leader of the movement and made it even more extreme, advocating a merciless struggle against the Social Democratic Party, campaigning for the expulsion of Jews to Palestine and the annexation of lands to the east of the Dnieper. Their views overlapped with Treitschke's but many others among Germany's ruling elite became convinced of the need for territorial expansion "when the time was ripe." Admiral Alfred von Tirpitz in particular argued that Germany's industrialization and overseas conquests were "as irresistible as natural law."[8] Not that Germany stood out too much in this regard—imperial belligerence was just as prevalent at the time in Britain, France, and Japan.

But there could be no doubting that Germany's military build-up was more impressive than anyone else's, the most awesome aspect being the rapid expansion of its navy after 1898, which under Tirpitz was transformed from the world's sixth largest to the second largest, after Britain's Royal Navy.[9] A final factor that set German militarism apart from anyone else's was her geographical location. She was, as David Calleo put it, "born encircled." Because Germany lay at the center of a continent, Germans were always prone to see themselves as encircled—and threatened—by France, Britain, Russia, and Austria-Hungary. This meant that Germany relied far more on statecraft than did other nations and, after Bismarck had been dismissed, that crucial ingredient was lacking.[10]

The Anti-Catholic Imagination

On July 19, 1870, France declared war on the North German Confederation. Only twenty-four hours before, the First Vatican Council had confirmed the proclamation of the Pope's "infallibility." For many in Germany this coincidence was just too much, and the speed of the German victory was therefore all the sweeter. Paris was taken and the German states united in an empire under the aegis of the Prussian king, now an emperor. Sybel spoke for many when he confided, in a letter to Hermann Baumgarten:

"How have we deserved God's grace to be permitted to experience such great and mighty things? And for what shall one live hereafter?"[11]

As Michael Gross has pointed out, Sybel and others soon found an answer. "They committed themselves now to a war against the Roman Catholic Church and with it the consolidation within Germany of modern society, culture and morality." Papal infallibility was provocative because it seemed to invite German Catholics to direct their allegiance to Rome rather than to the newly installed Kaiser.[12] Paul Hinschius, a Liberal deputy, said that the Vatican's proclamation was "nothing less than a death sentence" passed against the newly unified state.

That was going too far, but his words had resonance because this *Kulturkampf*, as it came to be called, had been building for some time. Catholicism in the land of Luther, Reformation, Protestantism, and Pietism came to stand as the enemy of liberalism, of belief in reform, above all of opposition to the cultivation of the human intellect and spirit in civic society through *Bildung*.

After the failed revolutions of 1848, liberalism had come under sustained threat in the reactionary 1850s, and the Catholic Church threw its weight in with the state to rub in the liberal defeat. From 1848 on, waves of missionaries swept across Germany, visiting thousands of towns and small villages from the Rhineland to the Baltic, and for more than twenty years initiated a counterrevolutionary, anti-Liberal, anti-Enlightenment onslaught. The missions—usually consisting of three Jesuit, Franciscan, or Redemptorist missionaries, though there could be as many as eight—concentrated on the smaller localities and between them, according to one account, mounted at least 4,000 interventions between 1848 and 1872.[13]

They were strikingly successful. Typically lasting for two weeks, the towns where they were located often swelled in size to double or even three times their normal population as pilgrims descended on the missions. The pilgrims would either be put up with locals or sleep in churches and even churchyards. Work in the fields was suspended, shops, theaters, and schools were closed. Congregations would be in their places from 4:00 or 5:00 in the morning. At Cologne, in 1850, when the mission arrived, as many as 16,000 people crowded into the cathedral.[14]

For two weeks, the missionaries held masses, heard confessions, and carried out exorcisms at an intense pace, though the chief attraction was invariably the sermon. These were held three times a day—at dawn, in the afternoon, and in the evening—and each lasted a full two hours. Two subjects appear to have been pre-eminent. One concerned the veneration of the Virgin Mother of Christ, whose Immaculate Conception had become dogma only in 1854. In adoring the exemplary nature of Mary's life, the missionaries railed against the sins of the contemporary world—alcohol consumption in the taverns, dancing, cards, gambling, sexual license, and the reading of political literature. Some sermons proved so popular that the police commissioner in Düsseldorf in 1851 had them printed as a pamphlet to encourage order in the city. The other theme—the more popular one—was on the fiery reality of hell. The Jesuits in particular specialized in bloodcurdling hell-sermons.[15] At Aachen, people were so keen to confess that brawls broke out as they fought to get to the confessional. Theaters were denuded of their patrons, inns stood empty. In one case the priest reported to his bishop that girls were even wearing their hats more modestly, "without bands and flowers."[16] Although the missions made a lot of noise, there is, says Michael Gross, no evidence that such "sins" as illegitimacy or alcoholism declined.

There was also a dramatic rise in the number of new monastic orders and religious congregations. In Cologne there were 272 monks and nuns in 1850; by 1872 that number had grown to 3,131. Four monasteries were built there before 1848; thirty-seven were built in the two decades afterward, and the figures were much the same in Paderborn, five before 1848, twenty-one between then and 1872.[17]

Although the authorities were sometimes concerned that the missions might provoke civil trouble, by and large they welcomed them because their conservative, antiradical aims suited the powers that be and, again for the most part, the sermons steered clear of blatantly political arguments. In May 1852 sixty-two deputies in the lower house of the Prussian Parliament formed the "Catholic *Fraktion*," the first sign of a Catholic Center Party that would come to full fruition with the founding of the empire in 1870. At the same time, worries did begin to be aired about the Jesuits, that they were a rogue "state-within-a-state" pushing anti-Prussian "ultramontanism" and reasserting Austrian influence.[18] These fears grew in the late 1850s as Protestant pastors found they had to work harder to educate their flocks about the differences between Protestant and Catholic. As a result, a Protestant revival took place in Germany toward the end of the 1850s and throughout the 1860s, an indirect and unintended consequence of the Catholic missions. Priests and pastors now saw themselves as being involved in a *Kriegszustand*, or state of war. Pastors argued that the Catholic campaign to improve morality was a sham, a Trojan Horse, and that they were just as keen to assault the virtues of the Enlightenment, "against Bildung and the humanity of our time."[19] Others argued that Catholicism was backward, whereas Protestantism was progressive and, yes, liberal. Treitschke, after Pius IX had issued the Syllabus of Errors in 1864, confessed, "what good luck it is to be Protestant. Protestantism has the capacity for endless, continuing Bildung."[20]

Feelings were just as strong against nuns and monks because the Kulturkampf was overlaid with a *Geschlechterkampf*, or a conflict between men and women. Rudolf Virchow, leader of the Progressive Party in the Prussian Parliament, and the man who actually coined the term "Kulturkampf," attacked the emergence of the women's movement and was joined by Sybel. The growth of female religious orders was singled out—these nunneries "were draining the marriage pool."[21] In the convent schools the girls were made to promise not to read Goethe or Schiller.

If there was a tipping point, it occurred in the early 1860s. Throughout the late 1850s, state after state had initiated policies that helped them join the Zollverein (the German Customs Union). In 1866 Prussia's decisive victory over Austria put an end to any idea of a *Grossdeutsch* solution to the German national question, a German state unified under the Hapsburgs. Moreover, Catholics—who had been in the majority in the German Confederation that had existed since 1815—now found themselves in a minority in the North German Confederation: 20 million Protestants to 8 million Catholics. The Protestants thought the victory a proof of their virility, Droysen characterizing it as "the triumph of the true German spirit over the false."[22] The "problem of Catholicism" became a major topic of debate, not helped by the Syllabus of Errors, which made it plain that the pope considered liberalism and Catholicism incompatible. Anti-Catholic literature, and anti-Catholic feeling in general, would soon reach hysterical proportions.[23]

By the early 1870s, then, several things came together—militant Catholicism (including better organization, politically), the women's movement, the demand for democratization, plus nascent socialism and even the fear of French revanchism, after the victories of 1870–71, to produce what Gross again has called a "meta-enemy."[24] All these factors were enemies allied against what was, after all, a *new* empire. And when, in October 1873, the pope claimed that everyone who had been baptized Christian "belonged" to him, all liberal fears were confirmed, and the campaign against the Catholics became urgent. According to the Catholic *Badische Beobachter*, a war was in the offing: "We have made peace with France; with Rome, we will never make peace."[25] That campaign was launched "in the name of German unity, the modern state, science, progress, Bildung and freedom." For many, nonetheless, this was the abandonment of liberal principles by the liberals.

The famous "pulpit paragraph" was passed in December 1871 and made public discussion of matters of state by clerics "in a manner endangering public peace" a criminal offense. In 1873 the first of the so-called May Laws was passed, stipulating that the appointment of all clerics be approved by the state.[26] A second law, passed the following July, banned the Society of Jesus and the Redemptorist and Lazarist orders from German soil. The following year a Court of Ecclesiastical Affairs abolished the authority of the pope over the Catholic Church in Prussia and a year after that abolished the state's subsidy to Catholic dioceses until their bishops agreed in writing to abide by all Kulturkampf laws. In all, 189 monasteries were closed and several thousand clerics banned. Twenty newspapers were closed down, 136 newspaper editors arrested, and 12 dioceses left without bishops. By 1876 1,400 Prussian parishes did not have priests.[27] Jews looked on nervously.

In cultural terms, the educated liberals of Germany had seen—or professed to see—in the rise of Catholicism the return of an ignorant, backward age, "indifferent or hostile to Bildung" and all that had helped raise Germany from what she had been in the mid-eighteenth century, to where she was in the 1870s, the country of von Liebig, Clausius, Helmholtz, Siemens, Heine, Koch, Zeiss, and Virchow. In winning the battle against the Catholics, however, they lost the fight with Bismarck. Their defeat was more far-reaching than their victory.[28]

The Uses and Abuses of Darwinism

Only in the late nineteenth century, with the advent of mass literacy, did science begin to have an impact on daily life. Never before or since has the prestige of science been so high and the interest of the layman so great.[29] In particular, Darwinism was a sensation in Germany. "Darwinism became a kind of popular philosophy in Germany more than in any other country, even England. Darwinism caught on rapidly in the German scientific community; indeed, Germany, rather than England, was the main centre of biological research in the late nineteenth century . . . not only was Germany the most literate of the major European countries, it also offered the richest environment for Darwinism to expand beyond the confines of science. Political liberalism had been thwarted in Germany in 1848, and Darwinism became a pseudopolitical ideological weapon for the progressive segments of the middle class."[30] Germans were aware that the *Naturphilosophen* had in many ways anticipated the idea of evolution, even if they had no understanding of natural selection.

In this atmosphere, a number of science popularizers emerged, of which the best known, and best remembered, were Ernst Haeckel, Carl Vogt, Ludwig Büchner, Carus Sterne, Edward Aveling, and Wilhelm Bölsche. Hundreds of books were published on Darwinism, and Bölsche was the single best-selling nonfiction author in the German language before 1933. In the process, Darwinism was changed and corrupted as people in all walks of life appealed to Darwin's authority. But in general, Germany's Darwinists sought to continue the Enlightenment tradition, to stamp on superstition, to inform and, in so doing, to liberate, as they saw it, and to continue the radical spirit of 1848.

Although literacy rates in Germany were much higher than anywhere else in Europe, mass reading didn't really catch on until the 1870s, when the price of books and newspapers dropped owing to the invention of new and more efficient printing presses. It was only after 1860 that the developments in science and technology, referred to earlier in Chapters 17–20, pp. 341–383, began to seep into popular consciousness.

A translation of *On the Origin of Species* was begun within a few weeks of the book's appearance in Britain and published in 1861; Darwin's collected works were published in 1875. And Darwinism, as Alfred Kelly says, "made rapid and deep inroads in the German scientific community. From the beginning it was broadly identified with progressive views." (Nipperdey says the German response to Darwin was "overwhelming.") As early as March 1861, Darwin wrote to his colleague Wilhelm Preyer, "The support which I receive from Germany is my chief ground for hoping that our views will ultimately prevail." At the end of 1899, the *Berliner Illustrierte Zeitung* asked its readers who they thought the greatest thinkers of the nineteenth century had been. Darwin came in third after Helmuth von Moltke and Kant, but *The Origin of Species* was voted the single most influential book of the century.[31]

Most histories remember Haeckel above all other popularizers, but it was Bölsche who was better known at the time, indeed his was a household name in millions of homes. The combined sale of his books by 1914 has been estimated at 1.4 million—Alfred Kelly says he was a "major cultural phenomenon." Bölsche was a founder of the Cologne zoo and a friend of Alexander von Humboldt, Carl Vogt, and Jacob Moleschott. His major work perhaps reads oddly now—*Das Liebesleben in der Natur: Eine Entwicklungsgeschichte der Liebe* (*Love-Life in Nature: The Story of the Evolution of Love*), three volumes, 1898–1901. Designed to reconcile Darwin with the Bible, it told the story of evolution from the perspective of sexual love and was a sensational success.[32] Sex, for Bölsche, was the ideal experience, a brief glimpse of eternity and the harmony that was the aim of evolution.

The rise of Darwinism took place, of course, at the same time as the Kulturkampf and, being allied with the forces of progress, it naturally came under attack from the reactionaries, who did not want the Protestant Church interfered with. The main battleground was the schools where, for the most part, Darwinism was excluded. In many places in Germany very little science was taught in the schools anyway (in some places, like Bavaria, it was simply not required), so Darwinism wasn't being singled out. Haeckel, on the other hand, thought that Darwinism ought to become the centerpiece of the school curriculum, and bitter battles were fought—in parliament, in the newspapers, in books—over this issue.

After its original enthusiastic reception, Darwinism in Germany developed in two ways. First, a variety of Social Darwinisms emerged, several examples of which are considered, pp. 428–29. The other way that

Darwinism developed was its conjoining with Marxism. The generation of German workers before World War I was extremely socialist in its outlook but, again according to Kelly, the tenets of basic Darwinism were much easier to understand than Marxism itself and so the latter became infused with Darwinist terminology, even to the point that most workers saw the future in evolutionary rather than revolutionary terms. Even Rudolf Virchow, noted liberal that he was, feared that Darwinism "could lead to socialism."[33] In 1899, A. H. T. Pfannkuche placed an advertisement in *Die Neue Zeit* asking the librarians of workers' libraries to send him lists of the most popular books and he published his findings as *Was liest der deutsche Arbeiter?* (*What Does the German Worker Read?*). Four out of the top ten books were books about Darwinism. Its appeal stemmed from the prestige of science at the time and its message of the inevitability of change.

Other scholars such as Hans-Günter Zmarzlik, Roger Chickering, and Richard Evans have added the important critique that, with the rise of Pan-Germanism and ideas about "racial hygiene" (see below), Social Darwinism in Germany veered from being a mainly left-wing concern to the right in and after the 1890s.[34]

The Fear of Degeneration

Kelly also observes that it was in the 1890s that Social Darwinism "began to undergo some ominous changes."[35] By 1890, there was a growing consensus—not least among medical men—that the industrial landscape of Europe was encroaching so quickly on what had gone before that a host of new disorders was being created in its wake: new forms of poverty, crime, alcoholism, moral perversion, and violence.[36] The concept of "degeneration," the very idea that the European population was no longer physically capable of supporting civilized life, had begun with the Italian doctor Cesare Lombroso, who espoused a theory that criminals were a special "atavistic" type, "criminaloids," throwbacks to primitive humanity.[37] But the man who made the most of this was Max Nordau, a socialist and committed egalitarian and a man who, in his fiction, called degeneracy the "malady of the century."[38] He and Ernst Haeckel were founder members of the National Peace League and the Society for Racial Hygiene. On top of its Italian beginnings, the French had built their own ideas of degeneracy, following the Franco-Prussian War, when the humiliation of defeat had shocked France's intellectual elite. But Nordau, a German-speaking Hungarian doctor and journalist, published *Entartung* (*Degeneration*) in 1892 and although it was close to 600 pages long, it became an immediate international best seller, translated into a dozen languages. Nordau argued that there was degeneration not just in people but in culture, "degenerates are not always criminals, prostitutes . . . lunatics; they are often authors and artists." He singled out Charles Baudelaire, Oscar Wilde, Henrik Ibsen, Leo Tolstoy, Émile Zola, Édouard Manet, and the Impressionists, who he thought painted the way they did as a result of nystagmus, a "trembling of the eyeball," which blurred and distorted their vision.[39]

Nordau thought the European aristocracy was beyond help, the only hope lying with the working class, whose self-confidence and vitality could be best ensured by physical exertion and strenuous outdoor exercises. His theories helped give rise to the mania for athletic clubs, the hiking and backpacking movements and bicycle races that engulfed Germany around the turn of the twentieth century. These activities overlapped with the Youth Movement, started in 1897 in Steglitz, a middle-class suburb of Berlin, under the charismatic leadership of Karl Fischer. Their hikes always took in difficult terrain and they soon embraced their own way of dressing, their own uniform. As well as physical exercise, their songs encouraged the experience of belonging. Their gatherings in uniform were theoretically illegal but that didn't stop them calling Fischer their "*Führer*" and greeting each other with "*Heil*."

The Aryan Mystique

The history of Germany is different from that of other European countries in so far as her textbooks invariably begin by concentrating on the expansion of the Germanic people in Italy, France, and Spain rather than just telling the story of Germans in Germany. This is reflected in the different names other nations give the Germans—they were "Saxons" to the Finns, "Niemcy" or "Swabians" to the Russians and Poles, "Germans" to the British, "Allemands" to the French, "Tedeschi" to the Italians, with the Germans themselves adopting the last root for their own "Deutsche."[40]

Is it this varied history that gave the Germans their concern with racial purity and the role of heroes in the achievement of that purity? All nations feel themselves special, but the feeling was particularly strong in Germany, where Friedrich Schlegel was the principal builder of the Aryan myth, laying the groundwork for others later to contrast the "old Latin race" with the "young Germanic race," so that Theodor Mommsen was required to warn against "the nationalist madmen who want to replace the universal Adam by a Germanic Adam, containing in himself all the splendours of the human spirit."[41]

The founder of modern racial theory, who anticipated the Aryan concept, was Christoph Meiners (1745–1810), a Göttingen professor who was the first to advance the theory that mankind had its origins in Africa, and who saw a sort of progress from orangutans to Negroes, to Slavs, to Germans. But his views were overtaken by those of the Romantics, who thought that, as the Schlegels put it, "Everything, absolutely everything, is of Indian origin" and Germany "must be considered the Orient of Europe."[42] Schopenhauer was sympathetic and it was around this time that the term "Aryan" began to be used, having been originally borrowed from Herodotus by the Frenchman Anquetil du Peyron to designate Persians and Medes. Whereas other countries (such as France) also had the Aryan myth, tracing many features of European cultural life (such as language) to "Indo-*European*" roots, the Germans referred to "Indo-*German*" traditions, seeing themselves as particularly strongly linked with that background, as a rival to the more classical Latin trajectory. They became, as the nineteenth century progressed, the quintessence of the white human race, the noblest of all races, the only ones capable of attaining a "higher spirituality." (Lorenz Oken went so far as to say that, since blacks could not blush, they could not express an interior life.)[43]

By 1860 the distinction between Aryans and Semites had become accepted in many quarters, right across Europe. The word was used by Darwin and by Nietzsche, in the novels of Gustav Freytag, and in the Proceedings of the Anthropological Society of London. The cult of "blond virility," later so popular, did not come into being, however, until 1870–71 and had its origins in the writings of the French anthropologist Armand de Quatrefages de Bréau, who, having witnessed the siege of Paris by the Prussians, in which civilians were bombarded, accused the Prussians of being Finns and Slavo-Finns, primitive inhabitants of northeastern Europe. This provoked a backlash in Germany, and it was as part of this backlash that Virchow's famous study of German craniology was undertaken. The theoretical background to the study was that a mainly blond and "dolichocephalic" (long-skulled) Aryan stock had migrated to Europe and mingled with the dark and "brachycephalic" (round-skulled) natives. Lasting more than ten years, involving more than 15 million schoolchildren, Virchow's study showed that the northeastern Prussians were blond, like the Finns, and that the southwestern Germans were not—they had been assimilated into the racial types of the Burgundians, the Franks, and the Goths. Both the Hapsburgs and the Hohenzollerns were of the northeastern type, so German honor was satisfied. But in general it was felt, as Ernst Renan wrote: "The inequalities of races is established . . ."[44]

Darwinism, when it arrived, provided a (spurious) basis for many of these theories, aided and abetted by the ideas of Herbert Spencer, who tried to amalgamate the theory of the conservation of energy with evolution and came up with the deadly phrases "the struggle for existence," and "the survival of the fittest," which at a stroke enabled people to combine militarism, nationalism, and cultural history into whatever miasma of beliefs they chose. Dr. Ludwig Woltmann (1871–1907) and Dr. Alfred Ploetz (1860–1940) were just two of the (predominantly but not exclusively German) scholars who subscribed to such views, going so far as to found secret Nordic societies and serve on the jury for the famous Krupp Prize in 1900, when 50,000 marks was offered for the best essay on state legislation in the field of heredity.

The culmination of the "Aryan epoch," as Leon Poliakov calls it, was perhaps 1889, when Max Nordau observed that "Darwin was well on the way to becoming the supreme authority for militarists in all European countries." Evolution, Nordau thought, acted as a cover for "natural barbarism," allowing people to cloak the "instincts of their inmost hearts" with the "last word of science." This is supported indirectly by the observation of Amos Elon that at the end of the nineteenth century the combination of material strength with cultural wealth in Germany "was unparalleled on the continent" and that although the pre-eminence of Jews in some cultural fields was "overwhelming," there was little if any "religiously Jewish" content in the work of, say, Albert Einstein, Sigmund

Freud, Gustav Mahler, Stefan Zweig, Franz Werfel, Edmund Husserl, Hugo von Hofmannsthal, or Paul Ehrlich.[45]

Darwin himself was not at all a "*terrible simplificateur,*" but his work was used in that way and, deliberately or otherwise, people missed the real lesson of Darwinism—that all of mankind is derived from the same source. The Aryan myth, incoherently, posited that the Germans had a heredity different from that of their fellow Europeans, indeed from all others.

The Hatred of Modernity

In the middle of all this, in 1890, Julius Langbehn published *Rembrandt als Erzieher* (*Rembrandt as Teacher*), in which he denounced intellectualism and science. Born in a small town in Schleswig, into a family of pastors, one of whom had studied under Luther, and the son of a philologist, Langbehn argued that it is art, not science or religion, that is the higher good, the true source of knowledge and virtue. In science, he maintained, the old German virtues of simplicity, subjectivity, and individuality were lost. Rembrandt, the "perfect German and incomparable artist," was pictured as the antithesis of modern culture and the model for Germany's "third Reformation," yet another turning inward (the first two had been sparked, he said, by Luther and Lessing). One theme dominated the entire book: German culture was being destroyed by science and intellectualism and could be regenerated only through the resurgence of art, reflecting the inner qualities of a great people, and the rise to power of heroic, artistic individuals in a new society. After 1871 Germany had lost her artistic style and her great individuals, and for Langbehn, Berlin above all symbolized the evils in German culture. The poison of commerce and materialism ("Manchesterism" or, sometimes, "*Amerikanisierung*") was corroding the ancient inner spirit of the Prussian garrison town. Art should ennoble, Langbehn said, so that naturalism, realism, anything that exposed the kind of iniquities that a Zola or a Thomas Mann drew attention to, was anathema.

He intended his book as a new bible for a new, reformed Germany.[46] The dominant theme was a hatred of science and, as we shall see, hatred itself is a unifying theme that runs through these few pages. His book also coincided with a wider-based critique of industrial society (Friedrich Nietzsche, William James, Henri Bergson, Wilhelm Dilthey, and Sigmund Freud) and this helps account for its appeal. His hatred was directed at, among others, Theodor Mommsen, who, like other professors, had "sacrificed his soul to his intellect . . . The professor is the German national disease." The underlying threat to Germany, Langbehn insisted, "was over-education."[47]

That Langbehn ignored politics was part of his attraction. In the Germany he appealed to, that ignorance was looked upon as high-mindedness. By means of a "flight into art," nationalism, faith, intuition, and philosophy could be mixed.[48] Art was particularly suited to Protestant nations, he said, because they were the most "inward."

The book was a sensation, presented by booksellers as "the most important work of the century" and warmly reviewed, despite its evident anti-Semitism, by Georg Simmel, Ludwig von Pastor, and Wilhelm von Bode, the latter Germany's (and the world's) pre-eminent Rembrandt scholar. Fritz Stern argues that 1890 marked a turning point in the cultural life of Germany. "The decade that followed witnessed a quickening of thought and hope, a new concern for the inner freedom of man, an anxious brooding on how this freedom could be realised . . . It was in the 1890s that cultural pessimism and anti-modernity became the twin resentments of the disaffected, conservative elements of imperial Germany."[49] Nietzsche saw where resentment would lead as a political force.

These modes of thought were by no means confined to Germany, even if they were most extreme there for a time.[50] Nor was the associated idea of eugenics, which found enthusiastic adherents in Britain and France, and almost as often among left-wing advocates as among right-wing. The main difference was between Britain and the United States, on the one hand, where the "soft" side of eugenics was preferred—government encouragement of selective breeding—rather than the "hard" variety favored on the Continent, which included forced abortion, sterilization, and euthanasia.[51]

The most vicious of the eugenicists was Alfred Ploetz, a doctor who grew up in Breslau and whose book *Die Tüchtigkeit unsrer Rasse und der Schutz der Schwachen* (*The Efficiency of Our Race and the Protection of the Weak*) Hitler would read as a young man in Vienna before World War I.

The following extract gives the flavor of Ploetz's argument: "Advocates of racial hygiene will have little objection to war since they see in it one of the means whereby the nations carry on their struggle for existence . . . In the course of the campaign it might be deemed advisable deliberately to muster inferior variants at points where the main need is for cannon fodder and where the individual's efficiency is of secondary importance." Ploetz was not an anti-Semite, however; he thought they were "racial Aryans."

After 1880, and especially after the Dreyfus trial in France in 1893, the Jews were increasingly identified as Europe's leading "degenerates." This was also a catalyst for the anti-Semitic political parties that were formed, and again not just in Germany (though Karl Lueger, mayor of Vienna, was especially vitriolic). There were over a hundred branches of the Society for Racial Hygiene in Germany in 1907, by which time a number of anthropologists and other scientists had formed the Ring der Norda, designed to cultivate Teutonic physical specimens. Max Sebaldt von Werth's many-volumed *Genesis* (1898–1903) and Jörg Lanz von Liebenfels *Theozoology* (1905) both claimed that the real "chosen people" of the Bible were Aryan-Teutons.[52]

The racial ideas of Joseph-Arthur de Gobineau passed into Germany via Richard Wagner. Wagner first became interested in Gobineau's work in 1876 while preparing for the first performances at Bayreuth (they met not long afterward). The composer was very taken with the French (self-appointed) aristocrat's theories, once describing Gobineau to his wife as "my only true contemporary." Wagner pushed Gobineau's ideas on his circle and they were taken up by two people in particular, Ludwig Schemann and Houston Stewart Chamberlain.

Schemann was struck by the similarity of Gobineau's ideas to those of a key figure in the ultranationalist German *völkish* movement, Paul Anton Bötticher, who wrote some fifty tracts under the name of Paul de Lagarde. Not unlike Langbehn, Lagarde believed that the German nation had a will of its own, "the expression of its *Seele*," or collective soul. That German soul, he further argued, was being destroyed by materialism, industrialization, and "middle-class greed." The real Germany, the Germany of rural customs and traditions of the common Volk, was being overwhelmed, progress was "a Trojan horse" hiding a soulless future of mechanization, liberal individualism, socialism, and, above all philistinism.[53]

Lagarde was viciously antimodern, seeing all about him, amid the fantastic and brilliant innovations, nothing but decay. A biblical historian (one of the areas where German scholarship led the world), he hated modernity as much as he loved the past.[54] He was also one of those calling for a new religion, an idea that, much later, appealed to Alfred Rosenberg, Hermann Göring, and Heinrich Himmler. Lagarde attacked Protestantism for its lack of ritual and mystery and for the fact that it was little more than secularism. In advocating a new religion, he said he wanted to see "a fusion of the old doctrines of the Gospel with the National Characteristics of the Germans."[55] To begin with he adopted the idea of "inner emigration," meaning people should find salvation within themselves, but then he advocated Germany's taking over all non-German countries of the Austrian Empire. This was because the Germans were superior and all others, especially Jews, inferior.

For Lagarde, the central aspect of Germanness was its Aryan heritage, an identity that went back to the "forest and bogs" of northern Europe and Scandinavia and provided an honorable and distinctive alternative to the classical Greek Mediterranean culture that so influenced Italy, Spain, and France. This identity and tradition, Schemann said, following Lagarde, survived in the völkish culture of Germany, a Pan-German feeling that was the sole bulwark against Europe's cultural, social, and racial "disintegration." Above all, Wagner's operas were seen as the authentic re-creation of the original Aryan myths. "Bayreuth became an annual festival where Aryan-Germans could participate in 'their primeval mysteries,' rediscover the origins of their *Kultur*, and be restored to spiritual health."[56] Lagarde's disillusion with Germany was aggravated by his visit to Britain, "where he thought he saw a unified people, a popular monarchy, and a responsible gentry—all things that Germany lacked."[57]

Lagarde's reputation, which has now all but vanished, was then very high. Thomas Mann called him a "*praeceptor Germaniae*" (the same term as had been attached to Treitschke), especially for those who were dissatisfied with their "humdrum existence in bourgeois society." In World War II, soldiers of the Third Reich were issued an anthology of Lagarde's work.[58]

Schemann's own writings were limited in their appeal, but not so those of Houston Stewart Chamberlain. Chamberlain, as his name implies, was born in England but grew up a Germanophile, married Wagner's daughter, becoming in the process more German than the Germans, an influential member of the Bayreuth circle. In 1899, long after Wagner's death, he published *Die Grundlagen des Neunzehnten Jahrhunderts* (*The Foundations of the Nineteenth Century*), a rambling survey of European history, the chief arguments of which were, first, that the European achievement was owed entirely to the Aryan race, a race that kept its identity against all the odds, and survived now as the Teutons, exhibiting great "physical health and strength, great intelligence, luxuriant imagination, untiring impulse to create."[59] Second, when Teutonic vitality was under threat, the main villain was invariably the Jews, Chamberlain claiming that the Jewish "race" was the degenerate result, he said, of "cross-breeding" between Bedouins, Hittites, Syrians, and Amorites from the Fertile Crescent in the Middle East. They were a "tainted race" who deliberately tried to sabotage and pollute the world that their "Teutonic superiors" had built.[60]

His book became part of the standard history curriculum in German schools, and we know that Hitler was introduced to his doctrines by Alfred Rosenberg and Dietrich Eckhart.[61] The two men met in 1927 when Chamberlain was quite old. Goebbels was there and later described the meeting, with Hitler and Chamberlain clutching each other's hands, the former telling the latter he was his "spiritual father." Chamberlain wrote to Hitler a few days later: "With one blow you have transformed the state of my soul. That Germany, in her hour of need, brings forth a Hitler—that is proof of her vitality. Now I will be able to sleep peacefully and I shall have no need to wake up again. God protect you!"

Chamberlain died before Hitler came to power, but Schemann lived on and, on his eighty-fifth birthday, received Germany's highest literary award, the Goethe medal, from the Third Reich.[62]

As Fritz Stern has said, the "rhapsodies of irrationality . . . illuminate the underside of German culture." Whatever their starting point, they all envisaged a *Germania irredenta*, a new destiny in which Germany, purged and disciplined, would emerge as the greatest power in the world. Above all there was the ideological attack on modernity, the dominance of *resentment* as the major psychological force.[63]

These issues touched the common man more in Germany than elsewhere, in particular the educated classes. It was an idealism that represented an attitude toward life, a set of sentiments and values, in which science and scholarship—however scientistic or tendentious—played an important role in adding a spurious force to the feeling. This idealism, with its emphasis on *Innerlichkeit*, on inwardness, did not encourage political involvement, a state of affairs reinforced by Bismarck's "semi-authoritarian" political regime. Norbert Elias has drawn attention to the division in late nineteenth-century Germany between, on the one hand, the *satisfaktionfähige Gesellschaft*, a society oriented around a code of honor, in which dueling and the demanding and giving of "satisfaction" occupied pride of place, which became brutalized, and, on the other, the educated middle class.

Nationalism, *Kultur*, and idealism fused into a cultural nationalism in which the German spirit was exalted above those of other nations, with an enthusiasm—even aggression—that had no parallel elsewhere. Elias has again shown how nationalism embodies a moral code, inegalitarianism, at variance with those of the "rising tiers" of society. This was so because the educated classes in particular—the academics, the bureaucrats, the professional people—were being rapidly overtaken by the new industrialists and were slipping down the power-pecking order. Having once been just below the old aristocracy, they were now just above the proletariat. "With a suddenness that has had no parallel, the industrial revolution changed the face and character of German society."[64] One revealing statistic is that, by 1910, Germany had almost as many large cities as the rest of Europe put together. This change to modernity was greater and quicker in Germany than anywhere else.[65]

23.

Money, the Masses, the Metropolis: The "First Coherent School of Sociology"

There is something rather fetid about many of the individuals in the previous chapter, with their tightly woven—even overwoven—works. Collectively, they seem to have lacked self-awareness of their own motives, to have shared a willful inability to face their own tendentiousness. At the same time, however, there was a raft of rather more serious thinkers in Germany concerned with a similar range of issues—in particular, the problems posed by industrialization, vast metropolises, and rapid technological innovation.

Some of these were writers. Following the failed bourgeois revolution of 1848, we saw in Chapter 14, p. 289, that some authors moved away from the developments of modernity and sought refuge in idyllic surroundings cut off from the mainstream, resulting in a special German path in literature, turning away from the harsh (mainly urban, industrial) realities of life to the inner concerns of *Bildung* (Adalbert Stifter, Gottfried Keller, etc.). In the latter half of the century, however, German writers did start to come to grips with at least *some* of the more contemporary issues. Among the first books of this kind were Gustav Freytag's *Soll und Haben* (*Debit and Credit*; 1856) and Friedrich von Spielhagen's *Hammer und Amboss*

(*Hammer and Anvil*; 1869), the very titles of which are suggestive and can be read as a sort of *Entwicklungsroman*, a parallel to a *Bildungsroman*, stories that lead to the eventual good fortune of the middle-class protagonist: instead of finding inner development outside society, Freytag's and Spielhagen's "heroes" find their niche *within* the business community.

After them, the two greatest Realist writers of the late nineteenth century were Wilhelm Raabe (1831–1910) and Theodor Fontane (1819–98), both of whom focused on the society—and morals—produced by capitalism. Raabe, very sympathetic to the aims of the French Revolution, modeled much of his work on that of Dickens, attempting to catch some of the British author's humor, but he never forgot that capitalism was his target and, in such books as *Abu Telfan* (1867), *Der Schüdderump* (*The Rumbledump*; 1870), and *Pfisters Mühle* (*Pfister's Mill*; 1884), about industrial pollution, he explored both the distortions produced by capitalism and the means of keeping alive our humanity in a capitalist world.

Theodor Fontane was a better writer, a man who turned to the novel only late in life, after a career that took in being an apothecary, a journalist (including a stint as a war correspondent), and a drama critic—and fighting in the 1848 revolution. In his novels he was particularly aware of the class basis of existence in Germany, and with him the German novel was reclaimed from its special path and joined the mainstream. Unlike many writers, he was generally in favor of capitalism, though he was not without sympathy for the Junker class. His main aim was to show the recklessness with which an "utterly conventional society holds sway over the lives of individuals." This had been true in the age of absolute state and it was, he insisted, *still* true. Society had changed less—hardly at all—in the moral grip in which its citizens were held. His most famous book in this regard was *Effi Briest* (1894), sometimes grouped with *Anna Karenina* (1878), and *Madame Bovary* (1857) in a trilogy of nineteenth-century marriages seen from the female point of view. Effi, daughter of a nobleman, is married off to a baron twice her age, who had courted her mother. Ignored by her husband, not accepted by the local aristocracy, she consummates a relationship with a married womanizer. Much later, the baron finds out. He kills the womanizer in a duel, divorces Effi, is awarded custody of their daughter, and Effi goes back home but only after her parents, who had rejected her, realize she is dying from tuberculosis. Everyone ends unhappily. Fontane's real target was the moral fog in Bismarck's Germany, its self-righteous emptiness, more destructive than creative.

It has to be said, however, that although Fontane and Raabe were Realist writers to a degree unknown earlier—in the age of Stifter and Keller, say—more enduring originality about modern society was shown at this time by a different cohort of men in a wholly different discipline: they delivered a set of analyses and warnings that, taken together, amount to what has been called the first coherent school of sociology.[1]

In Germany, and perhaps characteristically, sociology had its origins in philosophy and with the work of a man who, though he may not be as well known today as some of his contemporaries, deserves his place as the starting point because, even now, with all that has gone on in between, he was a very clear and sane observer, whose decisive clarity makes him a breath of fresh air.

"Thought Cannot Go Behind Life": The Birth of Human Studies

Wilhelm Dilthey (1833–1911) was fortunately placed. Born in 1833, he too was the son of a clergyman but one better off than most, being court preacher to the Duke of Nassau. His mother was the daughter of a conductor, and Dilthey inherited a strong interest in music, together with a dislike of the lower classes, in whom—rather like Marx—he had little confidence.[2]

Dilthey first pursued theology, at Heidelberg, but transferred to Berlin because of its greater sophistication, especially in music. He formed his own reading circle to explore Shakespeare, Plato, Aristotle, and St. Augustine and completed his doctorate with a thesis on Schleiermacher. After teaching spells at Basel, where he was a friend of Jacob Burckhardt, and at Kiel, where he began publication of his biography of Schleiermacher, the first installment in what would become "a history of the German spirit," he was offered the chair of philosophy at Berlin, which Hegel had once held. He remained there for the rest of his life.[3]

In the manner of the ancient Greeks, Dilthey looked about him and observed what he called "the enigmatic face of life." People, he thought, inevitably found themselves confronted by "circumstances they do not understand, seeing irrational forces and blind chance at work." Out of all this, he said, they need to manufacture a coherent picture of their circumstances, and they need ideals to strive for and principles by which to order their conduct. Somewhere inside them, Dilthey said, everyone has a "metaphysical impulse"—this is what gives rise to art, religion, and politics. "When the response is based on sustained and critical thought it becomes philosophy." Philosophy differs only in that it is—or should be—more abstract and closely reasoned than other activities. The history of philosophy, he thought, was a succession of ideas, none of which has proved wholly triumphant, and we should use this fact, this failure, as our guide. Worldviews or systems that have stood the test of time must contain some elements of truth, but each is limited "because the human mind is limited and conditioned by circumstances."[4] When worldviews or philosophical systems make exclusive claims, their credibility is dented.

This led Dilthey to deny metaphysics. For him, the rich and varied multifariousness of reality can never be adequately captured by any one conceptual approach. He was convinced that there is nothing to discover behind those experiences which, taken together, we call life. "Thought cannot go behind life," he insisted. Political decision making, for example, could never be explained by the elementary processes of formal psychology, such as reflex action. Instead, we can only analyze, in this example, the products of political thought. "Unmutilated experience" was for him the basic ingredient of human studies.

He insisted too that man's basic position in an evolved universe must be linked to his morality, but he also thought it important to say that man's nature is not fixed, that it too developed in the course of history and that, therefore, moralities also change and cannot be fixed either. This was another ingredient in human studies, a reorientation of disciplines that must be counted as Dilthey's most distinctive achievement.[5]

As part of this, he asked himself what difference man's possession of a mind had made to the world and arrived at a set of phenomena that, he said, have no equivalent elsewhere in nature. One is purposiveness. Man struggled with this notion, to free himself of the religious mantra that the whole of nature is purposive. Since Galileo, this project had largely succeeded, culminating in Baruch Spinoza. But Dilthey wasn't sure that man was any happier, or understood the universe any better, because of this. A second phenomenon was value. Man shares with other creatures an ability to respond to the environment around him, but only man makes judgments in the abstract as to what is good and what is not. Third, there are norms and rules and principles in human life, everything from high moral principles to traffic regulations that, importantly, differ from the laws of science. Human laws are conventions and therefore changeable. Finally, human life is aware of itself as historical. Nature itself is mindless: planets cool, glaciers melt, sea levels rise and fall, producing change that affects humans. Only because of consciousness and memory does the *cumulative* effect of successive events become important.[6]

All this led Dilthey to conclude, simply enough, that the world of the mind cannot be directly observed. Purposes, values, and norms cannot be seen, nor can history, since it exists only in the past. It follows that our knowledge of the mind can come from just two sources—inner experience, which tells us about purposes, values, and norms, and reflects on the past by means of memory; and communication, "without which the individual's knowledge would be minimal." It is through these phenomena, these ingredients, that meaning emerges, a meaning that cannot be observed.

None of this was radical, but its clarity—to repeat—was cleansing and made Dilthey's coining of the phrase "human studies" seem commonsensical. He split human studies into two, into historical and systematic disciplines. The historical disciplines included political history, economic history, intellectual history, and the history of science. By "systematic disciplines," he meant those activities—economics, sociology, psychology—that seek to explain phenomena by means of general laws.[7]

Dilthey also introduced, or reintroduced, the concept of understanding (*Verstehen*). He followed Kant in agreeing that we have evolved so as to understand the world. For him, there is a fundamental difference between sheer facts and true understanding.[8] Understanding is for him a specific capacity of the human intellect and should be regarded as such. This is shown in the failings, or fallings short, of specific systems, such as behaviorism. Such systems may tell us *something* about human behavior, human experience, but it is inevitably limited and can never approach total understanding. However many opinion polls were carried out, there could never be enough for total understanding. "Man does not discover what he is through speculation about himself or through psychological experiments but through history." The historical dimension—and the education on which it is based—underlines the need for interpretation and this, like understanding, to which it was linked, was another insight of Dilthey's.[9] Because the activity of the mind was the chief phenomenon of the human world, Dilthey said that understanding in that world was always more likely to resemble literary or legal interpretation than physics or chemistry. In saying this, Dilthey challenged head-on the claims of modern science to be the paradigm of all knowledge.

Dilthey therefore derived five important principles by which human studies were to be guided. First, individual cases are intrinsically interesting and it is beside the point to generalize from such cases "because their differences are just as important as their similarities and are due to their historical character." General laws, applicable within science, have no place in human studies. Second, the relationship of parts to wholes is different so far as people are concerned. The sense in which people are part of communities is nothing like pistons in a machine. Third, investigation must start at the level of complexity we find in nature. Nothing is gained by attempting to understand the poet's imagination through simple processes that may be observed in animals or small children. Fourth, we are free to switch disciplines "whenever that helps." Fifth, man is both subject *and* object. Circumstances have made him—he is an object. But he also knows himself and controls his actions.[10]

We are left with an important conclusion. Because assumptions and interpretations are involved, knowledge of the human world "can never be a kind of photograph of reality." It is always a construction, constantly under revision.[11]

Mental Life in the Metropolis

Despite Dilthey's very great influence, formally the first German sociologist of note was Wilhelm Heinrich Riehl (1823–97), whose study of the German peasant was especially interesting to foreigners, where such a way of life had died out owing to the industrial revolution fifty years earlier. The peasant's idea of constitutional government, he found, was very tenuous. He also identified important differences between Lower, Middle, and

Upper Germany (Middle Germany being more "individualistic" than the other two areas), and he conceived the notion of the "social philistine," the individual "indifferent to all social interests, all public life, as distinguished from selfish and private interests." This was not the "inwardness" of the Bildung classes but its petit bourgeois equivalent and much more problematic.

But Riehl is now overshadowed by a raft of bigger names, still read with profit today, of which the first was Georg Simmel (1858–1918). Born in the very center of Berlin, Simmel was in every sense a modern urban man. After he had read Simmel's first book, Ferdinand Tönnies wrote to a friend, "The book is shrewd but it has the flavour of the metropolis." As someone else said of him, "Simmel suffers from modernism."[12]

Simmel was the youngest of seven children. His father, a successful Jewish businessman who had converted to Christianity, died when Georg was still a boy, and a friend of the family, a music publisher, took over as his guardian. Simmel studied history and philosophy at the University of Berlin, where he was taught by a wide raft of luminaries—Mommsen, Treitschke, Sybel, Droysen, the physicist Hermann von Helmholtz, and the anthropologist Moritz Lazarus (who also taught Dilthey and William Wundt.)[13]

In 1885, he became a *Privatdozent* (an unpaid lecturer dependent on student fees) at the University of Berlin, giving courses on ethics, sociology, Kant, Schopenhauer, "The Philosophical Consequences of Darwinism," and Nietzsche. He was a superb performer, and his lectures became one of the socio-intellectual attractions of Berlin, attended not only by students but by the cultural elite.

Despite this success, despite the fact that the reception of his books brought him eminence throughout Europe and as far afield as Russia and the United States, where he was an advisory editor of the *American Journal of Sociology*, despite the fact that he had the friendship and support of leading academic figures like Max Weber, Edmund Husserl, and Adolf von Harnack, despite the fact that he was a close friend of both Rainer Maria Rilke and Stefan George, and that with Tönnies and Weber he was a cofounder of the German Sociological Society, Simmel was constantly rebuffed by university authorities when chairs became available. When he was finally made professor, at the University of Strasbourg, it was already

1914 and within a short time the university was closed and converted into a military hospital. He didn't let anti-Semitism get him down. His predicament was eased by financial security when his guardian left him a considerable fortune. Instead of growing bitter, he perfected his performance as a lecturer, "punctuating the air with abrupt gestures and stabs, dramatically halting, and then releasing a torrent of dazzling ideas." As one admiring American observer described it, "Simmel 'simmelifies.'"[14]

Capitalism as Prostitution

For the Paris exhibition of 1900, the American sociologist Lester Ward presented a report on the state of sociology on both sides of the Atlantic. After referring to the absence of any chairs of sociology in Germany at the time (a situation not rectified until after World War I), the report singled out Simmel who, it said, had been offering courses in sociology "for the last six years." Simmel was thus the first professional sociologist in Germany, his rise coinciding with what Thomas Nipperdey describes as the "collapse of philosophy."[15]

Simmel had first come up with what he called "a new concept of sociology" in the 1880s. Darwinism and Social Darwinism were in the ascendant, but Simmel in his lectures argued that no single element can be identified as decisive in the ceaseless interaction of society. For Simmel it was the "*interaction* of the parts" that counts. He thought there was a key difference—insufficiently appreciated—between "what takes place merely *within* a society as a framework and that which really takes place *through* society."[16] The latter was the concern of the sociologist.

He was particularly interested in the wider ways in which people in the new social circumstances (post–industrial revolution) organized themselves. Accordingly, in *Über sociale Differenzierung* (*On Social Differentiation*) his early contribution was to show that, as the social group to which the individual belongs gets larger (as in the great metropolises), the individual achieves greater moral freedom: "the purely quantitative enlargement of the group is merely the most obvious instance of the moral unburdening of the individual."[17] He also made the point, though, that much of this freedom is illusory, "since beneath the choice there is a relentless pressure, and the capacity to choose itself is a sign of rootlessness."[18]

He also enlarged the concept of collective responsibility—in a closely woven urban society, where people live cheek by jowl, we must all bear a share of the guilt for various forms of pathology, and this is not an easy thing to do. We are both more free and more responsible.[19] At the same time, with the stronger development of individuality that occurs in urban society, there is bound to be a weaker sense of group belonging. Simmel observed an "increase in nervous life" in the cities, brought about by the fact that a city has more differentiated social circles. This promotes superficiality and imitation—"one of the lower intellectual functions." The most obvious form of imitation is the phenomenon of fashion—not just in clothes but, for example, in musical consumption, a way of belonging *and* differentiating.[20]

The climax of *On Social Differentiation* was Simmel's differentiation between objective and subjective culture. By the former he meant books that had been published, paintings painted, operas performed—achievements "out there" to which individuals can relate and define themselves, responses they can share, benchmarks and canons that can be agreed upon (or argued about) and which mark where someone belongs and differs from his or her fellows. By subjective culture, Simmel meant mainly business culture, where he saw bankers, industrialists, entrepreneurs, and shopkeepers sharing much less, having far fewer benchmarks with which to compare themselves, leading more private (but not necessarily more intimate) lives. He thought such life impoverished but saw at the same time that small town life was "unbearable" for people who had lived in metropolises.

In May 1889 Simmel gave a paper on "The Psychology of Money."[21] Somewhat expanded, this was to become *Die Philosophie des Geldes* (*The Philosophy of Money*) which, when it appeared in 1900, was welcomed with great acclaim. Karl Joël declared it to be "a philosophy of the times," Max Weber thought its analysis of the spirit of capitalism was nothing less than "brilliant," and Rudolf Goldscheid suggested it formed "a very interesting correlate to Marx's *Capital*."

Simmel's argument owed something to Dilthey because he argued that "Money, like other phenomena, can never be grasped from a single science." Nevertheless, for Simmel money was important because it symbolized "the fundamental relatedness of social reality." The meaning of money, he said, is grounded not in production, as Marx would have it, but in *exchange*. Exchange, he insisted, is the source of value, embodying the process of sacrifice and gain. "Every interaction [in society] has to be regarded as an exchange: every conversation, every affection (even if it is rejected), every game, every glance at another person." Such interactions always involve the exchange of personal energy, and this, for Simmel, is what metropolitan life is about and why it is new.[22]

A money economy creates new dependencies, "especially upon third persons, not as persons but as representatives of functions." One consequence is that the personalities of those we now become dependent upon are irrelevant and far less tense emotionally. Money enables us to participate in a much wider range of relationships and associations but without any real emotional involvement, still less commitment. Money is both "disintegrating and isolating" but also "unifying," in that it brings together elements of a society that otherwise "would have no connection whatsoever." Simmel even compared the money economy to prostitution: "The indifference as to its use, the lack of attachment to any individual because it is unrelated to any of them, the objectivity inherent in money . . . which excludes any emotional relationship . . . produces an ominous analogy between money and prostitution."[23]

In the final chapter of *The Philosophy of Money*—which ranks as one of the first sociological analyses of modernity—Simmel tried to work out an updated theory of alienation. Money, he said, helps modern culture toward a "calculating exactness," and the ease of understanding of these monetary relationships eclipses all others, so that the individual's chances to be creative and develop in particular directions becomes restricted. Furthermore, it is in the nature of the production process that more impersonal objects are preferred "because they are suitable for more people and can be produced cheaply to satisfy the widest possible demand." In this way individual experience becomes flattened, intimacy is lost—*this* is the modern alienation and, for Simmel, the "tragedy of culture," compounded further by the loss of philosophical coherence brought about by scientific specialization.

In 1903 Simmel published his no less famous work, *Die Großstadt und das Geistesleben* (*The Metropolis and Mental Life*), later expanded into *Soziologie* (*Sociology*; 1908). Here he argued that great cities are not only characterized by far more social differentiation than before but also that there is a totally new phenomenon, the "indefinite collectivity" we refer to as crowds, which are characterized by "total indifference to one's fellow human beings."[24] This experience—unknown in traditional, rural communities or market towns—provokes "extreme subjectivism," he said. The individual's struggle for self-assertion, against "the pervasive indifference" of much metropolitan social interaction, leads to excessive behavior, "the most tendentious eccentricities, the specifically metropolitan excesses of aloofness, caprice and fastidiousness, whose significance no longer lies in the content of such behaviour, but rather in its form of being different, of making oneself stand out and thus attracting attention." Metropolitan life atrophies genuine individuality and replaces it with an artificial, contrived, calculated individuation. This too is a form of alienation.[25]

Both Georg Lukács and Walter Benjamin found *The Philosophy of Money* second only to Marx in importance. It also helped imbue writers like Houston Stewart Chamberlain and Oswald Spengler with "a passionate hatred" of the modern city, one of those themes taken up in 1920s Weimar Germany, paving the way for National Socialism.

Simmel was also an important influence on the Chicago school.

Two Types of Individualism

Ferdinand Tönnies conceived sociology as part of an already-existing "cognitive continuum" extending from geometry at one end to narrative history at the other. Much influenced by Hobbes and Hume, he thought sociology was in principle no different from linguistics, mathematical physics, or the theory of law—it was a new form of logic or epistemology, brought about by modern life.[26]

With the benefit of hindsight, we can say that Tönnies's background was a heavy influence on some of his concepts. Born in 1855, in the marshlands of east Schleswig, where his grandparents had included Lutheran pastors, Tönnies was ten when the family left their country home for the nearby town of Husum, where his father—formerly a farmer—became involved in banking. All his life, right through to Nazi times, Tönnies seems to have found adjustment to metropolitan mass culture difficult.

In his university education he followed the cosmopolitan model, studying at Strasbourg, Jena, Berlin, Leipzig, Heidelberg, Kiel, and Tübingen.[27] Between 1878 and the outbreak of World War I Tönnies visited England several times, and in 1905 went on to the United States. His attitudes to these countries was mixed. He found the contradiction of capitalism and poverty "hypocritical" but at the same time admired their constitutional liberties, a contradiction that was to remain at the heart of his theorizing.

After his first doctorate, in philosophy, he formed a friendship with Friedrich Paulsen, the philosopher and educator, and under his influence Tönnies began to investigate the pre-Kantians. In the course of this he encountered the works of Thomas Hobbes, which led to those visits to England, where he gained access to original Hobbes material in the British Museum, St. John's College, Oxford, and the country seat of the Duke of Devonshire, at Hardwick Hall in Derbyshire. He discovered Hobbes documents that had been overlooked by other scholars, enough material to publish four papers, helping to make his name both north and south of the Channel.

Hobbes led to an appraisal of Adam Smith, then to other economists, and this was the intellectual background that brought about Tönnies's first "sketch" of what would become *Gemeinschaft und Gesellschaft* (*Community and Civil Society*), in which he tried—seriously—to counterpose what he saw as two fundamentally contrasting models of human social organization. Tönnies formed the view that the traditional, small-scale village in which he had himself grown up was dying, and he felt the loss profoundly. At the same time, Bismarck's high-handed repression of all opposition produced in Tönnies "a growing disenchantment with the much vaunted achievements of the new imperial German Reich."[28]

There was, however, little sign as yet that his sketch would ever be expanded and he returned to England to work on Hobbes. It was only a series of mishaps, during the course of which his British publishers canceled Tönnies's projected volume on Hobbes, that caused him to return his attention to *Gemeinschaft und Gesellschaft*. It was 1887 and it was a fateful decision. There were to be eight editions of the book during his lifetime, the last in 1935, shortly before his death, and though it was ignored to begin with, it caught on in the run-up to World War I.[29]

The book was divided into three parts; the first compared small-scale "communities" and large-scale, market-based "civil societies." The second examined the way these two types affect how people think and behave. The third looked at how all this affected politics, government, and law. Tönnies's central point was that at one extreme, the will, or consciousness, was "natural," spontaneous and unreflecting (what he called *Wesenswille*); at the other extreme, the will was artificial, deliberative, limited to "rational calculation" (*Willkür* to begin with, later changed to *Kürwille*). For him, these were two kinds of freedom. One entailed "unselfconscious fulfilment of a function or duty within a predetermined social context," while the other, the "rational will," was more self-conscious but brought with it "unlimited choice" and therefore absolute "self-sovereignty." Tönnies thought both forms of will and freedom were latent in all human beings but tended to be expressed differently according to different social circumstances. Rational will was more characteristic of men than women, of adults than children, of city-dwellers than villagers, of traders than creative artists. He built on these differences to argue that there were in modern society two totally different types of human psyche. The "natural will" produced a "self" that was in harmony with its habitat and closely linked with others. In contrast, the "rational will" produced "subjects" (not selves) that invented their own identities and were in fact estranged from their natural selves, perceiving other people, moreover, as mere things or "objects."[30]

This dichotomy of the human psyche was closely linked with social and economic organization. The "organic" community (*Gemeinschaft*) was characterized by ties of kinship, custom, history, and the communal ownership of primary goods. The converse was "society" (*Gesellschaft*), in which "free-standing individuals interacted with each other through self-interest, commercial contracts, a 'spatial' rather than a 'historical' sense of mutual awareness, and the external constraints of formally enacted law." The dichotomy ran through everything. For example, in community material production was primarily for "use" not "gain." In society, in contrast, "all personal ties were subordinate to the claims of abstract individual freedom." In community, work and life were integrated in a "vocation" or "calling," while in society business produced profit that was then used to provide "happiness." "The entire civilisation has been turned upside down by a modern way of life dominated by civil and market Society, and in this transformation civilisation itself is coming to an end."[31]

Tönnies's arguments overlapped with at least some of the things the Social Darwinists and vulgar nationalists were saying. But his dichotomy didn't catch on until the international temperature was on the rise before World War I. After the war, Tönnies was much better appreciated, and not just in Germany. Even in Europe and North America, *Gemeinschaft und Gesellschaft* became a canonical text of classical sociology.

Tönnies always said that he emphatically *wasn't* comparing individualism and collectivism. Rather, there were two distinct forms of individualism, "the unself-conscious kind, which was created by and naturally flowed from *Gemeinschaft*, and the self-conscious kind which fostered and was manufactured by the culture of *Gesellschaft*." Most readers, however, interpreted his book as an attack on modernity.

Simmel and Tönnies are important in the way that they identify aspects of "cultural lag" at the end of the nineteenth century that would have long-lasting consequences for the "belated nation" that was Germany in the twentieth century (see Chapter 41, p. 757). As did Simmel, Tönnies identifies something that Riehl had categorized as "un-German" about modern society, and that was, through such writers as Julius Langbehn, Spengler, and Chamberlain, to drive conservative thinking into intellectual proximity with the Nazis, in the shape of the so-called Conservative Revolution (see Chapter 33, p. 611). As Keith Bullivant, historian of the Conservative Revolution, has pointed out, the basic ideas behind *Gemeinschaft und Gesellschaft* continued to play a central part in German thinking, even after the Second World War, as Chapter 9 of Ralf Dahrendorf's 1968 *Gesellschaft und Demokratie in Deutschland* clearly shows: it is titled "Gemeinschaft und Gesellschaft" (again, see Chapter 41).

Heroes versus Traders

At the time, Werner Sombart was probably even better known than Tönnies, and not just in Germany.[32] He has come to be regarded as the "reactionary modernist" par excellence, best known for *Der moderne Kapitalismus* (*Modern Capitalism*, in which he introduced the concept and the word "capitalism"), for *Die Juden und das Wirtschaftsleben* (*The Jews and Modern Capitalism*), and *Warum gibt es in den Vereinigten Staaten keinen Sozialismus?* (*Why Is There No Socialism in the United States?*). Many of the first generation of sociologists—Tönnies and Weber in Germany and Everett C. Hughes and Robert Park in America—regarded him as brilliant and original, though they had certain reservations, summed up by Joseph Schumpeter, who said that *Modern Capitalism* "shocked professional historians by its often unsubstantial brilliance."[33]

Der moderne Kapitalismus, published in two volumes in 1902 and reissued as a single work in 1916, was the book that made Sombart's name.[34] Many, though again not all, of his colleagues regarded this book as an immediate classic. In this work, he rejected Marx's base-superstructure theorem where, it will be remembered, productive forces comprise the fundamental layer of society, with an ideological architecture built on top. Sombart thought that the essence of capitalism was its spirit, in some ways anticipating Max Weber's ideas. More than any of the other sociologists at the time, Sombart's work overlapped with the lesser figures addressed in the previous chapter. Such issues as race, Judaism, Germanness, technology, Marxism, and nationalism are returned to time and again.

In their analysis of Sombart's career, Reiner Grundmann and Nico Stehr argue that he performed a volte face on two crucial issues, and this perhaps explains the conviction with which he wrote: he had the passion of a convert. In the first place, he began as a Marxist and an ardent socialist.[35] After the turbulent 1890s, however, Sombart became an equally ardent *anti*-Marxist, and one, moreover, with anti-Semitic overtones. His attitude to his own country changed too. Early on, he had had serious misgivings about where Germany was headed, but all that changed around 1910 and he developed a strident nationalism. In his book *Die deutsche Volkswirtschaft im neunzehnten Jahrhundert* (*The German Economy in the Nineteenth Century*), published in 1903, he began to argue that the national character of the German people was responsible for the spirit of capitalism. This somewhat muddy argument became clearer as Sombart began to distinguish between two types of capitalists: entrepreneurs and traders. By 1909, and then in *Der Bourgeois*, in 1913, he was writing that the entrepreneur is "quick in comprehension, true in judgement, clear in thought, with a sure eye for the needful . . . Above all he must have a good memory." Contrast that with his view of the trader, whose "intellectual and emotional world is directed to the money value of conditions and dealings, who therefore calculates everything in terms of money." This type, in particular, he said, was epitomized "by the Jewish species." He argued that there was indeed a "spirit of capitalism" that was based on a particular form of rationalism, best seen in the United States and England. Later he was more specific still. "The capitalist spirit in Europe was cultivated by a number of races, each with different characteristics of its own and that, of [these] races, the Trading peoples (Etruscans, Frisians and Jews) may be divided off from those we have termed Heroic . . . The Scotch, the Jews, the Frisians and Etruscans are trading peoples, the Celts and the Goths heroic people. Since the Jewish spirit is capitalistic, and the English are said to possess the capitalist spirit, they also possess the Jewish spirit." By Nazi times, in *Deutscher Sozialismus* (1937), he was able to write: "What we have characterised as the spirit of this economic age . . . is in many respects a manifestation of the Jewish spirit . . . which dominates our entire era."[36] Since he had been saying this long before Hitler, he came to see himself as the Third Reich's chief ideologue. The Nazis did not.

The "German Line" in Sociology

Jeffrey Herf (1984) characterizes Sombart, along with Ernst Jünger, Oswald Spengler, Hans Freyer, Carl Schmitt, Gottfried Benn, and Martin Heidegger, as one of the main architects of what he terms "reactionary modernism." The essence of reactionary modernism in Germany was support for industrial development combined with a rejection of liberal democracy, a view that appealed to the Nazi regime. Men like Ernst Jünger and Gottfried Benn were in favor of technological progress and even approved of some modernist aesthetic developments, but they eschewed many of those institutions that would act as checks and balances in political and social affairs. "The rural conditions of a natural, living *Gemeinschaft* constitute much more desirable conditions of social existence than those of the artificial *Gesellschaft*."[37] Sombart, in his concept of heroes and traders, aligned himself with such thinking, not keeping much of a distance, as Tönnies tried hard to do.

Stefan Breuer has identified what he calls a "German" line in sociology. Its main elements are (or were) a "romantic criticism of capitalist rationality, the utility principle, and the lament over the breaking up of community bonds." He includes Tönnies, Sombart, and even Simmel in this group, though he thinks that Max Weber stands apart. "Weber did not welcome the First World War as a chance for redemption from fragmentation and alienation by means of German heroism and, what seems more important, he defended liberal democracy and its institutions." More than anyone else, then, Sombart *felt* the loss of Gemeinschaft. And, more than anyone else, as Herf says, he tried "to identify the guilty party for ruining it."[38]

Economic Bildung

Max Weber (1864–1920) was also troubled by the "degenerate" nature of modern society. He was much influenced by Dilthey, Simmel, and Tönnies but differed from them in believing that what he saw around him was not wholly bad.[39] No stranger to the "alienation" that modern life could induce, he thought that group identity was a central factor in making life bearable in modern cities and that its importance had been overlooked. A tall, stooping man, for several years around the turn of the century he had produced almost no serious academic work (he was on the faculty at the University of Freiburg), being afflicted by a severe depression that showed no signs of recovery until 1904. Once begun, however, few recoveries can have been so dramatic.

Weber was that rare combination, being both very practical and very theoretical. He wrote on the practicalities of parliamentary government, attacked the "Bismarck legend" but at the same time gave much thought to the methodology of the social sciences. He was equally interested in religion, bureaucracy, the whole question of authority—why some people obey others—in urban societies and in the role of scholarship and universities in the modern world. He was convinced that we don't have to *be* Caesar to understand him, that "understanding" involves interpretation, which is a different form of explanation from causative explanation in the hard sciences, and he developed and contrasted this with his notion of "adequate causation." He thought that there were types of religion—ascetic, mystic, and prophetic (or savior) religions—the latter coming more into conflict with the world and the former showing less the tension that he thought always exists between "religiosity and cognition." He was impressed by Confucianism, which "knew no radical evil or salvation," its aim being "dignified acceptance of the world and graceful adjustment to it."

By far his most well-known work on religion and sociology was *Die Protestantische Ethik und der Geist des Kapitalismus* (*The Protestant Ethic and the Spirit of Capitalism*), the very opening of which shows Weber's way of thinking: "A glance at the occupation statistics of any country of mixed religious composition brings to light with remarkable frequency a situation which has several times provoked discussion in the Catholic press and literature, and in Catholic congresses in Germany, namely, the fact that business leaders and owners of capital, as well as the higher grades of skilled labour, and even more the higher technically and commercially trained personnel of modern enterprises, are overwhelmingly Protestant."[40]

That observation is, for Weber, the nub of the matter, the crucial discrepancy that needs to be explained. (Thomas Nipperdey says that certain Englishmen had pointed out this link before but, as their observations were confined to Britain, they had not attracted a lot of attention.)[41] Early on, Weber makes it clear he is not talking just about money. For him, a capitalist enterprise and the pursuit of gain are not at all the same thing. People have always wanted to be rich, but that has little to do with capitalism, which he identifies as "a regular orientation to the achievement of profit through [nominally peaceful] economic exchange." Pointing out that there were mercantile operations—very successful and of considerable size—in Babylonia, Egypt, India, China, and medieval Europe, he insists it is only in Europe since the Reformation that capitalist activity has become associated with the *rational organization of formally free labor.*[42]

Weber was also fascinated by what he thought to begin with was a puzzling paradox. In many cases, men—and a few women—evinced a drive toward the accumulation of wealth but at the same time showed a "ferocious asceticism." Many successful entrepreneurs actually pursued a lifestyle that was "decidedly frugal." Why work hard for so little reward? After much consideration, Weber thought he had found an answer in what he called the "this-worldly asceticism" of Puritanism, a notion that he expanded by reference to the concept of "the calling," a form of economic Bildung.[43] Such an idea did not exist in antiquity and, according to Weber, it does not exist in Catholicism either. It dates only from the Reformation, and behind it lies the (Pietist) idea that the highest form of moral obligation of the individual, the best way to fulfill his duty to God, is to help his fellow men, now, in this world. Whereas for Catholics the highest ideal was the purification of one's own soul through withdrawal from the world (as with monks in a retreat), for Protestants the virtual opposite was true.

Weber backed up these assertions by pointing out that the accumulation of wealth, in the early stages of capitalism and in Calvinist countries in particular, was morally sanctioned only if it was combined with "a sober, industrious career." Idle wealth that did not contribute to the spread of well-being, capital that did not *work*, was condemned as a sin. For Weber, capitalism, whatever it has become, was originally sparked by religious fervor, and without that fervor the organization of labor that made capitalism so different from what had gone before would not have been possible.

Weber also focused on bureaucracy and on science. There were two faces to bureaucracy, he said. Modern societies couldn't do without bureaucrats and he thought that Germans have displayed a better talent for rational administration than other nationalities, this having something to do with the idea of Bildung, which he thought was in decline in his day. Necessary as they were, bureaucrats, he thought, always risked stifling innovation, as had happened in China in medieval times, and that was one reason why Bildung was so important.

In his famous address, "Wissenschaft als Beruf" (Science as a Vocation), delivered in 1917, he found science hoist by its own petard, in that he thought originality could only come from increased specialization that, important as it was, was also a form of impoverishment, both for the individual scientist, who could never cultivate his "whole soul" in this way, and for the rest of us, in that through science we would become progressively disenchanted, magic would be removed from the world, as would meaning. Weber thought that scientific concepts, even when they weren't scientific pseudoconcepts, were bloodless abstractions incapable of capturing the reality of life. Science could not offer meaning, he said; it could offer nothing we can base our values on. We are therefore left to create our own values without ever being able, even in principle, to know that they are right. This is our predicament. His analysis is almost as bleak as Nietzsche's.

Weber wasn't as angered by modernity as some of his colleagues were.

He wasn't an unmitigated admirer, but he knew he had to be *involved*.[44] Perhaps that is why he has in general been more influential and is better remembered.

These last two chapters have marked the emergence of "*Kulturkritik*," a new philosophical and literary genre and an early indication of a pattern that would flourish in the twentieth century—the idea of cultural crisis and cultural pessimism. Involving warnings about the imminent cultural collapse of German *Kultur*, this was one of the ideas that helped propel the Conservative Revolution in Weimar Germany in the 1920s and made the rise of National Socialism possible.[45]

24.

Dissonance and the Most-Discussed Man in Music

The progression—Haydn, Mozart, Beethoven, Schubert, Schumann, Mendelssohn, Wagner, which had occupied the century from 1780 to 1880—might seem an unparalleled peak in musical history. But in the run-up to World War I there was another burst of creative energy in the German-speaking lands, which produced Johannes Brahms, Hugo Wolf, Johann Strauss I, Johann Strauss II, Richard Strauss, Gustav Mahler, Anton Bruckner, Max Reger, and Arnold Schoenberg. The well of musical genius seemed inexhaustible.

The career of Johannes Brahms (1833–97) overlapped that of Wagner and, while Wagner was alive, Brahms was the only composer who could stand comparison. But how different they were. Wagner changed everything, but Brahms in a strange way looked back. With him, the symphony as evolved by Beethoven, Mendelssohn, and Schumann achieved a splendid finale. "Brahms, like Bach, summed up an epoch." At the same time, many of the musical afficionados of Vienna were bitterly divided about his achievement. Mahler described Brahms as "a manikin with a somewhat narrow heart," while Hans von Bülow, showing the influence of the revolution in physics, put it this way: "*Brahms ist latente Wärme*"—Brahms is latent warmth.[1]

It is undeniable, however, that Brahms is still very much with us. His works have become—and remain—a lively part of the repertoire. The four symphonies, the four concertos (two piano, one violin, one double concerto) are now classics, together with the *Haydn Variations* and the *German Requiem*.[2] These have a greater appeal today than many of the more innovative works of Mendelssohn, Schumann, and Liszt.

Perhaps what most appeals is Brahms's sheer seriousness. From the word go, he set himself the task to write music that would be beautiful but would at the same time counter the self-centered flamboyance of Liszt and Wagner. This meant he became known as a "difficult" composer, even "a philosopher in sound." And how much did that reflect the fact that he was himself uncompromising and "difficult"? Prickly, oversensitive, cynical, and bad-tempered, he was as much feared and disliked as Hans von Bülow, who was notorious for his tempers and antagonisms. At one party in Vienna, it is said, Brahms left in a huff, grumbling, "If there is anyone here that I have not insulted, I apologise."[3]

Handsome in his youth, Brahms was slight, with fair hair and vivid blue eyes. He became much heavier as he grew older, his face framed with a beard "of biblical proportions."[4] The one extravagance he allowed himself was a collection of original music manuscripts, the jewel of which was Mozart's G Minor Symphony.

Born in Hamburg in 1833, Brahms was the son of a professional double bass player and was only six when it was discovered he had perfect pitch and a precocious musical ability. By ten he was giving piano recitals in public. Strangely, however, given that his father was a double bass player, Brahms was set to playing in waterfront bars and bordellos in the red-light district of what was, after all, a port city. This may have helped the family finances, but it left its psychological scars. Throughout his life Brahms seems to have been sexually comfortable only with prostitutes, and this almost certainly got in the way of his marrying.

By the time he was twenty, Brahms had written several works for the piano. They were monumental, with deep-sounding basses in the background, but they hardly let up—there was no spark to draw people in.[5] However, at this time his career as a pianist was going from strength to strength, and in 1853, on tour, he encountered Joseph Joachim (1831–1907), a young violinist already famous who was very much taken with Brahms's piano playing.[6] Joachim introduced Brahms to Liszt and, more consequentially, to Schumann, the latter making an entry in his diary for September 30, 1853: "Brahms to see me (a genius)." Schumann was in fact so impressed by the young man that, generous as ever, he wrote a lengthy piece about Brahms in the *Neue Zeitschrift für Musik*, referring to him as "a young eagle." The article would be the last Schumann wrote for the publication he had founded, but so intense was the attraction between the two men that Schumann insisted Brahms move in with him. (Brahms almost certainly fell in love with Clara, though for her part, when Schumann died, "she became a professional widow who wore mourning clothes all her life.")[7]

In 1862 Brahms traveled to Vienna, liked it, went back the following year—and stayed for the rest of his life. His decision was helped by his appointment as conductor at the Academy of Singing though he remained there for just two years, afterward concentrating on composing, with short breaks for concert tours.

Brahms's first truly famous work was not a piano piece, as might have been expected, but the *Deutsches Requiem*, and here there was a paradox. He himself was a freethinker, from Protestant Hamburg, writing in Catholic Vienna. The text of the Requiem is in German (not Latin) and is taken from the Lutheran Bible but bears no relation to any known liturgy; it is his own language and there is nothing nationalistic or political about it.[8] There is no mention of Christ. Yet when it was performed in 1868, first in Dresden (incomplete), then in Leipzig—the full version—it was a great success, the right blend of reflection and stirring choral harmonies. Brahms all but gave up touring as a pianist as music poured from his pen.

All composers in the nineteenth century were confronted—or *felt* they were confronted—with the monumental presence of Beethoven's Ninth Symphony, "a vast wall of invention and sound." (Even Wagner had been daunted.) Brahms's Symphony no. 1 in 1876 was several years in the making, and the composer was all too aware that Beethoven, at the age of forty-three, which Brahms then was, had produced eight of his nine symphonies. Brahms, who some people in Vienna were calling Beethoven's successor, had to be sure of his ground.

Harold Schonberg says that the C Minor Symphony—and perhaps its reception—seemed to "unlock" something in Brahms. He now entered upon a period of great creativity, producing masterpiece after masterpiece—the Violin Concerto in 1879, the B-flat Piano Concerto in 1881, the Third Symphony in 1883, the Fourth Symphony in 1885, the Concerto for Violin and Cello in 1887.[9] Between 1891 and 1894 he produced a raft of remarkable works for the clarinet—the Clarinet Trio and Clarinet Quintet (both in 1891), and two clarinet sonatas (1894)—written for another of his friends, Richard Mühlfeld, the first clarinetist of the Meiningen Orchestra, which played an important role in Brahms's career. Hans von Bülow, the conductor who had taken over in 1880, had turned it into "*the* precision instrument of European orchestras" and he now used it to become the greatest interpreter of Brahms.[10]

Not that this adulation seems to have had much effect on the composer himself. Brahms aged badly, growing more irascible, more sarcastic, more cynical, falling out with his erstwhile friends, with both von Bülow and Joachim. Yet toward the end, Brahms's music became very tender and relaxed. The D Minor Violin Sonata, the Clarinet Quintet, and his very last work, eleven choral preludes for organ, "have a kind of serenity unique in the work of any composer." At a time when the formidable operas of Wagner and the alarming modernist dissonances of Richard Strauss were the talk of Europe, the music of Brahms, the sound of the old, premodern, pre–World War I era, serenely slipped away.[11]

The Greatest Lieder Composer of All Time?

The year of Brahms's death, 1897, was a year of tragedy for Hugo Wolf (1860–1903). In that year, the man who many think was the greatest song composer of all time, was consigned to a lunatic asylum. We shall never know whether the cause was the syphilis he had contracted as a teenager or whether he was in any case temperamentally and constitutionally weak. A slim, aristocratic-looking man, often photographed in an elegant velvet jacket and flamboyant artist's tie, his dark eyes were always burning, always troubled. Yet within the space of a few years "this tortured creature left the world a legacy that carried the German art song to its highest point."[12]

A bohemian and a malcontent, he was yet able for a short time to achieve an intensity of feeling in his music unmatched by any of his contemporaries.[13] He wrote nearly 250 songs that, although they show a great affinity for the poetry on which they are based and are more original and more advanced harmonically even than Schubert, have an equilibrium belied by Wolf's own stormy life.

He started writing songs when he was in his teens, but his best work was done about a decade later. In the four years between 1888 and 1891 he produced more than 200 songs, two or three a day at times, using a variety of (often comic) poems by Eduard Mörike, Joseph von Eichendorff, Goethe, of course, and Gottfried Keller.[14] In 1897 Wolf collapsed into insanity and the last four years of his short life were spent, on and off, in an asylum—his song-writing career lasted just seven years.[15]

Despite the speed of their composition, the quality of Wolf's songs was immediately recognized. Leading singers took them up straightaway with Wolf, a good pianist if not in Brahms's class, accompanying them. The difference Wolf introduced lay in his use of melody. For him melody could contribute to the meaning of a poem, a good example being "Wer rief dich denn," in which the singer sings one thing while the accompaniment suggests that the words he is mouthing are false.[16]

In the first instance, Wolf's madness took the form of delusions that he had been appointed director of the Vienna Opera. From the asylum he wrote detailed plans for what he wanted to do now that he had succeeded Mahler.[17] He died in the asylum in 1903, his life never achieving the equilibrium of his best work.

"The Laughing Genius of Vienna"

Brahms and Wolf, though rivals (at least in the latter's eyes), were united in being serious composers. Harold Schonberg points out that in the nineteenth century three composers of "light" music have survived time and fashion "so triumphantly that they legitimately can be called immortals." The waltzes and Viennese operettas of Johann Strauss II, the opéra bouffe of Jacques Offenbach, not to mention the operettas of Arthur Sullivan, remain with us, as charming, pert, and inventive as ever.

The German form of light music, the waltz, originated in the 1770s from the *Ländler*, an Austro-German dance in three-quarter time. It took all Europe by storm, even though Vienna was always its "headquarters."

Michael Kelly, the Irish tenor who performed at the world premiere of Mozart's *Le Nozze di Figaro*, wrote in his memoirs (in 1826) that women "regularly waltzed from ten in the evening until seven in the morning," and that special rooms were prepared near the dance floor "for heavily pregnant ladies to give birth should the need arise." Schubert, Weber, Brahms, and Richard Strauss all wrote waltzes (there is even a waltz in Alban Berg's *Wozzeck*). But of course the waltz is forever associated with the name of Johann Strauss, father and son.[18]

Johann Strauss I was born in Vienna in 1804. By the age of fifteen he was a professional violinist, playing in a number of orchestras. In 1826, when he was still in his early twenties, he and one of his fellow violinists, a sensitive young man called Josef Lanner, formed their own dance group.[19] All went well until Strauss turned his hand to composing, when the two men fell out badly; Strauss stormed off to form his own orchestra, soon employing 200 musicians, servicing six balls a night. His compositions were also a success and he wrote one "hit" after another—the *Donaulieder* (op. 127), the "Radetzky-Marsch," and so on.

Despite the successes, Strauss did not want any of his six children to become a musician. Johann II, born in 1825, was therefore forced to take lessons in secret, at least until his father brought scandal on the family by moving out of the conjugal home to take up with another woman, with whom he had another four children. All Vienna was agog at the scandal but it was soon overtaken by another, namely that the elder Strauss saw his position threatened by his son, who began to outshine his father before he died in 1849.[20] Johann II was just nineteen when he decided to take on his father. He obtained a booking at Dommayer's Garden Casino and Restaurant in Vienna and, on that first evening at least, he took care to end the concert with his father's *Lorelei-Rheinklänge*.[21] But there were no secrets in gossipy Vienna, everyone knew about the divided nature of the Strauss household, and one Viennese newspaper summed up the situation with the headline "Good night, Lanner. Good evening, Father Strauss. Good morning, Son Strauss."

After Johann I died, Johann II merged his father's orchestra with his own. At the height of his fame, he had six orchestras, dashing from one to the other, appearing for one or two waltzes at each venue. Eventually he gave up this exhausting routine to devote himself to composing, leaving his brother Eduard to take the baton. Now began his great series of compositions, which included "Perpetuum Mobile," "Geschichten aus dem Wienerwald" (Tales from the Vienna Woods), "Kaiserwaltzer" (Emperor Waltz) and, not least, the "sound of the river"—"An der schönen blauen Donau" (On the Beautiful Blue Danube).[22] The orchestration and rhythms of these tunes have been much admired and mark them out "as much more than 'mere' dance music." No less a figure than Brahms understood this, autographing Frau Strauss's fan one evening with a few notes of the "Blue Danube," adding the words: "Alas, not by Johannes Brahms." Richard Strauss called him "the laughing genius of Vienna."[23]

Wagner's Natural Successor

Some people in Vienna referred to Richard Strauss (1864–1949) as "the third Strauss" but for others—modernists—he was without question the first, the only Strauss, "the most-discussed man of European music" between *Don Juan*, which had its premiere in 1886, and 1911, when *Der Rosenkavalier* was staged. His symphonic poems were considered "the last word in shocking modernism" while *Salome* and *Elektra* (1905 and 1909) provoked riots.

In his early years, an aura of nervous energy and scandalous sensation surrounded the tall, slim Strauss. It was not just the size of his orchestras that attracted and repelled people—often at the same time—but the fact that his music was, for many, painfully dissonant and, moreover, immoral. *Salome* set music to a text by Oscar Wilde, who had been sent to prison for his homosexuality.

Paradoxically, Strauss was himself a solid bourgeois, with a sober—even staid—private life. Alma Mahler was at the rehearsal of *Feuersnot* in 1901 and confided to her diary: "Strauss thought of nothing but money. The whole time he had a pencil in his hand and was calculating the profits to the last penny."[24] His wife, Pauline, was a grasping woman, once a singer, who would scream at her husband, when he was relaxing at cards, "Richard, go compose!" Their house at Garmisch had three separate doormats, on each of which Pauline insisted that the composer wipe his feet.

Until *Der Rosenkavalier* each of Strauss's works was different, one from another, exciting and electric. "Then he seems to have hit a wall." There is no shortage of critics and historians who dismiss *all* of Strauss's post-*Rosenkavalier* oeuvre as a regression, mechanically repetitive and lacking in innovation.[25] Ernest Newman was one of those: "A composer of talent who was once a genius," is how he summed up the predicament. After *Elektra*, Newman said, "the premiere of a Strauss opera was no longer an international event."

Strauss's father was a peppery, outspoken man—Franz was "the most celebrated horn player in Germany," who considered Wagner "subversive." Richard, born in Munich, was a prodigy, playing the piano at four and a half, the violin shortly afterward, composing at six. Richard's father, however, had no wish to form him into a second Mozart: it was accepted in the family that the boy would be a musician "but all in good time." In 1882 he attended the University of Munich though he never took a degree, then spent time in Berlin, playing at musical soirees and in due course met Hans von Bülow. He showed the celebrated conductor his Serenade for Winds in E-flat, op. 7, which was snapped up for the Meiningen Orchestra. In fact, Bülow so loved the Serenade that he commissioned another piece there and then—what became the Suite for Winds in B-flat, op. 4. This, too, impressed Bülow so much so that he appointed Strauss his assistant at Meiningen (it was by then 1885). It was a heady time for the young composer and might have led Strauss in a very different direction, except that in Meiningen he met Alexander Ritter, a violinist with the orchestra. Ritter had married Wagner's niece and introduced Strauss to Berlioz, Liszt, and Wagner himself, and it was the latter who encouraged Strauss to explore new forms of music. Until this point, his compositions had been along largely traditional, familiar lines. The break came in 1889, with *Don Juan*, his first tone poem. Premiered in Weimar on November 11, it was immediately obvious that a new voice had emerged.[26]

With *Don Juan*, Strauss staked his claim as Liszt's—and Wagner's—natural successor. The score stipulated a vast orchestra and was unprecedented in its difficulty. Its bold leaps and twists of sinuous melody were also new.[27] His standing as a conductor grew in parallel. In 1898 he was appointed conductor at the Royal Opera in Berlin, where he remained until 1918, at which time he took up the post as codirector of the Vienna Opera. Even Pauline was impressed.

He had his failures. "It is incredible what enemies *Guntram* [1894] has made for me," he complained in a letter. "I shall shortly be tried as a dangerous criminal."[28] In 1905, however, with *Salome*, he at last had an opera that galvanized the public as much as his symphonic poems had done, though the scandal it provoked had as much to do with its plot as his score. Who but the most puritanical would not want to see Salome make love to the severed head of Jochanaan, or be there as she took off her seven veils one by one?[29]

The opera was based on Oscar Wilde's play, which had been banned in London, but Strauss's score "added fuel to the fire." To highlight the psychological contrast between Herod and Jochanaan, Strauss employed the unusual device of writing in two keys simultaneously. The continuous dissonance of the score reaches its culmination with Salome's moan as she awaits execution. This, rendered as a B-flat on a solo double bass, registers the painful drama of Salome's plight: she is butchered by guards crushing the life out of her with their shields.

After the first night, opinions varied. Cosima Wagner was convinced the new opera was "Madness! . . . wedded to indecency." The Kaiser would only allow *Salome* to be performed in Berlin after the manager of the opera house shrewdly modified the ending, so that a Star of Bethlehem rose at the end, a simple trick that changed everything, and the opera was performed fifty times in that one season. Ten of Germany's sixty opera houses chose to follow Berlin's lead so that within a few months Strauss could afford to build his villa at Garmisch in the art nouveau style. Despite its success in Germany, it was banned outright in New York and Chicago (in the former city after one night). Vienna also banned the opera, but in Graz the opening night was attended by Giacomo Puccini, Gustav Mahler, and a band of young music lovers who traveled down from Vienna, including an out-of-work would-be artist named Adolf Hitler, who later told Strauss's relatives he had borrowed money to make the trip.

Despite the offense *Salome* caused in some quarters, its eventual success contributed to Strauss's appointment as musical director of the Hofoper in Berlin. He began work there with a one-year leave of absence to complete his next opera, *Elektra*. This was his first major collaboration with Hugo von Hofmannsthal (1874–1929), whose play of the same name, realized by that magician of the German theater, Max Reinhardt, Strauss had seen in Berlin. What appealed to him was its theme, so very different from the noble, elegant, *calm* image of Greece traditionally set out in the writing of Winckelmann and Goethe.

Elektra uses a larger orchestra even than *Salome*, 111 players, to produce a much more dissonant, "even painful experience." The original Clytemnestra was Ernestine Schumann-Heink, who described the early performances as "frightful . . . We were a set of mad women . . . There is nothing beyond *Elektra* . . . We have come to a full stop."[30]

Strauss and Hofmannsthal were trying to do two things. At the most obvious level, they were doing in musical theater what the Expressionist painters of Die Brücke and Der Blaue Reiter (see Chapter 27, p. 503) were doing in their art, using unexpected and "unnatural" colors, disturbing distortion and jarring juxtapositions to change people's perceptions of the world. Most scholars had inherited an idealized picture of antiquity from Winckelmann and Goethe, but Nietzsche had changed all that, stressing the instinctive, savage, irrational, and darker aspects of pre-Homeric ancient Greece (fairly obvious, for example, if one reads the *Iliad* and the *Odyssey* without preconceptions). But *Elektra* wasn't only about the past.[31] There can be little doubt that Hofmannsthal had read *Studies in Hysteria* and *The Interpretation of Dreams*. The presence of Freud's, and Nietzsche's, ideas on stage, undermining traditional understanding of ancient myth and the exploration of the unconscious world beneath the surface, did not make people content, but it made them think.

Elektra made Strauss think too, and he abandoned the discordant line he had followed from *Salome* to *Elektra*. In doing so he left the way open for others, of whom the most innovative would be Arnold Schoenberg (1874–1951).

Elektra had achieved one other thing. It had brought Strauss and Hofmannsthal together. For nigh on a quarter of a century, Strauss and Hofmannsthal collaborated—*Der Rosenkavalier* (1911), *Ariadne auf Naxos* (1912), *Die Frau ohne Schatten* (1919), *Arabella* (1933). Their most fruitful collaboration after *Elektra* was *Der Rosenkavalier*. After the dark depths of *Elektra*, Strauss thought a comedy was needed (he was not wrong) and Hofmannsthal provided the idea. Although there had been no widely acclaimed German comic opera since *Die Meistersinger* in 1868, *Der Rosenkavalier* had a difficult birth. Hofmannsthal was trying to lead Strauss to a different aesthetic, lighter, more sophisticated, with no burdensome psychological archetypes. "People do not die for love in Hofmannsthal's world."[32] Is this why, for many, Strauss appeared to stop after this opera?

In their day, three other German composers were overshadowed by Richard Strauss. Gustav Mahler was well known in his lifetime but chiefly as a conductor, as the main figure in what are remembered as the "golden years" of the Vienna Opera, the decade between 1897 and 1907. His symphonies were heard but not often. Anton Bruckner's following was even smaller. Max Reger had his followers and there was at least a vogue for his music after he died. But not until the 1960s did Bruckner and Mahler find any kind of general popularity.[33]

Bruckner, born in Ansfelden, Upper Austria, in 1824, studied at the monastery of St. Florian, known for its magnificent Altdorfer altarpiece. He became choirmaster and organist there, playing for its community of Augustine monks. He was close to being a peasant. Photographs show that his head was shaven, and accounts record that his clothes were homemade. Despite this and his rural accent, he was appointed teacher of organ and theory at the Vienna Conservatory in 1868 and made full professor not long afterward; he was also given parallel appointments at the University of Vienna. These were serious positions, such that leading conductors, Mahler among them, started to pay attention to his music. But Bruckner's problem was with the critics, not the conductors, in particular with Eduard Hanslick, Brahms's champion in the press. Bruckner always suspected that Brahms was the dark shadow behind Hanslick.[34]

Bruckner never lost his air of provincial awkwardness and, in the middle of his lectures, delivered in his peasant garb, he would stop everything and kneel to pray when the Angelus sounded. But he was sophisticated enough in music, and people were forced to acknowledge it. He loved slow, solemn, deliberate music, and became known in Vienna as the "Adagio-Komponist."[35] Critics said he composed the same symphony nine times, but the unhurried serenity of his music has helped it endure.

Mahler was the opposite. For many of *his* critics, his music is much too neurotic, though Mahler afficionados are, if anything, even more fanatical than those of Bruckner. Mahler, a patient of Freud's, was typical of one kind of Viennese, who took life seriously, anxious to make sense out of their circumstances.[36] The difference between Beethoven's struggles and

Mahler's has been well put: Beethoven was a titan and a heroic figure, Mahler was a "psychic weakling," a sentimentalist, a "manic-depressive with a sadistic streak," who took four-hour walks with Freud as a form of therapy. His orchestras respected him but rarely enjoyed the experience. His music has been attacked as "monotonous."[37]

Born in Kalist, Bohemia, in 1860, Mahler was master of all he surveyed for ten years in Vienna. To his credit, his approach—unpopular though it was—worked; under him the opera was revitalized and cleared of debt. Avant-garde productions proliferated and so did the furors. Mahler was pointed out in the street to visitors by cabdrivers who referred to him simply as "Der Mahler!" He was essentially a Romantic composer (especially in the Third and Eighth Symphonies), less harsh than Wagner and less progressive than Strauss.[38]

The Emancipation of the Dissonance and the Musical Equivalent of $E=mc^2$

Richard Strauss was ambivalent about Arnold Schoenberg. He thought he would be better off "shovelling snow" than composing, yet recommended him for a Liszt scholarship (the revenue of the Liszt Foundation was used to help composers or pianists). Born in 1874 into a poor family, Arnold Schoenberg—like Brahms and Bruckner—always had a serious disposition and was largely self-taught. A small, wiry man, "easily unimpressed," who went bald early on, Schoenberg was strikingly inventive—he carved his own chessmen, bound his own books, painted (Wassily Kandinsky was a fan), and built a typewriter for music.[39] In Vienna he frequented the cafés Landtmann and Griensteidl, where Karl Kraus, Theodor Herzl, and Gustav Klimt were great friends. He mixed with the philosophers of the Vienna Circle.[40]

Schoenberg's autodidacticism served him well. While other composers made the pilgrimage to Bayreuth, Schoenberg was more impressed by the Expressionist painters who were trying to make visible the distorted and raw forms unleashed by the modern world and analyzed and ordered by Freud. His aim was to do something similar in music. The term he himself liked was "the emancipation of the dissonance."[41]

Schoenberg once described music as "a prophetic message revealing a higher form of life toward which mankind evolves." However, he found his own evolution slow and painful. Though his early music owed a debt to Wagner, *Tristan* especially, it had a troubled reception in Vienna, its high seriousness out of place in a city that was, in Alex Ross's words, "forced to glitter."[42] In addition to his difficulties with the public, Schoenberg had problems in his private life. In the summer of 1908, the very moment of his first atonal compositions, his wife, Mathilde, abandoned him for a friend. Rejected by his wife, isolated from Mahler, who was in New York, Schoenberg was left with nothing but his music. But in that year he composed his Second String Quartet, based on the "esoteric and remote" poems of Stefan George.[43]

The precise point at which atonality arrived, his "voyage to the other side," according to Schoenberg himself, was during the writing of the third and fourth movements of the string quartet.[44] He was using George's poem "Entrückung" (Ecstatic Transport) when he suddenly left out all six sharps of the key signature. As he completed the part for the cello, he abandoned completely any sense of key, to produce a "real pandemonium of sounds, rhythms and forms." As luck would have it, the stanza ended with the line, "Ich fühle Luft von anderen Planeten" (I feel the air of other planets). It could not have been more appropriate. The Second String Quartet was finished toward the end of July. Between then and its premiere, on December 21, one more personal crisis shook the Schoenberg household. In November the painter for whom his wife had left him hanged himself after failing to stab himself to death. Schoenberg took back Mathilde, and when he handed the score to the orchestra for the rehearsal, it bore the dedication "To my wife."

The premiere of the Second String Quartet turned into one of the great scandals of musical history.[45] After the lights went down, the first few bars were heard in respectful silence. But only the first few. Most people who lived in Vienna then carried whistles attached to their door keys. If they arrived home late at night, and the main gates of the building were locked, they would use the whistles to attract the attention of the concierges. On the night of the premiere, the audience got out its whistles en masse. A wailing chorus arose in the auditorium to drown out what was happening onstage. Next day one newspaper labeled the performance a "Convocation of Cats," and the *New Vienna Daily* printed their review in the "crime" section of the paper.[46]

Years later Schoenberg conceded that this was one of the worst moments of his life, but he wasn't deterred. Instead, in 1909, continuing his emancipation of the dissonance, he composed *Ewartung*, a thirty-minute opera, the story line for which is so minimal as to be almost absent: a woman goes searching in the forest for her lover; she discovers his body not far from the house of the rival who has stolen him. The story does not so much tell a story as reflect the woman's moods—joy, anger, jealousy. In addition to the minimal narrative, it never repeats any theme or melody. Since most forms of music in the "classical" tradition usually employ variations on themes, and since repetition, lots of it, is the single most obvious characteristic of popular music, Schoenberg's Second String Quartet and *Ewartung* stand out as the great break, after which "serious" music began to lose the faithful following it had once had. It was fifteen years before *Ewartung* was performed.

Although he might be too impenetrable for most people's taste, Schoenberg was not obtuse. He knew that some people objected to his atonality for its own sake. His response was *Pierrot lunaire*, appearing in 1912. It features a familiar icon of the theater—a dumb puppet who also happens to be a feeling being, a sad and cynical clown allowed by tradition to raise awkward truths so long as they are wrapped in riddles. Out of this format, Schoenberg managed to produce what many people consider his seminal work, what has been called the musical equivalent of Picasso's *Les demoiselles d'Avignon* or Einstein's $E=mc^2$. *Pierrot*'s main focus is a theme we are already familiar with, the decadence and degeneration of modern man. Schoenberg introduced in the piece several innovations in form, notably *Sprechgesang*, literally song-speech in which the voice rises and falls but cannot be said to be either singing or speaking. Listeners have found that the music breaks down "into atoms and molecules, behaving . . . not unlike the molecules that bombard pollen in Brownian movement." Schoenberg saw himself more as an Expressionist, and he shared many of the aims of Kandinsky though some of his early atonal pieces have the sunny fog and silence of Caspar David Friedrich's landscapes.[47]

The first performance took place in mid-October in Berlin, in the

Choralionsaal on the Bellevuestrasse (destroyed by Allied bombs in 1945). Following the premieres of the Second String Quartet, the critics gathered, ready to kill off the clown. But the performance was heard in silence and, when it was over, Schoenberg was given an ovation. It was short, so many in the audience shouted for the piece to be repeated, and they liked it even better the second time. So too did some of the critics, one going so far as to describe the evening "not as the end of music; but as the beginning of a new stage in listening." Like it or not, Schoenberg had found a way forward after Wagner.

25.

The Discovery of Radio, Relativity, and the Quantum

Two upheavals took place in physics at the turn of the twentieth century. These were, first, the unexpected discoveries of x-rays, the electron, and radioactivity; and then, what some people regard as "the real revolution," the discovery of the quantum and the theory of relativity. As well as being possibly the greatest intellectual adventure of the twentieth century, this was also one of the most international—advances being made by New Zealanders, Danes, Italians, French, British, and Americans, besides Germans, many of whom, to begin with at least, behaved with a commendable sense of international camaraderie. So if this chapter concentrates on the German contribution, this is not in any way to belittle the contributions of others, which were vital.

Nevertheless, Amos Elon says that in the natural sciences there was at this time "talk of a new German 'Age of Genius,' second only to the era of Goethe, Schiller, Hegel and Kant," and Helge Kragh, in his study of twentieth-century physics, gives two tables which show how the Germans were at least ahead of others in their physics *institutions.*[1]

Physics Institutes and Faculty			
	No. of Institutes	Faculty, 1900	Faculty, 1910
Britain	25	87	106
France	19	54	58
Germany	**30**	**103**	**139**
United States	21	100	169

Physics Journals in 1900			
	Core Journal	Papers, 1900	%
Britain	*Philosophical Magazine*	420	19
France	*Journal de Physique*	360	18
Germany	***Annalen der Physik***	**580**	**29**
United States	*Physical Review*	240	12

Many new physics laboratories were built between 1890 and World War I, twenty-two in Germany, nineteen in the British Empire, thirteen in the United States, twelve in France.[2] The *Dictionary of Scientific Biography* lists 197 physicists who were twenty years old in 1900: 52 were German (and 6 Austrian), Britain came next with 35, France with 34, and the United States with 27.

It is not altogether clear *why* such attention was being paid to physics. When Max Planck started at the University of Munich in 1875, he was warned by his professor that his chosen field "was more or less finished and that nothing new could expect to be discovered."[3]

Waves Through the Air

But there *was* undoubtedly change in the air. Most physicists still clung to a mechanical view of the universe, even James Clark Maxwell, whose field theory found many supporters. This was accompanied by the rise to prominence of the idea of a universal ether as a "quasi-hypothetical," continuous, and all-pervading medium through which forces were propagated at a finite speed.[4] This helped people consider the possibility that the foundation of these forces was electromagnetic rather than mechanical. In this environment, new ideas began to proliferate—rudimentary notions of antimatter, for example, of extra dimensions, most important a new field of "energetics," put forward by the German physicist Georg Helm and his chemist colleague Ludwig Ostwald. In this view, energy, not matter, was the essence "of a reality that could be understood only as processes of actions." Energetics turned out to be important in that, although turn-of-the-century physics revolved around the two "upheavals" mentioned earlier, the first German name to shine was in a different but related field, where the "ether," electromagnetism, and, by implication, energy, were also important elements.

Heinrich Rudolf Hertz was born in Hamburg in 1857, the son of a Jewish lawyer who had converted to Christianity. Heinrich was a clever linguist, learning Arabic and Sanskrit but he also had a liking for the natural sciences and a facility for building experimental equipment, particularly in physics. He went to the university in Munich and afterward in Berlin where he studied under Gustav Kirchoff and Hermann von Helmholtz and also attended Treitschke's lectures.[5] His PhD dissertation in 1880 was so well received that he became Helmholtz's assistant, after which he was appointed a lecturer in theoretical physics at Kiel. Though distinguished enough as a university, Kiel was not very big and, unlike other institutions, had hardly any laboratory space—this was why *theoretical* physics flourished there, a relatively new discipline, as we have seen, in which Germany led the way. At Kiel, Hertz produced his first important contribution when he derived Maxwell's equations but in a way that was different from Maxwell's own and did not involve the assumption of an ether.[6] On the strength of this, Hertz was appointed the following year at the age of twenty-eight to the chair of physics at Karlsruhe, a much bigger, better-equipped university. There his first significant discovery was of the photoelectric effect, whereby ultraviolet radiation releases electrons from the surface of a metal (Einstein's explanation of the photoelectric effect, not his work on relativity, won him the Nobel Prize—see below). Hertz was becoming *the* theoretical physicist par excellence.

With his gift for manufacturing laboratory equipment, in 1888 he produced his most innovative device yet. The central element here was a metal rod in the shape of a hoop with a minute (3mm) gap at midpoint (not unlike a large key-ring).[7] When a sufficiently strong current was passed through the hoop, sparks were generated across the gap (he darkened the room to facilitate observation).[8] At the same time violent oscillations were set up in the rod forming the hoop. Hertz's crucial observation was that these oscillations sent out waves through the nearby air, a phenomenon he was able to prove because a similar circuit some way off could detect them. In later experiments Hertz showed that these waves could be reflected and refracted—like light waves—and that they traveled at the speed of light but had much longer wavelengths than light. Later still, he observed that a concave reflector could focus the waves and that they passed unchanged through non-conducting substances. These were originally called Hertzian waves, and their initial importance lay in the fact that they confirmed Maxwell's prediction that electromagnetic waves could exist in more than one form—light. Later they were called radio.

Asked by a student what use might be made of his discovery, Hertz famously replied, "It's of no use whatsoever. This is just an experiment that proves Maestro Maxwell was right—we just have these mysterious electromagnetic waves that we cannot see with the naked eye." Asked, "So what's next?" he answered, "Nothing, I guess." A young Italian, on holiday in the Alps, read Hertz's article about his discovery and immediately wondered whether the waves set off by Hertz's spark oscillator might be used for signaling. Guglielmo Marconi rushed back home to see whether his idea might work.[9] Had Hertz lived (he died from bone disease at thirty-seven) he would have been as surprised as anyone at the direction physics was about to take. Rollo Appleyard says he was in all respects "a Newtonian." [10]

A New Kind of Ray

In Chapter 17, p. 341, we saw that gases had claimed physicists' attention earlier in the century, at first for the light they threw on the conservation of energy, then for the statistical behavior of their atoms and molecules. As part of this, and especially with the growth of interest in electromagnetism and the possibility that the "void" between atoms was filled, as Maxwell said, with an electromagnetic field, a new specialism grew up that entailed the discharge of electricity in gases. As this interest developed, a new piece of apparatus was conceived, eventually known as the cathode ray tube. This was a glass tube with metal plates sealed into either end, and the gas sucked out, leaving a vacuum. If the metal plates were then connected to a battery and a current generated, the empty space, the vacuum inside the glass tube, glowed or fluoresced. This glow was generated from the negative plate, the cathode, and was absorbed into the positive plate, the anode.* The Berlin physicist Eugen Goldstein in 1876 was the first to label the new equipment "cathode ray" tubes.

William Crookes in Britain hypothesized in 1879 that cathode rays were a "fourth state of matter" (i.e., neither solid, liquid, nor gas), but that wasn't convincing and most physicists still asked what exactly cathode rays *were*. The situation remained both confusing and promising, and several physicists started to take a look. One was the professor of physics at the University of Würzburg, Wilhelm Conrad Röntgen.

Born in the Lower Rhine province, Röntgen grew up in the Netherlands before studying in Zurich under Clausius. At Würzburg, he started investigating cathode rays—in particular their penetrating power—toward the end of 1895. It was by then common to use a barium platino-cyanide screen to detect any fluorescence caused by cathode rays.[11] This screen was not really part of the experiment, more a fail-safe device should there be any anomalies. In Röntgen's case, the screen was some way from the cathode ray tube which was in fact covered with black cardboard and operated only in a darkened room. On November 8, 1895, now a famous date in the history of science, Röntgen noticed—to his surprise—that the screen, though a good distance from the tube, also fluoresced. This could not possibly have been caused by the cathode rays. But did that mean the apparatus was giving off *other* rays, invisible to the naked eye? He confirmed his results, noting also that the paper screen covered with barium platino-cyanide fluoresced "whether the treated side or the other be turned towards the discharging tube."[12]

When his discovery was published, it caused a great stir, well beyond the confines of professional scientists, and he was soon commanded to demonstrate his discovery to the Kaiser. But what exactly *were* the new

* This is also the basis of the television tube. The positive plate, the anode, was reconfigured with a glass cylinder attached, after which it was found that a beam of cathode rays passed through the vacuum toward the anode made the glass fluoresce.

rays?[13] In his follow-up studies, Röntgen found they had some of the properties of light, in that they followed straight lines, affected photographic plates, and were not interfered with by magnetic fields. At the same time, they were different from light and from Hertz's electromagnetic waves in not being either reflected or refracted. For as much as a decade, physicists worked fruitfully with x-rays, as they came to be called (*x* for unknown, though they are called *Röntgenstrahlen* in Germany), without understanding exactly what they were.[14]

Eventually, in the early twentieth century, it was shown that x-rays were a form of electromagnetic wave with an extremely short wavelength, but some confusion remained until the wave-particle duality was clarified with the advent of quantum mechanics (see Chapter 32, p. 595). In 1912 Max von Laue, a physicist who worked with Planck and Einstein, realized that, with their very small wavelength, x-rays could only be studied (i.e., reflected or refracted) by substances that had a very small grid structure, and that such spacings were to be found in the "inter-atomic distances between the ions of a crystal."[15] The experiment to test this prediction was carried out by the Munich physicist Walter Friedrich and his student Paul Knipping in the spring of 1912, with this collaboration constituting the first-ever "x-ray diffraction." This provided proof that x-rays are indeed electromagnetic waves and that their wavelength is very short indeed, somewhere near 10^{-13} meters. This set off a whole new area of research, the use of x-ray diffraction in crystallography, and came to play an important role in chemistry, geology, metallurgy, and, not least, biology. It was a key technique in the determination by James D. Watson and Francis Crick, in 1953, of the double-helix structure of the DNA molecule.

The Discovery of the Quantum

In 1900 Max Planck was forty-two.[16] He had been born into a very religious, rather academic family, and was an excellent musician (he had a harmonium specially built for him).[17] But science was Planck's calling and by the turn of the century he was near the top of his profession, a member of the Prussian Academy and a full professor at the University of Berlin, where he was known as a prolific generator of ideas that didn't always work out.[18]

In 1897, J. J. Thomson, who had followed Maxwell as director of the Cavendish Laboratory in Cambridge, England, had pumped different gases into the cathode ray tubes and at times surrounded them with magnets. By systematically manipulating conditions, he demonstrated that cathode rays were in fact infinitesimally minute *particles* erupting from the cathode and drawn to the anode. He discovered that the particles were lighter than hydrogen atoms, the smallest known unit of matter, and exactly the same *whatever* the gas through which the discharge passed. Thomson had clearly identified something fundamental, what is today known as the electron.[19]

Many other particles of matter were discovered in the years ahead, but it was the very notion of particularity itself that interested Max Planck, and in 1897, the year Thomson discovered electrons, Planck began work on the project that was to make his name. It had been known since antiquity that as a substance (iron, for example) is heated, it first glows dull red, then bright red, then white. This is because longer wavelengths (of light) appear at moderate temperatures, and as temperatures rise, shorter wavelengths appear. When the material becomes white-hot, all wavelengths are given off. Studies of even hotter bodies—stars, for example—show that in the next stage the longer wavelengths drop out, so that the color gradually moves to the blue part of the spectrum. Planck was fascinated by this and by its link to a second mystery, the so-called black body problem. A perfectly formed black body is one that absorbs every wavelength of electromagnetic radiation equally well. Such bodies do not exist in nature, though lampblack, for instance, comes close, absorbing 98 percent of all radiation.[20] According to classical physics, a black body should only emit radiation according to its temperature, and then such radiation should be emitted at every wavelength—it should only ever glow white. But studies of the perfect black bodies available to Planck, made of porcelain and platinum and located at the Bureau of Standards in Charlottenburg, showed that when heated, they behaved more or less like a lump of iron, giving off first dull red, then bright red-orange, then white light. Why?[21]

Planck's revolutionary idea first occurred to him around October 7, 1900. On that day he sent a postcard to his colleague Heinrich Rubens on which he sketched an equation to explain the behavior of radiation in a black body.[22] The essence of Planck's idea, mathematical only to begin with, was that electromagnetic radiation was not continuous, as Newton had claimed, but could only be emitted in packets of a definite size. It was, he said, as if a hosepipe could spurt water only in "packets" of liquid. By December 14 that year, when Planck addressed the Berlin Physics Society, he had worked out his full theory. Part of this was the calculation of the dimensions of this small packet of energy, which Planck called *h* and which later became known as Planck's Constant.[23] This, he calculated, had the value of 6.55×10^{-27} ergs each second (an erg is a small unit of energy). Planck had identified this very small packet as a basic indivisible building block of the universe, an "atom" of radiation, which he called a "quantum." It was confirmation that nature was not a continuous process but moved in a series of extremely small jerks. Quantum physics had arrived.

Or not quite. So many of the theories Planck had come up with in the twenty years leading to the quantum had proved wrong that when he addressed the Berlin Physics Society, he was heard in polite silence, and there were no questions. It took four years for the importance of his idea to be grasped—and then by a man who would create his own revolution: Albert Einstein.[24]

The "Annus Mirabilis" of Science

Germany, as we have seen, led the way in the tradition of theoretical physics—Clausius, Boltzmann, Hertz, Planck. But the most famous theoretical physicist in history was Albert Einstein, and he arrived on the intellectual stage with a bang. Of all the scientific journals in the world, the single most sought-after collector's item by far is the *Annalen der Physik*, volume XVII, for 1905, for in that year Einstein published not one but three papers in the journal, causing 1905 to be dubbed the annus mirabilis of science.

Einstein was born in Ulm, between Stuttgart and Munich, on March 14, 1879.[25] Hermann, his father, was an electrical engineer. Unhappy at school, Albert hated the autocratic atmosphere just as he hated the crude nationalism and vicious anti-Semitism. He argued incessantly with his fellow pupils and teachers, to the point where he was expelled. When he was sixteen, he moved with his parents to Milan, attended the University of Zurich at nineteen, and later found a job as a patent officer in Bern. Half educated and half-in and half-out of academic life, he began in 1901 to publish scientific papers.[26]

The first were unremarkable. Einstein did not, after all, have access to the latest scientific literature and either repeated or misunderstood other people's work. However, one of his specialities was statistical techniques—Ludwig Boltzmann's methods—and this stood him in good stead. More important still perhaps, the fact that he was outside the mainstream of science may have helped his originality, which flourished suddenly in 1905. Einstein's three great papers were published in March, on quantum theory, in May, on Brownian motion, and in June, on the special theory of relativity.[27] Though Planck's original paper had caused little stir when it was read to the Berlin Physics Society in December 1900, other scientists soon realized that Planck must be right: his idea explained so much, including the observation that the chemical world is made up of discrete units—the elements. Discrete elements implied fundamental units of matter that were themselves discrete. But at the same time, for years experiments had shown that light behaved as a wave.[28]

In the first part of his paper Einstein, showing early the openness of mind for which physics would become celebrated, made the hitherto unthinkable suggestion that light was *both*, a wave at some times and a particle at others. This idea took some time to be accepted, or even understood, except among physicists, who realized that Einstein's insight fit the available facts. In time the wave-particle duality, as it became known, formed the basis of quantum mechanics in the 1920s.*

Two months after this paper, Einstein published his second great work, on Brownian motion. When suspended in water and inspected under the microscope, small grains of pollen, no more than a hundredth of a millimeter in size, jerk or zigzag backward and forward. Einstein's idea was that this "dance" was owing to the pollen being bombarded by molecules of water hitting them at random. Here his knowledge of statistics paid off, for his complex calculations were borne out by experiment. This is generally regarded as the first proof that molecules exist.

It was Einstein's third paper, on the special theory of relativity, that

* If you have difficulty visualizing something that is both a particle and a wave, you are in good company. We are dealing here with qualities that are essentially mathematical, and all visual analogies will be inadequate. Niels Bohr, a Dane who was arguably one of the twentieth century's top two physicists, said that anyone who wasn't made "dizzy" by the very idea of what later physicists called "quantum weirdness" had lost the plot.

would make him famous.[29] It was this theory that led to his conclusion that $E = mc^2$. It is not easy to explain the special theory of relativity (the general theory came later) because it deals with extreme—yet fundamental—circumstances in the universe, where common sense breaks down. But a thought experiment might help. Imagine you are standing at a railway station when a train hurtles through from left to right. At the precise moment that someone else on the train passes you, a light on the train, in the middle of the carriage, is switched on. Now, if you assume that the train is transparent so you can see inside, you, as the observer on the platform, will see that by the time the light beam reaches the back of the carriage, the carriage will have moved forward. In other words, the light beam has traveled slightly less than half the carriage. However, the person inside the train will see the light beam hitting the back of the carriage at the same time as it hits the front of the carriage. Thus the time the light beam takes to reach the back of the carriage is different for the two observers. The discrepancy, Einstein said, can only be explained by assuming that the perception is relative to the observer and that, because the speed of light is constant, time must change according to circumstance. His most famous prediction was that clocks would move more slowly when traveling at high speeds. This anti-commonsense notion was actually borne out by experiment many years later. Physics was transformed.[30]

The Secret of Continuity and the Meaning of "Between"

In the late nineteenth century, Germany bred an extraordinary generation of "pure" mathematicians who were very concerned with ideas that, though extremely theoretical to begin with, would eventually prove fundamental and practical in equal measure.[31] Together with Planck, but operating from a very different starting point, they conceived the basis of what would, in the future, become the digital revolution.

As we have seen, Carl Friedrich Gauss, Bernhard Riemann, and Felix Klein had helped established Göttingen as the world capital of mathematics, though other German university towns—Heidelberg, Halle, and Jena—were close seconds. In these small, remote, and self-contained worlds, away from the teeming metropolises, mathematicians' minds were free and cleared to explore basic issues. And for many, number theory was the ultimate abstraction.

Richard Dedekind, born in 1831, was one of Gauss's last students at Göttingen and a pallbearer at his funeral.[32] Dedekind was a dedicated academic who never married and spent much of his time editing for publication one or two contributions by Gauss and the bulk of the papers by his other great teacher, Peter Lejeune Dirichlet, on differentiable functions and trigonometric series (Dirichlet, he liked to say, had made "a new man" of him).[33] This exercise had given Dedekind some ideas of his own, and though it was no more than a pamphlet when it was published in 1872, *Continuity and Irrational Numbers* soon became a classic, the best description to date of what mathematicians call the "numerical continuum" or the secret of continuity.

The "secret" of continuity is one of those issues that troubles mathematicians if no one else (though of course it was theoretically linked to quantum theory, which advocated that energy was not continuously emitted, as Newton had said). The problem of continuity becomes apparent once you try to grasp what it means to be "between." As early as the sixth century B.C., Pythagoras knew that fractions lay "between" whole numbers. Then irrational numbers quashed this thinking, their decimal representation just went on and on and on. The problem of "between" now had to be restated. If irrational numbers lay between whole numbers and rational fractions, how many numbers were there between, say, 0 and 1? More perplexing still, there seemed to be just as many numbers between 0 and 1 as there were between 1 and 1,000. How could that be?[34]

Dedekind's solution was as simple as it was elegant. In mathematical terms, he wrote, "if one could choose one and only one number, *a*, which divided all the others in the interval into two classes, *A* and *B*, such that all numbers in *A* were less than *a*, all in *B* were greater than *a*, while *a* itself could be assigned to either class, then the interval was continuous by definition." Dedekind had defined (the numerical meaning of) continuity by removing the concept of "between."

The concept of "between" borders on the philosophical, and brings to mind such concepts as "before" and "beyond," which had troubled Kant. This, in a sense, is what interested Dedekind's colleague, Georg Cantor.

Just as Dedekind had studied with Gauss, so Cantor had been a pupil of Karl Weierstrass in Berlin at least until 1866. Born in 1845 into a devout Lutheran family, Cantor was very interested in metaphysics and believed that when he made his major discovery—infinite cardinal numbers—they had been revealed to him by God.[35] A manic-depressive, he ended his days in an asylum, but between 1872 and 1897 he created the theory of sets and the arithmetic of infinite numbers.[36]

The paper that started it was entitled "On the Consequences of a Theorem in the Theory of Trigonometric Series." In this paper, with its jaw-breaking title, Cantor made the concept of "set" one of the most interesting terms in both mathematics and philosophy (he is generally regarded as the founder of set theory).[37] But it was his next step that took mathematicians by surprise (though in truth it was also a surprise that no one had noticed this before). The series, 1, 2, 3 . . . *n*, was an infinite set and so was 2, 4, 6 . . . *n*. But it followed from this that some infinite sets were larger than others—there are more integers in the infinite series, 1, 2, 3 . . . *n* than in 2, 4, 6 . . . *n*. Next came Cantor's proof that the infinite number of points on a line segment is equal to the infinite number of points in a plane figure. "I see it, but I don't believe it!" he wrote to Dedekind on June 29. It is as well to remember this sentence.

Not everyone thought that the revolution was anything of the kind. From Berlin, Cantor's former professor, Leopold Kronecker, attacked the new ideas, as did Hermann von Helmholtz and even Friedrich Nietzsche, who agreed that numbers, though necessary, were "fiction." By now both Gottlob Frege at Jena, upriver from Cantor at Halle, and the Italian Giuseppe Peano, were also grappling with the nature of number. Frege's answer was less complicated than Peano's (and Dedekind's) but used a special notation he had himself devised.[38]

Born in 1848, Frege is now known for two fundamental works, the *Begriffsschrift* of 1879 and *Die Grundlagen der Arithmetik* (*Foundations of Arithmetic*) of 1884, in which his basic idea was that language described logic much as mathematics does and that by comparing them, the essential elements of logic would become clear. This was an approach that interested another of Weierstrass's students, Edmund Husserl, of Halle, whose doctoral dissertation, *Über den Begriff der Zahl* (*On the Concept of Number*), was followed in 1891 with his ambitiously titled *Philosophie der*

Arithmetik (*Philosophy of Arithmetic*). Having used Frege's *Foundations of Arithmetic* in his own work, Husserl sent the *Philosophy* to Frege as a mark of respect.[39] Instead of trying to define a set mathematically, Husserl asked how the *mind* forms generalities—to convert multiplications into units in the first place. In other words, it was a philosophical or epistemological problem before it was a mathematical one. And he gave a Kantian answer. The continuum of real numbers, Husserl said, could never be made present to consciousness. Continuity was like space or time, or infinity, a creation of our minds. This was too much for Frege, who dismissed *Philosophy of Arithmetic* as a "devastation."[40]

The youngest of the second great generation of German mathematicians, David Hilbert (1862–1943), was in the Frege/Husserl mold in that he viewed mathematics philosophically. But he was also in the Cantor/Dedekind mold, being equally interested in the mathematics of sets.

Born in Königsberg, East Prussia, now Kaliningrad, he attended the Collegium Friedrichianum, the school Kant had himself attended 140 years before. Hilbert was a professor at Königsberg until 1895, when Felix Klein lured him to Göttingen. There he became a mentor to many other subsequently famous mathematicians, including Hermann Weyl, Richard Courant, and John von Neumann.[41]

Hilbert was interested in number and in the difference between intuition and logic. He thought that certain aspects of number (for example, order and some sets) were intuitive, and he wanted to define where logic took over from intuition. He became best known for his "exceptional" identification of twenty-three unsolved problems of mathematics, which he presented at the International Congress of Mathematicians in Paris in 1900, although these were, he said, just a sample of problems that remained to be discovered.[42] Later he became interested in what he called "infinite dimensional Euclidean space," later called a "Hilbert space," and he worked with Einstein on the final form of General Relativity, the so-called Einstein-Hilbert action.

Physics and mathematics had, in a sense, a conceptual overlap, both being concerned with the nature of continuity and particularity. That concern would help produce dividends in the digital revolution—but decades in the future.

26.

Sensibility and Sensuality in Vienna

In early September 1887, Arthur Schnitzler—physician, writer, amateur pianist—was out for one of his regular strolls in Vienna (he was such a familiar figure, says Clive James, that he was "practically part of the Ringstrasse"), when he encountered an attractive young woman who called herself "Jeanette." Socially they were poles apart. He was from the well-educated bourgeoisie, she was an embroiderer. Nonetheless, two days later she visited his rooms and they became lovers. Over the next months Schnitzler recorded in his diary every sexual encounter that took place. When the affair ended, acrimoniously, at the end of 1889, he was able to calculate that in the intervening two years they had made love 583 times. The exactitude was remarkable, but so was Schnitzler's potency: he had been away from the city many times and on occasions, to keep up his tally, he and Jeanette had performed five times a night.[1]

This admixture of scientific certitude, bravado, experimentation, and sexual license (in a pre-Salvarsan world) illustrates as well as anything the miasma of ideas swirling around in Vienna at the close of the nineteenth century. And in 1900, among German-speaking cities, Vienna certainly took precedence. If one place could be said to represent the mentality of continental Europe as the twentieth century began, it was the capital of the Austro-Hungarian Empire.

The Original Café Society

The architecture of Vienna played a crucial role in determining the unique character of the city. The Ringstrasse, a circle of monumental buildings that included the university, the opera house, and the parliament building, had been erected in the second half of the nineteenth century around the central area of the old town, in effect enclosing the intellectual and cultural life of the city inside a relatively small and accessible area. There had emerged the city's distinctive coffeehouses, an informal institution that helped make Vienna different from London, Paris, or Berlin.[2] The marble-topped tables were just as much a platform for new ideas as the newspapers, academic journals, and books of the day. These cafés were reputed to have had their origins in the discovery of vast stocks of coffee in the camps abandoned by the Turks after their siege of Vienna in 1683. Whatever the truth of that, by 1900 Viennese cafés had evolved into informal clubs where the purchase of a small cup of coffee carried with it the right to remain there for the rest of the day and to have delivered, every half hour, a glass of water on a silver tray. Newspapers, magazines, billiard tables, and chess sets were provided free of charge, as were pen, ink, and (headed) writing paper. Regulars could have their mail sent to them at their favorite coffeehouse; at some establishments, such as the Café Griensteidl, large encyclopedias and other reference books were kept on hand for writers who worked at their tables.[3]

Leadership as an Art Form

A group of bohemians who gathered at the Café Griensteidl was known as Jung Wien (Young Vienna). This group included Schnitzler; Hugo von Hofmannsthal; Theodor Herzl, a brilliant reporter, an essayist, and later a leader of the Zionist movement; Stefan Zweig, a writer; and their leader, the newspaper editor Hermann Bahr. His paper, *Die Zeit*, was the forum for many of these talents, as was *Die Fackel* (*The Torch*), edited no less brilliantly by Karl Kraus, more famous for his play *Die letzten Tage der Menschheit* (*The Last Days of Mankind*).[4]

The career of Arthur Schnitzler (1862–1931) shared a number of intriguing parallels with that of Freud. He too trained as a doctor and neurologist and studied neurasthenia. But Schnitzler turned away from medicine to literature, though his writings reflected many psychoanalytic concepts (he thought love affairs provided an education). His early work explored the emptiness of café society, but it was with the story "Lieutenant Gustl" (1901) and the novel *Der Weg ins Freie* (*The Road into the Open*; 1908) that Schnitzler really made his mark. "Lieutenant Gustl," a sustained interior monologue, takes as its starting point an episode when a "vulgar civilian" dares to touch the lieutenant's sword in the busy cloakroom of the opera. This gesture provokes in the lieutenant confused and involuntary "stream-of-consciousness ramblings" that in some ways prefigure Proust. In *The Road into the Open*, the dramatic structure of the book takes its power from an examination of the way the careers of several Jewish characters have been blocked or frustrated. Schnitzler indicts anti-Semitism, not simply for being wrong, but as the symbol of a new, illiberal culture brought about by a decadent aestheticism and by the arrival of mass society which, together with a parliament "[that] has become a mere theatre through which the masses are manipulated," gives full rein to the instincts, and which in the novel overwhelms the "purposive, moral and scientific" culture represented by many of the Jewish characters. Schnitzler was a committed realist who thought, for example, that "the battle between imagination and fidelity" was "a fact of life."[5]

Hugo von Hofmannsthal (1874–1929) went further. Born into an aristocratic family, his father introduced his son to the Café Griensteidl set when Hugo was quite young, so that the group around Bahr acted as a forcing house for the youth's precocious talents. In the early part of his career, Hofmannsthal produced what has been described as "the most polished achievement in the history of German poetry," but he was never totally comfortable with the aesthetic attitude and, noting the encroachment of science on the old aesthetic culture of Vienna, he wrote in 1905, "The nature of our epoch is multiplicity and indeterminacy. It can rest only on *das Gleitende* [the slipping, the sliding]." Could there be a better description about the way the Newtonian world was slipping after Boltzmann's and Planck's discoveries? "Everything fell into parts," Hofmannsthal wrote, "the parts again into more parts, and nothing allowed itself to be embraced by concepts any more."[6] Like

Schnitzler, Hofmannsthal was disturbed by political developments in the dual monarchy and in particular the growth of anti-Semitism. For him, this rise in irrationalism owed some of its force to science-induced changes in the understanding of reality; the new ideas were so disturbing as to promote a large-scale reactionary irrationalism. He therefore abandoned poetry at the grand age of twenty-six, feeling that the theater offered a better chance of meeting current challenges. Hofmannsthal came to believe that (in the Greek manner) theater could help to counteract political developments. His works, from the plays *Fortunatus and His Sons* (1900–01) and *King Candaules* (1903) to his librettos for Richard Strauss, are all about political leadership as an art form, the point of kings being to preserve order and control irrationality. Yet the irrational must be given an outlet, Hofmannsthal says, and his solution is "the ceremony of the whole," a ritual form of politics in which no one feels excluded.[7] His plays are attempts to create ceremonies of the whole, marrying individual psychology to group psychology in dramas that anticipate Freud's theories. As he put it, the arts had become "the spiritual space of the nation." Hofmannsthal always hoped that his writings about kings would help Vienna throw up a great leader, someone who would offer moral guidance and show the way ahead. The words he used were uncannily close to what eventually came to pass. What he hoped for was a "genius . . . marked with the stigma of the usurper," "a true German and absolute man," a "prophet," "poet," "teacher," "seducer," an "erotic dreamer."[8]

Just as Hofmannsthal's aesthetics of kingship and "ceremonies of the whole" were a response to *das Gleitende*, induced by scientific discoveries, so too was the new philosophy of Franz Clemens Brentano (1838–1917). Brentano was a popular man, with students—among them Freud and Tomáš Masaryk—crowding his lectures. A statuesque figure, he was frequently to be seen swimming the Danube. He published a best-selling book of riddles. Josef Breuer was his doctor.

Brentano's main interest was to show, in as scientific a way as possible, proof of God's existence. For him, philosophy went in cycles and there had been three—ancient, medieval, and modern—each divided into four phases: investigation, application, skepticism, and mysticism. These he laid out in the following grid:

	CYCLES		
PHASES	*Ancient*	*Medieval*	*Modern*
Investigation	Thales to Aristotle	Thomas Aquinas	Bacon to Locke
Application	Stoics, Epicureans	Duns Scotus	Enlightenment
Skepticism	Skeptics, Eclectics	William of Occam	Hume
Mysticism	Neoplatonists, neo-Pythagoreans	Lullus, Cusanus	German Idealism

This approach helped make Brentano a classic halfway figure in intellectual history. His science led him to conclude, after twenty years of search, that there does indeed exist "an eternal, creating and sustaining principle," which he called "understanding" (an echo of Kant). At the same time, his view that philosophy moved in cycles led him to doubt the progressivism of science. Despite this, his approach did spark two other branches of philosophy that were themselves influential in the early years of the twentieth century, Edmund Husserl's phenomenology and Christian von Ehrenfels's theory of *Gestalt*.[9]

Husserl (1859–1938) was born in the same year as Freud and in the same province, Moravia, as both Freud and Mendel. Like Freud he was Jewish, but he had a more cosmopolitan education, studying at Berlin, Leipzig, and Vienna. His first interests—as we have seen earlier—were in mathematics and in logic, but he found himself drawn to psychology. In those days, in the German-speaking countries psychology was usually taught as an aspect of philosophy, but it was growing fast as its own discipline, thanks in particular to the laboratory psychology pioneered by Wilhelm Wundt. Wundt (1832–1920), a prolific professor at Leipzig—his works run to 53,000 pages—had shared a laboratory with Helmholtz and tried all his life to introduce the methodology of experimental physiology into psychology. Contrary to Dilthey, Wundt firmly believed that our psychology would eventually be explained by small physical events, such as reflex arcs, observed in the laboratory.[10]

Husserl, who attended some of Wundt's lectures, is best understood as a post-Kantian, post-Darwinian, post-Nietzschean—and therefore post-Christian—philosopher, whose concern was to understand the phenomenon of existence, of being, in a nonreligious way. The key concepts for him, therefore, were consciousness, logic, and language. How are we to understand the phenomena *of* the world and the phenomena *in* the world? Through the mind we are conscious, the central psychological phenomenon of our existence; to what extent are the phenomena available to us through consciousness real—i.e., independent of our mind, our consciousness? Or do our minds in some way "intend" these phenomena? Does an apparently straightforward phenomenon like logic exist "out there" in the world, or is logic an "intention" of the mind? And how does all this relate to our use of and understanding of language? Is language an accurate reflection/description of phenomena and if so, how does the analysis of language help in understanding the world?

Husserl's big book on the subject, *Logische Untersuchungen* (*Logical Investigations*), was published in 1900 (volume one) and 1901 (volume two).[11] His main conclusion can be characterized as an updated Idealism, that there is a "tendency" of the mind to organize experience, to order consciousness. Husserl was not a great stylist, it has to be said, and many—especially in the Anglophone world—have difficulty following him. But one accessible way of approximating what he was getting at, in an admittedly simple and basic way, is via the well-known visual illusion that may be seen either as a candlestick (in black), or two faces opposing each other (in white). The fact that we switch—almost involuntarily—between these two perceptions means, for Husserl, that there is some organizing principle in our consciousness that can determine—or help determine—how we experience the world.

Husserl was fascinated by how one individual both changes and stays the same over time: what does it mean to have a continuous identity, and what does it mean to be part of a whole? He was convinced that there are entire areas of being, of consciousness, that science can never address, even in principle, and in this Husserl (who left a vast archive) is best understood now as the immediate father of the so-called continental school of twentieth-century Western philosophy, whose members were to include Martin Heidegger, Jean-Paul Sartre, and Jürgen Habermas. They stand in contrast to the "analytic" school begun by Bertrand Russell and Ludwig Wittgenstein, which is more popular in North America and Great Britain.[12]

Brentano's other notable legatee was Christian von Ehrenfels (1859–1932), the father of Gestalt philosophy and psychology. In 1897 Ehrenfels accepted a post as professor of philosophy at Prague. Here, starting with Ernst Mach's observation that the size and color of a circle can be varied "without detracting from its circularity," Ehrenfels modified Brentano's ideas, arguing that the mind somehow "intends Gestalt qualities"—that is to say, there are certain "wholes" in nature that the mind and the nervous system are "*pre*pared," *pre*disposed, to experience. Gestalt theory became very influential in German psychology for a time, and although in itself it led nowhere, it did lay the groundwork for the theory of "imprinting," a readiness in the neonate to perceive certain forms at a crucial stage in development.

The Pathologies of Science

Also prevalent in Vienna at the time were a number of avowedly rational but in reality frankly scientistic ideas, and they too read oddly now. Chief among these were the theories of Otto Weininger (1880–1903).

The son of an anti-Semitic but Jewish goldsmith, Weininger developed into an "overbearing coffee house dandy." He had a tendency to be withdrawn and taught himself more than eight languages before he left the university and published his undergraduate thesis.[13] Renamed *Geschlecht und Charakter* (*Sex and Character*) by his editor, the thesis was released in 1903 and became a huge hit. The book was rabidly anti-Semitic and extravagantly misogynist, Weininger putting forward the view that all human behavior can be explained in terms of male and female "protoplasm," which contributes to each person. A whole lexicon of neologisms was invented by Weininger to explain his ideas: idioplasm, for example, was his name for sexually undifferentiated tissue; male tissue was arrhenoplasm; and female tissue was thelyplasm. According to him, all the main achievements in history arose because of the masculine principle—all art, literature, and systems of law, for example. The feminine principle, on the other hand, accounted for the negative elements, and all these negative elements converge, Weininger says, in the Jewish race. Commercial success and fame did not settle Weininger's restless spirit. Later that year he rented a room in the house in Vienna where Beethoven died and shot himself. ("In a city that considered suicide an art, Weininger's was a masterpiece.")[14] He was twenty-three.

A rather better scientist, no less interested in sex, and the emergence of "sexual science," was the Catholic psychiatrist, Richard von Krafft-Ebing (1840–1902). His fame stemmed from a work he published in Latin in 1886, titled (in German) *Psychopathia Sexualis: eine klinische-forensische Studie*, quickly translated into seven languages. Most of the "clinical-forensic" case histories were drawn from courtroom records and attempted to link sexual psychopathology either to married life, to themes in art, or to the structure of organized religion. The most infamous "deviation," on which the notoriety of his study rests, was his coining of the term "masochism." The word was derived from the novels and novellas of Leopold von Sacher-Masoch, the son of a police director in Graz. In the most explicit of his stories, *Venus im Pelz*, Sacher-Masoch describes his own affair with a Baroness Fanny Pistor, during the course of which he "signed a contract to submit for six months to being her slave."[15]

Psychopathia Sexualis clearly foreshadowed some aspects of psychoanalysis. Krafft-Ebing acknowledged that sex, like religion, could be sublimated in art—both could "enflame the imagination."[16] For Krafft-Ebing, sex within religion (and therefore within marriage) offered the possibility of "rapture through submission," and it was this process in perverted form that he regarded as the etiology for the pathology of masochism.[17]

"Design Is Inferior to Art"

The dominant architecture in Vienna was the Ringstrasse. Begun in the mid-nineteenth century, after Emperor Franz Joseph ordered the demolition of the old city ramparts and a huge swath of space was cleared in a ring around the center, a dozen monumental buildings were erected in this ring over the following fifty years. They included the Opera, the Parliament, the Town Hall, parts of the university, and an enormous church. Most were embellished with fancy stone decorations, and it was this ornateness that provoked a reaction, first in Otto Wagner, then in Adolf Loos.

Otto Wagner (1841–1918) won fame for his "Beardsleyan imagination" when he was awarded a commission in 1894 to build the Vienna underground railway.[18] This meant the construction of more than thirty stations, plus bridges, viaducts, and other urban structures. Wagner broke new ground by not only using modern materials but *showing* them. For example, he made a feature of the iron girders in the construction of bridges. These supporting structures were no longer hidden by elaborate casings of masonry, in the manner of the Ringstrasse, but painted and left exposed. His other designs embodied the idea that the modern individual—living his or her life in a city—is always in a hurry, anxious to be on his or her way to work or home. The core structure therefore became the street, rather than the square or vista or plaza. For Wagner, Viennese streets should be straight and direct; neighborhoods should be organized so that workplaces are close to homes, and each neighborhood should have its own center, rather than there being just one center for the entire city.

Adolf Loos (1870–1933) was close to Freud and to Karl Kraus, editor of *Die Fackel*, and the rest of the Café Griensteidl set, and his rationalism was more revolutionary than Wagner's—he was *against* the Zeitgeist.[19] Architecture, he declared, was not art. "The work of art is the private affair of the artist. The work of art wants to shake people out of their comfortableness [*Bequemlichkeit*]. The house must serve comfort. The art work is revolutionary, the house conservative." Loos, who had lived in Chicago, extended this perception to design, clothing, even manners.[20] He was in favor of simplicity, functionality, plainness. He thought men risked being enslaved by material culture, and he wanted to reestablish a "proper" relationship between art and life. Design was inferior to art because it was conservative, and when he understood the difference, man would be liberated. "The artisan produces objects for use here and now, the artist for all men everywhere."[21]

Weininger especially, but Loos too, was carried away with rationalism. Both adopted scientistic ideas, but quickly went beyond the evidence to construct systems as fanciful as the nonscientific ideas they disparaged.

Nothing better illustrates this divided and divisive way of looking at the world in turn-of-the-century Vienna than the fight over Gustav Klimt's paintings for the university, the first of which was delivered in 1900. Klimt, born in Baumgarten, near Vienna, in 1862, was, like Weininger, the son of a goldsmith, but there the similarity ended. Klimt made his name decorating the new buildings of the Ringstrasse with vast murals.[22] These were produced with his brother Ernst, but on the latter's death in 1892 Gustav withdrew for five years, during which time he appears to have studied the works of James Whistler, Aubrey Beardsley, and Edvard Munch. Klimt did not reappear until 1897, when he emerged at the head of the Vienna Secession, a band of nineteen artists who, like the Impressionists in Paris and other artists at the Munich and Berlin Secessions (see Chapter 27, p. 503), eschewed the official style of art and instead followed their own version of art nouveau. (In the German-speaking lands this is known as *Jugendstil.*)

Klimt's style, bold and intimate at the same time (as photos show the man himself to have been), had three defining characteristics—the elaborate use of gold leaf (a technique learned from his father), the application of small flecks of iridescent color, hard like enamel, and a languid eroticism applied in particular to women. Klimt's paintings were not quite Freudian: his women were not neurotic, far from it. They were calm, placid, above all lubricious, "the instinctual life frozen in art." Nevertheless, in drawing attention to women's sensuality, Klimt hinted that it had hitherto gone unsatisfied. His women were presented as insatiable—here were women capable of the perversions reported in Krafft-Ebing's book, tantalizing and shocking at the same time. Klimt's new style immediately divided Vienna, but it quickly culminated in his commission for the university.[23]

Three large panels were asked for: *Philosophy*, *Medicine,* and *Jurisprudence.* All three provoked a furor but the rows over *Philosophy* came first. For this picture, the commission stipulated as a theme "The Triumph of Light over Darkness." What Klimt actually produced was a "deliquescent tangle" of bodies that appear to drift past the onlooker, a kaleidoscope of forms that run into each other, and all surrounded by a void. The professors were outraged, and Klimt was vilified as presenting "unclear ideas through unclear forms." Philosophy was supposed to be a rational affair; it "sought the truth via the exact sciences." Eighty scholars petitioned that Klimt's picture never be shown at the university. The painter returned his fee and never presented the remaining commissions. The significance of the fight is that it brings us back to Hofmannsthal and Schnitzler, to Husserl and Brentano. For in the university commission, Klimt was attempting a major statement. How can rationalism succeed, he is asking, when the irrational, the instinctive, is such a dominant part of life? Is reason really the way forward? Instinct is an older, more powerful force. It may be more atavistic, more primitive, a dark force at times, but where is the profit in denying it? This remained an important strand in German thought until World War II.

* * *

Theodor Herzl (1860–1904) was a Hungarian Jew who studied law at Vienna. As a pupil in his Gymnasium he had written a poem praising Luther as a champion of Germany and in Vienna he helped organize a German-National student group. A handsome man who wrote comedies, he became more successful as a journalist in the 1880s, contributing feuilletons to the *Wiener Allgemeine Zeitung* and the *Neue Freie Presse*. (The feuilleton was originally a French idea, an article—often below the fold on the front page of a newspaper—that eschewed "hard" news in favor of wittily written comment, bouncing off the news, posing awkward questions about what was, and was not, being revealed.) The crucial change in Herzl's life came in 1891 when he became the Paris correspondent of the *Neue Freie Presse* and arrived in France to witness a tide of economic anti-Semitism brought about by the Panama scandal.[24] Three years later, in 1894, Herzl was aghast when France formed an alliance with Russia at the very time pogroms were killing thousands of Ukrainian Jews. It was the indifference—not just of the West, but of Western Jews—to the fate of the eastern Europeans that caused him, in 1895, to publish his proposal for a Jewish state. Although he continued with journalism and wrote yet more plays, the rest of his career was devoted to realizing this one idea, that the governments of Europe should grant to a Jewish stock company sovereignty over a part of the colonial territory under their control to be turned into a refuge for any Jews who wished to take advantage of it. From 1896 until his death in 1904, Herzl organized a series of six world congresses of Jewry; among their aims was to persuade the sultan of the Ottoman Empire to release part of Palestine for the purpose of establishing a Jewish state. Failing that, Herzl would have accepted areas in Africa or Argentina (which many of his supporters would *not* accept).

Herzl knew he probably would not live to see his dream realized but never had any doubt that, one day, his vision would come about (this is evident in his copious correspondence). When he died, the Zionist bank in London, the Jewish Colonial Trust, had 135,000 shareholders, "then the largest number financing any enterprise in the world." More than 10,000 Jews from all over Europe attended his funeral in 1904.

The Primacy of Physics and Psychology

At the same time, there was in Vienna a strain of thought that was wholly scientific and frankly reductionist. The most ardent, and by far the most influential reductionist in Vienna was Ernst Mach (1838–1916). Born near Brünn, where Mendel had outlined his theories, Mach, a precocious and difficult child who questioned everything, studied mathematics and physics in Vienna. He made two major discoveries. Simultaneously with Breuer, but entirely independently, he discovered the importance of the semicircular canals in the inner ear for bodily equilibrium.[25] Second, using a special technique, he made photographs of bullets traveling at more than the speed of sound. In the process, he discovered that they create not one but two shockwaves, one at the front and another at the rear, as a result of the vacuum their high speed creates. This became particularly significant after World War II with the arrival of jet aircraft that approached the speed of sound, and is why supersonic speeds (on the Concorde, for example) were given in terms of a "Mach number."[26]

After these achievements, however, Mach became more and more interested in the philosophy and history of science. Implacably opposed to metaphysics of any kind, he dismissed as worthless concepts such as God, nature, soul, and "ego." All knowledge, Mach insisted, could be reduced to sensation, and the task of science was to describe sense data in the simplest and most neutral manner. This meant that for him the primary sciences were physics, "which provide the raw material for sensations," and psychology, by means of which we are aware of our sensations. For Mach, philosophy had no existence apart from science. An examination of the history of scientific ideas showed, he argued, how these ideas evolved. He firmly believed that there is evolution in ideas, with the survival of the fittest, and that we develop ideas in order to survive. For Mach, therefore, it made less sense to talk about the truth or falsity of theories than to talk of their usefulness. Truth, as an eternal, unchanging thing that just *is*, for him made no sense. The Vienna Circle was founded in response as much to his ideas as to Wittgenstein's.

The "Aryan Deficit" in Culture

All this was the Vienna that the young Adolf Hitler arrived in from Linz, where he had grown up, in 1907. It was bewildering. Brigitte Hamann tells us that in 1907 Vienna had 1,458 automobiles, which caused 350 accidents a year (overshadowed by the 980 accidents caused by horse-drawn carriages).[27] The Westbahnhof, the station where Hitler arrived, was lit by electric light, as were the city's ten inner districts. There were great battles in the newspapers of the time about the merits of modernism. For the opponents of modernism, the term "degenerate" (*entartet*) was a favorite form of abuse. Modernism in Vienna, at the time Hitler was there, was often referred to as "Jewish modernism," though this was clearly not true—Klimt, Oskar Kokoschka, Alban Berg, Otto Wagner, and Adolf Loos were not Jewish but the label suited the opponents of innovation. According to August Kubizek, it was in Vienna that Hitler began to reflect on how the "Aryans'" obvious educational deficit and lack of interest in culture could be reduced."[28]

The two best-known anti-Semites in Vienna (though there was no shortage) were George Schönerer, the leader of the Pan-Germans (see Chapter 22, p. 417), who lost his seat in the Austrian Parliament the year Hitler arrived, and Karl Lueger. The Pan-Germans pledged allegiance to their "Führer," sang "Schönerer songs," and wrote poetry in his praise. Full-page advertisements were taken out in the newspapers, with "HEIL TO THE FÜHRER" in the headlines.[29] Schönerer's early fight was against Russian Jews, fleeing the pogroms of the tsars, and he made a point of appropriating Wagner to the anti-Semitic cause.

Dr. Karl Lueger was Schönerer's archenemy, and their followers often feuded. Hitler, however, was impressed by Lueger even while he was a follower of Schönerer.[30] Lueger had been mayor of Vienna for ten years by the time Hitler arrived there. A handsome man, fond of the mayoral chain, he had, to his credit, masterminded the modernization of his city with efficiency and charisma, often attacking the local merchants for profiteering from their customers. "He knew how to turn disputes over such personal matters as milk prices and refuse disposal to his advantage."[31] And he raised anti-Semitism to an art form, becoming a superb, demagogic mass orator: the Jews, he insisted time and again, were to blame for most of the misfortunes of the Viennese (the Jewish population of the city had risen from 2,000 in 1860 to 175,300 in 1910).[32]

We must be careful, however, not to ascribe all of Hitler's characteristics to his time in Vienna. The Austrian capital was a sophisticated, cosmopolitan, multifaceted city which could be—and was—enjoyed in many different ways. The tumult of war and the divided and divisive landscape of the Weimar Republic still lay ahead. Still, there is no escaping the fact that Vienna in the first decade of the twentieth century provided the young Hitler with a whole raft of experiences that, almost certainly, he would never have had elsewhere.

27.

Munich/Schwabing: Germany's "Montmartre"

"Munich was radiant . . . Young artists with little round hats on their heads . . . carefree bachelors who paid for their lodgings with colour-sketches . . . Little shops that sold picture-frames, sculptures, and antiques there were in endless number . . . the owners of the smallest and meanest of these shops spoke of Mino da Fiesole and Donatello as though he had received the rights of reproduction from them personally . . . You might see a carriage rolling up the Ludwigstrasse, with such a great painter and his mistress inside. People would be pointing out the sight . . . Some of them would curtsy."

This is Thomas Mann, in "Gladius Dei," a story published in 1902 that in part compares contemporary Munich with quattrocento Florence.[1] Mann was himself just one of the artists drawn to Munich, where the beer was famous, the architecture and landscaping incomparable, the opera, theater, and university likewise renowned and where, the breweries apart, there was no real industry and the poverty associated with it.

Elsewhere in his work Mann was not always so positive about Munich's position but there is no question that, at the turn of the century, the arts community was integral to the city. A municipal committee set up in 1892 to inquire into the Sezessionist Controversy (which we shall come to) confirmed that "Munich owes its outstanding importance among

German cities to *art and artists*, however highly one may rate other factors in its development."[2] The poet Erich Mühsam described Schwabing, the cultural quarter of Munich, as Germany's "Montmartre." The Café Stephanie, known as Café Megalomania, was where the poets and artists met, played chess, borrowed money, and tried not to lust after Lotte Pritzel, "the most endearing amoralist ever known."[3]

The other jewel in some ways was the Neue Pinakothek, which Ludwig I (1786–1868) built for his collection of contemporary art (even today it houses only art produced since 1800). At his death in 1868 nearly half of his pictures were by non-German artists. Munich also boasted the Royal Academy of Fine Arts, "by mid-century the pre-eminent teaching institution in Central Europe." This drew students from all over the world. "An entire generation of American realists studied in Munich in the 1870s, as did many Scandinavian, Russian and Polish students."[4]

A final aspect of the Munich art world that was unique was its exhibition space. Unlike other cities in Germany, by mid-century Munich had two buildings large enough to host shows of significant size. The first was the Kunst- und Industrieausstellungsgebäude, used for exhibitions of local art, and second the Glaspalast, designed by August Voigt to house the Allgemeine Deutsche Industrie-Ausstellung (German industrial exhibition) of 1854 and subsequently used for major shows of both historical and contemporary art, including foreign artists.

These exhibitions were mounted by the Munich Artists Association (MAA), established in 1868, which held shows of both German and foreign artists every three years. To begin with they were insignificant affairs, but after it was granted a royal charter, the association began to use some very imaginative techniques to interest the general public in art. One device was a torchlight parade in which 800 artists walked through the streets of Munich behind a horse-drawn float showing four allegorical figures of genius. In 1892, just before the Munich Sezession occurred (see p. 505), the association consisted of 1,020 artists and was an eclectic mix of established—even famous—artists, together with their much less successful colleagues, and students. The association's regulations were strict, limiting the number of paintings any one artist could exhibit in their shows to three, but the exhibitions were marked by the same imaginative devices as the torchlight parade. Beer gardens were installed in the exhibitions, and lotteries were arranged with paintings and drawings as prizes.[5]

Following 1871, says Maria Makela, Munich entered a golden age for the arts. Besides its lower cost of living, Munich was seen as a more relaxed city than most, with looser morals, benefiting from its proximity to the Alps and Italy, and as a rail crossroads on the way to the Orient. By 1895 some 1,180 painters and sculptors were registered there, 13 percent of the total in Germany, and more than were registered in Berlin (1,159), which had a population four times that of Munich (Dresden had 314 registered artists, Hamburg 280, and Frankfurt 142). The English painter John Lavery, who visited Munich in the 1880s, remarked that painters there "had the status of generals."[6]

Despite these positive factors, by the early 1890s disagreements over the "equal rights for all" exhibition policy at the association were reaching critical dimensions and in February 1892 eleven artists announced they had established an informal club that would pursue its own goals outside the association. None of the eleven at that stage intended to resign from the MAA but their "Sezession" was soon supported by nine other artists, and this made them more optimistic that a rival organization could succeed. They announced their decision to mount their own exhibition and that it would counter the aims of the association, which, they said, had grown too unwieldy in size and was too dominated by mediocre artists. For the city fathers, the Sezession threatened the artistic unity of Munich so it was not universally popular and there was briefly a plan by the Sezessionists to hold exhibitions in Berlin. But a member of the city's board of works agreed to give the new group a plot of his own land for a five-year term. This kick-started the project, and the Sezessionists obtained funding to build their own gallery, an impressive achievement, making the Sezession a powerful presence in Munich even before its first exhibition opened in 1893.[7]

The "Apostle of Ugliness"

The best-known, and for many people the best of the Sezessionists in Munich, was Max Liebermann. Though his family of wealthy cotton manufacturers was Jewish, they might almost have been one of Max

Weber's Protestants in their devotion to hard work and thrifty simplicity (Liebermann's grandmother even did the laundry herself). Max's father naturally wanted his son to go into the family business, so the son received little encouragement when it became clear he wanted to be a painter.

Although he was eventually allowed to attend the Weimar Academy, Liebermann never rebelled against his bourgeois background. Indeed, his lifestyle was so like that of his businessman-father that Gerhart Hauptmann was once moved to remark: "How is it possible for such a philistine to paint such [beautiful] pictures!"[8]

In fact, Liebermann was someone who kept his distance, both in his life and in his art. He announced his approach while he was still a student at Weimar, with *The Goosepluckers*, a large canvas showing women plucking the feathers from geese. At one level this could be seen—and *was* seen—as social comment, working women being exploited to pluck down in order to "warm the affluent." On closer inspection, however, the women are shown enjoying their work, and they have a quiet dignity as they get on with it. The picture provoked widespread criticism, much of which missed the point, critics taking affront at what they took to be references to exploitation, rather than that there is dignity in honest toil.[9]

Between Weimar and the Munich Sezession, Liebermann made trips to the Netherlands and to France. In the former he was very impressed by the humane policies of the Dutch toward orphans and the elderly, and he made a number of pictures showing this side of Dutch life. Here too he focused on the quiet dignity among even the most unfortunate souls, observing that everyone is capable of reflection, thoughtfulness, even peace of mind. Only later did his style become lighter, using broader brushwork and a palette knife to create scumbled passages that gave his work an even lusher appeal. This too was new and, for many, unappealing. Liebermann became known as "the apostle of ugliness."

After his travels in the Netherlands and France, he returned to Germany and settled in Munich. He exhibited at the association but as the 1890s passed, Liebermann's style changed again. He started to collect paintings of the French Impressionists, and their lightness rubbed off on him. He used his distinctive grays more sparingly now, his pictures became more colorful, airy, and light—and he began to concern himself less with the world of the poor and unfortunate and more with the elegant world of the bourgeoisie.[10]

Liebermann eventually settled in Berlin, where he was one of those who helped found the Sezession there and became its president. His "Impressionist" works—of Beer Gardens and parks—lack the bite of his earlier work, but his brilliant technique was just as well suited to the more fashionable world that he now portrayed. Even here he succeeded in keeping his distance. His paintings still contain some sharp observation.[11]

The Other Dachau

Unlike Max Liebermann, Fritz von Uhde's interest in art was supported by his father. President of the Lutheran Church Council in Wolkenburg, Saxony, Uhde's father was himself a part-time painter and was married to the daughter of the general director of the Royal Museums in Dresden. Born in 1848, Udhe was thus encouraged to enter the Dresden Academy. Like Liebermann, he went to the Netherlands to paint *en plein air* to gain first-hand experience of the unusual light the Dutch landscape had to offer. This produced a looser, lighter style, which he used to depict mainly lower-class life in the 1880s. As part of this, Uhde and his fellow artists used the unusual landscape and light of the moorlands northwest of Munich, at Dachau. This served more or less as the German Barbizon, and was widely known as such before it became indelibly linked with the atrocities of Nazi Germany. The marshy topography and watery landscape of Dachau fascinated several other painters besides Uhde, in particular Adolf Hölzel and Ludwig Dill who, though not widely known in the twenty-first century, were both—in their landscapes—using the peculiar climate to grope their way toward semiabstract forms that anticipated what Kandinsky would realize a decade later.[12]

Franz von Stuck was a brooding sensualist from Tettenweis, a village in Bavaria; he was the son of a miller who had little feeling for art and assumed his son would take over the family business. Fortunately for the boy, his mother helped ensure that Franz was sent to a *Kunstgewerbeschule* in Munich where he learned the principles of design and architecture. In his early work he produced prints based on the work of some well-known Austrian and German artists—Max Klinger and Gustav Klimt among them.[13]

Stuck's own style began to emerge, well described as an amalgam of the erotic and the sinister, the naked female torso featuring in such works as *The Hunt* (1883), and *Sin*, which dominated the first Sezession show (in fact, according to Heinrich Voss, fully three-quarters of Stuck's paintings involve the erotic). While this sounds like a lot—while it *is* a lot—it was not so unusual in Europe at that time, when many artists—Fernand Khnopff, Paul Gauguin, Ferdinand Hodler—began to express visually their pent-up frustration with the repressions of "civilized" society.[14]

Stuck has also been seen as one of those leading the way to abstraction, in that the psychological component of his works is achieved by the juxtaposition of horizontal and vertical lines and forms, by strongly contrasting color schemes that, in effect, render the actual figures secondary to the overall effect. Kandinsky, who arrived in Munich in 1896, chose Stuck as one of his teachers, working hard for a year to be accepted into his class.[15]

Jugendstil: Reducing the Ugliness of Modernity

Richard Riemerschmid was born in Munich, studied painting at the Munich Academy, where his early works presented nature—as those of Caspar David Friedrich had done—as a form of religious substitute (trees with halos, for example), landscapes as profane altarpieces. He was condemned as blasphemous.

The crucial episode in his life turned out to be his marriage to the actress Ida Hofmann. After searching for furniture for the marital apartment and failing to find anything suitable, Riemerschmid designed some for himself. He hit upon a style in which the decoration used motifs taken from nature, with flowing lines that recalled leaves and fronds.[16] Others liked what they saw and he received a number of commissions, the motifs of which soon caught on. Other Germans—Bernhard Pankok, Hermann Obrist, and August Endell—all began to design a wide variety of objects (light fixtures, cooking utensils, even clothing), replacing an exclusive concern with the "fine" arts. Like other artists they felt that Germany's rapid industrialization and urbanization was robbing the world of something precious, where even Munich, much less industrial than other *Grossstädte*, had nearly tripled its population, from 154,000 to 415,500 between 1868 and 1896. Their idea was that the disagreeable aspects of modernity could be erased by the arts, that, in the words of Hermann Obrist, the ugliness and misery of modern life might be alleviated, "that life in the future will be less toilsome than now."[17]

The fact that the undulating organic lines of nature might recover, regenerate, *rejuvenate* what was in the process of dying gave rise first to the title of George Hirth's art nouveau journal, *Jugend*, and then to "Jugendstil" for the whole art form. The Munich Sezession played an important role in the dissemination of Jugendstil ideas: in 1899 it mounted its most important show, combining the fine and decorative arts, featuring entire living rooms, dining rooms and bedrooms, with entries from Scotland (Charles Rennie Mackintosh), France (René Lalique) and Russia (Peter Carl Fabergé)—everything from embroidered tablecloths to jewelry to framed mirrors.[18]

One final founding member of the Munich Sezession who soon went his own way, and made a name for himself in doing so, was Peter Behrens. Born in Hamburg in 1868, Behrens studied at that city's school for the applied arts before going on to the Karlsruhe School of Art and the Düsseldorf Art Academy. He was in Munich from 1890, working as a painter and graphic artist, an early advocate of Jugendstil, producing woodcuts, designs for bookbindings and other artifacts. In 1897, together with Hermann Obrist, Richard Riemerschmid, and Bernhard Pankok, he was one of those who founded the Vereinigte Werkstätten für Kunst und Handwerk, which produced handmade utilitarian objects as part of the general approach to reducing the ugliness of everyday life.

Behrens received his big break in 1906 when he secured his first commission from AEG.[19] He shared many friends with Walter Rathenau, and this may account for the commission. He was asked to design advertising material, after which Emil Rathenau chose Behrens as an artistic consultant on a wide range of projects, including the Turbinenhalle in Berlin, one of the first concrete and glass factories, housing for the factory's workers, and a number of electrical appliances, standardizing their components so as to make them interchangeable. He designed salesrooms, sale catalogs, even price lists, famously creating for the first time a "corporate image" for the company, which gave it an immediately recognizable identity.

A year later, together with Peter Bruckmann, Fritz Schumacher, and Richard Riemerschmid, he founded yet another organization, the Deutscher Werkbund.[20] This took its color from the British Arts and Crafts movement, and the aim was to produce everyday objects with standardized interchangeable parts that would be within reach of everyone's pocket, but of high quality like handmade goods, the underlying rationale being to remove the alienation from life. At much the same time, Behrens founded his own architectural and design practice in Berlin where, over the next few years, in the run-up to war, Walter Gropius, Ludwig Mies van der Rohe, and Le Corbusier also worked. Among the firm's architectural commissions were the German Embassy in St. Petersburg and the IG Farben Höchst headquarters in Frankfurt.[21]

Brothers Mordant and Melancholic

But there was more to Munich than its painters. The two Mann brothers (Heinrich was born in 1871 and Thomas four years later) were sons of a prominent grain merchant in the Baltic port of Lübeck. Heinrich developed faster than Thomas, qualifying for university at the age of eighteen, the same year that two of his stories were published in the local Lübeck press. He left school and worked in a Dresden bookshop, later transferring, in April 1890, to Berlin to work for the publisher Samuel Fischer.

After their father died in October 1891, and the family firm was liquidated, both brothers received a settlement that enabled them to embark on their literary careers without too much hardship. Two years later, Frau Mann moved to Munich with the three younger children. Having already published anonymously in his school magazine, *Der Frühlingssturm*, Thomas then published a novella and some poems under his own name in the avant-garde monthly *Die Gesellschaft* and, joining his family in Munich, started working in an insurance company.[22] At his mother's suggestion (she wanted him to be a journalist) he started attending lectures at the Technische Hochschule and, on the strength of that, and because he had received an encouraging note from Richard Dehmel, submitted a story, "Der kleine Professor," to the new quarterly Dehmel edited, called *Pan*.

Heinrich had by now accepted the editorship of a new magazine, *Das Zwanzigste Jahrhundert* (The Twentieth Century). The magazine was conservative and anti-Semitic, not qualities that would be long associated with Heinrich, but it was also polemical and anti-monarchist because the editors felt that Wilhelm II had "sold out" to capitalist and moneyed groups in Germany at the expense of the hard-working bourgeoisie.[23] Heinrich wrote a great deal for the magazine, from militarism (a bête noir of his) to anti-Semitism to Nietzsche, whom he found the most interesting modern philosopher.

For a time the brothers' careers ran more or less in parallel, for by now Thomas's new novella, "Der Wille zum Glück" (*The Will to Happiness*), was appearing in yet another new Munich magazine, the satirical weekly *Simplicissimus*. This publication was the brainchild of Albert Langen, the son of a rich industrialist who had originally started a publishing house and only turned to the magazine later. Just 15,000 copies of the first edition were sold but even so the magazine soon became the sharpest-tongued publication in all Germany. Very liberal, *Simplicissimus* constantly attacked the government of the Reich and supported the workers against the employers. The emperor accused the magazine of undermining Germany's international prestige, and in 1898 a lawsuit was brought against Langen, as publisher, Frank Wedekind, as writer, and Thomas Heine, a cartoonist. Langen fled to Switzerland, remaining in exile for five years, while Heine and Wedekind were imprisoned for six and seven months respectively.

This publicity only helped the magazine, sales soared to 85,000, which helped it attract other writers like Ludwig Thoma (himself imprisoned later) and Rainer Maria Rilke. Soon after Thomas Mann began writing for it, he was asked to join the staff. He was, as we would say today, a copy-taster, vetting the stories sent in for publication. In this way he met many writers, satirists, and cartoonists of the day.

It was about this time that differences between Heinrich and Thomas began to appear. Their only collaboration, a *Bilderbuch für artige Kinder* (*Picture Book for Good Children*), was produced in 1896–97, but with *Im Schlaraffenland* (*In the Land of Cockaigne*; 1900), and *Buddenbrooks* (1901), the divergence of the brothers became obvious. Thomas's *Buddenbrooks* was a long, beautifully written account of a declining bourgeois family, which owed as much to Thomas's reading—and appreciation of—Tolstoy, as to anything else. But the book was bleak. Thomas Buddenbrook and his son Hanno die at a relatively early age—Thomas in his forties, Hanno in his teens—"for no other very good reason than they have lost the will to live." Behind their fate lies the specter of Darwin, Schopenhauer, Nietzsche, nihilism, and degeneracy. Although it sold only slowly to begin with, it was critically well received, and eventually earned for Thomas enduring fame and the Nobel Prize.[24]

Heinrich's *Im Schlaraffenland*, on the other hand, has been described as "the first completely 'new' novel of the twentieth century in Germany," showing a debt to Balzac, Maupassant, and Zola. "Half my being consisted at this time," he said later, "of French sentences." The story concerned the temptations of an innocent young writer anxious to make his mark in fin-de-siècle Berlin. Into this apparently simple story line, Heinrich brought to bear his acid powers of observation, a mordant eye for the corrosive aspects of social climbing, lust for money, commercial deceit, and sham in all its guises. It was an angry book, "impudent," as his publisher described it, and cast in a style that hadn't existed before, at least not in Germany. Heinrich was close to a breakdown after producing it, but the book was unique in the history of the German novel, "the first major foundation of German expressionism."[25]

Whereas Heinrich became ever more acerbic, an ever louder critic of Wilhelmine Germany (as he was one of the first, much later, to predict the annihilation of the Jews by the Nazis), Thomas was more melancholic, more interested in the arts.[26] This affected his choice of follow-up to *Buddenbrooks*, a work in which, as Thomas later said, "I learned to use music to mould my style and form." This was *Tonio Kröger*, which he later described as the "dearest" of all his books and the most personal.[27] *Tonio Kröger* is about a young writer's struggle to find his true self as an artist, his disillusion and his comparison of the life of the bourgeoisie—to which he is also drawn—and that of an artist. This theme, the place of art in life, and its relation to "engagement" and politics, was to dog Thomas all through his career.

Between their first successes and the outbreak of World War I, both brothers wrote a great deal and both had one more major triumph before hostilities sent them their separate ways. In Heinrich's case it was *Professor Unrat*, published in 1905. This was about a small-town secondary-school teacher, Professor Rat, who is so loathed by his students that they nickname him "Unrat," altering the meaning of his name from "counsel" to "excrement."[28] Unbalanced by this, Unrat one evening follows a group of students to a shady nightclub near the town's port, a club called the Blue Angel, where he intends to expose them and ruin their careers. Instead, he falls for the nightclub singer Künsterlin Fröhlich. His obsession is now threatened with exposure by the very students he had intended to disgrace, and it is he who begins to sink ever lower in society. Dismissed by his school, he marries the singer and, with the aid of the gambling that takes place in the club, exercises a corrupting influence on the whole town.

Professor Unrat showed Heinrich at his bitter best but such was the nationalism in Germany at the time that his message was scarcely welcome. However, when the book was re-issued in the middle of the war it did much better, selling over 50,000 copies. It did even better in 1930 when *Der blaue Engel* (*The Blue Angel*) became one of the first sound films released in Germany, directed by Josef von Sternberg, with a script by Carl Zuckmayer and starring Marlene Dietrich as the nightclub singer.

Der Tod in Venedig (*Death in Venice*), published by Thomas in 1913, was much better received. Gustav von Aschenbach is a writer newly arrived in Venice to complete his masterpiece. He has the appearance, as well as the first name, of Mahler, whom Mann fiercely admired and who died on the eve of Mann's own arrival in Venice in 1911. No sooner has Aschenbach arrived than he chances upon a Polish family staying in the same hotel and he is struck by the dazzling beauty of the young son, Tadzio, dressed in an English sailor suit. The story follows the aging Aschenbach's growing love for Tadzio; meanwhile he neglects his work, and his body succumbs to the cholera epidemic encroaching on Venice. Aschenbach fails to complete his work and also, deliberately, fails to alert Tadzio's family to the epidemic so they might escape. The writer dies, never having spoken to his beloved.

Aschenbach, with his ridiculously quaffed hair, his rouge make-up, his elaborate and dated clothes, is intended by Mann to embody a once-great culture now "deracinated and degenerate." He is also the artist himself. In Mann's private diaries, published posthumously, he admitted to being erotically stirred by young, handsome men, though his 1905 marriage to Katja Pringsheim (daughter of a well-known professor at the University of Munich and herself the first female student at that institution) seemed happy enough. The horrors lurking beneath the surface of the story also call to mind the general climate of opinion in "civilized" Europe in the run-up to war.

Beer and Satire

Simplicissimus was not only the name of a satirical magazine. It was also adopted as the name for one of the new "cabarets" in Munich. The city had a long tradition of popular entertainment—there were, according to one account, nearly 400 folksingers performing in Munich in 1900.[29] Their entertainment was tied to the popular beer-drinking culture of the city.

But one man stood out: Frank Wedekind (1864–1918). His father was a doctor and his mother a singer and actress, and the family lived in Hanover. The doctor was a fierce democrat, had taken part in the 1848 revolution, and afterward escaped to America (his son's name was actually Benjamin Franklin Wedekind).[30] In America the doctor made a quick fortune in land speculation and it was there, in San Francisco, that he met his wife, twenty-three years younger than he. He returned to Germany but, dismayed by Bismarck's policies, immigrated to Switzerland and bought a castle in Lenzburg, where Frank grew up. He attended first the University of Lausanne and then the University of Munich but abandoned his legal and literature studies, taking a job as a publicity agent for the Swiss soup company Maggi. He visited London, Paris, and Zurich, where he met the Swedish playwright August Strindberg and had an affair—and a child—with his wife.

Back in Munich Wedekind led a bohemian life, producing work that often flirted with the censor, and sometimes went well beyond what even the liberal Munich authorities would allow. His first full-length drama of importance was *Frühlings Erwachen* (*Spring Awakening*; 1891), which appeared in book form and wasn't realized on stage until Max Reinhardt produced it fifteen years later. Its theme was adolescent sexuality, and most people found it far too obscene to be actually performed (a fourteen-year-old girl dies as the result of a botched abortion).

Wedekind, as we have seen, was one of the cofounders of *Simplicissimus*, with Langen. For his satire on the Kaiser he was forced into exile, then jailed, but in 1901 he paraded with a group of artists, writers, and students, denouncing censorship, after which eleven of the demonstrators established a cabaret called *Die 11 Scharfrichter* (The Eleven Executioners).[31] They rented a smallish room at the back of an inn (it seated only eighty people) and decorated it with paintings by their friends from *Jugend* and *Simplicissimus* as well as some instruments of torture—which appealed to Wedekind's love of the grotesque. In the cabaret, Wedekind sang his own songs and accompanied himself on the guitar.[32]

Among the many women in Wedekind's life, there was only one, Tilly Newes, the actress he married in 1906. Tilly starred in Wedekind's magnum opus, *Lulu*, which appeared in two parts, *Erdgeist* (*Earth Spirit*) in 1895, and *Die Büchse der Pandora* (*Pandora's Box*) in 1904.[33] She played Lulu to Wedekind's Jack the Ripper. Lulu is a wild, untameable, beautiful beast, the very embodiment of female sexuality—or what men would like female sexuality to be. "She was created to stir up great disaster," is how Wedekind himself described his greatest creation. She became even better known when Karl Kraus staged *Pandora's Box* privately in Vienna in 1905. Sitting in the sixth row was the composer Alban Berg (1885–1935). His postwar opera would introduce a whole new public to *Lulu*.

The Road to Abstraction

In 1896 the Russian Wassily Kandinsky inherited enough money from an uncle for him to become financially independent. His interest in art was kindled by seeing one of Monet's haystack paintings at an exhibition in Moscow and also by a performance of Wagner's *Lohengrin* at the Bolshoi which he later said he experienced mainly as "a series of wild lines and colours." He moved to Munich where he entered the private art school run by Anton Ažbe.

There he met Alexei Jawlensky and Marianne von Werefkin, fellow students at the school and, the following year, he visited the Sezession exhibition, encountering the works of Max Liebermann, Lovis Corinth, and Hermann Obrist. Kandinsky's own works were rejected by the Munich Art Academy but in 1900 he was accepted into Franz von Stuck's class, where Paul Klee was a fellow student. Stuck encouraged Kandinsky to work with strong light and dark contrasts, and he began a series of color works on black paper and some early woodcuts. In May 1901 he was instrumental in the founding of the Phalanx group with Waldemar Hecker and Ernst Stern. The Phalanx artists were—like the Sezessionists—opposed to old-fashioned and conservative art.

Kandinsky was by now integrated into the German art scene: he was friendly with Behrens and Obrist, his work showed the influence of Jugendstil, and he exhibited at the Berlin Sezession. On a visit to the Netherlands in 1904 with Gabriele Münter, a fellow Phalanx artist eleven years his junior, Kandinsky began to apply paint mainly with a palette knife and at the same time started to make notes about his new theories on color and form.[34] His first solo show took place in Munich in 1905, at the Galerie Krause. That year, too, he exhibited for the first time at the Salon des Indépendants in Paris, after which he and Gabriele moved to Sèvres, west of Paris, where he could observe the work of Henri Matisse, Paul Cézanne, Pablo Picasso, Vincent van Gogh, Paul Gauguin, Georges Rouault, Henri Rousseau, and Edvard Munch and where he formed a friendship with Gertrude Stein, who had a comprehensive collection of paintings by these masters. That year too his pictures were shown again in the Berlin Sezession, next to the painters of Die Brücke.

In 1908, on a cycling tour, Kandinsky and Gabriele discovered (or in his case rediscovered, because he had been there in 1904), the village of Murnau. Kandinsky explained its attractions in a postcard: "It is very, very beautiful . . . The low-lying and slow-moving clouds, the dusky, dark-violet woods, the gleaming white buildings, velvety deep roofs of the churches, the saturated green of the foliage, remain with me; I even dreamt of these things." The landscape around Murnau gradually became a decisive motif in Kandinsky's output, the colors growing more lively, ever more vivid as the forms began to dissolve.[35]

In 1909 Kandinsky and Gabriele, together with Alexei Jawlensky, Marianne von Werefkin, and the art historians Oskar Wittenstein and Heinrich Schnabel, founded the Neue Künstlervereinigung München (Munich New Artists' Association), the NKVM, with Kandinsky as chairman. The aim of the association read: "Our assumption is that artists, apart from impressions observed in the outside world, progressively collect experiences of the inner world," adding that the artist's task was to "free the line for the inner sound." That same year Kandinsky produced

Painting with Skiff, a work he later described, for the first time, as an "improvisation," subsequently defined as "mainly sub-conscious . . . impressions of an 'inner nature.'"[36] In that year too Gabriele Münter bought a house in Murnau, which they called The Russian House and where she and Kandinsky spent several months a year from then on. The following year he began to refer to his works as "C," for composition, followed by a number.

Beginning in 1910 Kandinsky embarked on a series of ten compositions, seven completed before 1914, which are now regarded as his most important paintings. His most significant relationship at this crucial time, after Gabriele, was with Franz Marc, who Kandinsky felt understood him instinctively, though he also came under the influence of Nietzsche.[37] In the exhibition organized that year by the NKVM his *Composition II* and *Improvisation 10* created a storm of protest, but Marc wrote a review about the "spiritualisation of the material world" and "an immaterial inner sensation which expresses itself through pictures." Kandinsky confirmed that this was his aim.

In 1911, in response to a Schoenberg concert, he painted *Impression III (Concert)*. Kandinsky later defined "Composition" as "planned and rationally structured 'Improvisations.'" But by now he was at variance with other members of the NKVM and, that year, when the association refused to show his *Composition IV* he resigned, together with Marc, Münter, and a few others and mounted a rival exhibition.[38]

The following year Kandinsky showed several works at the famous Cologne Sonderbund Exhibition and published extracts of *Concerning the Spiritual in Art* in *Camera Work*, a journal produced by Alfred Stieglitz, a photographer whose "291" gallery in New York specialized in contemporary art.[39] In February 1913 he took part in the New York Armory Show. It was in that year that he did further work on *Composition IV* and on *Bright Picture* and *Black Lines*, which he later described as "purely abstract pictures." It had been some time coming, but now abstraction had fully arrived.

Kandsinky was not German. The German role in the birth of abstraction was threefold—the intellectual freedom (relatively speaking) of Munich, the landscape around the city, which so inspired Kandinsky and

Münter, and the whole Germanic concern with the inner life, the new world of the sub- or unconscious, which so fascinated Kandinsky and many other artists, writers, and musicians of the time.[40] The unconscious sparked three artistic movements (at least) in the twentieth century—abstraction, Dada/Surrealism, and Expressionism. Each began in the German-speaking lands.

28.

Berlin Busybody

In 1871 when Germany was at last unified, following the victory over France by a Prussian-led coalition of German states, Berlin became the capital of the new nation.[1] It was not yet the city it would become. It did, however, celebrate the great victory with the largest military parade ever seen there. On Sunday June 16, 1871, in butter-bright sunshine, 40,000 soldiers wearing iron crosses paraded from the Tempelhof Field through the Brandenburg Gate to the royal palace on Unter den Linden. Eighty-one captured French flags, many in tatters, were carried down the route.

Among the dignitaries at the head of the parade was Helmuth von Moltke, carrying the field marshal's baton he had just been awarded, and Otto von Bismarck, who had been made a prince. Behind him came Germany's new Kaiser, Wilhelm I, "his erect posture belying his seventy-four years." More than one rider fainted from the heat but not the Kaiser's twelve-year-old grandson, also named Wilhelm, who, despite his withered left arm, contemptuously refused to acknowledge a spectator who called out to him as "Wilhelmkin."[2]

Not everyone agreed that Berlin should be the capital. The Kaiser himself (who was a reluctant emperor) would have preferred Potsdam, seat of Prussia's greatest king, Friedrich the Great. Non-Prussian Germans disliked Berlin's eastern orientation, fearing it was no more than a "colonial frontier city on the edges of the Slavic wilderness." Catholics thought it dangerously Protestant. Theodor Fontane considered it too commercial.[3] "The large city has no time for thinking and, what is worse, no time for happiness."[4] This ambivalence was reflected in the fact that the Reichstag was not awarded a building of its own until 1894, until then conducting its business in "an abandoned porcelain factory."[5]

At the time of the victory parade, the city's population was around 865,000. By 1905, it had passed 2 million, the growth coming mainly through immigration from the east, East Prussia and Silesia. Among the newcomers were many Jews from the Prussian provinces or from eastern Europe. In 1860 Berlin had 18,900 Jews, a figure that rose to 53,900 by 1880. Having been banned in their own countries from owning land or serving in the military, the *Ostjuden* were experts in commerce, finance, journalism, the arts, and law. The new metropolis was their natural habitat and from 1871 on Berlin became known as "Boomtown on the Spree," its expansion deriving from three other elements: the abolition of remaining internal tariffs; more liberal rules regarding banks and joint-stock companies; and a sudden infusion of reparations from France, no fewer than 5 billion gold marks. This translated into gold for every man, woman, and child. As David Clay Large puts it, "Imperial Germany was born with a golden spoon in its mouth."

This was reflected above all in Berlin. Within two years of unification, 780 new companies were established in Prussia, the country's greatest banks—the Deutsche, Dresdner, and Darmstädter (the "three Ds")—were installed there, along with the best of the country's newspapers.[6] Jews were prominent in this new liberal climate and, besides publishing, they took a full role in the rise of the department store (Wertheim, Tietz, and Israel), the stock market, and banking. After 1871, Jews controlled about 40 percent of all banks in the Reich, compared with a quarter owned exclusively by Christians.

Notable in this field was Gerson Bleichröder, Bismarck's personal banker and financial adviser. "Bleichröder's father, the son of a gravedigger, had managed to become the Berlin agent of the powerful Rothschild banking dynasty, thereby building a potent banking business of his own."[7] His astute advice made Bismarck "a respectable prince," and Bleichröder received the first hereditary title awarded to a Jew in the new Reich. (Yet

Bismarck told anti-Semitic jokes about Bleichröder behind his back, "as if half-embarrassed by the riches his *Privatjude* had earned him.")[8]

There were a number of attempts to make Berlin a rival to Paris or London in regard to its urban amenities. The fashionable residences on Unter den Linden, Berlin's most famous street, were replaced by shops, restaurants, and hotels. The Kaiser-Gallerie, a glass-covered shopping mall inspired by Milan's Galleria Vittorio Emannuelle opened in 1873, with fifty shops, Viennese-style cafés, and other entertainment facilities. New hotels kept pace, since Berlin was now attracting roughly 30,000 visitors a day, compared to 5,000 before unification.[9]

Circumstances were improving on the surface, but were less impressive beneath it. Berlin did not build a modern sewer system until the 1870s and was notorious for its smell. Only later would *Berliner Luft* (Berlin air) become a source of pride. Men smoked cigars constantly to avoid tasting the Berlin atmosphere. They also smoked while they ate, and peppered concerts and theatrical performances "with bodily sounds quite uncensored." There was an overwhelming deference to the military, such that a "merchant carrying a pile of hats would step off the sidewalk to allow a sergeant to pass."[10] Berlin's many beer gardens struck visitors as unfortunate, "raucous places where all social classes crammed together on benches." And there were numerous "dissolute dancing places," boasting "nudities in postures difficult to describe."[11]

On February 8, 1873, a National Liberal Reichstag deputy named Edward Lasker made a three-hour speech in parliament that attacked head-on imperial Germany's economic boom, in particular the railways, which, he said, were little more than a giant house of cards where corrupt officials were protecting get-rich-quick speculators. Some of the unpalatable facts that Lasker released in his speech sparked a wave of selling on the stock market and, when both the Vienna and New York markets crashed, a raft of bankruptcies followed. In 1874, 61 banks, 116 industrial enterprises, and 4 railway companies went under.

Although the laissez-faire liberals were blamed at first, fingers soon began to point at the Jews, since not a few of the leading liberals and bankers were Jews. This was when Heinrich von Treitschke published his article in the *Preussische Jahrbücher* in which he used the phrase, "The Jews are our misfortune." Even Theodor Fontane admitted that his avowed "philosemitism" was tested by these events.[12] Bismarck took firm action—but not against the Jews. The economic liberalism of the *Gründerzeit* was dispensed with, and a program of high tariffs and state subsidies introduced to protect hard-pressed manufacturers. Economic nationalism became the order of the day.

Anti-Semitism did not disappear. The term itself was coined at that time by a Berlin journalist named Wilhelm Marr, who recognized the change in public sentiment because anti-Jewish agitation in Tsar Alexander III's Russia had caused Jewish immigration to the German capital to increase rapidly. Although immigration became a major political issue, anti-Semitism was always a less disturbing force in Berlin than in Vienna. In Berlin there were countervailing voices to those of Treitschke and Marr—most important, a "Declaration of Notables" was issued, signed by university professors, liberal politicians, and a few progressive industrialists, that condemned anti-Semitism as "a national disgrace" and an "ancient folly."[13]

As the 1870s came to an end, the German economy was recovering, this "second industrial revolution" further aided by Germany's adoption of the gold standard and the introduction of a single national currency.* The infrastructure was overhauled, a horse-drawn train on rails being introduced in the 1870s, soon to be replaced by a steam railway system (the Ringbahn, or Circle Line) built on the course of the old city wall. Next came the Stadtbahn, or city railway, linking the center of Berlin with its suburbs. Alongside this, electric lamps were introduced on many of the main streets in the 1880s. Mark Twain visited Berlin in 1891 and found it to be "the German Chicago."[14] Julius Langbehn dismissed Berlin as the "epicentre of all modern evil," its nightlife the embodiment of sin.[15]

The Imperial Know-All

Kaiser Wilhelm II also disliked Berlin. For a start, it was far too free-

* The Goldmark was introduced in 1873, replacing the varied currencies of the different German states, most of which were linked to the Vereinsthaler, a silver coin of 16.6 grams; one Goldmark = three Vereinsthaler.

thinking and disrespectful of royalty. Nevertheless, Germany needed a capital to match his ambitions, and he therefore insisted that it must become "the most beautiful city in the world." To that end he made sure he had a finger in almost every pie: churches, prisons, barracks, hospitals—all bore the imprint of his vision, for good or ill.[16]

Under him, the most significant building to be constructed in Berlin was scarcely his favorite. The new Reichstag, started in 1884 and dedicated a decade later, was originally intended to be a simple affair on the Wilhelmstrasse. But politicians and architects alike argued that this would not reflect the "newly unified, glorious German nation, on the verge of taking over the leadership of Europe." The architect, Paul Wallot, charged with "capturing the German spirit in stone," produced a cross between the Paris Opera and a Palladian palazzo.[17]

Hardly better was the Siegesallee, an avenue built in 1901 in the Tiergarten and lined with marble busts of Hohenzollern heroes. The Kaiser was extremely fond of the Siegesallee and himself produced drawings for the figures, which should, he insisted, resemble his contemporary friends and supporters of royalty. Which is why the Elector Friedrich I, founder of the Hohenzollern dynasty, came to look like Philipp zu Eulenberg, the Kaiser's closest friend. Many thought the whole project embarrassing and nicknamed the street "Die Puppenallee"—the avenue of the dolls. The Kaiser's reputation also suffered: this was a man, Berliners quipped, "who could not attend a funeral without wanting to be the corpse."[18]

Kaiser Wilhelm II saw it as his duty—and his right—to be involved in all aspects of Berlin's artistic and intellectual life.[19] He saw himself as particularly suited to this because he felt he had a gift for drawing and for writing plays. He designed ships and produced a play of his own, titled *Sardanapal*, in which the central character is a king who sets fire to himself rather than be captured by the enemy. Visiting dignitaries would be forced to sit through performances of his play; among the dignitaries was his uncle, King Edward VII of England, who fell fast asleep until the raucous fire scene, when he suddenly awoke—and called for the fire department to be summoned. In artistic and cultural matters Wilhelm was a backward-looking archconservative, and his constant meddling eventually provoked a backlash.[20]

In the theater, matters came to a head in 1889. "[That year] was the year of the German theatrical revolution, just as 1789 was the year of the revolution of humanity," wrote Otto Brahm, founder of Berlin's Freie Bühne (Free Stage) movement, somewhat overstating his case to make his point. The Free Stage was a private club and not subject to censorship as public theaters were. For that reason Brahm felt able to produce Ibsen's *Ghosts*, otherwise banned because it dealt with syphilis. Emboldened, he next tried Gerhart Hauptmann's *Vor Sonnenaufgang* (*Before Dawn*), an exploration of everyday life among the working classes. Hauptmann (1862–1946), born in Silesia (now part of Poland), would win the Nobel Prize in 1912. He was one of the founders of Realism, again one of those terms that evoke little reaction nowadays, though it was very different in the Kaiser's Germany.[21] At the performances of *Before Dawn*, brawls were reported in the auditorium between the advocates and opponents of such modernism.[22]

Brahm was so encouraged by these reactions that he bought a public theater, the Deutsches Theater, and began mounting ever more political plays. The climax came in 1894 with a production of Hauptmann's *Die Weber* (*The Weavers*), a ferocious indictment of the social conditions condemning Silesian textile workers to extreme poverty in the 1840s.[23] The play was banned by the police on the grounds that "it was likely to stir up the lower orders." Judges overturned the ban after conceding that "the lower orders" could scarcely attend a play where the admission charge was well beyond their means. *The Weavers* (*De Waber* in the Silesian dialect) proved a great success.[24]

The Kaiser hated what Hauptmann represented. At the end of an evening, he thought, people should leave a performance "not discouraged at the recollection of mournful scenes of bitter disappointment, but purified, elevated, and with renewed strength to fight for the ideals which every man strives to realise." On the grounds that his plays contravened these self-evident rules, Wilhelm had Hauptmann arrested in 1892 "for subversion." The courts, to their credit, could find little reason to keep the writer in jail, so the Kaiser tried other methods of intimidation and vetoed Hauptmann's award of the Schiller Prize for dramatic excellence, giving it instead to one of his favorite hacks.[25]

Much the same happened with Max Reinhardt, a Jew from Austria who had arrived in Berlin at the turn of the century intending to be an actor. Reinhardt (1873–1943), born Max Goldmann, arrived when modern theater was taking off in all directions—Wagner, Zola, Ibsen, Strindberg. For him, Berlin was "Vienna multiplied by more than ten," as he wrote to a friend. Reinhardt never succeeded as an actor, but he turned out to be a brilliant director after he founded his own cabaret called *Sound and Smoke*. This brought commissions from the legitimate stage, in particular the Deutsches Theater, which he took over from Brahm in 1905 and where he changed the offerings from grim realism into a form of "magic and excitement."[26] He still kept the theater serious—everything from Sophocles to Büchner—but he introduced new lighting effects, new staging techniques, making the theater more *spectacular* than it had ever been in Berlin. (Marsden Hartley says Reinhardt probably handled the largest theatrical quantities outside wars, volcanic eruptions, or train crashes.)[27] There was nothing specific in Reinhardt's techniques that the Kaiser could object to—it was their very modernity that he didn't like. Therefore, unable to invoke any law, the Kaiser simply ordered his productions off-limits to the military. Childish to the end, when war broke out in 1914, Wilhelm rejected the playwright's offer to tour the Front with his company.[28]

Another of the great figures in the Berlin cultural world whose admirers did not include the Kaiser was the conductor Hans von Bülow. By the turn of the century, as we have seen, Berlin had long been a center of international respect so far as music was concerned. Ever since 1842, the royal orchestra, once led by Felix Mendelssohn, had had an excellent name. In the 1880s, a second, privately funded symphony orchestra had been established by a certain Benjamin Bilse. The former leader of a military band, he turned his new orchestra into a rival of the royal orchestra but was something of a martinet in style. In 1882, a group of his musicians who had grown weary of being treated in such a domineering way broke off to form a rival outfit, calling themselves the Berlin Philharmonic. Their early years proved difficult and they were forced to perform in a converted roller-skating rink. But in 1887 they came under the direction of Hans von Bülow. Not only was he a brilliant and charismatic conductor himself, who liked the classics and contemporary music equally, but he also had interesting friends. In 1889 he brought one of these, Johannes Brahms, to Berlin to conduct his D Minor Concerto. The occasion was a sensation.

The Kaiser, it will be no surprise to learn, hated modern music as much as he loathed modern art and modern theater.[29] He and Bülow clashed in particular over Wagner. Bülow was an experienced interpreter of Wagner, even though the composer had stolen his wife, Cosima, and the Berlin Philharmonic's offerings of Wagner had become a shining jewel of the Berlin opera scene, which had had little to show for itself since Giacomo Meyerbeer in the 1840s. Even so, many long memories in Berlin could not forget the composer's support for the revolution of 1848, and the Kaiser used this to put Wagner down. Soon after ascending the throne, he announced portentously that "Gluck is the man for me; Wagner is too noisy." He had much the same view of Richard Strauss, allowing Strauss to take over at the Royal Opera only because the composer promised he would make Berlin an even greater international center of music than it already was.[30] In fact Strauss continued to compose in the discordant way that the Kaiser hated. "I raised a snake in the grass to bite me," he growled, and told Strauss to his face that he considered his music "worthless."

In 1871, at the time of national unification, Berlin was a good way behind Munich in the world of fine art. Munich, as we have seen, had by far the largest community of painters and sculptors. In the 1880s and 1890s, however, as it became clear that Berlin was to be enriched by new monuments and museums, artists began migrating to the new capital. Here too, the Kaiser could never resist taking sides.

Until the influx, Berlin's best known artist was Adolf von Menzel, a native of Breslau who had lived in the Prussian capital since 1830. At first Menzel painted impressionistic depictions of Berlin's rougher edges, its dingy streets and archaic factories (Degas admired Menzel). In the 1870s, however, Menzel changed radically, turning instead to the history of the state and the monarchy.[31] For example, *The Flute Concert* and *The Round Table* treated the court of Friedrich the Great and did so reverentially; other compositions showed a straightforward adulation for Prussian *Macht*. The change achieved its aim, and Menzel duly became admitted to the court himself, this "unfortunately ugly painter" soon gracing the playgrounds of high society, which he now chronicled in loving detail. In 1905 the Kaiser marched in his funeral cortege.[32]

Not too dissimilar was Anton von Werner, whose vast, precise canvases were reproduced in German schoolbooks and became familiar as few art works are. His *Kaiser Proclamation in Versailles*, which depicted the emperor and his generals toasting the foundation of the German Empire in Louis XIV's Hall of Mirrors, was a gift to Bismarck. Appointed president of the Academy of Fine Arts in 1875, Werner subsequently became Wilhelm II's tutor, reinforcing the young Kaiser's instinctive loathing for modern art.

But though Wilhelm and Werner were a powerful minority, they were a minority nonetheless. As early as 1892 the Association of Berlin Artists invited Edvard Munch to exhibit his work there. Fifty-five paintings were planned and the conservatives duly incensed. Werner led the chorus, and the show was canceled. The old guard were not so successful with Max Liebermann. Liebermann's very Germanic humanity was self-evident in his pictures, but Wilhelm did not think that painting should make "misery even more hideous than it already is."[33] Accordingly, he did his best to keep Liebermann barred from official exhibitions. This did not stop the popular painter from exhibiting in private shows, and eventually he grew so popular that he was admitted to the official salon. In 1897 he won the Gold Medal, was elected to the Prussian Academy of Art, and appointed professor at the Royal Academy.

If the Kaiser lost (for the time being) to Liebermann, he won with Käthe Kollwitz. A jury recommended in 1898 that Kollwitz, a powerful, emotional artist, who lived in a Berlin slum, be awarded a gold medal for her cycle of etchings, *The Revolt of the Weavers*, based on Hauptmann's play. (The revolt of the Silesian weavers in the 1840s had great significance for the making of the German working class.) The Kaiser had to be consulted before the medal could be announced, and it was too much. "Please, gentlemen," he complained, "a medal for a woman, that's really going too far . . . Orders and honours belong on the chests of deserving men." Coming on top of the Munch affair, this was too much for the artists.[34] In that same year Liebermann and others announced the Berlin Sezession, modeled on the earlier reactions in Vienna and Munich. Their aim was to show art they thought worth showing and without interference. They obtained backing from wealthy collectors, not a few of whom were Jewish. The Cassirer cousins, Bruno and Paul, were the chief supporters—their gallery in Kantstrasse was a leading venue for modern art. For the Sezession, they built a new gallery.[35]

The Kaiser did not disappoint. All military officers were forbidden to attend the Sezession when in uniform, and Sezession members were banned from serving on juries of the salon. Sezession artists were likewise banned from showing at the 1904 St. Louis World's Fair. Subsequently, officials of the Cultural Ministry held out an olive branch, conceiving a plan for a Liebermann retrospective at the Royal Academy. But the Kaiser was having none of it. The painter, he said, "was poisoning the soul of the German nation." In fairness to the Kaiser, only three artists of the Berlin Sezession have stood the test of time, Liebermann, Walter Leistikow, and Lovis Corinth. It was Corinth who coined the term "Expressionist" for a predominantly German art form he was himself at odds with.[36]

The battles between the Kaiser and the artists went on. Die Brücke, a group of Expressionist painters, was founded in Dresden in 1905, and moved to Berlin five years later. Its spokesman, Herwarth Walden, launched a magazine and art gallery called *Der Sturm* which, on the brink of the war, was the heart and brains of the German avant-garde.[37] The art of Die Brücke was—much more than the Sezession—an urban art. The two most important figures were Ludwig Meidner and Ernst Ludwig Kirchner. Both were concerned with the side of Berlin that the Kaiser thought had no place in art—Meidner focused on suspension bridges, gas tanks, express locomotives, while Kirchner's contorted figures represented, as he put it, the raw energy in the city's streets and taverns, the new psychology Simmel had identified, the "so-called distortions" in his paintings "generated instinctively by the ecstasy of what is seen."[38] Static representation was impossible, he insisted, when the inhabitants of the city were in perpetual motion, "a blur of light and action. The city required of its artists a new way of seeing." The Kaiser, it goes without saying, was affronted.

While Berlin's reputation as a city of modern art was new, and far from settled, it had been a museum town since 1830, when Schinkel had designed the Altes Museum on the small spit of land in the middle of the Spree that soon became known as the Museuminsel (Museum Island). The Neues Museum was added in 1855, and the Nationalgalerie in 1876.[39] In this area, Wilhelm was lucky in having one of Europe's cleverest collectors and connoisseurs, Wilhelm von Bode, who took over as director of the new Kaiser-Friedrich-Museum when it opened on Museum Island in 1904.[40] Bode obtained for Berlin a raft of impressive old masters that included Rembrandt's *Man in a Golden Helmet* and Dürer's *Hieronymus*

*Holzschuher.** His success was underlined by the fact that his activities were spared the Kaiser's interferences. In fact, Wilhelm positively helped von Bode by awarding titles to those who offered works to the royal collections.

In contemporary art the familiar problems resurfaced. The director of the National Gallery, Hugo von Tschudi, was as accomplished a man as Bode, and an expert on French painting and contemporary art and sculpture.[41] The Kaiser, however, refused to give him the free hand he gave von Bode and, on one visit, noticed that some German works had been removed, their place taken "by pictures of modern taste, some of them of foreign origin." He insisted the originals be put back.

The Kaiser couldn't be everywhere at once, however, and Tschudi did find ways to acquire some contemporary masterpieces, including a Cézanne, making him the first museum director in the world to do so (at that stage, not even the French state had any Cézannes in its official collections).[42] Later, however, when Tschudi bought works by Eugène Delacroix, Gustave Courbet, and Honoré Daumier, the Kaiser exploded, complaining that Tschudi might "show such stuff to a monarch who understood nothing of art, but not to *him*." In 1908 Tschudi decamped to Munich to become head of the royal museums there.

Despite his conservative and retrograde taste in the arts, the Kaiser was nonetheless proud of the scientists and engineers who were building Germany's prosperity and, since he thought of himself as a man of the future, and believed that the *application* of new knowledge was the key to progress, he prevailed upon the University of Berlin at the turn of the century to recognize the graduates of the recently created *Realgymnasien*, which emphasized science at the expense of the humanities. For all his striking contradictions, the elevation in the status of Realgymnasien was conceivably the best thing the Kaiser ever did in cultural and intellectual affairs; it was built on in 1910 when, to mark the one hundredth anniversary of the founding of the Friedrich-Wilhelm-Universität, he announced a new institution for the natural sciences. The Kaiser Wilhelm Society, Germany's answer to France's Pasteur Institute and America's Rockefeller Institutes, was funded by private industry *and* the government, and would pay dividends in spades.[43]

* The Rembrandt was subsequently shown to have been produced, perhaps, by his assistants.

29.

The Great War between Heroes and Traders

In the early months of the Great War, the Viennese developed a set of symbolic acts that fortified them and helped them identify with the troops at the front. For instance, in the Schwarzenbergplatz, off the Ringstrasse, was a wooden statue, called the *Wehrmann im Eisen*, the "soldier in iron." Anyone who wanted to could buy a handful of nails—the profits from which went to benefit war widows and orphans—and hammer the nails into the statue, "covering him in iron, enveloping him in the collective strength of the Austrian *Volkskraft*."[1]

Support for the war was not, as Matthew Stibbe has recently revealed, quite as enthusiastic in Germany in 1914 as previously reported. Outside the main cities, and among the working class in particular, the mood, he says, was one of "resignation, indifference or passive acceptance" rather than aggressive nationalism. It was intellectuals who believed they were "called upon" to underpin the belligerence with a coherent philosophy, "which idealised the power conflict in terms of an alleged spiritual antithesis between German *Kultur* and political forms and those of its enemies." (Though Norbert Elias felt that Nietzsche, "almost certainly without being aware of it," in his book *Der Wille zur Macht* [*The Will to Power*], gave philosophical form to the belligerence of the Wilhelmine middle class.)[2]

For many, *Kultur* was the central factor in the war.[3] What these individuals meant by "Kultur" was the set of achievements represented by Goethe, Kant, and Beethoven—"high culture," art, music, literature, and scholarship together with "a set of collective virtues" (diligence, order, and discipline) regarded as characteristically German. Writers, historians, and philosophers on both sides of the political divide shared these views—Thomas Mann, Friedrich Meinecke, Ernst Troeltsch, Werner Sombart, Max Scheler, and Alfred Weber, to name only a few.

The Ideas of 1914

From the outbreak of war, there was in Germany a very public meditation on what was distinctive about German culture, with dichotomies being sharpened on a "polar opposition" between *Kultur* and *Zivilisation*. These polar opposites quickly came to a head, sparked by what the rest of the world saw as Germany's barbaric behavior during its conquests in Belgium and northeastern France, when the ancient library in the Belgian town of Louvain was burned and the cathedral in Rheims was badly damaged, and Belgian civilians in Dinant and elsewhere were massacred, in "retaliation" for alleged acts of sabotage. British and French academics led the cry that the best-known figures of culture and science in Germany must publicly distance themselves from Prussian militarism, but the effect was not what they anticipated. Whole swaths of German cultural and academic figures rallied behind the German war effort and, on October 4, 1914, a group of ninety-three of the most distinguished German scholars and artists issued the "Manifesto of the 93"—an "Appeal to the Cultural World" (Der Aufruf der 93, "An die Kulturwelt") in which they flatly refuted all charges of barbarianism in Belgium and insisted instead: "It is not true that the struggle against our so-called militarism is not also a struggle against our civilisation, as our enemies hypocritically pretend it is. Were it not for German militarism, German civilisation would have long since been extirpated from the earth. The former arose from the need to protect the latter in a country which for centuries has been afflicted by predatory invasions."[4]

The signatories to this appeal included the writers Richard Dehmel and Gerhart Hauptmann; the painters Max Klinger, Max Libermann, and Hans Thoma; the musicians Engelbert Humperdinck, Siegfried

Wagner, and Felix von Weingartner; the theater director Max Reinhardt; prominent academics such as Ernst Haeckel, Fritz Klein, the Nobel Laureate physicists Philipp Lenard, Richard Willstätter, and Max Planck; the future Nobel Laureate chemist, Fritz Haber; the theologian Adolf von Harnack; the economists Lujo Brentano and Gustav Schmoller; the philologists Karl Vossler and Ulrich von Wilamowitz-Moellendorff; the philosopher Alois Riehl; and the psychologist Wilhelm Wundt, together with the historians Karl Lamprecht, Max Lenz, Eduard Meyer, and Friedrich Meinecke. Even before this, a group of academics had renounced their honorary degrees awarded by British universities.[5]

All this may sound unreal now, and irrelevant, after the widespread horrors that happened later in World War I, and then in the 1930s and World War II, but the "Appeal" did reflect the views of educated people in the Germany of the time, that the war would bring about the country's rise to world power status and would therefore "go down in history as the German war." Following the "Appeal" there was a raft of speeches, books, and other events in the same vein. The Bund Deutscher Gelehrter und Künstler, which had its head office in Berlin, recruited 200 leading figures from the literary and artistic world, including Thomas Mann, to argue the intellectual case for war.[6] One of the main themes was the superiority of Germany's authoritarian constitution over the parliamentary regimes of the west.

These ideas remained important. Max Lenz, Otto von Gierke, Max Scheler, and Karl Lamprecht all advanced arguments for German "world leadership" and Lamprecht, one of the advisers to wartime chancellor Theobald von Bethmann-Hollweg, was, like others, not averse to playing the race card: "It is subjectively recognised and objectively proven that we are capable of the highest achievements in the world and must therefore be at least considered entitled to share in world rule . . ."[7] Lamprecht argued that the British were guilty of a sense of "innate superiority" that was too much: "For the other nations this [English] feeling [of superiority] is completely intolerable, and I dare say that the world cannot return to peace until this feeling has been replaced . . . by a more modest appraisal."

More impressive were the arguments of the generation of historians who comprised Max Lenz, Erich Marcks, Otto Hintze, and Hans Delbrück among others.[8] Their view—commonplace since the 1890s—was that the system of old European states, which existed in Ranke's day, would soon be replaced by a small number of world states (empires) in which the Germans would take their place as an equal. For them, the point of the war was to force Britain, the oldest of the established world powers, to surrender its pre-eminence and grant Germany equality.

The effect was twofold. It meant that Britain had to be seen as the instigator of the war, and it provided yet more justification for militarism. Even a moderate like Hans Delbrück, who later came to oppose government war policy, could write as follows in the early months of hostilities: "This nation is invincible . . . against that island nation [Britain] . . . [these] men of commerce, who merely hand out money, who send out mercenaries and mobilise the barbaric masses and think they are able to defeat us—it is these [men] who we need to be fighting against . . . with the certainty of our eternal inner superiority . . ."

Not everyone fell into this category. In 1915, for instance, Otto Hintze, Friedrich Meinecke, Hermann Oncken, and Hermann Schumacher got together to produce *Deutschland und der Weltkrieg*, aimed at counteracting the effects of English propaganda on neutral countries, in particular the United States.[9] They sought specifically to counter the British propagandists who had resurrected the French argument that there were two Germanies, the Germany of Goethe and Schiller and Beethoven, on the one side, and the Germany of Treitschke, Nietzsche, and General Friedrich von Bernhardi on the other. Hintze, Delbrück, and Meinecke stopped short of advocating the utter destruction of Britain, arguing instead for a "balance-of-power" and they thus seemed reasonable, certainly in comparison with everyone else.[10] In general, however, they were drowned out by more openly annexationist writers and speech makers.

Oswald Spengler, later well known as the author of *Der Untergang des Abendlandes* (*The Decline of the West*; see pp. 560–562), believed that Germany's decision "to challenge England for world domination" was a turning point in history. The fight with Britain was for him a crude Darwinian struggle between "English" liberalism, "with its emphasis on individual freedom and self-determination," and "Prussian" socialism, "with its emphasis on order and authority."[11] Elsewhere he confessed, "In the Germany which made its world position secure through technical skill, money and an eye for facts, a completely soulless Americanism will rule, and will dissolve art, the nobility, the Church . . . in a materialism such as only once has been seen before—in Rome at the time of the First Empire."[12]

As the war continued, and the stalemate grew staler, still the arguments kept coming. Even Max Weber, so sane in so many ways, said this in a speech in Nuremberg in August 1916: "It would be shameful if we lacked the courage to ensure that neither Russian barbarism, English monotony, nor French grandiloquence ruled the world. That is why this war is being fought." The historian Friedrich Meinecke went much further, claiming that the German nation as a whole "has a mission from God to organise the divine essence of man in a separate, unique [and] irreplaceable form. It is like a great artist, who, by means of his personal genius, creates something above his own personality . . . Only the Germans had managed to find the combination of *Innerlichkeit*, individual freedom and willingness to sacrifice selfish interests to the good of the whole that characterised their unique spiritual heritage." The philosopher Eduard Spranger wrote about the need to keep alive the German tradition of Bildung.[13]

Even as the war started to go against Germany, the cultural arguments remained strong. The philosopher Adolf Lasson insisted: "The whole of European culture, which is surely the only universal form of human culture, has gathered itself together like a focal point on German soil and in the hearts of the German people. It would be quite wrong to express ourselves on this point with modesty and reservation. We Germans represent the . . . highest of all that European culture has ever brought forth; upon this rests the strength and the fullness of our self-esteem."[14]

In his own wartime essay "Gedanken im Kriege" (Thoughts in War), Thomas Mann spoke of Germany's "indispensable role as missionary" in defending the unique status of German *Kultur* against the superficial, liberal *Zivilisation* of the West. And he went on, "It is not so easy to be a German . . . [It is] not so comfortable as it is to be English, and not at all such a distinct and cheerful thing as it is to live as the French do. This people has difficulty with itself, it finds itself questionable, it suffers from itself to the point of outright disgust; but . . . it is those that suffer the most that are of the most worth, and whoever would wish that German manners should disappear from the world in favour of *humanité* and *raison* is committing a sacrilege." Perhaps inevitably, at that stage anyway, he argued that Western-style democracy was not the German way. "[T]his most introspective of people, this people of metaphysics, of pedagogy and of music, is not a politically oriented, but a morally oriented people. And thus it has shown itself to be more hesitant and less interested in political progress towards democracy, towards parliamentary forms of government, and especially towards republicanism, than other [peoples]."[15]

Each of these critiques, beneath their contempt (or alleged contempt) for Britain and France (and America), exhibit a revulsion at the profound changes that industrial growth had wrought on society. Many across the world shared this view. Where the Germans differed, according to Roger Chickering and others, is that the educated class in particular believed that the state should intervene to "check the materialistic excesses of self-seeking minorities in the interests of the general good."[16]

Not for the first time, it is a relief to turn away from this airless atmosphere, which was, in any case—and not to mince words—wrong. To jump ahead of ourselves for a moment, in 1961 the German historian Fritz Fischer published his book *Griff nach der Weltmacht* (translated as *Germany's Aims in the First World War*; 1967). In the 1950s he had been given access to the East German archives in Potsdam, where he came across an "explosive" set of files that, he claimed, showed that imperial Germany had aggressive annexation plans before World War I, and that, among other things, in December 1912, at an infamous "war council," Wilhelm II and his military advisers "had made a decision to trigger a major war by the summer of 1914 and to use the intervening months to prepare the country for this settling of account."[17] Fischer claimed that there was a new kind of nationalism abroad in Germany from 1890 on that had racial overtones, that many of the country's historians and intellectuals supported the great expansion of naval hardware, that Nietzsche's "will to power" was a view agreed on by these very same people as an important psychological factor in modern life, that there was in imperial Germany very little difference between business interests and political interests, that Germany's main aim was to wipe out France and to keep Britain neutral. He further found that such a view was always unrealistic, that Germany initiated the arms race, that the Kaiser and his advisers came to the view that the time for diplomacy was over, believing "inter-racial conflict" was inevitable in the "settling of accounts."[18] Fischer also concluded that it was Germany who most seriously misjudged the fighting abilities of her enemies or potential enemies.

Fischer's work will be discussed more fully in a later section of the book (he was accused of "treason" by fellow German historians). For now we can confine ourselves to the remarks of Fritz Stern who, in commenting on Fischer's book, said that if one factor can account for World War I, it is the constant miscalculations of Germany's prewar policies, stemming from "a chronic blindness," a false estimation of themselves and of others, "a rare combination of *Angst*, arrogance and—in assessing the non-German world—political ignorance and insecurity."[19]

The Manifesto of the 93 provoked a fierce reaction in both France and Britain. French scholars were revolted by what they saw as the "intellectual servility, lack of objectivity, and craven spirit" of the scholars who signed the manifesto. Nevertheless, William Keylor concluded of the French academics that they too "rapidly abandoned their pre-war commitment to higher truths in the summer of 1914 and surrendered to the basest form of jingoist hysteria during the next five years."[20]

That was perhaps overstating the case. Three questions concerned the French: (1) What remained worthy of respect within German culture? (2) Did France owe more to nineteenth-century German culture than to ancient Greece and Rome? (3) Was German science related to German Kultur, or were the undoubted successes of German science rooted in the philosophical traditions of France and Britain?

At the center was the philosophy of Immanuel Kant. The conservatives and Catholics in France disparaged Kant because they regarded his ethics and epistemology as the foundations of "unrestrained individualism, subjectivism, and atheism." These, in their turn, were seen by the same conservatives as the bases of republicanism, fostering notions of rights and duties. Their opponents favored Kant because of his theory of moral obligation and individual responsibility, which, in the years before the war, had been made the cornerstone of (republican) civics in the French schools. Kant also lay at the heart of French theories about the "two Germanies." Living next door, the French had long been uneasily aware of the two faces of their neighbor—immensely cultured and inward, but at the same time militaristic and expansionist. This view had been sharpened in the wake of the Franco-Prussian War. In December 1870, E. Caro, writing in the *Revue des deux mondes*, had advanced the idea of the two Germanies, one "mystical and metaphysical," the other "materialistic and militaristic." Kant, he said, was the apotheosis of the former, the French defeat at Sedan of the latter which, in the end, had gained the ascendancy. This second tradition, said Caro, originated with Hegel.

As Martha Hanna has pointed out, at the beginning of the war there was also a belief, widely held in France as elsewhere, that science was "if not uniquely, then at least especially, a German enterprise." This had the unfortunate side effect that, after war broke out, science became suspect in France, a view reinforced when, in April 1915, the German army became the first to use poison gas. Science was now seen as "the regrettable product" of a materialist ethos. Hanna says French scientists worked hard to counter this belief, arguing that science was just as much a French and British activity as it was German.[21]

In Britain before the war there was widespread agreement about a "knowledge revolution" and the "institutionalisation of the German influence" in scholarship, and though some British academics, visiting Germany at that time, were repelled by the belligerent atmosphere, far more were attracted by the ideology of *Wissenschaft*, which was "virtually a way of life."[22] Stuart Wallace, in his study of British academics in World War I, published a list of fifty-six prominent British scholars who had studied in Germany, including Lord Acton, E. V. Arnold, James Bryce, H. M. Chadwick, William McDougall, A. S. Napier, W. H. R. Rivers, R. W. Seton-Watson, Henry Sidgwick, and W. R. Sorley. On August 1, 1914, the London *Times* carried a letter by nine scholars supporting Germany as the more civilized country in its struggle with Russia. After the invasion of Belgium this attitude was completely reversed (on August 29 the German Wolff news agency announced: "the ancient town of Louvain, rich in art treasures, no longer exists today") and in December 1914 the *Times* published a letter by A. H. Sayce, professor of Assyriology at Oxford, arguing that, in science, "none of the great names" was German, that apart from Goethe there were no great names in German literature, that Schiller was a "milk-and-water Longfellow" and Kant "more than half Scottish."[23]

Other scholars, writers, and artists, like the French, "perceived the war against Germany as a war . . . in defence of civilisation, against a barbarous, many-headed enemy." At one extreme were those ardent patriots who believed that not even Brahms should be played in wartime Britain. Like the French, British scholars described their horror at the intellectual subservience to the state shown by German academics, though they too helped fashion propaganda. Again like the French, British scholars had long admired German scholarship, but esteem for the German way of doing things quickly withered. Philosophers found it more of a problem. "Hegelianism had indelibly marked British Idealism, the most influential school of philosophical thought in Britain before 1914."[24]

The war infected scholarship in a different way where archaeology was concerned. The budget for the Deutsche Archaeologische Institut was *increased* between 1915 and 1916, excavations continued in Babylonia, at Tiryn, Dipylon, and Olympia and commenced at Laon, Arras, and Soissons in occupied France.[25] Attempts, ultimately unsuccessful, were also made to "corner" the excavation market in the Ottoman Empire. Because archaeology was so close to the Kaiser's heart, and several leading archaeologists were welcomed at his court, archaeologists, classicists, and philologists became a hot-bed of "monarchist nostalgia and apoplectic reaction" after the war.[26]

From Eden to Berlin

In America, the response was more measured than in France or Great Britain (the United States did not join the war as a belligerent until April 1917). Most notably, in 1915 two leading American intellectuals—John Dewey and George Santayana—both published their assessments of German philosophy and scholarship. Each was a short, pithy book.

John Dewey, then professor of philosophy at Columbia University, achieved a clear synthesis of German philosophy, linking the history of the country's thought to the war, in an analysis that still reads well and is all the more impressive for having been written nearly twenty years before the events that led to the Holocaust. The book began life as three one-hour lectures and was given the title *German Philosophy and Politics*. It was also, in part, a reply to General Friedrich von Bernhardi's book *Deutschland und der nächste Krieg* (*Germany and the Next War*), published in 1911, which had famously claimed: "Two great movements were born from German intellectual life, on which, henceforth, all the intellectual and moral progress of mankind must rest:—The Reformation and the critical philosophy . . . whose deepest significance consists in the attempt to reconcile the result of free inquiry with the religious needs of the heart, and thus to lay a foundation for the harmonious organisation of mankind . . . To no nation except the German has it been given to enjoy in its inner self 'that which is given to mankind as a whole' . . . It is this quality which especially fits us for leadership in the intellectual domain and imposes on us the obligation to maintain that position."[27]

Dewey's first point was that history has shown that to think in abstract terms is dangerous, "it elevates ideas beyond the situations in which they were born and charges them with we know not what menace for the future." He observed that British philosophy, from Francis Bacon to John Stuart Mill, had been cultivated by men of affairs rather than professors, as had happened in Germany (Kant, Fichte, Hegel). He thought there was always a connection between abstract thought and "the tendencies of collective life" and that the Germans "have philosophy in their blood."[28] In particular, he thought that Germany—and its well-trained bureaucracy—had "ready-made channels through which philosophic ideas may flow on their way to practical affairs," and that Germany differed from the United States and Britain in that this channel was the universities rather than the newspapers. He noticed a crucial difference, he said, in that whereas most nations are proud of their great men, "Germany is proud of itself for producing Luther . . . A belief in the universal character of his genius thus naturally is converted into a belief of the essentially universal quality of the people who produced him."[29]

Dewey attached most importance to the achievements of Kant and his idea that the two realms of science and morals are what matter most in life, that each has its own "final and authoritative constitution." The chief mark of "distinctively German civilisation," Dewey said, is its combination of "self-conscious idealism with unsurpassed technical efficiency and organisation . . . The more the Germans accomplish in the way of material conquest, the more they are conscious of fulfilling an ideal mission," so that the "distinguishing mark of the German spirit" is a supreme regard for inner truth and the inner meaning of things, "as against, say, the externality of the Latin spirit or the utilitarianism of Angla-Saxondom."[30] Dewey conceded: "It does seem to be true that the Germans, more readily than other peoples, can withdraw themselves from the exigencies and contingencies of life into a region of *Innerlichkeit*, which at least *seems* boundless; and which can rarely be successfully uttered save through music, and a frail and tender poetry, sometimes domestic, sometimes lyric, but always full of mysterious charm."[31]

A second achievement of Kant, said Dewey, after the separation of the realms, came in "the gospel of duty," Kant's idea of self-imposed duty as stern but noble, as a phenomenon that separates us from the animals. Equally important, however, Dewey felt that Kant had told us to do our duty without specifying what those duties were, or are.[32]

He also thought the distinction made in Germany between society and state was important, as was that between civilization and culture. "Civilisation is a natural and largely unconscious or involuntary growth. It is, so to speak, a by-product of the needs engendered when people live close together . . . Culture, on the other hand, is deliberate and conscious. It is a fruit not of man's natural motives, but of natural motives that have been transformed by the inner spirit . . . And the real significance of the term 'culture' becomes more obvious when [Kant] adds that it involves the slow toil of education of the Inner Life, and that the attainment of culture on the part of an individual depends upon long effort by the community to which he belongs."[33]

Dewey went on to examine the ideas of society and the state. In American and British usage, he said, "the state" generally refers to society "in its more organised aspects," one government agency or another. But in Germany, "the State, if not avowedly something mystic and transcendental, is at least a moral entity, the creation of self-consciousness operating in behalf of the spiritual and ideal interests of its members. Its function is cultural, educative . . . its purpose is the furtherance of an ideal community . . . Hence the peculiar destiny of the German scholar and the German State. It was the duty and mission of German science and philosophy to contribute to the . . . spiritual emancipation of humanity . . . The scholar . . . is, in a peculiar sense, the direct manifestation of God in the world—the true priest . . ."[34]

This was, one can now see, a perceptive analysis of the achievements of nineteenth-century Germany and a delicate and eloquent exploration of how Germany differed from, say, France, Britain, and the United States.

Dewey's final point was that there had been an unfolding of a great sequence in that same nineteenth-century Germany—1815, 1864, 1866, 1870–71—which he put alongside the fact that Germans had accepted the idea of evolution long before Darwin came up with natural selection. Added to this, "The very fact that Germany for centuries has had no external unity proves that its selfhood is metaphysical, not a gift of circumstance . . ."[35]

In contrast, a parallel and simultaneous work by George Santayana, while not devoid of numerous perceptive comments, was written with such sarcasm and bile that the author's attitude repeatedly got in the way of what he was trying to say. Santayana (1863–1952), born in Madrid, studied in Germany under Paulsen, and then became one of the best teachers Harvard ever had (his pupils included Conrad Aiken, T. S. Eliot, Robert Frost, Wallace Stevens, Walter Lippmann, Felix Frankfurter, and Samuel Eliot Morison). He retired back to Europe, from where he published *Egotism in German Philosophy* in 1916, covering much the same ground as Dewey—Kant, Fichte, Hegel—but adding Schopenhauer and Nietzsche for good measure.

Santayana didn't think much of any of them. He thought transcendental theory was a set of "desperate delusions," even that there was "something sinister" at work beneath it all, close to a (false) religion.[36] He admitted there was "an obvious animus pervading these pages, which it was a pleasure for me to vent."[37] He conceded that the direction in which German philosophy was profound was its inwardness, its consciousness of "inward light and of absolute duties"; but he thought this was egotistical, and egotism he defined as "subjectivism become proud of itself." He thought there was "something diabolical about its courage, something satanic in its courage," amounting to a "moral disease."[38] He thought German Idealism had inherited from Protestantism an earnestness and pious intention, that notions of the spirit or will resembled the notion of Providence. He dismissed Kant because he did not live up to his beliefs, that though he was mild-mannered himself, "his moral doctrine was in principle a perfect frame for fanaticism."[39]

Hegel, he thought, had retracted all belief in a real world and set in its place his knowledge of it—a "monstrous egotism" that enabled him to pass off the prejudices of his time and country "as the real thing." This was the wrong way round, making things conform to words, not words to things.[40] He felt there was something "immature" about German thought (Nietzsche in particular): "They have not taken the trouble to decipher human nature, which is an *endowment*, something many-sided, unconscious, with a margin of variation, and have started instead with the will, which is only an *attitude* . . ."[41] "Ideal" aims, he said, were not necessarily "higher" than personal ones, indeed they were more likely to be "conventional humbug." He pilloried the absurdity of Hegelianism by paraphrasing its message, saying that history had begun in Eden and had its end in Berlin. Nietzsche he dismissed similarly for using an abstraction, the Will to Power. But "what power would be when attained and exercised remains entirely beyond his horizon."[42] He described German philosophy as a work of genius but then qualified it in this way: "Idealism simply overlooks the all-important fact that our whole life is a compromise, an incipient loose harmony between the passions of the soul and the forces of nature."[43]

Better Fighters, Who Lost the War

So far then, in this account of World War I, the German genius has taken rather a battering. But there are other ways of looking at the events of 1914–18 and in his study of the German army and general staff, *A Genius for War: The German Army and General Staff, 1807–1945* (1977), Colonel Trevor Dupuy concluded "that the Germans, uniquely, discovered the secret of *institutionalising* military excellence." Particularly in World War I, Dupuy showed that although Germany was on the losing side, its defeat was due to superior numbers of the enemy and that, in most battles, and man-for-man, the Germans were better fighters.[44]

During World War I, the Germans mobilized about 11 million men and suffered almost exactly 6 million casualties. Against Germany alone, the Allies mobilized roughly 28 million men, more than two-and-a-half times as many, and casualties against Germany (ignoring Austria-Hungary, Turkey, and Bulgaria) totaled about 12 million. "Thus on average each mobilised German soldier killed or wounded slightly more than one Allied soldier; it took five Allied soldiers to incapacitate one German." On the other hand, the Germans were more often than the Allies in defensive positions, and experience shows that troops on the defensive have the advantage of position, fortification, and so on, with research confirming that defensive positions are, roughly speaking, 1.3 times as efficient as attacking ones. Taking this into account, Dupuy concluded that "there was an overall German superiority of 4 to 1 in inflicting casualties."[45] In a parallel study, Alexander Watson reported that the percentage of *Krankheiten des Nervengebiets* in the German west field army and nervous disorders in the British army, were 3.67 and 3.27 respectively, though the overall psychiatric casualties among the British battle injuries was 6.54 percent.[46]

In the end, the superior numbers of the Allies proved decisive (and the Germans were outperformed in the activities of intelligence and spying), but, man-for-man, German soldiers were better fighters.

Dupuy carried out his studies as part of the Historical Evaluation and Research Organization (HERO), looking at some sixty engagements in World War II, mainly in 1943 and 1944, later extended backward to World War I. He was particularly concerned to show that the Germans were not especially militaristic.[47] Between 1815 and 1945, Prussia and Germany participated in six significant wars (two of them minor), whereas during that time France was engaged in ten significant wars (six in Europe, four overseas), Russia fought thirteen wars (ten European), Great Britain fought seventeen wars (three in Europe, four in Africa, ten in Asia), and the United States was engaged in seven significant wars.[48] Dupuy's point was that there is no evidence that Prussians, or Germans, are excessively militaristic in any genetic or historical sense.

Instead, he argued, as Paul Kennedy argued (as discussed in Chapter 22, p. 417), that the Germans' superior fighting ability was due to their *institutionalization* of military excellence. These same characteristics, he said, would distinguish the German soldiers in World War II as well. "It was only Hitler who was rigid and inflexible."

Still on the technical war front, the Germans—surprisingly enough, given their level of technological and industrial development—were slower than the Allies in bringing in scientists to aid the war effort. They did do work on communications with submarines, and they developed some flame-throwing devices, but their tank experiments were too late to have any real impact on events. As the fighting turned against Germany, the enhancement of food production became a scientific priority. Two dark areas where Germany did lead the way were in the realms of chemical and aerial warfare. Under Fritz Haber, himself a future winner of the Nobel Prize, three other future laureates were pressed into service in the design and production of chlorine gas as a weapon—James Franck, Gustav Herz, and Otto Hahn, the future discoverer of nuclear fission.[49]

At the beginning of 1918, the Wehrmann in Vienna was abandoned. "The number of visitors grew smaller and smaller as the war stretched on, until finally nobody looked after him at all," reported one newspaper. Another noted, shortly after the war, that golden nails donated to the Wehrmann by Austria's allies had been stolen. "The last visitor was, then, a thief."

30.

Prayers for a Fatherless Child: The Culture of the Defeated

At no other time in the twentieth century has verse formed the dominant literary form, as it did in World War I (at least in the English language), and there are those, such as Bernard Bergonzi, whose words these are, who argue that English poetry "never got over the Great War."[1] It was no different in Germany where, according to one estimate, some 2 million war poems were written in the German language during the course of the war, and where in August 1914, 50,000 poems were written *every day*. Five hundred were submitted to newspapers every day and one hundred printed.[2] As with British war poetry, says Patrick Bridgwater, most of the German poems broke with tradition in that, until then, most war poetry had glorified war, in particular the heroic and chivalrous aspects of hand-to-hand combat. The advent of mechanized war changed all that.

German poets differed from their British equivalents in several ways.[3] Both Georg Heym and Georg Trakl wrote poems *about* war, or predicting war, well before hostilities broke out, as a test of heroic qualities. Once the fighting started, both Rainer Maria Rilke and Stefan George wrote verse ("hymns" in Rilke's case) about the war without actually seeing any action. George showed a marked indifference to the fighting, an indifference to others' suffering, revealing his conviction that modern war is "bestial rather than heroic," and most deaths had no dignity:

Zu jubeln ziemt nicht: kein triumph wird sein
Nur viele untergänge ohne würde. . .
Heilig sind nur die säfte
Noch makelfrei verspritzt—ein ganzer strom.

(There is no call for rejoicing: there will be no triumph, only many deaths without dignity . . . Holy alone the blood which is spilled innocently, a great river.)

Change came in the second winter of the war, with a more direct reaction to the horrors the poets were seeing around them, though it was slower in coming to the Germans than to the British.[4] Within this general picture, and against a background where, in an ideal world, perhaps, a dozen poets would be worth considering, three stood out—Georg Trakl, August Stramm, and Anton Schnack.

Trakl was Austrian, a man obsessed, as he put it himself, with his own "criminal melancholy."[5] He actually wrote very few war poems—just five—but they were all memorable. Trakl's gift was for very dense images, in the manner of Hölderlin, and his central message, once war had broken out, and once he could see what was happening, was that "the war may mark the end of man as a spiritual being":

Am Abend tönen die herbstlichen Wälder
Von tödlichen Waffen, die goldnen Ebenen
Und blauen Seen, darüber die Sonne
Düstrer hinrollt; umfängt die Nacht
Sterbende Krieger, die wilde Klage
Ihrer zerbrochenen Münder.
Doch stille sammelt im Weidengrund
Rotes Gewölk, darin ein zürnender Gott wohnt,
Das vergossne Blut sich. . .

Die heisse Flamme des Geistes nährt heute ein gewaltiger Schmerz,
Die ungebornen Enkel.

(In the evening autumn forests ring with deadly weapons, gold plains

> and blue lakes, over which the sun more darkly rolls; night embraces dying warriors, the wild lament of their broken mouths. Yet silently in the willow-grove a red cloud gathers, in which an angry god resides, shed blood gathers . . . a mighty grief today feeds the hot flame of the spirit, the unborn grandchildren.)[6]

Even when Trakl is describing red clouds, shed blood, and the hot flame of the spirit, his words are sparing; his effect is achieved by the cumulative nature of cool, honed images, icy and invigorating, stopping short of sentimentality.

The poems of August Stramm were much shorter than Trakl's, than anyone else's for that matter, making use of onomatopoeic and alliterative devices, neologisms and word layout, all designed to intensify the poetic experience, just as war intensified *all* experience associated with it.[7] Born in Münster, Westphalia, in 1874, Stramm was called up immediately on the declaration of war. He served on the Western Front to begin with, saw heavy fighting in northern France, and had won the Iron Cross by January 1915. In April he was transferred to the Eastern Front, where he again saw heavy fighting and was recommended for the Iron Cross (First Class). His publishers negotiated a release for him from the military, but he refused to take up the offer of "an alibi" and continued to serve. He had seen action seventy times when, on September 1, 1915, he was shot in the head in hand-to-hand fighting on the Rokitno marshes.[8]

Clearly a brave man, Stramm was nonetheless against the war and did not write a single chauvinistic poem even when hundreds of people around him were doing so. He wrote instead about how fear turns to courage, how ordinary law-abiding people are converted into murderers, and how—again—there is nothing heroic about modern warfare. Most of his poems were published in *Der Sturm* and collected after his death. This is "Schlachtfeld" (Battlefield), written in the autumn of 1914.

Schollenmürbe schläfert ein das Eisen
Blute filzen Sickerflecke
Roste krumen
Fleische schleimen
Saugen brünstet um Zerfallen.

Mordesmorde
Blinzen
Kinderblicke.

(Clod softness lulls iron off to sleep, bloods clot ooze patches, rusts crumble, fleshes slime, sucking ruts around decay. Child eyes blink murder upon murder.)[9]

The neologisms (bloods, not blood; rusts, not rust; fleshes, not flesh) emphasize that there are many men—not just the poet—who suffer, the lack of punctuation conveys the way everything in the battlefield is chaos, one thing running into another, just as dying can be achieved by oozing to death as by being killed outright in a flash. Iron sleeps, iron weapons can be killed as people can—there is no difference here in No Man's Land.[10]

In *"Angststurm" (Attack of Fear)*

Grausen
Ich und Ich und Ich und Ich
Grausen Brausen Rauschen Grausen
Träumen Splittern Branden Blenden
Sterneblenden Brausen Grausen
Rauschen
Grausen
Ich

(Dread. Me and me and me and me. Dreading roaring crashing dreading. Dreaming splintering burning dazzling. Dazzling star-shells roaring dreading. Crashing. Dread. Me.)[11]

This poem has been described as a chain of battle sounds and the reactions they provoke, themselves set out as a rattle, the "*au*" sound similar to a cry of pain.

Born in 1892 at Rieneck in Unterfranken, Anton Schnack produced a steady stream of verse from January 1917 onward, mainly in broken sonnet form, the most important examples of which were collected in his *Tier rang gewaltig mit Tier*, released in 1920, and consisting of sixty war poems. Generally regarded as the best single collection of war poems produced by a German poet, it has been compared with the works of the Britons Wilfred Owen and Isaac Rosenberg. The poems juxtapose original observations cheek-by-jowl with more ordinary, even banal, images, reminding us that "poetry is not an end in itself," that to cull beauty in such circumstances cannot be wholly appropriate, that the *accumulation* of images, of experiences, is as much an aspect of war as the vivid—short, intense—flashes and explosions of metaphor or simile.

"Im Granatloch" (In a Shellhole) tells about life in a temporary trench, which ends:

Was sang Ninette? . . . Leichtes, Südliches.—Weinen will ich, dass
ich lagere in Mord und Stürmen, im blauen Raketenmeer, im
Sausen des Windes,
Unter lärmenden Nachthimmeln, in grünen Wassern voll Schnecken
und roten Würmern, in Erwartung des Todes, faul und gross;
im Sterbeschrei der Pferde,
Im Sterbeschrei der Menschen, ich hörte Dunkle rufen aus Dunkelm,
Hängend in Drähten: so singen Vögel, die sterben wollen,
einsam, vertrauert, in Frühlingsjahren.
Und, über dem Rheine, weit, das schwerbestürmende, eines
vaterlosen Kindes. . .

(What was it Ninette used to sing? . . . Something gay, something southern.—I could cry that I am lying here amid murder and assaults, in a blue sea of rockets, in the wind's sighing, beneath turbulent night skies, in green waters full of snails and red worms, awaiting death, putrid and swollen, amid the dying screams of horses, amid the dying screams of men, I heard them, calling out of the dark, hanging in the wire; thus do birds sing who are ready to die, lonely, pining away, in the spring of their lives. And beyond the Rhine, far away, somebody opened a creaking door, and from the opening came prayer, the overwhelming prayer of a fatherless child . . .)[12]

Nothing chauvinist here, nothing about German "Kultur" and its alleged superiority. The overall tone is by no means bitter, but rather elegiac at the pity of it all.

Both Bertolt Brecht and Karl Kraus wrote bitter antiwar and savagely satirical poems toward the end of the hostilities. They were not always successful: satire, especially in such a context, risked being seen as "un-German."

Not only poets died. August Macke, the Blaue Reiter painter, was shot as the German forces advanced into France; Franz Marc was killed at Verdun; Max Planck lost one son (another, Erwin, was executed in 1945 for his part in the resistance against Hitler), as did the painter Käthe Kollwitz (she also lost her grandson in World War II); Oscar Kokoschka was wounded, and Albert Einstein ostracized. The mathematician and philosopher Ludwig Wittgenstein was interned in a Campo Concentramento in northern Italy, from where he sent Bertrand Russell the manuscript of his recently completed work, *Tractatus Logico-Philosophicus.*

The war produced many intellectual and cultural ramifications. Some of them took years to reveal themselves, but not all.

In film, although Germany had a strong industry in 1914, it was still dominated by output from abroad, France, America, and Italy in particular. At the outbreak of war, the importation of foreign films was stopped. Cinema audiences rose in the war—the combination of entertainment and newsreels proving irresistible, though film equipment was so bulky then that real action footage was rare and fiction films about war gradually took over. Documentaries were often more subtle, such as Ernst Lubitsch's film about the entry of women into male professions.

Theater was more lively, and more critical. Georg Kaiser's *Gas I* was not set anywhere specific, but its plot, pitting the manufacturer, who wants to cease production, against the military and industrial chiefs, who want ever-greater quantities turned out, was close to the bone. Ernst Toller, who had fought at Verdun and suffered a breakdown, wrote *Die Wandlung* (*The Transformation*), about the metamorphosis of an enthusiastic volunteer into an artist who leads an uprising (as Toller himself did in the Bavarian Soviet), and he also came too close to current events: his play could not be produced for some months. (Although it did appear late in 1919, while its author was still in prison—where the play had been written—serving a sentence for treason, for his part in the uprising.)

But it is Karl Kraus's *Die letzten Tage der Menschheit* (*The Last Days of Mankind*) that demands most attention. Written between 1915, when the first segment was published, and 1922, it featured a large cast, innumerable dialects, and it brilliantly and bitterly exposed the mendacities of the authorities, the dishonest jingoism of the media, in a world where all authority had broken down. Kraus's target, for which he often used actual documentation, was the "wilful romanticization of war," the "predatory greed" and "insatiable imperialism" that were for him the real driving force behind the war effort. It is now generally accepted that much of what Kraus put in his play was factually wrong, but dramatically it was the equivalent of Otto Dix's frightening automaton/cripples that became so familiar in the Weimar Republic, and his thesis is an anticipation of Hannah Arendt's idea about "the banality of evil."[13]

One of the other main changes wrought by the war lay in the field of psychiatry, where two developments overtook psychoanalysis.

By the time of the outbreak of the fighting, psychoanalytic societies existed in six countries and an International Association of Psychoanalysis had been formed in 1908. At the same time, the "movement," as Freud thought of it, had seen a number of other prominent figures emerge—most of them German-speaking—and suffered its first defectors. Alfred Adler, along with Wilhelm Stekel, left in 1911, Adler because his own experiences gave him a very different view of the psychological forces that shape personality. He conceived the idea that the libido is not a predominantly sexual force but inherently aggressive, the search for power becoming for him the mainspring of life and the "inferiority complex" the directing force that gives lives their shape. His phrase "inferiority complex" passed into general usage.

Freud's break with Carl Jung, which took place between the end of 1912 and early 1914 was more serious—and more acrimonious—than any of the other schisms because Freud, fifty-eight at the outbreak of war, saw Jung as his successor.[14] Although Jung was devoted to Freud at first, he had squabbled with other early analysts, and the break with the master came because, like Adler, Jung revised his views on two fundamental Freudian ideas. He thought that the libido was not, as Freud insisted, a solely sexual instinct but more a matter of "psychic energy" as a whole, a reconceptualization that vitiated the entire idea of childhood sexuality, not to mention the Oedipal relationship. Second, and perhaps even more important, Jung argued that he had discovered the existence of the unconscious for himself, independently of Freud.

He had discovered this, he said, when he realized that a woman he was treating, at Burghölzli hospital in Zurich, who was allegedly suffering from an untreatable mental illness, dementia praecox, had in fact killed her favorite child (by poisoning her with infected water) in order to free herself for a lover, the woman acting from an unconscious desire to obliterate all traces of her present marriage to make herself available for the man she really loved. Jung did not at first query the diagnosis of dementia praecox. The real story only emerged when he began to explore her dreams, prompting him to give her the "association test." This test, which subsequently became famous, was invented by Wilhelm Wundt (see Chapter 26, p. 489). The patient is shown a list of words and asked to respond to each one with the first word that comes into his/her head. The rationale is that in this way conscious control over unconscious urges is weakened. Using the test Jung revealed the woman's unconscious motives and was able to face her with the unpleasant truth. Within weeks, he claimed, she was cured.

There is already something defiant about Jung's account of his discovery of the unconscious, the Swiss implying he was not so much a protégé of Freud's as his equal. Soon after they met, they became very close and in 1909 traveled to America together. Jung was overshadowed by Freud in America, but it was there that he realized his views were diverging. As the years passed, patient after patient reported early experiences of incest, all of which encouraged Freud to lay even more emphasis on sexuality as the motor driving the unconscious. For Jung, however, sex was not fundamental—instead, it was itself a transformation from religion. When he looked at the religions and myths of other peoples around the world, as he began to do, he found that in Eastern religions (Hinduism, for example) the gods were depicted in temples as very erotic beings. For him, this frank sexuality was a symbol and one aspect of "higher ideas." Thus he began his examination of religion and mythology as "representations" of the unconscious "in other places and at other times."

The rupture with Freud first broke into the open in 1912, after they returned from America and Jung published the second part of *Wandlungen und Symbole der Libido* (translated as *Symbols of Transformation*). This extended paper, which appeared in the *Jahrbuch der Psychoanalyse*, was Jung's first and public airing of what he called the "collective unconscious." He concluded that at a deep level the unconscious was shared by everyone—it was part of the racial memory. Indeed, for Jung, that's what therapy *was*, getting in touch with the collective unconscious. The more Jung explored religion, mythology, and philosophy, the further he departed from Freud and from the scientific approach. He pointed first to the "extraordinary unanimity" of narratives and themes in the mythologies of different cultures. He next argued that "in protracted analyses, any particular symbol might recur with disconcerting persistence but as analysis proceeded the symbol came to resemble the universal symbols seen in myths and legends." Finally he claimed that the stories told in the delusions of mentally ill patients often resembled those in mythology.

Jung's other popular idea was the notion of archetypes, the theory that all people may be divided according to one or another basic (and inherited) psychological types, the best known being introvert and extrovert. These terms only relate to the conscious levels of the mind, of course; in typical psychoanalytic fashion, the truth is really the opposite—the extrovert temperament is in fact unconsciously introvert, and vice versa.

Although Jung's very different system of understanding the unconscious had first come to the attention of fellow psychoanalysts in 1912, it was only with the release of *Symbols of Transformation* in book form in 1913 (published in English as *The Psychology of the Unconscious*) that the split with Freud became public. Freud, while troubled by this personal rift, which also had anti-Semitic overtones, was more concerned that Jung's version of psychoanalysis was threatening its nature as a science.[15] Henceforth, Jung's work grew increasingly metaphysical, even quasi mystical, attracting a devoted but fringe following. From World War I on, the psychoanalytic movement was divided into two.

Thanks to World War I, however, psychoanalysis was undergoing another change—it was achieving respectability. Until the war, it had still been regarded as an exotic speciality—or worse, British doctors referring slightingly to Freud's "dirty doctrines." What caused a change was the fact that, on both sides in the war, a growing number of casualties were suffering from shell shock (or combat fatigue, or battle neurosis, or post-traumatic stress disorder, to use the terms now favored). There had been cases of men breaking down in earlier wars, but their numbers had been far fewer than those with physical injuries. What seems to have been crucially different this time was the character of the hostilities—static trench warfare with heavy bombardment, and vast conscript armies, containing large numbers of men unsuited for war. The considerable incidence of battle neurosis shook psychiatry and medicine as a whole.

Psychoanalysis was not the only method of treatment tried. Both the Allied and Central Powers found that officers were succumbing as well as enlisted men, in many cases highly trained and hitherto very brave men; these behaviors could not in any sense be called malingering. As one of Freud's biographers says, the Freudian age dates from this moment.

Early on it was discovered that men with neuroses could not be moved from the Front; if they were, they never came back and became a "pension burden." In Germany there were several other methods of so-called aggressive treatments that were tried. One involved "phoney operations" in which the patient was encouraged to believe his illness was somatic and surgically curable; another was an isolation technique, an attempt to "bore" the patient out of his symptoms by deprivation of food, light, and human contact. Work groups of neurotic patients were formed, where the life was more arduous than at the Front. The two most widespread techniques were Max Nonne's hypnotherapy in which it was suggested to patients that their symptoms were not real, and Fritz Kaufmann's "overpowering" electrotherapy method in which patients were told to expect painful electric shocks through their brains and body and were then "ordered" by a doctor and superior officer to lose their symptoms and get well. These techniques sound bizarre now but success rates of 90 percent or above were claimed, much reducing the pension burden.[16]

Tectonic Shifts

As the 1915 publication of Freud's *The Psychopathology of Everyday Life* in English in Britain shows, intellectual life did continue during the war, and it was not always disfigured by nationalistic, or chauvinistic, emotions.

Two other German ideas first saw the light of day in the 1914–18 period, both extremely influential and each having nothing to do with war.

Alfred Wegener, born in Berlin in 1880, was a meteorologist who received his PhD from the University of Berlin. An Arctic explorer, who was wounded in World War I, he first aired his ideas about "continental drift" in 1912 at a meeting of the German Geological Association at Frankfurt, but his full theory wasn't set down in book form, *Die Entstehung der Kontinente und Ozeane* (*The Origin of Continents and Oceans*), until 1915.[17] His idea, that the six continents of the world had begun life as one "supercontinent," was not wholly original—it had been aired earlier by an American, F. B. Taylor, in 1908—but Wegener collected much more impressive evidence than anyone else so that his theory, much ridiculed at first, eventually convinced most skeptics. In fact, with the benefit of hindsight one might ask why scientists had not reached Wegener's conclusion sooner. By the end of the nineteenth century it was obvious that to make sense of the natural world, and its distribution around the globe, some sort of coherent explanation was needed. For example, there is a mountain range that runs from Norway to north Britain and should cross in Ireland with other ridges that run through north Germany and southern Britain. In fact, it looked to Wegener as though the crossover actually occurs near the coast of North America, as if the two seaboards of the North Atlantic were once contiguous. Similarly, plant and animal fossils are spread about the earth in a way that can only be explained if there were once land connections between areas that are now widely separated by vast oceans.

Wegener's answer was bold. The six continents as they now exist—Africa, Australia, North and South America, Eurasia and Antarctica—were once one huge continent, one enormous landmass which he called Pangaea (from the Greek, for *all* the *earth*). The continents had arrived at their present position by "drifting," floating like huge icebergs.

The idea took some getting used to, but it could not go unexamined.[18] How could entire continents "float"? And on what? If the continents had moved, what enormous force had moved them? By Wegener's time, the earth's essential structure was known. Geologists had used analysis of earthquake waves to deduce that the earth consisted of a crust, a mantle, an outer core, and an inner one. The first basic discovery was that all the continents of the earth are made of one form of rock, granite. Around the granite continents are found a different form of rock—basalt, much denser and harder. Basalt exists in two forms, solid and molten (we know this because lava from volcanic eruptions is semi-molten basalt). This suggests that the relationship between the outer structures and the inner structures of the earth was clearly related to how the planet formed as a cooling mass of gas that became liquid, then solid.

The huge granite blocks that form the continents are believed to be about 50 kilometers thick, but below that, for about 3,000 kilometers, the earth possesses the properties of an "elastic solid," or semi-molten basalt. Millions of years ago, when the earth was much hotter than it is today, the basalt would have been less solid, and the overall situation of the continents would have resembled more closely an iceberg floating in the oceans. Even so, it took time for the notion of continental drift to be accepted—textbooks as late as 1939 were still treating it as "a hypothesis only." It was not until sea-floor spreading was confirmed in 1953, and the Pacific-Antarctic Ridge identified in 1968, that Wegener was finally vindicated.

The work that Ludwig Wittgenstein produced during the war was not a response to the fighting itself. At the same time, had Wittgenstein not been exposed to the real possibility of death, it is unlikely that he would have produced *Tractatus Logico-Philosophicus* when he did, or that it would have had quite the tone it did.[19]

Wittgenstein enlisted on August 7, the day after the Austrian declaration of war on Russia, and was assigned to an artillery regiment serving at Kraków on the Eastern Front. He later suggested that he felt the experience of facing death would, in some indefinable manner, "improve" him. On the first sight of the opposing forces, he confided in a letter: "Now I have the chance to be a decent human being, for I am standing eye to eye with death."

Wittgenstein was twenty-five when war broke out. His large family was Jewish, wealthy, perfectly assimilated into Viennese society. Franz Grillparzer was a friend of Ludwig's father, and Brahms gave piano lessons to both his mother and his aunt. The Wittgensteins' musical evenings were well known in Vienna: Gustav Mahler and Bruno Walter were both regulars, and Brahms's Clarinet Quintet received its first performance there. Margarete Wittgenstein, Ludwig's sister, sat for Gustav Klimt.

Ludwig was as fond of music as the rest of the family, but he was also the most technical and practical-minded. As a result, he wasn't sent to the Gymnasium in Vienna but to a Realschule in Linz, a school chiefly known for the teaching of the history master, Leopold Pötsch, a rabid right-winger who regarded the Hapsburg dynasty as "degenerate." There is no sign that Wittgenstein was ever attracted by Pötsch's theories, but a fellow pupil with whom he overlapped for a few months certainly was. His name was Adolf Hitler.

After Linz, Wittgenstein went to Berlin, where he became interested in philosophy. He also developed a fascination with aeronautics, and his father suggested he go to the University of Manchester in England, where there was an excellent engineering department. There he was introduced to Bertrand Russell's *Principles of Mathematics*. This book showed, or attempted to show, that mathematics and logic are the same, and for Wittgenstein the book was a revelation. He spent months studying the *Principles* and also Gottlob Frege's *Grundgesetze der Arithmetik* (*Fundamental Laws of Arithmetic*). In the summer of 1911 Wittgenstein visited Frege in Jena, and Frege was impressed enough by the young Austrian to recommend he study under Russell at Cambridge.[20]

Wittgenstein arrived there later in 1911, and by 1914 Luki, as he was called by then, began to form his own theory of logic. But, in the long vacation, he went home to Vienna, war was declared, and he was trapped. He proved brave in the fighting, was promoted three times, was decorated, but in 1918 was taken prisoner in Italy with half a million other soldiers. While incarcerated in a concentration camp, he concluded that the book he had just finished, during a period of leave, had "solved all the outstanding problems in philosophy" and that he would give up the discipline after the war and become a schoolteacher. He also decided to give away his fortune. He was as good as his word on both counts.

Wittgenstein had great difficulty finding a publisher for his book, which did not appear in English until 1922. But when it did appear, *Tractatus Logico-Philosophicus* created a sensation.[21] Many people did not understand it; others thought it stated the obvious. Maynard Keynes wrote to Wittgenstein, "Right or wrong, it dominates all fundamental discussions at Cambridge." In Vienna, it attracted the attention of the philosophers led by Moritz Schlick—a group that eventually evolved into the famous

Vienna Circle of logical positivists. Frege, whose own work had inspired the *Tractatus*, died without ever understanding it.[22]

Wittgenstein's major innovation was to realize that language has limitations, that there are certain things it cannot do and that these have logical and therefore philosophical consequences. Wittgenstein argues that it is pointless to talk about value—simply because "value is not part of the world." It therefore follows that all judgments about moral and aesthetic matters cannot—ever—be meaningful uses of language. The same is true of philosophical generalizations that we make about the world as a whole. They are meaningless if they cannot be broken down into elementary sentences "which really are pictures" of part of our world. Instead, we have to lower our sights, says Wittgenstein, if we are to make sense. The world can only be spoken about by careful description of the individual facts of which it is comprised. In essence, this is what science tries to do. Wittgenstein was saying that we can go no further than that. This is what he implied by his famous last sentence of the *Tractatus*: "Whereof one cannot speak, thereof one must be silent."[23]

One of the most influential postwar ideas in Europe was released in April 1918, in the middle of the Ludendorff offensive—what turned out to be the decisive event of the war in the West, when General Erich Ludendorff, Germany's Supreme Commander in Flanders, failed to pin the British against the north coast of France and Belgium and separate them from other forces, weakening himself in the process. In that month, Oswald Spengler, a schoolmaster living in Munich, published *Der Untergang des Abendlandes*, translated into English as *The Decline of the West.* He had actually written the book in 1914 and used a title he had first conceived even earlier, in 1912, but despite all that had happened he had changed hardly a word of his text, which he was to describe modestly ten years later as "*the* philosophy of our time."[24]

Spengler was born in 1880 in Blankenburg, southwest of Berlin, and grew up in a home steeped in "Germanic giants"—Richard Wagner, Ernest Haeckel, Henrik Ibsen, Friedrich Nietzsche, and Werner Sombart. For Spengler, there were two important personal turning points. He failed his doctoral thesis, which meant he became a writer, not an academic. And second, the Agadir incident in 1911, when Germany backed down after its cruiser, *Panther*, had sailed into the Moroccan port, bringing Europe to the brink of war.[25] Spengler felt this humiliation keenly and for some reason drew the conclusion that this marked the end of the realm of rational science that had arisen since the Enlightenment. It was now a time for heroes, not traders. He set to work on what would be his life's project, his theme being how Germany would be *the* country, *the* culture, of the future.

Spengler drew on eight civilizations to sustain his argument—the Babylonian, the Egyptian, the Chinese, the Indian, the pre-Columbian Mexican, the classical or Graeco-Roman, the West European, and the "Magian," a term of his own that included the Arabic, Judaic, and Byzantine. His main theme was to show how each of them went through an organic cycle of growth, maturity, and inevitable decline, one of his aims being to show that Western civilization had no privileged position in the scheme of things.[26] For Spengler, *Zivilisation* was not the end product of social evolution, as rationalists argued; instead it was *Kultur*'s old age. Moreover, the rise of a new *Kultur* depended on two things—the race and the *Geist* or spirit, "the inwardly lived experience of the 'we.'" For Spengler, rational society and science were evidence only of a triumph of the indomitable Western will, which would collapse in the face of a stronger will, that of Germany. Germany's will was stronger, he said, because her sense of "we" was stronger; the West was obsessed with matters "outside" human nature, like materialist science, whereas in Germany there was more feeling for the inner spirit—*this* is what counted.

The Decline was a great and immediate commercial success. Thomas Mann compared its effect on him to that of reading Schopenhauer for the first time, and Wittgenstein confessed himself "astounded" by the book. Elisabeth Förster-Nietzsche was so impressed that she arranged for Spengler to receive the Nietzsche Prize. This made Spengler a celebrity, and visitors were required to wait three days for an appointment to speak to him.

From the end of World War I throughout 1919, Germany was in chaos and crisis. Central authority had collapsed, revolutionary ferment had been imported from Russia (though Germany had helped export Lenin from Switzerland), and soldiers and sailors formed armed committees, called *Räte*, or "soviets." Whole cities were for a time governed at gunpoint. Eventually, the Social Democrats, the left-wing party that installed the Weimar Republic, had to bring in their old foes, the army, to restore order.[27] This was achieved but involved considerable brutality—thousands were killed.

Against this background, Spengler saw himself as the prophet of a nationalistic resurgence in Germany—he saw it as his role to rescue socialism from the Marxism of Russia and apply it in "the more vital country" of Germany.[28] A new political category was needed: he put Prussianism and socialism together to come up with National Socialism. This would lead men to exchange the "practical freedom" of America and England for the "inner freedom . . . which comes through discharging obligations to the organic whole." Among those impressed by this argument was Dietrich Eckhart, who helped form the Deutsche Arbeiterpartei (DAP) or German Workers Party (GWP), which adopted the symbol of the Pan-German Thule Society Eckhart had previously belonged to. This symbol of "Aryan vitalism," the swastika, now took on a political significance for the first time. Alfred Rosenberg was also a fan of Spengler and joined the GWP in May 1919. Soon after, he brought in one of his friends just back from the Front, a man named Adolf Hitler.

When war broke out, Thomas Mann—as we have seen—was as nationalistic as many others. He was not yet one of the giants of European literature, but he did have a growing reputation. He volunteered for the Landsturm, or reserve army, but the doctor who examined him was familiar with his work and, reasoning that he would make a greater contribution to the war effort as a writer rather than as a soldier, failed him physically for active service.

Like other intellectuals, Mann also saw the fighting as a clash of cultures, a battle of ideas. His first essay, "Thoughts in Wartime," written in August 1914, claimed he had seen the war coming, that Germany had been coerced into war by its "envious" adversaries and that, altogether, war was "a tremendous creative event," helping to stimulate "national unity and moral elevation."[29]

After "Thoughts in Wartime," Mann intended to spend the following months and years completing his next major work, *Der Zauberberg* (*The Magic Mountain*), which would contain much implied criticism of the corrupt prewar world that had led Europe to the abyss. But that was to reckon without his brother Heinrich. In his own lifetime, Heinrich would pass through the entire political spectrum, from (as we saw earlier) the editor of a racialist publication to become a supporter of Stalin. But in 1916, he published an essay on Émile Zola in a new dissident journal and in the essay there were a number of disparaging references to Thomas. Heinrich insisted that politics were important and he accused his brother of ignoring this dimension.[30]

So upset was Thomas by Heinrich's attack that he broke into work on *The Magic Mountain*, and devoted several months to a long essay, *Betrachtungen eines Unpolitischen* (*Reflections of a Nonpolitical Man*), which reached the bookshops just before the armistice in 1918. In this work, he discovered that he was more nationalistic than he had anticipated but, more important, he found he was a "profoundly apolitical being." This was not due to any failings in his education, he said, but as a matter of principle. Politics, he thought, "was not a fit occupation for aristocrats of the spirit." He therefore wrote about the war, in Walter Lacquer's words, with great confidence and "almost total abstraction."[31] "Mann thought of the war mainly as great drama, a conflict of ideas . . . He had attached certain attributes to the German spirit and also to the French, the Russian and the British; America had no civilisation and did not count."[32] His view, directly contradictory to Heinrich's, was that the "ultimate questions of mankind" could not be solved by politics.

A rambling but powerful (and in parts shrewd) critique of democracy, Mann pointed out its weakness and predicted it would not suit the Germans who, he thought, wanted and needed authority. He was dismissive, too, arguing that a democratic Germany would be "boring." Walter Lacquer again: "He lived to realise that the boredom of the 1910s was greatly preferable to the excitement of the 1930s."[33]

The Dada Virus

During the war many artists and writers retreated to Zurich in neutral (but German-speaking) Switzerland. James Joyce wrote much of *Ulysses* there; Hans Arp, Frank Wedekind, and Romain Rolland were also there. They met in the cafés of Zurich which for a time paralleled in importance the coffeehouses of Vienna at the turn of the century—the Café Odeon was the most well known. For many of those in exile in Zurich, the war seemed to mark the end of the civilization that had spawned them. It came after a period in which art had become a proliferation of "isms," when science had discredited both the notion of an immutable reality and the concept of a wholly rational and self-conscious man. In such a world, the Dadaists felt they had to transform radically the whole concept of art and the artist.

Among the other regulars at the Café Odeon were Franz Werfel, Alexei Jawlensky, and Ernst Cassirer, the philosopher. There was also an unknown German writer, a "Catholic anarchist" named Hugo Ball (1886–1927), and his girlfriend, Emmy Hennings (1885–1948).[34] Hennings was a journalist but also performed as a cabaret actress, accompanied by Ball on the piano. In February 1916 they opened a review or cabaret with a literary bent. It was ironically called the Café Voltaire (ironic because Dada eschewed the very reason for which Voltaire was celebrated) and occupied premises on the Spiegelgasse, a steep and narrow alley where Lenin lived.[35] Among the first to appear at Voltaire were two Romanians, the painter Marcel Janco and a young poet, Sami Rosenstock, who adopted the pen name of Tristan Tzara.[36] The only Swiss among the early group was Sophie Taeuber, Hans Arp's wife (he was from Alsace). Others included Richard Hülsenbeck and Hans Richter from Germany.

For a review in June 1916, Ball produced a program and it was in his introduction to the performance that the word "Dada" was first used. Ball's journal records the kind of entertainment at Café Voltaire: "Rowdy provocateurs, primitivist dance, cacophony and Cubist theatricals." Tzara always claimed to have found the word *Dada* in the Larousse dictionary, but whether the term ever had any intrinsic meaning, it soon acquired one, best summed up by Hans Richter. He said it had some connection "with the joyous Slavonic affirmative 'Da, da' . . . 'yes, yes' to life." In a time of war it lauded play as the most cherished human activity. Dada was designed to rescue the sick mind that had brought mankind to catastrophe and restore its health. Dada questioned whether, in the light of scientific and political developments, art—in the broadest sense—was possible. Rather than follow any of the "isms" they derided, Dada turned instead to childhood and chance in an attempt to recapture innocence, cleanliness, clarity—all as a way to probe the unconscious.

No one succeeded in this more than Hans Arp (1886–1966) and Kurt Schwitters (1887–1948). Arp produced two types of image. There were his simple woodcuts, toylike jigsaws; as children do, he loved to paint clouds and leaves in straightforward, bright, immediate colors.[37] At the same time he was open to chance, tearing off strips of paper that he dropped and fixed wherever they fell, creating random collages. Kurt Schwitters made collages too, but he found poetry in rubbish.[38] A cubist at heart, he scavenged his native Hanover for anything dirty, peeling, stained, half-burnt, or torn. Although his collages may appear to have been thrown together at random, the colors match, the edges of one piece of material align perfectly with another, the stain in a newspaper echoes a form elsewhere in the composition. The detritus and flotsam in Schwitters's collages were for him a comment, both on the culture that leads to war, creating carnage, waste, and filth, and uncomfortable elegies to the end of an era, a new form of art that was simultaneously a relic, a condemnation of that world, and a memorial.

Toward the end of the war, Hugo Ball left Zurich for the Ticino and the center of gravity shifted to Germany. It was in Berlin that Dada changed, becoming far more political. Berlin, amid defeat, was a brutal place. In November 1918, the month of the armistice, there was a general uprising, which failed, its leaders Karl Liebknecht and Rosa Luxemburg murdered. The uprising was a defining moment for, among others, Adolf Hitler, but also for the Dadaists.

It was Richard Hülsenbeck who transported "the Dada virus" to Berlin.[39] He published his Dada manifesto in April 1918, and a Dada club was established. Early members included Raoul Hausmann, George Grosz, John Heartfield, and Hanna Höch.[40] George Grosz and Otto Dix were the fiercest critics among the painters, their most striking image being the wretched half-human forms of the war cripple.[41] Their depiction of the prostheses with which these people were fitted made them seem half-human, half-machine, with an element of the puppet, the old order still in command behind the scenes.

The Stolen Victory

In explaining what came next, it is necessary to explore briefly the nature of Germany's defeat in World War I. Although by common consent all the European combatants had fought themselves to a standstill and Germany was, to all intents and purposes, the loser, it was possible for Germans to derive some—albeit grim—comfort from what had actually come to pass. As Wolfgang Schivelbusch puts it, in his book on *Die Kultur der Niederlage* (*The Culture of Defeat*), "Only America's sudden intervention, at the Allies's behest, had saved England and France from the coup de grâce of the spring invasion of 1918. Put simply, the Americans had stolen Germany's victory." To the Germans, the very fact that the Allies had "ridden America's coattails to victory" instantly converted the other European nations into "second-rate powers . . . Germany had not been subdued in and by Europe." This meant that Germany was a loser only in relation to America, that she, from now on, would be "the only serious participant in a future Europe-America duel."[42] Unlike the Triple Entente, Germany had fought the war, not with American aid, but using only its own resources. France was the real weakling and loser. Her old ideas of revanchism since 1870–71 had implied subduing Germany in a one-on-one contest, but France had suffered much damage and, by herself, would have been quickly overcome. After the war, General Ludendorff gave as his explanation for Germany's loss that she had not been defeated by the enemy but "stabbed in the back" by forces at home (*Dolchstosslegende*). The desire for a unanimity of national feeling, the *Burgfrieden* (literally "fortress-peace") remained strong.

There are two ways of looking at these remarks. One, that the German views about her 1919 predicament were real enough, sharpened by her defeat. Or that they were fantasies, ignoring the *Realpolitik* of the situation (*why* had America supported the Allies?). Either way, they played a part in what came next. As Norbert Elias has said, the defeat of 1918 had interfered with the whole process of Germany's "catching up."[43]

31.

Weimar: "Unprecedented Mental Alertness"

The old Vienna officially came to an end on April 3, 1919, when the republic of Austria abolished titles of nobility, forbidding the use even of "von" in legal documents. The peace left Austria a nation of only 7 million people with a capital that was home to 2 million of them. The years that followed brought famine, inflation, a chronic lack of fuel, and a catastrophic epidemic of influenza. Housewives were forced to cut trees in the woods, and the university closed because its roof had not been repaired since 1914. Coffee, historian William Johnston tells us, was made of barley, and bread caused dysentery. Freud's daughter Sophie was killed by the influenza epidemic, as was the painter Egon Schiele.

Freud, Hofmannsthal, Karl Kraus, and Otto Neurath all stayed on in Vienna and the Vienna-Budapest (and Prague) German-speaking axis did not disappear completely, still producing individuals such as Michael Polanyi, Friedrich von Hayek, Ludwig von Bertalanffy, Karl Popper, and Ernst Gombrich. But it wasn't the same, and these people came to prominence only after the Nazis forced them to flee to the West. Vienna was no longer the buzzing intellectual center it had been, its cafés no longer the warm, informal meeting place of a worldly elite. What Vienna offered now, as we shall see, was the occasional literary or scientific or philosophical firework, which had to shine the more brilliantly against a brighter firmament to the north. The city's thunder had been stolen, and her lightning too.

The First "Art Film"

Berlin was a different matter. Following World War I, Germany was turned almost overnight into a republic. The fact that this could happen at all shows what is often overlooked—that, as mentioned earlier, *some* parliamentary/democratic traditions had been established. Berlin remained the capital, but Weimar was chosen as the seat of the assembly after a constitutional conference had been held there to decide the form the new republic would take. The choice of Weimar was based partly on its reputation dating from the time of Goethe and Schiller, and partly on worries that the violence in Berlin and Munich would escalate if either of those cities was selected. (Hitler always hated the wit and cynicism of Berlin.) The Weimar Republic lasted for fourteen years, until Hitler came to power in 1933, "a tumultuous interregnum between disasters" which nevertheless managed to produce a distinctive culture both brilliant and singular, and despite the steady decay of the state's monopoly on violence during those years and which, as Norbert Elias has highlighted, was as much a part of Weimar culture as anything else. [1]

The period is conventionally divided into three clear phases. From the end of 1918 to 1924, "with its revolution, civil war, foreign occupation, and fantastic inflation, [there] was a time of experimentation in the arts." Expressionism dominated politics as much as it dominated painting or the stage. This was followed, from 1924 to 1929, by a period of economic stability, a relief from political violence, and increasing prosperity reflected in the arts by the Neue Sachlichkeit, the "new objectivity," a movement whose aims were "matter-of-factness," sobriety.[2] Finally, the period 1929–1933 saw a return to political violence, rising unemployment, and authoritarian government by decree; the arts were cowed into silence and replaced by propagandistic Kitsch.

After painting, the art form most influenced by Expressionism was film. In February 1920 a horror film was released in Berlin that was, in the words of one critic, "uncanny, demonic, cruel, 'Gothic,' a Frankenstein-type story filled with bizarre lighting and dark, distorted sets. Considered by many to be the first "art-film," *The Cabinet of Dr. Caligari* was a huge success, so popular in Paris that it played in the same theater every day between 1920 and 1927. But the film was more than a record breaker.

Caligari was a collaboration between two men, Hans Janowitz, a Czech, and Carl Meyer, an Austrian, who had met in Berlin in 1919. The film features the mad Dr. Caligari, a fairground vaudeville actor who entertains with his somnambulist, Cesare. Outside the fair, however, there is a second string to the story that is far darker. Wherever Caligari goes, death is never far behind—anyone who crosses him ends up dead. The darkest part of the story starts after Caligari kills two students—or thinks he has. In fact, one survives, and it is the survivor, Francis, who discovers that the sleepwalking Cesare is unconsciously obeying Caligari's instructions, killing on his behalf without understanding what he has done. Realizing he has been discovered, Caligari flees into an insane asylum, where Francis finds out that Caligari is also the *director* of the institute. Still, there is no escape for Caligari when his double life is exposed, he loses all self-control and ends up in a straitjacket.

This was the original plotline of *Caligari*, but before the film appeared it went through a drastic metamorphosis. Erich Pommer, one of the most successful producers of the day, and the director, Robert Wiene, actually turned the story inside out, rearranging it so that it is Francis and his girlfriend who are mad.[3] The ideas of abduction and murder are now no more than *their* delusions, and the director of the asylum is in reality a benign doctor who cures Francis of his evil thoughts.

Janowitz and Meyer were furious. In Pommer's version, the criticism of blind obedience had disappeared and, even worse, authority was shown as kindly, even safe. The irony was that Pommer's version was a great success, commercially and artistically, and film historians have often wondered whether the original version would have done as well. Perhaps there is a fundamental point here. Though the plot was changed, the *style* of telling the story was not—it was still expressionistic, a new genre.[4] Expressionism was a force, an impulse to revolution and change. But, like the psychoanalytic theory on which it was based, it was not fully worked out. The Expressionist Novembergruppe, founded in December 1918, was a revolutionary alliance of all artists who wanted to see change—Emil Nolde, Walter Gropius, Bertolt Brecht, Kurt Weill, Alban Berg, and Paul

Hindemith, among others. But revolution needed more than an engine; it needed direction. Expressionism never provided that. And perhaps in the end its lack of direction was one of those factors that enabled Adolf Hitler's rise to power.

It would be wrong, however, to see Weimar only as a temporary way station on the path to Hitler—it boasted many solid achievements. Not the least of these was the establishment of some very prestigious academic institutions, some of which are still centers of excellence even today. These included the Berliner Psychoanalytisches Institut—home to Franz Alexander, Karen Horney, Otto Fenichel, Melanie Klein, and Wilhelm Reich—and the Deutsche Hochschule für Politik (the German Institute for Politics), which had more than 2,000 students by the last year of the republic—its teachers included Sigmund Neumann, Franz Neumann, and Hajo Holborn. There was also the Warburg Institute of Art History, in Hamburg, with its impressive library. This library was the fantastic fruit of a lifetime's collecting by Aby Warburg, a rich, scholarly, and "intermittently psychotic individual" who shared Winckelmann's obsession with classical antiquity and the extent to which its ideas and values could be perpetuated in the modern world. The charm and value of the library was not just that Warburg had been able to afford thousands of rare volumes on many recondite topics, but the careful way he had put them together to illuminate one another: art, religion, and philosophy were mixed in with history, mathematics, and anthropology. The Warburg Institute would become the home of many important art historical studies throughout the twentieth century. In particular, Erwin Panofsky's way of reading paintings, his "iconological method," as it was called, would prove hugely influential after World War II.

Europeans had been fascinated by the rise of the skyscraper in America, but it was difficult to adapt on the eastern side of the Atlantic: the old cities of France, Italy, and Germany were all in place and were too beautiful to allow the distortion that very tall buildings threatened. But the new materials of the twentieth century, which helped the birth of the skyscraper, were very seductive and proved very popular in Europe, especially steel, reinforced concrete, and sheet glass. In the end, glass and steel had a bigger effect on European architects than concrete did, and especially on three architects who worked together with the leading industrial designer in Germany, Peter Behrens (see p. 509). These men were Walter Gropius, Ludwig Mies van der Rohe, and the Frenchman Charles-Édouard Jeanneret, better known as Le Corbusier. Each would make his mark but the first was Gropius. It was Gropius who founded the Bauhaus.

Influenced by Marx and William Morris, Gropius always believed, contrary to Adolf Loos, that craftsmanship was as important as "higher" art. Therefore, when the Grand Ducal Academy of Art, which had been founded in the mid-eighteenth century, was merged with the Weimar Arts and Crafts School, which had been established in 1902, Gropius was an obvious choice as director. The fused structure was given the name Das Staatliche Bauhaus Weimar, with Bauhaus chosen because it echoed the *Bauhütten*, medieval lodges where those constructing the great cathedrals were housed.[5]

The early years of the Bauhaus, in Weimar, were troubled and it was forced to move to Dessau, which had a more congenial administration. This seems to have brought about a change in Gropius himself.[6] He now announced that the school would concern itself with practical questions of the modern world—mass housing, industrial design, typography, and "the development of prototypes."[7]

After a lost war and an enormous rise in inflation, there was no social priority of greater importance in Weimar Germany than mass housing. And so Bauhaus architects were among those who developed what became a familiar form of social housing, the *Siedlung* or "settlement." Although the *Siedlungen* were undoubtedly better than the nineteenth-century slums they were intended to replace, the lasting influence of the Bauhaus has been more in the area of applied design. The Bauhaus philosophy, "that it is far harder to design a first-rate teapot than paint a second-rate picture," has found wide acceptance—folding beds, built-in cupboards, stackable chairs and tables, designed with mass-production in mind and with an understanding of the buildings these objects were to be used in. Bauhaus designers like László Moholy-Nagy never lost their utopian ideals.[8]

The Marriage of Freud and Marx

The catastrophe of World War I, followed by the famine, unemployment, and inflation of the postwar years, confirmed for many people Marx's theory that capitalism would eventually collapse under the weight of its own "insoluble contradictions."

In fact, it soon became clear that, despite the activities of theoreticians like Karl Kautsky (1854–1938), it wasn't communism that was appearing from the rubble of war in Germany, but fascism.[9] Some Marxists were so disillusioned by this that they abandoned Marxism altogether. Others remained convinced of the theory, despite the evidence. But there was a third group who wished to remain Marxists but felt that Marxist theory needed reconstructing if it were to remain credible. This group assembled in Frankfurt in the late 1920s and made a name for itself as the Frankfurt school, with its own institute—founded by a millionaire interested in Marxism—in the city. Thanks to the Nazis, the institute didn't stay there long, but the name stuck.[10]

The three best-known members of the school were Theodor Adorno (1903–69), a man who "seemed equally at home in philosophy, sociology and music," Max Horkheimer (1895–1973), a philosopher and sociologist, less innovative than Adorno but perhaps more dependable, and the political theorist Herbert Marcuse (1898–1979), who in time would become the most famous of all. Horkheimer was the director; in addition to his other talents, he was also a financial wizard who brilliantly manipulated the investments of the institute, both in Germany and afterward in the United States. In addition there were Leo Lowenthal, the literary critic, Franz Neumann, a legal philosopher, and Friedrich Pollock, who was one of those who argued—against Marx and to Lenin's fury—that there were no compelling reasons why capitalism should collapse.[11]

In its early years the school was known for its revival of the concept of alienation but the Frankfurt school developed this idea so that it became above all a *psychological* entity and one, moreover, that was not necessarily, or primarily, the result of the capitalist mode of production. Alienation, for the Frankfurt school, was more a product of all of modern life. This view shaped the school's second and perhaps most enduring preoccupation: the attempted marriage of Freudianism and Marxism. Marcuse took the lead to begin with, though Erich Fromm wrote several books on the subject later. Marcuse regarded Freudianism and Marxism as two sides of the same coin. Freud argued that repression necessarily increases with the progress of civilization; therefore, aggressiveness must be produced and released in ever greater quantities. So, just as Marx had predicted that revolution was inevitable, a dislocation that capitalism must bring on itself, so, in Marcuse's hands, Freudianism produced a parallel, more personal backdrop to this scenario, accounting for a build-up of destructiveness—self-destruction and the destruction of others.

The third contribution of the Frankfurt school was a more general analysis of the vital question of the day: "What, precisely, has gone wrong in Western civilisation, that at the very height of technical progress we see the negation of human progress: dehumanisation, brutalisation, the poisoning of the biosphere, and so on? How has this happened?" To try to answer this question, they looked back as far as the Enlightenment and then traced events and ideas forward to the twentieth century. They claimed to discern a "dialectic," an interplay between progressive and repressive periods in the West. Moreover, each repressive period was usually greater than the one before, owing to the growth of technology under capitalism, to the point where, in the late 1920s, "the incredible social wealth that had been assembled in Western civilisation, mainly as the achievement of capitalism, was increasingly used for preventing rather than constructing a more decent and human society." The school saw fascism as a natural development in the long history of capitalism after the Enlightenment, and in the late 1920s earned the respect of colleagues with its prediction that fascism would grow. The school's scholarship most often took the form of close readings of original material, from which views uncontaminated by previous analyses were formed. This proved very creative in terms of the understanding produced, and the Frankfurt method became known as critical theory. It was, in is way, an updating of the higher criticism.

The King of "Secret Germany"

The Psychoanalytic Institute, the Warburg Institute, the German Institute for Politics, and the Frankfurt school were all part of what Peter Gay has called "the community of reason," an attempt to bring the clear light of scientific rationality to communal problems and experiences. But not everyone believed that cold rationality was the answer.

One part of what became a campaign against the "cold positivism" of science in Weimar Germany was led by the *Kreis* (circle) of poets and writers that formed around Stefan George, "king of a secret Germany." In practice, the Kreis was more important for what it stood for than for what it produced (though a minority always had a high regard for George's poetry).[12] Several of its writers were biographers—and this wasn't accidental. Their intention was to highlight "great men," especially those from more "heroic" ages, men who had by their will changed the course of events. The most successful book of this genre was Ernst Kantorowicz's biography of the thirteenth-century emperor Frederick II (see p. 608). For George and his circle, Weimar Germany was a distinctly unheroic age; science had no answer to such a predicament, and the task of the writer was to inspire others by means of his superior intuition.

George never had the influence he expected because he was overshadowed by a much greater poetic talent, Rainer Maria Rilke. Born René Maria Rilke in Prague in 1875 (he Germanized his name only in 1897), Rilke was educated at military school. Early in his career, he tried writing plays as well as biography and poetry, but his reputation was transformed by *Fünf Gesänge* (*Five Cantos/August 1914*), which he wrote in response to World War I. Young German soldiers took his slim volumes with them to the Front, and his were often the last words they read before they died, making Rilke "the idol of a generation without men."

His most famous poems, the *Duineser Elegien* (*Duino Elegies*), were published in 1923, their strange mystical, philosophical, "oceanic" tone perfectly capturing the mood of the moment.[13] The bulk of the elegies were "poured out" in a "spiritual hurricane" in one week, between February 7 and 14, 1922.[14] After he had finished his exhausting week, Rilke wrote to a friend that the elegies "had arrived." In the poems he wrestles with the "great land of grief," casting his net over the fine arts, literary history, mythology, biology, anthropology, and psychoanalysis, exploring what each has to offer to help our suffering. The second elegy reads:

Earliest triumphs, and high creation's favourites,
Mountain-ranges and dawn-red ridges,
Since all beginning, pollen of blossoming godhead,
Articulate light, avenues, stairways, thrones,
Spaces of being, shields of delight, tumults

Of stormily-rapturous feeling, and suddenly, singly,
Mirrors, drawing back within themselves
The beauty radiant from their countenance.

Science, Modernity, and the Novel

Whereas Rilke shared with Hofmannsthal and Stefan George the belief that the artist can help shape the prevailing mentality of an age, Thomas Mann was more concerned, as Schnitzler had been, to describe that change as dramatically as possible. Though not as famous today (in Germany) as *Buddenbrooks*, *Der Zauberberg* (*The Magic Mountain*), published in 1924, did extremely well (it appeared in two volumes), selling 50,000 copies in its first year. It is heavy with symbolism (too heavy in translation), and the English translation has also succeeded in losing some of Mann's humor, not exactly a rich commodity in his work. Set on the eve of World War I, *The Magic Mountain* tells the story of Hans Castorp, "a simple young man" who goes to a Swiss sanatorium to visit a cousin who has tuberculosis (a visit Albert Einstein actually made, to deliver a lecture). Expecting to stay only a short time, he catches the disease himself and is forced to remain in the clinic for seven years. The overall symbolism is pretty obvious. The hospital is Europe, a stable, long-standing institution that is filled with decay and corruption. "Like the generals starting the war, Hans expects his visit to the clinic to be short, over in no time." Like them, he is surprised—appalled—to discover that his whole time frame has to be changed. Among the fellow inmates are rationalists, would-be heroes, and innocents. The inadequacies of science as a form of self-knowledge run through the book, Mann's goal being to sum up the human condition (at least, the Western condition), aware as Rilke was that a whole era was coming to an end, and that heroes were not the answer. For Mann, modern man was self-conscious as never before.

Thomas Mann has often been compared with Hermann Hesse—and as often contrasted. They were introduced in 1904 in Munich by Samuel Fischer, the publisher. They kept in touch all their lives, exchanging numerous letters, but only became real friends in the 1930s. Their careers had parallels and differences. Hesse cut himself off and remained more or less in one place in Switzerland; Mann moved on, and on. Mann was in favor of World War I, at least to begin with; Hesse opposed it, his "pacificist duet" with Romain Rolland during that time earning him as many enemies as friends. Both flirted with Jungian ideas and both won the Nobel Prize for Literature.

Hesse was a headstrong child and fiercely solitary, not much engaged with the world (other than his writing), and his first two marriages ended in failure.[15] He was a prolific author (seven novels, numerous volumes of poetry, 3,000 reviews; he edited fifty volumes of literary classics and wrote 35,000 letters). Many of his works are autobiographical, none more so than *Steppenwolf* (*Prairie Wolf*) which he began in the year *The Magic Mountain* appeared. It was published in 1928 and tells the story of Hans Haller (the same initials as Hesse himself), who leaves the manuscript of a book he has written to a chance aquaintance, the nephew of his landlady. When he reads the book (the book-within-the-book) the nephew finds that, magically, part of it is addressed directly to him. The book is about personality and human nature and whether we are one self or more than one and whether inner coherence is possible, even in principle.

It was a theme eerily paralleled (but in a very different way) by Robert Musil. His three-volume work, *Der Mann ohne Eigenschaften* (*The Man without Qualities*), the first volume of which was published in the same year as *Steppenwolf*, is for some people the most important novel in German written during the last century, eclipsing anything by Mann or Hesse.

Born in Klagenfurt in 1880, Musil came from an upper-middle-class family, part of the Austrian "mandinarate." He trained in science and engineering and wrote a thesis on Ernst Mach. *The Man without Qualities* is set in 1913 in the mythical country of "Kakania," clearly Austro-Hungary, the name referring to *Kaiserlich und Königlich,* or *K. u. K.*, standing for the royal kingdom of Hungary and the imperial-royal domain of the Austrian crown lands. The book, though daunting in length, is for many the most brilliant literary response to developments in other fields in the early twentieth century. There are three intertwined themes that provide a loose narrative. First, there is the search by the main character, Ulrich von . . . , a Viennese intellectual in his early thirties, whose attempt to penetrate the meaning of modern life involves him in a project to understand the mind of a murderer. Second, there is Ulrich's relationship (and love affair) with his sister, with whom he had lost contact in childhood. Third, the book is a social satire on Vienna on the eve of World War I.[16]

But the real theme of the book is what it means to be human in a scientific age. If all we can believe are our senses, if we can know ourselves only as scientists know us, if all generalizations and talk about value, ethics, and aesthetics are meaningless, as Wittgenstein tells us, how are we to live? asks Musil. At one point Ulrich notes that the murderer is tall, with broad shoulders, that "his chest cavity bulged like a spreading sail on a mast," but that on occasions he felt small and soft, like "a jelly-fish floating in the water" when he read a book that moved him. In other words, no one description, no one characteristic or quality, fits him. It is in this sense that he is a man without qualities: "We no longer have any inner voices. We know too much these days; reason tyrannises our lives."

Franz Kafka was also fascinated by what it means to be human and by the battle between science and ethics.[17] In 1923 when he was thirty-nine, he realized a long-cherished ambition to move from Prague to Berlin (he was educated in the German language and spoke it at home). But he was in the Weimar Republic less than a year before the tuberculosis in his throat forced him to transfer to a sanatorium near Vienna, where he died, aged forty-one.

A slim, well-dressed man with a hint of the dandy about him, he had trained in law and worked successfully in insurance. The only clue to his inner unconventionality lay in the fact that he had three unsuccessful engagements, two of them to the same woman.

Kafka is best known for three works of fiction, *Die Verwandlung* (*Metamorphosis*; 1916), *Der Prozess* (*The Trial*; 1925; posthumous), and *Das Schloss* (*The Castle*; 1926; also posthumous). But he also kept a diary for fourteen years and wrote copious letters. These reveal him to have been a deeply paradoxical and enigmatic man. He was engaged to the same woman for five years, yet saw her fewer than a dozen times in that period; he wrote ninety letters to one woman in the two months after he met her, including several of between twenty and thirty pages, and to another he wrote 130 letters in five months. He wrote a famous forty-five-page typed letter to his father when he was thirty-six, explaining why he was still afraid of him.

Although Kafka's novels are ostensibly about very different subjects, they have some striking similarities, so much so that the cumulative effect of Kafka's work is much more than the sum of its parts. *Metamorphosis* begins with one of the most famous opening lines in literature: "As Gregor Samsa awoke one morning from uneasy dreams he found himself transformed in his bed into a gigantic insect." If a man is turned into an insect, does this help him/us understand what it means to be human? In *The Trial,* Joseph K. (we never know his last name) is arrested and put on trial. Neither he nor the reader ever knows the nature of his offense, or by what authority the court is constituted, and therefore he and we cannot know if the death sentence is warranted. Finally, in *The Castle* K. (again, all we are told) arrives in a village to take up an appointment as land surveyor at the castle that towers above the village and whose owner owns all the houses there. However, K. finds that the castle authorities deny all knowledge of him, at least to begin with, and say he cannot even stay at the inn in the village. Characters contradict themselves, vary unpredictably in their attitudes to K., or lie. He never reaches the castle.

An added difficulty with interpreting Kafka's work is that he never completed any of his three major novels, though we know from his notebooks what he intended. He also told his friend Max Brod what he planned for *The Castle,* his most realized work. All three stories show a man not in control of himself, or of his life. In each case he is swept along, caught up in forces on which he cannot impose his will, where those forces—biological, psychological, logical—lead blindly. There is no development, no progress, as conventionally understood, and no optimism. It is bleak and chilling. W. H. Auden once said, "Had one to name the author who comes nearest to bearing the same kind of relation to our age as Dante, Shakespeare or Goethe have to theirs, Kafka is the first one would think of." Eerily, he also prefigured the specific worlds that were soon to arrive: Stalin's Russia and Hitler's Reich.

So too did Lion Feuchtwanger. It is a pity that he is less well known now than Mann, Kafka, Hesse, or even Musil. Quite apart from his books, his life was in some ways exemplary: he escaped *twice*, in two separate wars, as a POW.

Born in Munich in 1884, the son of a wealthy Jewish industrialist, and a frequent traveler, Feuchtwanger found himself in 1914, at the outbreak of war, in Tunisia, then a French possession. He was imprisoned as an enemy alien but escaped, returned to Germany, and enlisted. His first real success was *Jud Süss* (translated as *Power*), an exploration of anti-Semitism. Written in 1921, it wasn't published until 1925 because he couldn't find a publisher, though it became an immediate success.

His masterpiece was *Erfolg* (*Success*), 1930, a roman à clef about Weimar Germany, where Johanna Krain, a young woman, seeks to secure the release of her lover, Krüger, from prison. He is a museum curator who has offended the Bavarian authorities by showing two controversial paintings, one of which is a nude. Krüger is tried for adultery with the woman who posed for the painting and for breach of public morality. Several well-known figures and institutions are identifiable in the book (Hitler, Brecht, IG Farben). Krüger dies in prison but not before Feuchtwanger exposes and predicts all the corruptions and specious rationalizations of the world giving rise to the Third Reich.

Feuchtwanger's assets were seized by the Nazis, but he fled to France where, in 1940, he was held in a concentration camp. This time he escaped dressed as a woman and went first to Spain, then to the United States ("God's own country"), where he joined the growing band of talented exiles who would become known as "Hitler's Gift."[18]

In Chapter 29, p. 531, we saw that the experiences of men in modern technological warfare in the Great War were so extreme as to bring about a whole new set of psychological problems among the admittedly brave men in the trenches. Do these subconscious anxieties account for the delay in the appearance of so many war-related novels, a delay that occurred on both sides? Ford Maddox Ford's *No More Parades* was published in 1925, Ernest Hemingway's *The Sun Also Rises*, about an injured war veteran, appeared in 1926, while Siegfried Sassoon's *Memoirs of an Infantry Officer*, was not released until 1930. Between these last two, in 1928, there was the most successful of them all, at any rate commercially—*Im Westen nichts Neues* (*All Quiet on the Western Front*), by Erich Maria Remarque.

Christine Barker and R. W. Last argue that *All Quiet* is one of the most important books of the twentieth century, but neither one of the greatest, nor yet the best work of Remarque himself. Born in 1898 in Osnabrück, Remarque was dogged by controversy, in particular concerning what he actually did in World War I and whether he won the medals he said he did. He was never stationed at the Front, but it appears he did perform heroically, helping carry wounded soldiers out of danger.

After the war he began writing short stories and sketches, moved to Berlin in 1925, and worked as a journalist on *Sport im Bild* (*Sport in Pictures*).[19] *All Quiet* was written two years later, serialized in the *Vossische Zeitung* at the end of 1928 (eleven other papers turned it down) and appeared between hard covers in January 1929, where its overnight success altered Remarque's life forever.[20]

The novel tells the story of a class of young men who are sent to the war, are much depleted, and carry on the fight with older, more experienced men. There are many passages when the men/boys contemplate life back home, and the world of love which—for most of them—hasn't really opened up yet. The claustrophobia of war gradually closes in on the men as, out of the original eight school friends, only one remains. Remarque explores the different ways that the men/boys are alienated, as they try to grasp whether they are cowards or heroes, individuals or comrades-in-arms, proud combatants, or ashamed and disappointed. They come to realize how being in such a terrible war has cut them off from people who will never have this experience. Although the book has it share of clichés, some of the images have become famous, as that of the dying cigarette hissing on the lips of its already-dead owner.[21]

All Quiet was bleak, very bleak, but it stimulated a boom in war novels and provoked enormous controversy, for its writing style, its "defeatism," and its unpatriotic depiction of war. About a year after publication, by which time the book had sold close to a million copies and been translated (or was in the course of translation) into several languages, the Nazis turned on Remarque, making the book into a political issue because he had challenged the myth of individual heroism in armed conflict. The campaign was spearheaded by Goebbels himself and began when Hitler Youth disrupted the screening of the American film of the book in 1930.[22]

Remarque left Germany and eventually reached America. His bank account was seized, but he had wisely already moved most of his money, together with his collection of Impressionist and post-Impressionist paintings—Cézanne, van Gogh, Degas, Renoir. *All Quiet* was burned in the notorious book-burning in Berlin in May 1933, but in some ways Remarque had the last laugh. In America he went to Hollywood, where he formed firm friendships with Marlene Dietrich, Greta Garbo, Charlie Chaplin, Cole Porter, F. Scott Fitzgerald, and Ernest Hemingway. He even acquired the title "King of Hollywood" because of the number of films made from his books, including *The Road Back, Three Comrades, A Time to Love and a Time to Die, Heaven Has No Favorites,* and *Shadows in Paradise.* For Remarque, in these books nothing endures—each individual is alone, there will never be any clear answers to the problems and mysteries that trouble us; life may have its moments of extreme beauty and even happiness, but that is all they are, moments. Remarque said he believed that human nature in Germany was especially bleak and that it had been so since Goethe and *Faust.*[23]

A Disappearing Homeland Where Only the Children Are Innocent

While we may attach more importance now to completed or fully realized *books,* at the time of the Weimar Republic, two writers in Germany made more of a name for themselves from more traditionally ephemeral skills—satirical squibs in newspapers and magazines, songs and skits for cabaret, pointed book reviews, savage poetry, and whimsical—and not so whimsical—newspaper columns. In Berlin, in their way, Kurt Tucholsky (1890–1935) and Erich Kästner (1899–1974) were the equivalent of Karl Kraus in Vienna. They form a bridge between the full-time authors considered above and the full-time dramatists, considered next.

Both of them filled almost entire magazines (again like Kraus) using many noms-de-plume, both were much influenced by their mothers (though Tucholsky's relationship was strained), and both were turned into pacifists by their experiences in World War I.

In their writings, however, both were extremely combative. Tucholsky, born in Berlin but brought up in Stettin (now in Poland) was precocious, poking fun at Kaiser Wilhelm's taste in art when he was just seventeen. At twenty-three, he started writing for the theater magazine *Die Schonbühne,* later renamed *Die Weltbühne* (The World Stage), which he was subsequently to edit, and which under him became one of the most colorful Weimar journals. Tucholsky wrote all manner of articles—poetry, book reviews, lead articles, aphorisms ("Either you read a woman or you embrace a book"), even court reports—in the course of which he denounced the military, the judiciary, the censor, the bourgeoisie, and in particular the series of political murders carried out by the conservative revolutionaries, which he felt were inciting the "mob" in Germany into a mood where only the National Socialists would profit.

He badly wanted the Weimar Republic to succeed but, as Erich Kästner was to remark, Tucholsky was a "little, fat Berliner" who sought to "prevent a catastrophe with his typewriter." In 1924 he became Paris correspondent (and coeditor) of *Die Weltbühne*, paralleling the exile, and Francophilia, of his idol and fellow Jew, Heinrich Heine. From France he eyed Germany—and its changes—no less sharply, so much so that he was several times sued by victims who judged they had been libeled in his attacks.

His work culminated in *Deutschland, Deutschland über alles*, bitter social criticism, illustrated by John Heartfield, in 1929, which he nonetheless claimed was a work of love for his disappearing homeland. "They are preparing to head towards the Third Reich," he wrote prophetically.

His typewriter was less effective than he hoped, and in 1930 he moved permanently to Sweden. *Die Weltbühne* itself had come under increasing fire. Carl von Ossietsky, who had replaced Tucholsky as editor, had run an investigation in the periodical revealing the Reichswehr's illegal air rearmament and been imprisoned. Although he considered it, Tucholsky didn't return to Germany to support Ossietsky, and always regretted this failure (he had himself been indicted for writing a piece in which he declared that "soldiers are murderers"). He never believed—as others did—that Hitler's regime would disintegrate and, weakened by chronic sinusitis, he took an overdose of sleeping pills in December 1935. Before his death he had campaigned for Ossietsky to receive the Nobel Peace Prize, a campaign that was successful, but only a year after Tucholsky's suicide.

Kästner was born in Dresden, the son of a saddle maker and a hair stylist. He found his military training very brutal (he was only fifteen when World War I broke out), much preferring history, philosophy, literature, and theater, which he studied at the University of Leipzig after the war. He became a journalist on the *Neue Leipziger Zeitung*, using several pseudonyms, but was in Berlin from 1927, publishing poems, articles, and reviews in a variety of outlets, including the respected *Vossische Zeitung* and *Die Weltbühne*. He became a leading figure in the Neue Sachlichkeit (New Objectivity) movement which, despite its sober style, was a satirical force in Weimar.

The work for which he is most well known is *Emil und die Detektive* (*Emil and the Detectives*; 1928), which sold 2 million copies in Germany alone and set up a genre of children's detective stories. (R. W. Last described Kästner as "one of the greatest writers for children of all times.") But he wrote for adults too, to such effect that his works were banned in 1933 and burned along with those of Brecht, Joyce, Hemingway, and the Mann brothers. He was twice arrested by the Gestapo but stayed in Germany where he was eventually banned from writing.

Fabian, his best known work for adults, was originally titled *Der Gang vor die Hunde* (Going to the Dogs) and surprised many who only knew him through his *Emil* stories. Published in 1931, the novel is set in Berlin: "In the east resides crime, in the centre swindling, in the north misery, in the west lechery, and to all points of the compass destruction lurks." Fabian is employed as an advertising copywriter by a cigarette company, charged with conceiving slogans to perpetuate a corrupt, disabling system. Loose sexuality, fetid family relationships, and unemployment come together in a pessimistic satire condemning the Germany that would allow Hitler's rise, a world where, taking *Fabian* and *Emil* together, only the children are innocent.

At the same time Kästner breathed new life into poetry, much of which appeared in cabaret, becoming known as "public poetry," holding up a mirror to the age:

Was man auch baut—es werden stets Kasernen
Whatever is built—always turns into barracks.

The fact that Kästner wrote in many different forms was a reflection of his understanding that the situation in Weimar Germany was urgent. He was less interested in creating "art" than in having an effect on his readers. What he said about Tucholsky was equally true of himself: he wanted to prevent the cataclysm he saw coming with his typewriter. Although he (and Tucholsky) failed in that, as Clive James has said, the journalists in Weimar Germany nevertheless enriched German-speaking culture by saving it from the "stratospheric oxygen-starvation of the deliberately high-flying thesis."

A New Grammar for Music

Edgar Vincent, Viscount D'Abernon, the British ambassador to Berlin, described in his memoirs the period after 1925 as an "epoch of splendour" in the city's cultural life. Painters, journalists, and architects flocked to the city, but it was above all a place for performers. Alongside the city's 120 newspapers, there were forty theaters providing, according to one observer, "unparalleled mental alertness." It was also a golden age for political cabaret, satirical songs, Erwin Piscator's experimental theater, Franz Lehar's operettas, jazz, and Josephine Baker, though Harry, Count Kessler, in his diary, says he found her unerotic even when naked (he was homosexual).

Among this concatenation of talent, three figures from the performing arts stand out: Arnold Schoenberg, Alban Berg, and Bertolt Brecht. Between 1915 and 1923 Schoenberg composed very little, but in 1923 he gave the world what one critic called "a new way of musical organisation." Two years before, in 1921, Schoenberg, embittered by years of hardship, announced he had discovered "something which will assure the supremacy of German music for the next hundred years."[24] This was what became known as "serial music." Serialism is not so much a style as a "new grammar" for music. Atonalism, Schoenberg's earlier invention, was partly designed to eliminate the individual intellect from musical composition; serialism took that process further, minimalizing the tendency of any note to prevail. Under this system a composition is made up of a series from the twelve notes of the chromatic scale, arranged in an order that is chosen for the purpose and varies from work to work. Normally, no note in the row or series is repeated, so that no single note is given more importance than any other, lest the music take on the feeling of a tonal center, as in traditional music with a key. Its melodic line was often jerky, with great leaps in tone and gaps in rhythm. Huge variations were possible under the new system—including the use of voices and instruments in unusual registers. Rudolf Serkin spoke for many when he said he loved Schoenberg the man "but I could not love his music."[25]

The first completely serial work is generally held to be Schoenberg's

Piano Suite (op. 25), performed in 1923. Both Alban Berg and Anton von Webern enthusiastically adopted Schoenberg's new technique, and for many people Berg's two operas, *Woyzeck* and the "stately but brutal" *Lulu*, have become the most familiar examples of, first, atonality, and second, serialism. Berg began to work on *Woyzeck*, which was based on the short unfinished play by Georg Büchner (see pp. 301–303), in 1918, although it was not premiered until 1925, in Berlin.[26] Berg, a large, handsome man, had shed the influence of Romanticism less well than Schoenberg or Webern (which is perhaps why his works are more popular), and *Woyzeck* is very rich in moods and forms—rondo, lullaby, a military march, each character vividly drawn.[27] The first night, with Erich Kleiber conducting, took place only after "an unprecedented series of rehearsals," but even so the opera created a furor. It was labeled "degenerate," and the critic for *Deutsche Zeitung* wrote, "We deal here, from a musical viewpoint, with a composer dangerous to the public welfare." Yet it received "ovation after ovation," and other European opera houses clamored to stage it. Schoenberg was jealous.[28]

Lulu is in some ways the reverse of *Woyzeck*. Whereas the soldier was prey to those around him, Lulu is a predator, an amoral temptress "who ruins all she touches." Based on two dramas by Frank Wedekind (see Chapter 27, p. 503), this serial opera also verges on atonality. Unfinished at Berg's death in 1935, it is full of bravura patches, elaborate coloratura, and confrontations between a heroine-turned-prostitute and her murderer. Lulu is the "evangelist of a new century," killed by the man who fears her. The opera's setting was the very embodiment of the Berlin that Bertolt Brecht, among others, was at home in.

Like Berg, Kurt Weill, and Paul Hindemith, Brecht was a member of the Novembergruppe, founded in 1918 and dedicated to disseminating a new art appropriate to a new age. Though the group broke up after 1924 when the second phase of life in the Weimar Republic began, the revolutionary spirit survived. And it survived in style in Brecht. Born in Augsburg in 1898, Brecht was one of the first artists/writers/poets to grow up under the influence of film, and Charlie Chaplin in particular. Brecht was always fascinated by America and American ideas—jazz and the work of Upton Sinclair were to be other influences later.[29]

Bertolt (christened Eugen, a name he dropped) grew up in Augsburg as a self-confident and even "ruthless" child with, according to one observer, the "watchful eyes of a raccoon."[30] Initially a poet, he was also an accomplished guitarist, with which talent, according to some (like Lion Feuchtwanger) he used to "impose himself" on others, "smelling unmistakably of revolution." He collaborated and formed friendships with Karl Kraus, Carl Zuckmayer, Erwin Piscator, Paul Hindemith, Kurt Weill, Gerhart Hauptmann, Elisabeth Hauptmann, and an actor who "looked like a tadpole," Peter Lorre. In his twenties Brecht gravitated toward theater, Marxism, and Berlin.

War themes were not popular in the theater in Weimar Germany, and Brecht's early works, like *Baal,* steered well clear, earning him a reputation among the avant garde.[31] But it was with *Die Dreigroschenoper* (*The Threepenny Opera*) that he first found real fame. This was based on a 1728 ballad opera, *The Beggar's Opera,* by John Gay, which had been revived in 1920 by Nigel Playfair at the Lyric Theatre in London, where it ran for four years. John Gay's main aim had been to ridicule the pretensions of Italian grand opera, but after Elisabeth Hauptmann translated it for Brecht, he cleverly moved the action to Victorian times—nearer home—and made the show an attack on bourgeois respectability and its self-satisfied self-image.[32]

Rehearsals were disastrous. Songs about sex had to be removed because the actresses refused to sing them. The first night did not start well. The barrel organ designed to accompany the first song refused to function, and the actor was forced to sing the first stanza unaided (the orchestra rallied for the second verse, though "orchestra" is putting it a bit strongly—the show was scored for seven musicians playing twenty-three different instruments).[33] But the third song, the duet between Macheath and the Police Chief, Tiger Brown, reminiscing about their early days in India, was rapturously received. The opera's success was due in part to the fact that its avowed Marxism was muted. As Brecht's biographer Ronald Hayman put it, "It was not wholly insulting to the bourgeoisie to expatiate on what it had in common with ruthless criminals; the arson and the throat-cutting are mentioned only casually and melodically, while the well-dressed entrepreneurs in the stalls could feel comfortably superior to the robber gang that aped the social pretensions of the *nouveau riches.*" Another reason for the show's success was the fashion in Germany at the time for *Zeitoper,* opera with a contemporary relevance. Other examples in 1929–30 were Hindemith's *Neues vom Tage* (Daily News), a story of newspaper rivalry; *Jonny spielt auf,* by Ernst Krenek; Max Brandt's *Maschinist Hopkins;* and Schoenberg's *Von Heute auf Morgen.*

Brecht and Weill repeated their success with *Aufstieg und Fall der Stadt Mahagonny* (*The Rise and Fall of the City of Mahagonny*)—like *The Threepenny Opera,* a parable of modern society. The responses of audiences and critics were extreme either way and, as Weill put it, "Mahagonny, like Sodom and Gomorrah, falls on account of the crimes, the licentiousness and the general confusion of its inhabitants."[34] It was also epic theater, which for Brecht was central: "The premise for dramatic theatre was that human nature could not be changed; epic theatre assumed not only that it could but that it was already changing."[35]

The Nazis took increasing interest in Brecht and Weill. When the latter attended one of their rallies out of mere curiosity in 1929, he was appalled to hear himself denounced "as a danger to the country," together with Albert Einstein and Thomas Mann. He left hurriedly, unrecognized.

Second Only to Hollywood

When Remarque arrived in America, in Hollywood in particular, he probably felt more at home than he might have expected.[36] This chapter began with a description of a film, *The Cabinet of Dr. Caligari,* but although that was a seminal Expressionist work, it was a long way from being the only one. In fact, the 1920s, the years of the Weimar Republic, saw what was without question the golden age of German film, when German film rivaled Hollywood in its creativity and impact, when there was a golden generation of German film directors who have given us many of the most beautiful and important films ever made. Moreover, each of that generation—Fritz Lang (1890–1976), F. W. Murnau (1888–1931), Ernst Lubitsch (1892–1947), Robert Siodmak (1900–73), Billy Wilder (1906–2002), Otto Preminger (1906–86) and Fred Zinnemann (1907–97)—ended up in America. All but Murnau were Jewish, and all left either before Hitler came to power or not long afterward. It is safe to say that among them they helped create the cinema as we know it today.

Scholarship about Weimar cinema has been undergoing a certain amount of revision recently. Traditionally, film in the 1920s in Germany has been described as "Expressionist," as an art form where the emotion generated is more important than the accurate depiction of reality, where distortion is a favorite technique, together with exotic effects and characters, with fantasy and (frequently) horror being a dominant motif. While no one denies that these features did characterize German films in the early years of the Weimar period (or the theater of Reinhardt and Brecht, come to that), the general feeling now is that film was more influenced by art nouveau or the more mechanical modernism of Marcel Duchamp, Hans Richter, and Fernand Léger, rather than Kirchner, Klee, or Nolde. As the twenties passed, montage became a dominant technique, as in *Menschen am Sonntag* (People on Sunday, 1929), by Robert Siodmak, Billy Wilder, and Fred Zinnemann, and *Kuhle Wampe*, 1932, by Slatan Dudow and Bertolt Brecht.[37]

This was also the period when talking pictures eclipsed silent movies (1929), when cinema audiences went up everywhere, but in Germany, it was claimed, there was something else, something fundamental that helped explain the success of German films. It was shown at its clearest in a book published in 1930 by Siegfried Kracauer. He too went to America and would later write a seminal book about German film in Weimar times (*From Caligari to Hitler*), but his 1930 book was *Die Angestellten*, literally "The Employees" or "Workers," though it was more an examination of a new class that he thought had come into existence since World War I, more akin to what we would now call, in English, "white-collar workers."[38] Kracauer identified a rootlessness, a physical isolation, and an emotional insecurity that produced in this new group a longing for—and a love of—spectacle; as modern life became more streamlined and monotonized, so this new class developed in its leisure time what he called a "culture of distraction." Elizabeth Harvey likewise argues that mass media came of age in Germany in the Weimar years and that the change produced its effect more among the working class than the middle class and more among women than men.[39] Kracauer's book was the original model for a long line of popular sociological analyses that would dominate the Western world in the second half of the twentieth century, but he also sought to explain why there was such a demand for film in Weimar. More than sound, it was the distraction film offered, the mix of high and low culture that was, he said, the natural home of film. Going to the cinema, much more than the theater or the opera, offered an opportunity to experience what Hofmannsthal had called "ceremonies of the whole" of a kind and on a scale never before experienced, subverting and sabotaging the caste system that Bismarck's reforms had left intact.

Among the golden generation, who should take precedence? Probably Lubitsch. The son of a Jewish tailor, Ernst Lubitsch (1892–1947) was born in Berlin and, at the age of nineteen, joined Max Reinhardt's Deutsches Theater. He made his film debut a year later as an actor, but his main love was directing and, at the end of World War I, he had three hits, one after the other. The first was *Die Augen der Mummie Ma* (*The Eyes of the Mummy*; 1918), with Pola Negri in the starring role, followed by *Carmen* (*Gypsy Blood*), with the same star. Later that same year Lubitsch released *Die Austernprinzessin* (*The Oyster Princess*), a comedy of manners satirizing American foibles. In this film he first showed traces of what would become known as "the Lubitsch touch," gentle humor driven home by witty visual flourishes, contrasting with brief scenes—often a single shot—that summed up the characters' motivations and thereby explained the plot.

On the strength of these successes, Lubitsch went to Hollywood early, in 1922, and became known for two entirely different types of film—comedies, often absurd comedies, and grand historical dramas. He made a number of classic films toward the end of the silent era (*Lady Windermere's Fan* and *The Student Prince*), but when sound came along he responded with some of the earliest musicals—*The Love Parade, Monte Carlo, The Smiling Lieutenant.* In 1935 he was appointed production manager of MGM, becoming the only director to run a large studio. But the hits kept coming and in 1939 he directed Greta Garbo in *Ninotchka*, cowritten with Billy Wilder, and notable for the fact that, as the publicity for the film announced, "Garbo laughs!" Lubitsch had left Germany permanently when the Nazis achieved power, and he became an American citizen in 1936.

Fritz Lang (1890–1976) was born in Vienna, studied painting in Paris, and saw action in Russia and Romania in World War I, where he was wounded three times. He was in some ways the quintessential Expressionist director, who began by working for Erich Pommer's company and whose films in the pre-sound era were crammed with spies, dragons, historical heroes, master criminals, and tyrants. He loved big-budget epics and the special effects that Max Reinhardt had made so popular. Lang was probably the most famous film director in Weimar Germany, his work being likened to a cross between Franz Kafka and Raymond Chandler. In the silent era his biggest hits included *Metropolis,* the world's most expensive film when it was released, and *M,* a study of a child murderer (based on an actual case in Düsseldorf and starring Peter Lorre), who is tracked down and brought to justice by his fellow criminals. Many consider this Lang's masterpiece.[40] The famous story about Goebbels summoning Lang to his office, to tell him that his most recent film, *Das Testament des Dr. Mabuse* (*The Testament of Dr. Mabuse*), was being banned as an incitement to public disorder, *and* at the same time offering him the position as head of UFA, the German film studio, has been dismissed as apocryphal. What *is* true is that Lang, as a Jew, left Germany for Hollywood soon after, while his wife, Thea von Harbou, stayed behind, and joined the Nazi Party.

In America, Lang was employed by MGM and is held to have been at least partly responsible for the emergence of film noir (despite its French name), his most famous work of this kind being *The Big Heat,* starring Glenn Ford and Lee Marvin. But he worked with many stars, including Henry Fonda, Spencer Tracy, Marlene Dietrich, Barbara Stanwyck, Tyrone Power, and Edward G. Robinson, in films such as *You and Me,* 1938, with songs by Kurt Weill, and *Hangmen Also Die!,* cowritten with Bertolt Brecht. Toward the end of his life he returned to Germany.

Billy Wilder, born Samuel Wilder (1906–2002) was brought up in a part of the Austro-Hungarian Empire that is now Poland but studied at the University of Vienna, where he dropped out to become a journalist in Berlin. He started on the sports pages, changed to film reviewing, and acquired a taste for screenwriting. He collaborated with Edgar G. Ulmer, Robert Siodmak, Eugen Schüfften, and Fred Zinnemann on the 1929 film *Menschen am Sonntag* (People on Sunday), but left Germany via Paris (with $1,000 in his hatband) for Hollywood, in 1933 (traveling on a British ship, to learn English). In Los Angeles he shared an apartment with Peter Lorre. His family who remained in Germany—his mother, grandmother and stepfather—all died in Auschwitz.[41]

Wilder's first success was *Ninotchka,* written with Ernst Lubitsch, but he then went on to have the most illustrious career of all the golden generation, or at least, for most people, the most remembered. Among his films, we may mention *Double Indemnity* (1944), a murder plotted for the insurance money, *The Lost Weekend* (1946), an examination of alcoholism, *Sunset Boulevard* (1950), about an aging film star dreaming of a comeback, *Ace in the Hole* (1951), an attack on gutter journalism, *The Seven Year Itch* (1955), *Some Like It Hot* (1959), and *The Apartment* (1960), which are all so well known as to need no further comment. What is worth adding is that, in the course of these films, Wilder coaxed Oscar-winning performances out of leading actors in unlikely roles—William Holden, Fred MacMurray, and James Cagney as comedians. It was Wilder who paired Jack Lemmon with Walter Matthau, notably in *The Front Page*, 1974. Credited with breaking the bonds of Hollywood censorship in several of his films, he won six Oscars and was nominated another fifteen times. When he died a French newspaper titled its front-page obituary: "Billy Wilder Is Dead. Nobody's Perfect."

Erich Korngold (1897–1957) was not a director, but a composer, the son of a Jewish music critic from Brno in what was then Austria-Hungary and is now the Czech Republic. He studied music under Alexander von Zemlinsky, and both Strauss and Mahler liked his work—the latter referring to him as a "musical genius." Korngold moved to the United States in 1934 and composed many film scores, beginning with an adaptation of Mendelssohn's music for *A Midsummer Night's Dream*, which he wrote for Max Reinhardt's 1935 version of the Shakespeare classic. In 1938 he was asked to compose the music for an Errol Flynn film, *The Adventures of Robin Hood*; and while he was in Hollywood the *Anschluss* took place, so he stayed in California. His film credits include the music for *Deception*, starring Bette Davis, Paul Henreid and Claude Rains, *Anthony Adverse*, *The Private Lives of Elizabeth and Essex*, *The Constant Nymph*, and *Of Human Bondage*. He also composed a piano, violin, and cello concerto and a symphony and arranged operettas by Strauss and Offenbach. His oeuvre—very rich chromatically—is at last being treated seriously in the history of twentieth-century music, and his opera, *Die tote Stadt* (*The Dead City*; 1920), a big hit in the 1920s, has recently been revived in Bonn, Vienna, San Francisco, and London.[42]

And then there was *The Blue Angel*. Besides being, according to some, the first masterpiece of sound cinema and the first major German sound film, it brought together four unusual talents. The plot, it will be recalled from Chapter 27 (pp. 512–513), was based on Heinrich Mann's *Professor Unrat*, the schoolteacher who is ruined by his charges even as he tries to expose them, as he falls hopelessly in love with a nightclub singer. The film, released in 1930, was directed by Josef von Sternberg, cowritten with Carl Zuckmayer, and starred Emil Jannings and Marlene Dietrich.

Sternberg, originally Jonas Sternberg, without the "von," which was added by a Hollywood studio, was Austrian-Jewish, from Vienna, though he spent much of his childhood in New York City where his father was starting anew. Sternberg got a job repairing films and in that way wormed his way into the business. His early films attracted the attention of Charlie Chaplin, who invited him to Hollywood, where Sternberg made his name with a series of gangster movies (the 1920s were the era of Prohibition). On the strength of this he went to Germany in 1930 for *The Blue Angel*, produced in both German and English.

Carl Zuckmayer (1896–1977) was brought up in Mainz and saw action on the Western Front during World War I. In 1917 he published a collection of pacifist war poems. His first plays did not do well but in 1924 he became dramaturge at the Deutsches Theater in Berlin, alongside Bertolt Brecht. There, his play *Der fröhliche Weinberg* (*The Merry Vineyard*; 1925) won him the Kleist Prize. *The Blue Angel* was not his only great success of 1930, a year when he also won the Büchner Prize. After 1933, however, his plays were banned and he moved to Switzerland and then America. Although he did some work in Hollywood, he bought a farm in Vermont and, after World War II, became a cultural attaché to Germany, helping in the postwar investigations of war criminals. He wrote several other plays, which were successes in Germany, and in 1952 won the Goethe Prize.

The Blue Angel achieved part of its effect from Sternberg's lighting, which intensified the emotional impact, and owed much to Zuckmayer's writing, which had to underscore Heinrich Mann's text, in which Professor Unrat is transformed from a confident, if not entirely likeable full character, into a shell. Emil Jannings (1884–1950), who played Unrat, was at the time a much better known actor than Dietrich, with a remarkable voice and delivery.[43] A Swiss, at the time shooting began he had become the first winner of the Academy Award for Best Actor, for *The Way of All Flesh* and *The Last Command*, but at the invention of sound his thick German accent ruined him for Hollywood. During the Third Reich he appeared in several propagandistic films, including *Führerprinzip* (1937) and *The Dismissal of Bismarck* (1942), with Goebbels naming him an "Artist of the State" in 1941. On account of this, Jannings was forced to undergo de-Nazification after the war.

But of course what everyone remembers, or knows, about *The Blue Angel* is Marlene Dietrich (1901–92), her voice and her legs (showing stockings and garters, in one of the most famous film posters of all time). Born in Berlin-Schöneberg, the daughter of a police officer, she was not at all well known going into *The Blue Angel*. She had studied violin at school, had failed her audition for Max Reinhardt's drama academy, but had nonetheless appeared as a chorus girl and in walk-on parts in such plays as Wedekind's *Pandora's Box*. In *The Blue Angel* she played Lola, the nightclub singer, and what struck a chord was her smoky, world-weary singing voice, in particular the song that made her famous and is always associated with her, "Falling in Love Again." (Ernest Hemingway famously said, "If she had nothing more than her voice, she could break your heart with it.") On the back of the success of the film, Paramount marketed her as a German Garbo, and she appeared in her first American film, *Morocco*, also directed by Sternberg.[44]

She made many other films, opposite such stars as James Stewart and John Wayne and working with directors such as Billy Wilder, Alfred Hitchcock, and Orson Welles. She took an active role in World War II, being one of the first stars to help raise war bonds, making anti-Nazi records for the OSS, including "Lili Marlene," and singing for the troops under General Patton, even playing the saw.[45] After the war, with her film career stalled, she re-invented herself as a cabaret star under the direction of Burt Bacharach. Her return to Germany in 1960 had a mixed reception, but she was buried in Berlin not far from where she had grown up.

The demise of Professor Unrat set the scene for the demise of Weimar. *The Blue Angel* was banned in Nazi Germany.

32.

Weimar: The Golden Age of Twentieth-Century Physics, Philosophy, and History

In many areas of science, the wake of war lasted for years. In 1919 the Allies established an International Research Council, but Germany and Austria were excluded. Not until 1925 and the Locarno Pact was this rule relaxed but even then German and Austrian scientists turned down the olive branch. The frost existed on more informal levels too—Germans were banned from international science conferences, they were not offered visiting fellowships, and their research was not incorporated into the leading journals. Notably, the Solvay Conferences of physicists were without German participation until 1923.[1]

At much the same time, a new organization was established in the Weimar Republic, the Assistance Fund for German Science, which brought together the universities, the academies, and the Kaiser Wilhelm Societies. While the financial and organizational situation thus slowly improved, problems began to emerge at more personal levels. Einstein began to experience anti-Semitism and he was not the only one. Richard Willstätter had won the Nobel Prize for Chemistry in 1920 for his work on the understanding of chlorophyll but, before that, in World War I, he had invented a triple-layered gas-mask, and as a result had been awarded the

Iron Cross. Yet he found Munich, where he was professor, so anti-Semitic that in 1924 he resigned his position.[2]

Out of this unenviable situation, however, something rather striking occurred. The period between 1919 and 1932 would become the golden age of physics, particularly theoretical physics, and although it was very much an international effort, the centers of gravity in those years were three institutes, in Copenhagen, Göttingen, and Munich.

Niels Bohr's Institute of Theoretical Physics had opened in Copenhagen in January 1921, quickly followed, in 1922, by the award of a Nobel Prize. Just before World War I, Bohr had explained how electrons orbit the nucleus only in certain formations, which married atomic structure to Max Planck's notion of quanta. But, in the same year that he was awarded the Nobel Prize, Bohr also explained the fundamental links between physics and chemistry, showing that successive orbital shells could contain only a precise number of electrons, and introduced the idea that elements that behave in a similar way chemically do so because they have a similar arrangement of electrons in their outer shells, which are the ones most used in chemical reactions.

The Advent of Quantum Weirdness

One of the international galaxy of physicists who studied at Copenhagen was the Swiss-Austrian Wolfgang Pauli. In 1924 Pauli was a pudgy twenty-three-year-old, prone to depression when scientific problems defeated him. One problem in particular had set him prowling the streets of the Danish capital.[3] It arose from the fact that no one just then understood why all the electrons in orbit around the nucleus didn't just crowd in on the inner shell. This is what should have happened, with the electrons emitting energy in the form of light. What was known by now, however, was that each shell of electrons is arranged so that the inner shell always contains just one orbit, whereas the next shell out contains four. Pauli's contribution was to show that no orbit could contain more than two electrons. Once it had two, an orbit was "full," and other electrons were excluded, forced to the next orbit out. This meant that the inner shell (one orbit) could not contain more than two electrons, and that the next shell out (four orbits) could not contain more than eight. This became known as Pauli's exclusion principle, and part of its beauty lay in the way it expanded Bohr's explanation of chemical behavior. Hydrogen, for example, with one electron in the first orbit, is chemically active. Helium, however, with two electrons in the first orbit, is virtually inert (i.e., that orbit is "full" or "complete").

The next year, 1925, the center of activity moved for a time to Göttingen. Before World War I, British and American students regularly went to Germany to complete their studies, and Göttingen was a frequent stopping-off place. Bohr gave a lecture there in 1922 and was taken to task by a young student who corrected a point in his argument. Bohr, being Bohr, hadn't minded. "At the end of the discussion," said Werner Heisenberg later, "he came over to me and asked me to join him that afternoon on a walk over the Hain Mountain."[4] It was more than a stroll, for Bohr invited the young Bavarian to Copenhagen where they set about tackling yet another problem of quantum theory. According to this theory, energy—like light—was emitted in tiny packets, but according to classical physics it was emitted continuously. How could that be? Heisenberg returned to Göttingen enthused by his time in Copenhagen but also confused. And so, toward the end of May 1925, when he suffered one of his many attacks of hay fever, he took two weeks' holiday in Helgoland, a narrow strip of land off the German coast on the North Sea, where there was next to no pollen, and he cleared his head with long walks and bracing dips in the sea. The idea that came to Heisenberg in that cold, fresh environment was the first example of what came to be called quantum weirdness. Heisenberg formed the view that if something is measured as continuous at one point, and discrete at another, that is the way of reality. If the two measurements exist, it makes no sense to say that they disagree: they are just measurements.

This was Heisenberg's central insight, but in a hectic three weeks he went further, developing a method of mathematics known as matrix math, originating from an idea by David Hilbert, in which the measurements obtained are grouped in a two-dimensional table of numbers where two matrices can be multiplied together to give another matrix.[5] In Heisenberg's scheme, each atom would be represented by one matrix, each "rule" by another. If one multiplied the "sodium matrix" by the "spectral line matrix," the result should give the matrix of wavelengths of sodium's spectral lines. To Heisenberg's, and Bohr's, great satisfaction, it did: "For the first time, atomic structure had a genuine, though very surprising, mathematical base." Heisenberg called his creation/discovery quantum mechanics, though Nancy Thorndike Greenspan's recent biography of Max Born confirms how his role in the conception of the probabilistic nature of quantum waves, and of matrices themselves, was underacknowledged in the past by the likes of Heisenberg. Born won the Nobel Prize in 1954, but his contribution has now been properly positioned.[6]

The acceptance of Heisenberg's idea was made easier by a new theory of Louis de Broglie in Paris, also published in 1925. Both Planck and Einstein had argued that light, hitherto regarded as a wave, could sometimes behave as a particle. Broglie reversed this idea, arguing that particles could sometimes behave like waves. No sooner had he broached this theory than experimentation proved him right. The wave-particle duality of matter was the second weird notion of physics, but it caught on quickly and one reason was the work of the Austrian Erwin Schrödinger, who was disturbed by Heisenberg's idea and fascinated by Broglie's. Schrödinger added the notion that the electron, in its orbit around the nucleus, is not like a planet but like a wave. Moreover, this wave pattern determines the size of the orbit, because to form a complete circle the wave must conform to a whole number, not fractions (otherwise the wave would descend into chaos). In turn this determined the distance of the orbit from the nucleus.

The final layer of weirdness came in 1927, again from Heisenberg. It was late February, and Bohr had gone skiing in Norway. In his room high up in Bohr's institute, Heisenberg decided he needed some air, so he trudged across the muddy soccer fields nearby. As he walked, an idea began to germinate in his brain. Could it be, Heisenberg asked himself, that at the level of the atom there was a limit to what could be known? To identify the position of a particle, it must impact on a zinc-sulphide screen. This alters its velocity, meaning it cannot be measured at the crucial moment. Conversely, when the velocity of a particle is measured—by scattering gamma rays from it, say—it is knocked into a different path, and its exact position at the point of measurement is changed. Heisenberg's uncertainty principle, as it came to be called, posited that the exact position and precise velocity of an electron could not be determined at the same time (Heisenberg said: "To measure is to disturb," "*messen ist stören*"). This was certainly disturbing both practically and philosophically, because it implied that in the subatomic world cause and effect could never be measured. The only way to understand electron behavior was statistical, using the rules of probability. "Even in principle," Heisenberg was affirming, "we cannot know the present in all detail." Einstein was never happy with the basic notion of quantum theory, that the subatomic world could be understood only statistically. It remained a bone of contention between him and Bohr until the end of his life.[7]

Several physicists were not very happy with Einstein himself. These were the "anti-relativists," notably Philipp Lenard and Johannes Stark. Both Lenard and Stark were good scientists but, as the 1920s passed, they convinced themselves that relativity was a bogus Jewish science. Lenard, memorably described as having an "angry beard," was Hungarian but had studied in Germany under Heinrich Hertz and became his assistant.[8] He himself won the Nobel Prize (in 1905) for showing that cathode rays could pass through atoms, confirming how much atoms were made of empty space. Despite his experimental brilliance, however, Lenard was a great hater—he delivered a series of lectures in 1920 attacking relativity, although by then some of its predictions had been confirmed experimentally. And in 1929 he published a book of scientific biographies, designed to show that "Aryan-Germans" were a leading creative/innovative force and attributing other discoveries, by Jews and foreigners, to little-known, but always German, individuals.

Stark was another Nobel Prize winner, in 1919 for "the Stark effect," the influence of electrical fields on spectral lines. Surrounded by "Einstein lovers" at the University of Würzburg, he resigned his chair and was not to get another until the Nazis came to power.[9] But he wrote a book, *Die gegenwärtige Krise der deutschen Physik* (*The Contemporary Crisis in German Physics*), which argued that relativity was part of the cultural malaise then afflicting the Weimar Republic, followed by an article, "Hitlergeist und Wissenschaft" (The Hitler Spirit and Science), written jointly with Lenard, in the *Grossdeutsche Zeitung* in May 1924, in which they compared Hitler with the giants of science. This marked the emergence of "Deutsche Physik" (German physics), which eschewed relativity and quantum theory, arguing that they were too theoretical, too abstract, and "threatened to undermine intuitive, mechanical models of the world."[10]

Yet the fresh data that the new physics was producing had very practical ramifications that arguably have changed our lives far more directly than was at first envisaged by scientists mainly interested in fundamental aspects of nature. Radio moved into the home in the 1920s; television was first demonstrated in August 1928. Another invention using physics revolutionized life in a completely different way: this was the jet engine, developed almost simultaneously by the Englishman Frank Whittle and the German Hans von Ohain.

In the early 1930s, Ohain, a student of physics and aerodynamics at the University of Göttingen, had had much the same idea as Whittle. But whereas Whittle tried to enlist the aid of the British government, Ohain took his idea to the private plane-maker, Ernst Heinkel.[11] Heinkel, who realized that high-speed air transport was much needed, took von Ohain seriously from the very start. A meeting was called at Heinkel's country residence, at Warnemünde on the Baltic coast, where the twenty-five-year-old Ohain was faced by some of the plane-maker's leading aeronautical brains. Despite his youth, Ohain was offered a contract, which featured a royalty on all engines that might be sold.[12] This contract, which had nothing to do with the air ministry, or the Luftwaffe, was signed in April 1936, one month after Whittle concluded a deal for Power Jets, the company eventually formed in Britain between a firm of city bankers, the Air Ministry, and Whittle himself. Between the British company being formed, and Ohain's agreement, Britain's defense budget was increased from £122 million to £158 million, partly to pay for 250 more aircraft for the Fleet Air Arm. Four days later, German troops occupied the demilitarized zone of the Rhineland, thus violating the Treaty of Versailles. War suddenly became much more likely, a war in which air superiority might well (and did) prove crucial.

The intellectual overlap between physics and mathematics has always been considerable. In the case of Heisenberg's matrices and Schrödinger's calculations, the advances made in physics in the golden age involved the development of new forms of mathematics. By the end of the 1920s, the twenty-three outstanding mathematical problems identified by David Hilbert at the Paris conference in 1900 (see Chapter 25, p. 475) had for the most part been settled, and mathematicians looked out on the world with optimism. Their confidence was more than just a technical matter;

mathematics involved logic and therefore had philosophical implications. If mathematics was complete, and internally consistent, as it appeared to be, that said something fundamental about the world.[13]

But then, in September 1931, philosophers and mathematicians convened in Königsberg for a conference on the "Theory of Knowledge in the Exact Sciences," attended by, among others, Ludwig Wittgenstein, Rudolf Carnap, and Moritz Schlick. All were overshadowed, however, by a twenty-five-year-old mathematician from Brno (Brünn) whose revolutionary arguments were later published in a German scientific journal in an article titled "Über formal unentscheidbare Sätze der *Principia Mathematica* und verwandter Systeme" (On the Formally Undecidable Propositions of *Principia Mathematica* and Related Systems). The author was Kurt Gödel, and this paper is now regarded as a milestone in the history of logic and mathematics. Gödel was an intermittent member of Schlick's Vienna Circle, which had stimulated his interest in the philosophical aspects of science. In his 1931 paper he demolished Frege's, Russell's, and Hilbert's aim of putting all mathematics on irrefutably sound foundations, with his theorem that tells us, no less firmly than Heisenberg's uncertainty principle, that there are some things we cannot know. As John Dawson Jr. has written, Gödel's work raises "the spectre of unsolvability."[14]

His theorem is difficult. The simplest way to explain his idea is by analogy and makes use of the so-called Richard paradox, first put forward by the French mathematician Jules Richard in 1905. In this system integers are given to a variety of definitions about mathematics. For example, the definition "not divisible by any number except one and itself" (i.e., a prime number), might be given one integer, say 17. Another definition might be "being equal to the product of an integer multiplied by that integer" (i.e., a perfect square), and given the integer 20. Now assume that these definitions are laid out in a list with the two above inserted as 17th and 20th. Notice two things: 17, attached to the first statement, is itself a prime number, but 20, attached to the second statement, is not a perfect square. In Richardian mathematics, the above statement about prime numbers is not Richardian, whereas the statement about perfect squares is. Formally, the property of being Richardian involves "not having the property designated by the defining expression with which an integer is correlated in the serially ordered set of definitions." But of course this last statement is itself a mathematical definition and therefore belongs to the series and has its own integer, *n*. The question may now be put: Is *n* itself Richardian? Immediately the contradiction appears. "For *n is* Richardian if, and only if, it does *not* possess the property designated by the definition with which *n* is correlated; and it is easy to see that therefore it is Richardian if, and only if, *n* is not Richardian."

No analogy can do full justice to Gödel's theorem, but this at least conveys the paradox. It was, for some mathematicians, a profoundly depressing conclusion, for Gödel had effectively established that there were limits to mathematics and to logic—and it changed mathematics for all time.[15]

One place where such questions were frequently discussed was among a group in Vienna who, in 1924, began to meet every Thursday. Originally organized as the Ernst Mach Society, in 1928 they changed their name to the Wiener Kreis, the Vienna Circle.[16] Under this title they became what is arguably the most important philosophical movement of the last century. The guiding spirit was Moritz Schlick (1882–1936), Berlin-born who, like many members of the Kreis, had trained as a scientist, in his case as a physicist under Max Planck, from 1900 to 1904. The twenty-odd members of the circle that Schlick put together included Otto Neurath from Vienna, a remarkable Jewish polymath; Rudolf Carnap, a mathematician who had been a pupil of Gottlob Frege at Jena; Philipp Frank, another physicist; Heinz Hartmann, a psychoanalyst; Kurt Gödel, the mathematician we have just met; and at times Karl Popper, who became an influential philosopher after World War II. Schlick's original label for the kind of philosophy that evolved in Vienna in the 1920s was *Konsequenter Empirismus*, or consistent empiricism. However, after he visited America in 1929 and again in 1931–32, the term "logical positivism" emerged—and stuck.

The logical positivists made a spirited attack on metaphysics, against any suggestion that "there might be a world beyond the ordinary world of science and common sense, the world revealed to us by our senses." For them, any statement that wasn't empirically testable—verifiable, or a statement in logic or mathematics—was nonsensical. And so vast areas of theology, aesthetics, and politics were dismissed. There was more to it than this, of course. As the British philosopher A. J. Ayer, himself an observer of the circle for a short time (one of only two outsiders ever allowed, the other being W. V. O. Quine), described it, they were also against "what we might call the German past," the Romantic and to them rather woolly thinking of Hegel and Nietzsche (though not Marx).[17] Otto Neurath would hum "metaphysics" every time the circle strayed from the logical positivist path.[18] The American philosopher Sidney Hook, who also traveled in Germany at the time, confirmed the split, saying that the more traditional German philosophers were hostile to science and saw it as their duty "to advance the cause of religion, morality, freedom of the will, the *Volk* and the organic nation state." Ayer observed that there were more philosophical books published in Germany than in all other places put together.[19] The aim of the circle was to clarify philosophy, using techniques of logic and science.[20]

Thinking with the Blood

Much opposed to the Vienna Circle was a man ill at ease with the whole of Weimar culture, with modernity in general and Berlin in particular. Martin Heidegger was arguably the most influential and certainly the most controversial philosopher of the twentieth century. Born in southern Germany in 1889, he studied under Edmund Husserl before becoming himself a professional teacher of philosophy. His deliberate provincialism, his traditional mode of dress—knickerbockers—and his hatred of city life all confirmed his philosophy for his impressionable students. In 1927, at the age of thirty-eight, he published his most important book, *Sein und Zeit* (*Being and Time*). Despite the fame of Jean-Paul Sartre in the 1940s and 1950s, Heidegger was—besides being earlier—a more profound existentialist.

Being and Time is an impenetrable book, "barely decipherable," in the words of one critic. Yet it became immensely popular.[21] For Heidegger the central fact of life is man's existence in the world, and we can only confront this central fact by *describing* it as exactly as possible. Western science and philosophy have all developed in the last three or four centuries so that "the primary business of Western man has been the conquest of nature." Heidegger saw science and technology as an expression of the will, a reflection of this determination to control nature. He thought, however, that there was a different side to man, which he aimed to describe better than anyone else and which, he said, is revealed above all in poetry. The central aspect of a poem, said Heidegger, was that "it eludes the demands of our will . . . The poet cannot will to write a poem, it just comes." This links him directly with Rilke. Furthermore, the same argument applies to readers: they must allow the poem to work its magic on them. This is a central factor in Heidegger's ideas—the split between the will and those aspects of life, the interior life, that are beyond, outside, the will, where the appropriate way to understanding is not so much thinking as submission. This sounds not unlike Eastern philosophies, and Heidegger certainly believed that the Western approach needed skeptical scrutiny (he had a famous exchange with a Buddhist monk), that science was becoming intent on mastery rather than understanding. He argued—as the philosopher William Barren has said, summing up Heidegger—that there may come a time "when we should stop asserting ourselves and just submit, let be."[22]

What made Heidegger's thinking so immediately popular was that it gave respectability to the German obsession with unreason, with the rejection of urban rationalist civilization, with, in effect, a hatred of contemporary Weimar itself.[23] Moreover, it gave tacit approval to those movements associated with the idea of the *Volk*, then being spawned, that appealed not to reason but to heroes, that called for submission in the service of an alternative will to science, to those who, in Peter Gay's striking phrase, "thought with their blood." Heidegger did not create the Nazis, or even the mood that led to the Nazis. But as the German theologian Paul Tillich was to write later, "It is not without *some* justification that the names of Nietzsche and Heidegger are connected with the anti-moral movements of fascism and national socialism."

Martin Heidegger is remembered now as much for his involvement with the Nazis (see Chapter 34, p. 629) as for *Being and Time*. Much less well known are two other philosophers, one of whom, certainly, is every bit as deserving of attention as Heidegger. Max Scheler, who was born in Munich in 1874 and died in Frankfurt in 1928, was—like Wilhelm Dilthey—one of those Germans we know far too little about. One man who thought he was very important was Karol Wojtyla, Pope John Paul II, whose PhD thesis in 1954 bore the title "An Evaluation of the Possibility of Constructing a Christian Ethics on the Basis of the System of Max Scheler."

Scheler's father was a Lutheran pastor, his mother was Jewish. He tried medicine first, then philosophy and sociology under Dilthey and Simmel, taking his doctorate at Jena. He met Husserl early in the new century and married Märit Furtwängler, sister of the conductor, Wilhelm. Scheler settled in Cologne and Frankfurt, where he formed a circle with Ernst Cassirer, Karl Mannheim, and others.[24]

Scholarship about Scheler has been intensifying lately, not just because of the late pope's interest but also because his arguments have relevance in the animal rights debate and the abortion dispute.

Scheler is known for two main ideas. The first centers around the phenomenon of sympathy.* The fact that sympathy exists and we cannot *escape* it, is for Scheler proof of God's existence, that love is at the center of our existence, that the "heart," not the mind determines values, and not in a rational way—values can only be felt, as colors are "seen," without any rational explanation. The existence of sympathy means that each person is morally unique, and that—above all—we do not exist *with* others, we exist *toward* them: we should accept this and use it. His other idea was that there is an "*ordre du coeur*," a hierarchy of values, from high to low as follows: values of the holy; of the mind (truth, beauty, justice); of vitality and nobility; of utility; of pleasure. Scheler thought that the mistake in most systems of ethics was to elevate one value above all others, rather than recognize that this hierarchy exists and tempers all judgments. He thought that when human beings elevate a lower value over a higher one "disorders of the heart" occur. For Scheler, reason has little to say about value (he overlapped here with Wittgenstein); instead the "heart" governs our approach to life rather than our intelligence; experience is what counts, not will. Feelings and love have a logic of their own, he said, quite different from the logic of reason. He was asserting that there is a fundamental connection among all of us, and that the work we do to render that connection stronger and clearer is the way to contentment.[25]

Like Dilthey, Ernst Cassirer's main concern was to explore what was similar and what was different about the forms of knowledge we know as the sciences, on the one hand, and the humanities on the other, though he preferred the phrase "cultural sciences" instead of "humanities." Cas-

* In *Zur Phänomenologie und Theorie der Sympathiegefühle und von Liebe und Hass* (1913).

sirer was born in Breslau in 1874 into a cosmopolitan and well-off Jewish family. Another branch of the family lived in Berlin, where one cousin, Bruno, was a publisher and a second, Paul, a well-known art dealer.[26] In 1919 Cassirer was offered two professorships himself, one in Frankfurt and one in Hamburg. He opted for the latter, and taught there for several years, becoming rector in 1929, the first Jew to hold such a position.[27]

Cassirer's main book, *Philosophie der Symbolischen Formen* (*Philosophy of Symbolic Forms*), was a three-volume exploration of symbolic forms, in which he argued that moral experience and mathematical experience were essentially the same, exploring the basis that moral choices were as "necessary" as mathematical logic. To give some idea of Cassirer's highly technical work (which approach he regarded as inevitable in the modern world), he looked at Leibniz's and Newton's approaches to differentiation, as a way to understand change, on a graph and applied that to change in other fields, exploring whether change—in history, for example—could be understood in a similar, or equivalent, way.[28] Can other areas of life, outside mathematics, be regarded as "formal" in the same fashion? He also examined what implications Einstein's concept of relativity had for Kantian philosophy, in that Kant had said that our understanding of space was instinctive, or intuitive, when of course Einstein's concept of "curved" space was anything but. Cassirer's other important book was *Zur Logik der Kulturwissenschaften* (*The Logic of the Cultural Sciences*; 1942), which explored the similarities and differences between the natural sciences, mathematics, and aesthetics, in which he argued that "thing perception" (*Dingwahrnehmen*) is generally given pre-eminence over "expressive perception" (*Ausdruckswahrnehmen*), and this is why the natural sciences are usually felt to have "a more secure evidential basis."[29]

Cassirer was forced to leave Germany in 1933. After stints at Oxford and Göteborg, he transferred to Yale and Columbia (being spurned by Harvard because, as a young man, he had turned down a visiting professorship there, considering it "too remote"). In America he wrote two books in English, including *The Myth of the State*, a reply to a number of German (and National Socialist) writers, in which he sought to explain fascism as arising logically from the Platonic tradition in European thought. He died tragically young after a heart attack while walking in New York in 1945. He influenced Erwin Panofsky and Peter Gay, among others.

A Patriot without a Country

Weimar Germany was also blessed with what the French scholar Alain Boureau calls "a momentous generation" of historians: Ludwig von Pastor, Percy Schramm, Ernst Kantorowicz, Norbert Elias, and Gershom Scholem. Most were interested in the Middle Ages, Pastor as a Catholic version of Leopold von Ranke, who had of course written a seminal history of the popes in the early nineteenth century. Born in Aachen, Pastor's greatest success was to persuade Pope Leo XIII of his seriousness of purpose, so that the contents of the Vatican Library—hitherto closed—were opened to him. This led to his lifetime's work, his *Geschichte der Päpste seit dem Ausgang des Mittelalters* (*History of the Popes from the Close of the Middle Ages*), sixteen volumes that run from the Avignon Papacy of 1305 to Napoleon's entry into Rome in 1799. Unlike Ranke, Pastor eschewed the institutional changes and innovations and concentrated instead on the individual incumbents. His theme was that the weakness of the papacy reflected the "flaws" of the times, and that its shortcomings were not always the weaknesses they were made out to be, enabling the popes to retain power and influence longer than would have otherwise been the case. Owing to the unprecedented access he was given, his history superseded all others and is still regarded as a seminal work.

Like Pastor, Percy Schramm and Ernst Kantorowicz were interested in the Middle Ages, but there the similarity ceases. Schramm (1894–1970) served in the army in World War I, after which he studied history and art history at Hamburg, Munich, and Heidelberg. He is generally credited with making art history a much harder, more interesting, and more powerful discipline than it was originally, less dilettante-like, showing how, in his most important book, *Kaiser, Rom und Renovatio* (*Emperor, Rome and Renovatio*), the German emperors of the medieval period had used the symbolism of the Romans to underwrite their power. In World War II, Schramm volunteered for service and was made official staff historian for the German High Command Operational Staff. His book *Hitler als militärischer Führer* (*Hitler: The Man and the Military Leader*), published in 1963, stresses the good side of the Führer as well as the negatives.[30] Schramm saw a lot of Hitler and was close to General Alfred Jodl, acting as a witness in Jodl's support at the Nuremberg Trials after the war, and was removed from teaching. He was reinstated in the late 1940s, and his inside accounts of the high command have become required reading.

Ernst Kantorowicz had many of the same interests as Schramm and a not dissimilar approach but, being Jewish, his fate was very different. After four years in the army in World War I, he studied philosophy at Berlin. An extreme right-winger, he joined the militia that attempted to put down the Spartacist uprising, and became involved with the *Georgekreis*, the artists and intellectuals devoted to Stefan George (see 574). This group, elitist and culturally conservative, had a big influence on Kantorowicz's first important book, a biography of Frederick II, which examined the king's charisma and spiritual qualities, rather than getting involved in the minutiae of the institutions of his rule.

By then it was the 1930s and although Kantorowicz had been appointed professor at Frankfurt, he was forced out, moving to Oxford at first, like Cassirer, then on to Berkeley. There he became known for two things, for refusing to sign the oath of loyalty demanded by Senator Joe McCarthy (Kantorowicz resigned from Berkeley and moved to the Institute for Advanced Study at Princeton), and for his second masterpiece, *The King's Two Bodies*, which sought to explain the birth of the modern state as growing out of the medieval conceit that a king possessed two bodies, one that was human and died, and another that was immortal and passed "mystically" from monarch to monarch.[31]

The fourth of this great generation, Norbert Elias, was also Jewish. Like Schramm and Kantorowicz, he also volunteered to fight in World War I, serving as a telegrapher. His interest in the German Zionist movement brought him into touch with the likes of Erich Fromm, Leo Strauss, Leo Lowenthal, and Gershom Scholem.[32] He took courses at Heidelberg with both Karl Jaspers and Alfred Weber, later moving to Frankfurt to work under Karl Mannheim and be near the Frankfurt Institute. In 1933 he had to flee Germany before his thesis could be presented; he went first to Paris and then on to Britain in 1935, where he started work on his most important contribution, *The Civilising Process.* This appeared in 1939 but, because of other events, wasn't noticed until much later, and for that reason is discussed in Chapter 40, p. 743.[33]

Gershom Scholem (1897–1982), the youngest of the golden generation, was born in Berlin and studied mathematics, philosophy, and Hebrew at the university there, where he came into contact with Martin Buber, Walter Benjamin, and Gottlob Frege. Sympathetic to Zionism, he immigrated to Palestine in 1923, becoming in time head of the department of Hebrew and Judaica at the National Library of Israel.[34] He was interested in Kabbalah and mysticism, feeling that Judaism had mystical origins and could not be properly understood without that element.[35] He tried to construct a narrative of Jewish belief which concluded that, contrary to what many Jews believed, the ultimate form of their religion was not achieved until relatively recent times, the Middle Ages, when Maimonides attempted a final reconciliation between Jewish thought and Greek thought. This was an important achievement of theological scholarship though by now, back in the Weimar Republic, hardly anyone was listening.

33 ·

Weimar: "A Problem in Need of a Solution"

On October 28, 1929, the notorious stock market crash occurred on Wall Street, and U.S. loans to Europe were suspended. In the weeks and months that followed, and despite the misgivings of many, Allied troops withdrew from the Rhineland. In Thuringia Wilhelm Frick was about to become the first Nazi to be appointed minister in a state government, while in Italy Benito Mussolini was clamoring for the revision of the Versailles Treaty. In Britain in 1931 a National Government was formed to help balance the budget, and Japan abandoned the gold standard. There was a widespread feeling of crisis.

Sigmund Freud, then seventy-three, had more personal reasons to feel pessimistic. In 1924 he had undergone two operations for cancer of the mouth. After the operation he could chew and speak only with difficulty (the prostheses didn't work properly), but he still refused to stop smoking, probably the cause of the cancer in the first place.[1] At the end of 1929, as Wall Street was crashing, Freud delivered the most telling of his cultural critiques. *Totem und Tabu* (*Totem and Taboo*) and *Die Zukunft einer Illusion* (*The Future of an Illusion*) both had mixed receptions, but *Das Unbehagen in der Kultur* (*Civilization and Its Discontents*) was much more timely. There had been famine in Austria and attempted revolution and mega-inflation in Germany, while capitalism appeared to have collapsed in America. The devastation and moral degeneration of World War I was still a concern to many people, Hitler was on the rise. Wherever you looked, Freud's title fitted the facts.[2]

In *Civilization and Its Discontents,* Freud developed some of the ideas he had explored in *Totem and Taboo,* in particular that society—civilization—evolves out of the need to curb the individual's unruly sexual and aggressive appetites. He now argued that civilization, suppression, and neurosis are inescapably intertwined because the more civilization there is, the more suppression of the instincts is needed and, as a direct result, the more neurosis. Man, he said, cannot help but become more and more unhappy in civilization, which explains why so many seek refuge in drink, drugs, or religion. Given this basic predicament, it is the individual's "psychical constitution" that determines how any individual adjusts. For example, "The man who is predominantly erotic will give first preference to his emotional relationships with other people; the narcissistic man, who inclines to be self-sufficient, will seek his main satisfactions in his internal mental process." We are, he insisted, progressively more and more cut off—alienated—from each other. The point of his book, he said, was not to offer easy panaceas but to suggest that ethics—the rules by which men agree to live together—can benefit from psychoanalytic understanding.

Freud's hopes were not to be fulfilled. The 1930s, as we know now, were, as one historian put it, a "dark valley" ethically.[3] Not surprisingly, therefore, his book spawned a raft of others that, though very different, were all profoundly uneasy with Western capitalist society.

The book closest to Freud's was published in 1933 by the former crown prince of psychoanalysis, now turned archrival. Carl Jung's argument in *Modern Man in Search of a Soul* was that psychoanalysis, by replacing the soul with the psyche, only offered a palliative.[4] Psychoanalysis, as a technique, could only be used on an individual basis; it could not become "organized" and used to help millions at a time, such as, for example, Catholicism. And so, the "participation mystique," as the anthropologist Lucien Lévy-Bruhl called it, was a whole dimension of life closed to modern man. This lack of a collective life, ceremonies of the whole as Hugo von Hofmannsthal called them, was the main ingredient in neurosis, and the general anxiety.

For fifteen years, Karen Horney practiced in Weimar Germany as an orthodox Freudian analyst, alongside Melanie Klein, Otto Fenichel, Franz Alexander, Karl Abraham, and Wilhelm Reich at the Berlin Psychoanalytic Institute. Only after she moved to the United States, first as associate director of the Chicago Institute and then in New York, at the New School for Social Research and the New York Psychoanalytic Institute, did she find herself capable of offering criticism of the founder of the movement. Her book, *The Neurotic Personality of Our Time,* overlapped with both Freud and Jung but was also an attack on capitalistic society for the way it induced neurosis.

Horney's chief criticism of Freud was his antifeminist bias (her early papers included "The Dread of Women" and "The Denial of the Vagina").[5] She was also a Marxist and thought Freud too biological in outlook and "deeply ignorant" of modern anthropology and sociology. Horney took the line that "there is no such thing as a universal normal psychology." For her, however, two traits invariably characterized all neurotics. The first was "rigidity in reaction," and the second was "a discrepancy between potentiality and achievement." Horney didn't believe in the Oedipus complex either. She preferred the notion of "basic anxiety," which she attributed not to biology but to the conflicting forces of society, conflicts that act on an individual from childhood. Basic anxiety she characterized as a feeling of "being small, insignificant, helpless, endangered, in a world that is out to abuse, cheat, attack, humiliate, betray, envy." Such anxiety is worse, she said, when parents fail to give their children warmth and affection. Such a child grows up with one of four rigid ways of approaching life, which interfere with achievement: a neurotic striving for affection; a neurotic striving for power; neurotic withdrawal; and neurotic submissiveness.

The most contentious part of Horney's theory was her blaming neurosis on the contradictions of contemporary American life. She insisted that in America more than anywhere else there existed an inherent contradiction between competition and success on the one hand ("never give a sucker an even break") and good neighborliness on the other ("love your neighbor as yourself"); between the promotion of ambition by advertising ("keeping up with the Joneses") and the inability of the individual to satisfy these ambitions. This modern world, despite its material advantages, foments the feeling in many individuals that they are "isolated and helpless."

From Hegel to Hitler

In 1924, the year that tuberculosis killed Kafka, Adolf Hitler celebrated his thirty-fifth birthday—in prison. He was not sent back to Austria, but was in Landsberg jail, west of the Bavarian capital, serving a five-year sentence for treason and his part in the Munich Putsch of 1923. The trial was front-page news in every German newspaper for more than three weeks, and Hitler broke through to a national audience. During his time in prison Hitler wrote the first part of *Mein Kampf*, which helped establish him as the leader of the National Socialists, helped him lay the foundation of the Hitler myth, and helped him clarify his ideas.

Whatever his other attributes, Hitler certainly thought of himself as a thinker and an artist, with a grasp of technical-military matters, of natural science, and above all of history. He was transformed into the figure he became first by World War I and the ensuing peace, but also by the education he gave himself. The Führer's ideas, as revealed in his table talk during World War II, are directly traceable to his thinking as a young man.

The historian George L. Mosse has disinterred the more distant intellectual origins of the Third Reich, on which this section is chiefly based.[6] He shows how an amalgam of *völkisch* mysticism and spirituality grew up in Germany in the nineteenth century, in part a response to the Romantic movement and to the bewildering pace of industrialization, and was also an aspect of German unification. In addition to the influence of thinkers and writers who helped create this cast of mind—people like Paul de Lagarde and Julius Langbehn, who stressed "German intuition" as a new creative force in the world, and Eugen Diederichs, who openly advocated "a culturally grounded nation guided by the initiated elite"—there were nineteenth-century German books such as that by Ludwig Woltmann, examining the art of the Renaissance, identifying "Aryans" in positions of power and showing how much the Nordic type was admired even then. Mosse also emphasizes how Social Darwinism threaded through society and describes the many German attempts at utopias—from "Aryan" colonies in Paraguay and Mexico to nudist camps in Bavaria, which tried to put völkisch principles into effect.[7]

In his own book Hitler insists that while at school in Linz he "learned to understand and grasp the meaning of history." "To 'Learn' history," he explained, "means to seek and find the forces which are the causes leading to those effects which we subsequently perceive as historical events." One of these forces, he felt (and this too he had picked up as a boy), was that Britain, France, and Russia were intent on encircling Germany, and he thereafter never rid himself of this view. For him history was invariably the work of great men—his heroes were Charlemagne, Rudolf von Hapsburg, Friedrich the Great, Peter the Great, Napoleon, Bismarck, and Kaiser Wilhelm I. Hitler therefore was much more in the mold of Stefan George or Rainer Maria Rilke than that of Marx or Engels, for whom the history of class struggle was paramount. For Hitler, history was a catalog of racial struggles, although the outcome always depended on great men: "[History] was the sum total of struggle and war, waged by each against all with no room for either mercy or humanity."

Hitler's biological thinking, says Mosse, was an amalgam of Thomas R. Malthus, Charles Darwin, Joseph-Arthur de Gobineau, and William McDougall. "Man has become great through struggle. . . . Whatever goal man has reached is due to his originality plus his brutality . . . All life is bound up in three theses: struggle is the father of all things, virtue lies in blood, leadership is primary and decisive . . . He who wants to live must fight, and he who does not want to fight in this world where eternal struggle is the law of life has no right to exist."[8]

Hitler's biologism was intimately linked to his understanding of history.[9] He knew very little about prehistory but certainly regarded himself as something of a classicist, fond of saying that his "natural home" was ancient Greece or Rome, and he had more than a passing acquaintance with Plato.[10] Partly because of this, he considered the races of the East (the old "Barbarians") as inferior. Organized religion, Catholicism in particular, was also doomed, owing to its anti-scientific stance and its unfortunate interest in the poor ("weaklings"). For Hitler, mankind was divided into three—creators of culture, bearers of culture, and destroyers of culture—and only the "Aryans" were capable of creating culture. The decline of culture was always due to the same reason: miscegenation.[11] This helps explain Hitler's affinity for Hegel. Hegel had argued that Europe was central in history and that Russia and the United States were peripheral. Landlocked Linz reinforced this view. "Throughout his life Hitler remained an inland-orientated German, his imagination untouched by the sea . . . He was completely rooted within the cultural boundaries of the old Roman Empire." This attitude may just have led Hitler to fatally underestimate the resolve of that periphery—Britain, the United States, and Russia.

It is doubtful that Hitler was as well read as his admirers claimed, but he did know some architecture, art, military history, general history, and technology, and also felt at home in music, biology, medicine, and the history of civilization and religion. He sometimes surprised his listeners with his knowledge in a variety of fields. One of his doctors, for example, was once astonished to discover that the Führer fully grasped the effects of nicotine on the coronary vessels. But Hitler was largely self-taught, which had significant consequences. He never had a teacher to give him a systematic or comprehensive grounding in any field. Furthermore, World War I, which began when Hitler was twenty-five, acted as a brake (and a break) in his education. Hitler's thoughts stopped developing in 1914; thereafter, he was largely confined to the halfway house of ideas in Pan-Germany described in Chapter 22, p. 417.

We must be careful, moreover, not to pitch the Führer's thought too high.[12] As Werner Maser highlights in his psychohistory of Hitler, much of his later reading was done merely to confirm the views he already held. Second, in order to preserve a consistency in his position, he was required to do severe violence to the facts. Hitler several times argued that Germany had abandoned its expansion toward the East "six hundred years ago." This had to do with his explanation of Germany's failure in the past, and its future needs. Yet both the Hapsburgs and the Hohenzollerns had had a well established "*Ostpolitik*"—Poland, for instance, being partitioned three times.

Cultural Pessimism, Conservative Revolutionaries, Reactionary Modernism

The well-established German tradition of cultural pessimism had been continued in Weimar by Arthur Moeller van den Bruck. Fritz Stern describes Moeller van den Bruck as an outsider from his early years. Expelled from his Gymnasium in mysterious circumstances, he went into exile to escape military service, while the modest fortune he inherited "freed him from the obligation of steady employment."[13] He began his writing with a trilogy on modern German art but he finished only the first volume. After other books, on theater, he was finally forced into military service where, for a short time at least, he was branded a military deserter.[14] He did know some of the early figures of German Expressionism, notably Ernst Barlach and certainly, to begin with, he was not anti-Semitic. But his extensive time abroad seems to have produced in him an idealized image of Germany and his eight-volume history of the Germans, *Die Deutschen* (1904–10) was the first expression of his nationalism. After this, he made the fateful turn to meta-history, distinguishing in *Die Zeitgenossen* (six volumes) between the "young peoples" and the "old peoples," between French skepticism, English common sense, and Italian beauty, on the one hand, and German *Weltanschauung*, American will, and Russian soil on the other. It was at this time that he changed his name from Moeller-Bruck to Moeller van den Bruck.[15]

In *Die Zeitgenossen*, Moeller van den Bruck lamented the absence of great spiritual and artistic interpreters of modernity (except for Walt Whitman, "the hero of the modern world"), and he decried in particular the decline of German culture since unification, arguing that Germany had "too much civilization, not enough culture."[16] His own contribution was to edit the twenty-three-volume German edition of the works of Dostoevsky. He was also involved in the Juni-Klub, an active force in German intellectual politics whose members, even in those days, were called neo-Conservatives, and with a journal, *Gewissen* (Conscience), which had much the same aim. As Lagarde had said, liberalism was the enemy—more than ever in Weimar—in particular the enemy of *Innerlichkeit*, Bildung, and idealism.[17] No form of social harmony was possible with liberalism.

This was the (very rough) background to Moeller van den Bruck's celebrated work *Das dritte Reich* (*The Third Reich*; 1922), which, again as Fritz Stern has described it, "accidentally provided the National Socialist state with its historic name." The book was a passionate polemic, an attack on liberalism and social democracy in Germany, an attack on ideal types that existed nowhere other than in Moeller van den Bruck's imagination, in which "he reduced socialism to Marxism, Marxism to Marx, and Marx to Judaism." This was a new theme for Moeller van den Bruck, who had not been anti-Semitic to that point. But he now criticized the Jews as an uprooted, homeless people "who had no fatherland." His main argument was that "liberalism is the expression of a society that is no longer a community . . ." the pre-1914 Germans were "the freest in the world," liberalism was synonymous with reason, which was inferior to understanding.[18] Many Nazis (not least Hitler) did not embrace Moeller van den Bruck, but Goebbels did and after his suicide in 1925 the writer became a hero in right-wing circles.

And he was far from alone. In the Weimar Republic people with Moeller van den Bruck's way of looking at the world went by several names—cultural pessimists, conservative revolutionaries, reactionary modernists—all overlapping: figures such as Ernst Jünger, Edgar Jung, the post-Nietzschean, pre-existential philosopher Ludwig Klages, Stefan George, Oswald Spengler, Ernst Toller, Thomas Mann, who shared a view that what Germany needed was a spiritual revolution, that democracy was culturally unacceptable, that Weimar was a problem in need of a solution, and that a return to a "*völkisch*" community the ideal.

Thomas Mann and Oswald Spengler have already been introduced. Among the others, Ernst Jünger stands out. A man who would live to be 102 (born in 1895, he died six weeks before his 103rd birthday in 1998), his long life enabled him (as Hans Baumann, another long-lived writer, was to say later) to correct many of his mistakes. Jünger ran away from home to join the French Foreign Legion, then fought bravely on the Western Front, being injured fourteen times and winning the Iron Cross and Pour le Mérite (the "Blue Max") at twenty-three, one of the youngest-ever recipients. After the war he trained as an entomologist and, in 1922, published *In Stahlgewittern* (*Storm of Steel*), an unrestrained war memoir that does not shy away from the casualties of war but is at its most lyrical and enthusiastic when describing the fighting. It is now often contrasted with *All Quiet on the Western Front*, treating war as a near-mystical, elevating, "internal event." Like many—like the Freikorps, the private armies that sprang up in Germany after World War I to combat revolutionary tendencies—Jünger, at least then, wanted Germany reinstated to a position of supremacy. For him, the Weimar Republic was a pale alternative to the "real" Germany, democracy and liberalism the twin enemies of all that is noble in life. His career would go through several twists before World War II had come and gone; he was never a Nazi but in Weimar, as Keith Bullivant has observed, he was a vivid presence among the Conservative Revolutionaries.[19]

The stance of the Conservative Revolutionaries was as much aesthetic and cultural as political. It set them against such figures as Kurt Tucholsky, Alfred Döblin, the novelist and author of the picaresque *Berlin Alexanderplatz*, about a criminal who can't break free from the underworld, and against Walter Benjamin. Born in Berlin in 1892, the son of a Jewish auctioneer and art dealer, Benjamin was a radical intellectual, a "cultural Zionist" as he described himself (meaning he was an advocate of Jewish liberal values in European culture), who earned his living as a historian, philosopher, art and literary critic, and journalist. Of a slightly mystical bent, Benjamin spent World War I in medical exile in Switzerland, afterward forming friendships with Hugo von Hofmannsthal, Bertolt Brecht, and the founders of the Frankfurt school. In a series of essays and books—*Goethes Wahlverwandschaften* (*Goethe's Elective Affinities*), *Ursprung des deutschen Trauerspiels* (*The Origin of German Tragic Drama*), and "The Politicisation of the Intelligentsia"—he compared and contrasted traditional and new art forms, anticipating in a general way the ideas of Raymond Williams, Andy Warhol, and Marshall McLuhan. His approach was to try to understand these new forms, not condemn them.

In the view of the conservatives, the extraordinary Weimar culture that has been the subject of the previous chapters had as one of its main faults that it neglected or belittled the lower classes, for them the salt of the earth. This attitude combined to create another characteristic of the times—the anti-intellectual intellectual. This, in turn, with its inherent anti-Semitism, sparked a resurgence of Jewish culture—in particular what became known as the Lehrhaus Movement and the *Wissenschaft des Judentums*.[20] By no means all of the conservatives embraced the Nazi Party (Ernst Jünger, for instance) but the general climate of opinion generated by cultural pessimism, with chaos threatening in the background, did contribute to the Nazis' growing self-confidence.[21]

The Cultural/Intellectual Basis for National Socialism

During the Weimar years, as we have seen, there was a continual battle between the rationalists—the scientists and the academics—and the nationalists, the pan-Germans, who remained convinced there was something special about Germany, her history, the "instinctive superiority" of her heroes. In *The Decline of the West* Oswald Spengler had stressed how Germany was different from France, the United States, and Britain, and this view, which appealed to Hitler, gained ground among the Nazis as they edged closer to power. From time to time Hitler attacked modern art and modern artists but, like other leading Nazis, he was by temperament an anti-intellectual; for him, most great men of history had been doers, not thinkers. There was, however, one exception to this mold, a would-be intellectual who was even more of an outsider in German society than the other leading Nazis.

Alfred Rosenberg's family came from Estonia, which until 1918 was one of Russia's Baltic provinces. As a boy he was fascinated by history, especially after he encountered Houston Stewart Chamberlain's *Foundations of the Nineteenth Century* on a family holiday in 1909. He now had a reason to hate the Jews every bit as much as his experiences in Estonia gave him reason to hate the Russians. Moving to Munich after the Armistice in 1918, he quickly joined the Nationalsozialistische Deutsche Arbeiterpartei (NSDAP) and began writing vicious anti-Semitic pamphlets. His writing ability, his knowledge of Russia, and his facility with Russian all helped to make him the party's expert on the East; he also became editor of the *Völkischer Beobachter* (The People's Observer), the Nazi Party's newspaper. As the 1920s passed, Rosenberg, together with Martin Bormann and Heinrich Himmler, began to see the need for a Nazi ideology that went beyond *Mein Kampf*, and in 1930 he published what he believed provided the intellectual basis for National Socialism. In German its title was *Der Mythus des 20. Jahrhunderts,* usually translated into English as *The Myth of the Twentieth Century.*[22]

Mythus is a rambling and inconsistent book. It conducts a massive assault on Roman Catholicism as the main threat to German civilization—the text stretches to more than 700 pages. The third section is titled "The Coming Reich"; other parts deal with "racial hygiene," education, and religion, with international affairs at the end. Rosenberg argues that Jesus was not Jewish and that his message had been perverted by Paul, who *was* Jewish, and that it was the Pauline/Roman version that had forged Christianity into its familiar mold by ignoring ideas of aristocracy and race and creating fake doctrines of original sin, the afterlife, and hell as an inferno, all of which beliefs, Rosenberg thought, were "unhealthy."

His aim—and at this distance Rosenberg's audacity is breathtaking—was to create a substitute faith for Germany.[23] He advocated a "religion of the blood" which, in effect, told Germans that they were members of a master race, with a "race-soul." He quoted the works of the Nazis' chief academic racialist, H. F. K. Günther, who claimed to have established on a scientific basis "the defining characteristics of the so-called Nordic-Aryan race." As with Hitler and others before him, Rosenberg did his best to establish a connection to the ancient inhabitants of India, Greece, and Germany, and he brought in Rembrandt, Herder, Wagner, Friedrich the Great, and Heinrich the Lion, to produce an entirely arbitrary but nonetheless heroic history specifically intended to root the NSDAP in the German past. For Rosenberg, race—the religion of the blood—was the only force that could combat what he saw as the main engines of disintegration—individualism and universalism. "The individualism of economic man," the American ideal, he dismissed as "a figment of the Jewish mind to lure men to their doom."

Hitler seems to have had mixed feelings about the *Mythus*. He held on to the manuscript for six months after Rosenberg submitted it, and publication was not sanctioned until September 15, 1930, *after* the Nazi Party's sensational victory at the polls. Perhaps Hitler put off approving the book until the party was strong enough to risk losing the support of Roman Catholics that would surely follow publication. He was being no more than realistic. The Vatican was incensed by Rosenberg's argument and, in 1934, placed the *Mythus* on the Index of Prohibited Books. Cardinal Schulte, the archbishop of Cologne, set up a "Defense Staff" of seven young priests who worked round the clock to list the many errors in the text, which were published as anonymous pamphlets printed simultaneously in five different cities to evade the Gestapo. Rosenberg nonetheless remained popular with Hitler, and when the war began, he was given his own unit, the Einsatzstab Reichsleiter Rosenberg, or ERR, charged with looting art.

Although it was incoherent and arbitrary, the *Mythus* left no doubt as to what the Nazis thought was wrong with German civilization.

A Violent Adoration for "Everything German"

We conclude this chapter with a view of Germany from outside. It was written during the Weimar years but just before the Nazis came to real prominence. For that reason it deserves to be taken more seriously as a critique. It also overlapped with earlier critiques by other non-Germans, for example, John Dewey and George Santayana.

Julien Benda's book, *The Treason of the Learned*, first appeared in 1927. The learned, or "*clercs*" in French, were not only German but also French and this too makes its arguments worth listening to: he wasn't being narrowly nationalistic. Benda (1867–1956) came from a once-prosperous Jewish Parisian family whose firm had gone bankrupt during World War I. A prolific author of some fifty books, he was one of the defenders of Alfred Dreyfus and saw himself as a supreme rationalist in the French tradition, setting himself against the "intuitionism" of Henri Bergson. Benda's main argument in his book was that the nineteenth century had seen the growth of political passion out of all proportion to anything that had gone before. The emergence of a bourgeois class, he said, had spawned the development of class hatred and a rise in nationalist sentiment that he put down to democracy. As Herbert Read outlined it in the introduction to the English edition, "Nationalism has become a widely diffused, mystical sentiment, with the result that national passions devastate national life." Not least, the intensifying of Jewish nationalism had spawned a corresponding spread of anti-Semitism.[24] Benda insisted that political passions had become much more "emphatic" in the nineteenth century, in particular *national* passions, "not only as regards their material existence, their military power, their territorial possessions, and their economic wealth, but as regards their *moral* existence. With a hitherto unknown consciousness (prodigiously fanned by authors) every nation now hugs itself and sets itself up against all other nations as superior in language, art, literature, philosophy, civilization, 'culture.' Patriotism is today the assertion of one form of mind against other forms of mind."[25] It was, he added, impossible to "overstress" the novelty of this form of patriotism in history, inaugurated in Germany in 1813, and embodying three ideas—the movement against the Jews, the movement of the possessing classes against the proletariat, and the movement of the champions of authority against the democrats.[26]

Most of all, Benda saw a change in the behavior of intellectuals, creative people, scientists, and philosophers. Before the nineteenth century, he said, people of the character of Leonardo da Vinci, Goethe, Erasmus, Kant, Thomas Aquinas, Kepler, Descartes, Roger Bacon, Pascal, and Leibniz "set an example of attachment to the purely disinterested activity of the mind and created a belief in the supreme value of this form of existence." Now, he said, it was very different. "Today, if we mention Mommsen, Treitschke, Ostwald, Brunetière, Barrès, Lemaître, Péguy, Maurras, d'Annunzio, Kipling, we have to admit that the 'clerks' now exercise political passions with all the characteristics of passion—the tendency to action, the thirst for immediate results, the exclusive preoccupation with the desired end, the scorn for argument, the excess, the hatred, the fixed ideas."[27] In descending to the level of the rest of the public, Benda thought these men were betraying what they—or their predecessors—had stood for. They were not acting like Socrates or Jesus, but like the mob.

Benda was anxious to show that this betrayal had occurred not just in Germany—indeed, as a Frenchman his chief focus was the French, but he extended his arguments from France to Germany, Italy, Britain, and America, more or less in that order, and he thought that German intellectuals had been especially culpable in World War I, in particular in regard to the Manifesto of the 93 (see p. 532). "We know how systematically the mass of German teachers in the past fifty years have announced the decline of every civilization but that of their own race, and how in France the admirers of Nietzsche or Wagner, even of Kant or Goethe, were treated by Frenchmen . . ."[28]

Although he excoriated his fellow French in this regard, Benda did think that German intellectuals had "led the way in this adhesion of the modern 'clerk' to patriotic fanaticism." He thought it had begun with Lessing, Schlegel, and Fichte, who were "organising in their hearts a violent adoration for 'everything German,' and a scorn for everything not German. The nationalist 'clerk' is essentially a German invention."[29]

Although he blamed novelists, dramatists, and artists equally, he reserved particular venom for historians, "German historians of the past half century and the French Monarchists of the past twenty years." " 'A true German historian,' declares a German master, 'should especially tell those facts which conduce to the grandeur of Germany. '" The same scholar praises Mommsen (who himself boasted of it) for having written a Roman history "which becomes a history of Germany with Roman names." And the philosophers were hardly better. "Fichte and Hegel made the triumph of the German world the supreme and necessary end of the development of Being . . ."[30] Not even the French could compete here, he said. The German historians, says Numa Denis Fustel de Coulanges, onetime director of the French school in Athens, "urge their nation to be intoxicated with its personality, even to its barbarity. The French moralist does not lag behind . . ." What had ruined Germany in World War I, Benda felt, was that its material strength was not equal to the arrogance that had been bred by this intellectual nationalism. The Germans, too, he said, were responsible for the cult of the powerful state. The learned had *divinized* politics.[31]

The most important effect in all this, Benda thought, and it was a profound point, was that the military life and war, fought inevitably with nationalist aims in mind, became attached to morality rather than utility.[32] Courage, honor, and harshness came to be extolled by the learned—even, in the case of Nietzsche, cruelty ("Every superior culture is built upon cruelty"). Another cult, the cult of the will (of the successful will, of course), had arisen, supported by "everyone in Germany since Hegel" and by many in France "since de Maistre."[33]

All this, said Benda, was in the ascendancy, whereas the passion of the learned to *understand*, the desire to be universal or objective, had, since Nietzsche and Sorel, been derided. Several French writers had insisted, he said, that people interested in purely intellectual things were "inferior to soldiers . . . A whole literature has assiduously proclaimed the superiority of instinct, the unconscious, intuition, the will (in the German sense, i.e., as opposed to the intelligence) and has proclaimed it in the name of the practical spirit, because the instinct and not the intelligence knows what we ought to do—as individuals, as a nation, as a class—to secure our own advantage."[34] For him, this all amounted to a "prodigious" decline in morality, a "sort of (very Germanic) intellectual sadism."[35]

He concluded that the battle was over. "Today . . . humanity is national. The layman has won . . . The man of science, the artist, the philosopher are attached to the nation as much as the day-labourer and the merchant. Those who make the world's values, make them for a nation . . . All Europe, including Erasmus, has followed Luther." Then, on page 145 (and remember this was first published in 1927), "This humanity is heading for the greatest and most perfect war ever seen in the world, whether it is a war of nations, or a war of classes."[36]

These are not all of Benda's arguments and I have made the book seem to be more about Germany than it was.[37] He was no less harsh in his treatment of the French than of the Germans, but this makes his arguments less nationalistic than would have otherwise been the case, and therefore more equable. Nonetheless, Benda made it clear he thought this "treason" had originated in Germany and spread to other countries, notably France, from there. Several of his points—the divinization of politics, the elevation of the will, the understanding of war as an instrument of morality, not utility, the downplaying of objectivity—had an uncomfortable resonance with what came later.

PART V

Songs of the Reich: Hitler and the "Spiritualization of the Struggle"

34 ·

Nazi Aesthetics: The "Brown Shift"

On January 30, 1933, Adolf Hitler became chancellor of Germany. Six weeks later, on March 15, the first blacklist of artists was published. George Grosz, visiting the United States, was stripped of his German citizenship. The Bauhaus was closed. Max Liebermann (then aged eighty-eight) and Käthe Kollwitz (sixty-six), Paul Klee, Max Beckmann, Otto Dix, and Oskar Schlemmer were all dismissed from their posts as teachers in art schools. A few weeks later the first exhibition defaming modern art—called "Chamber of Horrors"—was held in Nuremberg, then traveled to Dresden and Dessau. These facts and events, and many others like them, are well known now, but they still have the power to shock. Four days before these dismissals took place, the Reich's Ministry for Popular Enlightenment and Propaganda was announced, with Joseph Goebbels as minister.

These brutal actions did not come out of the blue, however. Hitler had always been clear that if and when the Nazi Party formed a government, there would be "accounts to settle" with a wide range of enemies. Artists were foremost among these "enemies." In a 1930 letter to Goebbels, he insisted that when the party came to power, it would not be simply a "debating society" so far as art was concerned. The party's policies, laid out in the manifesto as early as 1920, called for a "struggle" against the "tendencies in the arts and literature which exercise a disintegrating influence on the life of the people." [1]

Some artists—like many scientists, philosophers, and musicians—seeing which way the wind was blowing, attempted to align themselves with the Nazis, but Goebbels was having none of it. For a time he and Alfred Rosenberg competed for the right to set policy in the cultural/intellectual sphere, but the propaganda minister sidelined his rival as soon as an official Chamber for Arts and Culture came into being under his control. Its powers were formidable—every artist was forced to join a government-sponsored professional body, and unless they registered, they were forbidden from exhibiting in museums and from receiving commissions. Goebbels further stipulated that there were to be no public exhibitions of art without official approval. In a speech to the party's annual meeting in September 1934, Hitler emphasized "two cultural dangers" that threatened National Socialism. On the one hand were the modernists, the "spoilers of art"—identified specifically as "the cubists, futurists and Dadaists."[2] What he and the German people wanted, Hitler said, was a German art that was "clear," "without contortion," and "without ambiguity." Art was not "auxiliary to politics," he insisted. It must become a "functioning part" of the Nazi political program. From May 1936 all artists registered with the Reichskammer had to prove their Aryan ancestry. In October that year the National Gallery in Berlin was ordered to close its modern art galleries, and in November Goebbels outlawed all "unofficial art criticism." From then on only the *reporting* of art events was allowed.

Some artists protested—Ernst Kirchner that he was "neither a Jew nor a Social Democrat," Max Pechstein that he had fought for Germany on the Western Front in World War I, that his son was a member of the SA, Emil Nolde that he had been a member of the Party since 1920—but it was all in vain. Some protested in their art—Otto Dix portraying Hitler as "Envy" in his 1933 painting *The Seven Deadly Sins*, and Max Beckmann caricaturing the chancellor as a *Verführer*, or "seducer." Many artists realized they had little choice but to emigrate, Kurt Schwitters to Norway, Paul Klee to Switzerland, Lyonel Feininger to the United States, Beckmann to the Netherlands, and Ludwig Meidner to Britain.[3]

As has been seen, the closure of the Warburg Institute in Hamburg actually preceded that of the Bauhaus, with the Frankfurt school the next to go. Most members of the school were not only Jewish but also openly Marxist. According to Martin Jay in his history of the school, its endowment was moved out of Germany in 1931, to the Netherlands, thanks to the foresight of the director, Max Horkheimer. Foreign branches had already been set up in Geneva, Paris, and London (the latter at the London School of Economics). Shortly after Hitler assumed power in March 1933, Horkheimer quietly crossed the border into Switzerland, only days before the school was closed down for "tendencies hostile to the state." The building on Victoria-Allee was confiscated, as was the library of 60,000 volumes. Only days after he escaped, Horkheimer was formally dismissed, together with Paul Tillich and Karl Mannheim. Horkheimer and his deputy, Friedrich Pollock, went to Geneva, and so did Erich Fromm. Offers of employment were received from France, initiated by Henri Bergson and Raymond Aaron. Theodor Adorno meanwhile went to Merton College, Oxford, where he remained from 1934 to 1937. Pollock and Horkheimer made visits to London and New York to sound out the possibilities of transferring there. They received a much more optimistic reception at Columbia University and so, by the middle of 1934, the Frankfurt Institute for Social Research was reconstituted at 429 West 117th Street. It remained there until 1950.

The migration of the Vienna Circle was perhaps less traumatic than that of other scholars. Thanks to the pragmatic tradition in America, not a few philosophers there were sympathetic to what the logical positivists were saying, and several of the Circle crossed the Atlantic in the late 1920s or early 1930s to lecture and meet like-minded colleagues. They were helped by a group known as Unity in Science, philosophers and scientists searching for the constancies from one discipline to another. This international group held meetings all over Europe and North America. Then, in 1936, A. J. Ayer, the British philosopher, published *Language, Truth and Logic*, a brilliantly lucid account of logical positivism that popularized its ideas still more in America, making members of the circle especially welcome there. Herbert Feigl was the first to go, to Iowa in 1931; Rudolf Carnap went to Chicago in 1936, taking Carl Hempel and Olaf Helmer with him. Hans Reichenbach followed in 1938, establishing himself at UCLA. A little later, Kurt Gödel accepted a research position at the Institute for Advanced Study at Princeton and so joined Einstein and Erwin Panofsky.

On May 2, 1938, Hitler signed his will. In it he ordered that, upon his death, his body was to be taken to Munich—to lie in state at the

Feldherrnhalle and to be buried nearby.* Even more than Linz, where he had been at school and grown up, Munich was home to him.[4] In *Mein Kampf,* Hitler described the city as "this metropolis of German art," adding that "one does not know German art if one has not seen Munich." Here the climax of his quarrel with the artists took place in 1937.

On July 18 that year, Hitler opened the Haus der deutschen Kunst, the House of German Art, in Munich, with nearly 900 paintings and pieces of sculpture by such Nazi favorites as Arno Breker, Josef Thorak, and Adolf Ziegler.[5] There were portraits of Hitler as well as Hermann Hoyer's *In the Beginning Was the Word*, a nostalgic view of the Führer "consulting his colleagues" during the early days of the Nazi Party. One critic, mindful that speculative criticism was now outlawed and only reporting allowed, disguised his criticism in reportage: "Every single painting on display projected . . . the impression of an intact life from which the stresses and problems of modern existence were entirely absent—and there was one glaringly obvious omission—not a single canvas depicted urban and industrial life."

On the day the exhibition opened, Hitler delivered a ninety-minute speech in which he reassured Germany that "cultural collapse" had been arrested and the vigorous classical-Teutonic tradition revived. Art was very different from fashion, he insisted. "Every year, something new. One day Impressionism, then Futurism, Cubism, and maybe even Dadaism." No, he insisted, art "is not founded on time, but only on peoples." Race—the blood—was all. What did it mean to be German? It meant, he said, "to be clear." Art is for the people, and the artist must present what the people see—"not blue meadows, green skies, sulphur-yellow clouds and so on." There can be no place for "pitiful unfortunates, who obviously suffer from some eye disease."[6]

This time there *was* criticism of a sort, albeit in a disguised way. The very next day, July 19, in the Municipal Archaeological Institute, across town in Munich, the notorious exhibition *Entartete Kunst* (Degenerate Art) opened. This displayed work by 112 German and non-German artists, 27 Noldes, 8 Dixes, 61 Schmidt-Rottluffs, 17 Klees, plus works by

* Later he planned to be buried in a crypt in "Germania" (Berlin) surrounded by his dead field marshals.

Gauguin, Picasso, and others. The paintings and sculptures had been plundered from museums all over Germany, and this exhibition surely ranks as one of the most infamous ever held. Even the Führer was taken aback by the way some of the exhibits were presented. Kirchner's *Peasants at Midday* was labeled "German Peasants as Seen by the Yids"; Ernst Barlach's statue *The Reunion*, which showed the recognition of Christ by Saint Thomas, was labeled, "Two Monkeys in Nightshirts."

If Hitler thought that he had killed off modern art, he was mistaken. Over the four months that *Entarte Kunst* remained in Munich, more than 2 million people visited the show, far more than the thin crowds that drifted through the House of German Art. This was small consolation for the artists, many of whom found the exhibition heartbreaking. Nolde wrote yet again to Goebbels, demanding that "the defamation against me cease." Beckmann was more realistic; on the day the show opened he took himself off into exile.

Yet another retroactive law, the degenerate art law of May 1938, was passed, enabling the government to seize "degenerate" art in museums without compensation. Some of the pictures were sold for derisory sums at a special auction held at the Fischer Gallery in Lucerne; there were even some that the Nazis deemed too offensive to sell and these were burned at a great bonfire in Berlin in March 1938.[7]

A different meaning of degeneration was fixed by Victor Klemperer (1881–1960), a Jewish professor of French literature at Dresden, who lived in Germany throughout the Third Reich, protected by friends. He kept a detailed account of the Nazis' use of language and showed, inter alia, how the word *Sturm* (storm), which had been the name of a (now banned) Expressionist art magazine, was appropriated as a hierarchical military term. *Schutzstaffel* (Protection Echelon) was soon reduced to abbreviations, SS and SA, "which became so satisfied with themselves that they were no longer really abbreviations at all," with official typewriters being fitted with special keys showing the angular SS character.[8] Elsewhere, among many other examples, sunshine became "Hitler weather,"* and Klemperer noted that in the university physics department the Hertz unit of frequency could not be referred to in that way because Hertz's father's family was Jewish. Just oc-

* "*Kaiserwetter*" is still used today, humorously.

casionally, he notes that the Nazi use of language gave him hope: whenever German troops were reported to be fighting "valiantly," they were losing.[9]

Verse "in the Key of Heroism"

In cultural terms, in its systems of thought, National Socialism was far more coherent than it is often given credit for. That was part of the problem. In their efforts to impose political synchronization, known officially as *Gleichschaltung*, the Nazis sometimes assumed, or determined, that there was more coherence than actually existed. Hitler's idea was that the prejudices of ordinary people could be molded into a Germanic worldview that would end the alienation felt by so many.

It is very doubtful if culture—any culture—can be prepackaged in this way for very long but certainly, so far as German culture in the Nazi period is concerned, the exercise showed that it can work in the short term.[10] Herman Burte, a playwright and poet and the author of an anti-Semitic novel, addressed an assembly of German poets in 1940 and, arguing that Hitler was essentially a poet-turned-statesman, asserted that all those involved in cultural creativity must use it to turn the energies of the German people to augment the "German mode of being." He thought the Führer was superior even to Goethe because Hitler understood the "organic" nature of the German people, the "primacy of the primordial German." Adolf Spemann, a Stuttgart publisher, lectured his fellow publishers that the age when they could release books that they might not agree with but thought intellectually or commercially worthwhile, was over. The publisher was no longer an "uninvolved cultural mirror," the "servant of the writer has been changed into a deputy of the state . . . literature is not to be separated from politics."[11] By 1937, between 50 and 75 percent of all books approved by the National Socialists were peasant novels, historical novels, and novels set in the "native landscape."[12]

According to Jay Baird in his study of Nazi heroes, three individuals in particular contributed significantly to National Socialist aesthetics, two of them writers—the poet Gerhard Schumann and the songwriter Hans Baumann. (The third, the film director, Karl Ritter, is considered on p. 639.)

Schumann, Baird says, was "a self-styled elitist," the son of a professor of education and a pious Christian. He joined the German youth movement and went on walking trips, exploring the landscape and ancient castles and churches. It was an idyllic upbringing but then he was pitched into "the bleeding cities of Weimar" and was "shocked into poetry."[13] In the early 1930s he produced his first collection, *Die Lieder vom Reich* (*Songs of the Reich*), giving voice to his longing to sabotage the "foreign ideologies" that had overtaken Germany and for a strong "Führer personality" to take the helm and save the nation.

. . . as he arose the halo of the chosen one
shone round his head. And as he descended
He called the torch illuminating the night.

Millions silently revered him . . . [14]

At the University of Tübingen he found the old, aristocratic, class-based emphasis on classical scholarship out of date with what was required in industrial Germany and as a result, in November 1930, at the age of nineteen, he joined the NSDAP. When Hitler assumed office, Schumann was taken on as a party writer, a position of some prestige. Here he became more infused with Nazi ideology as applied to the arts.[15] The so-called *Asphaltliteraten*, the central approach of the Weimar years, was now outdated, and Schumann was attracted instead by the idea that the new verse was to be "struck in the key of heroism."

A prolific author (he published nineteen volumes of verse, as well as journalism and two plays), one of his specialities as a party writer, and then as *Präsidialrat* of the Reich Chamber of Music, when he was only twenty-four, was composing scripts for the great ceremonial occasions. His poems had such titles as "Germany, You Eternal Flame," "The Purity of the Reich," and "Hitler."

In one will all the towering force
of millions living and dead. . .

In one hand the brotherly greeting
of millions of outstretched hands. . .

With the thunderous power of all the bells
his voice is ringing over the world.

And the world will hear.

He celebrated the *Anschluss* with a poem that Hitler so loved that he insisted it be broadcast time and again.[16]

Schumann saw action in France and Russia in World War II and won the Iron Cross. Despite being severely injured on the Russian front, he volunteered again as soon as he was fit, and continued writing poems that, as he put it, in wartime were really "prayers disguised as poems." His most famous, "Soldier's Prayer," was scored by Eugen Papst and sung far and wide, especially on ceremonial occasions:

O God, we are not much with words.
But please hear our prayer now:
Make our souls firm and strong.
The rest we'll do ourselves.[17]

"Do Not Count the Dead"

Hans Baumann, "the troubadour of the Hitler Youth," was nineteen in 1933. An innocent from the Bavarian forest, he enjoyed a meteoric career in the Third Reich, aged only thirty-one when it ended. As a boy he was known as "Happy Hans" and in his memoirs described his mother as "the best mother in the world." When his father came back from World War I, he brought some old hand grenades which Hans called "my first friends."[18]

This idyllic childhood was tempered by the inflation and unemployment in Weimar Germany and some of Baumann's early poems, written when he was just fourteen, have titles like "Unemployed" and "Four Flights Up," about living in tenement slums.

As a Catholic, however, he became active in the Catholic youth movement, and this gave rise to some of his early songs, the most famous of which, "Morgen gehört mir," "Tomorrow Belongs to Me," was popularized in the 1972 film *Cabaret*. This song helped Baumann's rise to fame. He had written it when he was eighteen, studying for a career as a teacher at a Jesuit academy in Amberg. The priest in charge was so taken with Baumann's songs that he approached the Catholic publishing house Köselverlag in Munich, and they brought out a collection in 1933. "Tomorrow Belongs to Me" became famous among the Catholic youth movement long before it was taken up by the Hitler Youth.

Baumann, however, became convinced that Hitler was a savior, and he composed more than 150 songs reflecting that view. These songs got more aggressive as the thirties passed, especially in regard to the East. In the *Blitzkrieg* years he could be shockingly cavalier about war.

Despite the trembling of brave men,
despite the distress in my heart
Afire with sorrow, I will raise the banner,
with hands that are no more.

Baumann liked to say it was a privilege to be living in Germany's era of greatness and, in a series of speeches, quoted these lines from Hölderlin:

The battle is ours! Hold high the banner,
O Fatherland, and do not count the dead!
Beloved nation, not one too many
has died for you.[19]

But Baumann changed. He seems to have had second thoughts around 1941–42, when he began to advocate more charity on Hitler's part toward Germany's enemies. This culminated in his play *Alexander*, which Gustaf Gründgens (see p. 648) snapped up for production in Berlin. A great success, with Gründgens in the title role taking twenty-five curtain calls on the opening night, its plot drew not-so-subtle parallels between Alexander the Great and Hitler. Alexander exclaims "Let us be contemptuous of the earthbound," but he also says, "I am victorious because I love." Again, Baumann stressed charity among the victors, perhaps feeling that was as much as he could get away with in the Third Reich. Goebbels had the play closed after two nights.

Baumann's second thoughts gathered pace, aided by the fact that his brother, an artillery captain in Kiev, had seen terrible things there, and his wife, who had been an entertainer on the Eastern Front, had also seen and heard about heartbreaking atrocities. In the army himself now, in the East, Baumann directed a program for cultural understanding for German soldiers, where even Russian works were performed and Russian guests allowed, even members of the Resistance. After the war Baumann became internationally acclaimed as a writer of children's books, his work translated into a score of languages and winning prizes in West Germany, Italy, and the United States. Toward the end of his life (he died in 1988 in Murnau, where Kandinsky had spent so much time), Baumann said, "The great thing about having a long life is that one can correct his mistakes."

In an age before television, Goebbels well understood the power of radio. He inherited a system of centralized control but even so his propaganda ministry acquired all the shares in the National Broadcasting Company, which exercised an influence on other—more peripheral—broadcasting outfits. Pressure was put on electrical companies to produce cheap radio sets—so all citizens could own one—and to construct them so that they could not receive foreign broadcasts.

Hitler moved in on the filmmakers as soon as he moved on the artists. One of Goebbels's first initiatives when he was appointed propaganda minister was to call together Germany's most prominent filmmakers and show them Sergei Eisenstein's *Battleship Potemkin,* his 1925 masterpiece that commemorated the revolution and was both a work of art and a piece of propaganda. "Gentlemen," Goebbels announced when the lights came on, "that's an idea of what I want from you." The minister wasn't looking for obvious propaganda; he was clever and knew better. But films must glorify the Reich: there was to be no argument about that. At the same time, he insisted that every cinema must include in its program a government-sponsored newsreel and, on occasion, a short documentary.[20]

By the outbreak of war, Goebbels's newsreels could be as long as forty minutes, but it was the documentaries that had the most effect. Technically brilliant, they were masterminded by Leni Riefenstahl, an undistinguished actress in the Weimar years who had reinvented herself as a director and editor. The best was *Triumph des Willens* (*Triumph of the Will*; 1935), commissioned by Hitler himself as a record of the first party convention at Nuremberg in 1934. Sixteen camera crews were involved and when it was shown, after two years of editing, the film had a mesmerizing effect. The endless torch-lit parades, one speaker after another shouting into the microphone, the massive regularity of Brownshirts and Blackshirts absorbed in the rhetoric and then bellowing "Sieg Heil" in unison, were hypnotic.

Almost as clever was *Olympia*, which Goebbels ordered to be made about the 1936 Olympic Games, staged in Berlin. It was there that the modern Olympic Games emerged, thanks to the Nazis. They implemented the idea of the "torch run," whereby a flaming torch was carried by runners from Greece to Berlin, arriving in time to open the games in style.

For Riefenstahl's film of the games she had the use of eighty cameramen and crew, and shot 1.3 million feet of film, eventually producing, in 1938, a two-part, six-hour movie with soundtracks in German, English, French, and Italian. She ennobled good losers, supreme winners, and dwelled on fine musculature, particularly that of Jesse Owens, the Negro athlete from the United States who, to Hitler's extreme displeasure, won four gold medals. Some of *Olympia*'s sections, particularly those dealing with platform diving, are unsurpassingly beautiful. But Riefensthal was not the only "heroine" of Nazi cinema: Kristina Söderbaum, Lilian Harvey, and Zarah Leander were all fêted.[21]

After the war started, Goebbels used all the powers at his command to make the most of propaganda. Cameramen accompanied the Stuka bombers and Panzer divisions as they knifed through Poland—but these documentaries were not only used for audiences back home. Specially edited versions were shown to government officials in Denmark, the Netherlands, Belgium, and Romania to underline "the futility of resistance."

The Song of the Stukas

Goebbels used to say that victory makes its own propaganda, while defeat "calls for creative genius," a sentiment that sums up the career of the film director Karl Ritter.[22] A pilot in World War I, Ritter became one of the top two or three film directors in the Third Reich.

After World War I Ritter tried his luck as an artist in Munich, and it was there that he first heard Hitler speak. Devastated by the outcome of the war, Ritter found Hitler's message more than congenial and joined the NSDAP in 1925. He found his way into films via poster design and public relations and in 1933, UFA, one of Germany's leading film companies (which would soon become *the* leading film company) offered him work as production director of *Hitlerjunge Quex* (*Hitler Youth Quex*), arguably the first Nazi film of any consequence. It concerns a youth who is torn between loyalty to his Communist father and his growing belief in the Hitler Youth movement. He is killed in the course of street disturbances, which the Nazis had made such a feature of their activities in the late 1920s and early 1930s.

Ritter went on to produce a series of films which, until Stalingrad, were increasingly influential. *Urlaub auf Ehrenwort* features a handsome junior officer in World War I and his unit as they are transported by train back to Berlin from one area of fighting before going on to another, where they will almost certainly meet their death. When the unit arrives in Berlin, the men have five hours to wait for their next train. Most of the soldiers come from Berlin and although leave is forbidden, the junior officer allows them to go home after they have given an undertaking they will all be back at the station in time for their train. The film then follows the challenges they face—pacifists and communists taunting them along the way, unfaithful wives, wives who have taken their husband's jobs, wayward children, illness. All the men save two are back at the station in time, and the others catch up with the train when it makes its first stop.

Non: Pour le Mérite, possibly Ritter's finest film, was a biography of a World War I pilot, made with the cooperation of the air ministry. Having downed an English ace, who is not wounded, the German pilots honor him—for his bravery—over dinner. Elsewhere, a German pilot withholds fire when he realizes the English pilot's guns have jammed. The film then shows the still-young pilots going to seed in the Weimar Republic, their talents and war record ignored, only to be redeemed with the arrival of Hitler.[23]

During the Blitzkrieg years, Ritter shifted his focus to World War II itself. The 1941 *Stukas* is devoted to comradeship in the flying corps and featured a very popular song, "Stukalied."[24] The film also depicted Prussian mothers who, "instead of sadness and melancholy" because their sons have been killed, feel "pride and a sense of fulfilment that [their boys] had the privilege of dying a heroic death."[25] The Stukalied ends:

We're not afraid of hell and never relent,
Till the enemy is destroyed,
Till England, till England, till England is crushed,
The Stukas, the Stukas, the Stukas.

Ritter was captured by the Russians, escaped, and returned to Germany. After being "de-Nazified," he emigrated to Argentina.[26]

Hitler's assault on music and musicians was—in its aims at least—no less severe than his attack on artists and publishers, but what transpired was more complex. The modernist repertoire was purged from early on in 1933, with "degenerate" composers such Arnold Schoenberg, Kurt Weill, Hanns Eisler, and Ernst Toch and conductors including Otto Klemperer and Hermann Scherchen, expelled. A commission of leading musicians—including Max von Schillings and Wilhelm Furtwängler—was set up in Berlin in June 1933, their task to supervise and censor the programs of music performed in the capital.[27]

Richard Strauss was treated gingerly and so was Furtwängler. Strauss's collaboration with Stefan Zweig, the Jewish writer who wrote the libretto for his opera *Die schweigsame Frau* was not stopped, though performances were forbidden shortly after the premiere, and Strauss was eventually asked to resign his post as director of the Reichsmusikkammer (RMK), "on grounds of age and ill health." Furtwängler was forced to resign his post as vice president of the RMK in protest against the regime's treatment of Hindemith, but he wasn't further harassed, not then anyway.[28]

Goebbels's propaganda ministry did not have its own music division until 1936, and only then was pressure stepped up. Just 2 percent of German music was of Jewish origin but the prominence of Jewish musicians, such as Schoenberg, Klemperer, Kurt Weill, and Hanns Eisler helped the Nazis spread the message that there was a conspiracy to debase what the average German saw as a national treasure—the tradition of German music. Jews were banned from various musical organizations but given their own self-financing cultural outfit, the Kulturbund deutscher Juden. After Kristallnacht, Jewish musical publishers were closed down or "Aryanized."[29]

Goebbels was, however, in general careful about precipitate dismissals. He was told early on that it would take time to replace dismissed Jewish soloists with "Aryan" substitutes and so he held on until they were ready before insisting on change. In the case of the Berlin Philharmonic Orchestra, Goebbels was forced to go especially slowly because the orchestra was private, its musicians unaffected by the April 1933 Civil Service Law that removed all Jews from public (civil) service and dismissed artists from their museum and art school posts. Instead, Goebbels arranged for the orchestra to be starved of funds until it was on the verge of bankruptcy. Then he rode to the rescue, guaranteeing funds but at the price—eventually—of the dismissal of all Jews and enemies of the regime. In the end, the RMK expelled more musicians than did any other branch of the RKK as these figures by Erik Levi show:

Number of Expulsions Carried Out by RKK			
Film	750	Press	420
Theater	535	Music	2,310
Writing	1,303	Art	1,657

Again, to begin with (fateful words), the Kulturbund provided the Nazis with propaganda in that it enabled them to insist there was plenty of work for Jewish musicians in Germany, and statistics appeared to bear this out: between 1934 and 1938 the Kulturbund gave 57 opera performances and 358 concerts, playing before 180,000 people in Berlin, Frankfurt, Cologne, Hamburg, and Munich. It was prohibited from performing *Fidelio* but otherwise gave a "standard" repertoire. The situation deteriorated first in the provinces and came to a head on Kristallnacht, November 9, 1938, after which the opera section was closed down, though concerts were given until September 1941.[30]

One interesting difference between art and music was that although paintings could not be "Aryanized," there were attempts to make this happen with music. The regime invited contemporary composers to produce replacements for Mendelssohn's ever-popular music for *A Midsummer Night's Dream* (forty-four scores were produced, though none really caught on). Both Schubert and Schumann had set to music poems by

Heine, who was Jewish, and there was widespread debate about the propriety of performing such "hybrid" creations. In the case of Mozart's *Così fan tutte*, *Le Nozze di Figaro*, and *Don Giovanni*, in which the librettos had been written by the baptized Jew Lorenzo Da Ponte and translated into German by the Jewish conductor Hermann Levi, the NSKG commissioned Siegfried Anheisser to provide a new "Aryan" retranslation. By 1938 his translations had been adopted in 76 of Germany's 85 opera houses.[31] There were several other examples of this procedure.

It wasn't only Jews who were targets. The contemporary music festival at Baden-Baden was stopped, and the experimental Kroll Opera in Berlin abolished in 1931, even before the Nazis came to power, but partly thanks to their agitation.*[32] Early in 1933 Nazi protests against contemporary music were stepped up, together with a ban on the broadcasting of jazz, viewed as a degenerate example of "Negro culture."† Censorship was temporarily relaxed in 1935–36, ahead of the Berlin Olympics, when a larger than usual number of foreigners—especially Americans—were in Germany, but was resumed soon afterward. An *Entarte Musik* exhibition was held in Düsseldorf in May 1938, the brainchild of Adolf Ziegler, a major feature of which was photographs of composers—Schoenberg, Hindemith, Webern—who were considered to have a destructive influence on German music. There were six booths where, at the touch of a button, visitors could hear examples of Hindemith, Weill, Ernst Krenek, and others.[33]

The dissonant music of Richard Strauss escaped censure but not that of Schoenberg's disciples Webern and Berg. *Lulu* was performed in Berlin in November 1934 and provoked such a scandal that no other work by Berg was ever performed in the Third Reich. The attitudes of the Nazis toward Paul Hindemith had been hostile even before 1933, not just because of his modernist music but because of his links to Bertolt Brecht. But, as professor of composition at the Berlin Hochschule für Musik since 1927, he occupied a high profile position in which he exerted an influence

* After the burning of the Reichstag, the premises of the Kroll Opera were used for the Parliament.

† Although jazz was banned inside Germany, swing was allowed in propaganda broadcasts to other nations, to attract listeners. There was also a "Propaganda Cabaret," led by Lutz Templin. In addition Göring established a small number of dance orchestras to play late-night concerts on the radio, and to go on tours.

over a whole generation of composers, and not just German ones. As an "Aryan" and Germany's next most prominent composer after Strauss, he had influence, the more so when Strauss nominated him to be part of the inner council of the RMK in November 1933. In February 1934, a concert to unveil the RMK was held, billed as "The first concert in the Reich," featuring works by Strauss, Hans Pfitzner, Siegmund von Hausegger, and Hindemith, who conducted the Berlin Philharmonic in a performance of his *Concert for Strings and Wind*, originally written to mark the fiftieth anniversary of the Boston Symphony Orchestra in 1930. A month later Furtwängler conducted the same Berlin orchestra in the first performance of Hindemith's new work, *Mathis der Maler*. This was well received, and former critics believed he had purged himself of the "ugly stains of the past." *Mathis* was quickly performed all over Germany.

Then denunciations started appearing, accusing him of "cultural bolshevism," of identification with the Jews, and "atonality." A timely invitation came from the Turkish government to establish a music school in Istanbul, and Hindemith accepted.[34]

In the first few months of taking office, forty-nine out of Germany's eighty-five opera houses had seen a change of senior personnel. Despite their replacement by administrators and musicians more amenable to the National Socialists, Erik Levi reports that the high standards of performance that had been maintained before 1933 were sustained.[35] Furthermore, the number of musicians contracted to German theaters increased as follows:

Season	Singers	Chorus	Orchestra
1932–33	1,859	2,955	4,889
1937–38	2,145	3,238	5,577

Even after the outbreak of war, says Levi, opera activity continued to be intensive, many theaters maintaining a "substantial" repertoire. The Nazis poured vast resources into Bayreuth, offering cheap tickets to armaments workers and war veterans.[36] German opera companies visited occupied territories until 1942.

Contemporary composers who declared support for the regime were assiduously promoted. Max von Schillings's operas received 117 performances in the 1933–34 season, compared with 48 and 24 in the previous two seasons. Much the same happened with Hans Pfitzner, whose performances increased from 46 in 1931–32 to 130 in 1933–34.[37] There was also a posthumous renaissance of operas by Siegfried Wagner, Richard's son, who had died in 1930 after composing even more operas than his father. At the same time, contemporary operas by foreign composers were discouraged. According to Levi, 170 or so new German operas were performed in the Third Reich. Once the war had started, the number of new operas reaching the German stage showed no reduction until the 1943–44 season, with between sixteen and twenty premieres a year. Wagner actually suffered a *decline* in popularity throughout the 1930s, performances dropping from 1,837 in 1932–33 to 1,154 in 1939–40, while those of Verdi and Puccini rose.[38]

There were 181 permanent orchestras working in Germany in 1940, according to the RMK. From the time the Nazis came to power, and if we ignore the Jewish experience, orchestral musicians in Germany experienced an upturn in their fortunes. Standards remained just as high as in the Weimar period, this having to do with both an outstanding generation of conductors (Furtwängler, Erich Kleiber, Bruno Walter, Karl Böhm, Otto Klemperer, Hans Knappertsbusch, Hermann Scherchen, many of whom had to leave Germany later) and the growth of commercial recording firms, which helped establish German orchestras as preeminent throughout Europe.[39]

The Berlin Philharmonic Orchestra fought hard to keep its Jewish instrumentalists, even giving them their own solo pieces during the early months of the new regime. Furtwängler fought in their corner too, arguing at least to begin with that race had nothing to do with ability. As we have seen, however, the orchestra eventually had to succumb, driven into near bankruptcy by Goebbels. Even so, the orchestra's level of quality was maintained, conductors from abroad were still invited, and tours all over Europe were made in the 1930s. Throughout the period, contemporary music accounted for about a third of the repertoire.[40]

The Brown Shift in Theater

Germany—and Berlin in particular—had been renowned in the Weimar years for the virility of its theater. Though Berlin theater held on to its strengths for a while, elsewhere in the country the decline was swift after the Nazis took power. Performances of Goethe and Schiller continued, but for the rest the content was soon reduced to light opera and the works of mainly now-forgotten playwrights, often performing modern plays about peasant themes—*and* in peasant dialect.[41] Theater also underwent what was called a "*braun* shift"—all aspects being politicized, taking their color from the brown shirts of the Storm Troopers. Both Hitler and Goebbels believed, or said they believed, that it was the theater in the Weimar period that had abused German culture the most.

The first National Socialist premiere took place in 1927, in Cologne, with Hans Johnst's *Thomas Paine*, about the nationalist American revolutionary, "forgotten by his country as he sits in a French Republican prison." But by May 1933 the change that was to come was heralded. On the sixth of the month, as minister president of Prussia, Göring took personal control of state and city theaters and, two days later, Goebbels addressed a special meeting of producers at the Hotel Kaiserhof in Berlin on "the tasks of the German theater." He insisted that "outmoded art forms" would be eliminated in favor of the new *Volk* art—political, patriotic, and "consistent with the philosophy of the ruling party."[42] "German art of the next decade will be heroic; it will be like steel; it will be romantic, non-sentimental, factual; it will be national with great pathos and at once obligatory and binding, or it will be nothing." On August 21, 1933, Goebbels announced the creation of the office of Reichsdramaturg to advise his ministry on the theater, under Dr. Rainer Schlosser, a former critic. By then the new administration had given some hope to the theater profession (provided it kept in line) by supporting twelve new theaters (on to a total of 248 across the country).

The earliest chance the Nazis had to show off their taste was at the First National Socialist Theater Festival, held in Dresden from May 27 to June 3, 1934. The brainchild of the Propaganda Ministry (known then as the Promi), it was graced by Hitler himself, who decided to attend at the last minute. By this time, the theaters were more or less firmly under Goebbels's control and he identified three Berlin houses in particular to be the showcase of his new policy, and new *Intendants*, or directors. These were Count Bernhard Solms for the Volksbühne (formerly associated with Piscator), Heinz Hilpert for the Deutsches Theater (Reinhardt), and Walter Brügmann for the Theater des Volkes (the former Grosses Schauspielhaus of Reinhardt). Ten days before the Dresden Festival, Goebbels announced the Unified Theater Law, which stipulated that all theaters—private and public—must abide by the Nazis' racial and artistic aims, that theaters had only one obligation, to be "conscious of national responsibility" and that otherwise artistic freedom "will not be altered in any way."[43] Theaters were to be licensed.

One of the arguments Goebbels had with what he called the "modernist big-mouths" was that they often showed strife *within* Germany (Hauptmann's *The Weavers*, for instance) and this would no longer be tolerated. The program of the Dresden Festival therefore included works by Kleist, Schiller, Ibsen/Eckart (*Peer Gynt*), Goethe, and Shakespeare. It was in effect a "safe" performance, not ramming propaganda down people's throats too hard.

The Berlin Grosses Schauspielhaus itself, as a building, was huge and had a varied history, being at one stage the home of a circus, at another the venue where Robert Koch's international congress dealing with tuberculosis had been held in 1890, and at another where the workers congregated on hearing of the death of Lenin in 1924.[44] Its great days were inaugurated with the arrival of Max Reinhardt (see pp. 524–525), in whose time the theater was equipped with a *Kuppelhorizont* (sky dome) for spectacular staging effects and a revolving floor. Financial difficulties forced Reinhardt to sell this theater even before Goebbels forced his hand with the others.

The very size of the theater was attractive to the Nazis. This "sense of massiveness" was always important for them as they attempted, one way or another, to create what Hofmannsthal had identified as "ceremonies of the whole" back in turn-of-the-century Vienna (see pp. 491–492). In the Grosses Schauspielhaus, productions were always grand, massive, achieving their effects by enormous choruses, hundreds of musicians, vast troops of dancers, real cocks crowing, and real dogs barking. The aim, says Yvonne Shaffer, was to elevate and unite the audience "in a mystical union," but from 1936 to 1940 the Grosses Schauspielhaus, now renamed the Theater des Volkes, became the home of operetta. It was argued that operetta could "lead the uninitiated to an appreciation of opera." Believing that the contentment of the workers was "a task of military importance," the point of operetta was that it returned audiences to life as it was in the good old days, which the Nazis promised to bring back.[45]

The most imaginative and versatile actor/director of those times was Gustaf Gründgens (1899–1963), who could sing and dance as well as act, who had worked all over the country with his wife, Erika Mann, her brother Klaus (children of Thomas), and Pamela Wedekind (daughter of Frank) and had a famous cabaret, the "Review of Four."[46] Gründgens, though married, was a homosexual, but this did not deter Göring from appointing him *Intendant* at the Schauspielhaus am Gendarmenmarkt and instructing him to attract the biggest names. Gründgens obliged, surrounding himself with actors like Werner Krauss, Emil Jannings, and Emmy Sonnemann (who Göring was courting just then). Most importantly, Gründgens employed Jürgen Fehling (1885–1968), who had been a big name in Weimar times, directing plays by several playwrights the Nazis had banned, though he also directed plays by those they approved.

By no means an ideal director from the Nazis' point of view, Fehling's ability protected him. Together with Heinz Hilpert (1890–1967), who had a long association with Reinhardt in many modernist Weimar productions, they managed to keep legitimate theater unpolitical and maintain its excellence. These men maintained some integrity, mainly by producing only German classics and Shakespeare, and keeping works by playwrights who adapted to the Nazis, such as the Austrian, Richard Billinger, winner of the 1932 Kleist Prize. A major scandal was caused when Gründgens allowed Fehling to stage *Richard III* in 1937. Werner Krauss, as Gloucester, hobbled around the stage, with a clubfoot, "an apparent Goebbels take-off." Equally bad, the murderers of Clarence appeared on stage "in brown shirts and jackboots, bearing a distinct similarity to Nazi storm troopers." To cap it all, when Gloucester became king, "a phalanx of eight men in black uniforms accented with silver bijouterie accompanied him; their resemblance to Hitler's SS was both immediate and frightening."[47]

Göring, who saw the play, wanted Fehling dismissed, but Gründgens refused and, for once, the Reichsmarschall was defeated. But Fehling went on to direct Nazi plays. As elsewhere, didactic plays became obligatory and the classics were reinterpreted to support Nazi dogma. [48]

35 ·

Scholarship in the Third Reich: "No Such Thing as Objectivity"

In the first two years after the Nazis took power, at least 1,600 scholars, some 32 percent of the total of 5,000 university teachers, were dismissed, on either political or racial grounds. By the end of 1938, Germany—including Austria—had lost 39 percent of its university teachers, with Berlin and Frankfurt hardest hit, closely followed by Heidelberg.[1] The first years of the Third Reich comprised the era of the worst student radicalism, which often disrupted the classes of faculty members deemed undesirable but, as Steven Remy has shown in his study of Heidelberg, many faculty members silently acquiesced in the purges. Remy says that American intelligence reports compiled in 1945 identified fifteen Heidelberg professors as informants on their colleagues during the years of the Third Reich. There were protests but they were few and far between. Alfred Weber, the sociologist brother of Max Weber, resisted a plan to fly the swastika over public buildings, including his own institute, was pilloried in the local press for doing so, and forced to resign his position not long after.

Only a few academics joined the Nazi Party before Hitler actually took power, including the Nobel Prize–winning physicist Philipp Lenard, but once the Nazis were in office "the encomiums began." In April 1933 the Association of German Universities issued a statement in support of the "new German Reich," and in November 700 out of a total of some 2,000 full professors signed a document of support to "Adolf Hitler and the National Socialist state." Hundreds of professors now joined the party.[2]

Remy says that at Heidelberg there was a rash of publications in support of the National Socialists, most of them decrying Weimar as "weak," "foreign," and "un-German" and welcoming what many liked to call the "national revolution . . . that embodied a mixture of continuity with Germany's past and the radical, youth-oriented element of the National Socialist 'movement.'" The sociologist Arnold Bergstraesser blamed democracy for failing to produce "that social and political unity which is necessary in order to overcome a crisis like the world crisis of 1929," arguing that one of the aims of National Socialism was to establish "a real unity between State and society." The main idea now, he said, was "not to allow the existence of any sphere apart from the State." When he attended conferences across the Channel, the theologian Martin Dibelius made it his business to spread before the British, "the wonder of German unity" and "the cleansing of moral life" taking place under the Nazis.[3]

Jurists did what they could to offer legal justification for the new Nazi laws. In general their view was to advocate the importance of "German common law" and to reject the concept of "individual rights protected by law."[4] Walter Jellinek (himself Jewish) praised the new Nazi laws for overcoming class, regional, and religious differences. "The individual . . . owes all his dignity of being human only to his subordination to the state." In 1934 Jellinek argued that the political power concentrated in Hitler's hands was no bad thing: "It must not be forgotten that a voluntary restriction of supreme power resides in the German word Führer, the ideological content of which can hardly be translated into a foreign language."[5]

That so many established scholars were dismissed created boom conditions for younger, more compliant colleagues. Many of them, on the radical right, saw such advance as little more than their entitlement. The sociologist Carl Brinkmann, for example, now began opening his lectures with the words "finally, we can speak freely." There also emerged a group of more senior figures who began to work on the shape of the universities under National Socialism. These included Ernst Krieck, Alfred Bäumler, Adolf Rein, Hans Freyer, and Martin Heidegger. Krieck, a professor of pedagogy and in 1933 rector of the University of Frankfurt, called for an overhaul of the university, a "levelling of its hierarchical structure" and the "total focus of its research and teaching on the ideological goals of the state." Walter Gross, chief of the Office of Racial Politics, charged with "heightening ethnic consciousness," worried that many scholars who claimed to endorse National Socialism actually withheld "inner support" by taking "refuge" in "apolitical" research projects.[6] He also realized that biologists had failed to identify Jewish blood by physiological traits, which meant that a switch to cultural stereotypes was needed.

Of all those involved in shaping the universities, none was more interesting or controversial than the philosopher Martin Heidegger, whose relationship with, and treatment of, Hannah Arendt was notorious. As a young philosophy student of eighteen, Hannah Arendt (1906–75) arrived in Marburg in 1924 to study under Martin Heidegger (1889–1976), then arguably the most famous living philosopher in Europe, and in the final process of completing his most important work, *Being and Time*, which appeared three years later. When Arendt first met Heidegger, he was thirty-five and married, with two young children. Born a Catholic and intended for the priesthood, he developed instead into a charismatic philosophy lecturer.

Arendt came from a very different background—an elegant, cosmopolitan, totally assimilated Jewish family in Königsberg. The love affair between Heidegger and Arendt is now well known. Each transformed the other but in 1933 their lives turned dramatically in different directions. He was made rector of the University of Freiburg, and rumors soon reached her that he was refusing to recommend Jews for positions and even turning his back on them. At Heidegger's rectorial address, he made a very anti-Semitic and pro-Hitler speech, which was reported all over the world. Arendt, now in Berlin and married to a man who, as she later admitted, she did not love, moving among the likes of Adorno, Marcuse, and Fromm, was deeply upset and confused by Heidegger's behavior. To make matters worse, Bertolt Brecht, persecuted as a Communist and forced to flee the country, left behind his address book, containing Arendt's name and phone number. She was arrested and spent eight days in jail being interrogated. As soon as she was released, she left Germany and settled in Paris. She and Heidegger would not meet for seventeen years.

Heidegger played a crucial role in Germany. As a philosopher, he gave his weight to the Third Reich, helping develop its thinking, which grounded

Nazism in history and the German sense of self. In this he had the support of Goebbels and Hitler. As an academic figure he played a leading role in the reorganization of the universities, the chief "policy" under his regime being the removal of all Jews. Through Heidegger's agency both Edmund Husserl, the founder of phenomenology, and his own professor, Karl Jaspers, who had a Jewish wife, were forced out of their university posts. Arendt later wrote that "Martin murdered Edmund."

Ernst Krieck considered himself a more important National Socialist philosopher even than Heidegger, but the influence of the philosophers was limited in the Third Reich: Hitler, Goebbels, and the other leaders were more interested in practical matters than abstract theorizing—they regarded academics in general as an "elite group of opportunists." The common theme that emerged from the speeches and publications by Krieck, Heidegger, and others, in 1933 and 1934, was that "research and teaching must serve the German '*Volksgemeinschaft*,' and not some abstract notion of 'objective truth' or knowledge for its own sake." At Göttingen, Paul Schmitthenner, the architect and arch-opponent of Walter Gropius, argued openly that universities must become "political universities," research being supported only if it served "the state and the people." Hans Frank, the Reichsminister and governor-general of occupied Poland, insisted that "The categorical imperative of action in the Third Reich is this: act in such a way that the Führer, if he knew of your action, would approve of it."[7]

The "German Spirit" in Scholarship

In the summer of 1936 Heidelberg celebrated its 550th anniversary, an event that Hitler deemed to be of "national" significance.[8] The Heidelberg celebrations offered an opportunity to set out what, exactly, was meant by the term "German scholarship" in the context of the Third Reich, what Reinhard Heydrich called "the spiritualisation of the struggle."

At Heidelberg, this had already begun the previous year when the physics institute had been renamed after Philipp Lenard, the prominent advocate of "Aryan physics." What Lenard meant was that "German natural sciences" differed from "Jewish science," the former consisting of "observation and experimentation and not excessive theorising and reliance on abstract mathematical constructions," unlike, say, relativity theory. Speeches at the anniversary ceremony carried this further, with Ernst Krieck arguing that in the nineteenth century science had been "shattered into a 'heap' of disconnected specialities that ultimately did not serve the people."[9]

The "Aryan" physicists tried to extend their influence, taking over various scientific journals and creating their own Deutsche Mathematik, devoted to Aryan mathematics. (Not everyone toed the line by any means, and Walter Bothe's research into nuclear physics—ignoring what the "Aryans" had to say—led to the creation of Germany's first cyclotron.) Their actions were attacked in the British (but also international) journal *Nature* to such an extent that, at the end of 1937, the Reich Education Ministry forbade subscriptions to *Nature* in Germany.[10]

The Heidelberg celebrations featured—besides a congratulatory telegram from Hitler—several talks about the new intellectual climate. The Reichserziehungsminister (education minister), Bernhard Rust, argued that the national and racial background of individual scholars could not help but shape their scholarship, that there was no such thing as purely "objective science," which was a "Jewish-Marxist" idea, and that this realization had "transformed the inner life of the German people," helping them to forge an "organic unity" between Wissenschaft and the Volk. Krieck spoke in much the same vein but added that "it can be fully demonstrated . . . that . . . every worthwhile achievement in the sphere of the natural sciences, no less than in the sciences of culture, has been intimately bound up with the . . . racial characteristics of the people concerned."[11]

All this was underlined by a number of new institutes and seminars that were created at Heidelberg in the years before the outbreak of World War II, focusing on the military and political preparedness of Germany. Schmitthenner, who would become rector of Heidelberg in 1938, described himself as a "soldier, politician, and scholar" (i.e., in that order). He founded a seminar on the history of warfare, styling himself as a "frontier professor." The Social and Economic Faculty focused on "spatial" research, *Raumforschung*, developing ideas of Werner Sombart that "economic life in Germany rested on two pillars, race and space." The departments of classics, theology, languages, and literature were all affected—race was seen as a "determinant" of language, and people's ability to absorb the Christian message was likewise regarded as a function of "blood, soil and race." History teaching was reorganized around a new set of "key dates" or turning points, such as March 1912, when the last legal bans against Jews in Germany were removed.[12]

Steven Remy sums up the "German Spirit" in scholarship as follows: it was fundamentally opposed to traditions of teaching and research throughout the Western world, in that (1) it rejected "objectivity," (2) it denied that scholarship served intangible notions of truth for truth's sake, insisting instead that German scholarship must serve the "Volk," (3) it opposed "hyper-specialization," and (4) race was a central concept, that representatives of "inferior" races, like the Jews, "were incapable of examining the natural world honestly and accurately."[13]

Biological Dogmatics: "The Language of Our Age"

Just as Steven Remy has reconstructed the Nazification of one university, so James R. Dow and Hannjost Lixfeld have explored the Nazification of a single discipline—folklore.[14]

There was in Germany a strong interest in folklore going back to Herder and the Grimm brothers. In fact, in 1940, Thomas Mann identified what he viewed as the fundamental difference between German culture and that of the West. "Whereas British and French writers produced art rooted in social and political reality, the Germans had dedicated themselves to the 'pure humanity of the mythical age,' which was based in nature itself rather than the circumstances of any historical era."[15] This interest remained strong, and in the Weimar years a number of international conferences were held at the German Central Work Station for Folk and Cultural Landscape Research, while in 1926 the first issue of the journal *Volk und Rasse* appeared. What appealed to people about Volk culture was that it was organic, traditional, the exact opposite of the industrial worker. "He [the industrial worker] works with dead tools on dead material . . . His work pace is determined not by the sun, the season, the weather, but by a machine which goes at its own pace summer and winter, day and night. His work is measured exactly in millimetres and kilograms, measurements that have no relationship to life."

On this view, only *Volk* culture could give people a sense of fulfillment, all others would become "an un-German greenhouse plant."[16]

Although the National Socialists appropriated the general approach of the folklorists, folklore had a wider appeal than that. Kurt Huber argued for a "resurrection" of folk culture to counter "the national loss of instinct by Germans, completely miseducated by too much humanism . . ." Wolfgang Emmerich examined Gottfried Benn's notion that "peasantry is an inner attitude, not a line of business."[17] The remains of German mythology, said another scholar, "were a force of secret resistance to bourgeois civilisation." Herman Wirth, the founder of Ancestral Inheritance Inc., was filled with an unshakable belief in the "continuity of the early Stone Age religion and world view." This was all a form of *biological dogmatics.* Max Hildebert Boehm reworked sociologist Wilhelm Riehl's four S's—*Stamm, Siedlung, Sprache, Sitte* (tribe, settlement, language, custom) "into the language of our age"—blood, soil, folk-nation, and folk order (*Blut, Boden, Volkstum, Volksordnung*). Above all there was the "ennobling" of the Social-Darwinistic mythos of "fateful struggle" for survival, an emphasis on the dark and tragic (the twilight of the gods) and the defamation of foreign influences. High culture, in this view, was "like a slut" that ran after every foreign influence that came its way. In 1928 William Stapel said that anyone who wanted to experience Germanness must have "lived in German forests, he must have courted German girls, and must have done German farming and German handiwork."[18]

Hermann Strobach noted that the League of German Societies for Folklore reacted immediately and "conspicuously" to the advent of the Nazis into power and at its conference in October 1933 there were lectures on "National Socialism and Folklore" and "The Sociopolitical Task of Folklore." At its conference the following year, the folklorists sent a telegram to Hitler, vowing that they would work toward "strengthening and increasing the Germanness of our people."[19]

Christoph Daxelmüller says that Jewish *Volkskunde* had developed since 1898 thanks to the cooperation of both east and west European scholars, with organizational centers for the "Science of Jewishness" in Hamburg and Vienna, and that there was also in Berlin an Academy for the Science of Judaism, and a Teaching Institute for Judaic Studies, plus a Jewish-Theological Seminar in Breslau. There was also a Society for Jewish Folklore. All of these were closed down and their assets—such as their books—destroyed or scattered. These closures were countered by the opening of several institutes "For the Study of the Jewish Question." The main one, in Berlin, was run by Wilhelm Ziegler, who became the "Jewish expert" for the Propaganda Ministry. Alfred Rosenberg's Institute for Research on the Jewish Question, founded in Frankfurt am Main in 1941, annexed the Bibliothek Rothschild from Paris, the Alliance Israélite Universelle, and the Librairie Lipschutz.[20] This institute saw itself as the coordinating center for a "unified solution" to the Jewish question.*

In the spring of 1934 a new journal appeared, *Volkstum und Heimat* (Folk-Nation and Homeland), which was the organ for the Der Reichsbund Volkstum und Heimat (Reich Union for Folk-Nation and Homeland). By the time the journal appeared, there were some 10,000 politically coordinated societies, about 4 million people, who were members of the Reich Union, though it was an "organization" with a difference. Specialisms were eschewed, instead "authentic, characteristic and valuable fellows," whose enthusiasm for local culture would impress itself on followers, were sought to lead local groups by example. Festivals were planned where local groups would parade showing their local attributes—workers with spades on their shoulders, "maidens" pulling plows, farmers with sowing bags wrapped around them. The job of the union was to glorify peasant life and rural values.

Anna Oesterle has examined the Office of Ancestral Inheritance, the Ahnenerbe, and its effect on folklore scholarship. It appears to have been a cesspit of rivalries and jealousies, involving itself in folksong research and religious folklore; it began as a private outfit, started with private money, called the Intellectual History Association.[21] Many of those involved in the early days were academics schooled in the Indo-Germanic tradition, and their main interests, to begin with, were ancient intellectual history. Herman Wirth, for instance, had an interest in the "strengthening of genuine German spirituality," hoping for "a rebirth of the Nordic race and the freeing of humanity from the curse of civilisation."

Himmler, who was the undisputed leader of the Ahnenerbe, though he often had to quibble with either Rosenberg or Göring, gradually took the SS further into academic life via the Ahnenerbe.[22] One central concern of

* Its library eventually contained half a million books, all looted.

Himmler's was German origins. In his mind there were two elements, a concern to trace the "Nordic" ancestry of the Germans and a profound interest in the so-called Aryan race in Central Asia, which he felt held the key to ancient religions and mythology and constituted the "founder" race of the Teutons. He enlisted the aid of more or less eminent—and more or less opportunist—anthropologists, ethnologists, orientalists, runologists, philologists, heraldry experts, and archaeologists, in a number of well-funded expeditions in Finland, Iceland, Mesopotamia, the Canary Islands (which Himmler thought formed the southern edge of the lost continent, Atlantis), the Andes (where Himmler thought the civilization had been founded by Aryans), and even Tibet. They studied Bronze Age rock carvings and Paleolithic caves, excavated ancient graves, collected different types of tents, coins, skulls, and needles, took endless photographs, recorded folk stories and dialects, and in Tibet worried they were being spied on.[23] They all fell ill on Christmas Day[24] before entering the "white- and wine-coloured walls" of Lhasa.[25] They made plaster casts of things they couldn't take away and studied ancient rites with a view to introducing them to Germany to replace traditional Christianity. They published a monthly magazine, *Germanien*, but, that apart, it is difficult to know what to make of these activities since few projects were properly completed or incorporated into Himmler's overall view because, during the war, under the guise of "repatriating" goods that were of help in understanding the history and continuity of "Germanness," the SS forces (and sometimes the Gestapo) carried out innumerable art and cultural robberies that were shameless in their extent. This process started in Lithuania and Estonia, then spread to Poland and even France. As part of the "cover" for this, Himmler declared himself to be Reichskommissar für Festigung deutschen Volkstums (Reich Commissioner for Solidifying the German Folk-Nation).

The Nazi Conception of Science

In a survey conducted in 2007 by *Stern* magazine, one in four Germans believed that Nazi rule "had its good points." One of those was "a high regard for the mother," and another was the *Autobahn* (highway) system. Hitler is often credited with starting Germany's impressive *Autobahnen*, but in fact the first such roads were conceived during the Weimar era. Similarly, Hitler was not quite the scientific enthusiast he is sometimes made out to be. Walter Dornberger, one of the physicists leading the army's ballistic missile program at Kummersdorf West in Berlin (well ahead of everybody else's), said later that the Führer never really grasped the significance of missile technology when it was explained to him. At his trial after the war, Albert Speer confirmed that Hitler had some pretty strange views in regard to technology. He was against the machine gun because "it made soldiers cowardly and made close combat impossible."[26] In 1944, when the Luftwaffe high command wanted to use the Me-262 jet as a fighter, Hitler insisted it should only be used as a bomber (he wanted to be able to bomb New York eventually), arguing that aerial combat in jets had a dangerous effect on the brain. He was unable to grasp the revolutionary nature of nuclear physics and distrusted German attempts to build an atomic bomb because it was based, he said, "on Jewish pseudo-science." (Before he had taken power, he promised he would reduce the amount of science pupils had to learn at school.) One of his enthusiasms was for Hanns Hörbiger's theory of "glacial cosmogeny," which claimed that the universe was formed from ice. Hitler believed that "progress" in science had led man to believe, mistakenly, that he could master nature. Instead, he said, he believed in an "intuitive acquaintance with the laws of nature."[27]

As with the artists, so the dismissal of scientists began almost immediately after Hitler became chancellor, in the spring of 1933. For the most part, one would think that science—especially the "hard" sciences of physics, chemistry, mathematics, and geology—would be unaffected by political regimes. It is, after all, generally agreed that research into the fundamental building blocks of nature is as free from political overtones as intellectual work can be. But in Nazi Germany nothing could be taken for granted. Some Jewish academics were exempt for a while, if they had been employed before World War I, or had fought in the war, or had fathers or sons who had done so. But such exemption had to be applied for and Hans Krebs, who was to win the Nobel Prize in 1952 for his discovery of the citric acid cycle, wrote a memoir in which he described how, all of a sudden, in the laboratories of the Freiburg hospital where he was then working, people who had shown only the mildest interest in Hitler, "were suddenly to be seen in the uniforms of Nazi organisations."[28] The situation was quickly polarized.

Einstein's persecution had begun early. He had come under attack largely because of the international acclaim he received after Arthur Eddington's announcement in November 1919 that he had obtained experimental confirmation for the predictions of general relativity theory. Einstein had some support—the German ambassador in London in 1920 warned his Foreign Office privately that "Professor Einstein is just at this time a cultural factor of first rank . . . We should not drive such a man out of Germany with whom we can carry on real cultural propaganda." Yet two years later, following the assassination of Walther Rathenau, the foreign minister, unconfirmed reports leaked out that Einstein was also on the list of intended victims, and he was described as an "evil monster."[29]

When the Nazis achieved power ten years later, action was not long delayed. In January 1933 Einstein was away from Berlin on a visit to the United States. Despite facing a number of personal problems, he made a point of announcing he would not return to his positions at the university in Berlin and the Kaiser Wilhelm Gesellschaft as long as the Nazis were in charge.[30] The Nazis repaid the compliment by freezing his bank account, searching his house for weapons allegedly hidden there by Communists, and publicly burning copies of a popular book of his on relativity. Later in the spring, the regime issued a catalog of "state enemies": Einstein's picture headed the list, and below the photograph was the text, "Not yet hanged." He eventually found a berth at the newly established Institute for Advanced Study at Princeton. When the news was released, one newspaper in Germany ran the headline: "Good News from Einstein—He Is Not Coming Back." On March 28, 1933, Einstein resigned his membership in the Prussian Academy of Sciences, preventing any Nazi from firing him, and was distressed that none of his former colleagues—not even Max von Laue or Max Planck—made any attempt to protest his treatment. He later wrote: "The conduct of German intellectuals—as a group—was no better than the rabble."[31]

Einstein was by no means the only famous physicist to leave Germany. Some 25 percent of the pre-1933 physics community was lost, including half its theoretical physicists and many of the top people in quantum or nuclear physics. In addition to Einstein and Franck, there were Gustav Hertz, Erwin Schrödinger, Victor Hess, and Peter Debye, all Nobel Prize winners, plus Otto Stern, Felix Bloch, Max Born, Eugen Wigner, Hans Bethe, Dennis Gabor, Georg von Hevesy, and Gerhard Herzberg, as well as the mathematicians Richard Courant, Hermann Weyl, and Emmy Noether, described by Einstein as the best female mathematician ever. Roughly one hundred world-class colleagues found refuge in the United States between 1933 and 1941, and Leo Szilard worked hard in Britain to set up the Academic Assistance Council to provide jobs for displaced academics. According to John Cornwell, the German physics community did not shrink in absolute numbers because there were plenty of people to replace those dismissed, "but the quality of the scientists declined and basic research stagnated."[32]

Max Planck tried to put in a good word for Fritz Haber, who had been forced to resign his post as president of the Kaiser Wilhelm Institute (KWI) for Physical Chemistry. Planck went to see Hitler and, according to what he himself later wrote, said that "there were different sorts of Jews, some valuable and some valueless for mankind" and that one had to make distinctions. Hitler rebuked him, saying, "That's not right. A Jew is a Jew; all Jews cling together like burrs."

For scientists only slightly less famous than Einstein or Haber, the attitude of the Nazis was sometimes difficult to anticipate. Karl von Frisch was the first zoologist to discover "the language of the bees," by means of which bees inform other bees about food sources through dances on the honeycomb. Frisch's experiments caught the imagination of the public, and his popular books were best sellers. This cut little ice with the Nazis who, under the Civil Service Law of 1933, still required Frisch to provide proof of his Aryan descent. The sticking point was his maternal grandmother, and it was possible, he admitted, that she was "non-Aryan." A virulent campaign was therefore conducted against Frisch in the student newspaper at the University of Munich, and he survived only because there was in Germany in 1941 an outbreak of nosema, a bee disease, killing several hundred thousand bee colonies. This seriously damaged fruit growing and at that stage Germany had to grow its own food. The Reich government concluded that Frisch was the best man to rescue the situation.

According to recent research, about 13 percent of biologists were dismissed between 1933 and the outbreak of war, four-fifths of them for "racial" reasons. About three-quarters of those who lost their jobs emigrated, the expelled biologists on average proving considerably more successful than their colleagues who remained in Germany. The subject suffered most in two areas: the molecular genetics of bacteria, and phages (viruses that prey on bacteria).

By one of those quirks of statistics we now know that doctors were more enthusiastic Nazis than members of any other profession—44.8 percent of German doctors joined the NSDAP.[33] It helped that, in the Weimar Republic, the doctor/patient ratio had been confined to 1 in 600. As Jewish doctors were expelled (some 2,600 by 1939), their non-Jewish colleagues were more in demand than ever.

In fact, the focus of modern historical research has shifted from a concentration on the small minority of medical men who specialized in "racial science" to the broader picture of whether Germany's doctors as a whole modernized too quickly in a "hard" scientific sense, at the expense of proper professionalization, involving greater ethical and socializing training. Thanks to widespread social insurance, Weimar Germany had a surfeit of doctors, 13 percent of them Jewish. In 1933, 36 percent of medical students were Jewish and so, when the racial laws came in, non-Jewish doctors and medical students had every reason to be grateful to the Nazis. As things now stand, the question as to whether German doctors were less "socialized" than in other countries has not been settled, but undoubtedly doctors were overrepresented in the NSDAP.[34]

Psychoanalysis came under attack because it was seen as a "Jewish science." The Berlin Psychoanalytic Society was purged of its Jewish members, and the leadership passed to M. H. Göring, cousin of Reichsmarschall Hermann. He let it be known that one of the basic texts of psychoanalysis in the Third Reich would be *Mein Kampf.* The German Society for Psychotherapy was renamed the International General Medical Society for Psychotherapy under its new president, Carl Jung, though Adlerians were equally strongly represented.[35] Jung was later to argue that he did all in his power to help Jewish colleagues, but Freud had long suspected him of anti-Semitism (see pp. 553–556) and, certainly, in his theoretical work, Jung targeted Freud, arguing that the founder's "soulless materialism" was a reflection, in part, of his Jewishness. Julius Streicher joined in.[36] The Nazification of psychology was completed with the takeover of six of fifteen full professors of psychology at German universities, in which chairs were occupied by Jews, and a purge of the German Society for Psychology and the Berlin Psychological Institute, where the offices of the director were ransacked for evidence of "treason" (nothing was found but the director,

Wolfgang Köhler, one of the founders of the Gestalt school, subsequently resigned, his life having been made intolerable). It was a rude shock too when, in October 1933, psychoanalysis was banned from the Congress of Psychology in Leipzig. Psychoanalysts, in increasing numbers, looked to the United States.

American psychologists were not especially favorable to Freudian theory—William James and pragmatism were more influential. But the American Psychological Association did set up a Committee on Displaced Foreign Psychologists and by 1940 was in touch with 2,169 leading professionals (not all psychoanalysts), 134 of whom had already arrived in America: Karen Horney, Bruno Bettelheim, Else Frenkel-Brunswik, and David Rapaport among them.

Freud was eighty-two and far from well when, in March 1938, Austria was declared part of the Reich. Several sets of friends feared for him, in particular Ernest Jones in London but even President Franklin Roosevelt asked to be kept informed. William Bullitt, U.S. Ambassador to Paris, was instructed to keep an eye on "the Freud situation," and he ensured that staff at the consul general's office in Vienna showed "a friendly interest" in Freud. Ernest Jones hurried to Vienna, having taken soundings in Britain about the possibility of Freud settling in London, but when he arrived, Jones found Freud unwilling to move. He was persuaded only by the fact that his children would have more of a future abroad.

Before Freud could leave, his "case" was referred as high as Himmler, and it seems it was only the close interest of President Roosevelt that guaranteed his ultimate safety. The Nazis insisted that Freud settle all his debts before leaving and sent through the exit visas one at a time, with Freud's own arriving last. When his papers did at last materialize, the Gestapo also brought a document, which he was forced to sign, which affirmed that he had been properly treated. He signed, adding, "I can heartily recommend the Gestapo to anyone."

In 1934, Bernhard Rust, the Third Reich's education minister, asked David Hilbert, the mathematician, how Göttingen—the home of Gauss, Riemann, and Felix Klein, and a world center of mathematics for 200 years—had suffered after the removal of Jewish mathematicians. "Suffered?" Hilbert famously replied. "It hasn't suffered, Minister. It doesn't exist any more!"[37]

* * *

After Hitler's inquisition had become plain for all to see; emergency committees were set up in Belgium, Britain, Denmark, France, Holland, Sweden, and Switzerland, of which two may be singled out. In Britain the Academic Assistance Council (AAC) was formed by the heads of British universities, under William Beveridge of the London School of Economics. By November 1938 it had placed 524 persons in academic positions in 36 countries, 161 in the United States. Not only mathematicians were helped, of course. A group of refugee German scholars established the Emergency Society of German Scholars Abroad. This sought to place colleagues in employment where it could, but also produced a detailed list of 1,500 names of Germans dismissed from their academic posts, which proved useful for other societies. The Emergency Society also took advantage of the fact that in Turkey, in spring 1933, Atatürk reorganized the University of Istanbul as part of his drive to Westernize the country. German scholars (among them Paul Hindemith, as we have seen, and Ernst Reuter, later mayor of West Berlin during the blockade) were taken on under this scheme and a similar one, in 1935, when the Istanbul law school was upgraded to a university. These scholars established their own academic journal since it was so difficult for them to publish either back home or in Britain or the United States. The German journal in Turkey lasted for only eighteen issues, which are now collectors' items. It carried papers on anything from dermatology to Sanskrit.[38]

A more enduring gift from Hitler was a very different periodical, *Mathematical Reviews.* The first issue of this new journal went largely unremarked when it appeared—most people had other things on their minds in 1939. But, in its quiet way, the appearance of *MR,* as mathematicians soon began calling it, was both dramatic and significant. Until then, the most important mathematical periodical, which abstracted articles from all over the world, in dozens of languages, was the *Zentralblatt für Mathematik und ihre Grenzgebiete,* launched in 1931 by Springer Verlag in Berlin. In 1938, however, when the Italian mathematician Tullio Levi-Civita, a board member and Jewish, was dismissed, several members of the international advisory board resigned. An article in *Science* reported that papers by Jews now went unabstracted in the *Zentralblatt* and American mathematicians, watching the situation with alarm, considered buying the title. Springer wouldn't sell but suggested two editorial boards, which would have produced different versions of the journal, one for the United States, Britain, the Commonwealth, and the Soviet Union, the other for Germany and nearby countries. American mathematicians were so incensed by this insult that in May 1939 they voted to establish their own journal.[39]

As early as April 1933, officials at the Rockefeller Foundation began to consider how they might help individual scholars. Funds were found for an emergency committee, but it had to move carefully; the Depression was still hurting, and jobs were scarce. In October that year, Edward R. Murrow, vice chairman of the committee, calculated that upward of 2,000 scholars, out of a total of 27,000, had been dropped from 240 institutions. That was a lot of people, and wholesale immigration not only risked displacing American scholars but also might trigger anti-Semitism. In the end the emergency committee decided its policy would be "to help scholarship, rather than relieve suffering." Thus they concentrated on scholars whose achievements were already acknowledged, the most well-known beneficiary being Richard Courant from Göttingen. Fifty-one mathematicians were eventually brought to America before the outbreak of the war in Europe in 1939; by 1945 the total was just under 150.[40] Every scholar, whatever his or her age, found work. Put alongside the 6 million Jews who perished in the gas ovens, 150 doesn't sound like much, yet there were more mathematicians helped than any other professional group.[41]

To give a complete picture of German scholarship, however, we now need to describe three areas where, despite the crude anti-Semitism, despite the poor grasp of scientific principles and method which the likes of Hitler, Himmler, and Rust displayed, German scholarship did well. Rocket and jet technology were not the only fields where the Germans were strong.

The first was the Nazi war on cancer. Because the Germans, as we have seen, led the way in regard to coal-tar derivatives—the basic process that had been behind the dye industry and then the pharmaceuticals industry—they were also aware that cancer was associated with many of the new products. By the same token, having invented x-rays, the Germans also noted—as early as 1902—a link between them and cancer, in particular leukemia.

These results produced in Germany, first, a campaign to *prevent* cancer—anti-smoking campaigns started there well before anyone else had the idea, and men were warned to check their colons "as often as they checked their cars." As early as 1938 German scientists had found a link between twelve different types of cancer and asbestos.[42] Richard (later Sir Richard) Doll, the Englishman who worked on the link between smoking and cancer in the 1950s, studied in Germany in the 1930s and was shocked to find Jews depicted as "cancers" and Nazi storm troopers as x-rays targeting these "tumours."

German scientists were among the first to explore the links between cancer and diet, especially food additives, and they were also the first to promote natural foods, in particular whole wheat bread (white bread being denounced as a "French revolutionary invention").[43] The role of alcohol in causing cancer was suspected, but the Nazis went much further than anyone else in concentrating on the role of tobacco: smoking was banned in public areas, cigarette advertising was banned, "non-smoking" carriages were established on trains. The shine is taken off these stories when it is realized that tobacco consumption in Germany grew every year after Hitler came to power, so that in 1940 it stood at twice the level of 1933. It only fell in 1944, possibly due to rationing.

"White Jews"

In physics where, save for Lenard and Starck, Germany had a record second to none, much the same process repeated itself as had happened over biology. Just as Frisch had come under pressure because one of his grandparents may have been "non-Aryan," so Werner Heisenberg came under similar pressure because he refused to recognize that "Jewish physics" (i.e., relativity theory) must be wrong or degenerate or both. He, Laue, Planck, and Walter Nernst refused to sign a manifesto organized by Starck pledging loyalty to Hitler. Each of these Nobel Prize winners insisted that physics had nothing to do with politics.

Then, in 1935, Arnold Sommerfeld, sixty-six, was preparing to leave his position as professor at Munich after nearly thirty years (he had taken over from none other than Ludwig Boltzmann). Heisenberg was the natural successor, but he was held to be too much in thrall to the "Jewish spirit" in physics and so became one of the first to be referred to as a "white Jew." He was attacked in the Nazi press and although he was supported by the Göttingen Academy of Sciences, he didn't get the job, which went to a much less able man. In his memoirs he passed over this incident, saying that so many friends and colleagues had to suffer so much worse.[44]

As the 1930s passed, physics began to take on an almost apocalyptic significance. In 1933, as Hitler came to power, Einstein was not the only German physicist abroad in the United States. So too was Otto Hahn, lecturing at Cornell.[45] That left Lise Meitner in charge at the KWI for Chemistry back in Berlin. She was Jewish but Austrian and so, for the time being, did not come under the racial laws.[46] She watched as former colleagues were dismissed or left on their own initiative, including Otto Frisch, her nephew, with whom she had often played piano, who was dismissed from his post in Hamburg, and Leo Szilard, Hungarian-Jewish, who left for England "with his life's savings hidden in his shoes."[47] It was to be Szilard, famously, who, when safely in London that September and crossing Southampton Row at some traffic lights, had the idea of a chain reaction, which would be self-sustaining and produce an explosion. (He patented the idea and assigned it to the British Admiralty on condition that it be kept secret.) Meanwhile Enrico Fermi, an Italian physicist based in Rome, had, without knowing it, split the uranium atom, though it was two Germans, Ida and Walter Noddack, at Freiburg-im-Breisgau, who first recognized this.

In 1936 Hahn and Meitner were nominated for the Nobel Prize by Max Planck, Heisenberg, and Laue, apparently in an attempt to protect their Jewish colleagues.* But when the *Anschluss* occurred in March 1938, Meitner and Hahn's protection was removed at a stroke. Carl Bosch, who had worked on the nitrate-fixation process with Haber, managed to get Meitner permission to travel and she went to the Netherlands, with just two suitcases and a diamond ring Hahn had given her so she would have something to sell.[48]

The climax of physics was achieved by these personalities in the run-up to and in the immediate wake of the outbreak of war. In Berlin, Otto Hahn found that if he bombarded uranium with neutrons he repeatedly got barium. In a letter he shared these bewildering results with Meitner, now in exile in Göteborg. As luck would have it, Meitner was visited that

* After Carl von Ossietsky was awarded the Nobel Peace Prize in 1936, no German was allowed to accept any of the Nobel Prizes.

Christmas by her nephew Otto Frisch, also in exile, with Bohr in Copenhagen. The pair went cross-country skiing in the woods, which were covered in snow. Meitner told her nephew about Hahn's letter, and they turned the barium problem over in their minds as they moved between the trees. Until then, physicists had considered that when the nucleus was bombarded, it was so stable that at most the odd particle could be chipped off. Now, huddled on a fallen tree in the Goteburg woods, Meitner and Frisch wondered whether, instead of being chipped away by neutrons, a nucleus could in certain circumstances be cleaved in two.

They had been in the cold woods for three hours. Nonetheless, they did the calculations before turning for home. What the arithmetic showed was that if the uranium atom *did* split, as they thought it might, it could produce barium (56 protons) and krypton (36)—56+36=92. As the news sank in around the world, people realized that, as the nucleus split apart, it released energy, as heat. If that energy was in the form of neutrons, and in sufficient quantity, then a chain reaction, and a bomb, might well be possible. But how much U_{235} was needed?[49]

The pitiful irony of this predicament was that it was still only early 1939. Hitler's aggression was growing, but the world was, technically, still at peace. The Hahn/Meitner/Frisch results were published openly in *Nature,* and thus read by physicists in Nazi Germany, in Soviet Russia, and in Japan, as well as in Britain, France, Italy, and the United States. The problem that now faced the physicists was: how likely was a chain reaction? America, with the greatest resources, and now the home of so many of the exiles, was a nonbelligerent after war broke out in Europe. How could she be persuaded to act? It was only after Frisch and Rudolf Peierls, two German exiles now working at the University of Birmingham in England, and walking the blacked-out streets of the city at night, calculated (in a three-page paper), that about one kilogram of uranium was sufficient (as opposed to 13–40 tons, according to earlier calculations), that movement began.[50] Mark Oliphant, Frisch and Peierls's professor at Birmingham, traveled to America and persuaded the Americans to explore whether a bomb could be built. Without informing Congress, President Roosevelt, motivated by a letter from Einstein (drafted by Szilard) found the money "from a special source available for such an unusual purpose." Thus German-Jewish physicists played a full role in bringing into existence the bomb that would end the war. [51]

The Concept of the Political

Carl Schmitt has been widely acclaimed as being "among the two or three most original political theorists of the twentieth century," yet his public enthusiasm for the Nazis, his anti-Semitism, and his "obdurate refusal" to recant after 1945 has put him in the same doghouse as Martin Heidegger.

Born in 1888, in Plettenberg, Westphalia, the son of a small businessman, Schmitt, like Heidegger, grew up in a provincial Catholic home. As a student, he tried his hand at satire—he was famously set against all aspects of modern culture. By 1914 he was a civil servant and did not volunteer until 1915, securing a desk job, though he later reminisced about falling from his horse—an episode never corroborated. Despite his hatred of modern culture, Schmitt much enjoyed the artistic Schwabing area of Munich, mixing with Expressionist painters and Dada artists and corresponding with Eugenio Pacelli, later Pope Pius XII. He trained in the law and attended Max Weber's lectures, but when revolution broke out in Munich in the wake of World War I, he abandoned both the bohemian life and the church and turned to teaching.

He now began his more formal, more systematic criticisms of democracy. Human history, Schmitt insisted, originated with Cain and Abel, not with Adam and Eve. Politics, for Schmitt, is located in concrete power struggles rather than abstract ideas. He had a love of conflict, and in 1932, when there was a reactionary coup d'état in Prussia, Schmitt defended the coup as a counsel in court, attracting the admiration of Göring, who thereafter became his protector.[52] Schmitt joined the Nazi Party as April turned into May 1933 and supported the notorious book burning of May 10. This support is generally taken as helping Hitler's bid for respectability.

In *Der Begriff des Politischen* (*The Concept of the Political*), published in 1932, Schmitt provides an amalgam of Heidegger and Nietzsche and pits liberalism, on one side, against extreme right and left on the other. Schmitt's essential point is that we achieve our political identity through conflict, intense conflict, even fatal conflict. The experiences of "we" and "our" (reminiscent of Spengler) are central to politics (a celebration of the "whole" again), and its clearest defining process is by struggle, by fighting for what "we" believe in. Liberalism and democracy can never do this because compromise is the defining factor of liberal democracies and their product is always shifting. Because of this, Schmitt thought that people in liberal democracies never know who they truly are and can never take full responsibility for their lives. Political resolution, he thought, cannot be brought about by reason, only by "blood and soil." Furthermore, he thought it was dangerous to base political aims on some ideological abstraction with a claim to universal moral principles. That never works because it is always overtaken by events.

Schmitt was a controversial figure in the 1930s and is still so. He was captured by the Americans in 1945, interned for more than a year, and never had another university job. But he was visited by, and/or praised by such varied luminaries as Ernst Jünger, Alexandre Kojève, Walter Benjamin, and various members of the Frankfurt school. Leo Strauss, a fellow political theorist and, like Schmitt, a friend of Martin Heidegger, renewed the focus on Schmitt in postwar America, where he was an exile. We shall meet Strauss, who was Jewish, again in Chapter 39, p. 713.[53]

"Scientific" Notions of "Germanness"

Götz Aly and Susanne Heim have identified what they claim is a new science, a new academic speciality, which emerged in Germany in the late 1930s, and all thanks to the Nazis: demographic economics. This was based, they say, on the development of a new concept among town planners, geographers, economists, and demographers and was applied particularly to areas of eastern and southeastern Europe. That concept was "rural overpopulation," held to account for low productivity and a lack of purchasing power. It applied especially in Poland.[54]

The concept was developed, say Aly and Heim, primarily by the Reichskuratorium für Wirtschaftlichkeit or RKW (Reich Board for Industrial Rationalization), a large, thorough outfit that, as an example, commissioned no fewer than 1,600 secret reports from the Kiel Institute of World Economic Studies as an aid to planning the war. It was these regional planners, statisticians, and agronomists—many of whom were initially cool toward the new regime, but whose careers were rapidly advanced as a result of the many dismissals—who did so much to make respectable a policy that started out as mere prejudice on the part of Hitler, Himmler, and others.

Poland had been singled out as a "population problem" as early as 1935 in a study by Dr. Theodor Oberländer at the Institute of East European Economic Studies in Königsberg, who argued that its system of smallholdings was chronically inefficient and ripe for an agrarian revolution "on the Russian model."* This analysis was later widened, in 1939, by the social historian Werner Conze, into a "demographic structural crisis in eastern Central Europe."[55] A corollary to the theory of overpopulation was the idea of "optimum population size," the size that allowed the maximum possible return to be extracted from the economic resources of a region.[56] Using such reasoning, the academics calculated that somewhere between 4.5 million and 5.83 million Poles, "every second person in Polish agriculture . . . represented nothing but dead ballast." So began the idea that a reduction in population numbers would help improve the efficiency of those areas that eventually came under German influence and control, together with the idea that enforced deportation would help ensure "social peace."

A second concept was spearheaded by the Reichskommissar für die Festigung deutschen Volkstums, or RKF, the Reich Commissioner for the Strengthening of German Nationhood. This decided which minority ethnic groups were capable of "Germanization" and which weren't. Himmler decided that one-eighth of the Polish population could be "Germanized" ("There are still a few Goths left in the Caucasus and the Crimea," he said in 1942) and the population divided into:

a. Full-fledged Germans;
b. Persons of German origin who must be taught to become full-fledged Germans again, who therefore possess German nationality but not, initially, the rights and status of full Reich citizenship; this category was to be deported to Germany for Germanization;
c. Valuable members of the dependent minority races, and German renegades who "possess German nationality subject to revocation";
d. Foreign nationals who do not possess German nationality.

* Oberländer later was made a federal minister in one of Adenauer's cabinets.

> These comprised 8 million, out of which 1 million were chosen in advance (and arbitrarily) to be included in category C.

Himmler had another system, also dividing people into four classes, of which the most startling was Class 3, members of minority races who had married Germans and shown themselves prepared to "conform to German notions of orderliness" and to "show a willingness to better" themselves. This then was the German (or at least the SS) idea of Germanness.[57]

The classification systems were more than theoretical exercises: the long-term plan was to likewise classify Polish land so that only the "deserving classes" who wished to better themselves would be given the best soil, the plan being that the proportion of Germans working on the best agricultural land in the East would be raised from 11 to 50 percent.[58] Their homesteads would be the first to receive electricity and they were to be organized as the demographers saw fit—villages of 400–500 were judged the most efficient and cohesive. This really was a re-creation of the *Volk*. The figures amassed were designed to help the Government General produce a Polish petit bourgeoisie in place of the Jews, whose businesses had been closed or ransacked, the aim being that Poland would become "a purely German country within the space of fifteen to twenty years."[59] To this end the ghettoes were as assiduously studied as other aspects of Polish society. The number of workers and the number of dependents were calculated, set against the minimal amount of nutritional requirements and the cost. Aly and Heim show that these calculations were made and remade until they showed a heavy loss, one that could in no way be recouped, and then the fate of the ghettoes was sealed, on economic if not on racial grounds. As one RKW report put it, "Conditions of undernourishment could be allowed to develop without regard for the consequences." Much the same arguments were used in Bulgaria, Romania, and Yugoslavia.

The project of the economists and demographers that turned out to be the most ghoulish, and important in its wider significance, was that which showed the Nazi leadership that mass murder would not be "significantly detrimental" to public morale.[60] This began, according to Aly and Heim, in a project by a Professor Karl Astel, head of the Thüringischen Landesamtes für Rassewesen (Thuringian Regional Office of Racial Affairs). As part of this project, an epidemiological study of the mentally ill was carried out, which came up with a figure of 65,000–70,000 mentally ill people who were to be eliminated on economic grounds. A list was drawn up and a program known as Aktion T4, after the address of its offices at Tiergartenstrasse 4 in Berlin, established. Aktion T4 calculations were designed to show how much money the Reich would save by not having to support mentally ill individuals. As part of this, in 1939 Hitler's personal doctor, Theo Morell, prepared a document in which he quoted from a survey of parents of severely handicapped children, carried out in the 1920s, in which they had been asked "purely hypothetical" questions as to whether they would consent to a painless procedure to cut short the life of their handicapped offspring. The vast majority had answered "yes," a minority adding that although they did not want to decide the fate of their own children, they would be quite prepared for the doctors to make the decision. Some had even suggested that the doctors do it and then tell the parents their child had died of an illness. On this basis, say Aly and Heim, the decision was made by the Nazi leadership to carry out the murder of German mental patients (and Hitler certainly knew about it).[61] It was done in secret, but the secret was allowed to leak out to see what the reaction of the patients' relatives would be. On April 23, 1941, an official report concluded that "in 80 percent of cases, relatives are in agreement, 10 percent speak out against, and 10 percent are indifferent." Nor was there any opposition from within the bureaucracy. "This was a lesson of fundamental importance for the organizers of the 'final solution of the Jewish question,'" say Aly and Heim. "It convinced them that cover names would not be questioned, but would on the contrary be gratefully accepted, indeed expected, as an invitation to denial and moral indifference."

The eagerness with which so many scholars embraced National Socialism and their ideas is still shocking after all this time. It cannot be explained simply by the fact that so many junior figures were given early promotion after the Jewish seniors had been dismissed, exiled, or deported. Many senior colleagues—Martin Heidegger, Philipp Lenard, Ernst Krieck, Paul Schmitthenner—were equally enthusiastic supporters of the National Socialists. This amounts to yet another "Traihison des Clercs" but on a much bloodthirstier scale than ever before.

36.

The Twilight of the Theologians

When he was a boy of six, Adolf Hitler was for a short time a choirboy at the Benedictine monastery at Lambach in Austria. What he loved most, he said later, was "the solemn splendour of church festivals." By the time he reached Munich in 1919, as a thirty-year-old ex-soldier, such religious feelings as he still had were a long way from Catholicism. By now Hitler was caught up with a *völkisch* sentiment, shaped by such people as Paul de Lagarde, whose version of Christianity was described earlier, in Chapter 22, p. 417, a bastardization of faith in which it was asserted that Catholicism and Protestantism were "distortions" of the Bible, brought about mainly by St. Paul, who, Lagarde insisted, had "Judaised" Christianity.[1]

Many crude books circulated in the Vienna of Hitler's day, one with the title *Forward to Christ! Away with Paul! German Religion!* Here too the argument was that the "poisoner Paul and his Volk" were the "arch-enemies of Jesus" who "had to be removed from the entrance to the kingdom of God" before "a true German church can open its doors." The difficulty of Jesus' being Jewish was circumvented in various ways, either by making him "Aryan" or, in the case of Theodor Fritsch, by arguing that Galileans were in fact Gauls, who in turn were German. (He claimed to have demonstrated this philologically.) All this became a central element in Hitler's own view of Christianity, but on top of that he claimed to see in

Jesus a mirror image of himself, "a brave and persecuted struggler against the Jews."

Despite all this, Hitler was not anxious for the fledgling Nazi movement to antagonize established religion. Dr. Artur Dinter, a former scientist and dramatist whose daughter died tragically in childhood, called for "a German national church" that would counter modernism, materialism, and the Jews, "much as Jesus had done" (his *Richtrunen* were intended to replace the Ten Commandments). Hitler dismissed him, writing to Dinter, who had joined the National Socialists before Hitler and held Party Card Number 5, that he would not waste time on a "religious reformation"; he would steer clear of religious issues, "for all time to come."[2]

The Theological Renaissance

As we shall see, he didn't stick to his word. When the Nazis did achieve power, their relationship with religion would remain troublesome. In some ways their religious views were simplistic, in other ways cynical and manipulative. Hitler himself seems to have had a vague notion of a "sacred universe," but above all, in purely intellectual terms, the Nazis largely ignored the fact that, just then, Germany was undergoing a renaissance in religious thought.[3] It is a fact largely overlooked that just as the Germans had produced a "golden generation" in physics, philosophy, history, and film as the 1920s turned into the 1930s, there was a similar cohort of very creative individuals in theology. According to Alistair McGrath, writing in 1986, modern German theology has an "inherent brilliance" but since World War I, "the equivalent of a theological iron curtain appears to have descended upon Europe, excluding ideas of German origin from the theological fora of the English-speaking world."[4]

The renaissance in theological thought had been sparked by Ernst Troeltsch, and by Adolf von Harnack, professor of church history at Giessen, whose book *Das Wesen des Christentums* (*The Essence of Christianity*; 1900), tried to go beyond all the historical criticisms that had accreted during the nineteenth century.

Ernst Troeltsch (1865–1923) was probably the first sociologist of religion. His main work, *Die Soziallehren der christlichen Kirchen und Gruppen* (*The Social Teaching of the Christian Churches*), was an attempt to bring a sociological understanding to the phenomenon of religion and to Christianity in particular. Troeltsch was affected by the same cultural pessimism as Werner Sombart, and he thought that the main source of alienation was the strong central state, which, however necessary, had helped to define modern social relations in economic terms, and was not what many people wanted, interfering as it did with their fulfillment and satisfaction. He hoped that a sociological understanding of religion might help form a state–church harmony through which many people could adjust their lives in the modern world.

His main point, after a historical survey, espoused both Dilthey and Simmel, arguing that Christianity cannot be looked at only from the vantage point of the committed Christian, that there are other ways of looking at the church and that these other ways have to be considered, and argued with, if religion is to survive.

He also noted that the social position of the church affected its attitude toward reform, that at some times church membership overlapped with the political classes more than at other times when, fairly naturally, it was less radical. He foresaw problems with Catholicism, with any church which claimed that Natural Law existed before the state, and therefore before any other forms of law. Troeltsch ended his survey in the eighteenth century, arguably selling himself short, because the nineteenth century saw some epic battles between Catholicism and Protestantism (not least in Germany), between Catholicism and secularism, Catholicism and science. But the specter of a theologian treating the church not as a solely theological entity but as a sociological one was new and was, until World War I, very influential.[5]

Harnack concentrated more on the Gospels, which were for him evangelistic rather than historical documents and, whatever historical details did or did not stand up, described the impression Jesus made upon his disciples, an impression they felt the need to transmit and *this* was the Gospels' main essence and purpose. In this view, the entire "Life of Jesus" movement was to be seen as a blind alley, which proved an immensely popular interpretation.[6] Harnack's book was translated into more languages than any other book except the Bible, and according to Paul Tillich, "Leipzig railway station was jammed by freight trains carrying Harnack's book all over the world."[7]

Troeltsch and Harnack's near contemporary was Rudolf Steiner (1861–1925), possibly one of the most indefatigable people who ever lived. Born in Croatia, the son of a telegraph operator on the Southern Austrian Railways, Steiner studied mathematics, physics, and chemistry at the Technische Hochschule in Vienna. He made his mark so quickly that even before he graduated he was recommended as the editor of a new edition of Goethe's works. On the strength of this, in 1896 Elisabeth Förster-Nietzsche asked Steiner to put her brother's archive in order, and the young man was very moved by his meeting with the now-catatonic philosopher.

Steiner's subsequent life was spent trying to bring together the worlds of science, literature, the arts, and religion into one spiritual synthesis. He founded journals and schools and built two "Goethaneums"—auditoria where people could "experience" lectures about the spiritual life among like-minded souls ("ceremonies of the whole" again), and he established a cult.[8] He advocated his view of the "Threefold Social Order," in which he maintained that the economic, political, and cultural aspects of society should be independent but equally important. For this, Steiner was attacked by Hitler himself.[9]

Steiner died exhausted at the age of sixty-four, but he left a considerable legacy—900 Waldorf (Steiner) Schools and a number of firms (including banks) and charitable societies operating on his principle that our aim should be "the higher life," a moral concern for others and an attempt to grasp the spiritual dimension, by which he meant, specifically, the Second Coming of Jesus, which he did not believe would be physical, but "etheric," only becoming apparent through communal life.[10]

Although he was nowhere near as worldly as Steiner and did not have the same kind of practical innovative genius, Karl Barth (1886–1968) is widely regarded as the greatest Protestant theologian of the twentieth century, and possibly the greatest since Luther himself.[11] Born in Basel, where his father, Fritz, was a minister and professor of New Testament and Early Church History, Barth studied at the universities of Bern, Berlin, Tübingen, and Marburg. At Berlin he attended Harnack's seminars, and it was there that he first encountered the ideas of liberal theology (mainly the search for the historical Jesus), which he would eventually rebel against.[12] After his studies he returned to Switzerland as a pastor.[13]

In World War I Barth was much disturbed by the Manifesto of the

93 ("Among whom I was horrified to discover almost all my hitherto revered theological teachers"), which he believed was a betrayal of Christian principles.[14] He came to believe that the Higher Criticism in Germany, although it had been responsible for many of the new scholarly techniques, nevertheless missed the point. The concern with Jesus as a historical figure obscured Jesus as the revealed word of God. Mankind no longer consulted the Bible in the way that its compilers intended it to be read.

In the midst of war, Barth reexamined the scriptures and, in particular, in 1916, began a careful examination of Paul's Letter to the Romans. This proved of great significance for Barth, and in 1922 he published *Römerbrief* (*The Epistle to the Romans*), the main message of which was, as Paul himself had said, that God saves only those people who "trust not in themselves but solely in God."[15] This led to Barth's central, seminal view, what he called "the Godness of God," that God "is wholly other," totally different from humans.[16] It was this idea that brought Barth to the attention of other theologians and many of the faithful. In the year that he published *The Epistle to the Romans*, he, together with a number of other theologians, including Rudolf Bultmann, who is considered next, started a journal, *Zwischen den Zeiten* (Between the Times), which formed the main outlet for what became known as "Crisis Theology" (the "crisis" being World War I and the "sinfulness," the very great distance from God of which it was evidence). *Zwischen den Zeiten* remained a powerful force until it was closed down in 1933.[17]

Barth's developed view was that the Bible is not a revelation from God but *a human record* of that revelation. God's single revelation occurred in Jesus Christ, meaning that we can approach God, or make ourselves available to be approached by Him, only by learning from and emulating Jesus, and we must do this for ourselves.[18] Barth was also essentially optimistic for mankind, saying that although individually we may turn away from God (his definition of sin), we are "powerless to undo what Christ has done."

Such was the impact of Barth's theology that, by the time the Nazis came to power in 1933, he was a public figure. He then emerged as one of the leaders—if not *the* leader—of church opposition to the National Socialists, expressed in the so-called Barmen Declaration of 1934.[19] In the previous April, the "Evangelical Church of the German Nation" (*Deutsche Christen*) had been created under Nazi influence and published its guiding principles, which made anti-Semitism a central plank of this new religion and forbade marriage between "Germans and Jews."[20] In reply, Barth was one of those founding the so-called Bekennende Kirche (Confessing Church), which rejected the attempt to set up an exclusively German church. In May 1934 representatives of the Confessing Church met at Barmen and delivered a declaration, based on a draft that Barth had prepared, in which they rejected the "false doctrine" that "there could be areas of our life in which we would belong not to Jesus Christ but to other lords." Barth himself refused to take the oath of unconditional allegiance to Hitler, was dismissed, and returned to Basel where he continued to speak out in support of the Jews.[21]

Much influenced by Karl Barth, and a member of the Confessing Church, Rudolf Bultmann (1884–1976) also shared a profound belief in the importance of the New Testament as witness to the hope God gave to the world—to us—in the person of Jesus Christ.[22] This, for him, made Christianity *the* form of faith that God intended for us. Born in Wiefelstede, the son of a pastor, Bultmann grew up in Oldenburg and studied at Tübingen, Berlin, and Marburg. He then lectured at Breslau and Giessen before returning to Marburg as a full professor in 1921, remaining there until he retired in 1951.

Die Geschichte der synoptischen Tradition (*The History of the Synoptic Tradition*), published in 1921, reflected his fascination with the Higher Criticism but also his belief that by clearing away all the historical accretions, we are better able to know the real Jesus, what Bultmann called the *kerygmatic* Jesus, the Jesus as revealed in his teaching, which is what matters more than historical details about his life.[23] Besides Barth, Bultmann was much influenced by his friend and Marburg colleague, Martin Heidegger, whose brand of existentialism, then being worked out in *Being and Time* (1927), was a secular, philosophical equivalent of what the philosopher was trying to say—namely, that there are four main categories of human existence: first, man has a relationship to himself (in the way that we say someone is "at one" or "at odds" with himself); second, man is a possibility, rather than a predetermined actuality; third, every man's experience is unique and defies classification; and four, man exists in the world, is caught up in it.[24] Bultmann saw about him the anxieties of the modern world—especially strong in the wake of World War I—and he observed what he described as a "flight" into business, money-making, social status-seeking, and the enjoyment of ephemera, what amounted to him as a world without God.[25]

In *Das Evangelium des Johannes* (*The Gospel of John*; 1941)—which proved to be a best seller—his argument was that a close reading between the lines revealed that John's Gospel was very different from the other three (now generally accepted) and that there was within it a series of signs that help us to know how to live once we have stripped away accretions derived from Jewish apocalyptic traditions and Gnostic redemption myths. This was what became famous as Bultmann's "demythologizing" of the Bible. He thought that the Gospel of St. John was intended for a largely gentile audience (not the Jewish Christians in Jerusalem, which other scholars claimed in regard to the other gospels) and that the Gospels were less books to be read than *preached* as sermons, to be heard and fired up by. Bultmann thought that the Resurrection was a metaphor, for the fact is that faith will always rise up again, however dead the world may seem. The German repudiation of the "Quest of the historical Jesus" reached its zenith in the theology of Rudolf Bultmann.[26] He never amended his views to accommodate the Nazis, but he steered clear of politics and kept out of the Führer's sights.

The third man in this renaissance of theology was Paul Tillich, born in 1886 in Brandenburg in eastern Germany and the son of yet another Lutheran pastor. He served as a chaplain in the German army throughout World War I, then taught theology at Berlin, Marburg (where he met both Bultmann and Heidegger), Dresden, Leipzig, and finally Frankfurt, where he became part of the Frankfurt school.[27]

As the 1920s passed, Tillich became an increasingly outspoken socialist, publishing *Die sozialistische Entscheidung* (*The Socialist Decision*), an examination of the relationship between religion and politics. Unfortunately, it was released in 1933, was quickly suppressed, with copies confiscated and burned by the newly appointed Nazis. Tillich himself was dismissed (his name was on the first list of those suspended from university teaching, dated April 13, 1933, along with Max Horkheimer, Paul Klee, and Alfred Weber), but as luck would have it Reinhold Niebuhr, a prominent socialist and a professor of practical theology at the Union Theological Seminary in New York, and himself the son of an emigrant German pastor, was in Germany just then, and he and the seminary's president, Henry Sloane Coffin, invited Tillich to join the Seminary. Tillich and his family emigrated soon after.

The Defining Edge of Evil

Dietrich Bonhoeffer (1906–45), in fact almost the entire Bonhoeffer family, must rank among the most courageous people in all Germany, a reproach to those who say there was no chance of resistance in the Third Reich, and to those who say that there were no good Germans.

Bonhoeffer was born in Breslau, he and his twin sister, Sabine, being two of eight children born to Karl and Paula (von Hase) Bonhoeffer. Karl was a leading psychiatrist, a professor at the University of Berlin, though an empiricist, not a Freudian.[28] Dietrich's brother, Walter, was killed in World War I; his sister Christel married Hans von Dohnanyi and became the mother of Christoph von Dohnanyi, the conductor, and Klaus von Dohnanyi, a mayor of Hamburg. Although his father was a psychiatrist, Dietrich was drawn to the church and studied at Tübingen and then Berlin. He took his doctorate when he was only twenty-one but was forced to wait until he was twenty-five before being ordained.[29] He spent a year at the Union Theological Seminary in New York, during which time he formed an enviable collection of African-American spirituals.[30]

He returned to Germany in 1931 where, together with 2,000 Lutheran pastors, he helped organize the Pastors' Emergency League in opposition to the state church controlled by the Nazis. This was the organization that evolved into the Confessing Church under Barth's leadership. At the outbreak of World War II, Bonhoeffer joined the Resistance, in particular a small group of senior officers in the *Abwehr*, Military Intelligence, intent on assassinating Hitler. Bonhoeffer was arrested in April 1943 after money used to help Jews escape to Switzerland was traced to him. He was tried and hanged (naked) on April 9, 1945. His brother Klaus and his brothers-in-law Hans von Dohnanyi and Rüdiger Schleicher, both of whom had been active in the Resistance, were shot elsewhere later that month.[31]

Before his arrest, Bonhoeffer composed his most important book, *Ethik* (*Ethics*), a profound work but one that shows the scars of the time and may be seen as a cross between Pietism and existentialism.[32] For Bonhoeffer, the task in life is to become a responsible person, modeled on the example of Jesus Christ, but always realizing that it is *action* that counts, that it is how we *participate* in life in regard to good and evil that determines who we are and how Christian we may regard ourselves. Bonhoeffer thought that what really matters is how we *confront* evil on those occasions in our life when it really matters. Moral choices only matter when they are real and immediate—their reality and immediacy are their defining attributes, "urgency is invariably the defining edge of evil." In any given situation, Bonhoeffer says, there is a right thing to do. We can recognize this in two ways—by asking what Jesus would do and by asking if our immediate concern is with the other person or with ourselves. There are risks in this approach—no outcome is ever guaranteed, either in the short run or the long run, but once we start thinking in terms *other* than the immediate, we are rationalizing our behavior to avoid feeling bad about ourselves and are potentially aligning ourselves with evil. To act responsibly is to act against evil without thought of the consequences.

Albert Schweitzer was not "just" a theologian. He was a philosopher, a doctor, a musician, and a missionary. He won the Goethe Prize for his writings and the Nobel Peace Prize for the activity, achievement, and example of his whole life.

Born in 1875 in Kaisersberg, he was brought up in the village of Günsbach, in Alsace, when it was German (the village became French after World War I). Schweitzer's father was a pastor, but the whole family seems to have been musicians and he was taught to play the organ at home. He studied in Paris and at Tübingen, writing his PhD thesis on Kant's religious views before becoming a pastor at a church in Strasbourg.[33] In 1905 he answered an invitation from a missionary society in Paris that was looking for a doctor. He studied medicine and eventually left for the Gabon in West Africa, where he ran a hospital. Later in life Schweitzer achieved fame in equal measure as a medical missionary and as an organist, but theologically he is best known for two things—his examination of *The Quest of the Historical Jesus* (*Von Reimarus zu Wrede*) and his philosophy/theology of "reverence for life," most notable perhaps because he practiced so well what he preached.[34]

In *The Quest* (1906) he did two things. He brought to an end (for a time at least) the great desire by historians to winnow the historical accretions away from the record of Jesus; Schweitzer argued that these exercises tell us more about the historians than about Jesus, and he made a convincing case that the actual Jesus, the historical Jesus if you like, was a figure who expected the imminent end of the world. Schweitzer's scholarship was convincing and, together with a later book, *Mystik des Apostels Paulus* (*The Mysticism of Paul the Apostle*; 1930), which ascribed much of the biblical view to Paul, his argument is still that subscribed to by many theologians and biblical historians.

Since Gabon was French, Schweitzer was interned in France during World War I, and afterward traveled across Europe, becoming better known, before returning to his hospital in Lambaréné which, over time, became celebrated. Like one or two others, Schweitzer, now equally well known for his music as for his missionary work and his theology, widened his interests and, after the invention of the atomic bomb, campaigned against it. He received the 1952 Nobel Peace Prize (in 1953) in recognition of his "reverence for life."[35]

Martin Buber (1878–1965) was born in Vienna but brought up in Lvov, the grandson of a renowned Jewish scholar who was also a successful investor in mines and banking. Martin underwent a religious crisis as a young man and came under the influence of Kant, Kierkegaard, and Nietzsche. At the University of Vienna he studied philosophy, art history, philology, and German studies, then went on to Leipzig, Berlin, and Zurich, coming under the successive influence of Stefan George, Wilhelm Wundt, Georg Simmel, Ferdinand Tönnies, and Wilhelm Dilthey.[36] Regaining his belief, he joined the Zionist movement and, in 1902, took up the editorship of *Die Welt*, the main journal of the Zionists. Unlike Theodor Herzl, who was a friend, he valued a return to the Holy Land more for its spiritual possibilities than for its political advantages and later withdrew to devote himself to his writing, later still cooperating with Franz Rosenzweig in the House of Jewish Learning and on a new German translation of the Bible.

Buber's greatest book, published in 1923, was *Ich und Du* (*I and Thou*), in which he argued that there are two modes of being, the dialogue and the monologue, and that the central element in life, the "premise of existence," is the *encounter*. He argued that the relationship *between* people is the central fact of life, that mutuality, exchange, meeting, is the central aspect of experience, in which satisfaction and meaning are to be found. He felt that modernity had induced more *Ich-Es* ("I-It") relations, basically monologues, and he went on to argue that "I-Thou" relations were needed in order to help people know how to have a relationship with God. This led him to favor the Hasidic tradition, where life was lived as a community.

Buber was given an honorary professorship at Frankfurt in 1930 but resigned when Hitler came to power. The Nazis banned him from teaching, and he left Germany in 1938 for Jerusalem, where he eventually became a professor at the Hebrew University. After the war he was awarded the Goethe Prize, the Erasmus Prize, and the Israel Prize.

The Nazi Form of Christianity

For a while after taking office, Hitler was careful to offer some comfort to the churches. He confided to Goebbels that the best way to treat them was to "hold back for the present and coolly strangle any attempts at impudence or interference in the affairs of state . . ."[37] In reality the Führer was contemptuous of the Lutheran clergy, "insignificant little people . . . They have neither a religion they can take seriously nor a great position to defend, like Rome."[38]

Catholicism, however, as this comment makes clear, was a different matter. Hitler recognized the Catholic Church's institutional force and, despite the fact that Pope Pius XI had condemned Mussolini's species of fascism in 1931 as "pagan worship of the state," the Führer signed a *concordat* with the Vatican two years later. On the Vatican side, the agreement was chiefly the work of Cardinal Eugenio Pacelli, the Vatican secretary of state and the future Pius XII, who had been nuncio in Munich in the 1920s and had lived in Berlin. Pacelli managed to retain autonomy for the German see and some control over education, at the price of diplomatic recognition for the new regime.*

The Nazis moved swiftly in religious education. New regulations stipulated that all parents must enroll each of their children in religious instruction. Seven Catholic feast days were sanctioned as public holidays, and Nazi Party members who had left the church were ordered to rejoin. Until 1936 the German army stipulated that every serving soldier must belong to either the Catholic or Evangelical denominations.[39]

* His controversial career falls outside the scope of this book.

But a lot of this, in retrospect, can be seen as tactical maneuvering. Many thought that the real founding moment of Nazism was the Nietzschean "death of God."[40] More recently, however, Richard Steigmann-Gall has shown how the National Socialists—Hitler as much as anyone—never really followed through on their earlier intentions, in particular the much-advertised attempt to introduce, or reintroduce, "pagan" ideas. Instead, the original plan of the Nazis in the religious sphere was for a concept expressed as "positive Christianity," which had three key ideas: "the spiritual struggle against the Jews, the promulgation of a new social ethic, and a syncretism designed to bridge the confessional divide between Protestant and Catholic."

Hitler, like many leading Nazis, held the view that Jesus was not a Jew and that the Old Testament should be discarded from Christian teaching.[41] The second aspect of Positives Christentum (Positive Christianity), its social ethic, was embodied in the phrase "public need before private greed." This perhaps inevitably glib slogan enabled the Nazis to present themselves in an ethical-moral light in regard to their supervision of the economy. They could advertise as one of their main aims the desire to end class strife in Germany and create, or more properly *re*-create, a "People's Community," an organic, harmonious whole.[42] The final element of Positive Christianity, the attempt to create a "new syncretism," was in some ways the most important aspect, because many leading Nazis viewed the divide between Catholics and Protestants as the greatest stumbling block to the national unity they needed. Himmler put this view most clearly when he said, "We have to be on our guard against a world power which makes use of Christianity and its organisation to oppose our own national resurrection by methods of which we're everywhere conscious." He added that he was anti-clerical but not anti-Christian.[43] The elevation of the *Volk*, the community, as a mystical, almost divine entity, was the main device to overcome sectarian divide and, at the same time, a political maneuver to combat the rival analyses of Marx and the materialist economists of the West.[44] More than theology, or paganism (which many leading Nazis, despite Himmler, thought was laughable), Positive Christianity stressed *active* Christianity—helping the *Volk*, preserving the sanctity of the family, keeping healthy, practicing anti-Semitism, getting involved in the Winter Relief program to feed the poor—rather than reflection. Indeed, these activities seemed designed to *prevent* contemplation, and again this suggests that the Nazis' real worry over Christianity was that it represented the most powerful force that might be used against them.

Nazi Theologians

By no means all religious leaders were as courageous as Barth or Bonhoeffer, and some theologians accommodated themselves—and their theology—to the new regime. Robert Ericksen has studied three who, he says, advanced theology at the time but now seem little more than opportunists.

Gerhard Kittel was professor of New Testament Theology at Tübingen. Born, like the other two, in 1888, he was the son of a famous Old Testament scholar, Rudolf Kittel. Gerhard joined the National Socialists in May 1933 and his main theological contribution concerned the Jewish background to the origins of Christianity. He wrote that Jesus, "if he was a Galilean," might have had "a couple of drops" of non-Jewish blood in his veins.[45] Over time, throughout the Weimar years, Kittel became more and more anti-Semitic to the point where, in June 1933, he delivered a public lecture in Tübingen titled *The Jewish Question* in which he considered "what should be done with the Jews in Germany?" He ruled out extermination, but on the grounds of expedience—"It has not worked before and it will not work now." He rejected Zionism and assimilation and, on account of the Diaspora, opted for "guest status," the forcible separation of the Jews from the people with whom they lived, including the prohibition of "mixed" marriages.[46] The theological basis for his argument was the "transition" that took place in Jewry between 500 B.C. and 500 A.D., since which time the race had "degenerated." The Diaspora, he said, made the Jews a "perpetual problem" for their neighbors, one consequence being that they were always "trying for World Power." His lecture caused an uproar and a heated exchange with Martin Buber. Theologically, Kittel was looking for a "spiritual basis" for anti-Semitism. He was imprisoned in 1945.

Paul Althaus (1888–1966) achieved distinction as a Lutheran scholar at Göttingen. He believed God speaks to man through nature and history, and he derived a concept of *Ur-Offenbarung* or natural revelation, one element of which may be summed up by saying that "God created and approves the political status quo."[47] God's will, according to Althaus, equals the situation at any given moment, and obedience toward God means accepting one's allotted position in life "as handed down by years of tradition." A second element was his idea of *Ordnung*, order, a primary element of which was the *Volk*. The *Volk*, he said, were ordained by God according to a mysterious process: "We have no eternal life if we do not live for our *Volk*."[48] A third element was a return to Luther's idea of the two realms, the kingdom of God, ruled by love, and the kingdom of man, ruled by the sword. All these came together, he said, in a great German "turning point," the National Socialist *völkisch* movement. Althaus was removed from teaching at Erlangen at the end of the war but reinstated a few months later.

Emanuel Hirsch was also the son of a pastor. He studied with both Althaus and Paul Tillich and theologically was much exercised by what he saw as the crisis of modernity. His philosophy, first revealed in his book, *Deutschlands Schicksal* (Germany's Fate), published in 1920, was that revelation teaches us universal values and "internal certainty." Two of the "certainties" that Hirsch identified were the failure of rationalism and the evolution of the state. These were not particularly new ideas in a German context, as will be apparent, but they had never had a theological stamp of approval before. In particular, he thought that "Germany can now create a new form of authoritarianism in which people freely give their obedience to the state so long as the state properly represents the *Volk*." Christianity, he said, fitted admirably into "the German concept of leader and follower."[49]

As Robert Ericksen has observed, Kittel, Althaus, and Hirsch were not isolated or eccentric figures. They were probably typical of many others who held their tongues.

Despite such eloquent and sophisticated ideas and rationalizations for Nazi practices from the Protestant theologians, assaults on Christianity grew in intensity as Nazi confidence solidified.[50] Having begun by insisting on religious instruction, attendance at school prayers was later made optional, and religion was dropped as a subject from school-leaving examinations. Then priests were forbidden to teach religious classes. In 1935, by Bryan Moynahan's count, the Gestapo arrested 700 Protestant pastors for condemning Nazi neopaganism from the pulpit. In 1937 the Gestapo declared that the education of candidates for the ministry of the Confessing Church was illegal and Martin Niemöller, its leading light, was condemned to a concentration camp, refusing the offer of release because it required his collaboration.[51] (The medical orderly in Sachsenhausen found him to be "a man of iron.")*

In 1936 the assault on Catholic monasteries and convents was begun—they were accused of illegal currency trading and sexual offenses. In that year too, the Nuremberg rallies took on an aura of paganism, where the songs—or hymns—were redolent pastiches of traditional Christian worship:

Führer my Führer
Thou hast rescued Germany from deepest distress
I thank thee for my daily bread
Abide thou long with me, forsake me not
Führer my Führer, my faith and light.

This was built on by the Nazi-backed German Faith Movement, one of its aims being to "dechristianize" rituals and festivals. At weddings, for example, the bride and groom would be blessed by "Mother Earth, Father Sky and all the beneficent powers of the air," with extracts from Nordic sagas being read out. The celebration of Christmas—the word itself being replaced by "Julfest," yuletide—was exchanged for a "festival of the winter solstice" held on December 21. The cross was never abolished; attempts *were* made in 1937 to take it out of school classrooms, but the measure had to be rescinded (perhaps confirming Himmler's view that Christianity was the only force powerful enough to threaten Nazi aims). The Vatican complained formally to Berlin on almost a monthly basis, but the regime took next to no notice.

From Hitler's point of view, probably his greatest achievement was in nullifying the oppositional potential that the church—had it so minded—could have mustered.

---

* A celebrated poem is attributed to Niemöller, though he himself always said he could not remember when he had first used the famous words, which read: "In Germany, they came first for the Communists, And I didn't speak up because I wasn't a Communist/And then they came for the trade unionists, And I didn't speak up because I wasn't a trade unionist/And then they came for the Jews, And I didn't speak up because I wasn't a Jew/And then . . . they came for me . . . And by that time there was no one left to speak up." Niemöller was in Sachsenhausen and Dachau from 1937 to 1945 but survived. Following a meeting with Otto Hahn, he began campaigning for nuclear disarmament and in 1961 became president of the World Council of Churches.

37 ·

The Fruits, Failures, and Infamy of German Wartime Science

Shortly after war broke out in September 1939, Paul Schmitthenner, the rector of the University of Heidelberg, announced that the university would become "der Waffenschmied der Wehrmacht des Reiches," the armorer of the Reich's army. His rhetoric was supported by the powers-that-be in that, throughout the war, academic research—and not just "hard" science—was well funded. The budget of the Education Ministry rose from 11 million Reichsmarks in 1935 to 97 million in 1942. The research budget of the Interior Ministry likewise rose from 43 million Reichsmarks in 1935 to 111 million in 1942, while the funding for the Kaiser Wilhelm Society jumped from 5.6 million Reichsmarks in 1933 to 14.3 million in 1944.[1]

The universities were temporarily closed toward the end of 1939 but Heidelberg was one of those allowed to reopen in January 1940, when courses picked up where they had left off, offering "*Frontkursen*" and research to support hostilities. The language and literature seminar was recast "to strengthen the nation's intellectual and spiritual powers of resistance" and courses about Britain reworked to explain why she was "the great enemy." In politics and history, courses were introduced that highlighted the links between geopolitics, war, and race, such as "East Asia as Living Space," "Foreign People's Economics," and "The Nature of Journalism Abroad." Not to be left out, the theology faculty offered "War and Religion in the History of Germany Piety."[2] In early 1940, Paul Ritterbusch, the rector of the University of Kiel, masterminded a series of sixty-seven publications on war-related issues, funded by the Deutsche Forschungsgemeinschaft (DFG). Titles included *Great Britain: Hinterland of World Jewry* and *Economic Liberalism as a System of the British Worldview*. In a sense this was a re-run of some of the arguments broached in World War I, that Britain was shallow and hypocritical, that the English national character knew no calling higher than exploitative profit making. Another theme was Alsace which, it was claimed, was a much more successful culture when it was German than when it was French.

There was not much resistance inside the universities, save for the Weisse Rose (White Rose) group in Munich. This very small group—its core consisted of five students and a philosophy professor, who distributed six different leaflets in 1942 and 1943 calling for the overthrow of Hitler—was eventually found out, and all were beheaded by the Gestapo. The text of their sixth leaflet was smuggled out of Germany, and copies dropped by Allied aircraft later in 1943. At Heidelberg there was a group of between thirty and seventy professors (membership varied), led by Alfred Weber and his sister-in-law Marianne Weber, many of whom had been dismissed but kept working and met to exchange ideas.[3] Some were conservative rather than radical, but they formed what was later called "*Resistenz*," a somewhat elusive term that involved refusing to accept Nazi ideology without publicly criticizing the regime's policies and could hardly be said to involve any kind of bravery. Many were reduced to what Leo Strauss later called "writing between the lines."[4]

Despite Leo Szilard's warnings, on March 18, 1939, the French scientists, the Joliot-Curies, insisted on publication in *Nature* of their observation that nuclear fission emitted on average 2.42 neutrons for every neutron absorbed, meaning that energy was released in sufficient quantities to maintain a chain reaction. In Germany, the article was read by Paul Harteck, a thirty-seven-year-old chemist at the University of Hamburg and an expert on neutrons. Harteck immediately recognized the implications of the paper, and he approached the weapons research office of the German Army Ordnance, to say that a weapon of mass destruction, derived from uranium fission, was a distinct possibility. After the war, as John Cornwell tells the story, Harteck said it was only the "opportunistic quest for scarce funding" for research that caused him to set the ball rolling, rather than belligerence.[5] Whatever the truth of that, the effect was the same.

In any case, by then Werner Heisenberg had already discussed the possibilities of an atomic bomb, and Abram Esau, a physicist in Bernhard Rust's Department of Education, had called a meeting to set up a "Uranium Club," prompted by physicists at Göttingen who also saw the potential of nuclear power in uranium.[6] A second, more important meeting was held at the office of Army Ordnance in Berlin in September 1939, the month the war began, at which Werner Heisenberg, Otto Hahn, Hans Geiger, Carl Friedrich von Weizsäcker, and Paul Harteck were all in attendance.[7] The "club" discussed nuclear power in general, in addition to its use as a weapon, and as a result the KWI for Physics, in Berlin, was requisitioned for war work.

This sounds decisive, but the uranium team in Germany was never to exceed a hundred members, compared with tens of thousands in the Manhattan Project at Los Alamos in the United States. Whereas Germany had the largest supply of uranium reserves—at the Joachimsthal mines in now-occupied Czechoslovakia—it had no cyclotron for the study of the properties of nuclear reactions.

The development of the German bomb—or rather the pace at which its research developed—has been the subject of much controversy (not least in Michael Frayn's play *Copenhagen*). The Germans concentrated in the beginning on isotope separation though later they explored plutonium. These approaches involved the participation of two equally controversial and even mysterious characters, Werner Heisenberg and Fritz Houtermans. Heisenberg would become notorious for his meetings with Niels Bohr in 1940, in which the two men fenced over how far each side had gone in the race to produce a bomb, and for whether or not Heisenberg was letting Bohr know that Germany *was* exploring nuclear power and whether he, Heisenberg, was trying to slow down developments. Houtermans was a brilliant physicist who in 1941 confirmed that a chain reaction was possible using plutonium, element 94. Houtermans had trained at Göttingen but, having socialist sympathies, had gone to work in the Ukraine, despite Stalin's purges. There he had been arrested as a German spy, had given a false confession and been imprisoned.[8] Released during the brief Hitler-Stalin pact, he returned to Germany where he was suspected of still having leftist leanings but was given a job by Max von Laue, helping another physicist, Manfred von Ardenne, study chain reactions. After reading the Joliot-Curie paper in *Nature* in 1939, Houtermans concluded that plutonium *was* a possibility for a chain reaction and was so disturbed when his calculations confirmed this that he smuggled a message out of the country with a refugee bound for America. Although he didn't know how far the Allies were advanced in *their* bomb, his message had two elements—to "hurry" and to say that Heisenberg was trying to slow things down.[9]

Given the actions of these two men, it is perhaps not surprising that the German bomb project was not successful, though there were other reasons for the failure. After the occupation of Norway, the world's only heavy water plant, at Vermork, became available to the Germans, but when they tried to transport a large load south, the Danish Resistance (at the request of British Intelligence) blew up and sank the ferry carrying the canisters (several Danes lost their lives in the sinking).[10] When Speer took over as armaments minister (from Fritz Todt, who had been killed in a mysterious airplane crash), he told Heisenberg he could have a cyclotron bigger than anyone else; Heisenberg replied that the Germans were so inexperienced that they would first need to learn on a smaller one.[11] Nor could they use the one in occupied Paris, he said, because the conditions of secrecy hampered their work, all of which has been used to suggest he was deliberately sabotaging Germany's nuclear program.[12] Speer eventually decided, or said he decided (after the war) that Germany could not have a bomb before 1947.[13] Since it was consuming chromium ore at such a pace that the war could not be prosecuted beyond January 1946, and since the sums needed were also draining its rocket program (much nearer to Hitler's heart), Germany's attempt to construct an atom bomb was abrogated in the autumn of 1942.[14]

Not all the German innovations in the war were as fruitless as the atom bomb. They had in the Me-262 the world's first operational jet fighter, though in too few numbers to materially affect the course of the war. The brilliant marine engineer Helmut Walther designed the technology for

Germany's diesel submarines, which could achieve speeds underwater of 28 knots when conventional submarines were capable of barely 10 knots.[15] Here too they were too late to have an effect. The Germans had their own form of radar and for a time appeared to be ahead of the Allies. And they had Enigma, their typewriter-like code machine, which, in its more developed forms, could produce 159 billion billion different ways of setting a message (159 followed by 18 zeroes). Its main effect on history is perhaps that the British device invented to cope with the codes the Germans sent led to the Colossus, in effect the world's first computer.

The Germans also looked at computers, or at least very advanced calculating machines. They were mainly the work of Konrad Zuse, an engineer who, as early as 1932, used a binary—on/off or 1/0—scheme, with holes punched in paper and pins that could be locked in place, to develop a machine that performed the "tremendous numbers of monotonous calculations necessary for the design of static and aerodynamic structures" in the aircraft industry.[16] His machines, which had an electrochemical memory, were used by the aircraft industry to solve simultaneous equations associated with metals stress but were never employed in cryptanalysis and never developed further because the war ended too soon.

Then there was rocket science. Under the 1919 Versailles Treaty, Germany was forbidden from forging large guns and had been famously forced to scuttle her fleet in Scapa Flow. Her army could not exceed 100,000 men, tanks and submarines were forbidden, and guns were not to exceed 105 mm. However, she secretly began to rearm, building ships under cover in Holland and Japan, and it was in these highly clandestine circumstances that her rocket development project took shape. The leading light was Wernher von Braun, though it was Hermann Oberth, a German-speaking Romanian, who in his 1923 book, published in Germany, first outlined how the problems of space flight might be solved by the use of alcohol and liquid oxygen, and liquid hydrogen and liquid oxygen, in a multistage rocket.[17] Building on this, the motor car manufacturer Fritz von Opel financed a number of tests for rocket-driven road vehicles, all of which experiences came together in the army after 1933. Experiments were run on the remote island of Borkum in the North Sea, where the A2 rocket was developed, the prototype for the V1 and V2 rockets (V for *Vergeltungswaffen* = revenge weapons) that were to torment London toward the end of World War II. More was spent on this weapon in World War II than on anything else except the Manhattan Project for building the atomic bomb in the United States.

The rocket development facility was established at Peenemünde in 1936 where at first Göring hoped for a rocket-assisted aircraft—it was only later that the pilotless plane, or drone, was conceived. The Germans were ambitious, seeking a missile that would have a range of 160 miles, with a payload of one ton, traveling at five times the speed of sound (at a stage when no rocket had ever broken the sound barrier).[18] Three wind tunnels were built at Peenemünde to help scientists reproduce the conditions of supersonic flight.

Hitler had high hopes for the pilotless rockets, in particular that they would terrify Londoners and encourage the British government to sue for peace. The pilotless rockets did see the light of day, roughly 11,000 of them being launched on Britain, of which about 3,500 landed in London or in the south of England (the rest crashed on the way or strayed off course), killing 8,700 and injuring three times as many. But the "terror effect" never materialized. In 1945, 118 German rocket scientists surrendered to the Americans and were taken en masse to Fort Bliss in Texas as part of the secret Operation Paperclip, to use their skills to develop the U.S. space program. In addition to von Braun, Ernst Stuhlinger's expertise proved crucial to the American effort to launch its first satellite after Russia shocked the world by putting *Sputnik I* into orbit in October 1957.

Poison gas was also developed. As early as 1936 Dr. Gerhard Schrader, working on insecticides at IG Farben, had discovered a substance he named Tabun, which attacked the human nervous system, disrupting a neurotransmitter that controls the muscles, causing the victim to choke to death. A year later his team came up with an even more powerful substance, isopropyl methylphosphorofluoridate, or "sarin," which causes coma, nosebleeds, loss of memory, paralysis, trembling, and many other symptoms. Sarin was named in "honor" of its discoverers, Gerhard **S**chrader, **A**mbros, **R**üdiger, and Van der **Lin**de. Both substances were produced at a factory built in Silesia, where 12,000 tons of Tabun were found by the Allies in 1945. The Germans never used these agents, fearing they would provoke a deadly chemical war.[19] (Hitler had himself been exposed to poison gas in World War I, and this may have been a factor.)

By the 1930s, certainly by 1939, IG Farben was the largest company in Europe and the fourth largest in the world, behind General Motors, U.S. Steel, and Standard Oil. The company had followed the nineteenth-century successes in dyestuffs and pharmaceuticals by remaining at the forefront of "big chemistry," meaning that by the time the Nazis achieved office, Germany led the world in synthetic fuels. It cost ten times more to produce than fossil fuels, but synthetic fuel was popular with the Nazis because its production was under their direct control and production figures could be kept secret. The same arguments applied to synthetic rubber, which IG Farben also manufactured on a grand scale. When the war began, Germany was threatened with a shortage of rubber, and IG Farben was pressed into service. Famously, the company settled on Auschwitz as the site for its rubber plant, apparently quite independently of Himmler's decision regarding the siting of concentration camps.

Over the years of war, IG Farben's use of forced or slave labor in the production of synthetic fuel and rubber plants, plus substances like nitrogen, methanol, ammonia, and calcium carbide, rose from 9 percent in 1941 to 30 percent four years later.[20] This resulted in twenty-four members of the board of IG Farben being tried at Nuremberg. Five were found guilty of "slavery and mass murder" and received between six and eight years imprisonment.[21] President Dwight Eisenhower wanted the company broken up, but in fact it was folded back into three of the old companies, Bayer, BASF, and Höchst. In 1955, Friedrich Jaehne, who had been sentenced to a year and half at Nuremberg, was elected chairman of Höchst. A year later, Fritz ter Meer, also convicted of plunder and slavery at Nuremberg, was elected chairman of the supervisory board of Bayer. The Auschwitz plant continues to this day.

Part of the chemists' job was to investigate and systematize the science of mass murder—the design of ovens, the invention of more "efficient" gases, the ordered disposal of the "remains." Eleven million Jews, according to Adolf Eichmann, were killed in the death camps, though a more widely accepted figure is six million. Kurt Prüfer, an engineer who designed furnaces for Topf and Son of Erfurt, was an important figure here.

Though the sheer numbers of people gassed still have the power to astound us, it is the actions of the biologists that, even after all this time, must count as the greatest betrayal of the long tradition of German genius. Using newly opened archives in Berlin and Potsdam, Ute Deichman and others have shown that some 350 qualified doctors or university professors of medicine were involved in concentration camp experiments.*

Professor Heinrich Berning of the University of Hamburg used Soviet prisoners of war for famine experiments, carefully noting what happened as they starved to death. At the Institute for Practical Research in Military Science, experiments were carried out on cooling, using inmates from Dachau. The ostensible reason for this research was to study the effects of recovery of humans who suffered frostbite, and to examine how well humans adapted to the cold; some 8,300 inmates died during the course of these "researches." In the experiments on yellow cross, otherwise known as mustard gas, so many people were killed that after a while no more "volunteers" could be found with the promise of being released afterward. August Hirt, who carried out these "investigations," was allowed to murder 115 Jewish inmates of Auschwitz at his own discretion to establish a "typology of Jewish skeletons." Homosexuals were injected with hormones to see if their behavior was changed.[22]

Among the eminent scientists who are now known to have conducted unethical research (to put it no more strongly) are Konrad Lorenz, who went on to win the Nobel Prize in Physiology or Medicine in 1973, and Hans Nachtsheim, a member of the notorious Kaiser Wilhelm Institute for Anthropology and Genetics in Berlin. Lorenz's best-known work before the war was in helping to found ethology, the comparative study of animal and human behavior, where he discovered an activity he named "imprinting." In his most famous experiment he found that young goslings fixated on whatever image they first encountered at a certain stage of their development. Lorenz had read Spengler's *Decline of the West* and was not unsympathetic to the Nazis.[23] In that climate he began to conceive of imprinting as a disorder of the domestication of animals and drew a parallel between that and civilization in humans: in both cases, he thought, there was "degeneration." In September 1940, at the instigation of the Party and over the objections of the faculty, he became professor and director of the Institute for Comparative Psychology at the University of

* Which means, says John Cornwell, one in every 300 of the German medical community.

Königsberg, and from then until 1943 Lorenz's studies were all designed to reinforce Nazi ideology. He claimed, for instance, that people could be classified into those of "full value" (*vollwertig*) and those of "inferior value" (*minderwertig*).[24]

Conferences were held to broadcast the findings of research carried out on concentration camp inmates, including one where 1,200 people in Dachau were deliberately exposed to mosquitoes (with a small box containing the insects strapped to their hands), or else injected with the glands of mosquitoes, to study the effects of malaria, then said to be threatening German troops in Africa. These experiments were carried out under the direction of Dr. Klaus Schilling, emeritus professor of parasitology at Berlin and at one time director of the Malaria Commission of the League of Nations. At least thirty people died as a direct result of these experiments.

The Kaiser Wilhelm Institute for Anthropology and Human Genetics was founded in 1927 at Berlin-Dahlem on the occasion of the Fifth International Congress for Genetics, held in the German capital. The institute, and the congress, were both designed to gain international recognition for the study of human inheritance in Germany because, like other scientists, its biologists had been boycotted by scholars from other countries after World War I. The first director of the institute was Eugen Fischer, the leading German anthropologist, and he grouped around him a number of scientists who became infamous.[25] Nearly all of them supported the racial-political goals of the Nazis and were involved in their practical implementation—for example, by drawing up expert opinions on "racial membership" in connection with the Nuremberg laws.[26] There were also extensive links between the institute's doctors and Josef Mengele in Auschwitz. The institute was dissolved by the Allies after the war.

One aspect of the role played by science in the Third Reich that has not received due attention is the polycratic nature of Hitler's dictatorship, especially in the war years. Rather than being a tightly controlled regime, as the Nazis themselves advertised, so many people were clamoring for Hitler's attention and approval, and authority was divided so much, that rivalries for the Führer's favor created bottlenecks and gaps in lines of command that played havoc with the German war effort. The Vengeance rockets illustrate this point. Hitler was convinced they would wreak havoc in London, whereas in fact their chief effect was to draw resources away from aircraft production. This could have been predicted—it *was* predicted—but no one dared say it. As Norbert Elias has observed, this meant there were far more pressures on Hitler than may have appeared from the outside.

This is underlined, in a way, by Trevor Dupuy's figures about the fighting ability of German soldiers in World War II. Dupuy conducted the same exercise for the second war as he did for the first (see pp. 543–544), comparing the combat effectiveness of the fighting men on both sides. His conclusion, which he wrote most reluctantly, because of his contempt for the Hitler regime, was that "the Germans consistently outfought the far more numerous Allied armies that eventually defeated them."[27] Across seventy-eight engagements for which figures were available, the average number of opponents a soldier killed was as follows:

		Allies	Germans	German preponderance
Attack:	Successful	1.47	3.02	2.05
	Failure	1.20	2.28	1.90
Defense:	Successful	1.60	2.24	1.40
	Failure	1.37	2.29	1.67
Average:		1.45	2.31	1.59

With the German preponderance so marked, and extending consistently across many battles, one may ask why they lost. The answer lies in part in Hitler's polycratic style of government, but also in the fact that, ultimately, the Allied numerical and matériel strength was just too much. Therein, perhaps, lies a profound truth. If you want to win, you need friends. Making friends was the one thing the Nazis were not good at.

38.

Exile, and the Road into the Open

Between January 1933 and December 1941, 104,098 German and Austrian refugees arrived in America, of whom 7,622 were academics and another 1,500 were artists, journalists specializing in cultural matters, or other intellectuals. The trickle that began in 1933 swelled after Kristallnacht in 1938, but never reached a flood. By then it had become difficult for many to leave, and anti-Semitism, and anti-immigrant feeling generally in America, meant that many were turned away.

Other artists and academics fled to Amsterdam, London, or Paris. In the French capital Max Ernst, Otto Freundlich, and Gert Wollheim formed the Collective of German Artists, and then later the Free League of Artists, which held a counter-exhibition to the Nazi Entartete Kunst (Degenerate Art) show in Munich. In Amsterdam Max Beckmann, Eugen Spiro, Heinrich Campendonck, and the Bauhaus architect Hajo Rose formed a close-knit group, for which Paul Citroen's private art school served as a focus. In London such artists as John Heartfield, Kurt Schwitters, Ludwig Meidner, and Oskar Kokoschka were the most well known in an intellectual community of exiles that was about 200 strong, organized into the Free German League of Culture by the Artists' Refugee Committee, the New English Arts Club, and the Royal Academy.

A German Academy of the Arts and Sciences in Exile was established, intended as a form of resistance to Hitler. Thomas Mann headed the literary section (in America) and Freud the scientific division (in London).[1]

In Germany itself, artists such as Otto Dix, Willi Baumeister, and Oskar Schlemmer retreated into what they called "inner exile." Dix hid away at Lake Constance, where he painted landscapes; that, he said, was "tantamount to emigration." Karl Schmidt-Rottluff and Erich Heckel removed themselves to obscure hamlets, hoping to escape attention. Ernst Kirchner took his life.

The immigration to the United States was the most important and significant. As a result, the landscape of twentieth-century thought was changed dramatically. It was probably the greatest transfer of its kind ever seen (see Chapter 39, p. 713).

In addition to the artists, musicians, and mathematicians who were brought to America, scholars were also helped by a special provision in the U.S. immigration law, created by the State Department in 1940, which allowed for "emergency visitor" visas, available to imperiled refugees "whose intellectual or cultural achievements or political activities were of interest to the United States."[2] Max Reinhardt, the theater director, Stefan Zweig, the writer, and Roman Jakobson, the linguist, all entered the United States on emergency visas.

Of all the schemes to help refugees whose work was deemed important in the intellectual sphere, none was so extraordinary, or so effective, as the Emergency Rescue Committee (ERC) organized by the American Friends of German Freedom. The Friends had been formed in America by the ousted German socialist leader Paul Hagen (also known as Karl Frank), to raise money for anti-Nazi work. In June 1940, three days after France signed the armistice with Germany with its notorious "surrender on demand" clause, requiring France to hand over any non-French person to German authorities, the committee's members held a lunch to consider what needed to be done to help threatened individuals in the new, much more dangerous situation. The ERC was the result, and $3,000 was raised immediately. The aim, broached at the lunch, was to prepare a list of important intellectuals—scholars, writers, artists, and musicians—who were at risk and would be eligible for special visa status. One of the committee's members, Varian Fry, was chosen to go to France to find as many threatened intellectuals as he could and help them to safety.

Fry, a slight, bespectacled Harvard graduate, had been in Germany in 1935 and seen what Nazi brutality was like. He spoke German and French and was familiar with the work of living writers and painters. Fry arrived in Marseille in August 1940 with that $3,000 in his pocket and a roster of 200 he had memorized, judging it too dangerous to carry written lists. These names had been collected in an ad hoc way. Thomas Mann had provided the names of German writers at risk, Jacques Maritain a list of French writers, Jan Masaryk the Czechs. Alvin Johnson, president of the New School of Social Research in New York City, submitted names of academics, and Alfred Barr, director of the Museum of Modern Art in New York, supplied the names of artists.

Fry soon grasped that not all the people on his list were in mortal danger. The Jews were, as well as the more outspoken, long-standing political opponents of Nazism. At the same time, it became clear that if many of the very famous, non-Jewish "degenerate" artists were protected by their celebrity in Vichy France, there were far more lesser-known figures who *were* in real danger.[3] Without referring back to New York, therefore, Fry changed the policy of the ERC and set about helping as many people as he could who fell within the ambit of the special visa law, whether they were on his list or not. He set up his own clandestine network using the French underground, which transported selected refugees out of France into Portugal, where, with a visa, they could sail for America. He found a "safe house," the Villa Air Bel, north of Marseille, and there he equipped his refugees with false documents and local guides who could lead them via obscure and arduous pathways across the Pyrenees to freedom.[4] The best-known figures who escaped in this dramatic fashion included André Breton, Marc Chagall, Max Ernst, Lion Feuchtwanger, Konrad Heiden (who had written a critical biography of Hitler), Heinrich Mann, Alma Mahler-Werfel, André Masson, Franz Werfel, and Wilfredo Lam, the Cuban painter. Fry helped around 2,000 individuals, ten times the number he had been sent to look for.

Alvin Johnson at the New School took ninety scholars to create a University in Exile, where the faculty included Hannah Arendt, Erich Fromm, Otto Klemperer, Claude Levi-Strauss, Erwin Piscator, and Wilhelm Reich. László Moholy-Nagy re-created a New Bauhaus in Chicago, and other former colleagues initiated something similar in what became

Black Mountain College, in the wooded hills of North Carolina. At one time or another its faculty included Joseph Albers, Willem de Kooning, Ossip Zadkine, Lyonel Feininger, and Amédée Ozenfant. After the war the college was home to a prominent school of poets, and it remained in existence until the 1950s. The Frankfurt Institute at Columbia University and Erwin Panofsky's Institute of Fine Arts at New York University were also started and staffed by exiles.

Once the Nazis took power, there was never much doubt that Arnold Schoenberg would have to leave. He had converted from Judaism to Christianity early in life, but that never made any impression on the authorities, and in 1933 he reverted to being a Jew. In the same year he was blacklisted as a "cultural Bolshevik" and dismissed from his Berlin professorship. He moved to Paris, where he was penniless and stranded. Then, out of the blue, he received an invitation to teach at a small private conservatory in Boston, founded and directed by Joseph Malkin, a Russian exile who had been solo cellist with the Berlin Philharmonic. Schoenberg accepted, arriving in America in October 1933. The first music he wrote in Boston was a light piece for a student orchestra, but then came the Violin Concerto, op. 36. Not only was this his real American debut, it was also his first concerto. Rich and passionate, it was—for Schoenberg—fairly conventional in form, though it demanded phenomenally difficult fingerwork from the violinist.[5]

Béla Bartók, Darius Milhaud, and Igor Stravinsky all followed to the United States in 1939–40. Many of the virtuoso performers, being frequent travelers as a matter of course, were already familiar with America, and America with them: Artur Rubinstein, Fritz Kreisler, Efrem Zimbalist, and Mischa Elman all settled in America in the late 1930s.

The only American rival to New York as a base for exiles in wartime was Los Angeles, where the roster of famous names living in close proximity (close in Los Angeles terms) was remarkable. Apart from Schoenberg (who had moved there from Boston), that roster included Thomas Mann, Bertolt Brecht, Lion Feuchtwanger, Theodor Adorno, Max Horkheimer, Otto Klemperer, Fritz Lang, Artur Rubinstein, Franz Werfel and Alma Mahler-Werfel, Bruno Walter, Peter Lorre, and Heinrich Mann, not forgetting the non-Germans Sergei Rachmaninoff, Igor Stravinsky, Man Ray, and Jean Renoir.[6]

"A New Type of Planned Order" . . . with "the Worst on Top"

It was perhaps only natural that a war in which very different regimes were pitched against one another should bring about a reassessment of the way people govern themselves. Alongside the scientists and generals and code breakers trying to outwit the enemy, others devoted their energies to the only marginally less urgent matter of the rival merits of fascism, communism, capitalism, liberalism, socialism, and democracy. This brought about one of the more unusual coincidences of the century, when a quartet of books was published during the war by exiles from the old dual monarchy, Austria and Hungary, looking forward to the type of society humanity should aim for after hostilities ceased. Whatever their other differences, these books had one thing in common to recommend them: thanks to paper rationing, they were mercifully short.

Karl Mannheim's *Diagnosis of Our Time* appeared in 1943. Mannheim was a member of the Sunday Circle who had gathered around George Lukács in Budapest during World War I, and included Arnold Hauser and Béla Bartók. Mannheim had left Hungary in 1919, grown up with the traditional German understanding of Bildung, studied at Heidelberg, and attended Martin Heidegger's lectures at Marburg.[7] He was professor of sociology at Frankfurt from 1919 to 1933, a close colleague of Adorno, Horkheimer, and the others, but after Hitler took power he moved to London, teaching at the London School of Economics (LSE) and the Institute of Education.

Mannheim took a "planned society" completely for granted.[8] For him the old capitalism, which had produced the stock market crash and the Depression, was dead. "All of us know by now that from this war there is no way back to a laissez-faire order of society, that war as such is the maker of a silent revolution by preparing the road to a new type of planned order." He was equally disillusioned with Stalinism and fascism. Instead, according to him, the new society after the war, what he called the Great Society, could be achieved only by a form of planning that did not destroy freedom, as had happened in the totalitarian countries, but that took account of the latest developments in psychology and sociology, in particular psychoanalysis. Mannheim believed that society was ill—hence "Diagnosis" in his title. For him the Great Society was one where individual freedoms were maintained, but informed by an awareness of how societies operated and how modern, complex, technological societies differed from agricultural peasant communities. He therefore concentrated on two aspects of contemporary society: youth and education on the one hand, and religion on the other. Whereas the Hitler Youth had become a force of conservatism, Mannheim believed youth was naturally progressive if educated properly. He thought pupils should grow up with an awareness of the sociological variations in society, and the causes of them, and should also be made aware of psychology, the genesis of neurosis, and what role psychology might play in the alleviation of social problems. He concentrated the last half of his book on religion because he saw that at bottom the crisis facing the Western democracies was a crisis of values, that the old class order was breaking down but had yet to be replaced by anything else systematic or productive. While he saw the church as part of the problem, he believed that religion was still, with education, the best way to instill values, but that organized religion had to be modernized—again, with theology being reinforced by sociology and psychology. Mannheim thought that postwar society would be much more informed about itself than prewar society. He acknowledged that socialism had a tendency to centralize power and degenerate into mere control mechanisms, but he was a great Anglophile who thought that Britain's "unphilosophical and practically-minded citizens" would see off would-be dictators.

Joseph Schumpeter had little time for sociology or psychology. Insofar as they existed at all, for him they were subordinate to economics. In his wartime book *Capitalism, Socialism and Democracy*, Schumpeter declared himself firmly opposed to John Maynard Keynes, to Marx, and to Weber up to a point. It is not hard to see why.[9] Educated at the Theresianum, an exclusive school in Vienna reserved for the aristocracy, Schumpeter was there by virtue of the fact that his mother had married a general after his father, an undistinguished man, had died. As a result of his "elevation," Schumpeter was always rather self-consciously aristocratic; he would appear at university meetings in riding habit and inform anyone who was listening that he had three ambitions in life—to be a great lover, a great horseman, and a great economist. After a number of adventures in Egypt and Austria, Schumpeter eventually made his way to Harvard, "where his manner and his cloak quickly made him into a campus figure." All his life he believed in "an aristocracy of talent."

Schumpeter's main thesis was that the capitalist system is essentially static. For employers and employees as well as for customers, the system settles down with no profit in it, and there is no wealth for investment.[10] Workers receive just enough for their labor, based on the cost of producing and selling goods. Profit, by implication, can only come from innovation, which for a limited time cuts the cost of production, until competitors catch up, and allows a surplus to be used for further investment. Two things followed from this. First, capitalists themselves are not the motivating force of capitalism; the impetus comes instead from entrepreneurs who invent new techniques or machinery by means of which goods are produced more cheaply. Schumpeter did not think that entrepreneurship could be taught or inherited; it was, he believed, an essentially "bourgeois" activity, and bourgeois people acted not out of any theory or philosophy but from pragmatic self-interest. This flatly contradicted Marx's analysis.[11] The second element of Schumpeter's outlook was that profit, as generated by entrepreneurs, was temporary. Whatever innovation was introduced would be followed up by others in that sector of industry or commerce, and a new stability eventually achieved. This meant that capitalism was inevitably characterized by cycles of boom and stagnation. As a result, Schumpeter's view of the 1930s was diametrically opposite to Keynes's (that economies could spend their way out of recession). Schumpeter thought the Depression was to an extent inevitable, a cold shower of reality. By the time the war began, he had developed doubts that capitalism could survive. He thought that, as a basically bourgeois activity, it would lead to increasing bureaucratization, a world for "men in lounge suits" rather than buccaneers. In other words, it contained the seeds of its own ultimate failure, an economic success but not a sociological one.

If Mannheim took planning for granted in the postwar world, and if Schumpeter was lukewarm about it, the third Austro-Hungarian, Friedrich von Hayek, was downright hostile. Born in 1899, Hayek came from a family of scientists, distantly related to the Wittgensteins. He took two doctorates at the University of Vienna, became a professor of economics at the LSE in 1931, and acquired British citizenship in 1938. He loathed Stalinism and fascism equally, but he was much less convinced than the others that the same centralizing and totalitarian tendencies that existed in Russia and Germany couldn't extend eventually to Britain and even America. In *The Road to Serfdom* (1944), he set out his opposition to planning and linked freedom firmly to the market, which, he thought, helped produce a "spontaneous social order." He was critical of Mannheim, regarded Keynesian economics as "an experiment" that in 1944 had yet to be proved, and reminded his readers that democracy was not an end in itself but "essentially a means, a utilitarian device for safeguarding internal peace and individual freedom." He acknowledged that the market was less than perfect, but he reminded his readers that the rule of law had grown up at the same time as the market, in part as a response to its shortcomings: the two were intertwined achievements of the Enlightenment.[12] His reply to Mannheim's point about the importance of having greater sociological knowledge was that markets are "blind," producing effects that no one can predict, and that that is part of their point, part of their contribution to freedom, the "invisible hand" as it has been called. For him, therefore, planning was not only wrong in principle but impractical.[13] Hayek then went on to produce three reasons why, under planning, "the worst get on top." The first was that the more highly educated people are always those who can see through arguments and don't join the group or agree to any hierarchy of values. Second, the centralizer finds it easier to appeal to the gullible and docile; and third, it is always easier for a group of people to agree on a negative program—on the hatred of foreigners or a different class, for example—than on a positive one. He conceded that the tendency to monopoly needed watching and should be guarded against, but he saw a greater threat from the monopolies of the labor unions under socialism.

As the war was ending, a fourth Austro-Hungarian, Karl Popper, released *The Open Society and Its Enemies.* Born in Vienna in 1902, Popper flirted with socialism, but Freud and Adler were deeper influences, and he attended Einstein's lectures in Vienna. He completed his PhD in philosophy in 1928, then worked as a social worker with children who had been abandoned after World War I, and as a teacher. He came into contact with the Vienna Circle, and was encouraged to write.[14] His first books, *Die beiden Grundprobleme der Erkenntnistheorie* (*Two Fundamental Problems of the Theory of Knowledge*) and *Logik der Forschung* (*The Logic of Scientific Discovery*), attracted enough attention for him to be invited to Britain in the mid-1930s for two long lecture tours.[15] But when Moritz Schlick was assassinated in 1936 by a Nazi student, Popper, who had Jewish blood, accepted an invitation to teach at the University of Canterbury in New Zealand. It was in the Southern Hemisphere that he produced his next two books, *The Poverty of Historicism* and *The Open Society and Its Enemies,* many of the arguments of the former title being included in *The Open Society.*[16]

The immediate spur to that book was the news of the *Anschluss.* The longer-term inspiration arose from the "pleasant sensation" Popper felt on arriving for the first time in England, "a country with old liberal traditions," as compared with a country threatened with National Socialism, which for him was much more like the original closed society, the primitive tribe or feudal arrangement, where power and ideas are concentrated in the hands and minds of a few, or even one, the king or leader. Popper, like the logical positivists of the Vienna Circle, was profoundly affected by the scientific method, which he extended to politics. For him, this meant that political solutions were like scientific ones—they "can never be more than provisional and are always open to improvement." This is what he meant by the poverty of historicism, the search for deep lessons from a study of history, which would provide the "iron laws" by which society should be governed. Popper thought there was no such thing as history, only historical interpretation.

This led Popper to the most famous passage in his book, the attack on Plato, Hegel, and Marx. (The book was originally called *False Prophets: Plato, Hegel, Marx.*) Popper thought that Plato might well have been the greatest philosopher who ever lived but that he put the interests of the state above everything, including the interpretation of justice. Popper was attacked for his dismissal of Plato, but he clearly saw him as an opportunist and as the precursor of Hegel, whose dogmatic dialectical arguments had led, he felt, to an identification of the good with what prevails, and the conclusion that "might is right." Popper thought this was simply a mischaracterization of dialectic. In reality, he said, it was merely a version of trial and error, as in the scientific method, and Hegel's idea that thesis generates antithesis was wrong: thesis, he said, generates modifications as much as it generates the opposite to itself.[17] By the same token, Marx was a false prophet because he insisted on holistic change in society, which Popper thought had to be wrong simply because it was unscientific—it couldn't be tested. He himself preferred piecemeal change, so that each new element introduced could be tested to see whether it was an improvement on the earlier arrangement. Popper was not against the aims of Marxism, pointing out, for example, that much of the program outlined in the *Communist Manifesto* had actually been achieved by Western societies. But that was his point: this had been achieved piecemeal, without violence.

Popper shared with Hayek a belief that the state should be kept to a minimum, its basic raison d'être being to ensure justice, to ensure that the strong did not bully the weak. He disagreed with Mannheim, believing that planning would lead to more closure in society, simply because planning involved a historicist and holistic approach, which went against the scientific method of trial and error. This led Popper to consider democracy as the only viable possibility because it was the only form of government that embodied the scientific, trial-and-error method and allowed society to modify its politics in the light of experience, and to change government without bloodshed.

The coincidence of these four books by Austro-Hungarian émigrés was remarkable but, on reflection, perhaps not so surprising. There was a war on that was being fought for ideas and ideals as much as for territory. These émigrés had each seen totalitarianism and dictatorship at close hand and realized that even when the war with Germany and Japan ended, the conflict with Stalinism would continue.

Erwin Schrödinger, the physicist who conceived the idea that the electron orbited the nucleus as a wave (see p. 598) and had been awarded the Nobel Prize in 1933, found a different way into the open. He left Germany in the year he won the Nobel, dismayed by the Nazis. He had been made a Fellow of Magdalen College, Oxford, and taught there before moving on to Ireland where, in 1943, he gave a series of well-received lectures, "What Is Life?" in which he considered how a physicist might define life. In these lectures he looked at the chromosome from the point of view of physics and showed that the gene must be "an aperiodic crystal . . . a regular array of repeating units in which the individual units are not all the same." He explained that the behavior of individual atoms could be known only statistically and therefore for genes to act with the very great precision with which they did act, they must be of a certain minimum size—which he calculated—with a minimum number of atoms. He concluded that the gene must consist of a long, highly stable molecule that contains a code. In 1943, most biologists were ignorant of the latest physics but among those who read Schrödinger's book based on his lectures and were excited by its arguments were Francis Crick, James Watson, and Maurice Wilkins.[18]

Walter Benjamin's road into the open turned into disaster. In 1933 he fled Germany for Paris where he worked for the Frankfurt Institute and published some of his most influential work, notably "Das Kunstwerk im Zeitalter seiner technischen Reproduzierbarkeit" (The Work of Art in the Age of Mechanical Reproduction), an argument brilliantly deconstructed by Clive James.[19] In this Benjamin argued that art from antiquity to the present has its origin in religion and that even secular work keeps to itself an "aura," the possibility that it is a glimpse of the divine. As Hofmannsthal, Rilke, and José Ortega y Gasset had argued before him, this implied a crucial difference between the artist and the nonartist. In the era of mechanical reproduction, however, this tradition, and the distance between artist and nonartist breaks down. Benjamin thought this was a good thing, and his view was to prove persuasive among postmodernists—that mass-produced entertainment can address the psychological problems of society at large. But he did not live to see what became of his idea. As the Nazis advanced on Paris, he headed south, planning to take advantage of the passage over the Pyrenees put in place by Varian Fry and others. By 1943, Benjamin thought that he had the necessary paperwork—he had a U.S. emergency visa and a Spanish transit visa. But then he found he also needed a French exit visa and, already exhausted as the result of a heart condition, the whole enterprise proved too much; he took his own life.

What are we to make of Ernst Jünger's road after 1933? In 1930 he had published *Über Nationalismus und die Judenfrage* (*On Nationalism and the Jewish Question*), in which he had condemned the Jews as a threat to the unity of Germany, followed in 1932 by *Der Arbeiter* (*The Worker*), which called for a totally mobilized society run by "warriors-workers-scholars." But Jünger began to have his doubts about Hitler's Reich, publishing *Auf den Marmorklippen* (*On the Marble Cliffs*) in 1939, which expressed some of these doubts, albeit metaphorically, and he still served in World War II as an army captain. In

Russia in 1942, however, his reputation was such that the generals there admitted to him the terrible atrocities taking place. For a time, Jünger comforted himself that all sides in the war were equally barbaric but eventually he came to see that the Germans were far worse. He was then fortunate to be stationed in Paris where he mixed with—and to an extent was able to protect—the likes of Pablo Picasso and Jean Cocteau. If that redeemed him somewhat, so too did the inspiration he seems to have offered to anti-Nazi conservatives in the German army who mounted an unsuccessful attempt (one of several) on Hitler's life in 1944. Nevertheless, after the war Jünger was banned from publishing for several years for not sufficiently resisting the Nazis.

Gottfried Benn fared little better. In the years before 1933 he had not been as bellicose as Jünger, enjoying a good reputation as a distinguished poet and an accomplished doctor. Born in Manfeld, he studied theology at Marburg before taking a medical degree in Berlin. In World War I he served as a military doctor in Belgium, though by then he had already published his first collection of Expressionist poems, *Morgue*, concerned with the decay of the body. After World War I, Benn came to loathe Weimar, in particular its liberal culture, and he railed against what he saw as the nihilism of the republic and the role of intellectuals in that process. After 1933, he agreed with the Nazis' attempt to "re-awaken" Germany and broadcast over the radio his apparently new view that "intellectual freedom was an anti-heroic ideology," mocking those authors in exile—in particular Thomas Mann and his son Klaus—who, he said, "had missed the chance to experience the concept, so alien to them, of *Volk*, rather than think about it in the abstract."[20]

However, Benn's enthusiasm for the Nazis did not outlast the Nacht der langen Messer (Night of the Long Knives; June 1934), and he broke with the regime, retreating into what he called "aristocratic inner emigration." He was attacked in the press, forced to resign from the Reich Chamber of Writers and forbidden from publishing. After the war he resumed both writing and his medical practice. To begin with his works were banned by the Allies but in 1951 he won the Büchner Prize. In his autobiography, *Doppelleben* (*Double Life*; 1950), he included a letter Klaus Mann had sent to him from France, which showed, he admitted, that Mann had judged the prewar situation better than he had. By then it was too late. Klaus Mann committed suicide in May 1949.[21]

PART VI

Beyond Hitler: Continuity of the German Tradition under Adverse Conditions

39 ·

The "Fourth Reich": The Effect of German Thought on America

When the American philosopher Allan Bloom first went to college, at the University of Chicago in the mid-1940s, just after World War II had ended, one of the things he soon noticed was that "American university life was being revolutionised by German thought." At that time, in Chicago anyway, Marx was revered, he said, but the two thinkers who generated the most enthusiasm were the sociologist Max Weber and the psychoanalyst Sigmund Freud, who in turn, as Bloom put it, had both been profoundly influenced by Friedrich Nietzsche. Between them (plus Georg Simmel and Ferdinand Tönnies later on), Bloom argued, "Freud and Weber are the immediate source of most of the language with which we are so familiar . . . part of that great pre-Hitler German classical tradition . . . Equality and the welfare state were now part of the order of things, and what remained was to complete the democratic project. Psychotherapy would make individuals happy, as sociology would improve societies."[1]

What we were witnessing, Bloom insisted, was an Americanization of German pathos that the Americans were not aware of. Americans, he said, were now straining in a search for inwardness, but the main effect of German thought on America (and perhaps, by extension, on the rest of the West) was its historicism, its rejection of universality and cosmopolitanism in favor of a culture rooted in a nation's history and achievements. "Our intellectual skyline has been altered by German thinkers even more radically than has our physical skyline by German architects."[2]

Henri Peyre agreed. Peyre (1901–88), professor of French at Yale, was one of five people who contributed to a series of lectures at the University of Pennsylvania in 1952 on the topic of "the cultural migration." The others were Franz Neumann on the social sciences, Erwin Panofsky on the history of art, Wolfgang Köhler on psychology, and Paul Tillich on theology. Peyre, in speaking of the effect of immigrants on literature in the United States, said it was already clear that they constituted "one of the most vigorous elements in present-day American intellectual life, around periodicals like the *Partisan Review* and *Commentary*." Particularly in their capacity for work and their concern for intellectual values, he said, they had transformed many university departments, with American pragmatism and fondness for factual empiricism being strengthened by "German patience" and the Germans' habits of collection of data, adding "those exiles from Germanic lands have enabled American speculation in many fields to leap forward with unheard-of boldness." He concluded: "Philosophy has invaded many academic curricula; psychological or sociological generalisations fascinate college youth. Tocqueville . . . wisely remarked that 'the Americans are much more addicted to the use of general ideas than the English.' In several respects, American intellectual life is today closer to the German than to the British." In fact, he said, the British contribution to American intellectual life was "surprisingly far behind" the German contribution.[3]

The De-Provincialization of the American Mind

The pithiest way to show how German refugees affected American life is to give a list of those whose intellectual contribution was such as to render their names, if not household words, then at least eminent among their peers: Theodor Adorno, Hannah Arendt, Rudolf Arnheim, Erich Auerbach, Paul Baran, Hans Bethe, Bruno Bettelheim, Arnold Brecht, Bertolt Brecht, Marcel Breuer, Hermann Broch, Charlotte and Karl

Bühler, Rudolf Carnap, Lewis Coser, Karl Deutsch, Marlene Dietrich, Alfred Döblin, Peter Drucker, Alfred Eisenstaedt, Hanns Eisler, Erik Erikson, Otto Fenichel, Ernst Fraenkel, Erich Fromm, Hans Gerth, Felix Gilbert, Kurt Gödel, Gottfried von Haberler, Eduard Heimann, Ernst Herzfeld, Julius Hirsch, Albert Hirschman, Hajo Holborn, Max Horkheimer, Karen Horney, Werner Jaeger, Marie Jahoda, George Katona, Walter Kaufmann, Otto Kirchheimer, Wolfgang Köhler, Kurt Koffka, Erich Korngold, Siegfried Kracauer, Ernst Krenek, Ernst Kris, Paul Oskar Kristeller, Fritz Lang, Paul Lazarsfeld, Kurt Lewin, Peter Lorre, Leo Lowenthal, Ernst Lubitsch, Heinrich Mann, Klaus Mann, Thomas Mann, Herbert Marcuse, Ernst Mayr, Ludwig von Mises, Oskar Morgenstern, Hans Morgenthau, Otto Nathan, Franz Neumann, Erwin Panofsky, Wolfgang Panofsky, Erwin Piscator, Karl Polanyi, Friedrich Pollock, Otto Preminger, Fritz Redlich, Max Reinhardt, Erich Maria Remarque, Hans Rosenberg, Arnold Schoenberg, Joseph Schumpeter, Alfred Schutz, Hans Simons, Leo Spitzer, Hans Staudinger, Leo Strauss, Leo Szilard, Edward Teller, Paul Tillich, Eric Voegelin, Kurt Weill, René Wellek, Max Wertheimer, Billy Wilder, Karl Wittfogel, Hans Zeisel, Heinrich Zimmer, Fred Zinnemann. This list is, of course, nowhere near exhaustive.[4]

Brecht described their experiences this way:

Hounded out by seven nations,
Saw old idiocies performed,
Those I praise whose transmutations
Leave their persons undeformed.

But who could remain undeformed, even when there were so many Germans in Washington Heights in New York that it became known as the "Fourth Reich"? Most of the refugees arrived during the 1930s, in the Great Depression, when unemployment was high and the general mood was not especially favorable to newcomers, however distressing their circumstances. Even so, they made their world. The Deauville restaurant on East Seventy-third Street in New York, the Éclair on West Seventy-second, the Café Royale on the Lower East Side, or the Blue Danube in

Hollywood, operated by Joe May, a Berlin director down on his luck, became homes-from-home, as close to the old life as they could find.* [5]

Most who were to become famous (and "de-provincialize" the American mind, in Anthony Heilbut's phrase) were under forty. They were flexible, but even so it wasn't always easy. One historian found American students disappointing—"They're so unequipped. I've never had one student from whom I learned a thing."[6] More than one remarked that "Americans were the kindest of people and the dullest."[7] Paul Lazarsfeld found German more precise than the "commercial discourse" of America. Theodor Adorno and his colleagues found American popular culture to be uncritical, a latent form of propaganda for commercial society.[8]

The Golden Age of Psychoanalysis

Probably the greatest single influence that the refugee Germans had in America, certainly in the more immediate aftermath of war, was in the realm of psychology in general and psychoanalysis in particular. Social psychologists like Kurt Lewin (1890–1947) and Gestalt psychologists like Wolfgang Köhler (1887–1967), Kurt Koffka (1886–1941), and Max Wertheimer (1880–1943), all exerted some influence, in the case of Lewin on such people as Margaret Mead and Ruth Benedict, and in the case of the Gestaltists on such behaviorists as Edward C. Tolman and the humanist psychologist Abraham Maslow.[9]

Freud had formed a rather warped view of the United States after his visit there early in the century (he thought America was "a mistake") but despite this, psychoanalysis had become popular in the United States even without fresh German input. Between the wars it had been incorporated into the medical establishment, in contrast to Europe where "lay analysts" were far more common, and this may have had something to do with its high prestige. Between 1940 and 1960 the membership of the American Psychoanalytic Association grew fivefold, making this, says Lewis Coser, "surely the golden age of psychoanalysis in America."[10] Coser puts this down

* Anthony Heilbut, *Exiled in Paradise: German Refugee Artists and Intellectuals in America from the 1930s to the Present* (University of California Press, 1983 and 1997, with a new postcript) is by far the most enjoyable book about refugees in America, beautifully written and by turns funny and moving.

partly to "America's more optimistic temper," but whatever the reason, the insights of several of the refugee psychoanalysts passed into the language.

Born in Frankfurt in 1902, Erik Erikson was abandoned by his Danish father and brought up by his Jewish stepfather. Taunted at school as a "Jew" yet treated suspiciously at his local synagogue because of his blond Danish looks, he was uncertain of his identity from an early age, and this may have shaped his work. He taught art at a school in Vienna and was drawn into a circle that included Anna Freud, who analyzed him. He moved to America after Hitler invaded Austria, his way smoothed by the analyst Hanns Sachs, who helped him secure a position at the Harvard Medical School and Massachusetts General Hospital as a child analyst. There he encountered Margaret Mead, Ruth Benedict, Gregory Bateson, and Kurt Lewin.[11] His anthropologist friends suggested that some of the generalizations he was prone to make about childhood did not apply across all cultures and so, acting on these criticisms, he visited a Sioux reservation in South Dakota, where he observed child-rearing practices. These inquiries led to his groundbreaking book *Childhood and Society*, written after he had moved to California (via Yale) and in which introduced his concepts of "ego identity" and "identity crisis."[12] He compared Americans with Germans, placing the appeal of Nazism in the cradle of the German family, where the son was set against the father, unlike in America, where fathers and sons "are friends," united against the wife-mother who "incarnates" social authority. This was why one's occupation was so important to Americans, he said—it was the American way of overcoming the dominance of the mother.[13]

Bruno Bettelheim arrived in 1939 after a traumatic year in Dachau and Buchenwald. He had studied philosophy and psychology in Vienna under Karl Bühler, though he too was influenced by Anna Freud. He found work at the University of Chicago and soon became the director of the university's school for disturbed children. His best-known books are *The Informed Heart* (1960), *The Empty Fortress* (1967), and *The Uses of Enchantment* (1976). In these works he drew on his treatments of disturbed children, but also on his experience of concentration camps and, as a Jew, of being a victim of anti-Semitism.[14] The books were thus an amalgam of clinical detail, contemporary history, and social criticism, his main argument being that modern mass society fails to take account of the unconscious and nonrational aspects of our make-up and that this leads people into either the extremes of crime, cruelty, and brutality, or else into ill health—mental and physical—suicide, or other forms of self-harm. He even thought the mentally ill had no place in American society—a chill echo of Nazi Germany.[15] The autistic child, for example, cannot "reach" adulthood, but is "held back by its own prison guards," Bettelheim going so far as to identify the autistic child's parents with Nazi guards.[16] In *The Uses of Enchantment* he examined children's classic fairy tales, concluding that they introduce children to the sometimes harsh world of adult reality, that they too have an unconscious aspect, the symbols of which help us understand the problems of children growing up.[17]

Erich Fromm probably enjoyed the largest readership among the lay (non–psychoanalytically trained) public. Born in 1900 in Frankfurt am Main, he was brought up in a strict Orthodox Jewish tradition and studied with, among others, Gershom Scholem. Fromm himself planned to become a rabbi but, while studying philosophy, sociology, and psychology at Frankfurt, and then at Heidelberg, he was drawn to what Scholem called the "torapeutic" sanatorium, a clinic where the psychoanalyst Frieda Reichmann combined teachings of the Torah with Freudian therapy.[18] Fromm did more than study with Reichmann—he married her, before associating, as we saw in an earlier chapter, with Adorno and Horkheimer at the Frankfurt Institute for Social Research. He moved to America in 1938 along with most of the other members of the institute.

His most famous book, *Escape from Freedom*, appeared in 1941 and it too may be seen as an attempt to marry Marx with Freud. Accepting the theory of the "oral," "anal," and "genital" stages of human development, and combining that with the concept of "social character" built up by Wilhelm Reich and Otto Fenichel—Marxist psychoanalysts he had met in Berlin—Fromm argued that, contrary to what Freud said, character was partly determined by class structure and socioeconomic conditions.[19] He distinguished, for instance, between the "hoarding orientation" of nineteenth-century merchants with their predisposition to punctuality, a saving mentality, and orderliness, and the "marketing orientation" of the twentieth century. It was Fromm who identified what he called the sadomasochistic or "authoritarian" personality, which he had observed first in Weimar Germany.[20] Such people respect the strong and loathe the weak, and Fromm thought this might help explain fascism. His Frankfurt colleagues took up the theme of the authoritarian personality in a more sociological context (see below).

Fromm's later books *Man for Himself* (1947) and especially *The Sane Society* (1955) became works of social criticism—applying a mix of clinical detail and contemporary observation, as Bettelheim was doing and as would become a familiar form of literature in the West from the 1960s on—which castigated modern culture, especially its greed, competitiveness, lack of moral backbone, and loss of community.[21] This was, in its way, a return to German cultural pessimism. Together with Hannah Arendt and Herbert Marcuse, Fromm was adopted as a guiding light by the students of the 1960s (again, see below).

Wilhelm Reich and Fritz Perls may be considered together as they are often regarded as joint initiators of the "sexual revolution" that began in the 1960s and gathered pace through the 1970s and into the 1980s.[22] This is hardly true of Reich and is based chiefly on his "invention" of the "orgone" box, a telephone booth–shaped instrument, wooden on the outside, with a metal lining which he claimed—fraudulently—had therapeutic properties. Reich began as a serious Freudian (he analyzed Perls) and, like so many others, attempted a marriage of Freud and Marx in the interwar years in Germany. His 1933 book, *Die Massenpsychologie des Faschismus* (*The Mass Psychology of Fascism*) was timely, so much so that with his background and interests he could not stay in Germany.[23] He reached America via Denmark, settling in the Forest Hills suburb of New York, where he underwent a profound reversal of feeling—from pro-Communism to virulent anti-Communism, and a growing paranoia (it was a paranoiac time). This gradually got the better of him (to the point where he included flying saucers among his enemies), and he was eventually convicted of fraud after the Food and Drug Administration took out an injunction against a shipment of "orgone accumulators," which he claimed attracted "orgone energy" (basic to life) and focused it on the body inside the box. In March 1957 he was sent to prison.

Fritz Perls studied dramatic direction in Berlin with Max Reinhardt, later becoming interested in Gestalt psychology, which, he thought, was "the next step after Freud." It was Perls's approach that produced the Esalen Institute in the 1960s and that, via "est," evolved into the "human potential" movement in the late 1970s, the basic idea of which was to "unlock previously blocked psychic energies," mainly by sexual and sensual liberation—hot tubs in the open air, nudity, drugs, the breaking of taboos. This may be seen as a form of post-Freudian *Bildung*. *Games People Play*, by Eric Berne, another émigré, was a best seller in 1964 and explored similar issues.

Heidegger's Children

After psychoanalysis, the area where German thought has had most influence in the United States is politics or, more accurately, political science—i.e., political theorizing rather than practical politics.[24] The first figure here is Hannah Arendt, but Richard Wolin has reminded us in his book *Heidegger's Children* (2001) how many of Heidegger's students became influential after the war on both sides of the Atlantic: besides Arendt, we may include Herbert Marcuse, Leo Strauss, Karl Löwith, Hans Jonas, Paul Tillich, and Hans-Georg Gadamer.[25]

As outlined earlier, Arendt arrived in New York in 1941 via Paris. She moved in the milieu that surrounded such small magazines as *Commentary* and the *Partisan Review*, later for a time becoming a professor—at Princeton, the University of California, Chicago, and, perhaps inevitably, the New School for Social Research. The New School had been founded in 1919 by a group of scholars linked to the *New Republic* magazine, and early professors there had included John Dewey and Thorsten Veblen. In 1933, to help support refugee scholars, the school founded the University in Exile, which evolved into the Graduate Facility of the New School for Social Research.

From 1945 until 1949 Arendt worked on the first of several major books, *The Origins of Totalitarianism*, which appeared in 1951 and had an enormous impact on its American audience and made her famous. The book attempted to come to grips with the events that led up to World War II and examined in particular how a small group—the Jews—became the catalytic agent for the Nazi movement, the world war, and the death factories.[26] She drew parallels between communism and fascism, arguing that although they were intended to lead mankind into a glorious future by eradicating class differences, they had instead produced only atomization, alienation, and homelessness. The point of mass society, she argued, was that instead of creating "a higher form of human community," it produced isolation and loneliness which, she insisted, were the common ground of terror, and the cold and inflexible logicality of bureaucracy, leading to the executioners. There was no role for heroism, she said, and this absence helped to crush man's soul. One of her main points in *Origins* was that there had been an alliance in Germany in the 1930s between the educated middle classes and the "mob," and this was one of the main reasons why what happened happened.[27]

The Origins of Totalitarianism offered no solutions to the problems it described and diagnosed, though her next book, *The Human Condition* (1958), argued that the main aspects of political life were structure and action, and that in the modern world these two entities had all but disappeared in the highly administered politics of modern society—no one had the power to alter the structure of public life and to *act* on it. This turned out to be an important message, and Arendt's books became influential texts in regard to the revolutionary student movements of the 1960s and helped to cohere the aims of the so-called alternative culture.

In some ways, however, Arendt—along with Brecht, whose art she admired as much as she loathed his politics—was the great nonsentimentalist, the writer who, even more than Thomas Mann, kept her individuality intact, uncontaminated by fame and by her status as a Jew. Even though she was a German victim of the Holocaust, she was never sentimental about it, and she distrusted inwardness—for her, public action in a public space was the only guarantee of honesty or authenticity in human affairs, and so the political, defined in that way, took priority. (Jewish émigrés, she liked to say, had "committed no act.") Private life, she insisted, was the great aim but, increasingly in the modern world, it was a luxury.[28] The real battle in the modern world, she felt, was not between classes but between the increasingly "totalitarian fictions" of all-powerful government and the "everyday world of factuality" in which we live.[29]

In her book *Eichmann in Jerusalem*, about the trial of Adolf Eichmann, in 1962, the "mastermind" of Auschwitz, who had been captured by Israeli special forces in Argentina, where he was hiding, and smuggled to Israel, she was unflinchingly unsentimental about the man on trial and about the Holocaust and Jewish behavior in resisting persecution. The book created a furor but she held to her view, that evil was banal, that it is where nihilism ends.[30]

Herbert Marcuse became, for a brief time, the most famous of the Frankfurt school theoreticians, though he was by then already elderly. Born in 1898 into a middle-class Jewish family, he had not been much involved in politics until the revolutionary movements that emerged in Germany in the wake of World War I (he was part of a "worker's council" in Berlin). But even then his involvement did not last long and he moved to Freiburg to study under Heidegger and Husserl. He broke with Heidegger as the latter's flirtation with the Nazis began to show itself, and joined Horkheimer, Adorno, and Fromm at the Frankfurt Institute.[31]

He was in America from the late 1930s, where he obtained a position at the new Brandeis University near Boston. In the postwar era, he became one of the main critics of the world he saw around him, a world of increasing uniformity, consensus, and order, all subsumed under what was for him a tyrannical rubric of "progress." This led to the first of two books by Marcuse that captured the public imagination. The first, *Eros and Civilization*, was published in 1955 and was intended as a liberating text in which Marcuse used Freud to modify Marx for the modern world. It became popular especially among the hippies of the counterculture.[32] His argument was that modern men and women need to educate their *desire*, that Marx says nothing about this, that modern conformist society kills the aesthetic and sensual side to life, that this is a form of repression, and that society needed to be based as much on the pleasure principle as on economic principles.[33]

Marcuse made this more explicit in *One-Dimensional Man* (1964), with its famous concept of "repressive tolerance" where, he said, in modern society, even the language of tolerance and liberation is used to keep people from being liberated. The world, the American world in particular, was one-dimensional in that there was only one way of thinking that was now regarded as legitimate—technological rationality, which perpetuated itself in science, in the universities, in industry and commerce. His remedy was "the great refusal," the "negation" of the reality that technological rationality has foisted on us. This stifling world, he said, needed to be replaced by imagination, art, and "negative thought."

Compared to these, the influence of yet another student of Heidegger, Leo Strauss, was less diffuse, more immediate, and by no means to everyone's taste. In 2004 Anne Norton wrote an impassioned attack on the charismatic Strauss and the many Straussians whose particular cult of militant conservatism she held responsible for taking over George W. Bush's White House and oversimplifying the world of politics into a crass battle between righteous Americans and a series of "convenient enemies."[34]

Born in Kirchhain in Germany in 1899, Strauss arrived in the United States in 1938, where he joined the University in Exile at the New School. His interests, to that point, were Spinoza, Maimonides, and Carl Schmitt, the National Socialist scholar, in particular his book *The Concept of the Political* (see pp. 668–669).[35] Strauss studied at the universities of Frankfurt, Marburg, Berlin, and Hamburg and his PhD dissertation was sponsored by Ernst Cassirer. After that he spent a year at Freiburg under Husserl and Heidegger. A formal and timid man, Strauss's thought "was marked by his revulsion from the preponderant tendencies of the modern age . . . His conservatism rested on his conviction that modern trends of thought, be they positivistic or historicist, were inimical to—nay, destructive of—what he cherished as being the perennial values."[36] He thought that these perennial values were the qualities that had distinguished the educated middle class in Nazi Germany and that modern "fads" had undermined the timeless qualities of the Greeks and "opened the floodgates to a nihilism of values of which the Nazi movement was the most extreme outcome." Science, neo-Kantianism, and the modern behavioral sciences all contribute to a contemporary nihilism and this, Strauss was convinced, was the modern predicament, out of which a way must be found. He doubted the redemptive power of politics and never once wrote about American thought,[37] becoming known for his doubts about technological supremacy and for his dogmatic stance on these issues.[38] He lamented the decline of religion "as the only means of keeping the mob in check" and, according to Arendt (when they were both students of Heidegger), he "concurred with fascism in every respect except its anti-Semitism."* His many students became very influential.[39]

Among political *practitioners*, as opposed to theoreticians, the most colorful is without question Arnold Schwarzenegger. Born in Thal bei Graz in Austria in 1947, he arrived in the United States in 1968, already famous for being voted the best-built man in Europe (and going AWOL

* Allegedly, he asked her out, to which she replied, "I don't go out with Nazis."

during basic training in the Austrian army in order to enter competitions). In Los Angeles he trained to be an actor, at the same time continuing to work on bodybuilding, but his acting career blossomed after *Conan the Barbarian* was a hit, most notably with the three science-fiction *Terminator* films. He was elected governor of California in October 2003, though for the time being at least his political achievements have been overshadowed by those of Henry Kissinger, conceivably the most conventionally successful of all German émigrés to the United States.

Born Heinz Alfred Kissinger in 1923, in Fürth, Bavaria, his Jewish family moved to New York in 1938. While still in college he served as a German interpreter in the Counter Intelligence Corps. After the war he forged an academic career at Harvard, specializing in foreign policy (and nuclear weapons in particular), becoming a consultant to various prestigious bodies and to Nelson Rockefeller, then governor of New York, who sought the Republican nomination for president in 1960, 1964, and 1968. After Richard Nixon became president in 1968 he made Kissinger national security adviser and then secretary of state, which office he continued to hold when Gerald Ford took over as president after Nixon resigned.

Kissinger was a very controversial secretary of state, a "power cynic" to some, pursuing *Realpolitik* and becoming the dominant force in American foreign policy between 1969 and 1977. His involvement in the Vietnam War, the carpet bombing of Cambodia, the Indo-Pakistan War of 1971, and the botched CIA intervention in Chile, when the Marxist Salvador Allende was assassinated after being legitimately elected president, brought Kissinger robust criticism, and later in his life there were repeated attempts in several countries to have him arraigned on war crimes charges. At the same time, he helped negotiate the end to the Yom Kippur War in 1973, the withdrawal of American forces from Vietnam, and, with Nixon, pursued the policy of détente, a relaxation of relations with the Soviet Union and China, which helped him win the 1973 Nobel Peace Prize.[40]

Damaged Lives?

Theodor Adorno was possibly the most arrogant and at the same time the most angry émigré in the United States. Brecht thought Adorno was "pompous and elusive, austere and sensual" and Anthony Heilbut concluded that his disdain for American culture "bordered on the pathological." He nevertheless remains a figure to be reckoned with.

The Frankfurt Institute had moved from Columbia to California in the early 1940s, in an attempt to fortify the declining health of Max Horkheimer, its director (though Horkheimer didn't die until 1973, a decade after he and the institute returned to Germany). It was an ironic destination, Los Angeles being the capital of the entertainment industry that attracted the brunt of Adorno's disdain.[41] Yet although Adorno's criticisms of American society and culture were clearly over the top at times, many of his points were well taken. His criticisms may be divided into those he advanced while he was in America, and those he made after his return to Germany. In *Dialectic of Enlightenment* (1944), which he co-authored with Horkheimer, the two men argued that Enlightenment led inexorably to totalitarianism, "everything can be illuminated in order to be administered."[42] Cultural life in capitalist society, in particular, they said, is as much a prison as a liberation, "style"—in fashion as in art—is a phony form of individualism, brought about by the need for commerce to maximize profits, trivializing experience.

This was followed by a more influential—but far more prosaic—work, *The Authoritarian Personality* (1950). The book was conceived as early as 1939, as a joint project with the Berkeley Public Opinion Study and the American Jewish Committee, to investigate anti-Semitism. It was the first time the institute had used a quantitative approach, and the results of their "F" (for fascist) scale "seemed to warrant alarm . . . Anti-Semitism turned out to be the visible edge of a dysfunctional personality revealed in the many 'ethnocentric' and 'conventional' attitudes of the general American population, as well as of a disquietingly submissive attitude towards authority of all kinds." The book concluded with a warning that fascism rather than communism was the chief threat facing America in the post-war world, that fascism was finding "a new home" on the western side of the Atlantic, and that bourgeois America and its great cities were now "the dark heart of modern civilisation." It was an arresting thesis, especially against the backdrop of the McCarthy shenanigans. But it was immediately attacked by fellow social scientists, who disassembled its findings. By then, however, the unsubstantiated phrase "the authoritarian personality" had caught on.

After he returned to Germany, Adorno wrote three more reflective works. In 1949, in *Minima Moralia: Reflexionen aus dem beschädigten Leben* (*Minima Moralia: Reflections from Damaged Life*), he examined—again—how capitalism and marketing vulgarized experience, where the media attach almost equal weight to all events, the political being regarded as no more important than, say, the death of a soap-opera character.[43] This he saw as a form of psychological damage, so too with television and film, where sentimental music often does the thinking for the audience, so that its responses are conditioned not by the objective situation but by the way a score is manipulated. Direction, presentation, staging, on this account, are a form of coercion, of bullying even, rather than a form of enlightenment or education.[44]

However, the main center of the social sciences, for the German refugees in the United States, was not the Frankfurt Institute but, as should already be clear, the New School for Social Research in New York, where the University in Exile had become the Graduate Faculty of Political and Social Science. Alvin Johnson, one of the founders of the New School, had encountered—and been impressed by—many German scholars in the course of editing Columbia's *New International Encyclopedia* and the *Encyclopedia of the Social Sciences.* He personally raised money for the exiles and, from his encyclopedia activities, knew who was likely to need help.

Exiles began arriving in 1933 and the graduate school soon acquired its own dean, Hans Staudinger, once a distinguished German civil servant and secretary of state in the Prussian Ministry of Commerce. Two journals were conceived, *Social Research* and *Zeitschrift für Sozialwissenschaft*, the latter being published until the outbreak of World War II and indicating, says Lewis Coser, that the scholars were not too concerned to build strong bridges to their new location.[45] In a variety of fields, such as phenomenology or econometrics, the New School offered pioneering courses. A number of members of the faculty—among them Hans Speier and Gerhard Colm, who had served in government institutions such as the Office of War Information during the war—helped to conceive the postwar German currency reform that was to prove such a success.

Paul Lazarsfeld (1901–76) stood out as the man most responsible for the introduction of sociological survey research into America. Born in Vienna to a mother who was a psychoanalyst, he arrived in America in 1933 and, following developments in Austria and the outlawing of the socialist party, he extended his stay, then made it permanent. His first eye-catching study was a survey of the effects of radio on American society, and this brought him into contact with the Harvard social scientist Hadley Cantril, who offered Lazarsfeld a job as director of the Office of Radio Research at Princeton, an outfit that moved to Columbia in 1939 and evolved five years later into the now-famous Bureau of Applied Social Research.[46] This bureau institutionalized a new approach to research, studying "aggregate behavior," how people make up their minds when voting, why people don't vote, why they buy some things and not others. Overall, as Coser has pointed out, Lazarsfeld revealed gradually the latent social structure of American life, new ways of understanding how people are grouped, beyond class structure, which was to have a significant effect on market research and political focus groups.[47] Lazarsfeld's influence was felt on a generation of famous American social scientists, including Seymour Martin Lipset, Alvin Gouldner, David Riesman, and Robert Merton.

A New Stage in Industrial Civilization

Among a whole raft of economists who were influential after their arrival in the United States (Fritz Machlup, Gottfried von Haberler, Alexander Gerschenkron, Paul Baran, Karl Polanyi, Fritz Redlich), there were a handful whose names became very familiar. Ludwig von Mises (1881–1973) was known for his work in his native Austria when he arrived in 1940 and became a guest of the National Bureau of Economic Research, eventually becoming a visiting professor at the Graduate School of Business Administration at New York University. He had begun by being interested in business cycles, which stimulated in him a belief in strict laissez-faire. While Keynesian economics held the day after World War II, Mises's approach was not popular but in the 1970s, as he was joined in his views by Hayek and Milton Friedman, he was listened to more and more.

Oskar Morgenstern, born in Görlitz, in 1902, and Albert Hirschman, born in Berlin in 1915, both became important figures soon after leaving Hitler's Germany. Morgenstern was a consultant to the Rand Corporation, the Atomic Energy Commission, and the White House, and Hirschman was at first the principal assistant to Varian Fry, helping refugee intellectuals and artists escape over the Pyrenees; he later served on the Federal Reserve Board.[48] Hirschman was the author of several books, including *Private and Public Happiness* (1982), an original work on how these two are—and are not—related. But his most influential work is probably his first book, *Strategy of Economic Development*, in which he pointed out that many other economic theorists had selected one or another overriding factor as the main determinant of economic performance—be it natural resources, capital, entrepreneurship, or creative minorities. Usually, as one determinant was adopted, the others were jettisoned, and Hirschman thought it time to acknowledge that such an approach was inadequate, that monocausal explanations explained nothing. Instead, we should acknowledge that economic development depends on discovering resources and abilities that "are hidden, scattered or badly utilised."[49] He has now become among the most cited of social scientists.

Peter Drucker (1909–2005) was the best known of three German-speaking refugees who were interested above all in consumer behavior (the others were George Katona and Fritz Redlich). He taught at Bennington College before becoming professor of management at New York University and much enjoyed making management a speciality and consumer behavior the focus of rational research. His books reflect that interest: *The End of Economic Man* (1939), *The Future of Industrial Man* (1941), *The Concept of the Corporation* (1946), and *Management: Tasks, Responsibilities, Practices* (1974). Drucker's main aim was to help people adjust to the modern world by emphasizing the difference—too often not appreciated—between nineteenth-century entrepreneurial capitalism and modern, postindustrial, managerial capitalism. Lewis Coser called him "a Max Weber for managers," except that whereas Weber was gloomy about "instrumental reason," Drucker thought it was the main means to salvation in the modern world.[50] He also thought that business promotes tolerance—because blacks and women are customers too. In 1980, *The Unseen Revolution: How Pension Fund Socialism Came to America* surged into the best-seller lists as yet another good-news message that the American way combined capitalism and socialism almost without knowing it. Drucker and Lazarsfeld were two German Dr. Panglosses among the many other cultural pessimists.

* * *

Among German philosophers in America, the greatest success story is that of Rudolf Carnap (1891–1970). As already noted, the members of the Vienna Circle were among the first refugees to arrive in the United States and were well received because their attempt to put an end to metaphysics found resonance with American pragmatists such as John Dewey and Willard van Quine.[51]

Born in Barmen in northwestern Germany, Carnap came from a family of deeply religious Protestant weavers.[52] After his father died when he was still quite young, Carnap was taught by his mother before studying mathematics, philosophy, and physics at Freiburg and at Jena, where he studied under Gottlob Frege.[53] He was drafted in 1917 and stationed in Berlin where revolution broke out the following year, a development Carnap welcomed. He retained his socialist beliefs all his life and for him, and others like him, the Weimar years were exciting. His main aim, like that of the others who made up the Vienna Circle, was "the final overthrow of all metaphysical speculation, all references to transcendent entities, and their replacement by resolutely this-world empiricism, informed by the symbolic logic of Frege and Russell."[54] He was also opposed to the specifically German idea that there is a fundamental divide between the natural sciences and the *Geisteswissenschaften*, the social sciences and the humanities. Instead, he held that there are only two types of knowledge—the purely formal and the empirical. This approach led to Carnap's best-known work, *Der logische Aufbau der Welt* (*The Logical Structure of the World*; 1928), which effectively sums up the aims of the Vienna Circle. He was offered a position at the University of Vienna in 1926 by Moritz Schlick, and together they set about forming the circle.[55]

They enjoyed rapid success but, as we have seen, since many of them were Jewish, the advent of the Nazis forced them abroad. Charles Morris, at Chicago, who had spent several years at the German university at Prague, and Willard van Quine, at Harvard, sponsored Carnap. He obtained a position at Chicago, where he taught until 1952.

In America, Carnap produced *Logical Foundation of Probability* (1950), which had a big influence on Nelson Goodman and Hilary Putnam. In 1953 Carnap accepted the chair at UCLA that his friend and Berlin colleague Hans Reichenbach had held.[56] Between them, Carnap and Reichenbach did much to establish logic and linguistics as integral to American philosophy. In 1971 van Quine described Carnap as "the dominant figure in philosophy from the 1930s onward."[57]

Paul Tillich's journey from Heidegger's Germany to Union Theological Seminary in New York has already been outlined. Once in America, although it took him a while to learn English properly and to disengage himself from Germany, and though he always retained a strong German accent, he became a prolific author, achieving fame well beyond theological and philosophical circles, most of all with *The Courage to Be* (1948).[58] Many former Marxists were becoming disillusioned, especially as Marxism manifested itself in East European and Chinese Communism, and the secular world that became especially visible after World War II seemed to many devoid of meaning, even as prosperity blossomed. Tillich proposed a "spiritual cure" for "uneasy souls."[59] He personally found the non-authoritarian, even anti-authoritarian ethos of America very attractive, and the theology he offered in *The Courage to Be* was a form of religious existentialism that arose from this absence of authoritarianism—people could find God wherever they looked, it was the *looking* that counted. After he retired from the Union Theological Seminary in 1955, he became a professor at Harvard and then moved on in 1962 to Chicago where, for the last three years of his life, he was professor of theology and, as Lewis Coser says, "an American institution."[60]

The successes of Tillich did not go unnoticed by Peter Berger, who was born in Vienna in 1929 and immigrated to the United States after World War II. Berger was one of the first to notice that religion was *not* declining as the secular social scientists had predicted, and he cannily argued that in an increasingly globalized world the experience of faith was changing: it was no longer taken for granted when people were growing up; and more and more individuals searched for a personal religious preference. This was an early sighting of what became known as "expressive individualism."

The Bias in History

German history was not well established at American universities before World War II.[61] This provided opportunities for the roughly three dozen historians who found refuge in America, among them Hajo Holborn,

Hans Rosenberg, Felix Gilbert, Paul Kristeller, Hans Baron, and Ernst Kantorowicz.[62] The most important (and the most "imposing," according to Coser) was Hajo Holborn (1902–69), who taught for many years at Yale, becoming the only refugee historian to be elected president of the American Historical Association.[63]

Historians were in a special position in Germany, as already noted. Germans were responsible for the very concept of historicism which, among other things, meant that historians were taken seriously. By the time of the Nazi takeover, almost all the professors there had been trained either by or in the tradition of Sybel, Treitschke, or Droysen (see Chapter 21, p. 401), and so all were in the Prussian mold who looked back more or less fondly to Bismarck and the Wilhelmine Reich "when the professoriat had been considered an essential pillar of the Prussian and German political establishment."[64] On the other hand, the refugees were usually younger, mainly in their thirties when they emigrated, and in many cases were students of the intellectual historian Friedrich Meinecke at the University of Berlin. Meinecke was unusual in that, while he had his traditionalist side (and had signed the Manifesto of the 93 in 1914), he made an accommodation with the Weimar Republic (he famously described himself as a monarchist at heart but a republican by virtue of reason) and this set him implacably against the Nazis.

Hajo Holborn was a Berliner whose father was a well-known physicist, highly political and deeply liberal. This rubbed off on the son, an influence reinforced when he studied under Meinecke, who stimulated in Holborn a lifelong interest in the history of ideas. His first book was a study of Ulrich von Hutten, Luther's close friend, in which Holborn argued that the Reformation and the history of humanism were parallel—but separate—intellectual movements, and that the conservative strand of German thought, culminating in Bismarck, was not as directly related to Luther as it suited the conservatives to say. His subtext was that German historiography had a right-wing bias that had caused German history to be misunderstood. From Berlin, Holborn moved to Heidelberg, where he unsuccessfully attempted to resist the Nazis' interference in history teaching and historical understanding.

He was barely thirty when he arrived at Yale.[65] There he worked on two major books, *The Political Collapse of Europe* (1951), which had an impact on American foreign policy in the 1950s and 1960s, and his three-volume *A History of Modern Germany* (1959–69), for many years the standard work. Here he showed how the founding idealism of German thought had become obsolete in the contemporary world. During the war he worked in the Office of Strategic Services as chief of the research and analysis branch, responsible for scrutinizing Nazi policies and drawing up plans for postwar Germany.[66] In 1946 he became an adviser to the U.S. Department of State and wrote a book on the American military government in Germany that had a major impact on postwar political organization. Later he became an unofficial mediator between the United States and the Federal Republic. His students included Leonard Krieger and Charles McClelland.[67]

In some ways the most successful—and successfully *adjusted*—émigré historian was Fritz Stern. He wasn born in 1926 in Breslau, where his father was a doctor and an enthusiastic member of the *Bildungsbürgertum*, who numbered Fritz Haber among his friends and who became Fritz Stern's godfather. The family emigrated in 1938, relatively late, when Fritz was already twelve. He avoided the admonition from his father to become a scientist, choosing history instead. Stern's life, as a professor at Columbia, and his father's, touched many of the well-known German émigrés—the Mann brothers, the Werfels, Einstein, Marcuse, Max Wertheimer at the New School, Felix Gilbert, Hans Jonas—and he formed friendships with many artists and scholars, among them Allen Ginsberg, Lionel Trilling, Kurt Hahn, Ralf Dahrendorf, Hajo Holborn, Tim Garton-Ash, and David Landes. His students included Peter Novick (see Introduction, p. 1) and Jay Winter, whose *Sites of Memory, Sites of Mourning*, is a brilliant and moving exploration of war memorials.[68]

In his work, Stern concentrated on two themes, German history in the run-up to World War II, and American historiography, especially in regard to Europe and Germany. He researched a variety of figures, including Lagarde, Langbehn, and Moeller van den Bruck, as well as scientific figures such as Haber, Einstein, and Planck. He also devoted fourteen years to a detailed study of the relationship between Bismarck and Bleichröder. This took him back to Germany (including East Germany) in search of correspondence, and made him well known in the German corridors of power. Many of his conclusions have been incorporated into this book.

Besides the books he produced, Stern served on several American-

German committees and academic and diplomatic bodies, making him a trusted expert on German-American relations, psychological as well as policy-oriented. He took part in a number of celebrated confrontations, including the *Historikerstreit*, the Fritz Fischer controversy, Goldhagen's *Hitler's Willing Executioners*, and the role of culture in the German self-image. When Richard Holbrooke was appointed ambassador to Germany by President Bill Clinton in the mid-1990s, Holbrooke took Stern with him as an adviser. A friend and/or colleague of Henry Kissinger and Chancellors Helmut Schmidt, Helmut Kohl, and Willy Brandt, Stern attended the notorious meeting at Chequers to advise Prime Minister Margaret Thatcher on what a reunified Germany would mean. Stern has had fingers in many pies on both sides of the Atlantic, and was asked by Joschka Fischer, Germany's foreign minister, to take part in the historical commission to investigate the ex-Nazis in the postwar German foreign service.[69] He also helped Arthur Schlesinger in the Kennedy White House.

Stern agreed with Hajo Holborn that German Idealism was a crucial factor underlying everything, and that partly because of it, "The split between Germany and the West will of necessity always be an important theme for historians."[70]

Stern successfully managed to be both German and American, never completely comfortable with what he called the "European arrogance" of Hannah Arendt. In his work (which I have leaned on heavily) he did a lot to explain how the "mood" that helped create National Socialism came about, especially in regard to German elites, but he concluded that, ultimately, the horror would never be explained fully.

The "native tongue" of art history, according to one American scholar, "was German."[71] Although there is a measure of truth in this, and although the most influential art historians of the postwar world were, arguably, Erwin Panofsky in the United States, and Ernst Gombrich in Great Britain, it is not true to say that there was no art history in either country until the refugees arrived. The first chair in art history in Germany, in Göttingen in 1813, did long predate any such position in America or Britain, yet the discipline had been organized since at least the 1920s, and Panofksy himself called that period a golden age in art historical scholarship.

The Museum of Modern Art, the Institute of Fine Arts at New York

University (IFA), and the Princeton Institute for Advanced Study (IAS) had all opened their doors not too long before the refugees arrived, and all were well endowed. Among the art historians, the biggest impact, without question, was in New York where, at the Institute of Fine Arts, director Walter Cook invited several renowned academics to take up positions—Erwin Panofsky, Walter Friedländer, Max Friedlander, Richard Krautheimer, Rudolf Wittkower, Richard Ettinghausen, Karl Lehmann, Ernst Kris, and Rudolf Arnheim. As Cook liked to say, "Hitler is my best friend; he shakes the tree and I collect the apples."[72]

Panofsky was no stranger to America—he had been teaching on and off at the Institute of Fine Arts since 1931 and held joint appointments there and in Hamburg, alternating terms between Germany and the United States. When the Nazis came to power, he simply stayed in America, eventually joining the IAS at Princeton while continuing to teach at the IFA.[73] Between eighty and one hundred refugee art historians arrived in America as a result of National Socialism, and through them many colleges began to offer art history courses.[74]

The Origins of Pop Art

As for the artists themselves, Joseph Albers, Hans Hofmann, and George Grosz became teachers as well as painters and the first two, in particular, exerted a major influence in that way. Albers taught at Black Mountain College where Robert Rauschenberg was among his students. Hans Hofmann, who had a Jewish wife and was on a visit to America when Hitler came to power, simply extended his stay, teaching for a time at the Art Students League in New York before opening his own school. There he became what one historian described as the most influential art teacher of his generation, teaching—among others—Helen Frankenthaler, Alan Kaprow, Louise Nevelson, and Larry Rivers.[75] He himself became a leading member of the school of Abstract Expressionists and may well have invented "action painting," splashing pigment on canvases as early as 1938, several years before Jackson Pollock. George Grosz, arguably the most famous of the three when he arrived in America, had the unhappiest—and least successful—time. He too taught at the Art Students League before opening his own school but he seemed overly keen to adopt the

American way. He became an illustrator for popular magazines and did a stint at *Esquire*.[76]

Unlike these three, Richard Lindner, Hans Richter, and Max Ernst never intended to stay in America, an attitude that limited their engagement with their temporary home. Lindner, however, is generally credited with being the founder, or at least one of the founders, of Pop Art. He described himself as "a tourist" in America, but a friendly tourist, whose visitor status meant that he saw New York "better than anyone who was born there," which is why the everyday paraphernalia of modern American life fascinated him so much.[77]

Among the German émigré photographers, illustrators, and cartoonists there were Robert Capa, Alfred Eisenstaedt (the famous "Kiss," showing a sailor embracing a total stranger on V-E day), Philippe Halsman, Lotte Jacobi, and Andreas Feininger. Otto Bettmann founded the Bettmann Archive, a celebrated library of historical photographs. They formed an exiles' community with art dealers and publishers, people such as Karl Nierendorf, who specialized in the Expressionists but gave Louise Nevelson her first exhibition, Samuel Kootz (Hofmann and Picasso), Curt Valentin (Lipschitz, Beckmann, Henry Moore), and Hugo Perls (Chagall, Calder). Kurt and Helen Wolff published Kafka and Kraus in Munich before the war and Heinrich Mann, Erwin Panofsky, Robert Musil, and Franz Werfel afterward in New York under the Pantheon imprint. They collaborated with the Mellons, Paul and Mary, who had both been patients of Carl Jung. Their Bollingen Press was designed to introduce Jung's ideas to America, though Jung's interest in the East and in mysticism and Oriental religion meant that other titles, such as the *I Ching*, also appeared under this imprint. Schocken Books and the New American Library were also started by German émigrés, as was Aurora Press, which originally published books only in German, to be read by prisoners of war. The name was devised by Brecht, to symbolize a new dawn but also the boat that fired a shot over the tsar's palace.[78]

Lenin Overboard

Music and musical theater had crossed the Atlantic—both ways—long before 1933. American jazz had sailed east and Wagner in particular went west, to be reinterpreted all over again in the New World.[79] Of the three great theatrical figures of prewar Germany—Max Reinhardt, Erwin Piscator, and Bertolt Brecht—Reinhardt fared worst in America, his Broadway productions failing and his one film, *A Midsummer Night's Dream*, proving a commercial flop, though as Anthony Heilbut says he drew brilliant performances from James Cagney (Bottom) and Mickey Rooney (Puck). In Los Angeles Reinhardt was forced to open his own school, where William Wyler and William Dieterle taught directing and Erich Korngold taught composing.

Erwin Piscator was much more successful, though it was hard at first. When he first arrived in America, he directed the Dramatic Workshop at the New School, which eventually closed, as did two of the other ventures he was involved with, the President Theatre and the Rooftop Theatre. Despite this, the list of writers and actors who studied under Piscator is second to none—Harry Belafonte, Marlon Brando, Tony Curtis, Ben Gazzara, Walter Matthau, Arthur Miller, Rod Steiger, Tennessee Williams, and Shelley Winters. In the so-called New Drama, Piscator introduced to the American stage Sartre, Kafka's *The Trial*, and the music of Hanns Eisler.[80] Though he liked America, he was appalled by the activities of McCarthy's House Un-American Activities Committee and, after being subpoenaed to appear in 1951, returned to Germany, where he directed Rolf Hochhuth's *Der Stellvertreter* (*The Deputy*), his assault on the Vatican, followed by Heinar Kipphardt's *In der Sache J. Robert Oppenheimer* (*The Case of J. Robert Oppenheimer*), a vivid modern American tragedy.

Bertolt Brecht was in America for six years after exile in Prague, Vienna, Zurich, Russia, and Denmark. Ironically, although he had been much interested in, and influenced by, American culture, especially popular culture (such as jazz), in the 1920s and early 1930s, Brecht's time in the United States was not especially happy, though he did produce one masterpiece, *Der kaukasische Kreidekreis* (*The Caucasian Chalk Circle*; 1948).[81] His swagger had less meaning in the land of swagger, his interest in popular culture was less controversial, indeed it was the orthodoxy, and his hatred of the cult of popularity, his loathing for sentimentality—equal to Hannah Arendt's—meant that although he didn't care how he went down, his was hardly an attitude guaranteed to produce success. He thought America was the most "vital spot" on earth but also the site of the "ultimate horrors of capitalism."[82]

Brecht was as much an anarchist as a Marxist at heart, but he wasn't without his realism either. He threw his copy of Lenin overboard during the crossing to America, knowing it might interfere with his acceptance. And he journeyed to California because Feuchtwanger assured him it was easier to make a living there. He made one Hollywood picture, *Hangmen Also Die,* not a success at the box office and he never really attempted to assimilate, not believing it possible.[83] He liked the "grace and generosity" of ordinary Americans but not their lack of dignity. As soon as he could, he went back to Germany, to the East.

When Thomas Mann arrived in America in 1938, he was acclaimed as the world's greatest living novelist, invited to dine at the White House and, along with Einstein, awarded an honorary degree from Harvard. He quickly became a public figure, and it was Mann who, in November 1941, was given the distinction of broadcasting, over the BBC (from the USA), the first news about the Holocaust.[84]

To begin with he adored America, in particular the leadership of President Franklin Roosevelt, although Mann's family gave cause for despair. He had arrived with his wife, Katja, and their six children, plus his brother Heinrich and his wife. Thomas's son Klaus, Heinrich's wife, and two sisters all committed suicide during or soon after the war.

Yet Thomas managed to produce a body of work while he was in America. In 1948 he published *Doctor Faustus,* arguably his masterpiece, about the life of a German composer, modeled on Schoenberg, who wasn't entirely happy with Mann's treatment. Leverkühn, the composer, is nihilistic, a man who concludes a Faustian bargain, contracts syphilis after a visit to a brothel, and destroys his lovers. There is more than a touch of Nietzsche in Leverkühn and there are several allusions to a figure who might or might not be Hitler; high art itself comes under cynical scrutiny, as do the Frankfurt school and the twelve-tone system itself. There is a more than passing reference to a true community of artists—the redemptive community that has so obsessed Germans of all stripes.

Mann always remained a serious writer, continually upset by the glibness and what he saw as the unreflective nature of American culture and public life, in particular the American tendency to oversimplify—a dangerous stance, he felt, both fostering and deepening the Cold War that emerged after 1945. He was proud of having become a citizen of Roosevelt's

America but became appalled by the "barbarous infantilism" of American life and by the country's decision, as he put it, "not to lead the world but to buy it." In June 1952, he emigrated again, back to Europe, to Switzerland, a German-speaking country that wasn't Germany. "America was as glad to see him go as it had been pleased to receive him years before."

In one way and another, because the language barrier proved too difficult, because the social-political-intellectual climate was so different in the New World, because they rarely had the popular successes of the filmmakers, most of the émigré writers returned to Europe when they could: Brecht, Bloch, Frank, Mehring, Döblin, Mann as we have seen, Remarque, and Zuckmayer. Several of them got only as far as Switzerland, a German-speaking country but not the country they had left. It could be said that Hitler had all but broken the spirit of an entire generation of writers, or at the least had deformed them out of recognition. In that sense, the monster had won. Anthony Heilbut quotes a letter Mann wrote to a friend shortly before he left America: "We poor Germans! We are fundamentally lonely, even when we are 'famous'! No one really likes us."[85]

The same could not be said of the scientists. Among the émigrés there were nineteen Nobel Prize winners, which shows their caliber and underlines Hitler's extraordinary decision to let them—even encourage them—to go. And to an extent reinforces the sentiment of Sir Ian Jacobs, Churchill's wartime military secretary, that "the Allies won the [Second World] War because our German scientists were better than their German scientists."

The role of Leo Szilard in conceiving the notion of a self-sustaining chain reaction, which would make an atomic explosion possible, has already been introduced. After he emigrated to America, he joined with two others, Albert Einstein and Hans Bethe, in trying to convince the country's military and political authorities not to pursue the atomic (and then the nuclear) bomb to its logical conclusion. Another émigré physicist, Wolfgang Pauli, refused point blank to have anything whatsoever to do with an atomic bomb project, while a fifth, the Hungarian (but German-speaking), Edward Teller, took a diametrically opposing view, and became a celebrated "hawk."

Bethe and Teller were close friends when they arrived in America, going mountain climbing together with their wives and jointly renting a home.[86] But the bomb came between them, as it divided émigré physicists in general: John ("Jancsi") von Neumann, whose ideas had been crucial to the speed with which the atomic bomb had been produced at Los Alamos, took Teller's side, and Victor Weisskopf took Bethe's.[87] The crunch came in 1953 over the Oppenheimer affair when J. Robert Oppenheimer, once the director of America's atomic research program, was charged with disloyalty, as shown by his opposition to A-bomb research (a charge that was dismissed), and with attempting to shield a left-wing friend.[88] Teller gave evidence against Oppenheimer. In response, Bethe wrote a paper—not declassified until 1982—which argued that the delays in the Los Alamos H-bomb project resulted more from miscalculations by Teller than from Oppenheimer's political doubts.

Bethe was taken on to the President's Science Advisory Committee and well into the sixties had a voice in tempering Teller's more belligerent approach (Peter Goodchild subtitled his 2004 biography of Teller *The Real Dr. Strangelove*).[89] In America the arguments for and against the atomic and nuclear bombs have been unusually closely associated with émigré physicists.

"Das Reich der Zwei"

Fortunately, that is not all the émigré scientists have been noted for. In the postwar years a friendship formed between Albert Einstein and Kurt Gödel, both of whom were fellows at the Institute for Advanced Study at Princeton.[90] Einstein was, of course, by far the better known of the two men, but it was Gödel who, at that stage, was doing the more significant work. (Einstein used to tell visitors to the IAS that he went into the institute simply to "have the privilege of walking home" with Gödel.) The younger man had never enjoyed good health and he would, soon enough, suffer a breakdown. But for a while their walks together endured and they laughingly described their time at Princeton as "Das Reich der Zwei," the Reich of two. [91]

Gödel's new thinking about relativity imagined a breaking out of our notions of time, as if certain limits we accept in common sense are no longer true. Einstein had introduced the concept of space-time, that it was all one entity, and that it could be curved or twisted. Gödel now imagined (or rather, worked out mathematically) that if the universe were rotating, as he calculated it was (this is now called a "Gödel universe"), then space-time could become so greatly warped or curved by the distribution of matter that were a spaceship to travel through it at a certain minimum speed (which he calculated), time travel would be possible.[92]

The idea of time travel naturally catches the eye. But Gödel was not a frivolous man—far from it—and his aim was deeply philosophical, an attempt to understand time in a post-Einstein world, his fundamental point being that the world was/is a space, not a time.[93] This is clearly not an easy concept to grapple with, and for many years Gödel and his new theory were ignored. But there were signs of increasing interest at the turn of the twenty-first century (he died in 1978) as his ideas show some overlaps with string theory.

The Return—and Americanization—of *Bildung*

Looking back to the beginning of this chapter, Allan Bloom's comments about the German influence in American cultural life sounded more than a little reminiscent of a plea for a return to Bildung, a rounded, humanistic education, harking back to the Greek and Latin classics, as the best (and first) that have been thought, written, painted . . . and so on. This is not so surprising, since Bloom was himself taught by the German exile Leo Strauss. Bloom argued that Americans were now looking for fulfillment by Freudian means rather than the more traditional educational route, and this was a cause of his pessimism, for he didn't see how it could succeed.

His views, and the book based on them, created a great furor, especially in the universities, where opinion was sharply divided as to whether he had a point: that, as he said, the big issues facing mankind have not changed and that many of the "new" ideas "discovered" by the social sciences were in fact introduced a long time ago, in ancient Greece ("the best model"), and subsequently by mainly German thinkers—Hegel, Kant, Nietzsche, Weber, Husserl, and Heidegger; or, as his critics alleged, that he was out of date and that culture should now be regarded, as one critic put it, as "a kind of ethnic carnival."

The furor has rumbled on since then, never really going away, and one eventual result was a conference held at Bard College in New York in August 2002, titled "Exile, Science, and Bildung," attended by scholars from the United States, Canada, the United Kingdom, France, Germany, and Hungary. The thrust of the conference was that the debate over Bildung had entered America in 1930 with the publication of Abraham Flexner's book, *Universities—American, English, German*, and they wanted to see how it had fared. Flexner, secretary of the General Education Board, the Rockefeller Foundation's first educational philanthropy, argued that neither American nor English institutes of higher education were really any more than secondary schools, while "Germany alone, building on the historic initiatives of Wilhelm von Humboldt, knew genuine universities."[94] Bildung was the chief purpose of a university.

The aim of the Bard conference was to examine the careers, publications and friendships of a raft of German émigré scholars in America, to see to what extent they brought with them the German idea of Bildung. The figures examined in the conference included Thomas Mann, Lázsló Moholy-Nagy, Erwin Panofsky, Paul Lazarsfeld, Ernst Cassirer, Theodor Adorno, Max Horkheimer, Siegfried Kracauer, Karl Mannheim, and Paul Oskar Kristeller—many of the names considered in this and previous chapters.

The conference found that, once in America, many of the émigré scholars lost their obsession with Bildung. Whether the experience of exile was just too much, whether America was just too different, no one could say. At the same time, aspects of the Bildung culture were incorporated into American life but Americanized in the process, in three areas in particular. First, the idea of a *critical* sociology, rather than number-crunching, seems to have gathered pace as a result of the Frankfurt Institute's work in America. This has not hindered number-crunching sociologists, but postwar critical sociology in the United States did flourish as a result of what Adorno, Horkheimer, and others brought to the table.

Second, was the German challenge to the empirical tradition. Their philosophers understood that there were "impersonal forces beyond their control that governed their fate." This was not just people like Hegel and Nietzsche, but Heidegger too, whose attitude that we should "submit" to the world as it is, that we should "care" for it rather than try to control it, was an ethical stance not exactly congenial to American materialist thought, but was an attitude that flowed from the Bildung approach and would grow in importance as the postwar decades passed.

But the Bildung concept with the greatest resonance in America, and the concept that was most Americanized, was the notion of "self-realization." At the same time that Bildung self-realization entered the American vocabulary, so did Freudianism and *its* concept of individual self-realization. Psychoanalysis, the psychological approach, was, as might be expected, deeply personal and individualistic. Its moral content was confined implicitly to the doctrine that a (mentally) healthy citizen is a better citizen than an unhealthy one. It did not explore, as traditional Bildung, traditional self-realization did, what it meant to be a good citizen in a moral or political sense or what it meant to be "a cultural-ethical personality." So, yes, Bildung had arrived in America, but it was in an impoverished form.

We shall return to these matters in the Conclusion. For now we may say that the relatively small number of German émigrés who lived in exile in the United States had an influence out of proportion to their size, but that they were themselves much influenced (and in some cases defeated) by the exile experience.

It was somewhat different in Great Britain.

40.

"His Majesty's Most Loyal Enemy Aliens"

While 130,000 German émigrés settled in the United States, the equivalent figure for Britain was roughly 50,000.[1] Proportionately, this was a bigger figure, since America had four times the population of Britain and ten times the landmass. Other than that, however, it is not easy to compare the experiences of the émigrés in the two countries. Daniel Snowman, who interviewed many of the more prominent German émigrés who settled in Britain, found that they were characterized by two qualities: they had been raised in homes that were "steeped in music," learned from their mothers, and in which Bildung, emanating from their fathers, was taken very seriously.[*2]

The country they arrived in was not especially interested in German culture. Most educated Britons directed their attentions to France and, to an extent, Italy. Christopher Isherwood, who spent time in Berlin and Hamburg in the 1930s, said he was attacked by his friends who deplored his interest in Germany "and wished that I went more often to France . . . the France of Proust and the French Impressionists." German art and culture had been anathema since the First World War, though before it the likes of

* *The Hitler Emigrés*, 2002, a book as enjoyable for its rich footnotes as for its text. I have relied heavily on Mr. Snowman's account.

the composers Edward Elgar (1857–1934) and Donald Tovey (1875–1940), had felt the need to receive recognition in Germany *before* Britain.

As with those émigrés who arrived in America early, so there were some who arrived in Britain in the early 1930s. Carl Ebert and Rudolf Bing were among the first, the former the *Intendant* at the theater in Darmstadt (who resigned after an ignominious meeting with Göring), the latter the manager of the Charlottenburg Opera in Berlin. They would collaborate in helping the Glyndebourne Opera get off the ground and, in 1947, Bing went on to found the Edinburgh Festival, though he "hardly knew where Scotland was."[3] Walter Gropius was another early arrival; he had visited Britain in 1934 for an exhibition of his work at the Royal Institute of British Architects and was invited back later in the year when he began a number of collaborations, notably Impington College north of Cambridge. Rudolf Laban's Labanschule in Stuttgart transferred first to Paris, in 1937, and then on to Dartington in Devon, where the arts community of Leonard and Dorothy Elmhirst was making waves, and where Laban's former pupil, the choreographer Kurt Jooss (see below), was already ensconced.[4] Alexander Korda, Hungarian by birth but German-speaking and Berlin-trained, had left the German capital in 1926, destined for Hollywood, where he failed to thrive, and so had turned up in London where his fortunes were transformed, and he became one of the most successful film producers of all time. Another Germanophile Hungarian émigré, Emeric Pressburger, who had worked alongside—if not actually *with*—Robert Siodmak, Billy Wilder, and Carl Meyer (creator of *Caligari*), was forced to leave in the spring of 1933. In England, Korda introduced him to an aspirant director Michael Powell, and an enduring partnership was born. Ernst Gombrich, the future art historian, left Vienna for Britain in 1935, joining painters such as Oskar Kokoschka and Kurt Schwitters.

Britain had not been overwhelmed with immigrants. During the first year of the Hitler regime, about 2,000 refugees arrived, rather less than those who chose France (21,000), Poland (8,000), and Palestine (10,000). But, as the 1930s darkened, Britain—with the United States—became the preferred destination, the more so when, in 1938, the British agreed to accept "shiploads" of minors from Germany-Austria. Daniel Snowman says that the *Kindertransport* (Children's Transport), as it was known, began in December 1938 and continued until the outbreak of war the following September, during which time 10,000 young people, three-quarters of them Jewish, found sanctuary in the United Kingdom.

For academics, the Academic Assistance Council was set up (later the Society for the Protection of Science and Learning), the brainchild of Leo Szilard, William Beveridge, director of the London School of Economics, and Lionel Robbins, a colleague there. This organization, which found rooms above the Royal Society in Burlington Gardens, helped people such as Karl Mannheim, Max Born, Hans Krebs, and Rudolf Peierls.[5] By 1992, no fewer than seventy-four refugees, or children of refugees, had become Fellows of the Royal Society and a further thirty-four were Fellows of the British Academy. Sixteen had won Nobel Prizes and eighteen had been knighted.[6]

The Jewish refugees tended not to settle in the traditional East End areas of London—instead a new synagogue was established in Swiss Cottage (the Belsize Square Synagogue). Many of the refugees, once they had obtained employment, naturally tried to help their fellows. They ran into a problem collectively when, in the spring of 1940, with invasion imminent, thousands of refugees were interned on the Isle of Man. Among those interned were Max Perutz, Stephen Hearst, Hans Schidlof, Hans Keller, Kurt Jooss, Sebastian Haffner, Kurt Schwitters, and Claus Moser.[7] The only piece of good news was that the talent on the Isle of Man ran so deep that excellent courses in everything from Chinese theater to the Etruscan language were available. The exiles joked that they were "His Majesty's Most Loyal Enemy Aliens."[8]

Some of the refugees were let out after only a few weeks (Claus Moser, for example), and the boards that were set up (staffed by such figures as Ralph Vaughan Williams, who headed a committee for assessing refugee musicians) mostly looked kindly on the internees. Lucie Rie, the potter, was let out to do fire-watching, Nikolaus Pevsner helped clear bomb damage, and Hans Schidlof trained as a dental mechanic. But not all worked on menial tasks. Rudolf Peierls, Klaus Fuchs, and Joseph Rotblatt went to America to be part of the Manhattan Project, while Ernst Gombrich, Martin Esslin, and George Weidenfeld worked for the BBC monitoring services. Stephen Hearst and Charles Spencer were given the chance to use their linguistic skills, interrogating prisoners of war.[9]

Alex Korda came into his own during the war, producing anti-Nazi films, such as *The Lion Has Wings*, about the invincibility of the RAF, begun even before war was declared, and Pressburger produced *49th Parallel*.[10] Martin Miller polished his (subsequently famous) impersonations of Hitler and founded the *Laterndl*, an Austrian theater-in-exile specializing in cabaret (then hardly known in Britain) and located in Westbourne Terrace, off Notting Hill. There was also the Blue Danube Club and the FDKB, or Freier Deutscher Kulturbund, the Free German League of Culture, an umbrella organization to support writers, actors, musicians. and scientists, and whose founding members included Stefan Zweig (who was never happy in Britain), Berthold Viertel, Fred Uhlman, and Oskar Kokoschka.[11]

As people and institutions were evacuated from London, Oxford became a center of refugee life: Rudolf Bing lived there, as did the composer (and Schoenberg pupil) Egon Wellesz (who had an honorary degree from Oxford), the poet Michael Hamburger, the philosopher Ernst Cassirer, and Nicolai Rubinstein, who lectured on Renaissance history. Paul Weindling, in his study of German academics in Oxford in World War II, identified more than fifty refugees, a large proportion of whom were medical men: a Berlin-Oxford axis had existed before the war and personal contacts were good. Cambridge had its share of émigré scientists—Hermann Blaschko, Hans Krebs, Rudolf Peierls, Max Perutz; it was also home to the LSE in exile, where Friedrich von Hayek was now ensconced. Hayek was by now identified as the chief rival to John Maynard Keynes, also a Cambridge man. Despite their differences, the two men became firm friends. After the war, the LSE became a distinguished home for distinguished refugee scholars: Claus Moser, John Burgh, Ralph Milliband, Ernest Gellner, Peter Bauer, Hilde Himmelweit (Hans Eysenck's assistant), Bram Oppenheim, and Michael Zander.

One can't ignore the problems. The painters Ludwig Meidner and Kurt Schwitters admitted that they never felt fully appreciated in Britain, Elias Canetti never got over how philistine he thought Britain was (though C. V. Wedgwood translated him and Iris Murdoch hailed his brilliance).* Claus Moser told Daniel Snowman that his parents "never recovered the élan of their earlier lives," living stoically in a semi-detached

* Canetti won the Nobel Prize for Literature in 1981.

in Putney, very different from the dazzling world of Berlin in the Weimar Republic.[12]

As the fortunes of war turned, and in the wake of war, opportunities did begin to appear. Walter Goehr, a musician who had performed all manner of jobbing roles, formed his own orchestra, the London Philharmonic, and asked Richard Tauber, a conductor as well as a tenor, to take the baton, which he did to great success. Kurt Jooss was asked to produce a new version of *The Magic Flute* for the New Theatre, and he choreographed a new ballet, *Pandora*, in 1944.

The story of Ernst Gombrich underlines the fact that one of the greatest émigré influences on British cultural life (together with science and music) was in publishing. Gombrich's writing career had begun back in 1934–35, in Vienna, when Walter Neurath had asked him to prepare a history of the world for children. This led to an idea for a history of art, which appeared as *The Story of Art* in 1950, by which time author and publisher had both relocated to Britain. The book was a sensation, still in print fifty years—and 6 million copies—later, probably the most successful art book of all time. Gombrich became Slade Professor of Art at Oxford and published two other seminal books, *Art and Illusion*, about the psychology of art, and one on decoration, *The Sense of Order*. In 1959 he became director of the Warburg Institute, a post he held until he retired in 1976, by which time he had been knighted.[13]

The following list of names reinforces the point about the émigré impact on publishing: George Weidenfeld, Tom Maschler, Walter Neurath, Paul Hamlyn, Peter Owen, Andre Deutsch, Paul Elek, Robert Maxwell. Stanley Unwin was partly responsible for ensuring that the Phaidon Press moved safely to Britain after he bought all the Phaidon stock, technically "Aryanizing" the company, which became a "subsidiary" of the British firm.

Probably the biggest publishing success story is that of George Weidenfeld. After the war, Weidenfeld's first aim was to start a magazine, a combination as he saw it of the *New Yorker*, the *New Republic*, and the *New Statesman*, to be called *Contact*. He commissioned articles from Bertrand Russell, Ernst Gombrich, and Benedetto Croce, but *Contact* was not the success it might have been, and the turning point in his fortunes occurred when he was having lunch with Israel Sieff, one of the directors of

Marks & Spencer. After lunch Sieff took Weidenfeld to his firm's Marble Arch store, where he showed his guest a counter where children's classics, produced in America, were "flying off the shelves." He invited Weidenfeld to do the same and sell direct to Marks & Spencer.

Weidenfeld quickly produced a series of such familiar out-of-copyright titles as *Treasure Island*, *Black Beauty*, and *Grimm's Fairy Tales*; books took over from magazines and in 1949 Weidenfeld & Nicolson was born. Among Weidenfeld's other notable publishing coups were Nabokov's *Lolita* (after much controversy), Isaiah Berlin's *The Hedgehog and the Fox*, and the memoirs of David Ben-Gurion, Golda Meir, Abba Eban, Moshe Dayan, and Shimon Peres, not to mention several of the books covered here, such as Ernst Nolte's *Three Faces of Fascism* and Ralf Dahrendorf's *Democracy in Germany* (see Chapter 41, p. 757). His ninetieth birthday, in September 2009, hosted in Geneva for him by the internationally acclaimed architect, Lord (Norman) Foster, was attended by 300 distinguished friends, including the deputy prime minister of Israel, 10 ambassadors, assorted celebrities from the media and publishing and, in Weidenfeld's own words, "some Hapsburgs."

Almost as impressive as Weidenfeld's achievement in publishing was that of Nikolaus Pevsner. He arrived in Britain in 1936 and worked for a furniture designer. During the war he took over as (stand-in) editor of the *Architectural Review*, taught part-time at Birkbeck College at the University of London and published a book on European architecture. In 1955 he gave the Reith Lectures on the BBC and chose as his theme "The Englishness of English Art." But two other projects probably had more impact. One was the result of a conversation with Allen Lane, the man who conceived Penguin Books, for a "Pelican History of Art," a mammoth multivolume series surveying the development of art across the world; and a second series, which Pevsner researched and wrote himself, about the most important and beautiful buildings in Britain.[14] It took thirty years, but his survey is still a monument. The Neuraths were at the center of a circle that included Henry Moore, Barbara Hepworth, and Ben Nicholson.

A later achievement was that of historian Eric Hobsbawm, who grew up in Vienna, moving to Berlin in his early teens. He was attracted to communism not only as an alternative to capitalism but as an alternative to Zionism, for which he had little sympathy. In Britain he became an impressive teacher of history at Birkbeck College, helped to found what became a very influential journal, *Past and Present*, wrote a number of books about the underclasses (*Primitive Rebels*, 1959, *Labouring Men*, 1964) and a very popular, synthesizing tetralogy, *The Age of Revolution*, *The Age of Capitalism*, *The Age of Industry*, and, most recently, *The Age of Extremes*.[15]

Both Karl Popper and Friedrich von Hayek had continued their attacks on socialism and historicism begun in the 1940s. In 1959 Popper published *The Logic of Scientific Discovery*, in which he set out his view that the scientist encounters the world—nature—as a stranger, and that what sets the scientific enterprise apart from anything else is that it only entertains knowledge or experience that is capable of falsification. For Popper this is what distinguishes science from religion or metaphysics; it is the very embodiment of an "open" society. Hayek left Britain for the University of Chicago in 1950 and could have been considered just as easily in the chapter on emigrants to the United States, though later still he went back to German-speaking Europe, to Salzburg and then Freiburg. In 1960, at the height of the Cold War, he published *The Constitution of Liberty* in which he extended his argument beyond planning—the focus of his earlier book—to the moral sphere. His argument now was that our values have evolved just as our intelligence has and that the evolved rules of justice *are* liberty. The concept of "social justice," which would become so popular in the 1960s, and "the Great Society" was and is a myth. Being evolved, law is "part of the natural history of mankind"; it is coeval with society and, therefore and crucially, it antedates the emergence of the state. The imposition of "social justice" is an unwarranted (and unworkable) interference with natural processes. Neither Popper nor Hayek were cultural pessimists in the traditional German fashion, but they were recognizably Darwinian in their approach. In 1974, Hayek won the Nobel Prize for Economics and in 1984 he was made a Companion of Honour in Britain. Popper was knighted in 1965.

A publishing-literary-historical venture of a different kind was the Holocaust Library put together in London by Alfred Wiener. A Berliner by birth, he fought in World War I and won the Iron Cross. Always mindful of the threat of National Socialism, from as early as 1928 he set about documenting its activities but in 1933 he was forced to flee, first to Amsterdam, then to London. After the war he established the Wiener Library, one of the major resources in documenting the Holocaust.[16]

Leo Baeck (1873–1956) was a similar figure, even more impressive in some ways. Born in Lissa, now in Poland but then in Germany, he studied philosophy in Berlin with Wilhelm Dilthey and became a rabbi. In 1905, in response to Adolf von Harnack's *The Essence of Christianity*, he published *The Essence of Judaism*, which mixed neo-Kantianism with Jewish ideas and the success of which made him something of a hero to his fellow Jews in Germany.[17] He acted as an army rabbi during World War I and thereafter performed in one capacity or another as a guardian of the Jewish community, remaining in Germany, serving on committees and bodies designed to protect Jewish interests. Eventually, in 1943, he was deported to Theresienstadt, where he became a member of the Council of Elders; he was still there when the camp was liberated by the Russians in May 1945. After the war he transferred to London where he published a second book, *This People Israel*, which further enhanced his standing. In recognition of his role during catastrophic times, the Leo Baeck Institute for the Study of the History and Culture of German-speaking Jewry was established in 1955. He died a year later. There are now Leo Baeck centers in Melbourne and Toronto and Leo Baeck Institutes in London, Jerusalem, and New York.

Émigré journalists in Britain crowded most around the BBC and, in print, David Astor's *Observer*, which published Sebastian Haffner, Arthur Koestler, Richard Löwenthal, Ernst ("Fritz") Schumacher, and Isaac Deutscher. The other strength of the BBC, besides journalism and plays, was, of course, music, and here the impact of émigrés was as substantial as it was in science, publishing, and social/political theory. In the postwar years, up to the 1970s, three Viennese émigrés had a disproportionate effect on the output of the broadcasting corporation: Hans Keller, Martin Esslin, and Stephen Hearst.

Born in Vienna just after the Great War had ended, Keller escaped from Austria after the *Anschluss* and made his way to Britain, where he had a sister. An admirer of George Gershwin as much as Claude Debussy, and deeply influenced by Freudian psychology, he began to make a name as a music critic and was an early admirer of Benjamin Britten. In 1959 William Glock was made controller of music at the BBC, and he soon afterward recruited Keller; both men shared a love of Haydn and Mozart (the latter not universally admired at that stage) but were also intent on promoting modern and contemporary music.[18] Through them musical appreciation in Britain achieved a sophistication it had never had before. His biographer described Keller as the "musical conscience" of British broadcasting.[19]

Esslin was doing much the same in BBC drama. Born in Budapest but raised in Vienna, he received a typically German education aimed at Bildung (Latin at eleven, Greek at twelve, philosophy not long after). The woman his father married after his own mother died gave Esslin an exposure to—and a passion for—Wagner, but she also gave him a puppet theater. Indirectly, this introduced him to Hauptmann, Schnitzler, and Brecht, about whom he was to write several well-received books and on the strength of which he was taken on by the BBC, becoming head of drama in 1963.[20]

As controller of BBC 2, Stephen Hearst was in charge of Britain's most far-reaching cultural institution. An Anglophile, while at school in Vienna he had staged a performance of *The Importance of Being Earnest* in English. He fled Austria the day after the *Anschluss*, joining the BBC after Oxford and always regarded himself as lucky, in the sense that the 1960s and early 1970s were the "heroic age" of arts and culture on British television, the time of Kenneth Clark's *Civilisation* and Alistair Cooke's *America*.[21]

Still in the realm of music, we have already encountered Rudolf Bing, starting the Edinburgh Festival in 1947 (he would later go on to be director of the Metropolitan Opera in New York). In that same year, 1947, Karl Rankl was appointed musical director of the Royal Opera in London. He was not the first choice—the post had been discussed with Eugene Goossens and Bruno Walter, but they were too demanding. Though Rankl was not as well known as the others, he was not without experience, having studied with Schoenberg and Klemperer in Berlin and held positions in Graz and Prague. He escaped to Britain in 1939 and was interned. It was a while before Rankl hit his stride. During the war, the opera house in Covent Garden had been let to the Mecca Café Ltd. as a palais de danse for troops home on leave, while the ballet had toured the country to great success. Rankl built the opera company solidly, and although it did not overtake the ballet in the public mind during his tenure, it would eventually do so.

For many people, for many years, the most obvious German pres ence on the British musical scene were four men known to initiates as the "Wolf Gang" and, more formally, as the Amadeus String Quartet. Their first concert was given at the Wigmore Hall on January 10, 1948, appropriately with Mozart's D Minor Concerto, K. 421. Three members of the quartet had become friendly only in exile in Britain. Norbert Brainin and Siegmund Nissel were violinists, and Hans Schidlof was a violist; all were of a similar age, and all had been students of Max Rostal; and it was another Rostal pupil who introduced them to the fourth member of the quartet, the cellist Martin Lovett. Their first performance, as the "Brainin String Quartet," was given at Dartington, at Imogen Holst's invitation in the summer of 1947.[22] Holst was impressed by the performance, and they eventually decided to use Mozart's middle name; that first London concert, at the Wigmore Hall, was so well received that offers from the BBC were soon followed by others, including a tour of Germany itself, in 1950, followed by a contract with Deutsche Grammophon not long after.

Claus Moser originally wanted to be a pianist but it didn't work out. Born in Berlin in 1922, he moved with his family to Britain (to Putney) in 1936, where he attended the LSE. After internment on the Isle of Man, he returned to the LSE, eventually becoming professor of social statistics (1961–70). Prime Minister Harold Wilson made him director of Britain's Central Statistical Office, an institution that had earlier turned him down because he was an enemy alien. This most loyal of all enemy aliens was knighted in 1973 and made a life peer in 2001, but this wasn't all: in his time Moser has been president of the Royal Statistical Society, warden of Wadham College, Oxford, pro-vice chancellor of Oxford, chairman of the British Museum Development Trust, and president of the British Association for the Advancement of Science. He was a governor of the Royal Academy of Music, a member of the BBC Music Advisory Committee, a trustee of the London Philharmonia Orchestra, and chairman of the Royal Opera House at Covent Garden. He became, in effect, a one-man establishment.

Hardly less multitalented was Ronald Grierson. Born Rolf Hans Griessmann in Nuremberg in 1921, educated at the Lycée Pasteur in Paris, he moved to London in 1936, and went to Balliol College, Oxford, but was interned before joining the army and seeing action in, among other places, North Africa, where he was mentioned in dispatches. After the war he was assigned to the Control Commission for Germany where his most delicate task was to persuade Konrad Adenauer out of his sulky retirement (Adenauer having been dismissed as mayor of Cologne). Later in the 1940s Grierson served at the fledgling United Nations, and at the European Commission in Brussels in the 1970s. He was a director of S. G. Warburg, chairman of the General Electric Company, and in 1984 took over as chairman of the South Bank Centre, the arts complex on the south side of the Thames in central London that houses the National Theatre, the National Film Theatre, the Royal Festival Hall, and the Hayward Gallery. He found contemporary music by far the most contentious art issue he had to face: the Centre was charged by the government with presenting "challenging music," but halls were often more than half empty. He was knighted in 1990.[23]

Among the many German-born scholars who settled in Britain and made a name there we may include Max Born, George Steiner, Rudolf Wittkower, Edgar Wind, Marie Jahoda, Max Perutz, Peter Pulzer, and Richard Wollheim. Probably the most well known was Ludwig Wittgenstein, whose *Philosophical Investigations*, his second masterpiece after the *Tractatus*, was published in 1953, two years after his death from cancer at the relatively young age of sixty-two. In this book one of his main arguments was that many philosophical problems, as construed, are in fact false problems, mainly because we are misled by language. For Wittgenstein, the concept of mind was unnecessary, and we need to be very careful how we think about the "brain." It is the *person* who feels hope or disappointment, not his or her brain. Talk of "inner" and "outer" in regard to mental life is, for Wittgenstein, only metaphor.

Wittgenstein's book was part of the attack on Freud that was growing in the late 1950s and 1960s. Freud himself had died in London in 1939, soon after arriving from Vienna with his family. After Sigmund died, his daughter Anna, who had trained with her father, set up a Hampstead War Nursery, and later another clinic, to examine the effects of wartime stress (including being orphaned) on children.[24] It was now that she came into conflict with another German-speaking female child psychoanalyst, Melanie Klein.

Klein was also Jewish and had been in psychoanalysis with both Sandor Ferenczi and Karl Abraham.[25] Born in Vienna in 1882, she too was interested in children and was invited to London by Ernest Jones. Klein had a rather unsatisfactory personal life, but she did have a sensitivity toward children and was the first to observe that a way into the thinking of disturbed infants can be through play, in particular their treatment of toys.[26] This gave rise to her theory of object relations, which states that the ego settles into a characteristic way of facing the world, and that this inflexibility is the cause of many problems.

She and Anna Freud had a long-running battle about the inner lives of children.[27] Anna Freud discerned distinct developmental stages that children go through that affect the presentation of symptoms, whereas Melanie Klein saw mental life as, in general, an oscillation between depressive and manic phases.[28] The two were never reconciled, and the British Psychoanalytic Society remains formally split in its training division into Kleinian, Anna Freudian, and Independent sections.

Norbert Elias (1897–1990) was introduced earlier. In Germany, he moved in a circle that included Erich Fromm, Leo Strauss, Leo Lowenthal, and Gershom Scholem, but the real influence on his life was Karl Mannheim, whose assistant he became at the University of Frankfurt. In 1933, when Mannheim's institute was closed by the Nazis, Elias moved to Paris where he began his most well-known book, *The Civilising Process*. In 1935 he immigrated to Britain and met up with Mannheim, again becoming his assistant, this time at the LSE. By the outbreak of war, he had finished his magnum opus but was interned on the Isle of Man. The big break in his career did not come until much later, in 1969, with the republication of *The Civilising Process*.[29] This traces the development in Europe of various forms of behavior—sexual behavior, table manners, bodily functions, forms of speech, and the relations between servants and their masters. Elias used documents, memoirs, and paintings as sources to show how etiquette at court spread out, how shame and repugnance developed and widened, and how self-restraint began to be praised as an aspect of democracy. His approach—once ignored—was now welcomed as central to the way psychology and the social sciences were developing, and the book was described by Richard Sennett as "Without doubt the most important piece of historical sociology since Max Weber."[30]

Like Elias, Ernest Gellner (1925–95) taught at the LSE and at Cambridge. He grew up in what Kafka called tri-cultural Prague, where he attended the English-language grammar school, a prescient move on the part of his father, for the family moved to Britain in 1939, and Ernest won a scholarship to Balliol College, Oxford. Before taking his degree he went to serve with the 1st Czechoslovak Armoured Brigade, taking part in the siege of Dunkirk. After Balliol, he moved to the LSE, subsequently becoming professor of philosophy, logic, and scientific method.

He made his name with *Words and Things* (1959), a clever critique of Wittgenstein, Gilbert Ryle, and other linguistic philosophers who, he thought, had been sloppy about their own methods. Ryle was so incensed he refused to have the book reviewed in *Mind*, the journal he edited. Bertrand Russell wrote to the London *Times* to complain, and the row went on for weeks. Among Gellner's other books were *Plough, Sword and Book* (1988), in which he argued that there have been three great phases in history—hunting and gathering, agrarian production, and industrial production—and that these fit with the three great classes of human activity: production, coercion, and cognition. Probably his most important work after *Words and Things* was *Nations and Nationalism* (1983). Gellner had moved to Cambridge, and into social anthropology, in the 1960s, and he made it his business to study societies other than those in the West.[31] After his retirement in 1993, he returned to Prague to head up a new Centre for the Study of Nationalism, funded by George Soros as part of the new Central European University. A mountaineer and enthusiastic beer drinker, his writing style was inimitable: "Dr J. O. Wisdom once observed to me that he knew people who thought there was no philosophy after Hegel, and others who thought there was none before Wittgenstein; and he saw no reason for excluding the possibility that both were right."

Although Kokoschka and Schwitters were the most famous mature painters to seek exile in Britain, Frank Auerbach, who was only eight when war broke out, has probably become the most well-known contemporary painter among émigrés, and one whose scumbling and heavy impasto technique is closest to the German Expressionist tradition. Born in Berlin, he was sent to Britain in 1939 by his parents, both of whom died in a concentration camp.[32] In England he was sent to Bunce Court, a school for refugees in Kent run by Anna Essinger, a Quaker with a Jewish heritage (the school had itself transferred from Herrlingen in the Swabian Jura district). Auerbach was sponsored by the writer Iris Origo.

Since his parents were dead, Auerbach stayed in Britain after the war, studying under David Bomberg, and becoming known for his scenes of industrial—or at least urban—inner London. Regarded as the most exciting "British" talent since Francis Bacon, Auerbach was given an Arts Council retrospective in 1978 and a major retrospective at the Royal Academy in 2001; he represented Britain in 1986 at the Venice Biennale, where he shared the Golden Lion prize with Sigmar Polke.[33] It was reported in 2003 that he had turned down a knighthood.

While none of the German émigrés in Britain earned the worldwide fame of, say Thomas Mann, Albert Einstein, Billy Wilder, Marlene Dietrich, Hannah Arendt, or Herbert Marcuse, they seemed overall happier in Britain than their counterparts did in the United States. Fewer went back to Germany, and they integrated themselves into British life more smoothly and completely, eventually occupying the higher echelons of the traditional "establishment"—the BBC, Oxford, and Cambridge, and major cultural institutions such as Covent Garden and the British Museum. Was this because Britain, being a European country, was easier for them to understand? Was it because they had the opportunity to fight in some fashion, or were closer to the fighting, which helped them adjust? There is no American book like Helen Fry's *The King's Most Loyal Enemy Aliens*, about Germans who fought for Britain, though undoubtedly many Germans in America did a lot for the war effort. Was the experience of internment in some way cleansing, too, in the sense that, although it was unpleasant while it lasted, it was a communal experience, people could see that—from a British point of view—it wasn't entirely unreasonable and, important psychologically, that when it ended it was over? Many émigrés in Britain were there by 1940 and shared the darkest days with their hosts—did the experience of "coming through" affect their subsequent adjustment and loyalties?

We can never be sure. What *is* certain is that the émigrés were much more influential than most Britons recognize.

41.

"Divided Heaven": From Heidegger to Habermas to Ratzinger

After the guns fell silent in Europe in May 1945, Germany was a wasteland, with millions of homeless and displaced people. George Orwell was in Cologne in March 1945 and wrote that to walk through the ruined city "is to feel an actual doubt about the continuity of civilisation."[1] Walter Gropius returned to Berlin on a visit in August 1947 and found the city little more than "a corpse"; he recommended that the Americans build a new capital at Frankfurt am Main.[2] Others felt the rubble should be left *as* rubble, a "monument to the obsoleteness of the Third Reich." New housing was needed on an unprecedented scale—according to one account, some 6.5 million units had been destroyed.[3] As Wolfgang Schivelbusch says, it was not just housing that was needed but a new vision. Did the country's architects and planners re-create and restore what had been destroyed, or did they start afresh?

They did both. In some areas—Munich, Freiburg, and Münster—they reconstructed what had been lost. In others—Düsseldorf, Hamburg, Cologne, and Frankfurt am Main—they started anew. Everywhere, however, they made use—ironically enough—of Albert Speer's plans drawn up for the "Working Staff for the Reconstruction Planning of Bombed Cities."[4] Besides houses, theaters, concert halls, universities, and sports stadia all rose from the rubble, some more pleasing to the eye than others, such as the Expressionist Town Hall at Bensberg (Gottfried Böhm, 1962–67), or the Philharmonic Hall in Berlin itself (Hans Scharoun, 1956–63).[5] The most remarkable building was Mies van der Rohe's glass and steel Neue Nationalgalerie, with austere lines where the paintings were displayed underground.[6]

Amid this rubble and reconstruction, Berlin's intellectual life enjoyed a short, sharp revival immediately after the end of the war as exiles returned, people who had been hiding underground dared to show their faces, and the Allies encouraged cultural life as something that was easier to rebuild than the fabric of bricks and stone. Brecht's pithy aphorism for what went on was "Berlin: an etching of Churchill's according to an idea of Hitler's."[7]

The Kulturbund zur demokratischen Erneuerung Deutschlands (Cultural Alliance for the Democratic Revival of Germany), shortened inevitably to the Kulturbund, was licensed by the Russians and soon had a membership of 9,000, showing—if nothing else—the appetite for culture in the ruined city. At first the Kulturbund thought that Thomas Mann might become its figurehead, but he had been attacked in an open letter accusing him, in effect, of watching the war from a "comfortable" distance, and he therefore spurned all overtures. Instead, Gerhart Hauptmann, then aged seventy and living in Silesia, was approached, and he agreed to be honorary president. The Kulturbund mounted a series of concerts and lectures on contemporary music, and founded an organization to promote the sciences and humanities. But then it attracted unwelcome attention from the British authorities, who viewed it as a Russian-inspired communist organization, and it was closed.[8]

Newspaper, radio, and film initiatives had to be licensed by the occupying powers, and one of the most interesting developments here was the decision to bring back Erich Pommer, the original force behind the Universum Film AG (UFA) productions of *The Cabinet of Dr. Caligari*, *The Testament of Dr. Mabuse*, *Metropolis*, and *The Blue Angel*. He returned in July 1946 and was welcomed, in his own words, "like the coming of the Messiah." A residence was made available for him in Berlin and he was given a personal servant. Hollywood objected vociferously—German films had been their main rival up until 1933—but agreement was eventually reached, both sides accepting the fact that their chief objective now would be to combat Soviet Cold War propaganda.[9]

All of these initiatives were interrupted when, on July 24, 1948, the Soviet Union halted all road and rail traffic between Berlin and the West. This resulted in the famous airlift (*Luftbrücke*), which endured until September 30, 1949.* By then the Cold War was firmly in place, culminating in the summer of 1961 when the Berlin Wall was constructed. An important issue in postwar German culture was the process of *Vergangenheitsbewältigung*—overcoming (or coming to terms with) the past. This process was not helped by Chancellor Konrad Adenauer's policy of employing former high-ranking Nazis in positions of authority if he felt they could help administer what was to become known as the "economic miracle" of the 1950s and 1960s and/or take part in the Cold War. Arguably the most disgraceful element in this direction was the fact that *both* the German and American governments knew that Adolf Eichmann had been living in Argentina since 1952 under the name Ricardo Clement and shielded him in case he made public information he possessed about figures such as Hans Globke, the author of a commentary on Hitler's Nuremberg race laws and then a senior figure under Adenauer. The Israelis did not capture Eichmann until 1962.

In these and other ways the exigencies of the Cold War continually interfered with the process of Vergangenheitsbewältigung.

Robert Conquest has made the point that the non-Nazi world was hindered in coming to grips with "Hitlerism" by the Soviet presence at Nuremberg. "It seems anomalous that one of the states passing judgement over Nazi Germany, as an aggressor, should itself have been expelled from the League of Nations six years previously on that charge."[10] However, the first full-length book that tried to ensure that Germans *were* forced to come to terms with their past did appear much sooner than anyone anticipated. Max Weinreich's *Hitler's Professors:The Part of Scholarship in Germany's Crimes against the Jewish People* appeared in March 1946, less than a year after the end of the war. Weinreich was born in Latvia in 1893 and studied German philology at Berlin and Marburg. After completing a doctorate on Yiddish, he eventually became director of the institution that was to evolve into the YIVO Institute for Jewish Research in Vilna. Weinreich was at a conference in Brussels when Poland was annexed and,

* The success of the American-led airlift means that, among Germans, West Berliners have generally been more supportive of the United States than have their countrymen elsewhere.

with difficulty, made his way to the United States. There, when he learned that Vilna had fallen under Soviet control as part of the Nazi-Russian partition of Poland, he set about re-creating YIVO in New York. These experiences made it natural for him to focus on the scholars under Hitler who had lent their good name and imprimatur to Nazi genocidal policies. In his book, Weinreich used 2,000 wartime publications, many of them secret until that point, and roamed among another 5,000 articles from within the Third Reich to identify, for example, "large-scale experimentation" as one of these policies, exposing how the Nazis made a science of the ghetto, how they had developed their concepts of "folk" and "space," the developments in "racial science," the scientific aspects of the death factories, and many of the matters discussed in Chapter 35, p. 649, of this book, on "Nazi Scholarship." Since the Wall came down in 1989, many other scholars have added to what Weinreich initially reported, but he set the scene, and his book is today rightly regarded as a classic.

Siegfried Kracauer also spent the war years in New York, where he met up again with his Weimar Republic colleague Theodor Adorno. After his groundbreaking study of 1930, *Die Angestellten* (The Salaried Classes), covered in Chapter 31, p. 567, he had—being Jewish—moved to Paris in 1933 and then on to the United States, where he worked at the Museum of Modern Art, sponsored by Guggenheim and Rockefeller fellowships. This led to his groundbreaking book of film criticism, *From Caligari to Hitler: A Psychological History of the German Film* (1947), in which he sought—and found—parallels between film, history, and the politics of the Weimar period that to an extent explained, he thought, the advent of Hitler. Kracauer, and later Lotte Eisner in *The Haunted Screen* (1955), which examined the aesthetics of cinema in the Weimar Republic, both argued that the background threat in the Weimar films is chaos (represented in *Caligari* by the circus), in which a tyrannical figure (Caligari) is redemptive.[11] Kracauer also looked at the other main films of the period, *M*, *Metropolis,* and *The Blue Angel*, and he widened his argument that the "screen of Weimar" was a ground on which the "German catastrophe" could be understood. In particular, he saw slapstick comedy as a metaphor for flirting with power and danger in which the comedian always escapes the grip of power but by chance alone; he retains his liberty, but the threat remains. Kracauer's book also became a classic, though the chance discovery of the original script of *Caligari* and other recent initiatives of scholarship have called into question its main theme as to whether films can be said to hold quite such a straightforward link to the imagination, and to politics, as he said.[12]

The German Syndrome

In 1961 the Hamburg historian Fritz Fischer published a book dealing with Germany's aims in World War I. This was referred to in Chapter 29, p. 531, when it was explained that, according to Fischer, at an infamous "war council" in December 1912, Kaiser Wilhelm II and his military advisers "had made a decision to trigger a major war by the summer of 1914 and to use the intervening months to prepare the country for this settling of account." More than that—and this is what makes the book important in the context of this discussion—Fischer suggested lines of continuity between German aims in the two world wars. This was too much for some fellow historians. Gerhard Ritter, for example, "angrily denied" the possibility of comparisons between Bethmann Hollweg and Hitler, between German foreign policy before 1914 and in the 1930s, between Bismarck's Imperial Germany and Hitler's Third Reich.[13] Fischer played up the role of the actors in the drama at the expense of anonymous economic and social forces and in so doing set alight a debate inside Germany about its past that, until then, had received more attention from Germans in exile, mainly across the Atlantic in America.

Within Germany, birth date came to matter. Those born in or after 1929 were regarded as innocent, part of the *weisse Jahrgänge*, the "white generation." Günter Grass (born in 1927), Martin Walser (1927), and Kurt Sontheimer (1928) thus formed part of the Third Reich, however minimally, but not Jürgen Habermas, Ralf Dahrendorf, and Hans Magnus Enzensberger (all born in 1929). The sociologist Helmut Schelsky was generally held to be correct when he identified these latter as a "skeptical generation," the first perhaps to overcome the traditional German "chasm" that had pitted a realm of "pure" culture against the "shallow and sordid world of politics."[14] In many people's minds, the Third Reich—at least the 1933–42 period—was still associated with "good times" and separated from the Holocaust and its "discovery," 1941–48, which was negative and traumatic.[15]

Several other studies on topics not unrelated to the Fischer affair appeared at the time. Among them were Wilhelm Röpke's *Die deutsche Frage* (1945), Leonard Krieger's *The German Idea of Freedom* (1957), Franz Neumann's *The Democratic and the Authoritarian State* (1957), Wolfgang Mommsen's *Max Weber und die deutsche Politik* (1959), Helmuth Plessner's *Die verspätete Nation* (1959), Friedrich A. von Hayek's *The Constitution of Liberty* (1960), Fritz Stern's *The Politics of Cultural Despair* (1961), Gerhard Ritter's *Das deutsche Problem* (1962), and Hermann Eich's *The Unloved Germans* (1963). Georg Lukács's *Die Zerstörung der Vernunft* (1962), and Ralf Dahrendorf's *Gesellschaft und Demokratie in Deutschland* (1965) stand out, however.

Lukács's book, which has been translated as *The Destruction of Reason*, looked at "the path to Hitler in philosophy" and was among the first to tread what would become a well-known path, from Ludwig Gumplowicz and Houston Stewart Chamberlain to Wilhelm Dilthey, Ferdinand Tönnies, Max Weber and Oswald Spengler to Max Scheler, Martin Heidegger, Karl Jaspers, and Carl Schmitt. Lukács was among the first to remark on Germany's "delayed" status in capitalist development, on the *misère* among German intellectuals in the late nineteenth and early twentieth centuries and their cultural pessimism. He attributed this to, first, the Idealism of Kant, "which gave intuition a good name," culminating, he said, in a form of vitalism that, among other things, prevented the far more rational and scientific Marxism from taking hold in Germany, where the "class struggle" was different from that elsewhere. Philosophically, Germany embraced the Goethe-Schopenhauer-Wagner-Nietzsche route (Lukács reserved special scorn for Nietzsche) rather than the more "enriching" path of Lessing, Heine, Kant, Hegel, Feuerbach to Marx and Engels. Lukács claimed to see irrationalism taking hold in the United States after World War II and this, combined with his argument that Leninism-Marxism was a "higher intellectual stage," at a time when Stalin's Great Terror was becoming known, somewhat vitiated his arguments.

Dahrendorf's book, translated as *Society and Democracy in Germany*, on the other hand, was less polemical and had the merit of considering most of the arguments in the other titles and of using recent sociological and survey work to confirm or contradict their theses.[16]

Born in Hamburg in 1929, Dahrendorf was the son of a Social Democratic member of parliament in the Weimar Republic. His education at the University of Hamburg bridged the traditional and the modern—he read classical philology and sociology. He took his PhD in 1956 at the London School of Economics, and thereafter his career straddled the academic world and practical politics. In 1969–70 he became a member of the German parliament and was subsequently appointed a commissioner to the European Commission in Brussels. After that, he returned to the LSE as director and then became warden of St. Anthony's College, Oxford, taking British citizenship and being elevated to Britain's House of Lords. He died in 2009.

In his book, Dahrendorf set out to answer what he called "the German Question": Why is it that so few in Germany embraced the principle of liberal democracy? He went on: "There is a conception of liberty that holds that man can be free only where an experimental attitude to knowledge, the competition of social forces, and liberal political institutions are combined. This conception has never really gained a hold in Germany. Why not? That is the German Question."

He began by highlighting some important differences between Germany's industrialization and the parallel (or not so parallel) process in other countries. He noted, for instance, that German industrial enterprises tend to be much bigger than those in Britain (three times as big in terms of capitalization), and that one result of this was that, "instead of developing it, industrialisation in Germany swallowed the liberal principle." This led him to conclude that, "Contrary to the beliefs of many, the industrial revolution is not the prime mover of the modern world at all."[17] The industrial sector was so large in Germany that it formed an alliance with the state, and "there was no place in these structures for a sizeable, politically self-confident bourgeoisie."[18]

He traced the origin of the idea, first popularized by Tönnies, that the original human *Gemeinschaft* is threatened by an artificial *Gesellschaft*, and found it unlikely that "a sweet community of minds" ever existed.[19] In a section headed "The nostalgia for synthesis," he argued that the Germans harbored different attitudes toward conflict than did some other modern nations, and that "different attitudes to conflict imply different interpretations of the human condition."[20]

In a section on universities, he drew parallels between science and politics—both were open-ended and their direction could (and should) not be forced along any single path—their unexpectedness was part of their point. Central to this was the experimental attitude, which he felt at times was in Germany compromised by the idea of *Wissenschaft*, embracing scholarship in general, in which philosophical speculation was at the center of things. He felt it symptomatic of Germany that the experimental sciences eventually left the universities at the turn of the twentieth century and found their home in the Kaiser Wilhelm Societies. Instead, the "inner freedom" of scholarly inquiry triumphed in the universities, which, he said, allowed the scholar to work in peace. On the other hand, "the experimental sciences necessarily require the political freedom that permits publicity and exchange." He concluded there were two notions of science, the experimental and the German, and this to him was crucial. "Knowledge by conflict corresponds to government by conflict . . . At any given time, a lively conflict of minds provides the market of science with the best possible result of knowledge."[21] On the other hand, "knowledge in the sense of both speculation and understanding [the German way] does not require debate." This resulted, he said, in a particularly German idea of truth, not one battered out by public debate, after a set of experimental results, but instead a way to "certain knowledge" available "at least for the chosen few" (i.e., the experts).

In politics, he drew attention to Max Weber's idea of charismatic leadership—a popular idea in Germany—as, once again, a process of harmony rather than competition or conflict.[22] He noted that in Germany, the "intellectual upper class" was more highly regarded than the "economic upper class" but that it was precisely these people who had undergone "inner emigration" during the Third Reich—again an internalization of their opposition, rather than public statement or action.[23]

He found the idea that the German was unpolitical to be untrue, in the sense that the public's participation in general elections had risen steadily from around 50 percent in 1871 to around 88 percent in 1961, and a recent survey had shown that nearly two-fifths of students had a "conscious commitment to politics." But he found the other three-fifths more interesting. The survey showed that they lived very different lives from their more committed colleagues—they valued their family life, their privacy, their detachment from the public virtues. From this he concluded that, in the early 1960s, "the political socialisation of the German is incomplete . . . Democratic institutions are accepted; but they remain external, distant, ultimately irrelevant . . . The German is unpolitical because the political is unimportant for him; he is authoritarian because he would much prefer not to be drawn out of the 'freedom' of his four walls."[24]

All of which comprised for Dahrendorf the German syndrome. He further justified his analysis by referring to the suddenness with which the German voters turned to the National Socialists (2.6 percent of the vote in 1928, 43.9 percent in 1933): the syndrome produced an explosive mix, the "extremism of the centre." This, together with the failure of a counter-elite to emerge to challenge the National Socialists, was both the explanation for the rise of Hitler, and a diagnosis of the German Question. Had a proper liberal elite existed in Germany, he said, the Nazis might have been stopped.

Dahrendorf made quite a bit of universities in the development of modern scholarship, science, and the division between the private world and the public. The role of Germany's academics in its modern history was the specific subject of Fritz Ringer's 1969 book, *The Decline of the German Mandarins: The German Academic Community, 1890–1933*.[25]

Ringer (1934–2006), born in Germany, immigrated to the United States in 1947, graduated from Amherst in 1956, received his PhD from Harvard in 1961, and became a professor at the University of Pittsburgh. Part of his argument was preempted by Frederic Lilge's much shorter work, *The Abuse of Learning: The Failure of the German University* (1948). Lilge, a professor at Berkeley, argued that the flowering of German humanism under the influence of Wilhelm Humboldt and others such as Schelling, Fichte, F. A. Wolf, and Schleiermacher had been all too brief, that the idealism and strength of German scholarship had begun to falter as early as 1837 when a group of seven academics at Göttingen (the Göttingen Seven), who regarded themselves as "the conscience of the country" and who protested the abolition and alteration of the Constitution of Hanover, had been dismissed. He thought the development of science, the introduction of laboratories, and the increasing specialization they represented further sabotaged the original idea of humanistic scholarship, that "hard" scientists soon developed a contempt for Idealism and Idealists, resulting in the isolation of science from philosophy, "and it remained a powerful discord in German intellectual life for the rest of the century." This isolation was all the greater after 1870–71, when governments everywhere saw the value of science in a military context.[26] In the end, Lilge felt, research became just an occupation, making reflection difficult and turning scholarship into drudgery, which could do no more than energize the small coterie of specialists to which they appealed. Lilge believed this was one of the reasons the ideas of people like Paul de Lagarde, Julius Langbehn, and Oswald Spengler caught on—these were men outside the universities with big, consistent systems of thought that appealed to people's need for coherence, if not certainty.[27] With the successful advent of science, Bildung had been forgotten.

Lilge's short book was pithy, but its arguments were a little too neat and left out a great deal. Fritz Ringer's book, though it had a similar theme, was altogether more convincing. Building on Julien Benda's arguments in *The Treason of the Learned* (see Chapter 33, p. 611), on Lilge, and on Dahrendorf, Ringer began by emphasizing that the non-noble bureaucrat in Prussia "represented an extreme which was equalled nowhere else in Europe." Learning, in Prussia, he said, was an ideal that could function as an "honorific substitute" for nobility of birth in a context where "learning means spiritual 'cultivation.'"[28] The administrative and professional classes drew together in the nineteenth century to produce "a kind of intellectual and spiritual aristocracy," involving not only specialized knowledge but also "general cultivation," to define a distinctive elite.[29]

Ringer's main theme, however, was that this elite began to decline in importance—both socially and intellectually—after 1890 and was in crisis in the 1920s, just before the advent of Hitler. The "Mandarins," as he called the joint community of bureaucrats and professoriate, saw itself being overtaken by the new financial and entrepreneurial groups, so that a new disillusioned, displaced alliance was formed in Germany between the rentiers, professional people, academics, and the artisans and petty clerks. This was shown most clearly by the striking statistic that, in 1913, the German higher official earned seven times as much as an unskilled laborer, and in 1922 only twice as much.[30] Ringer traced the Mandarin tradition, via the Aufklärung, Pietism, the concept of Bildung ("the single most important tenet of the mandarin tradition"), the humanism of the Humboldtian university, Idealism and the historicist tradition, the difference in meaning between Wissenschaft and science, all of which underlined that a university education was intended to be "spiritually ennobling rather than a narrowly utilitarian influence."[31]

Gradually, however, the cultivated elite began to assume a more defensive position, which made them more and more conservative. They became more concerned to defend German cultural traditions, especially as World War I approached, "feeling that a counterweight to the English was needed in this field."[32] They felt that national greatness came through cultural creativity and "could see no point in material prosperity, if it interfered with these objectives, if it did not create the preconditions for the fullest possible self-development of the individual."[33] There was a persistent interest among the mandarins in a strong presidency at the head of the Republic, because even after World War I—indeed, *especially* after it—they looked for a leader who would return Germany to "a natural aristocracy based on culture and capability, intellect and spirit," to counter the "shallowness" of materialist, interest politics.[34]

The consequence of these forces was that the German universities, especially in the 1920s, "became strongholds of right-wing opposition to the new regime." In particular, anti-modernity and anti-Semitism joined forces.[35] There was a return in the 1920s to the attack on specialization and empiricism and a search for synthesis, unity in scholarship, and a form of thinking that was neither Marxist nor socialist but "German socialist," in effect "a metaphysics of reaction" with vague new concepts such as *Volk* and *Reich*, in which one important element was "voluntary submission to the community." The German university professors, said Ringer, felt themselves involved in a genuine tragedy in which "Geist and its representatives had lost control of society."[36] No one knew, he said, how this division between Geist and politics had come about, but it generated a kind of self-pity among the mandarins, which often turned to hysteria and sometimes to hate. Ringer thought that intellectuals in France and elsewhere also agonized over these problems, but that the general anxiety was at its most intense in Germany. One of the problems was that so many of the mandarins—Scheler, Meinecke, and Spranger, for example—felt that only a small minority, an elite, was capable of benefiting from and expressing the great tradition. As Karl Jaspers put it, "all standards had been sacrificed in an effort to accommodate a mass of mediocre minds."[37]

These efforts at bringing knowledge together in the search for meaning, for overall coherence, became known as the "Synthesis Movement."[38]

The result, Ringer affirms, was a certain uncomfortable form of anti-intellectualism in Weimar Germany, at least in some quarters, and this allowed the National Socialists to profit. Because they had lost the economic fight, and then the fight for the hearts and minds of the general population, the mandarins put up little resistance to the Nazis. German Idealism had been weakened by the onslaught of materialism and positivism and specialization, and purely technological thinking "had destroyed the link between knowledge and cultivation." Ringer concluded that "Hitler's hordes" had only completed (and accelerated) something that was happening anyway.

Norbert Elias, in a series of publications set out over the decades but centered on 1969, argued that the inherent conflict in Germany concerned the *satisfaktionfähige Gesellschaft*, a term difficult to translate but referring to a society oriented around a code of honor in which dueling and the demanding and giving of "satisfaction" took pride of place. The effect of this, he says, was to brutalize a large section of the middle class, setting them against the Bildungsbürgertum and converting them from a humanist orientation to a nationalist orientation. They formed the bulk of the World War I officers who established the paramilitary Freikorps in the Weimar Republic, producing an insistent background noise of violence which helped to destabilize the 1920s.

The long-term effects of this, he felt, were manifold: their very existence, and the role they played, made a positive self-image of middle-class Germans very difficult to maintain; with their greater concern for honor rather than for morals, conscience formation in Germany was weaker than among her neighbors; there was a more pronounced gap among Germans than among other Europeans between ideals and identity, making them more prone (as Ringer and others had said) to self-pity. Later, partly because of this, they tolerated the unrealistic plans and policies of National Socialists more than would otherwise have been the case.[39]

Theodor Adorno, as we have seen, was as much concerned with the shortcomings of America as of Germany, though he did become a major force in political thinking in Germany after he returned there. He, and people like Karl Löwith (see p. 772), Franz Neumann, and Arnold Bergstraesser were distinguished by the term "rémigré" or "rémigrant."[40] But other major figures from before the war were still alive and still involving themselves in the broader areas of philosophy, the humanities, and social criticism in regard to Germany and the modern condition.

Having a Jewish wife, Karl Jaspers had not fared well under the Nazis. In September 1937 he was dismissed from his post, his attempts to move to Oxford, Paris, and Basel all fell through, and in 1943 he was banned from publishing anything at all. Matters looked up after the war ended, and he figured prominently on the Allies "White List" of people untainted by links to the Nazis. He was made one of the professors charged with reopening the University of Heidelberg (described by one American observer as "once famous, now notorious," because it was "still infested" with Nazis) and now began his most creative period of writing, not just in philosophy but also in politics.[41] Jaspers laid great store by civic morality, arguing that a liberal humanistic education was the best means of disseminating democratic ideas throughout Germany. He remained firmly against the rehabilitation of professors who had a history of Nazi affiliation, and his writings and radio broadcasts became a major force at the time.[42]

To an extent, therefore, Jaspers adopted a classic British or French view of political liberty.[43] In theory he wanted to see this model imported into his native Germany but he must have had doubts that this would happen, because in 1948 he accepted a professorship at the University of Basel and became a Swiss national, saying "he felt he was breathing again for the first time in fifteen years."[44] That year too he published *Der philosophische Glaube angesichts der Offenbarung* (translated as *Philosophical Faith*), a complicated but influential work in which he claimed that the "evidence" produced by faith (through revelation, say) is always likely to be paradoxical and uncertain. Therefore, dogmatism is unconvincing in religious belief, and a critical form of philosophy can be an important help in adding to what religions have to say; Jaspers saw it as one of philosophy's main aims to update and relate theology to present circumstances. This led him into confrontation with both Karl Barth and, more especially, Rudolf Bultmann.[45] Jaspers also returned to Marx's point about the role of the educated bourgeois elite. Marx had criticized this elite for making culture its refuge at the expense of politics. Jaspers argued that societies where the educated bourgeois elite see their role undermined will always be "inherently unstable" and that this segment of the body politic has a primary role to play in upholding democratic culture.

Jaspers's erstwhile friend Martin Heidegger had been through a much rougher time since the end of the war. In 1946 he had been banned from teaching by the Allies (this lasted until 1949), and his two sons were still in Russian captivity.[46] One bright spot for him was Hannah Arendt, who visited Heidegger in 1950 and again two years later. She eventually found it in her heart to forgive him and, from 1967 on, they met every year until his death in 1976.[47]

Heidegger's postwar career in philosophy embraced three subjects—humanism, the nature of thinking, and the issue of technology.[48] There was also a very public exchange with Theodor Adorno, who criticized Heidegger in no uncertain terms.

In a 1946 essay, *Über den Humanismus* (*On Humanism*), Heidegger was again the unyielding—and unrepentant—critic of "reason" and "modernity," still in the name of "Being" and "poesis," a stance that dovetailed well with the emerging postwar climate of opinion in the West, especially America, where many people were disillusioned with the modern world, "with its attendant horrors and catastrophes." As often as not, Heidegger's views were broadcast by French followers, such as Sartre and Jacques Derrida.[49] This approach formed a major plank in the postwar development of postmodernism.

The Bavarian Academy of Fine Arts invited Heidegger to give lectures there from the early 1950s and the lecture of 1953 became famous. His subject was "Die Frage nach der Technik" (The Question of Technology), and the hall was packed with Munich's intelligentsia—Werner Heisenberg, Ernst Jünger, and José Ortega y Gasset. Rüdiger Safranski says it was probably Heidegger's greatest success in postwar Germany, and he was given a standing ovation. By the time of this lecture, as again Safranski points out, there was widespread anxiety about the threat of a technological society, not only in Germany but especially so there.[50] Alfred Weber's *Der dritte oder der vierte Mensch* (*The Third or Fourth Man*) came out in the same year with a horrific vision of a future robotic world, and Friedrich Georg Jünger (Ernst's brother) had released *Die Perfektion der Technik* (*Perfection of Technology*), in which he argued that technology had already changed mankind, that technological man was locked in an irreversible exploitation of the earth that would eventually destroy us. Shortly afterward, Günther Anders published *Die Antiquiertheit des Menschen* (*The Obsoleteness of Man*), in which he too argued that technology had to be deliberately curbed or it would destroy us.

Heidegger saw technology as a vicious circle: technology breeds more technology, it "challenges" nature, and people live in a *Gestell*, or "frame," of technology. In doing so, we lose elements of freedom. With technology so rampant and so ever-present, the original experience of Being, says Heidegger, is lost. We cannot let nature "be," we are less able to submit, to surrender to that experience of being; the "releasement towards things" is simply unavailable in a technological society: the poetic experience of the world is sidelined and overwhelmed by technology.

This was reinforced by Heidegger's views on America. The United States had often been the object of German thought. For Heine, America was the symbol of all that Romanticism detested. After a visit across the Atlantic, Nikolaus Lenau, sometimes called the German Byron, described the country as disfigured by its politics, with its culture imposed from outside. Nietzsche expected America to spread a spiritual emptiness (*Geistlosigkeit*) over Europe and neither Moeller van den Bruck nor Spengler cared much for it, though Ernst Jünger admired America's ability to involve all the country in World War I.[51] As we have seen, Freud thought America "a mistake" (whatever that might mean). For Heidegger, America was the symbol of the crisis of our age, "which is also the deepest crisis of all time." It represents the greatest alienation of man, his profoundest loss of "authenticity," and it was the supreme impediment to spiritual reawakening. America reduced everything to its lowest common denominator, all experience to routine—all was trivialized and rendered bland. Americans, said Heidegger, were "totally oblivious" to "man's encounter with Being."[52] After the first space probes, Heidegger wrote that "there is no longer either 'earth' or 'heaven,' in the sense of poetic dwelling of man on this earth." The age of technology is our fate and America the home of this "catastrophe."[53]

Heidegger had written about America before the war, and essentially his arguments hadn't changed, only been updated. And this was partly Theodor Adorno's point in his celebrated attack on Heidegger, published as a pamphlet in the mid-1960s and titled *Jargon der Eigentlichkeit* (*The Jargon of Authenticity*). There was, it has to be said, a wider context to

Adorno's uncompromising stance. The developing Cold War in the 1950s had undoubtedly played into the hands of former Nazis. Chancellor Adenauer was keen to have the distinction between "politically impeccable" and the "not-so-impeccable" abolished, and in 1951 a law was passed that allowed "compromised" people to again hold public office, while the Loyalty Law of 1952 helped ensure that some people persecuted under the Nazi regime were now removed from public posts on suspicion of being Communists. Adorno and Horkheimer again became the subject of anti-Semitism.

The Jargon of Authenticity was important because Adorno felt that the whole idea of the authentic in Heidegger—the rural allusions, the emphasis on the *Volk*, the hatred of modernity as artificial—was phony, a form of "sacred gibberish . . . devoid of content . . . except self-idolisation."[54] When people use words like "authentic," Adorno said, they make them sound as if they meant something "higher" than what they had actually said, and Heidegger was especially guilty of this.[55]

Heidegger never replied to Adorno. The latter had scored some telling points, but they hardly affected his target's long-term reputation. That had more to do with what Richard Wolin has called *Heidegger's Children*. Of these, we have already met Hannah Arendt, Hans Jonas, Herbert Marcuse, and Leo Strauss, all of whom stayed in the United States after their emigration. The other two were Karl Löwith, who returned to Germany in 1952, and Hans-Georg Gadamer, who never left it.

After his return, Löwith was made a professor at Heidelberg and became chiefly known for three highly original works, *From Hegel to Nietzsche*, an account of the fragmentation of German philosophy, *Meaning in History*, about the relationship between modern philosophy and its theological predecessors, and *Max Weber and Karl Marx*, describing the emergence of sociology. Throughout his works, Löwith argued that the twentieth-century disasters were originally shaped in the middle of the nineteenth century "as the educated elite decisively turned their backs on the classicism of Goethe and Hegel. Increasingly, they grew impatient with values that were 'timeless' or that transcended the finitude of human temporal existence. Nature and the heavens ceased to be the touchstone for value and meaning, instead, 'man' became the measure." In Löwith's view, Europe's descent into nihilism culminated in there being "no constraints" upon the "sovereignty of the human will." Nietzsche's "will to power" was for Löwith "an amoral excess." Instead of Marx or Nietzsche, he preferred Heidegger and his advocacy of Stoicism, "acquiescence to fate."[56]

Löwith studied with Husserl before Heidegger, under whom he examined the role of intersubjectivity in the formation of the self. In his dissertation he argued that the "I" is primarily formed and shaped by a world of human intimacy, what he called "the co-world." According to Löwith, interpreting Heidegger, "Human beings are not 'rational animals' but instead ecstatic 'shepherds of being.'" Scientific thinking, seeking to control the world, is a decline from this original feeling of ecstasy.[57]

The career—and thought—of Hans-Georg Gadamer (1900–2002) was very different from that of the rest of Heidegger's children. Born in Marburg and the son of a pharmacology professor, Gadamer studied at Breslau before returning to Marburg after World War I. There his early teachers were Paul Natorp and Nicolai Hartmann, but it was Heidegger who exerted the most influence and, for a time, Gadamer worked as Heidegger's assistant.[58] "I always had the damned feeling that Heidegger was looking over my shoulder," he said later.[59]

During the 1930s and 1940s, Gadamer accommodated himself, first to National Socialism and then, briefly, to Communism. He was never a member of the NSDAP, and he seems to have kept his head down, though he was later criticized for being "too acquiescent." At the end of the war he received an appointment at Leipzig and, having been found untainted by Nazism by the American occupation forces, was made rector of the university. But Communist East Germany was not to his liking and he left, eventually succeeding Karl Jaspers in Heidelberg in 1949. To an extent, while there he tried to aid the rehabilitation of Heidegger. In 1953, with Helmut Kuhn, he founded *Philosophische Rundschau*, a highly influential journal, though Gadamer did not become known outside his own professional circle until the publication in 1960 of *Wahrheit und Methode* (*Truth and Method*). This established him in the eyes of many as one of the most important thinkers of the twentieth century.[60]

One of his starting points in *Truth and Method* was a series of lectures Heidegger gave in 1936 (not published until 1950), titled "Der Ursprung des Kunstwerks" (The Origin of the Work of Art). Here Heidegger introduced his concept of the "event" of truth, the "unconcealment" of truth, an idea that contrasts with the notion of truth as "correctness," usually taken to mean some sort of correspondence between a statement and the world. Works of art have a coherence and within that coherence "a" truth is revealed, stemming from disclosure, but it is disclosure that is an interpretation, which can never be total or truly objective. We play a part in whatever we choose to understand as a truth. The influence of Kant is clear.

Gadamer took these ideas much further in *Truth and Method.* He said that our involvement in the event of truth is always based on our prejudices, "anticipatory structures" that we have within us and that allow, or determine, that our understanding of any truth event will be grasped in a certain way, together with "the anticipation of completeness," another neo-Kantian notion that involves the presupposition "that what is to be understood constitutes something that is understandable, that is, something that is constituted as a coherent, and therefore meaningful, whole."[61] At the same time, history also plays a part in our understanding, says Gadamer. We are "embedded" in our particular history and cannot escape its effects. Understanding also needs another to be certain it is not mere subjectivism; new meaning emerges not by access to some "inner realm" but by the "fusion of horizons" (*Horizontverschmelzung*).[62]

Gadamer therefore concluded that the humanities, the *Geisteswissenschaften*, could never achieve the methodological footing of the "sciences of nature," that such an attempt was misguided. He even thought that the natural sciences claimed too much for their method, that understanding was an ongoing process with no final completion, a stance that marked Gadamer as in the same mold as the later Wittgenstein and Thomas Kuhn. Eventually, he arrived at a model for understanding that said it was like a "conversation": it takes place in language and each brings his or her understanding to the conversation or negotiation.

A final aspect was his exploration of culture, in particular "the relevance of the beautiful," in which he considered "art as play, symbol and festival."[63] He thought that the meaning, or role, or function of art often got lost in the modern world, and that play—the activity of disinterested pleasure—was also overlooked.[64] The symbolic role of art was to open up for us "a space in which both the world, and our own place in the world, is brought to light as a single but inexhaustibly rich totality," where we can "dwell" out of ordinary time. The disinterested pleasure we take in art is an aid to escaping ordinary time and moving into "autonomous time." The final quality of the successful artwork, as festival, also takes us out of ordinary time and opens us up "to the true possibility of community."[65]

Gadamer engaged in two famous debates, with Jacques Derrida and with Jürgen Habermas, on whether we can ever transcend history and how this affects criticism of contemporary society, whether such criticism can ever be truly objective (and therefore what validity it can have). The debate with Derrida was inconclusive, but as a result of the other debate, Gadamer and Habermas became good friends, and the former helped secure the latter's appointment to a professorship at Heidelberg.

The Achievements of Reflection

Habermas, however, was much more interested in politics than Gadamer, and more critical of his own country. Born in 1929 in Gummersbach, Habermas was the son of the chairman of the Cologne Chamber of Industry and the grandson of a pastor.[66] The Nuremberg trials had a big effect on the teenage Habermas, and he became especially critical of his own country, in particular its scholars. He studied philosophy at Göttingen from 1949 to 1954 and was alarmed to observe that most of the professors made no allowance in their teaching for the events of 1933–45. Accordingly, he first put pen to paper in a critique of Heidegger and his failure to repudiate the ideas of Hitler. An interest in Marxism led Habermas to both Lukács's *Geschichte und Klassenbewusstsein* (*History and Class Consciousness*) and Horkheimer and Adorno's *Dialektik der Aufklärung* (*Dialectic of Enlightenment*), his first encounter with the critical school with which he was himself to be so much identified. He taught at Heidelberg before taking a chair in philosophy and sociology at Frankfurt in 1964; in 1971 he took up a position at the Max Planck Institute in Starnberg, near Munich. During those years he became well known internationally as a theorist of the student protest movement.

Habermas's written output has been prodigious—politics, philosophy, the evolution of society, the role of religion and the social sciences in modern life, Freud, even the role of child psychology in civic life. But his most innovative and enduring contributions have been in the realm of critical theory and "communicative action."

The aim of critical theory, for Habermas, is to facilitate the understanding of communicative action—the way different aspects of society link with each other, often in unconscious and unintended ways, so as to enable cultural evolution, a key idea.[67] He began, in *Theorie und Praxis* (*Theory and Practice*), with the observation of four historical developments that rendered Marxism obsolete. The most important of these is that the state is no longer separated from the economy as it was in the days of laissez-faire capitalism, but plays a crucial role in regulation and enablement, meaning that the functioning of the state now requires careful critical attention. A second crucial observation is that rising standards of living in advanced societies have changed modes of oppression in ways not foreseen by Marx but still not appreciated by those undergoing that oppression. The new constraints are psychological and ethical rather than economic and it is in this realm that his theory of communicative action in particular applies.[68] More especially, Habermas thinks that the advent of the welfare state in capitalist societies renders true human emancipation far more difficult, that science and technology condition the way we think without our—for the most part—knowing it.

Habermas lies in the tradition of the Frankfurt school in trying to provide a marriage of Marx and Freud. In his case, he sees the Freudian method of psychoanalysis as not just a favored method but as a metaphor for what he would like to see in the wider society, where deep reflection reveals the many hidden constraints acting (usually unconsciously) on individuals, leading to self-insight and emancipation. These, his considered reflections about reflection, are the subject of *Erkenntnis und Interesse* (*Knowledge and Human Interests*). Only in such a fashion, he says, can we rediscover "ways to live together in harmony and mutual dependence, while respecting individuals' autonomy, but without sacrificing the advances of modern technology."[69] Habermas has never been anti-science as so many postmodernists have been. For him, we have to find ways of "sustaining [a] moral community in the face of rampant individualism."[70] This depends on people's being able to communicate effectively with each other.

Habermas believes that in crucial ways the modern period differs from all previous periods. In particular the concept of reason has been distorted by the advance of science. The agenda of the Enlightenment philosophes was to develop critiques that would assess and criticize the prevailing assumptions of an age; empirically grounded and leading to greater freedoms, these critiques amounted to forms of reflection that expanded human self-awareness. Science, however—and here he agreed with Weber—offered instrumental reason, reason as a way of controlling and manipulating nature.[71] Traditional scholarship, on the other hand, he defined as human emancipation through enhanced capacities for reflection, and this is where the cultural sciences came in, to make us more aware of the achievements of reflection.[72]

Living without Consolation

We have today, therefore, a very different and more pervasive form of "false consciousness" from that which Marx introduced: we are living in a thoroughly distorted version of reality or, as Habermas puts it, "systematically distorted communication." In fact, this is now the accepted state of affairs, in which we all know, at some level, that facts and values "cannot be accepted uncritically as 'givens,'" nothing we are told can be accepted at face value: late capitalism thrives on marketing and public relations, so that we are surrounded in the mass media by acts of communication that say one thing and mean another—not completely another, but with an agenda of their own, unspoken but present.[73]

Habermas argues that the solution is "an ideal speech community," in which politics is taken out of the hands of the "experts" and some sort of "public sphere" is created in which a consensus can emerge based on mutual concerns.[74] The natural home for this might be the university (though Habermas also considered consciousness-raising groups), but so far, it is fair to say, such a mechanism has not emerged. The contemporary university, says Habermas, has reverted more to the eighteenth-century idea of teaching institutions than homes of critical reflection.

On top of which, he says, all the scientific progress in the world has done little to advance our understanding of suffering, grief, loneliness, and guilt, the traditional concerns of religion. Having destroyed the basis of faith, the sciences have done nothing to provide a replacement and we must "resign ourselves to living without consolation."

The "Caesura" of 1968

Habermas was a prominent figure in the impressive survey published by Konrad Jarausch in 2006, with the title, in English, *After Hitler: Recivilising Germans, 1945–1995.* This book was published in German as *Die Umkehr: Deutsche Wandlungen, 1945–1995* (literally, The Turning Back: German Transformations, 1945–1995).[75] Both titles were controversial in their different ways.

Jarausch, Lurcy Professor of European Civilization at the University of North Carolina at Chapel Hill, and also director of the Center for Research in Contemporary History at Potsdam in Germany, identified three well-defined periods as of crucial importance for the post–World War II history of Germany. The first was the immediate postwar period, which saw the disarmament and demilitarization of Germany, the dissolution of Nazi institutions and the prohibition of Nazi propaganda, together with the decentralization of the economy to eliminate the potential for war.[76] He examined what this involved (80,000 Nazi leaders arrested, 70,000 Nazi activists dismissed, 3,000 German companies dismantled), together with the population's gradual acceptance of their "partly active, partly passive" participation in the genocide of the Jews, the retreat from nationalism ("the collapse of the nation as a reference point") and the origin of the idea of a "postnation nation."[77] He observed that "radical nationalism" was more "deeply anchored" in German culture than National Socialism, and that the privations people suffered in that period (a time when "all bellies disappeared") provoked feelings of self-pity "for their newfound role as victims" that helped transform "the formerly aggressive nationalism" into a defensive, "residual sense of nationality . . . Though German identity had been badly damaged by the crimes of the Nazis, it did not disappear entirely but, rather, transformed its character into a 'community of fate.' "[78]

He observed a remarkable period of economic growth throughout the 1950s (an average annual increase, thanks to certain Keynesian measures, of 8.2 percent). However, it wasn't until the 1960s, after the "relative stabilization" of the Adenauer period, and the Americanization of values and behavior, owing to prolonged occupation and "some intelligent exchange schemes," that the breakthrough to a modern civil society occurred.[79]

Jarausch then identified 1968—the year of the student revolts in Poland, Berlin, New York, and Paris, of the Soviet invasion of Czechoslovakia, and the beginning of the end of the war in Vietnam—as a caesura "that requires a cultural approach [as opposed to a political one] to be better understood." Jan-Werner Müller agreed, arguing that this time brought about "new structures of feeling," and Dirk van Laak even went so far as to assert that the 1960s were a threshold of change as much as the 1920s were.[80] A motivating force here was the young generation that had grown up since the war and was more willing—much more willing—to examine the involvement of their parents' cohort in National Socialism than were the parents themselves. The particular circumstances of Germany, therefore, sharpened this generational divide and had important cultural consequences. In particular, Jarausch identified the emergence of a "critical public sphere" and a new professional ethos "that favoured contemporary criticism over approval of government policies." A critical discourse emerged in the 1960s, he says, that advocated a broader social self-determination. Habermas, in *Der Strukturwandel der Öffentlichkeit* (*The Structural Transformation of the Public Sphere*), argued that public discussion was "a crucial precondition for civic freedom." This was, perhaps, old news in other Western countries but, says Jarausch, in Germany older authoritarian thinking was still widespread, and many people were still reluctant to engage in politics. But the events of 1968, he insisted, and the emergence of a critical public sphere, marked the internalization of democratic values and behavior, at least among the educated middle classes. Habermas agreed: he called the movement of the "68ers" the "first fairly successful German revolution, while Elias described it as an important break in the 'chain of generations' and the final stage in German 'catching up' with the West." (It has to be said that many older Germans reject this picture, insisting that there *were* early exposés of Nazi wrongdoing, which they faced head-on. One book often cited is Eugen Kogon's *Der SS-Staat: Das System der deutschen Konzentrationslager*, published in 1946.)[81]

Protest, Jarausch says, was centered on the "inner emptiness" of consumer society, and Herbert Marcuse's book, *One-Dimensional Man*, with its concept of "repressive tolerance" became a key text.[82] The at times violent confrontation took a full decade to subside, culminating in the

"German autumn" of 1977, with the murder of the president of the League of Employers, the liberation of a hijacked plane in Mogadishu, and the "controversial suicides" of Ulrike Meinhof, on May 8, 1976, and Andreas Baader, on October 18, 1977. The strategy of confrontation had failed, says Jarausch, and most of the confrontationists, including Joschka Fischer, "found their way back to the constitutional state." But that could not disguise the fact that German society had changed fundamentally: even though the power structure of Germany had not been changed by the events of 1968 and the decade afterward, there had been an "anti-authoritarian transformation of values." There was also the beginning of a change from the concept of German identity as a negative one—as shown in the works of Günter Grass, Heinrich Böll, Rolf Hochhuth, and Peter Weiss (see Chapter 42, p. 789)—and a turn to greater internationalism (or, more accurately, *non*-nationalism; this generation was no longer ashamed to be German, but it was not necessarily international in outlook either).[83] Nevertheless, as Klaus Schönhoven put it, "there was more air to breathe."

Jarausch's analysis is important for its emphasis on the different views of themselves that Germans have as compared with what most of their near neighbors understand about them. It is not the whole picture, however. There was in Germany, according to one researcher, "historical illiteracy on a staggering scale." In one survey, published in 1977 and titled *What I Have Heard about Adolf Hitler*, children described him variously as Swiss, Dutch, or Italian; he was a professor, a leader of the East German Communist Party; he had lived in the seventeenth—or the nineteenth—century. Rolf Hochhuth's *Die Juristen* (*The Lawyers*), premiered in 1978, concerned a real-life attorney who was *still* denying his war crimes.[84] This was the time when Chancellor Helmut Kohl used the phrase "the blessings of late birth" to describe a generation that could have had no role in the Nazi evil. Though this had an element of truth, it was also less than the whole truth, given the historical illiteracy already displayed.*[85]

Jarausch's third period in Germany's transformation centers around Unification Day (Der Tag der deutschen Einheit), October 3, 1990,

* The German phrase, "Gnade der späten Geburt," was originally coined by Günter Gaus and had an easier reception.

though he spent some time examining the change of thinking in East Germany in the 1980s as a precursor to investigating whether "a middle course of democratic patriotism" would now be possible in what the theologian Richard Schröder called the "difficult fatherland." Jarausch thought that there was, even then, "a continuing weakness in the newly emerging structures of [German] civil society."[86] Despite everything, Jarausch found that critical minorities had gradually developed in the GDR, and that there had been "a retreat into private life," a *dacha* culture, the cultivation of a conscious double life "that meant conformity in public and defiance in the private sphere." This was in some ways a disturbing parallel to the "inner emigration" of the Nazi years but, says Jarausch, it contributed to the stability of the East German state because it diverted the dissatisfaction with the regime inward.[87] Nevertheless, in a small way the idea of a "negotiation society" was established.

A "Normal" Germany

The most important psychological/intellectual change provoked in Germany by reunification, when in the words of Rolf Hochhuth "the German clock struck unity" (and which was feared by some, such as Günter Grass and Margaret Thatcher, who believed there might be a return to aggressive German nationalism), has been the search "for a post-national self-understanding conditioned by the Holocaust," a search for the extent to which Germany is now—and can ever be—"normal."[88]

What did that entail? The changes in the East in the early 1990s were breathtaking although for a time many people retained "a wall in the head." There remained a great divide between "Ossis" and "Wessis," and the exposure of the widespread Stasi collaboration produced a depression in many Easterners. Later, though, in the latter half of 2000, there emerged an "Ostalgie", a nostalgia for the East among its former inhabitants, in particular a nostalgia for once-familiar products and brands, for Florena soap, Konet Foods, and even the Ampelmännchen, the chirpy man who was East Germany's green traffic light symbol which, Martin Blum says, have now been elevated to cult status among the young. These products, where they can be found, are not consumed as such, but left in living rooms, intact in their packaging, to serve as a challenge to the supposed superiority of Western consumer culture.[89] More formally, the Dokumentationszentrum Alltagskultur der DDR (Dok) in Eisenhüttenstadt maintains changing exhibitions on the material culture of the German Democratic Republic.

In these debates, four people have stood out: Karl Heinz Bohrer, Hans Jürgen Syberberg, Botho Strauss, and Martin Walser. Bohrer, professor of German literature at the University of Bielefeld, a journalist, and the editor of *Merkur*, the "German Journal for European Thought," argued that reunification was necessary so that the two "partial nations" could come together to "remember together" and establish a common memory; he wrote that reunification, involving reconciliation with each other, could become a reconciliation with the past, so that Germany's "soul" could find peace and the nation as a cultural phenomenon be reformed.[90] Only now, at last, could there be a modern German nation.[91]

Syberberg, well known for his films, *Hitler, ein Film aus Deutschland* (*Hitler, a Film from Germany*; 1977) and *Parsifal* (1982), which treated "irrationalism, music and Romanticism as the core of German identity and intellect," has also written a number of books, in which he argues that the core of German identity was lost after World War II and that the void was filled by foreign, largely American, culture.[92] His arguments are starkest in *Vom Unglück und Glück der Kunst in Deutschland nach dem letzten Krieg* (*On the Misfortune and Fortune of Art in Germany after the Last War*; 1990).[93] Originally from East Germany, Syberberg examined German identity and aesthetics in the light of reunification. The book consciously embraced German exceptionalism, referring back to the German tradition of pessimistic anticapitalism and to a form of anti-American consumerism, arguing that art and aesthetics are the primary sphere of human existence, "that all other spheres are secondary." The most tragic victim of the Nazi era, on this account, was not the Jews but art itself. Hitler, for Syberberg, was the culmination of modernization, the embodiment of the dark side of the Enlightenment, and he argued that the instrumental rationalism identified by Weber has foisted an ugliness and inhumanity on the world, above all a meanness that was incarnated in "the Bonn democracy of money." For him, Germany was a unique province on the map of "European authenticity . . . the home of a new depth that has to be rediscovered."[94]

Botho Strauss is also a writer, of plays mainly, but his controversial essay, "Anschwellender Bocksgesang" ("Goat Song, Swelling Up"), was published to acclaim and controversy in 1993. "The song of the goat" was of course the original meaning, in Greek, of "tragedy," and so Strauss too, like Syberberg, like Nietzsche in fact, was giving voice to a longing for the primitive power of art rather than instrumental reason.[95] "The increasing volatility and unpleasantness of German life," Strauss suggested, "came from the feeling that an entire way of life had reached its unnatural limit, and that it was impossible to go on with the thoughtless, smug, wasteful materialism of the West German past." Strauss mourned the loss of what he saw as the most valuable part of Germany's cultural heritage, its irrationalism as a critique of economic utilitarianism and materialism. "We know nothing about the face of the future tragedy. All we can hear is the sound of the mysteries growing stronger . . ."[96]

In May 2002, Joschka Fischer, by then Germany's foreign minister, wrote an article in the *Frankfurter Allgemeine Zeitung* in which he explored the idea of "normality" in the German context. He introduced the thought that there were two important meanings of "normality" for Germans. One related to Germans and Jews and here, he said, he favored a return to normality. Germany, as many people do not realize, he pointed out, was by then made up of *one third* of immigrants who had arrived since 1945, including 3 million Turkish Muslims and 150,000 Soviet Jews who, in the 1990s, joined some 28,000 mainly elderly Jews already living in Germany. Since the year 2000 German nationality requirements had shed their "blood and soil" criteria, citizenship being extended now to individuals born in Germany to a parent who had resided there for more than eight years. All this pointed to an increasing German accommodation with "foreignness."

At the same time, Fischer resisted the concept of normality that reflected the efforts by conservatives to "draw a line" under Germany's Nazi past and "reinstate a positive German national identity." Like Fischer, Habermas was also against this.[97]

But some Germans thought obstacles were being put in the way of their return to "normality" (seven-tenths of all Germans were born after World War II). In 1998 the novelist Martin Walser was awarded the prestigious Peace Prize of the German Booksellers Association, and in his acceptance speech—another controversy— he questioned what he saw as an increasing emphasis on the Holocaust in the 1990s, remarking that he himself had begun to "look away" when "constantly subjected to media images of Germany's shame."[98] He rejected the new Holocaust Memorial in Berlin as "an instrumentalisation of our shame for present purposes" and though he himself would "never leave the side" of the guilty he insisted that a private conscience, a "redemptive individuality," was more important and relevant than "constant public preoccupation." Although many in his audience that day agreed with him (Gerhard Schröder and Joschka Fischer were seen nodding their heads, and he was given a standing ovation), one man, Ignatz Bubis, president of the Central Council of Jews in Germany, subsequently attacked Walser for being anti-Semitic in such terms that the Schröder government was forced to acquiesce in building the Holocaust monument. The philosopher Hermann Lübbe made a parallel point to Walser when he said that collective *kommunikatives Beschweigen* (communicative silence) about the past in the early Federal Republic had enabled West Germany to evolve into a functioning democracy by providing stability.[99] The argument continued (it was too acrimonious to call it a debate). In 2007, Saul Friedländer, the eminent historian of the Holocaust, was the recipient of the Peace Prize and in an interview he gave at the time of the ceremony claimed that Walser's speech of 1998 was typical of a recurrent German tendency to end concern about the Holocaust.

Karlheinz Stockhausen created another furor when, at a press conference in Hamburg on September 16, 2001, he described the events of 9/11 as "the greatest possible work of art which ever existed," an action that "accomplished things beyond music . . . for ten years people practice incessantly and absolutely fanatically for one concert and then they die. That is the greatest work of art one can imagine in the whole cosmos."[100] Subsequently Stockhausen's concerts were canceled and there were calls for him to be placed in an asylum. But as Klaus Scherpe has pointed out, in German literary history there are many fictions of a cataclysmic America, even of New York being destroyed as a symbol of modernization (in the work of Max Dauthendey, Bernhard Kellermann, Gerhart Hauptmann).

Habermas, who himself received the German Booksellers Peace Prize in October 2001, barely a month after the 9/11 attacks, saw parallels between religious fundamentalism and Nazism. He thought we should not attribute either to "others," or to "barbarians," but should recognize that both were the "fruits" (my word, not his) of modernity, that both represented the dark side of the Enlightenment. This is a bleak way for the Germans to achieve "normality," to recognize that others may have joined them in regard to the commission of atrocities, and not everyone is likely to accept such reasoning anyway. But Habermas was surely right in pleading for "a permanent deconstruction of essential and dogmatic beliefs."[101]

Finally, two other surveys suggest that a different kind of Germany is at last emerging, that a new phase in its postwar history is now under way. A. Dirk Moses's study was referred to in the Introduction, when it was explained how he has offered a "generational" model for understanding how Germans have coped—or failed to cope—with the legacy of National Socialism. For Moses, the "Forty-Fivers," the people who were born in the late 1920s and were on the verge of adulthood in 1945, had been socialized by National Socialism, knew almost nothing of Germany before that time, and did not feel in any way responsible for the atrocities because they were too young. Nonetheless they formed a "silent majority," at least until 1968, helping Germany on its road to becoming a Federal Democracy and to achieving a stability which, they judged, involved shielding their parents, who *were* responsible for the atrocities. This is why, as the psychoanalysts Alexander and Margarete Mitscherlich argued in their study *Die Unfähigkeit zu trauern* (*The Inability to Mourn*; 1967), this generation of Germans was frozen in "psychic immobilism" as regards the past. Moses divided the Forty-Fivers into "non–German-Germans" and "German-Germans," the former wanting to hurry Germany into becoming a Western democracy on the American/British/French model, and the latter wanting to retain much of the traditional flavor of the pre-1933 Germany. This division, Moses said, formed a kind of "culture war" in Germany throughout the post–World War II period, further delaying its "long road west." The generation of 1968 certainly attacked the Forty-Fivers, German-German and non–German-German alike, and this too formed part of the culture wars. But even Moses found, at the end of his study—and this is an important point—that the "fourth generation of Germans after the Holocaust" had at last begun to place trust in the country's institutions, that the memories of the atrocities "are of increasingly less existential significance for the youth of the twenty-first century," and the Berlin Memorial to the Murdered Jews of Europe "less a stigma or stigmata than a lucrative tourist attraction, an object of indifference, or a de facto playground for children." A parallel point was made by Max Hastings about the study by the Potsdam Institute of Military History, *German Wartime Society 1939–1945*, published only in July 2008 (and referred to in the Introduction), when he observed that its young scholars, all born well after the war, have been able at last to face—and tell—the unvarnished truth, "that almost every German was aware" of what happened to the Jews *and* moreover believed that they "deserved their fate." As Hastings also said, this study was a tribute to a new generation of Germans ready to pass judgment on their parents with a rigor few others could manage.

If Moses and Hastings are right, and this fourth generation *is* ready and able to look about itself without "psychic immobilism," may that have something—everything—to do with the fact that the whole truth about wartime Germany has at last been admitted?

Pastor, Professor, Pope

On April 19, 2005, Joseph Alois Ratzinger was elected pope, at the age of seventy-eight, in succession to John Paul II. Born in 1927 in Marktl am Inn, Bavaria, Ratzinger is the ninth German pope, but the first since the Dutch-German Adrian VI (1522–23). His father was a police officer, but both Joseph and his brother Georg knew they wanted to enter the church from a very early age. In 1939 Joseph enrolled in the seminary at Traunstein and, at more or less the same time, became a member of the Hitler Youth, as all fourteen-year-old boys were required to do.[102] Ratzinger's family, however, was opposed to Hitler, the more so when, in 1941, one of Joseph's cousins, also fourteen, who had Down syndrome, was murdered by the Nazis as part of their eugenics program. In 1943, Joseph was drafted into the anti-aircraft corps though poor health kept him from active service. In 1945, as the war was ending, he deserted his post and returned home just as American troops established their local headquarters in the Ratzinger family house. He was interned for a few months as a POW and, on release, re-entered the seminary, again with his brother.

They were ordained in 1951 and then began Joseph's glittering academic career, which saw him become a professor first at Freising College, then at the University of Bonn, then at Tübingen, where he was a colleague of Hans Küng's and locked horns with other leading theologians such as Edward Schillebeeckx and Karl Rahner.[103] During the Second Vatican Council (1962–65) he served as theological consultant to Josef Frings, a reform-minded cardinal of Cologne, afterward, in 1969, helping to found the distinguished theological journal *Communio* (now published in seventeen languages). In 1977 he was made archbishop (and cardinal) of Munich and Freising and four years later Pope John Paul II made him Prefect of the Congregation for the Doctrine of the Faith, formerly known as the Holy Office, the historical Inquisition.

Ratzinger has published many books and, although a conservative and traditionalist on many matters, has not been afraid to engage in full debate with contemporary philosophers, social critics, and academics, both religious and secular. His works abound with references to classical Greek thought, to Nietzsche and Heidegger and, for example, to the more recent works of Jean Lyotard, Leo Strauss, Alasdair MacIntyre (research professor of philosophy at Notre Dame University, Indiana), Nicholas Boyle (professor of German Literary and Intellectual History at Cambridge, whose distinguished two-volume biography of Goethe and his age was discussed in Chapter 4), and Jürgen Habermas, with whom Ratzinger coauthored *Die Dialektik der Säkularisierung* (*The Dialectics of Secularisation*) in 2007.

Ratzinger's theological and philosophical priorities are recognizably in the German tradition, showing a concern with the theological implications of what Kant, Dilthey, Max Weber, and Dietrich Bonhoeffer have offered—that is, a grappling with what we can know, the dangers of making partial points of view absolute guides, and above all the dangers of what Weber called "instrumental reason," scientific reasoning used to control the world rather than to enjoy it.[104] A close student of Augustine as well as of Thomas Aquinas, Ratzinger was also much exercised by the events of 1968, which for him, as for others, marked the real arrival of postmodern society and relativism, with its ideas of culture as a "carnival," where all worldviews have equal validity.[105]

For Ratzinger, the central event in modern history was the Enlightenment and this, he says, occurred—and could only have occurred—in a Christian environment like Europe. The development of reason, which the Enlightenment philosophes made so much of, is itself an aspect of revelation and so the world can never be enjoyed to the full so long as we maintain a division between faith and reason—this division being our central predicament. Ratzinger believes that the mystery of the Trinity is there to help us grasp the reality of the modern—but also ancient—trinity: the links between beauty, goodness, and truth. These links, he says, show us that there *are* eternal and timeless values, and the way God lets us know that this is true is through the phenomenon of hope, a gift. Nietzsche said that hope was the last joke that God played on mankind, but Ratzinger insists that hope is one of those "memories of God" that is etched within all of us.

Only Christianity—Catholicism—can offer the right mix of faith and reason, says Ratzinger, for it alone is responsible for recognition of the perceived dichotomy, it alone has established the traditions—the intellectual traditions (Augustine, Aquinas) as well as the liturgical traditions—that will help us "encounter Jesus," a central element in his thought.[106]

Christianity, for Ratzinger, is *the* master story; for him the postmodern world is quite at sea in asserting that master narratives are wrong in principle and often dangerous in practice, and he insists that the "evolutionary ethos" of the contemporary world means that our only choice is between Christianity and nihilism. His answer to the nihilists is that they must first realize that they "need to be given something."[107] And he uses here, both as evidence and analogy, marriage. We all feel we need love, but we have no control over erotic love—we have no control over who we fall in love with, we experience it as a bolt from the blue, and for Ratzinger it is a gift of God. Erotic love inevitably fades, however, and, with help from the church, the *community* of the church, the tradition of the church, erotic love is turned into something else. "The erotic dimension of love, which does not ask my permission to happen, is fulfilled only in the agapic dimension of gratuitous self-giving."[108] The phenomenon of agape, self-giving, for Ratzinger helps our ascent to the divine, invests us with a "spiritual chivalry." "Thus today we often see in the faces of the young people a remarkable bitterness, a resignation . . . The deepest root of this sorrow is the lack of any great hope and the unattainability of any great love." As Roger Cohen has said, in a sense, Pope John Paul II overcame Europe's physical division. "It could be that Pope Benedict XVI overcomes the continent's historical wound."[109]

42.

Café Deutschland: "A Germany Not Seen Before"

In 1967, two German psychoanalysts, Alexander and Margarete Mitscherlich, published *Die Unfähigkeit zu trauern* (*The Inability to Mourn*), an investigation into Germany's collective long-term reaction to the collapse of the Third Reich and the subsequent horrific revelations about the Holocaust. They concluded, controversially, that Germany was still gripped by a "psychic immobilism." It was frozen emotionally, having "deliberately forgotten" its excesses. The enormity of the collective crime, they argued, was such that, for the Germans to admit culpability, and their "narcissistic attachment" to Hitler and his ideology, would entail guilt and shame on a scale so overwhelming that "the self-esteem needed for continued living" would be simply unattainable. Instead, they concluded, Germans needed to view themselves as victims, especially those known as "Forty-fivers," who achieved maturity "between fascism and democracy," and had become known, in a book by Helmut Schelsky, as *Die skeptische Generation* (*The Sceptical Generation*).[1] The Mitscherliches also insisted that the problem of psychic immobilism had persisted right through the 1950s and into the 1960s.[2]

Their study was important in itself (we explored in Chapter 41 how it fits in with other analyses of Germany's post–World War II intellectual life). Those conclusions do, however, help inform us about the pattern of German literature since World War II, an art form that is, arguably, the most articulate aspect of Germany's intellectual and moral life in the contemporary world.

We find that new elements we shall identify are underpinned by two traditional, all too familiar, concerns. To begin with, we may say that whereas the main body of English literature can best be described as "elegant entertainment" (Keith Bullivant's phrase), modern German literature, as in the United States, has had a much closer relationship with contemporary political and social developments, it has been *engagé* in the best sense, or has tried to be. This raises once more the specter of Henry Sidgwick reaching for the term "prig" but, as we shall presently see, it hardly applies here. Second, we may say also that contemporary German literature is marked (bedeviled?) by that familiar clash, already encountered far more than once, between realism and "inwardness," an insistence that *Innerlichkeit* is the true realm of literature, that it is the "intuitive wisdom" of the poet or novelist, as opposed to knowledge derived from rational thought processes, that really matters, and that everything else, especially realism, is trivial (*Trivialliteratur* in German).[3]

For some time after 1945, the sheer ubiquity of the physical rubble (*Trümmer* in German) imposed itself on the imagination of writers, although *Trümmerliteratur*, as it is called, did not produce much in the way of lasting achievement. Beyond this, is it a surprising thing to say that there was no immediate radical break in German literary life after 1945, no innovative explosion or brilliant caesura? One might have expected such a watershed until one remembers that the great books about World War I took years to appear (see Chapter 31, p. 567). The delay after World War II was even greater. Careful literary reconstruction in the 1960s showed that even the achievements of the so-called Gruppe 47 (Group 47), founded in 1947 and supposed to be representative of young German writers, were much overrated.

There was really no such thing as *Stunde Null*, or a zero hour, as the phrase implied, and the untidy truth is that writers like Ernst Jünger and Gottfried Benn were still alive after 1945 and still rejecting the modernist movement as a "false straitjacket" imposed on man by the Enlightenment—false because "it took no account of human nature." Those authors still focused on that "inner world" that we have chronicled throughout modern German history. This wasn't the only factor militating against change, however. Many other writers who tried to work out their individual responses to the new circumstances in which they found themselves in the immediate aftermath of war, and who were often ex-soldiers and/or prisoners of war, were, among other things, hindered by the American obsession with establishing collective guilt. In some ways, this was another straitjacket.

As the Mitscherliches also showed, Germany—both Germanies—tried hard to "shrug off" their old identities and aligned themselves instead with the victors, the Soviet Union or America (as indeed did Japan). They both embarked on "a mindless labor" of reconstruction, which created in the West Konrad Adenauer's "economic miracle" and, in the East, "the most successful economy of the Soviet bloc."[4]

In these confused circumstances, three authors emerged who were not especially young (all being "Forty-fivers," members of the skeptical generation), but who were the first to come to grips with the immediate German past in the postwar world.

First came a series of warnings and protests from Heinrich Böll (1917–85). Born in Cologne and wounded four times in the war, Böll felt the moral failures of the Nazi years very keenly (he resisted joining the Hitler Youth), and he devoted much energy to chronicling the chaos and brutality, the black market, the hunger, the homelessness. In *Wanderer, kommst du nach Spa. . .*(*Traveller, if You Come to Spa*; 1950), a fatally wounded schoolboy-soldier is taken to an emergency operating theater, which turns out to be the very school he had left only six months before. There, amid the rubble of destruction, he recognizes a Greek epigram scrawled on a blackboard in his own handwriting. It is not just his death we are being shown, but that of Bildung too.[5]

Böll's main warning to his fellow Germans was that "affluence may bring forgetfulness." In books that included *Billard um halb zehn* (*Billiards at Half Past Nine*; 1959), and *Ansichten eines Clowns* (*Views of a Clown*; 1963), he explored how the Adenauer doctrine of "business as usual" led to morally damaging clashes within families, with the younger generation set against the older. This conflict would be reinforced later on in the raft of *Vaterromane*, the so-called father novels of the late 1970s.

Böll's work culminated in *Die verlorene Ehre der Katharina Blum oder Wie Gewalt entstehen und wohin sie führen kann* (*The Lost Honour of Katharina Blum or How Violence Develops and Where It Can Lead*; 1974), in which he attacked the Axel Springer press for advocating authoritarian tactics by the government security services against students and left-wingers who, they felt, were "preparing the way mentally" for the terrorists who so disfigured Germany in the 1970s.[6] Certainly, Böll's books do appear to dwell on the pathologies of capitalism, and for this reason his work was always very popular in Eastern Europe. Böll, who won the Nobel Prize for Literature in 1972, gave sanctuary to Alexander Solzhenitsyn when he was expelled from the Soviet Union, and Steve Crawshaw says *Katharina Blum* and the film that was made of it were a turning point, after which the silence on talking about the past, and the fact that many ex-Nazis were *still* in positions of power, was at last overcome.* (This was already 1974, reinforcing the Mitscherliches' argument.)

Günter Grass (b. 1927), the second of the triumvirate, who won the Nobel Prize for Literature in 1999, is best known for *Die Blechtrommel* (*The Tin Drum*; 1959), part of what is sometimes called his Danzig Trilogy, completed by *Katz und Maus* (*Cat and Mouse*; 1961), and *Hundejahre* (*Dog Years*; 1963), all dealing with the rise of Nazism in and around Danzig. On the surface, *The Tin Drum* follows the life of Oskar Matzerath, who decides at the age of three to stop growing and then proceeds to drift through life armed only with a tin drum. Nothing seems to touch him in any way, not even the most farcical and terrible absurdities of Hitler's Reich, although he does end his days in a mental asylum, where he composes his memoirs.[7] Underneath, however, the book is a satire on what Grass sees as the self-righteous (priggish) tradition of the Bildungsroman. The book achieves its effect stylistically by contrasting the childlike understanding revealed in the narrative with Grass's super-sophisticated language, intended as a metaphor for Germany's postwar predicament: technical virtuosity alongside an underformed morality. One scene describes a fashionable restaurant that only serves onions. People relish the

* When Crawshaw visited the Georg Eckert Institute for Schoolbook Research in Braunschweig, he found the way the "vast bulk" of textbooks avoided confronting the truth "remarkable." With these, he said, the tone changed earlier, after the Eichmann trial in 1961.

onions so that, in the "tearless [twentieth] century" they may experience crying. This "underformed" morality would obsess Grass in the years that followed.[8]

Both Böll and Grass saw themselves as moral guardians in postwar Germany.[9] Both were involved politically with the peace movement, but the shine was rather taken off Grass's role when it was revealed in 2006 that he had himself been a member of the Waffen-SS.[10] Not everyone accepted that it was simply a case of a misguided teenager "doing his duty."

Among his critics was the third of the triumvirate, Martin Walser (b. 1927), who was himself in the Wehrmacht and may have joined the Nazi Party during the war. Walser was a much wittier writer than the other two, with a caustic turn of phrase, but not so well known outside Germany. His dominant theme to begin with was the effects of the rat-race on middle-class employees (*Halbzeit* [*Halftime*; 1960] and *Das Einhorn* [*The Unicorn*; 1966]), though later and more interestingly he explored the psychological consequences of living in a divided country (as in his novella *Dorle und Wolf*, translated as *No Man's Land*; 1987). Walser didn't want to look back, not in public: the present-day problems of Germany were too pressing. His most well-known work in the English-speaking world is *Ein fliehendes Pferd* (*Runaway Horse*; 1978).[11]

Mourning Becomes Oedipus

Böll and Grass in particular, then, had at last begun to come to grips with the Third Reich, stimulating an appetite that would finally germinate in the 1960s and 1970s. But their collective achievement went wider than that. In the words of Keith Bullivant, "Gone at last was the ultimate concern with a transcendental world, gone the allegorical, mystical treatment of the great questions of life without real regard for those of the day."[12]

By this time, however, none of them were young men and so they did not involve themselves with the student radicalism of the 1960s to anything like the same extent as, say, Max Frisch, Peter Schneider, and Peter Weiss. This younger generation was also much affected by the (Baader-Meinhof) terrorism of the early 1970s, mainly for the way in which it provoked (and therefore revealed) the *still*-authoritarian instincts of the government agencies. Schneider's *Lenz* (1973) and Frisch's *Stiller* (1964) in particular explore the authoritarian ground *before* the internalization of democratic values around 1968 referred to in the previous chapter. This culminated in the late 1970s in what became known as the "German autumn," when the chief prosecutor, a high-ranking banker, and the head of the Federation of German Industry were all assassinated. The gap between the writers and the government widened in times of terrorism. Writers accused the state of restricting civic rights, and politicians repeated their claim that the writers were offering "mental support to anarchism."

But not all German literature can be snugly fitted into this pattern of "coming to terms with the past." The critics Hans Magnus Enzensberger and Walter Jens both drew attention to the greater readiness, from about 1970 on, of German writers to be more aware politically—in the broadest sense (civil rights, U.S. nuclear missiles on German soil, secularization), important aspects of this "new realism" being an emphasis on language (*Sprachrealismus*), documentary narrative (concrete realism), and the emergence of the women's movement in Germany, as elsewhere. Probably the most significant names here were Ingeborg Bachmann (1926–73) and Elfriede Jelinek (b. 1946), both Austrian and both Catholic. In *Die Klavierspielerin* (*The Piano Player*; 1983), and in particular *Lust* (1989), Jelinek exposed the allegedly modern worlds of film and media where women are still treated in the same old way, as often as not as sex objects.[13] Jelinek's style is deliberately deadpan in order to confront the reader, especially the male reader, with how pornography strikes women.[14]

The *Vaterromane*, or "father novels," which faced the burden of the parental Nazi past in a kind of collective Oedipal revolt, formed a distinct subgenre of German literature in the late 1970s. The (delayed) timing of these works rather reinforces the Mitscherliches' conclusions.[15]

The Alliance of Writers and Readers in the GDR

To the east was another Germany, the Deutsche Demokratische Republick (DDR) or German Democratic Republic (GDR), which underwent its own trajectory in which religious instruction was abolished in schools, censorship was imposed and relaxed, imposed and relaxed, limiting subject matter and creating eventually an implicit alliance of readers and writers who learned to create and assimilate a series of codes, allegories, and subtexts that, up to a point, the authorities would accept. As a result the GDR became known as "*Leserland*," the Land of Readers.[16] There was an initial optimism, with authors such as Anna Seghers presenting the brand-new socialist state as a viable alternative to the Third Reich, even daring to compare it with Goethe's and Schiller's Weimar.[17] This didn't last. From the late 1950s on, in such works as Bruno Apitz's best seller about Buchenwald, *Nackt unter Wölfen* (*Naked amidst Wolves*; 1958), Hermann Kant's *Die Aula* (*The Lecture Hall*; 1965), Ulrich Plenzdorf's *Die neuen Leiden des jungen W.* (*The New Sufferings of Young W.*; 1972), and in several titles by Christa Wolf, in particular *Kassandra* (1983), writers considered what was being lost, aesthetically, because so much of the communist state was centralized. Here, as later, artistic license was allowed where direct criticism was not.[18]

It was to be expected that women writers would shine in East Germany because it was (officially, at any rate) committed to the equality of the sexes, backed up by laws governing such things as maternity leave and day care. But that overlooks the fact that East Germany occupied much of the territory of what had been Prussia, where most men were conservative traditionalists. This made conflict over sex roles more ingrained there than anywhere else. And it was these factors, or some of them, which conspired to produce Christa Wolf, a writer who, perhaps, could not have existed anywhere else. Her *Nachdenken über Christa T.* (*Reflections on Christa T.*; 1968), is the pivotal work, a spare narrative in which the eponymous Christa T. goes in pursuit of a form of Bildung, a fuller realization of her self than is allowed by the drab confines of the communist world she inhabits, and where the terminal illness of her close friend became a metaphor for the GDR.[19] Most controversial of all was *Was bleibt* (*What Remains*), a book that Wolf wrote in 1979, but that was not released until 1990. It gives a semifictional account of her surveillance by the Stasi and her adjustment to that life. However, because the text was only released after the *Wende* (the "turn" or reunification of Germany), Wolf was attacked for not being brave enough to "go public" from *within* the GDR.[20]

A number of books, as Martin Swales has pointed out, specifically addressed the psychological and cultural differences between East and West

Germany; among them are Uwe Johnson's *Zwei Ansichten* (*Two Views*; 1965), Peter Schneider's *Der Mauerspringer* (*The Wall Jumper*; 1982), and Thorsten Becker's *Die Bürgschaft* (*The Pledge*; 1985).[21] Johnson, who settled in England, showed in his books how there can be no simple, single truth about character, an argument (or message) with more resonance in Germany than elsewhere.

One of the biggest cultural differences between East and West, as will already be clear, was freedom of expression, a matter on which the GDR shot itself in the foot in 1976 when it stripped Wolf Biermann of his citizenship, "because he told too many uncomfortable truths." Biermann told these truths in poems and songs, and he and the other *Liedermacher*, or Song Makers, were an important dimension in politicizing young people on both sides of the Wall in the late 1970s and 1980s. Biermann's expulsion provoked a wave of immigration of literary figures to the West and was followed by the passing of the "Lex Heym," a law criminalizing any material that "might" damage the state—this was after Stefan Heym had released his novel *Collin* in the West in 1979, in which he attacked the corruption of Stalinism in the 1950s.[22] Others kept up the pressure. In books such as Peter Handke's *Die Stunde der wahren Empfindung* (*The Hour of True Feeling*; 1975), Christoph Meckel's *Suchbild: Über meinen Vater* (*Picture Puzzle: About My Father*; 1980), and Volker Braun's *Hinze-Kunze-Roman* (*Every Tom, Dick, and Harry Novel*; 1985), the dominant theme was the links between political restrictions and a deadened psychology.

That makes it seem as though criticism of East Germany from within was—if not widespread—at least active. This did not stop Günter Kunert from arguing that the GDR would have collapsed much earlier if so many writers had not constantly legitimized it, and Peter Schneider agreed. After 1989, incredible as it may seem now, more than one East German writer went on record as regretting the demise of "a utopian alternative" to the federal/capitalist state. No less telling was Christa Wolf's argument, in an interview she gave in 1990, that "the great ideologies had not only become more and more dubious, but also less important, no longer offering a guide as to moral values or behaviour."[23] Such views were mocked mercilessly by Enzensberger for their anguish over lost "fundamental experiences" and "slow-moving Sundays."[24]

The Dimensions of German Suffering

Once again we should remind ourselves not to shoehorn postwar German writing into one or two simple patterns ("We didn't just have autumn and winter," said the East German actress Corinna Harfouch, "we had spring and summer too").[25] Here we must mention a breed of angry writers, people such as the Austrians Thomas Bernhard, Felix Mitterer, and Gerhard Roth, and the Swiss Peter Bichsel. Bernhard died in 1989, just as unification occurred, but not before he (and others like him) had published a raft of books denouncing his country as "a vile place," a cold and isolated "sump of immorality" that had never addressed its past. His titles reflect his verdict—*Der Keller: eine Entziehung* (*The Cellar: A Withdrawal*; 1979); *Die Kälte: eine Isolation* (*The Coldness: An Isolation*; 1981); and *Auslöschung: ein Zerfall* (*Extinction: A Degeneration*; 1986).[26]

Toward the end of the century, three authors emerged who paralleled Böll, Grass, and Walser. Bernhard Schlink's best-known book is *Der Vorleser* (*The Reader*; 1995), set in the 1950s and telling the story of a young teenager, later a law student, who has an affair with Hanna, an older woman who is an uneducated tram conductor. Only after we are well into the book do we discover that Hanna has a secret past as a guard in a concentration camp. By the time of the revelation, we have responded to her as a sympathetic character, but Schlink's underlying point is that Hanna admits her guilt and that in Germany, in the immediate postwar years, it was easier for those guilty of lesser crimes to concede their culpability than it was for those responsible for much greater wrongdoing. For Schlink, distinguishing between the greater and lesser evils is an important element in overcoming the past. In its way, this brings the Mitscherlich argument full circle, that the greater the crimes a war criminal has committed, the more likely he or she is to be mired in psychic immobilism. The character of Hanna, based in part on a real woman, was deconstructed by Tom Bower in the London *Sunday Times*, showing how such a figure—unable to read and write—could not have existed in the Third Reich.[27]

W. G. Sebald, in *Austerlitz* (2001), his last and best-known book, tells the story of a child evacuated from Prague to Wales in 1939 who goes in search of his past. This journey, which at one stage includes a single sentence ten pages long, describing a visit to the concentration camps, what is left of them, brings him face to face, in a calm tongue reminiscent of Goethe, with *German* suffering, a theme Sebold returned to in *Luftkrieg und Literatur* (translated as *On the Natural History of Destruction*; 1999), where he gives a vivid description of the firestorm unleashed by the Allies over Hamburg in 1943. This theme was explored even more starkly in Jörg Friedrich's *Der Brand* (*The Fire*; 2002), six hundred pages about the imbalance in the number of dead in the bombing raids. In particular, 80,000 died in *two* night raids on Hamburg and Dresden, more deaths from bombing than in the entire United Kingdom during the whole war. Overall, 600,000 Germans died from the bombing, more than ten times the British deaths. Friedrich was not trying to excuse the Nazis, nor being "dangerously soft"—he had, in 1984, in *Die kalte Amnestie* (*Cold Amnesty*), cataloged how the West German establishment "remained infected" by Nazism in the years after the war. But he did draw attention to the fact that the Allied mass killing brought no military gain.[28]

In the same year, Günter Grass published *Im Krebsgang* (*Crabwalk*). This again dealt with memory and the question of German suffering through the sinking of the passenger liner *Wilhelm Gustloff*, torpedoed by a Soviet submarine in January 1945, with the loss of 9,000 passengers. This made it the largest-ever maritime disaster, with losses six times those of the *Titanic*.[29]

As Steve Crawshaw has said, these authors are not attempting to "airbrush" unpleasant facts out of the picture: German suffering and Jewish suffering are not equal. "They are, however, both real . . . Germany is sometimes seen as unchanging. In reality, it is a nation of crabwalkers—moving, more rapidly than Germany itself sometimes seems to notice, towards the future and towards the creation of a Germany that we have not seen before."[30]

Volker Weidermann, editor of the arts and literature section of the *Frankfurter Allgemeine Sonntagszeitung*, has become something of an authority on recent literary history in Germany. In 2008 he published *Das Buch der verbrannten Bücher* (*The Book of Burnt Books*), an account of the auto da fé on May 10, 1933, when, prompted by students, the Nazis burned the works of ninety-four German and thirty-seven foreign writers. Weidermann tracked down the surviving German authors to rescue several from oblivion. Two years earlier, he had published *Lichtjahre* (*Light

Years), in which he identified the latest raft of German writers worth reading, among them Ingo Schulze, *33 Augenblicke des Glücks* (*33 Moments of Happiness*; 1995), Thomas Brussig, *Helden wie wir* (*Heroes Like Us*; 1995), and Thomas Meinecke, whose work tries to bridge pop culture and high seriousness.

Weidermann also devotes space to Walter Kempowski, who was born in Rostock, arrested in 1948 for smuggling documents to the West that showed the Russians were breaking strategic agreements with the Americans, and sentenced to twenty-five years' hard labor. Released after eight years, Kempowski, whose novels were popular in the 1970s, set about chronicling Germany's twentieth-century tragedy via stories of ordinary people, amassing an archive of 8,000 diaries and 300,000 photographs. He also wrote a series of novels going back in German history, the account of which was for him far from heroic. Perhaps because of this, recognition was slow in coming but, at his death in 2007, his reputation was rising.[31]

Poetry, Silence, and Intimacy After Auschwitz

In 1949 Theodor Adorno famously announced that to write poetry after Auschwitz "is barbaric," but in one sense in Germany there was more of a need for poetry after the war than ever before. The nature of guilt, of grief, of shame, is a private as well as a public matter, and their expression in intimate terms has been as much a feature of postwar German poetry as has anger at what was done in Germany's name.

Early on, Wolfgang Weyrauch identified a need for what he called *Kahlschlag*, a "clearing of the terrain," a purging of the language, disposal of the "rubble" of the past and a need to invent a language cleansed and freshened but at the same time worthy. This is the context for the first truly successful German poem realized after Auschwitz, Günter Eich's strikingly simple "Inventur" (Inventory; 1948), notable for its "bald stocktaking of existence" in which deliberately flat language is used as a form of cleanliness, discarding the traditions of meter, rhythm, and metaphor, poetry without the furniture of poetry.[32]

In the 1950s Gottfried Benn's *Statische Gedichte* (*Static Poems*) became known to the public. He had continued to write in secret, although forbidden from doing so by the Nazis, and the collection had circulated privately. In one of the best, called "Farewell," Benn conceded that, early in the 1930s, he had betrayed "my word, my light from heaven" and that redemption was impossible: "There are only two things: emptiness and the constructed self." As Nicholas Boyle sums it up, "If the self has become pure construction, not made out of interactions with its past experiences or with a given world, there is no place for poetry as it had been practised in Germany from Goethe to Lasker-Schüler."[33] In his 1951 essay *Probleme der Lyrik* (*The Problems of Poetry*), Benn outlined his view that art obeys its own rules and that poetry in particular should aim for intimacy and privacy so as to remain beyond the reach of politics; poetry, he said, is a closed world with its own rules, and in remaining so, it becomes redemptive.[34] This hermeticism proved influential for writers such as Paul Celan, Ingeborg Bachmann, and Rose Ausländer.[35]

Celan's best-known work, "Todesfuge" (Death Fugue), was written in the year war ended, when the full extent of the Holocaust was becoming known. The poem, which begins "Schwarze Milch der Frühe" ("Black milk of daybreak") and climaxes with the terrible phrase "der Tod ist ein Meister aus Deutschland" ("Death is a master from Germany"), is a commemoration of the death-camps, its title reflecting (and anticipating) the dangerous ambiguity that would come to afflict memory of the crime: a fugue is a piece of music, a work of art, but also a flight, an avoidance, a psychological illness, and (it can be) an escape.[36] Celan's style gradually became terser—he called this "straightening," *Engführung*. Later still, he argued that true poetry reflects a natural "tendency towards silence," and this too may be understood as poetry after Auschwitz. Celan, who was Jewish, committed suicide in 1970.

Not everyone shared this tendency toward silence. In his *verteidigung der wölf* (*defense of the wolves*; 1957), Hans Magnus Enzensberger became both the natural successor to Brecht (who had died the year before) and the leader of a school which adapted Brecht's conviction that a poem should be an "object for use" (*Gebrauchsgegenstand*). Enzensberger's work is characterized by anger and aggression and urges greater political awareness among his readers, the very opposite of the aim of Benn and Celan.[37]

A Deformed Reality

On the other side of the Iron Curtain, inside the GDR, the early environment for poetry seemed encouraging once it had shaken free of its much-derided "tractor verses" ("Boy meets tractor" is how Adorno satirized it), and when J. R. Becher, a poet of no mean talent, was made first minister of culture in 1949.[38] But he was soon displaced and it was in any case to reckon without Brecht himself who lost little time producing a series of snapshots ("Der Rauch" [Smoke], "Der Radwechsel" [Changing the Wheel], and "Böser Morgen" [Bad Morning]), a succession of unpalatable truths about life in the East. After Brecht's death in 1956, Günter Kunert did his best to fill the gap, taking aim at the bureaucracy and, in doing so, helping to provoke a younger generation of poets at what was by now the J. R. Becher Institute for Literature in Leipzig.[39] Volker Braun was (and is) the best of this school, his finest work juxtaposing the intense personal discomforts unique to the GDR alongside the avowed utopian assertions of the state. He was joined by Sarah Kirsch and Wolf Biermann.[40]

In theory at least, the East German bureaucracy encouraged these voices, especially so in Erik Honecker's notorious "no taboos" speech delivered in 1971. But the inherent tension could not be disguised for long, and in 1976 Biermann was, as mentioned earlier, expelled. His expulsion was too much for many people, and Günter Kunert, Reiner Kunze, and Sarah Kirsch all followed him west.[41]

After that, the only way was up. The liberalization known as *glasnost*, in the 1980s, played a part, but so too did the new generation of poets "born into" socialism, who had very little expectation of the GDR as a form of utopia. For writers such as Heinz Czechowksi, Ernst Moritz Arndt, and Rainer Erb it was simply a deformed reality, and they plucked up the courage to say so. Wolfgang Emmerich went so far as to claim that "a major legacy" of the GDR would be its lyric poetry. This recalls Anna Akhmatova's claim that the "lyrical wealth" of Russia could not be destroyed by the Stalinists.

East and West German poets were in any case drawing closer together in the 1980s, both turning away from socialism (known in the GDR as *Abscheid nehmen*, or "taking leave of a disappearing world") and both showing the traditional German anxiety about the relentless march of technology.

Enzensberger stood out here too in his long narrative poem, "Der Untergang der Titanic" (The Sinking of the Titanic; 1978). Joachim Kaiser observed that in fact German literature had no need of reunification: "Its profound communality was never broken . . . only endangered."[42]

Amid the explosion of verse that erupted during and after the euphoria of the *Wende* of 1989 (a turning point that Peter Schneider thought was "intellectually comparable" to 1945), Volker Braun's "Nachruf" (Obituary) stood out as a paradigm of the 1990s, mourning the dead utopian dreams of the GDR but, more than that, the lives who suffered for those ideals. "Obituary" formed a kind of brackets with "Inventory."

Reunification also saw the emergence of younger poets, many living in Berlin, such as Barbara Köhler and Durs Grünbein, whose *Grauzone morgens* (Grayzone in the Morning; 1988), was rapturously received, though *Porzellan: Poem vom Untergang meiner Stadt* (Porcelain: Poem on the Death of My City; 2005), about the firebombing of Dresden, received more mixed reviews.[43]

Bearing in mind where we started this chapter—with the Mitscherliches' conclusions—this brief survey of postwar literature confirms that the realm of mourning *has* now been explored by German authors. Not necessarily in a way that will please everyone, but if outsiders are ever to understand modern Germans, it is to their imaginative writers that we must first turn. Modern German literature goes far beyond "elegant entertainment."

Theater as Culture, Not Entertainment

In the realm of theater—to include opera and dance—the names of Brecht, Piscator, Reinhardt, Laban, and Jooss led the world up to and across World War II. Immediately after the war, Piscator interested himself in a new form of theater that was to resonate across Europe, especially in Britain in, for example, the works of David Hare. Notable examples of this "documentary theater" were Rolf Hochhuth's *Der Stellvertreter* (*The Deputy*; 1963), which tackled the unfortunate role of Pope Pius XII in doing next to nothing to stop the Holocaust, and *Die Juristen* (*The Lawyers*; 1979), which heaped such odium on Hans Filbinger, the prime minister of Baden-Württemberg, that he was forced to resign.

Whereas Piscator had been the equal of Brecht in the Weimar years and was more successful than him in the United States, in Germany it was Brecht who developed a clear edge. As both a playwright and director of the Berliner Ensemble from 1949 until his death in 1956, Brecht's innovative stagecraft was so powerful as to be felt right across Europe. His often austere sets distilled the drama, reinforced by his concept of *Verfremdung* (alienation), an attempt to make the familiar unfamiliar, so audiences *experienced* alienation, and were not simply passive onlookers.

But Brecht's was by no means the only tradition. The most notable alternative was the theater of "ordinary people." Known as the *Volksstück* ("people's play," following Lessing), it concentrated on postwar German working-class life, the main authors being Martin Sperr (1944–2002), Rainer Werner Fassbinder (1945–82), and Botho Strauss (b. 1944).[44] Nor can we ignore the tradition of Büchner and Wedekind, of Max Reinhardt and Fritz Lang—experimentation and spectacle of a type that is almost unthinkable outside Germany.[45] Outstanding among these plays are Peter Weiss's overwhelming drama located in a lunatic asylum but depicting events of the French Revolution, *Die Verfolgung und Ermordung Jean Paul Marats durch die Schauspielgruppe des Hospizes zu Charenton unter Anleitung des Herrn de Sade* (*The Persecution and Assassination of Jean Paul Marat as Performed by the Inmates of the Asylum of Charenton under the Direction of the Marquis de Sade*, usually known simply as *Marat/Sade*; 1964). The genre survives in Heiner Müller's seven-and-a-half hour *Hamlet-Maschine*, produced at the Deutsches Theater in Berlin in 1990.

Since then, German-language theater has boasted two innovative playwrights difficult to categorize other than to say that they exhibit late Expressionist tendencies and other modernist influences, first identified by the critics Marcel Reich-Ranicki and Helmut Heissenbüttel. The Swiss Friedrich Dürrenmatt (1921–90) is best known for *Der Besuch der alten Dame* (translated as *The Visit*; 1955), which Michael Patterson and Michael Huxley say is the most widely performed postwar drama in Germany. The main character is a wealthy woman, much face-lifted, who visits her hometown to seek revenge on the boy—now an old man—who had seduced her and denied he was the father of her child, when they were both young.

Since Dürrenmatt's death in 1990, his mantle has been taken over by Peter Handke. He had already made his mark in the 1960s with the *Sprechstücke* (speak-ins) *Publikumsbeschimpfung* (*Offending the Audience*) and *Selbstbezichtigung* (*Self-Accusation*), which presented an empty stage in plays without apparent plot, characterization, or (at times) dialogue. Handke's reservations about language link German theater to philosophy—Wittgenstein, for example—and find final form (thus far) in his play *Die Stunde da wir nichts von einander wußten* (*The Hour We Knew Nothing of Each Other*; 1992). The play is set in a town square in which, for one hour and forty-five minutes, two dozen actors playing close to 400 roles pass across the space, but nobody speaks.[46] The director Peter Stein has also been creating waves.

Just as Dürrenmatt's *The Visit* is the most-performed German play since World War II, so Kurt Jooss's *Der grüne Tisch* (*The Green Table*) is the most-performed dance-drama, apparently "staged by more dance companies than any other work in the modern repertoire."[47] Jooss died in 1979 and since then his pupil Pina Bausch (1940–2009) developed what she and others call *Tanztheater*. Building on the tradition of Rudolf Laban (1879–1958) in Munich, John Cranko (1927–73) in Stuttgart, and Mary Wigman (1883–1973) in Berlin, it is a mixture of narrative, expressionism, and sheer sensuality.[48] Her breakthrough came in 1971 with a commission from the Wuppertal Theater for *Aktionen für Tänzer* (*Actions for Dancers*), followed by a series of works that quickly became modern classics, the best being *Café Müller*, in which the chairs that virtually fill the stage are moved around at breakneck speed: only very fit, highly coordinated dancers could perform this piece without catastrophe.

Which underlines the fact that, again as Patterson and Huxley point out, theater by and large represents "culture" rather than "entertainment" in Germany. There is no equivalent to "show-business" in the German language, and there is no equivalent of Broadway in German theater, though there are some theaters—*Boulevardtheater*—that specialize in comedies. In Germany theater audiences by and large expect a range of serious high-culture plays reflecting the European heritage. This requires subsidies that far outweigh those available elsewhere. The figures provided by Patterson and Huxley show that state and municipal subsidies to German theaters at the turn of the twenty-first century stood at roughly *seven times* the amount of public funding the United States provides for all the arts, while the Berlin Opera House alone receives almost as much as the British Arts Council spends on *all* the theaters it supports. German theaters tend to stand in their own grounds and the interiors are more elaborate and the productions more ambitious, as a result of which the role of the theater in the cultural life of Germany has been more important than it has elsewhere.

In East Germany, in the 1980s, as with poetry, theater became a debating chamber for politics, albeit at one remove, via coded references, as mentioned in the previous section.[49] In fact, in the 1980s only the churches and the theaters provided a space for political debate.

"The Second Flowering of German Film"

In film, as with the novel and poetry, the landscape of "rubble" was an early theme (Roberto Rossellini's *Germania anno zero*, 1947) but this was not a genre that flourished. More successful was Wolfgang Staudte's *Die Mörder sind unter uns* (*The Murderers Are Among Us*; 1946), in which a traumatized doctor tries to come to terms with the war and at the same time to bring his former commanding officer to justice.

In East Germany filmmaking was controlled by the party, with DEFA (Deutsche Film-Aktiengesellschaft), maintaining a virtual monopoly over production (not so different from the UFA during the Nazi era). The early films often had capitalism in their sights, though other genres included the "anti-fascist film," the doctrinal film, to promote the image of the ruling SED party (Sozialistische Einheitspartei Deutschlands), with a lot of money also going into children's films (likewise a form of indoctrination). The former UFA studios, in Potsdam-Babelsberg, were now used by DEFA; their most memorable anti-Nazi film was *Sterne* (Stars; 1959) in which an infantryman falls in love with a Jewish woman who is about to be sent to Auschwitz.

Then there was the *Gegenwartsfilm*, which was a deliberate throwback to the "proletarian films" of the Weimar Republic. In such productions as Slatan Dudow's *Unser täglich Brot* (*Our Daily Bread*; 1949), directors tried to show the real conditions existing in East Germany, but they were overtaken by events, in particular the Hungarian uprising in 1956, after which stricter censorship was introduced, until the closing of the border, in 1961, when topical issues were again allowed. Konrad Wolf's *Der geteilte Himmel* (*The Divided Heaven*; 1964), based on a novel by Christa Wolf, looked at the division of Germany from an Eastern point of view and stimulated mildly critical films by younger directors, examining such themes as the conflict between the generations and corruption in the legal system. This relative freedom came to an end again with the Eleventh Plenum of the Central Committee of the SED in 1965, as a result of which no less than half a year's production was binned.[50] These so-called *Verbotsfilme* (forbidden films) were vivid evidence that a new generation had become critical of the old. Matters changed yet again after Erich Honecker's notorious speech in 1971, already referred to, in which he condemned taboos in the arts. One immediate consequence was Heiner Carow's *Der Legende von Paul und Paula* (*The Legend of Paul and Paula*; 1972), a big success, based on the popular novel by Ulrich Plenzdorf. This explored a couple's search for individual happiness within the East German system, in the process introducing sexuality as a means of achieving freedom and modernity. Frank Beyer's *Jakob der Lügner* (*Jacob the Liar*; 1974), also based on a novel, told the story of a Jew in wartime who pretends he has a radio and makes up the "news" for his friends to keep their spirits high. It was the only DEFA film nominated for an Academy Award.[51]

In West Germany, though "rubble" films were also produced immediately after the war, the Germans' own work was overwhelmed by a flood of American films, usually dubbed. Otherwise, the period was characterized by *Heimatfilme*—safe, escapist romances that raised no ghosts.

There are two views about the effect of American film on Germany. One is that they trivialized German culture but another is that Germans willingly embraced American culture as a means of breaking with the Nazi past.[52]

In fact, so weak was the (West) German film industry at this time that, in 1961, at the Berlin Film Festival, the Federal Film prize was not awarded. This gesture seems to have had some effect and only a year later, at the Oberhausen Short Film Festival, a group of twenty-six young directors signed the founding document of Das neue Kino, the New German Cinema. Known as the Oberhausen Manifesto, the document resulted in an organization to subsidize new films by young directors, the Kuratorium junger deutscher Film. But in practice it wasn't until the late 1960s and 1970s that German film experienced its real renaissance and, like most renaissances, this had to do with a constellation of genuine "stars" maturing at much the same time.

In 1968, Werner Herzog released his first feature, *Lebenszeichen* (*Signs of Life*), the first in a series of works inhabited by bizarre loners and outcasts but redeemed with caustic humor. Herzog's films seek to show, he says, "ecstatic truth" rather than the "accountant's truth" of *cinéma verité*. They explore solitude, inner states, "inner landscapes"—Caspar David Friedrich is a favorite artist. "Tourism is a sin," says Herzog and the twentieth century a "catastrophic mistake." His loathing of "technological civilization" is reminiscent of Heidegger's, though he lives in—and loves—Los Angeles for its "collective dreams."[53]

At much the same time, Rainer Werner Fassbinder produced his trilogy of gangster movies, also examining the inner worlds of loneliness and isolation (*Liebe ist kälter als der Tod* [*Love Is Colder than Death*] introduced Hanna Schygulla to the world). Herzog and Fassbinder were quickly followed in 1970 by Wim Wenders and his celebrated road movies, also inhabited by rootless, haunted loners, most notably *Summer in the City*.[54]

Directors such as Fassbinder, Alexander Kluge, Margarethe von Trotha, Volker Schlöndorff, and Reinhard Hauff were by no means blind to the evils of National Socialism but, like Martin Walser, they preferred to deal with late twentieth-century issues such as immigrant workers (Fassbinder's *Angst essen Seele auf* [*Fear Eats the Soul*; 1974]), and terrorism (the collective film *Deutschland im Herbst* [*Germany in Autumn*], and Reinhard Hauff's *Messer im Kopf* [*Knife in the Head*], both 1978). The division of Germany was left largely unexplored by this generation of directors, though Hans Jürgen Syberberg did confront Nazism, in the four-part *Hitler, ein Film aus Deutschland* (*Hitler, a Film from Germany*; 1977).[55]

Government support for the film industry fell away in the 1980s, but talent had begun to flow. This period saw the release of Edgar Reitz's *Heimat*, 1984, an eleven-part chronicle of life in the fictional Hunsrück village of Schabbach, which was well received when shown on television both in Germany and elsewhere, and Wim Wenders's *Der Himmel über Berlin* (*Wings of Desire*; 1987). Partly written by Peter Handke, this tells the story of two angels—unseen to everyone but children—who wander through Berlin listening to everyday people and their problems. It won a prize at the Cannes Film Festival. The 1980s were also the years of several very good documentaries, notably Hartmut Bitomsky's *VW-Complex*, and a raft of films by new women directors among whom Margarethe von Trotha was prominent.

The fall of the Berlin Wall in 1989 stimulated the production of many films, the most unexpected of which was the reunification comedy, inaugurated by Helmut Dietl's *Schtonk!*, 1992, a satire on the "Hitler Diaries" fiasco, and Christoph Schlingensief's *Das deutsche Kettensägenmassaker* (*The German Chainsaw Massacre*; 1990), a vicious parody of consumerist culture, in which a mad Wessi family seeks out Ossis and, using chainsaws and axes, turns them into sausages. In *Good-Bye, Lenin!* (2003), Christiane, who has lived a near-normal life in East Germany, suffers a heart attack and goes into a coma on the very day that the first big antigovernment protest occurs in October 1989. She doesn't regain consciousness for several months, by which time the GDR is about to disappear. The doctors warn that any shock might kill her, so her children are forced to pretend—hilariously—that East Germany still exists. They bring back the old furniture, which they had in the meantime replaced, and concoct "broadcasts" that "explain" some of the changes (the government has generously allowed Wessis to flee east, as refugees from capitalist imperialism). In *Das Leben der Anderen* (*The Lives of Others*), written and directed by Florian Henckel, which won the Oscar for Best Foreign Film in 2007, a Stasi officer gradually loses sympathy with the regime he is part of. Though he tries to help some of the people he has under surveillance, he can do nothing to prevent the various levels of corruption from combining to produce a tragedy that ends with the suicide of a woman he has himself, inadvertently, helped to trap.

Three things stand out about German television culture. One, it is very popular: ZDF, or Zweites Deutsches Fernsehen, the Second German Television Channel, is the largest TV station in Europe. Two, internationally it has had much less impact than German music, painting, dance, or film. And three—perhaps most interesting of all—there has been much more controversy about the impact of television culture in Germany than elsewhere. Helmut Schmidt, when he was chancellor of Germany, stigmatized cable television as "more dangerous than nuclear power." Several professional critics such as Günther Anders, Hans Magnus Enzensberger, and Jürgen Habermas all agree in seeing television as a cultural "black hole."

The Dominance of Darmstadt in Music

Music, we must never forget, had not suffered in the Third Reich as had other activities such as painting and scholarship. Despite what had happened, postwar Germany beyond the rubble still boasted—incredibly—some 150 opera houses and orchestras, unparalleled conservatories for musical education, an undiluted musicological tradition producing musical scholarship of unequaled quality and originality, and a larger number of specialized periodicals devoted to music than in any other country.[56]

In West Germany composition and musical production quickly regained their former position once the "economic miracle" had begun to exert itself. In 1948 Richard Strauss remarked "I have outlived myself," but it wasn't true—the next year he produced what would turn out to be his ever-popular *Vier letzte Lieder* (Four Last Songs).[57] The Berlin Philharmonic, under Wilhelm Furtwängler from 1947 to 1954, and then under Herbert von Karajan, a child prodigy on the piano, quickly recovered its pre-Goebbels pre-eminence and the Deutsche Grammophon Gesellschaft was likewise reinvigorated.[58] Karajan's Nazi past dogged him (he had been a party member since 1935), his favored status emphasized by the fact that when he had married his second wife, in 1942, and she turned out to be one-quarter Jewish, the NSDAP had made her one of Germany's five "honorary Aryans." Several musicians, such as Isaac Stern and Itzhak Perlman, refused to play with Karajan because of his Nazi past, and in 1946 he was banned from conducting by the Soviet occupation authorities.

Did the pressure begin to get to him? He sought psychoanalytic help from Carl Jung but in 1948 Karajan helped build the newly formed Philharmonia Orchestra in London and in 1955 he was appointed musical director for life of the Berlin Philharmonic, in succession to Furtwängler, two years later becoming artistic director of the Vienna State Opera.[59] He was also intimately involved with the Salzburg Festival, and over the next three decades, as "the genius of the economic miracle," discovered several artists (Anne-Sophie Mutter, Seiji Ozawa) and became the top-selling classical music recording artist of all time, with some 200 million records sold.

Given that the Nazis had been so opposed to the work—as well as the person—of Schoenberg and his pupil Anton von Webern, it was all but inevitable that their technique should become almost a new orthodoxy in the 1950s, boosted by the annual summer course in composition that was begun at Darmstadt. Three brilliant young composers, Bernd Alois Zimmermann (1918–70), Hans Werner Henze (b. 1926), and Karlheinz Stockhausen (1928–2007), each of whom had studied at Darmstadt, emerged in the 1950s. The critic Erik Levi, best known for his book on music in the Third Reich, has described Zimmermann's opera *Die Soldaten* as "the most significant German opera since Berg's *Lulu*."[60]

Henze was a fervent Schoenberg enthusiast who nonetheless removed himself to Italy to keep himself open to other musical innovations. Two of his operas, *Elegie für junge Liebende* (*Elegy for Young Lovers*; 1961), and *Die Bassariden* (*The Bassarids*; 1966), with a libretto by the English/American poet W. H. Auden (and premiered by Karajan), were immediately recognized as successful marriages of music and drama. Later, Henze became involved with the student revolutionary movement of the late 1960s, and his compositions acquired a more strident edge, looking across the world to the music of Castro's Cuba and back to Kurt Weill.[61]

Stockhausen, the third of the postwar young Turks, was the most radical. He famously pioneered electronic music, experimented with indeterminacy (or chance), as devised by the American composer John Cage, becoming a cult figure in the 1970s, not least among certain rock musicians.[62] Stockhausen's influence declined in the 1980s, his prominence not helped by the fact that, after 1977, he concentrated on a Wagnerian-like cycle of seven sacred operas, each one representing a day in the week. Known collectively as *Licht* (Light), and lasting for twenty-nine hours, it had not been staged in its entirety at the time of Stockhausen's death in 2007. One of the logistical difficulties is that, at one point, a chamber orchestra is directed to play from above the opera house in helicopters.

Music students still flock to Germany today to study and play.

Overcoming the Past in Painting

The art form that has adjusted best to Germany's Nazi past is painting and sculpture. The first figure to consider is Alfred Otto Wolfgang

Schulze (1913–51), better known as Wols. Born in Berlin, he studied the violin but preferred instead the Bauhaus under Moholy-Nagy; he moved to Barcelona in 1933 (where he refused to be called up) and then to Paris where he made a living as a photographer until he was interned in 1939 and began to paint. He suffered in great poverty during the war but was befriended and supported by Jean-Paul Sartre, though this couldn't prevent him from drinking himself to an early grave at the age of thirty-eight. Wols's pictures, formally called Tachiste, show openly the scars of his own life and his nation around him—they are, as one critic said, "eruptions of blood-red and black," reminiscent in form of carcasses, suppurating wounds, and the insect life that feeds on those wounds. There is no redemption in Wols's work.[63]

The main general phenomenon immediately after the war was the revival of abstraction. This was marked by the rehabilitation of artists like Klee and Kandinsky but it had one unfortunate side effect, forcing neglect on some German artists, which eventually provoked a group of painters in Düsseldorf in 1957 to mount a series of exhibitions in the studio of Otto Piene. Düsseldorf had emerged as the leading school, with artists from East Germany such as Gerhard Richter and Sigmar Polke moving there in the late 1950s and early 1960s to create their blend of kitsch and Pop Art, known satirically as Capitalist Realism. (Richter's father-in-law had been responsible for overseeing the mass sterilization of women, including the painter's own aunt, under the Nazis.)

The group around Piene, much influenced by Yves Klein and his doctrine that art is about ideas rather than any one view of reality, took the name Zero.[64] Their objective was to "strike out reality," notably by denying form, producing images (often in monochrome, or as explorations of whiteness) that had no physical presence other than sheer energy. Piene's 1963 *Venus of Willendorf* is the defining work here, along with Günther Uecker's *Hunsrückenstrasse*, a whole street painted entirely white. The developments of the Zero group were accompanied by an important private initiative that turned into one of the more successful elements of the late twentieth-century art world. This was the establishment of Documenta in Kassel, which has evolved into one of the great forums for contemporary art.

Josef Beuys's Dialogue with Time

All this was overshadowed by the advent of Joseph Beuys, who stands apart (and, for many people, above) all else in German postwar art. Beuys, born in Krefeld in 1921, never deviated from his conviction that his artistic aim was to find a new visual language that would come to terms with the war and at the same time find a way forward that did not ignore all that had happened.

The work of art, Beuys believed, exists in "eternal time, historical time, and personal time."[65] Having himself been shot down over Russia as a Luftwaffe pilot in the Second World War, he was treated for frostbite by his Russian captors, who used felt and fat, which became the materials Beuys used in (some of) his art, fused with other, less personal substances.[66] He felt the spectator should be aware of what these materials meant to the artist, adding a level of consciousness to the aesthetic experience: the artist is a person with a past, part of the national past. His famous piece *Strassenbahnhaltestelle* (*Tram Stop*) fuses his own experience (as a boy he used a tram stop near an important monument), with the national past, featuring railway lines to remind the viewer what railways were used for in Nazi Germany. *But*, his lines were slightly curved, to hint at progress, a way forward, and *up*. In experiencing the present-day beauty of his sculptures, Beuys is saying, we must relive past events—this is his dialogue with time.[67]

Beuys exerted an influence through his pupils, notably Jörg Immendorf and Anselm Kiefer, though a younger generation—Markus Lüpertz, Georg Baselitz, and A. R. Penck—has reacted against the high-culture associations that even Beuys's work displayed. Kiefer's materials include sand, straw, and burnt wood, and they often superimpose one simple image floating above the landscape below—denser, more damaged, more chaotic. This, for Kiefer, is the level of shame.[68] Baselitz, influenced by Munch, is a painter of monumental images that are, in his own words, an attempt to create an "aggressive disharmony" of color, though he too incorporates his own experiences—of the Wall and of the rebellions of 1968—into his work. Penck's stick figures mix cave painting and graffiti, the lurid, the sensual, the abuse and pathology of intimacy, which was, for him, the Nazis' greatest crime.[69]

In the works of the latest generation, Rainer Fetting, Helmut Middendorf, and Jiri Georg Dokupil, all high-culture trappings have been abandoned in favor of rock culture "at its most frenzied," so that there is an (all too postmodern) ironic distance between the artists and the issues depicted.[70] Here the iconic work is Immendorf's great series of *Café Deutschland* paintings, produced in the 1970s, in which the main elements of German political history take place in discotheques and druggy, cross-dressing cabarets, even on the flight decks of airplanes. Painting, in the words of Irit Rogoff, has become (and this too is a Germany we haven't seen before) "raucously informal."[71]

Culturally and intellectually, the biggest development since the Wall came down has been the collapse of East Germany. East German individuals still shine in painting, in film, and in literature, and there are opera houses and artists' colonies in such cities as Dresden and at the Baumwollspinnerei in Leipzig, with strong choirs in Leipzig, Dresden, and Berlin. But during the years of a divided Germany few outside the GDR understood how rotten the infrastructure was, that the trade currents within Comecon would simply disappear, that there was nothing to build on. In the Meissen factories the skill of the people remained, but after the Wende *all* the machines had to be thrown out. Even the vineyards in East Germany needed new plants. The Dresden camera maker, Pentacon, had a 10 percent share of the world market in the Communist era, with 5,000 employees in 1990; within a year just 200 were left.[72] The Berlin-Brandenburg Academy of Natural Sciences and Humanities, which had 24,000 members in the old East Germany (many of them "parked" there, as a form of hidden unemployment), now has 200 members and 175 research fellows.[73]

Among former East German scholarly projects, the definitive edition of the works of Marx and Engels has been taken over by the Berlin-Brandenburg Academy, as has the Goethe Lexicon, a dictionary and analysis of the language of Goethe. The scholarly strengths of the GDR were mainly in the realms of mathematics and computer science, molecular biology, pharmacology and energy, and, in the humanities, in Oriental and antique languages (where Manfred Bierwisch was well known); but most of these areas have now collapsed and almost all academic journals have been discontinued (whereas publishing in Germany overall has been growing at 1.9 percent annually).

In science, since 1945 German-born physicists, chemists, and medical doctors have won twenty-five Nobel Prizes, in addition to the Nobel Prizes for Literature won by Böll, Grass, and Herta Müller (in 2009) and Willy Brandt's Peace Prize in 1971. This success in science is due largely to the Max Planck Society: there are seventy-six Max Planck Institutes scattered over Germany, one in the Netherlands, two in Italy, one in Florida, and a sub-institute in Manaus, Brazil. Current research strengths lie in turbulence studies, superconductivity, quantum optics, quantum Einstein gravity, and evolutionary biology. The Max Planck Institute for the History of Science, in Berlin-Dahlem, is returning to traditional German concerns about the nature of knowledge, conducting research projects on The History of Laboratory Sciences, The Rise and Decline of the Mechanical World View, and an investigation of the links between knowledge and belief.

Between 1999 and 2004, the Max Planck Society investigated the role of the Kaiser Wilhelm Institutes in the National Socialist era, focusing on the continuity or discontinuity of scientific activity, the extent to which science was used as a "legitimation" for the regime's policies, which experts knew what and when, racial hygiene, military research, *Ostforschung* (research on the east), and *Lebensraumforschung* (research on living space). No detailed results have yet been published.[74] Several universities now teach science courses in English because that is the language of science.

Elsewhere in Berlin, the city has become Europe's biggest concentration of contemporary architecture. The first buildings to rise from the rubble of 1945 were considered at the beginning of the previous chapter. Hans Scharoun, also mentioned in the previous chapter, continued to shine in the 1960s.[75] His twin apartment blocks at Stuttgart Zuffenhaussen, known locally as Romeo and Juliet (1954–59), lean this way and that in a configuration that would be made popular worldwide by Frank Gehry.[76] Much influenced by Scharoun was Günter Behnisch who, with Fritz Auer and Frei Otto, designed the celebrated Olympic Games Complex in Munich (1965–72). Their soaring "tent roof" over and above the Stadium, Olympic Hall, and Olympic Pool would find an echo across the world as far afield as the Barbados airport. In the 1980s, a raft of museums was erected throughout West Germany, notably in Frankfurt and notably its Museum of Modern Art, designed by the Austrian Hans Hollein (1985, 1987–91).[77]

Reunification created architectural opportunities on an unprecedented scale. Initially the main aim was to rebuild central Berlin, which had been in the East. Among the early completed or renovated buildings are the Reichstag, with Norman Foster's glass cupola, the Federal Archive, the DG Bank (Frank Gehry), the library of the Freie Universität (Norman Foster), the Holocaust Memorial (Peter Eisenmann), and the Potsdamer Platz, an entire area that has been rebuilt since 1995.

In some ways the most effective—and most beautiful—architectural project (or is it sculpture?), which also presents the face of a Germany we haven't seen before, are the *Stolpersteine*, the "stumbling stones" of Gunter Demnig. These stones, set into the pavement, in Cologne to begin with, are slightly raised cobblestones located outside houses where murdered Jews once lived. A brass plate is nailed to each stone containing basic details: "Here lived Moritz Rosenthal. b. 1883. Deported 1941. Lodz. Died 28.2.1942." The first stones Demnig installed were illegal, but the idea caught on and in 1999 he was officially approved. More than a thousand stones are now in place in Cologne and in several other cities.[78]

In 1999, Dietrich Schwanitz, a historian and philosopher, who had studied in London and Philadelphia as well as Freiburg, and was then a professor of English literature at the University of Hamburg, published *Bildung*, "a handbook" that was essentially a device to address what he saw as a crisis in German education, by reintroducing a "canon" of works that would teach students to be at home in culture, to understand why they should know "Shakespeare, Goethe, and van Gogh," to have a conversation with history, to grasp the "great European narratives" that have brought us to this point. He thought that, in Germany at least, there had developed a disruption between school life and university life and that Bildung was the best way to bridge it. He somewhat spoiled his argument by asserting that Bildung was also a game, with snobbish elements, but perhaps he felt that such glosses were necessary in the contemporary world so as to "sell" the idea. But he discussed Humboldt, Hardenberg, Herder, and Hegel in a spirited attempt to turn the clock back that, to an extent, succeeded in the sense that as this book went to press his *Bildung* was in its twenty-second printing.[79]

* * *

When we look at the way German culture has "come back" since the years of National Socialism, at the very great depth and variety of German postwar poetry, at the country's serious theater, at the high ambition of its dancers, at its continued dominance in musical composition, performance, and scholarship, at its second film renaissance, at its preference for art over entertainment, at the bitter debate about the deleterious effects of popular culture in general and television in particular, we realize that High Culture is the culture of the educated middle class and that that whole constellation of ideas and concepts is more deeply rooted in Germany than elsewhere, even now, and after all that has happened.

Conclusion

German Genius: The Dazzle, Deification, and Dangers of Inwardness

"The finest characteristic of the typical German, the best-known and also the most flattering to his self-esteem, is his inwardness."

—Thomas Mann

In January 1939, W. H. Auden, the English poet, arrived in America. He had emigrated, he said, because it was easier there to "live on one's wits." One of the most famous homosexuals of the twentieth century, Auden was married at the time: in 1936 he had wed Erika Mann, Thomas Mann's daughter, in order to provide her with a British passport and escape from Nazi persecution ("What are buggers for?" he asked). In the United States he saw a lot of the Manns—he was an editorial adviser to Klaus Mann's magazine, *Decision*, he visited Thomas and Katja in California, and at the house they rented on Rhode Island from Caroline Newton, a rich East Coaster who had been psychoanalyzed by both Sigmund Freud and Karen Horney. He met Wolfgang Köhler, one of the originators of Gestalt psychology, who he described as "a great man with quite a lot of neuroses." He moved in a German/German-friendly world. But then Auden, so unusual in so many ways, was unusual in the sense that, unlike many other educated English people and Americans, of his and other times, he had long been fascinated by German culture. He had spent several months in Germany in 1929, some of them with his friend Christopher Isherwood; they had collaborated on a play, Auden had written a handful of verse in (poor) German, started work on his long poem *The Orators* and, as he put it in his Berlin journal, spent "my substance on strumpets, and taking part in the White Slave traffic." This sexual underworld notwithstanding, he recognized that Germans had in many ways defined the age in which he lived. He dedicated a poem, "Friday's Child," to Dietrich Bonhoeffer, and another, written after the death of Sigmund Freud, in 1939, emphasized the nature and extent of his influence:

For one who lived among enemies so long;
If often he was wrong and at times absurd,
To us he is no more a person
Now but a whole climate of opinion.

Under whom we conduct our differing lives. . .

Two years later, in 1941, Auden went on to describe Franz Kafka as "the artist who comes nearest to bearing the same kind of relationship to our age that Dante, Shakespeare and Goethe bore to theirs." [1]

Despite the horrors of National Socialism, Auden remained involved in German culture and ideas.[2] In New York he used to visit a German-language cinema in Yorkville, he collaborated with Brecht, formed a close friendship with Hannah Arendt, and was fascinated by another German psychiatrist in exile in the United States, Bruno Bettelheim, and what he had to say about autism, believing that his own allegedly partly autistic childhood and his calling as a poet were related.

In 1959, after buying a house at Kirchstetten, outside Vienna, Auden became increasingly drawn to Goethe (describing himself as a "minor Atlantic Goethe"). He composed a series of "prose meditations on love," called "Dichtung und Wahrheit" (Poetry and Truth) after the title of Goethe's autobiography, a short while later taking on a translation of the

German genius's Italian travel book. He collaborated with Hans Werner Henze on the opera *The Bassarids*, which many think is Henze's masterpiece. Auden's interment took place in the local church at Kirchstetten; the music at the ceremony was Siegfried's funeral march from *Götterdämmerung*.

Auden remained close to German culture, and German ideas, despite everything that happened in the first half of the twentieth century, and in this too Auden was, for a well-known Anglo-American, unusual if not unique. But, as should now be clear, he was not wrong in following the path that he did. As he himself might have put it, the climate of opinion under which we live our differing lives is, much more than we like to think, German.

With the exception of market economics and natural selection, the contemporary world of ideas is one that, broadly speaking, was created by, in roughly chronological order, Immanuel Kant, Georg Wilhelm Friedrich Hegel, Karl Marx, Rudolf Clausius, Friedrich Nietzsche, Max Planck, Sigmund Freud, Albert Einstein, Max Weber, and two world wars. The ideas of another German, Gregor Mendel, are gaining ground fast at the start of the twenty-first century—it has now been shown that genes govern all manner of behaviors, from certain forms of violence to depression and promiscuity—but they do not cohere together into the overall picture as created by these other German geniuses.

Along with his fellow German-speaker, Adolf Hitler, Karl Marx probably had a more direct effect on the recently completed twentieth century, and the shape of the contemporary world, than any other single individual. Without him there would have been no Lenin, no Stalin, no Mao Zedong, and few if any of the other dictators who disfigured those times. Without him there would have been no Russian Revolution, and without World War II (or Max Planck and Albert Einstein), would there—could there—have been a Cold War, a divided Germany? Would decolonization have occurred in the way that it did, would there have been an Israel where it is, the Middle East problem that there is? Would there have been a 9/11? Ideas don't come any more consequential than Marxism.

In his biography, *Das Kapital*, the British writer Francis Wheen asserts in his final sentence that Marx "could yet become the most influential thinker of the twenty-first century." He quotes a series of figures who one would normally take to be right-wing, conservative big-business men—the exact opposite of Marxist—who have come back to Marx and even to Rosa Luxemburg. It is not just that what Marx had to say about monopolization, globalization, inequality, and political corruption sounds so pertinent after 150 years, but that we take so much of Marx for granted now, without most of us even knowing it. We accept, implicitly, that economics is the driving force of human development; we accept, implicitly, that the social being determines consciousness; we accept, implicitly, that nations are interdependent; we accept, implicitly, especially in the realm of the environment, that capitalism destroys as it creates. After the credit crunch and stock market collapse of 2008, the sales of *Das Kapital* rose markedly, especially in Germany.

Wheen's point, and the argument of the people he quotes, is that these matters have become ever more visible since the Berlin Wall came down, and the socialist "alternative" to capitalism collapsed. Did the existence of two Germanies, and the rivalry they stimulated, keep capitalism seemingly more healthy than the alternative, than would otherwise have been the case? Either way, Germans and Germany are at the center of the argument.

The Most Consequential Contemporaries of the Modern World

Sigmund Freud's influence was less catastrophic than Marx's, but no less consequential. There are two ways of looking at Freud's legacy. One is to consider him on his own, to outline the specific ways in which psychoanalysis has affected all our lives; the other is to consider him together with his contemporaries, Nietzsche and Max Weber. Both approaches will be attempted here since this is the only way that the full impact of this cohort of German thinkers can be appreciated.

Alfred Kazin, the American critic, maintained in an essay he published in 1956 to mark the one hundredth anniversary of Freud's birth that "Freud has influenced even people who have never heard of him."[3] Kazin thought that, at mid-century in America, "to those who have no belief, Freudianism sometimes serves as a philosophy of life."[4] He thought that at "every hour of every day now," people could not forget a name, feel depressed, or end a marriage without wondering what the "Freudian" reason might be. He thought that the novel and painting (Thomas Mann, T. S. Eliot, Ernest Hemingway, William Faulkner, Pablo Picasso, Paul Klee, Expressionism, Surrealism, Abstraction) had been reinvigorated by the Freudian knowledge that "personal passion is a stronger force in people's lives than socially accepted morality" and that the "most beautiful effect" of Freudianism was the increasing awareness of childhood "as the most important single influence on personal development."[5] He thought the insistence on personal happiness—the goal of psychoanalytic therapy—was the most revolutionary force in modern times, a modern form of self-realization.

Another aspect of Freud's legacy is that we are now, to use Frank Furedi's phrase, taken from Philip Rieff's book *The Triumph of the Therapeutic*, living in a "therapeutic society." In the therapeutic society, as Furedi puts it, "there is an inward turn . . . The quest for personal self-understanding through the act of self-reflection is one of the legacies of modernity . . . the self acquires meaning through the experience of the inner, emotional life . . ."[6] Especially among those who are no longer religious, there is a widespread belief in an alternative self, somewhere within, and with it goes the essentially therapeutic belief that, if we can only "get in touch" with this inner, alternative (better and "higher") self, we can find happiness, contentment, fulfillment. The "soul" has been secularized.

Not everyone has been so sanguine about Freud. Richard Lapierre thought that "the Freudian ethic," as he called it, was responsible for many of the discontents and false pathways of modern society. "In the Freudian concept, man is not born free with the right to pursue life, liberty and happiness; he is shackled by biological urges that can never be freely expressed and that set him in constant and grievous conflict with his society."[7] Lapierre thought Freudianism had been responsible for "the permissive home," "the progressive school," "the condoning of crime," and "the maternalization of politics" (now called single-issue or identity politics), none of which he cared for.

Christopher Lasch, himself a psychoanalyst, was still more caustic. He said frankly that we now have what he called a culture of narcissism, economic man (Marxist man) having given way to psychological man. He too said we have entered a period of "therapeutic sensibility": therapy, he argued, had established itself "as the successor to rugged individualism and to religion." This new narcissism means that people are more interested in personal change than in political change, that encounter groups and other forms of awareness training have helped to abolish a meaningful inner private life—the private has become public in "an ideology of intimacy." This makes people less individualistic, less genuinely creative, and far more fad- and fashion-conscious. It follows, says Lasch, that lasting friendships, love affairs, and successful marriages are much harder to achieve, in turn thrusting people back on themselves, when the whole cycle recommences. Modern man, Lasch concluded, was actually *imprisoned* in his self-awareness. He longs for "the lost innocence of spontaneous feeling. Unable to express emotion without calculating its effects on others, he doubts the authenticity of its expression in others and therefore derives little comfort from audience reactions to his own performance."[8]

There is no shortage of evidence for the startling penetration of the "therapeutic sensibility" in our society. A troop of Brownies in California has its own stress clinic for eight-year-olds; a primary school in Liverpool, England, gives its stressed children aromatherapy. In 1993 British newspapers used the word "counseling" 400 times in a year; by 2000 it had risen to 7,250; some 1.2 million counseling sessions take place each month in Britain. Recently, the Archbishop of Canterbury himself claimed that therapy was "replacing Christianity" in Western countries, that "Christ the Saviour" is becoming "Christ the counsellor."[9]

If that all seems rather a lot to lay at Freud's door—well, we are not done yet. Freud must also be understood in the context of his German-speaking contemporaries Friedrich Nietzsche and Max Weber.

The "Entrance Door" to Modern Thought

Nietzsche's most well-known—some might say notorious—aphorism is "God is dead." One of his most important achievements, along with Max Weber, was to *think through and confront* the implications of that sentiment, to work out in what he saw as terrifying detail the consequences of modernity, a world of vast populous cities, mass transport, and mass communications, in which the old certainties had been dissolved, where the comforts and consolations of religion had disappeared for many people, and in which science had acquired an authority that was, in his view, as arid and empty as it was impersonal and impressive. It is in this sense that Martin Heidegger called Nietzsche the "culmination" of modernity—i.e., Nietzsche felt the loss of whatever had gone before more keenly than anyone else, and he described that loss in more vivid hues.

Formally, Nietzsche's influence is second only to that of the Greeks and Kant, and maybe even that doesn't do him justice. Until, roughly speaking, the Second World War, his influence was primarily literary and artistic. Robert Musil regarded Nietzsche's thought "as one of the great events of the twentieth century."[10]

Beyond art, Anatoly Lunasharski and Maxim Gorky tried to construct a "Nietzschean Marxism" in Russia but that did not outlast the rise of National Socialism and *their* appropriation (and inversion) of some of his themes (the last thing Nietzsche was, was an anti-Semite). But as the twentieth century lengthened, Nietzsche's relevance became clearer. Stephen Aschheim, in his study of the Nietzsche legacy in Germany, lists books detailing the philosopher's influence in Italy, "Anglo-Saxony" (Britain, the United States), Spain, Austro-Hungary, and Japan, and on the Catholic Church and Judaism. Karl Jaspers saw Nietzsche as "perhaps the last of the great philosophers of the past" and, as Ernst Behler has noted, divided the intellectual history of the West into two periods: "one marked by the domination of the logos and the admonition 'Know Thyself,' culminating in Hegel; the other characterized by a radical disillusionment with the self-confidence of reason, the dissolution of all boundaries, and the collapse of all authority, a period that began with Kierkegaard and Nietzsche." Together with Marx, Behler said, "They stand at the entrance door to modern thought."[11]

For Heidegger, Nietzsche's philosophy "is the completion of Western metaphysics"; with the interpretation of Being as the will to power, he "realised the most extreme possibility of philosophy."[12] Among modern philosophers, Nietzsche's influence has been reflected most keenly in the work of Michel Foucault, Gilles Deleuze, Richard Rorty (who characterizes the entire present age as "post-Nietzschean"), Alexander Nehamas, Eugene Fink, and Jacques Derrida. As Bernd Magnus and Kathleen Higgins put it, "Nietzsche's influence has become unavoidable in our culture."[13]

That culture is "modernity" and in his quest to understand and explain modernity, Nietzsche in effect tells us that the search for "absolute truth, universal values and complete liberation" is impossible.[14] Our profound psychological/philosophical condition in the modern world, says Nietzsche, is that we long to believe the old, the traditional certainties, but we cannot, we are trapped on the far side of scientific discoveries that destroy the old beliefs while replacing them with—nothing. Progress, philosophical progress, has reached an impasse: " . . . it is the disorganising principles that give our age its character."[15]

Nietzsche called this condition, the absence of any moral purpose to the world, any direction, "nihilism," and it had three—at least three—important consequences: there is no meaning to events, we lose faith that anything is to be *achieved*, or *can* be achieved; there is no coherent pattern in history; and there is nothing universal that we can all agree upon or aspire to. Our world is motivated mainly by our own inner psychological needs, rather than by any "truth" (a meaningless and malleable commodity, the only purpose of which is to enhance our feeling of power). He thought that our main psychological need was just that—the celebrated "will to power" and, for himself, felt that the only basis for any judgment, now that all other bases had disappeared, was the aesthetic one.

Even in making aesthetic judgments, since we have no grounds for agreement in any "deep" or universal sense, because there is no longer any basis for meaning, the only criterion by which originality or creativity or beauty may be judged is by their "newness." Even here, however, newness will be obsolete more or less immediately because it can have no meaning over and above the fact that it is new. This applies to changes in ourselves as much as in conventional works of art or developments in history or fashion. There can be no direction in our personal development, only meaningless change, change for the sake of it.

This is, needless to say, arguably the bleakest analysis of the human condition there has ever been, and Nietzsche intended it as such. ("I am by far the most terrible human being that has existed so far," he said in a famous passage. "This does not preclude the possibility that I shall be the most beneficial.") He thought there was no escape, that he was—we are—living at a unique time in history, when a sea change in philosophy and psychology was taking place, a "new man" was being born. It was this chilling message that echoed down the twentieth century and was only slightly alleviated by what Max Weber observed.

Just as Nietzsche's most famous aphorism was that "God is dead," so Weber's was that we now live in a world that is in a state of *Entzauberung*, that is "disenchanted." Weber made two main claims about modern life. One, its discontents were brought about, as Lawrence Scaff glosses it, by capitalism, technology, economic rationalism, and the institutionalization of instrumentalism—in other words, the main aim now is to *control* the world in an abstract, intellectual manner rather than to enjoy it in an aesthetic or sensual way. The modern condition is that we have to choose between knowledge that in Weber's words is "untimely and troubling" or undergo a "sacrifice of the intellect" as when we embrace a religious faith or a closed philosophical system like Christianity, Marxism, or Hegelianism.[16] We believe we can master all things by calculation—there is now a "romanticism in numbers"—and that science can preserve life. At the same time science cannot "answer whether the quality of the life preserved is worth having."[17] The idea of a "unified self" simply lies beyond our grasp in the modern world.[18]

Weber's other argument was that modernity involved a heightened preoccupation with the "inner self" that leaves us having to create our own ideals and values "from within our chests . . . We cannot read the *meaning* of the world in the results of its investigation, no matter how perfect, but must instead be in a position to create that meaning ourselves . . . therefore the highest ideals, which move us most powerfully, are worked out for all time only in struggle with other ideals, which are just as sacred as ours are to us."[19] Only in the West, he said, has mankind developed the idea of understanding himself in a *universal* way, in other words on principles that apply to all human beings at all times—this is essentially what science aims at. In other cultures, people do not have this aim, they are content to explain themselves to themselves as they are, at their particular point in history and location in the world. Why should the West be so concerned about it, and doesn't it condemn us to an empty, cold existence? The result, he said, for many—for most—people was that the only meaning in life was the pursuit of pleasure, entertainment, self-gratification, or money. In America, he said, capitalism, devoid of any religious or ethical meaning, had acquired the character of sport, and this had replaced the search for salvation. We are burdened by a surfeit of knowledge that doesn't tell us how to live or what living is for.

The final nail in the coffin that is modern culture was for Weber the fact that most people worked too hard and too long, so that they had no time—and no inclination after a day's work—to come to grips with the modern condition and sort out, for themselves, how best to *experience* the world, to address the question, "What Comes Next?"[20]

There is clearly a heavy overlap between Weber and Nietzsche. Each had a common core of things to say about the terrors built into the modern world, and each amplified what the other was saying. Weber was, if anything, marginally less pessimistic than his near contemporary. His way of writing implied that the modern world could at least be renounced, whereas Nietzsche, by and large, thought there was nothing to be done. Heidegger's concepts of "submitting" to the world as it is, or "caring" for it, rather than controlling it, take up Weber's challenge, as did Marcuse and his idea of "the great refusal."

But it should now be plain to what extent we are indeed living in a post-Nietzschean, post-Weberian nihilistic world—for example, in the realm of contemporary high art, where the only criteria by which it is now judged is by newness, where the big auctions have all the qualities of a game, and collecting has become for so many a form of salvation. The world of fashion, where again the defining criterion is sheer newness, is another nihilistic aspect of the modern world. In all these realms, money is a prominent feature.

But, in a sense, these are peripheral. To what extent, we may ask, were the terrible brutalities of the twentieth century carried out by nihilists who, since they could see no moral purpose to the world, could see no objections to the cruelties they inflicted? Hannah Arendt said that terror lay at the root of totalitarianism, and nihilism is surely the greatest terror there is.

Furthermore, beyond the nihilistic horrors of Fascism and Stalinism and Maoism, there is another way our lives have been affected—and are *still* affected—by the cold, empty, bleak landscape Nietzsche and Weber identified. This takes us back again to Freud. The vast majority of people have almost certainly never read Nietzsche or Weber. But, just as Alfred Kazin said that people who had never heard of Freud had nevertheless been influenced by him, so the same is true of those who have never heard of Nietzsche or Weber.

Despite the great economic earthquake that took place in the late summer and early autumn of 2008, we still live in a world of unprecedented prosperity and comfort—at least, many of us do in the West. Even the worst off in the developed world are cushioned from absolute material degradation by a welfare state. And yet—and it is revealing that it is a commonplace to say so—we are surrounded by criminal violence, drug abuse, child abuse, high-school massacres, gangland vendettas, piracy on the high seas, organized prostitution, and sexual slavery. There are more people in prison and in mental hospitals than ever before, vandalism is widespread, and alcoholism is rampant. It is not too much to say that these are all responses, however inchoate, to the nihilistic existential landscape of modern life, by people who, though they may never have read Nietzsche or Weber, nevertheless recognize, or experience, or feel themselves trapped in the empty, cold, bleak terrain these German speakers identified. The incoherence of their response is part of the condition.

This surely helps explain why Freud has had the impact he has. In recent years he has come under sustained attack, justifiably so, for fabricating evidence, falsifying his early "cures" and being, generally, wrong. But, in the context of this discussion, that is to misconstrue him. Contemporaneously with Nietzsche and Weber (and too little has been made of this), as they were diagnosing the predicament of modern life, Freud was finding, or inventing, or stumbling across, a solution to that predicament. Psychoanalysis, therapy, "talking cures," are misconceived if they are understood simply or mainly as a way to treat neurosis and other forms of mental illness (and this is why, generally speaking, they have been judged a failure in that regard). What Freud set in train with *The Interpretation of Dreams*, published in the very year Nietzsche died, was a method by which people could use their individual histories so as to reconstruct *meaning* into their lives, a way—however tendentious, hypothetical, abstract, clinically suspect—they could relate to the fragmentation and sheer *emptiness* of the modern world around them. The fact that therapy is so much a part of our lives (even very young lives) emphasizes that we are inhabiting a Nietzschean, nihilistic world.

The First XI of Modern History

To repeat: Kant, Humboldt, Marx, Clausius, Mendel, Nietzsche, Planck, Freud, Einstein, Weber, Hitler—for good or ill, can any other nation boast a collection of eleven (or even more) individuals who compare with these figures in regard to the enduring influence they have had on modern ways of thought? I suggest not. But the German genius is not just a matter of numbers. In the Introduction, several pages were devoted to a question that many people have found—and still find—fascinating, obsessive even, namely whether German history went through a *Sonderweg*, a special path which, *necessarily*, was destined to result in the horrors and excesses of National Socialism and the Holocaust. To my knowledge, no one has explored in any scholarly way, and in an overall sense, whether there is a systematic relationship between political history and cultural history. However, looking at modern German *culture*, as this book has been designed to do (using "culture" in its Franco-Anglo-American sense, rather than the German sense of *Kultur*), and bringing that cultural history right up to date—looking at its achievements *after* the Holocaust as well as before, beyond Hitler in both directions—one may conclude that there were several features of that culture that may be construed as, if not *necessarily* leading to catastrophe, then at least helping to explain why what happened in Germany happened there when it did.

Of course, no explanation is ever a complete explanation. But the argument here is that there were five distinct yet interlocking aspects of modern German culture that, as a group, accounted for both its dazzling brilliance and its shocking demise.

An Educated Middle Class

Conventional wisdom, especially conventional wisdom since Marx, has it that societies are most usefully understood as being divided into three levels or classes: the aristocracy, the middle classes, and the proletariat or working class. It should now be clear, however, that the *educated* middle class has very little in common with the rest of the middle class and certainly, in Germany, may historically be considered a separate entity. Indeed, the educated middle class is middle class only in the classically

Marxist sense that its mode of production is that of neither the aristocracy nor the laboring classes. Yet the educated middle class—inhabiting the world of scholarship, the arts and humanities, science, the legal, medical, and religious professions—has very little in common with, for example, organized labor, shopkeepers and retailers, industrialists, or financiers, either in terms of motivation, aspiration or, indeed, everyday interests and activities. These differences were more marked in the nineteenth century, but it is clear from what has gone before that Germany was the first country to boast an educated middle class of any size and that this was all important for its emergence as a great power.

A few statistics will underline this. Prussia enforced school attendance for children between the ages of seven and fourteen from the 1820s (in Britain children were not compelled to go to school until 1880) and by the 1890s had two-and-a-half times as many university students in proportion to population as did England.[21] We saw in Chapter 22 how in the late nineteenth century, illiteracy in the German army was much lower than among Italian or Austro-Hungarian soldiers, 1 in 1,000, as opposed to 330 in 1,000 among Italians, and 68 in 1,000 among Austro-Hungarians. In another chapter we saw that, in Germany, in 1785 there were 1,225 periodicals published, compared with 260 in France. In 1900 Germany had 4,221 newspapers, France roughly 3,000 (and Russia 125).[22] In the early nineteenth century, when England had just four universities, Germany had more than fifty. James Bowen, in his three-volume history of Western education, points out that Germany took the lead in the establishment of scientific societies in the early nineteenth century, published the greatest number of journals in the vernacular, and became the leading language of scientific scholarship.[23] In 1900 illiteracy rates in Germany were 0.5 percent; in Britain they were 1 percent and in France 4 percent. By 1913 more books were published annually in Germany (31,051 new titles) than in any other country in the world.[24]

For what it is worth, the Germans are still ahead in some familiar ways, even though many of them don't think so (see the comments of Dietrich Schwanitz, p. 815). In a survey reported in 2006, it was found that the average brain size of northern and central Europeans was larger than that of southern Europeans (1320 cc compared with 1312 cc). This translated into higher intelligence, with Germany and the Netherlands coming in on top (107 IQ points), Austria and Switzerland at 101, while Britain (where the research was done) scored 100, and France 94.[25]

It was the educated middle class that made the exciting advances in scholarship that so attracted academics from abroad (especially from America), that rendered the bureaucracy of the ever-coalescing German state so efficient and creative and led to the groundbreaking scientific achievements of the second half of the nineteenth century, that transformed Germany economically, and on which so much of modern prosperity—not just in Germany—is based. The rise—and then the fall—of the educated middle class is central to what happened in Germany and still has a contemporary relevance.

The development of modern scholarship, the concept of Bildung, and the innovation of the research-based university were seen at the beginning of the nineteenth century in Germany as a form of moral progress. Education was not simply the acquisition of knowledge but looked upon as a process of character development during the course of which a person would learn to form critical judgments, make an original creative contribution, *and* learn about his or her place in society with its duties, rights and obligations. Education as Bildung involved a process of *becoming*, a form of secular perfection or salvation that was, for the educated middle class, the very *point* of life in a world between doubt and Darwin.

The educated middle class had essentially taken over and expanded the role occupied in earlier times by the clergy and would remain the most important and innovative element in Germany in the century that lasted from 1775 to 1871. Toward the end of that time, the situation began to change and grow more complex, as is also discussed.

"Inwardness"

It seems clear that the Germans were (still are?) a more "inward" people than others—the French, British, or Americans, for example (though as Gertrude Himmelfarb notes, the Enlightenment in England "throve *within* piety"). The Germans certainly seem to have seen themselves for the most part in this light, as the lines from Thomas Mann quoted at the head of this Conclusion confirm.[26]

The combination of Lutheranism and Pietism was a starting point here, both being more concerned with inward conviction than with outward displays of religiosity. Another factor is that Germany's centers of learning—its universities—came on stream between the advent of doubt and the arrival of Darwin's theory of natural selection, when the theological understanding of man was under severe threat and Darwin's biological understanding not yet available. This state of affairs applied not just in Germany, of course, but it was stronger there than anywhere else, for several reasons. Many people became convinced that, if traditional notions of God were under threat, there must be some other purpose to life, some other teleology and, as we saw in Chapters 2 and 5, pp. 65 and 135, the Germans embraced an evolutionary form of teleological biology and at the same time the great systems of speculative philosophy came into being: Kant's Idealism, Fichte, Hegel, Naturphilosophie, Marxism, and Schopenhauer. The era between doubt and Darwin was the great period of speculative philosophy, and many people thought that Kant in particular had devised a new way of looking inward, of observing new structures of our minds.

The reading revolution interacted with this. Reading was a much more private—and therefore inward—activity than the most popular cultural activity that had preceded it, dancing and singing (see pp. 55–58). Given that Germans read more than anyone else because they were more literate, this too added to their inwardness.

Romanticism and music were still other aspects of inwardness. Listening to "the inner voice" was one of the main aims of Romanticism, one of the principles it adopted from "inner" Oriental religions, in which the artist, who creates from within, is the most advanced type of human being. Kant's instinct and intuition, Schopenhauer and Nietzsche's will, Freud and Jung's "unconscious" are all "inner" entities, inner concepts, as is the "second self," locked within, waiting to be released.

Schelling thought that music—music, the German art form par excellence—was the "innermost" of the arts, and the repeated and long-term association between German poetry and German music, finding expression in Schubert, Schumann, and Hugo Wolf, only adds to all this. At the turn of the nineteenth century, E. T. A. Hoffmann thought that music provided an entrance to a "separate realm, beyond the phenomenal"; as we saw in Chapter 6, p. 153, the symphony was regarded as an aspect of philosophy precisely because of its ability to penetrate inward, beyond words.

As we have also seen in earlier pages, for Wilhelm von Humboldt, Bildung was education through the humanities as the true path to inner freedom. The main aim of the German *Aufklärer* was the *Bildungstaat*, a state where the ideal was to "enrich the inner life of man." Suzanne Marchand has noted that F. A. Wolf's pursuit of philological expertise "contributed to the turning-inward of the university community after 1800" and that this was an "important innovation in scholarship." Kandinsky and Franz Marc confirmed that what they were trying to do, as abstract painting was born, was to give "impressions of an inner nature," "immaterial inner sensations." According to Erica Carter there was a "post-1968 interiority" brought about by the changes induced in that year of revolution, changes that were more psychological in Germany than they were elsewhere.[27] Martin Walser, in the words of Jan-Werner Müller, epitomized a "German form of interiority," the opposition of the "authentic private self and an untainted *Innerlichkeit* versus a superficial, even hypocritical public sphere," when he famously claimed that "poetry and inwardness" provided escape routes from the "inauthentic world of opinions," which usually led to a form of self-righteousness, part of the "entertainment industry."[28] Psychoanalysis, Expressionism in painting and film, the very concept of alienation in all its guises, the inward journey of the heroes in that uniquely German form of the novel, the Bildungsroman, the very dichotomy of "heroes *versus* traders," all these emphasize the inwardness of the German, the German way of life, and the traditional German set of values. Both Karl Jaspers and Günter Grass referred to Herder's "other, greater, deeper Germany"—i.e., the *Kulturnation*.[29] Martin Walser claimed that because of their "religious, inward-looking piety," Germans found it difficult to "act politically, like Englishmen."[30] Karl Heinz Bohrer thought that the most urgent task of reunification was to recover Germany "as a spiritual-intellectual possibility."[31] Even the events of 1968, according to Jan-Werner Müller, were a mixture "of Marxism and psychoanalysis."[32]

Inwardness comes with consequences, of course, as does everything. Karl Heinz Bohrer derided "Protestant inwardness," this "power-protected inwardness," arguing that it resulted in a form of provincialism and a neglect of national identity that was "likely to breed nationalist violence" and contributed toward its "belatedness."[33] Perhaps the most fateful consequence of inwardness was the concept of Bildung itself. Gertrude Himmelfarb is one of several historians who have commented on the similarity of Adam Smith's "invisible hand" and Hegel's "the cunning of reason." But whereas the invisible hand enables man to embrace an open-ended future, "reason" in Germany became embedded in Bildung, which idealized a distant Greek state 2,500 years in the past and out of which, ultimately, the disease of cultural pessimism emerged. However much we may aspire to high scholarship and the ideal of the well-rounded man, in Germany the shadow of Bildung was in the end the more powerful force.

The stereotypes we have of other people are too often crude and, almost by definition, overly simple, and they add to our problems rather than ease them. In the German case their stereotypes *about themselves* have been part of the problem too.

Bildung

Bildung is in some ways the primary achievement of educated inwardness—indeed, it could be held to be the natural end product. Goethe, it will be recalled, said specifically that the purpose of life when there is no God (this was after he lost his faith in the summer of 1788) is to *become*, to become much more than one was. "The ultimate meaning of our humanity is that we develop that higher human being within ourselves . . ." (see p. 120). Kant thought the difference between animals and man was that man can set himself goals and "cultivate the raw potentialities of his nature." In creating the very idea of purpose within us, he felt, we "enlarge" ourselves and those around us. This is inwardness, Bildung, and community (see p. 836) all in one.

William Bruford traced the idea of Bildung in novels all the way through the nineteenth century into the twentieth—Adalbert Stifter, Nietzsche, Thomas Mann in *The Magic Mountain*—and in the middle of the 1900s Karl Mannheim described Bildung as "the tendency toward a coherent life-orientation, the development of the individual as a cultural-ethical personality." He thought sociological study could add to our understanding of Bildung. Fritz Ringer described Bildung as "the single most important tenet of the mandarin tradition" and Christa Wolf, in *Nachdenken über Christa T*, explored the meaning and possibility of Bildung in Communist East Germany. In America, Allan Bloom's book, *The Closing of the American Mind*, was essentially a plea for a return to this German ideology. Bildung suited the educated middle classes—it made education the central aspect, the most important purpose of life in a post-Christian world. Of course the educated middle class, by definition, had privileged access to it. Bildung defined them and their difference from others. In 1968 there was a campaign in Germany for "Bildung für alles."

There was a crucial role here for pastors's sons, something else distinctive to Germany. As will have been noted, many of Germany's thinkers, right up until contemporary times, have been the sons and/or grandsons of pastors—Samuel Pufendorf, Gotthold Lessing, J. M. R. Lenz, Christoph Wieland, Friedrich Schelling, Friedrich and August Wilhelm Schlegel, Friedrich Schleiermacher, Johann Herder, Karl Schinkel, Johann Christian Reil, Rudolf Clausius, Bernhard Riemann, Theodor Mommsen, Jacob Burckhardt, Gustav Fechner, Heinrich Schliemann, Julius Langbehn, Wilhelm Wundt, Friedrich Nietzsche, Wilhelm Dilthey, Ferdinand Tönnies, Max Scheler, Karl Barth, Rudolf Bultmann, Paul Tillich, Albert Schweitzer, Emanuel Hirsch, Martin Niemöller, Gottfried Benn, Carl Gustav Jung, Jürgen Habermas (and not overlooking Angela Merkel, who is the *daughter* of a pastor). Besides being inward, many of these individuals had lost their own faith but nonetheless could not help but be influenced by their fathers; in many cases the secularization of salvation, of perfection, was part of their inheritance and achievement. The metaphor of salvation was difficult to lose. Many German professors retain the aura of the pastor even today.

Not all the influence of Bildung was good. Fritz Ringer concluded that in Germany the classical idea of the humanists became "entangled" with political conservatism and social snobbery.[34] This was to have profound consequences.

Research, the PhD, Scholarship, and Modernity

Research was not a German invention. As early as the twelfth century, Robert Grosseteste, bishop of Lincoln and chancellor of Oxford University, had conceived of the experiment as a way to further knowledge. But the important—the significant—achievement of the German universities of the late eighteenth and early nineteenth centuries was to *institutional-*

ize research, at the University of Berlin and subsequent universities modeled on it. In particular, the concept of the modern PhD is a German idea—and this is, conceivably, after Idealism, Marxism, and Freudianism, the most influential German innovation of modern times but much less appreciated.

This may seem an excessive claim to make, but the habit of having a well-educated young adult, usually in his or her mid- to late twenties, spend three or more years examining in detail a very specific aspect of the world about us, for little money but instead for love of the subject and, no less important, the honor of putting the letters "Dr." before one's name, setting one slightly apart (and above) and being accepted as part of the professoriate, has had an extraordinary effect on our times. It means that, at relatively little expense, we know our world in far more detail than anyone before, say, 1780, could ever have imagined.

The *institutionalization* of research released an entirely new activity on the world that many people who were not geniuses were nevertheless very good at. Modern democracies are characterized by entire new industries, each with their own talents—advertising and marketing, film directing, sports, journalism. Research was one of the first and by far the most important because so much else is based on it.

A third aspect of the institutionalization of research is that it has been a factor in the differentiation and fragmentation of the world. A direct effect of the PhD has been the proliferation of new disciplines, not just in the sciences (though proliferation has been especially strong there) but in the humanities, too, and in the social sciences. The fragmentation of the world and the specialization of scientists are issues of modernity that have especially taxed German writers, philosophers, and artists.

A fourth effect is that research is now a rival form of *authority* in the world—a rival, that is, to tradition, to religion, and to political experience. Almost all government policies and the practices of large industrial and commercial corporations are undertaken now only after assiduous research exercises. Moreover, many of us are more comfortable with this form of authority than any other because, provided the research methodology is sound, it tends to meet both rational and moral criteria. The fact that this authority is *impersonal* is both a strength and a weakness. It is fairer, but perhaps alienating.

In fact, research is now so important in our lives—as it has been for decades, if not for over a century—that it should really take its place alongside urbanization, industrialization, and the development of the mass media as a defining phenomenon of modernity itself.

The Longing for a Redemptive Community

This theme runs through modern German philosophy, literature, social science, history, art, and politics. It overlaps with, and is associated with, a similar longing for the "whole."

Kant was obsessed with the relationship between the whole and its parts, the meaning of organic unity; singing in choruses was understood by Goethe to be appropriate training for citizenship. Hofmannsthal believed that the ultimate theatrical experience was the "ceremony of the whole." The point—and the tragedy—of mass society, Hannah Arendt said, was that instead of creating "a higher form of human community," it produced isolation and loneliness, which, she insisted, was the common ground of terror and the cold and inflexible logicality of bureaucracy, leading to the executioners. For Max Weber there was no salvation in the modern world other than "the sentiment of community."[35] Wagner wanted to create the *Gesamtkunstwerk*, the whole artwork; Friedrich Meinecke advocated the formation of "Goethe communities" that would renew devotion to the "German spirit"; Gestalt psychology was an entire system built around the perception of "naturally occurring" wholes; and Ferdinand Tönnies and Werner Sombart wrote books about community and its redemptive possibilities.[36] Ernst Kantorowicz, the historian, identified what he specifically called the redemptive community. The National Socialists had their concept of the "community of fate." Goebbels insisted that the purpose of radio in the Third Reich was to "solidify the community," and Hitler spoke of the "Volkswagen community," a shared freedom of the new Autobahnen that brought people together in the enjoyment of new technological achievements. Redemptive communities are a specific feature of Thomas Mann's *Dr. Faustus*. Martin Walser upholds an ideal of "non-alienated, to some extent communal subjectivity."[37] This is why he thought the conscience should be a private affair "so that the newly united national community could be reconciled with itself."[38]

In Germany scholars themselves were part of their own redemptive community, more so than scholars elsewhere. Not only were many of Germany's thinkers the sons of pastors—growing up in a nineteenth-century background where the pastor was the very center of the community—but it was commonplace for scholars in Germany (in marked contrast to other countries) to attend three or four universities during the course of their training—it was a privilege built into the system. It naturally follows that the sense of an academic community, a redemptive community of scholars, a union of the educated middle class, was much stronger in Germany than anywhere else. Gadamer, in his exploration of the "relevance of beauty," thought that art festivals "take us out of ordinary time" and open us up to "the true possibility of community." For Habermas the central problem of modern life is how we find ways to "sustain a moral community in the face of rampant individualism."

These five elements were each important in themselves. If they weren't unique to Germany, they were more developed there, of longer standing, taken more seriously. But so far we have only considered them separately. As with Nietzsche, Weber, and Freud, they are much more potent, and more revealing, when considered together as an interlocking dynamic system.

Nationalist Cultural Pessimism

It should perhaps come as no surprise that, in the wake of the advent of doubt, when people began to lose their faith, two things happened. First, we see the rise of the (more secular) educated middle class, taking over some of the functions formerly served by the clergy. This change was eased in Germany by the fact that so many of the new thinkers were themselves the sons of pastors—they represented this change perfectly. It was helped too by the reading revolution, occurring at exactly the same time, and which, as Benedict Anderson has shown us, helped to generate the phenomenon of an ideal community—the very educated middle class we are considering, who thought of themselves for the first time *as a group*. Simultanously, and secondly, it was natural for this group of people to try to replace religious ideas with something else. Here, again, two things happened. One was the third revival of Greek (pagan) antiquity, thanks to Winckelmann, and the second was the arrival and achievements of Kant and other speculative philosophers. It was only natural, in the circumstances, for theology to be replaced by speculative philosophy in the era between doubt and Darwin. The successes of these developments led to the resurgence of German culture and intellectual life in general, to the concept of Bildung, of education as cultivation, essentially a secular form of salvation, and to inwardness as a way of approaching the truth—not just in Idealistic philosophy, but in Romanticism and in music. All this may be characterized as the growth of inwardness.

Alongside this rise of inwardness went the other main achievement of the educated middle class, the invention of modern scholarship and in particular the institutionalization of research. The fundamental significance of this in the transition to modernity was mentioned above, but there was another way in which research was of profound importance for the educated middle class in Germany. Research began as a tool of the early scholarly specialities—predominantly the humanities such as classics, philology, and history. But, beginning in the 1830s and 1840s, particularly with the growth of modern (cell) biology and in physics (the discovery of the conservation of energy), it began increasingly to be applied in the "hard" sciences. This change was all-important.

Whereas research in the humanities was institutionalized in 1809–1810 at the University of Berlin, the great commercial and industrial laboratories got under way in Germany, as we saw in Chapter 18, p. 355, only in the late 1850s and the 1860s. In the first place, this change contributed to the decline in status of the traditional scholars in, for example, classics, history, and literature. And with the rise of the hard sciences, a wedge was driven between the humanities on one side and these sciences on the other, creating a divide that—although it occurred in other countries (such as Great Britain)—was nowhere near as wide (or in time as bitter) as it was in Germany, where different terms, *Kultur* and *Zivilisation* and *Wissenschaft* and *Bildung*, were introduced to encapsulate the division. This was exacerbated in the late nineteenth century when scientific research moved out of the universities into the independent Kaiser Wilhelm Societies. The division, and the loss of status of the humanities that went with it, itself produced an effect on scholarship.

Now began the great age of nationalist cultural pessimism, with the works of Heinrich von Treitschke, Johann Gustav Droysen, Paul de Lagarde, Julius Langbehn, and Max Nordau culminating in Werner Sombart's *Heroes versus Traders*, and Oswald Spengler's *Decline of the West*. So far as the traditional scholars were concerned, these jeremiads described something all too real—their world *was* declining: the sciences *had* appropriated the idea and practice of research and, by the time Germany became a unified country in 1871, science was well on its way to producing the array of hi-tech products that would create modern mass society, discussed in Chapters 17–20 and 25 of this book, in which traditional areas of scholarship would feel increasingly peripheral. Cultural pessimism, and the reasons for it, have been a major topic for German writers and academics ever since, and still are. This also helps explain the pronounced conservative streak in German thought, not to mention the growth of anti-Semitism in the later nineteenth century.

A further consequence of the advent of doubt—and again this applies especially in Germany, with its tradition of Pietism—was the growth of the idea of a redemptive community. Helping people in *this* life was a natural ethic to emerge from the collapse of the idea of a future state, the Afterlife, so integral to Christianity. After the death of God, community—the basis of living together with other people—was perhaps the only ethical space left to explore. Germany—the land of Pietism and of 300 small independent states, the *Kulturnation* before it was a territorial nation—was a natural home for such an idea.[39] A concern with the redemptive powers of community runs across German scholarship, culture, and politics throughout the modern period.

The redemptive community and cultural pessimism are related, of course, the former being seen usually as a "cure" for the latter. Cultural pessimists, for the most part, seek a return to an earlier, more ideal form of community. (The idea that there was ever a golden age of communal life, before modernity took hold, is savaged in Michael Haneke's film *The White Ribbon*, which won the 2009 *Palm d'Or*.)

The German literature of cultural pessimism—though it typified a tradition of overarching syntheses, was not the only form of scholarly analysis in those years. In contrast to the speculative systems of Fichte, Hegel, Marx, Schopenhauer, and, to an extent, Nietzsche, the philosophies of Dilthey, Simmel, and Scheler were much more modest, more commonsensical, and all the more refreshing and instructive for that. But the overwhelming reality is that, in the face of the advances being made by science, especially in the forty to fifty years before World War I, the educated middle classes in Germany, the *traditionally* educated middle classes, the "Bildung classes" as we can call them, suffered two crucial setbacks, setbacks that were exacerbated in the 1920s in the Weimar Republic. First, they lost status and influence, finding their traditional intellectual interests downgraded and marginalized in the newer, mass urban spaces, and then, in the great inflation, they found their economic interests decimated. Second, in Germany in particular, the traditionally educated Bildung class found itself estranged from—and replaced by—the *scientifically* educated middle class. This was of crucial importance because, when it came to the crunch, when the Nazis began to flex their muscles, there simply was not in Germany a critical mass of educated people in positions of power and responsibility to provide any real resistance.

T. S. Eliot provided an appropriate framework in his short book *Notes Towards the Definition of Culture* (1948) when he said that the most important purpose of culture lies in its impact on politics. The power elite needs a culture elite, he said, because the culture elite is the best antidote, providing the best critics for the power brokers in any society, and that criticism pushes the society forward and prevents it from stagnating and decaying. For Eliot, within any one culture, the higher, "more evolved" levels positively influence the lower levels by their greater knowledge of, and use of, *skepticism* (and you cannot be properly skeptical unless you have knowledge to be skeptical *with*). For Eliot, that is what knowledge and education are *for*. In Germany, with the benefit of hindsight, we can see that that didn't happen.

This surely provided the subtext and the context of the Weimar years. In 1914 the Manifesto of the 93 had proclaimed that the war was being fought to defend the ideals of German culture. The war was then lost, and there was a surrender of nerve and of will. Spengler in 1918 and Moeller van den Bruck in 1922 continued with their versions of cultural pessimism, emphasizing that the war had solved nothing. The great inflation of 1923–24 seemed to confirm those worries while the riotous culture then in vogue—cabaret, Expressionism, especially in the new art form of film, surrealism, the subversive world of Brecht, Schoenberg, and Richard Strauss, the slip-sliding world of Pauli's exclusion principle, Heisenberg's uncertainty principle, and Gödel's limits to what we can know—had collapsed traditional ideas and pushed the classics-loving Bildung classes further and further to the periphery, even as Max Weber told them these new sciences could never tell them how to live. Wolfgang Schivelbusch, in his investigation of what he terms the "cultures of defeat," shows how many of the German postwar observers dated the origins of the catastrophe (the lost war) to the founding of the empire; they wanted a return not to the prewar world but to a *pre-1871* world, the world created by the Bildung classes, a universal world of "spiritual substance" that had, they felt, been destroyed by materialism, mercantilism, and science, which had caused Germany "to lose its soul."[40]

This was the context for what Hannah Arendt said when she argued that what happened in Germany in the 1920s and 1930s was a temporary alliance of the educated elite with the mob. She also noted that the First World War was itself "the true father of a new world order," the "constant murderous abitrariness" being "the great equaliser" that broke down the classes and transformed them into "the masses."[41] This, she felt, had created a "community of fate" in which the aim, going forward, was to do something "heroic or criminal" in which both the mob and the educated elite could express their "frustration, resentment and blind hatred, a kind of political expressionism . . ."[42] This *collective bitterness*, she said, was the "pre-totalitarian atmosphere" in which the ultimate end was the death of respectability, in which the difference between truth and falsehood "ceases to be objective and becomes a mere matter of power and cleverness."[43] Julien Benda agreed and so did Niall Ferguson. Benda thought that a barbaric nationalism had been sparked in Germany, initiated by its intellectuals. In *The War of the World* (2006), Ferguson wrote: "An academic education, far from inoculating people against Nazism, made them more likely to embrace it."[44]

None of that need *necessarily* have led to the horrors of 1933–45, but what we can now say is that the crucial failure in Germany in those years and in the years immediately before, was first and foremost among the educated middle class, *precisely because* they alone possessed the education needed to exercise skepticism and forestall mob action and behavior. Hannah Arendt said, much later, that only educated people can have a private life, and that fits together nicely with Eliot's argument about skepticism being the great aim of education that we must never forget—it provides people with enough of a private space for them to develop a healthy skepticism. People without a private life soon become a mob, where everything that matters, or seems to matter, takes place on the streets.

That is all in the past. I do not mean only that the betrayal of Germany's Bildung class took place more than seventy years ago. I also mean that such a betrayal could not take place again. How can we be sure? Because for once Germany has fashioned its own democratic revolution, albeit one that—surprising as it may seen—has gone very largely underappreciated by the world outside.

In 1945 Germany once more had a revolution imposed on it from above, just as it had in 1848 and 1871, only this time it came not just from above but from outside. The occupying powers imposed a political and legal structure on postwar Germany. But, and this is the crucial point, a point that many outside Germany still do not grasp (with the Germans themselves failing to see why outsiders do not appreciate this profound truth): the social revolution of 1968, particularly in West Germany, *was a much bigger set of events there than anywhere else.*

Konrad H. Jarausch has chronicled this change, which he describes as nothing short of a "caesura."[45] He argues that, despite the successful establishment of democratic institutions in Germany after the war, authoritarian thought patterns "tended to persist," and it was not until the 1960s that the "modernisation deficit" (Ralf Dahrendorf's phrase) was overcome. A crucial factor here, he says, was the "generational rebellion of 1968," when the younger cohort turned on its parents for the acquiescence in horror they had shown in the Third Reich (the "Brown Past") and for their inability to face their guilt; and only then, in 1968, did Germans start to internalize democratic values, develop a "counter-elite" and demand self-government and "democratic counterpower."[46] Jan-Werner Müller essentially agreed when he described the events of 1968 as a mixture of "Marxism and psychoanalysis."[47]

The substance of this change was explored in Chapter 41; here we need only add two key points. One, that henceforth Germany had a critical, *skep-*

tical public, an entity that had been common enough in, say, Great Britain, France, or the United States for generations, but that had now finally arrived in Germany. And two, that there began a concern with the quality of life, with culture, and with the environment in particular. This would lead in time to the formation of the Green Party and bring about a sea change in the political life of the Federal Republic.[48] The Germans had turned away from an emphasis on inwardness—perhaps no bad thing. In Heinrich Winkler's words, the country had completed its "long road west." The studies of Dirk Moses and the Potsdam Institute of Military History, referred to in the Introduction and in Chapter 41, suggest that the above analyses are correct, that the process is maturing, and that the fourth postwar generation has adjusted to the terrible German past and has the courage to face up to the fact that "almost everyone" in the Third Reich knew what was going on.

It may not be that we shall ever know how Hitler came about, but acknowledging how widespread the knowledge of the crimes was is clearly a significant advance.

In June 2006, Thomas Kielinger, London correspondent of *Die Welt*, wrote an article in the London *Daily Telegraph* in which he took his hosts to task. The constant "harping on" about a "happily extinct" Germany by the British was no longer funny, he said. "The funny side escapes us if the Germany of the Nazis is confused with the Germany of today . . . There is a distinct fire break in our minds about then and now; between the swastika'd pariahs and the country we have rebuilt, 'with liberty and justice for all'—including the liberty to mock ourselves for the past descent into hell. By contrast, for too many Britons, the old adversary has become frozen in time, encapsulated in 1945 like an insect in amber . . . Germany has moved on, with a vengeance."[49] I would add this: when you talk to Germans at length, many of them will admit, after a time, that they are not yet completely at ease with themselves. At the same time a recent biography of the last Kaiser has been well received. The Germans are changing more than the British are and more than the British think the Germans are. Of course, one might argue that in Germany there is more to change, and maybe more need of change. But Germany is not as static in its attitude to the Third Reich and World War II, as Britain thinks it is, or as Britain (or France, or America, to a lesser extent) is itself.

The German Ideology and the Future of Human Nature

The German genius is alive and well. It has been a curious journey in some ways, unreal at times—or it has felt that way. Despite the long night between 1933 and 1989, contemporary German artists can bear comparison with the best of other countries, its filmmakers are enjoying a resurgence, even in English-language countries (*Goodbye Lenin!*, *The Lives of Others*, which won an Oscar), its novelists are coping with the dominance of the English language better than most (W. G. Sebald, Bernhard Schlink, Daniel Kehlmann, and Günter Grass, still), and its composers and choreographers continue to shine. More names could have been mentioned—such as Hans J. Nissen, whose team of archaeologists has done so much to illuminate the ancient civilizations of Mesopotamia, at least until the Gulf Wars, and the new Leipzig school of painters who kept alive the tradition of figurative art. Germany's scientific community, though it has not yet returned to its position of pre-eminence of 1933, when it had won more Nobel Prizes than scientists from Britain and America put together, nonetheless *has* returned to prize-winning ways—with Nobels in 1995, 1998, 2000, 2001, 2005, and two in 2008. In Europe, Germany leads the table for patent registration with almost three times the number of its next rival, France. In tables drawn up in 2008 of the leading nations in physics, Austria and Germany came fifth and sixth respectively, behind Switzerland (top), Denmark, and the United States, but ahead of England, France, and Russia.[50] In engineering the only non-American institutes in the top twenty worldwide in 2008 were number 15, the Max Planck Society, 16, the Eidgenössische Technische Hochschule (ETH) in Zurich, and 20, the Technical University of Denmark (institutes in France and Britain did not feature).[51]

Despite this, in the spring of 2008 yet another historical controversy erupted in Germany, this time about the planned reintroduction of the Iron Cross medal for military bravery. Notwithstanding the long and colorful provenance of this award (see pp. 213–214), with the well-to-do in the Napoleonic Wars wearing iron jewelry because they had donated their gold to the war effort, it was felt by the government in Berlin that the Iron Cross was still too closely linked to the Nazis, and the reintroduction was canceled.

How long must this attitude persist? As this book has tried to show, we owe a great deal to the Germans and, as the Iron Cross incident highlights, there is much more to German history than 1933–45. Let us, therefore, end on an equally controversial note and consider what we might *learn* from one of the most contrary philosophers of the twentieth century—Martin Heidegger. Yes, he was a Nazi. Yes, he betrayed his Jewish lover Hannah Arendt, and in cowardly fashion. Yes, in a sense, as she herself said, Heidegger "murdered" his Jewish colleague, Edmund Husserl. But, as the twenty-first century gets into its stride, there are two important areas (at least) where the German philosophical tradition—the German ideology, as the French scholar Louis Dumont calls it—may come back into focus and have much to teach us. Many non-Germans find the Idealist cast of mind—if not Kant then certainly Fichte, Hegel, Husserl, and Heidegger—obscure and vague, using a language ("Being," "authenticity," "releasement") that is alien to, and uncomfortable in, the empirical tradition: they recall Wickham Steed's crack about the Germans diving deeper but coming up muddier. At the same time, the German ideological antipathy to technology and its advances can seem (again, to the empirical Anglophone mind) altogether unreal, a plaintive, overtheoretical, and thinly abstract opposition to inevitable "progress."

And yet, as shown by Jürgen Habermas, who is surely the most interesting post–World War II example of the German philosophical genius, recent developments in the world of biotechnology suggest that Heidegger, duplicitous, self-serving, and unapologetic as he was for his Nazi involvement, may have had a point all along: he perceptively anticipated the threat that technology would ultimately pose and, moreover, he did so in the very language that we now need to contemplate if we are to consider seriously where we are headed.

In his book *Die Zukunft der menschlichen Natur* (*The Future of Human Nature*; 2003), Habermas reflects on and himself anticipates the new forms of "damaged life" (Adorno's phrase) that we may be about to inflict on ourselves. He notes that recent developments in biotechnology allow, or will very soon allow, prenatal genetic intervention, giving parents the choice to select not only for characteristics they don't want their children to have (major handicaps, "negative eugenics") but also for characteristics (eye color, hair color, sex, higher intelligence, musical ability) that they *do* want their children to have—"positive eugenics." Habermas cautions us that here a line may be being crossed, a Rubicon he calls it, with profound implications for our understanding of freedom, and that it requires a *philosophical* resolution, not a technical-scientific-psychiatric one.[52]

In the future, children of one generation will be given characteristics by another generation (their parents') that are irrevocable. What, he asks, will this do to an individual's understanding of him- or herself, his or her sense of—as Heidegger put it—*being*? For Habermas, this new technology blurs the line between the "grown" and the "made," between chance and choice, all of which are essential ingredients in who were are, who we feel ourselves to be. For Habermas, if these processes are allowed to continue, future generations risk becoming *things* rather than beings. Any new generation will, to an extent, have been selected by its parents' generation and will, to that degree, be less free. As he puts it, again using Heideggerian language, the ethics of "successfully being oneself" will have been compromised. The inviolability of the person, "which is imperative on moral grounds and subject to legal guarantees," for him is something we can never "dispose over."

For Habermas, not only does this pose a threat to our essential "sense of being," it poses a threat to our capacity to see ourselves as equally free and autonomous as the next individual, to the idea of "anthropological universality," that man is everywhere the same.[53] For Habermas the evolution of the species is a matter for nature; to intervene in this process at the very least marks a new epoch in the history of mankind and perhaps something much worse.[54] Evolution, he insists, should not be a matter of "bricolage," however well-intentioned parents may be.

His worry is that such intervention amounts to nothing less than a third "decentration" of our worldview, after Copernicus and Darwin, so that a person's sense of "I" and his or her understanding of "we" would be changed irrevocably, with incalculable consequences for our shared moral life.[55] People, he warns, may feel that they are no longer "ends in themselves," no longer irreplaceable, no longer so completely at home in their bodies, no longer have the same relationship to such emotions as shame or pride, not weighing human life in the same fashion, no longer having equal respect for each other. Most fundamentally, Habermas worries that for genetically preprogrammed people the initial conditions of identity-formation will have been altered and the "subjective qualification essential for assuming the status of a full member of a moral community" will be affected beyond recall.[56] "The technicisation of 'inner nature' constitutes something like a transgression of natural boundaries."[57]

Habermas does wonder whether he is being oversensitive here. To an extent, genetic intervention already exists, in China, where the one-child-per-family policy has resulted in a heavy preponderance of male children and entire villages where there are no partners for young men. This has produced social problems but, so far as we know, no clinical-psychiatric epidemics. But Habermas feels this is a special, atypical case, where the whole individual has been chosen, so there is no specific intervention regarding identity.

He urges that our attitude to "being" is a complex philosophical issue and points to our ethical behavior in regard to corpses and dead fetuses. We insist on their dignified disposal—they are more than inert matter to us; they were *beings*—grown, not made—and therefore they are not *things*.

We don't *have* bodies, he concludes, we *are* bodies, and this typcially Heideggerian distinction is all-important. We stand on the verge of a major transformation in the understanding of human nature and the way we *choose* to go forward (since nothing is inevitable, however blinded we may be by the notion of "progress") is a *philosophical* matter, not a scientific-psychiatric-technical one.

There is of course a further ironical contextual level to all this. Given the notorious eugenic policies carried out in the Third Reich, Habermas, as a German identifying the future risks of genetic preprogamming, has a redemptive quality. Habermas is himself naturally aware of this context, quoting Johannes Rau, president of the Federal Republic of Germany, in 2001: "Once you start to instrumentalise human life, once you start to distinguish between life worth living and life not worth living, you embark on a course where there is no stopping point."[58]

Genetic preprogramming is not the only philosophical problem we face as a matter of urgency. As global warming starts to lay waste our planet, as the rain forests and ice caps shrink together, as inland seas disappear, as terrorists threaten nuclear annihilation, as genocide and famine continue to ravage Africa, as India and China begin to run out of water, does it not ring ever more true that Heidegger had a profound point (and wasn't being merely "priggish") when he said we should stop trying to exploit and control the world with our technological brilliance? Is this not a form of hubris that will in time destroy all we have, should we not instead learn to accept the world, to submit—without interference—to the pleasures nature has to offer, to enjoy them as poets enjoy them, and should not our main stance now, our first and only priority, be to *care* for the world?

Heidegger was caught up in what he saw as the redeeming energies of National Socialism and about that he was wrong, very wrong. And yet, despite the undoubted advantages that science and capitalism have wrought, they now seem incapable of rectifying the ravages they have also brought about. Hannah Arendt counseled us to be adult, and part of her own achieved adulthood was that she forgave Heidegger and in that sense redeemed him. Can we not do the same and learn from him (and her), despite what went before?

For that matter, is Germany itself always to remain unredeemable? Perhaps Norbert Elias was correct in saying that the country cannot move ahead until a convincing explanation for the rise of Hitler has been given. Yet Heidegger was prescient and was part of a recognizable line of German thinkers, from Kant through Fichte, Hegel, Schopenhauer, Nietzsche, Gadamer, and Habermas himself, who remained and remain skeptical of modernity (there's that word again, "skepticism"), who remind us that human nature—life itself—is as much about pride, shame, independence, coherence, and respect for others and for ourselves, for morality, for our "inner environment," for autonomy, intuition, and disgust, as it is about money, the markets, the profit motive, and the hard drives of technology. Germany is not only a "belated" nation in terms of modernity; it is also a reluctant nation and maybe there is a lesson in that reluctance. If science and capitalism—the market—cannot prevent the degradation of our environment, our very world, indeed if they are now the primary ingredient *in* that devastation, then only a change within us, a change of *will*, can do it. The way out of our dilemma, the Germans tell us, is not technical or scientific, but *philosophical*.

There must never come a time when a *Schlußstrich*, a final line, is drawn under Germany's past, when the events of 1933–45 become just another episode, another catastrophe mothballed in the chain of history. Gerhard Schröder had it right when he said, "We cannot emerge from our past so easily. Perhaps we should not even wish to."[59]

Germany should not wish, or seek, to leave its past behind. But embracing this view, as Beuys showed, as Gunter Demnig and his "stumbling stones" show, as Habermas and Ratzinger show, Germans do not need to remain *chained* to their past forever. All Germans, as Steve Crawshaw phrases it, are not "umbilically linked" to Hitler. The German past consists of much more than the events of the Third Reich and, as this book has tried to show, still has a lot to teach us.

The German predicament is not easy and the arguments in this book will not please everyone. It is to those who find it difficult to move beyond Hitler that *The German Genius* is dedicated.

Appendix

Thirty-five Underrated Germans

I do not mean to suggest for a minute that the names mentioned below are unknown. They are not. Indeed, to many specialists they include some of the very finest minds of their—or any—day. The point of this appendix, rather, is to underline in a vivid way one of the main arguments of the book—namely, that because two world wars have interfered with our view of the past, these German names are *generally* less well known than they deserve to be, and that they are worthy of being appreciated by a much wider public.

Many of the scientists, for example, are easily on a par with Freud, Mendel, and Einstein in regard to their influence on our lives. Several of the philosophers, though they cannot perhaps match Hegel, Nietzsche, and Schopenhauer, *are* the equal of Wittgenstein, Bertrand Russell, Henri Bergson, William James, and John Dewey, whose names are virtually household words. Writers and mathematicians have also suffered.

Wilhelm von Humboldt (1767–1835)

As is made clear throughout the main text (but in Chapter 10 and the Conclusion in particular), Wilhelm von Humboldt was responsible for the concept of the modern university, for the institutionalization of research, for much of modern scholarship and, indirectly, for the rise of modern science. He should now be given full credit for being one of the most important creators of modernity.

Alexander von Humboldt (1769–1859)

Alexander von Humboldt was at one stage the most famous man of science in the world, with more than a dozen geographical features (and one on the moon) named after him. In 1859, his obituary occupied the whole of the front page of the *New York Times*. At the same time, as Stephen Jay Gould also said, he then became the "most forgotten" man of science. His expeditions, his identification of new scientific fields of inquiry, and his active encouragement of so many younger colleagues mark him out as one of the great figures from the heroic age of nineteenth-century discovery. It is time that the reversal in his fortunes was itself reversed.

Caspar David Friedrich (1774–1840)

While Friedrich perhaps had the German vice of being a very theoretical painter, he was technically brilliant, foreshadowing many modern movements, such as Surrealism and the great American landscapists. He deserves to be as well known as, say, J. M. W. Turner, John Constable, and Salvador Dalí.

Carl Friedrich Gauss (1777–1859)

Of course, Gauss is well known to mathematicians and scientists, but his wide-ranging achievements, and his invention of mathematical *imagination*, which ensured he was the precursor of Einstein, really mean that he should join the exalted pantheon of mathematical geniuses, alongside Archimedes, Euclid, Copernicus, and Newton.

Karl Schinkel (1781–1841)

Every bit as distinguished as Christopher Wren, Paul Nash, James Barry, and Georges-Eugène Haussmann, a painter and designer as well as an architect, Schinkel is nevertheless often described as an "architects' architect" who should be a publicly recognized architect as well. Berlin is unthinkable without him.

Ludwig Feuerbach (1804–72)

Feuerbach deserves to be better known if only for his seminal influence on such diverse figures as Karl Marx and Richard Wagner. But his work on Christianity, his realization that God is as much created by us as we are by Him, makes him as important to our intellectual history as, say, Baruch Spinoza or Giambattista Vico.

Jan Evangelista Purkyně (1787–1869), Karl Ernst von Baer (1791–1876), Friedrich Wöhler (1800–82), Justus von Liebig (1803–73), Matthias Jakob Schleiden (1804–81), Theodor Schwann (1810–82), Rudolf Virchow (1821–1902), August Kekulé (1829–96), Robert Koch (1843–1910), Paul Ehrlich (1845–1915)

This constellation of names constitutes possibly the biggest black hole in the intellectual history of the West. Though they are well enough known to specialists, none of these names comes close to, say, Freud, Mendel, or Einstein as scientists whose name-recognition among the general public is near universal; on the contrary, the general public remains largely unaware of their achievements, either individually or collectively. Yet each had a profound effect either on our understanding *of* nature, or on our relationship *with* nature, or on the substances and structures and processes of life itself, or on our understanding of disease, its treatment, and its control. Without their achievements, modern life would be unthinkable and unbearable.

Friedrich Engels (1820–95)

In a sense, everyone who has heard of Karl Marx has heard of Engels. And yet, in typing out these sentences on a laptop, the Microsoft Word spell-checker recognizes Marx, who is not underlined in red, but not Engels, who is. There is no such thing as Engelism, as there is Marxism. As joint author of

The Communist Manifesto and editor of volumes two and three of *Das Kapital*, Engels's influence is great, but his own books deserve to be better known: they are more wide ranging, more learned, and more fun than Marx's. Engels's achievement as "the most educated man in Europe" is deserving of much wider appreciation, not least because he was amazingly prescient.

RUDOLF CLAUSIUS (1822–88), LUDWIG BOLTZMANN (1844–1906), HEINRICH HERTZ (1857–94), HERMANN VON HELMHOLTZ (1821–94), WILHELM RÖNTGEN (1845–1923)

This constellation comprises another intellectual black hole, yet the tradition of *theoretical* physics, one of the great adventures of the twentieth century, had its origins in these figures in Germany in the nineteenth century. This was, as the main text shows, a very international field, though the Germans led the way.

These two nineteenth-century scientific "black holes"—in biology and in physics—had a more direct effect on our lives than did the earlier scientific breakthroughs of the much better known Kepler, Copernicus, Galileo, and Newton.

WILHELM DILTHEY (1833–1911)

One of the common stereotypes of Germans, most particularly their philosophers, is that they are a theoretical, abstract people, who love overarching, all-embracing systems. Dilthey gives the lie to this; he is a man who showed how far it is possible to go with common sense.

HUGO WOLF (1860–1903)

Many of the cognoscenti regard Wolf, quite simply, as the greatest song composer of all time, who "carried the German art song to its highest point." A rebel, a bohemian, and a malcontent whose productive life occupied an intense three years when he wrote more than 200 songs, set to the words of Goethe, Keller, and others, and who ended his life in an asylum, Wolf surely awaits discovery by a Hollywood film director who sees in his art and life a modern tragedy of epic dimensions.

GEORG SIMMEL (1858–1918)

Taught by several of the fetid nationalist historians of his day (Treitschke, Sybel, Droysen), but also by the more open-minded Helmholtz, Simmel became more highly regarded abroad (especially in Russia and the United States) than in his own country, certainly among anti-Semitic university authorities. But he was among the first to identify the new moral conditions brought about by modernity, the conundrum that we are both more free and more responsible. He was the first to forecast that modern life would be "more nervous" and that "the lower intellectual functions would be promoted."

ROBERT MUSIL (1880–1942)

For some, *The Man without Qualities* eclipses anything that Thomas Mann or Hermann Hesse wrote and is the most remarkable response to developments in other fields in the early twentieth century. If all we can know about ourselves is what scientists tell us, if ethics and values are meaningless, how are we to live? Musil brilliantly exposes the central dilemma of modern life.

MAX SCHELER (1874–1928), RUDOLF BULTMANN (1884–1976), KARL BARTH (1886–1968), DIETRICH BONHOEFFER (1906–45)

The theological renaissance in Germany at the beginning of the twentieth century is the third intellectual black hole that should be better appreciated. In the wake of the "death of God," identified by Nietzsche, and the "disenchantment" of the world, as described by Max Weber, these other German theologians/philosophers produced a more cogent and coherent response to the "crisis conditions" than anyone else. The fact that two late twentieth-century popes, Jean Paul II (Karol Wojtyla) and Benedict XVI (Joseph Ratzinger) have picked up on their ideas, shows how these (Protestant) thinkers have been readily assimilated *within* the Catholic Church, if not yet outside it.

LION FEUCHTWANGER (1884–1958)

A man who escaped as a prisoner of war *twice* is clearly out of the ordinary and very brave. Bravery runs through his masterpiece, *Success*, in which such "characters" as Hitler and IG Farben are identified and excoriated. Thankfully, Feuchtwanger escaped to America. Had he not escaped, and instead perished, he would probably be more well known now than he is.

Karl Jaspers (1883–1969)

Jaspers's identification of the "axial age," of the origins of modern spirituality right across the world at more or less the same time (Isaiah, Confucius, the Buddha, Plato), mark him as one of the great synthesizers in history, explaining our world in a way that is every bit as fundamental as, say, John Dewey or William James, and possibly more so.

Heinrich Dreser (1860–1924), Arthur Eichengrün (1867–1949), Felix Hoffmann (1868–1946)

More than 40,000 tons of aspirin are now produced every year, more than a century after the drug's invention. This is certainly one measure of its impact. Having early on shown its efficacy in cases of pain control, migraine, rheumatoid arthritis, fever, and influenza, as well as in the control of various veterinary diseases, in the last decades of the twentieth century the drug was found helpful as an anti-blood-clotting agent, and effective in the prevention of angina, heart attack, and strokes. This is more than enough to suggest that the names of Dreser, Eichengrün, and Hoffmann are engraved on any role of honor. They have certainly helped mankind much more than, for example, the far better known Carl Jung, who would surely lead any list of the most *over*rated German-speakers.

Notes and References

When two dates are given for a publication, the first refers to the hardcover edition, the second to the paperback edition. Unless otherwise stated, pagination refers to the paperback edition. All translations are from the German unless otherwise indicated. Every attempt has been made to trace the names of translators; the author would be grateful to hear from readers who can fill in the gaps that, inevitably, remain. The *Dictionary of Scientific Biography*, referred to throughout these notes and references, was originally published in 16 volumes between 1970 and 1980 under the auspices of the American Council of Learned Societies by Charles Scribner's Sons in New York, with Charles Coulston Gillispie as editor in chief. The index may be found in volume 16. Several supplements were published up until 1990. In 2008 a *New Dictionary of Scientific Biography*, 8 volumes, was published in Detroit, also by Charles Scribner's Sons and also under the auspices of the ACLS, with Noretta Koertge as editor in chief. This later venture, however, is nowhere near as comprehensive as the original *DSB*. In the references that follow, volume numbers given in roman numerals refer to the original *DSB*, while volume numbers given in Indo-Arabic numbers refer to the *NDSB*. Readers will thus be able to see for themselves which scientists were not included in the latest compendium. Some of the omissions are surprising.

Introduction: Blinded by the Light: Hitler, the Holocaust, and the "Past That Will Not Pass Away"

1. These matters were discussed repeatedly in British newspapers, but many were collected together in John Ramsden's *Don't Mention the War* (London: Little, Brown, 2006), p. 393.
2. Ramsden, *Don't Mention*, p. 392.
3. Ibid., p. 413.
4. Ibid., p. 394.
5. Ibid., p. 411.
6. Ibid., p. 412.
7. Ibid., p. 364.
8. *Times Higher Education Supplement*, February 2, 2007, p. 6.

9. Ibid.
10. *Daily Telegraph*, May 8, 2005, p. 18.
11. Ibid.
12. Ramsden, *Don't Mention*, p. 402.
13. Ibid., p. 417.
14. *International Herald Tribune*, April 22, 2005.
15. D. D. Gutenplan, *The Holocaust on Trial: History, Justice and the David Irving Libel Case* (London: Granta, 2001).
16. Peter Novick, *The Holocaust and Collective Memory* (London: Bloomsbury, 2000), p. 2.
17. Novick, *Holocaust*, p. 69.
18. Ibid., p. 105.
19. Ibid., p. 65.
20. Ibid., p. 144.
21. Ibid., p. 164.
22. Ibid., p. 202.
23. Ibid., p. 232.
24. Norman G. Finkelstein, *The Holocaust Industry: Reflections on the Exploitation of Jewish Suffering* (London: Versa, 2000), passim.
25. Charles Maier, *The Unmasterable Past: History, Holocaust and German National Identity* (Cambridge, Mass.: Harvard University Press, 1988), p. 55.
26. Ibid., p. 56.
27. Richard J. Evans, *In Hitler's Shadow: West German Historians and the Attempt to Escape from the Nazi Past* (London: Tauris, 1989), p. 13.
28. Mary Fulbrook, *German National Identity after the Holocaust* (Cambridge, U.K.: Polity, 1999), p. 36.
29. Maier, *Unmasterable Past*, p. 101.
30. Ibid., p. 54.
31. Wulf Kansteiner, *In Pursuit of German Memory: History, Television and Politics after Auschwitz* (Athens, Ohio: Ohio University Press, 2006), pp. 54–56.
32. London *Daily Mail*, February 15, 2007, p. 43.
33. Steve Crawshaw, *An Easier Fatherland: Germany and the Twenty-First Century* (London: Continuum, 2004), p. 199.
34. Peter Watson, "Battle over Hitler's Loot," London *Observer Magazine*, July 21, 1996, pp. 28ff.
35. Pierre Péan, *A French Youth: François Mitterrand, 1934–1947* (Paris: Fayard, 1994).
36. Henry Rousso, *The Vichy Syndrome: History and Memory in France since 1944* (Cambridge, Mass.: Harvard University Press, 1991), passim.
37. Michael R. Marrus and Robert O. Paxton, *Vichy France and the Jews* (New York: Basic Books, 1981).
38. See Marrus and Paxton, *Vichy France*, pp. 341 ff. for what Vichy knew about the Final Solution.
39. See, for example, Lee Yanowitch, "France to Boost Efforts to Restore Nazi-looted Property to Jews," *Jewish News Weekly*, December 4, 1998.
40. *Times* (London), October 13, 2007, p. 52.
41. Richard J. Evans, *Rereading German History: From Unification to Re-unification, 1800–1996* (London: Routledge, 1997), pp. 149 ff.
42. Daniel Jonah Goldhagen, *Hitler's Willing Executioners* (New York: Random House, 1996), p. 77.
43. Ibid., p. 465.
44. Evans, *Rereading German History*, pp. 155ff.
45. Fritz Stern, *The Politics of Cultural Despair: A Study in the Rise of the German Ideology* (Berkeley and Los Angeles: University of California Press, 1961/1974), p. 202. A comparison of anti-Semitic acts and attitudes toward Jews in the popular press of Germany and four European nations (France, Great Britain, Italy, and Romania) from 1899 through 1939 demonstrates that Germans, before 1933, were among the least anti-Semitic people. William I. Brustein, *Roots of Hate:Anti-Semitism in Europe before the Holocaust* (Cambridge: Cambridge University Press, 2003), Chapter 6. Until that point, no census in Germany had gathered data on ethnicity. Quoted in Claudia Koonz, *The Nazi Conscience* (Cambridge, Mass.: Belknap

Press of Harvard University Press, 2003), p. 9. Fritz Stern also tells us that Arthur Moeller van den Bruck, whose book *The Third Reich* was a major work of cultural pessimism in the Weimar Republic, helping to create the mood in which the National Socialist Party could thrive, showed no sign of anti-Semitism in his many books published before World War I (see Chapter 33).

46. Norman G. Finkelstein and Ruth Bettina Birn, *A Nation on Trial: The Goldenhagen Thesis and Historical Truth* (New York: Holt), 1998.
47. Fritz Stern, *Einstein's German World* (Princeton, N.J.: Princeton University Press, 1999), pp. 276–278.
48. Finkelstein and Birn, *Nation on Trial*, p. 139.
49. Richard J. Evans, *Rereading German History*, p. 164. Goldhagen also avoids saying just what anti-Semitism means. As Clive James points out in his essay on the Austrian Jewish dramatist Arthur Schnitzler, "If he encountered anti-Semitism in grand [Viennese] drawing rooms, there were few grand drawing-rooms he could not enter." Clive James, *Cultural Amnesia: Notes in the Margin of My Time* (London: Picador, 2007), pp. 684–705.
50. Stern, *Einstein's German World*, p. 287.
51. Crawshaw, *Easier Fatherland*, p. 144.
52. Kansteiner, *In Pursuit*, pp. 104, 109, 116 and 210. A. Dirk Moses, *German Intellectuals and the Nazi Past* (Cambridge: Cambridge University Press, 2007), especially pp. 55–73. Alexander and Margarete Mitscherlich, *Die Unfähigkeit zu trauern: Grundlagen kollektiven Verhaltens* (Munich: Piper, 1967). Ralf Blank *et al.*, *German Wartime Society 1939–1945: Politicization, Disintegration, and the Struggle for Survival*, trans. Derry Cook-Radmore (Oxford: Clarendon Press, 2008). Max Hastings, "Germans Confront the Nazi Past," *New York Review of Books*, February 26–March 11, 2009, pp. 16–18.
53. Evans, *In Hitler's Shadow*, p. 12. Leopold von Ranke, "Die grossen Mächte," in the same author's *Preussische Geschichte*, ed. Willy Andrews (Wiesbaden, 1833), vol. 1, p. 16. I thank Werner Pfennig for this reference.
54. Maier, *Unmasterable Past*, p. 103.
55. Evans, *In Hitler's Shadow*, p. 13.
56. David Blackbourn and Geoffrey Eley, *The Peculiarities of German History: Bourgeois Society and Politics in Nineteenth-Century Germany* (Oxford: Oxford University Press, 1984), passim. Maier, *Unmasterable Past*, p. 107. In February 1871, three weeks before the Proclamation of the German Empire in Versailles, Benjamin Disraeli, at that time leader of Britain's opposition, said in the House of Commons that German unification would be "a greater political event than even the French Revolution" and that the European balance of power "is completely destroyed with no new one in sight." Walter Dussman, "Das Zeitalter Bismarcks," in *Handbuch der deutschen Geschichte*. (Frankfurt am Main: Akademische Verlagsgesellschaft Athenaion, 1968), vol. 2, part 2, p. 129. Richard Münch also compared the development of the Enlightenment in the United States, the United Kingdom, France, and Germany and concluded that four elements are decisive for what is modern, despite many differences: rationalism, activism, individualism, universalism. See his *Die Kultur der Moderne*, 2 vols. (Frankfurt am Main: Suhrkamp, 1986). I thank Werner Pfennig for this reference.
57. Evans, *In Hitler's Shadow*, p. 17.
58. Ibid., p. 141.
59. Maier, *Unmasterable Past*, p. 161.
60. Ibid., p. 168.
61. Crawshaw, *Easier Fatherland*, p. 202.
62. Nicholas Boyle, *Goethe: The Poet and the Age*, vol. 1, *The Poetry of Desire (1749–1790)* (Oxford: Clarendon Press, 1991), p. 4.
63. Wolf Lepenies, *The Seduction of Culture in German History* (Princeton, N.J., and Oxford: Princeton University Press, 2006), p. 4.
64. Ibid. The full differentiation in Germany is between *Wissenschaft* (scholarship), *Kunst, Kultur, Lebensart* (*feine Lebensart*), and *Zivilisation.*
65. Ibid., p. 6.
66. Ibid., p. 5.
67. See also Fritz Stern, *Five Germanies I Have Known* (New York: Farrar, Straus and Giroux, 2006), p. 16.
68. See also Lepenies, *Seduction of Culture*, p. 24.

69. Gordon Craig, *The Germans* (New York: Meridian, 1991; reprint, originally Putnam, 1982), pp. 214–218.
70. Lepenies, *Seduction of Culture*, pp. 17–19 and 28.
71. Ibid., pp. 27–29.
72. Ibid., p. 73.
73. T. S, Eliot, *Notes Towards a Definition of Culture* (London: Faber & Faber, 1948/1962), p. 31.
74. Fritz Stern, *Einstein's German World*, p. 3.
75. Keith Bullivant, *Realism Today: Aspects of the Contemporary West German Novel* (Leamington Spa/Hamburg/New York: Oswald Wolff, 1987), p. 158. Georg Lukács, *German Realists in the Nineteenth Century*, trans. Jeremy Gaines and Paul Keast, edited and with an introduction and notes by Rodney Livingstone (London: Libris, 1993), p. 168.

Chapter 1: Germanness Emerging

1. James Gaines, *Evening in the Palace of Reason* (London: HarperCollins, 2005), p. 5.
2. Jan Chiapusso, *Bach's World* (Bloomington: Indiana University Press, 1968), p. 37. Albert Schweitzer, *J. S. Bach*, trans. Ernest Newman. 2 vols. (London: Breitkopf & Kärtel, 1911).
3. Gaines, *Evening*, p. 7.
4. Robert Eitner, "Johann Gottfried Walter," *Monatshefte für Musikgeschichte* 4, no. 8 (1872): 165–167. Quoted in Gaines, *Evening*, p. 8.
5. Gaines, *Evening*, p. 9.
6. Ibid., back cover.
7. Karl Hermann Bitter, *Johann Sebastian Bach*. 2 vols. 2nd ed. (Berlin: W. Baensch, 1881), vol. 2, p. 181.
8. Gaines, *Evening*, p. 237.
9. Boyle, *Goethe*, vol. 1, p. 9.
10. Steven Ozment, *A Mighty Fortress* (New York: HarperCollins, 2004), p. 125.
11. Many of Pufendorf's works have been translated into English. See Ian Hunter, *Rival Enlightenments: Civil and Metaphysical Philosophy in Early Modern Germany* (Cambridge: Cambridge University Press, 2001), pp. xvii and 148–196.
12. Ozment, *Mighty Fortress*, p. 126.
13. Ibid., p. 27.
14. Richard L. Gawthrop, *Pietism and the Making of Eighteenth-century Prussia* (Cambridge: Cambridge University Press, 1993), pp. 1 and 2.
15. Ibid., p. 9.
16. Ibid., p. 10.
17. For Pietist conversion narratives and links to Puritanism, see Gisele Mettele, "Constructions of the Religious Self: Moravian Conversion and Transatlantic Communication," *Journal for Moravian History* 2 (2007). Also personal communication from the author. Gawthrop, *Pietism*, p. 12.
18. Johannes Wallmann, *Philipp Jakob Spener und die Anfänge des Pietismus* (Tübingen: J. C. B. Mohr [Paul Siebeck], 1970), p. 300. Martin Brecht, "Philipp Jakob Spener, sein Programm und dessen Auswirkungen," in *Der Pietismus vom siebzehnten bis zum frühen achtzehnten Jahrhundert*, vol. 1 of *Geschichte des Pietismus: im Auftrag der Historisches Kommission zur Erforschung des Pietismus*, ed. Martin Brecht. (Göttingen: Vandenhoeck & Ruprecht, 1993), p. 315. This is a magisterial four-volume history of Pietism.
19. Martin Brecht, "August Hermann Francke und der Hallische Pietismus," in Brecht, *Pietismus*, vol. 1, pp. 440–539.
20. Gawthrop, *Pietism*, p. 94.
21. Ibid., pp. 143–144. The Franckesche Stiftungen still exist in Halle and have been revitalized since reunification. I thank Werner Pfennig for this information.
22. Ibid., p. 145.
23. Wallmann, *Philip Jakob Spener*, pp. 89–90 and 94–95.
24. Wolf Oschlies, *Die Arbeits- und Berufspädagogik August Hermann Franckes (1883–1727): Schule und Leben im Menschenbild des Hauptvertreters des halleschen Pietismus* (Witten: Luther-Verlag, 1969), p. 107. Gawthrop, *Pietism*, p. 160.

25. Gawthrop, *Pietism*, p. 183.
26. Martin Brecht, "Der Hallische Pietismus in der Mitte des 18. Jahrhunderts—seine Ausstrahlung und sein Niedergang," in Brecht, *Pietismus*, vol. 2, pp. 319–357. Gawthrop, *Pietism*, p. 198.
27. Gawthrop, *Pietism*, p. 213.
28. Ibid., p. 221.
29. Hartmut Rudolph, *Das evangelische Militärkirchenwesen in Preussen: Die Entwicklung seiner Verfassung und Organisation vom Absolutismus bis zum Vorabend des ersten Weltkrieges* (Göttingen: Vandenhoeck & Ruprecht, 1973), p. 22. Gawthrop, *Pietism*, p. 225.
30. Gawthrop, *Pietism*, p. 228.
31. Terry Pinkard, *German Philosophy 1760–1860: The Legacy of Idealism* (Cambridge: Cambridge University Press, 2002), p. 5.
32. Gawthrop, *Pietism*, p. 241.
33. Ibid., p. 268.
34. Charles E. McClelland, *State, Society, and University in Germany, 1700–1914* (Cambridge: Cambridge University Press, 1980), p. 28.
35. Ibid., p. 199.
36. G. von Selle, *Die Matrikel der Georg-August-Universität zu Göttingen, 1734–1837.* 2 vols. (Hildesheim and Leipzig: A. Lax, 1937), vol. 1, p. 14.
37. McClelland, *State, Society*, p. 37.
38. Thomas Howard, *Protestant Theology and the Making of the Modern German University* (Oxford: Oxford University Press, 2006), p. 110.
39. Emil F. Rössler, *Die Gründung der Universität Göttingen: Entwürfe, Berichte, und Briefe der Zeitgenossen* (Göttingen: Vandenhoeck & Ruprecht, 1855), p. 36. McClelland, *State, Society*, p. 42.
40. McClelland, *State, Society*, p. 45.
41. Howard, *Protestant Theology*, pp. 116–117.
42. Ibid., p. 119.
43. Ibid., p. 87.
44. Ibid., p. 55.
45. William Clark, *Academic Charisma and the Origins of the Research University* (Chicago: University of Chicago Press, 2006), p. 174.
46. Ibid., p. 8.
47. Ibid., p. 237.
48. Ibid., p. 60.
49. McClelland, *State, Society*, p. 96.
50. Clark, *Academic Charisma*, p. 19.
51. Thomas Ahnert, *Religion and the Origins of the German Enlightenment: Faith and the Reform of Learning in the Thought of Christian Thomasius* (Rochester, N.Y.: University of Rochester Press, 2006). See also Hunter, *Rival Enlightenments*, and Howard, *Protestant Theology*, p. 26.
52. Clark, *Academic Charisma*, p. 211.
53. T. C. W. Blanning, *The Power of Culture and the Culture of Power: Old Regime Europe, 1660–1789* (Oxford: Oxford University Press, 2002), p. 242.
54. W. H. Bruford, *Culture and Society in Classical Weimar, 1775–1806.* (Cambridge: Cambridge University Press, 1962), p. 1.
55. Blanning, *Power of Culture*, p. 133.
56. Ibid. Jürgen Habermas, *The Structural Transformation of the Public Sphere: An Inquiry into a Category of Bourgeois Society*, trans. Thomas Burger with the assistance of Frederick Lawrence. (Cambridge, U.K.: Polity Press, 1989), p. 72.
57. Pinkard, *German Philosophy*, p. 7.
58. Blanning, *Power of Culture*, p. 144. See also the table on p. 145.
59. Ibid., p. 150.
60. Ibid., p. 159. Habermas, *Structural Transformation*, pp. 25–28.
61. For a note on how Leibniz converted Latin words to German ones, see Christian Mercer, *Leibniz's Metaphysics: Its Origin and Development* (Cambridge: Cambridge University Press, 2001), p. 278, note 46. See also G. W. Leibniz, *Discourse on Metaphysics and Other Essays*, ed. and trans. David Garber and Roger Ariew (London: Hackett, 1991); and *Protogaea*, trans. and ed. Claudine Cohen and Andre Wakefield (Chicago: University of Chicago Press, 2008).
62. Eric Blackall, *The Emergence of German as a Literary Language, 1700–1775* (Cambridge: Cambridge University Press, 1959), p. 69.

63. For Schiller's views of Thomasius, see his letter to Goethe, May 29, 1799, in S. Seidel, ed., *Briefe der Jahre 1798–1805*, vol. 2 of *Der Briefwechsel zwischen Schiller und Goethe* (Munich: Beck, 1985). See also Ahnert, *Religion and the Origins of the German Enlightenment*, and Blanning, *Power of Culture*, p. 201.
64. Blanning, *Power of Culture*, p. 239.
65. Ibid., p. 201.
66. Benedict Anderson, *Imagined Communities: Reflections on the Origin and Spread of Nationalism* (London: Verso, 1983), p. 41. Habermas, *Structural Transformation*, pp. 18–19.
67. Anderson, *Imagined Communities*, p. 49.
68. Ibid., p. 82.
69. Blanning, *Power of Culture*, p. 161.
70. Ibid., p. 162.
71. Ibid., p. 176.
72. Ibid., p. 180.
73. Ibid., p. 243.
74. For Frederick's theories of government, see Reinhold Koser, *Geschichte Friedrichs des Grossen*. 3 vols. (Berlin, 1925); also the catalog for the 1981 Berlin exhibition *Preussen: Versuch einer Bilanz*. 5 vols. (Reinbek bei Hamburg: Rowohlt, 1981); Theodor Schieder, *Frederick the Great*, ed. and trans. Sabina Berkeley and H. M. Scott (Harlow, U.K., and New York: Addison Wesley Longmann, 2000). This latter is generally regarded as the most recent substantive contribution to scholarship on Friedrich the Great and the Prussia of his time. See also Blanning, *Power of Culture*, p. 131.
75. Blanning, *Power of Culture*, p. 132. The library was later auctioned off by order of Friedrich Wilhelm I.
76. Schieder, *Frederick the Great*, p. 37.
77. G. P. Gooch, *Frederick the Great* (New York: Dorset Press, 1990), p. 140.
78. See also Schieder, *Frederick the Great*, p. 257.
79. Ibid., Chapter 9, "Philosopher-King" discusses many of Friedrich's books, pp. 233–267. The king's intellectual work is also examined in Friedrich Meinecke's *Machiavellism*, English translation published by Manchester University Press, 1957, pp. 275–310.
80. Blanning, *Power of Culture*, p. 219.
81. Ibid., p. 222.
82. Ibid., p. 228.
83. Schieder, *Frederick the Great*, pp. 43–44. Blanning, *Power of Culture*, p. 141.

Chapter 2: *Bildung* and the Inborn Drive toward Perfection

1. See, for instance, John Redwood, *Reason, Ridicule, and Religion, 1660–1750* (London, Thames & Hudson, 1976), p. 150; Karen Armstrong, *A History of God from Abraham to the Present: The 4000-Year Quest for God* (London: Heinemann, 1993), p. 330; Richard Popkin, *The Third Force in Seventeenth-Century Thought* (Leiden: Brill, 1992), pp. 102–103.
2. Peter Hanns Reill, *The German Enlightenment and the Rise of Historicism* (Los Angeles: University of California Press, 1975), p. 31.
3. Moses Mendelssohn, *Über die Empfindungen* (Berlin: Bey Christian Friedrich Voss, 1755), p. 52. Also Reill, *German Enlightenment*, p. 43.
4. Half a century later, Johann Christoph Gatterer, one of the first historians to hold a chair at Göttingen, still mocked the Chinese: "They wish to appear far older than they are . . . They play with millions of years as children play with balls." Reill, *German Enlightenment*, p. 78.
5. Johann Salomo Semler, *Beantwortung der Fragmente eines Lebens, Beschreibungen berühmter Gelehrter* (Leipzig, 1766), vol. 2, p. 290. Hartmut Lehmann, *Der Pietismus*, in Etienne François and Hagen Schulze, eds. *Deutsche Erinnerungsorte* (Munich: Beck, 2001), vol. 2, pp. 571–584.
6. Reill, *German Enlightenment*, p. 82.
7. Johann David Michaelis, "Schreiben an Herrn Professor Schlötzer die Zeitrechnung von der Sündflut bis auf Salomo betreffend," in *Zerstreute kleine Schriften*. 2 vols. (Jena, in der akademischen Buchhandlung, 1794), vol. 1, pp. 262ff.
8. Reill, *German Enlightenment*, pp. 78–79.
9. Reill, *German Enlightenment*, pp. 92–93.

10. Reill, *German Enlightenment*, p. 220.
11. Ibid., p. 90.
12. Ibid.
13. Johann David Michaelis, *Mosaisches Recht*. 6 vols. (Frankfurt, 1770–1775), pp. 88ff.
14. August Ludwig von Schlözer, *Allgemeine nordische Geschichte: Forsetzungen der Allgemeinen Welt-Historie durch eine Gesellschaft von Gelehrten in Teutschland und Engeland ausgefertiget.* Part 31 (Halle, 1771), p. 263.
15. August Ludwig von Schlözer, *Verstellung der Universal-Historie.* 2 vols. (Göttingen, 1772–1773), vol. 2, p. 273.
16. Ursula Franke, *Kunst als Erkenntnis: Die Rolle d. Sinnlichkeit in d. Ästhetik d. Alexander Gottlieb Baumgarten* (Wiesbaden: Steiner, 1972). Reill, *German Enlightenment*, p. 56.
17. H. R. Schweizer, ed., *Theoretische Ästhetik: Die grundlegenden Abschnitte der "Aesthetica"* (Hamburg: Felix Meiner Verlag, 1983). This comprises edited extracts from the original Frankfurt (1750–1758) edition.
18. Reill, *German Enlightenment*, p. 61.
19. Ibid., p. 202.
20. Ibid., p. 62.
21. Ibid., p. 65. Later, Martin Heidegger would argue that it is possible to philosophize in only two languages: Greek and German.
22. Ibid.
23. Isaak Iselin, *Über die Geschichte der Menschheit.* 2 vols. (Basel, 1768), vol. 1, pp. 7–8.
24. Reill, *German Enlightenment*, p. 217.
25. Ibid., p. 219.
26. Ibid., p. 214.
27. Walter Hofer, *Geschichtsschreibung und Weltanschauung: Betrachtungen zum Werk Friedrich Meineckes* (Munich: Oldenborg, 1950), pp. 370f. The standard work on Sturm und Drang in English is Roy Pascal, *The German Sturm und Drang* (Manchester: Manchester University Press, 1953).
28. Reill, *German Enlightenment*, p. 98.
29. Ernst Mayr, *The Growth of Biological Thought* (Cambridge, Mass.: Belknap Press of Harvard University Press, 1982), p. 36.
30. Reill, *German Enlightenment*, p. 126.
31. Ibid., p. 50.
32. Ibid., p. 104.
33. I thank Werner Pfennig for this information.
34. Reill, *German Enlightenment*, p. 156.
35. Ibid., p. 157.
36. Ibid., p. 187.
37. Ibid., p. 158.
38. Ibid., p. 106.
39. Ibid., p. 130.
40. Timothy Lenoir, *The Strategy of Life: Teleology and Mechanics in Nineteenth-Century German Biology* (Chicago: University of Chicago Press, 1989), pp. 17–18. Johann Blumenbach's *The Institutions of Physiology* was translated into English by John Elliotson in 1817, printed "by Bensley for E. Cox," and dedicated to HRH Prince Augustus Frederick, Duke of Sussex.
41. Lenoir, *Strategy of Life*, p. 19.
42. Ibid., p. 20.
43. Ibid.
44. Ibid., p. 26.
45. Ibid., pp. 36–37.
46. Ibid.
47. Ibid., p. 43.
48. Ibid., p. 48.
49. Ibid., p. 50.
50. Ibid., p. 197.
51. Ibid., p. 202.
52. Ibid., p. 310.
53. Ibid., p. 343.

54. Ibid., p. 391.
55. Peter Hanns Reill, "History and the Life Sciences in the Early Nineteenth Century," Chapter 2 of George G. Iggers and James M. Powell, eds., *Leopold von Ranke and the Shaping of the Historical Discipline* (Syracuse, N.Y.: Syracuse University Press, 1990), pp. 21ff.
56. Wilhelm von Humboldt, *Gesammelte Schriften, Preussischen Akademie des Wissenschaften*, ed. Albert Leitzmann (Berlin, 1903–06) vol. 1, p. 262.
57. Lenoir, *Strategy of Life*, p. 27.
58. In this, he was much influenced by his brother, Alexander. See Ilse Jahn, *Dem Leben auf der Spur: Die biologischen Forschungen Alexander von Humboldts* (Leipzig, Jena, Berlin: Urania-Verlag, 1969), pp. 40ff.
59. I have used mainly Paul Sweet, *Wilhelm von Humboldt*. 2 vols. (Ohio State University Press, 1980), in this case, vol. 2, pp. 394ff. But see also Wilhelm von Humboldt, "Essai sur les langes du nouveau continent," *Gesammelte Schriften*, vol. 3, pp. 300–341; and Clemens Menze, *Humboldts Lehre* (Ratingen bei Düsseldorf: Henn, 1965).
60. Friedrich Schiller, *Die Briefwechsel zwischen Friedrich Schiller und Wilhelm von Humboldt*, ed. Siegfried Seidel (Berlin: Aufbau, 1962), 2 vols; Clemens Menze, *Wilhelm von Humboldt und Christian Gottlob Heyne* (Ratingen bei Düsseldorf: Henn, 1966); Aleida Assmann, *Arbeit am nationalen Gedächtnis: Eine kurze Geschichte der deutschen Bildungsidee* (Frankfurt am Main: Campus, 1993). I am grateful to Dr. Gisele Mettele for drawing this reference to my attention.
61. Pinkard, *German Philosophy*, p. 7.
62. Thomas Albert Howard, *Protestant Theology and the Making of the Modern German University* (Oxford: Oxford University Press, 2006), p. 7; Blanning, *Power of Culture*, p. 205.

Chapter 3: Winckelmann, Wolf, and Lessing: The Third Greek Revival and the Origins of Modern Scholarship

1. For the Basel of Burckhardt's day, see Lionel Gossman, *Basel in the Age of Burckhardt: A Study in Unseasonable Ideas* (Chicago: University of Chicago Press, 2000).
2. Peter Burke, Introduction to *Jacob Burckhardt: The Civilisation of the Renaissance in Italy*, trans. S. G. C. Middlemore (London: Penguin, 1990), p. 12.
3. For Burckhardt's conception of cultural history, see Felix Gilbert, *History: Politics or Culture? Reflections on Ranke and Burckhardt* (Princeton, N.J.: Princeton University Press, 1990), especially Chapters 4 and 5.
4. Christopher Charles Parslow, *Rediscovering Antiquity: Karl Weber and the Excavation of Herculaneum, Pompeii and Stabiae* (Cambridge: Cambridge University Press, 1995), pp. 85, 177, and 215–232.
5. E. M. Butler, *The Tyranny of Greece over Germany: A Study of the Influence Exercised by Greek Art and Poetry over the Great German Writers of the Eighteenth, Nineteenth, and Twentieth Centuries* (Boston: Beacon Press, 1958), p. 12.
6. Henry Hatfield, *Aesthetic Paganism in German Literature: From Winckelmann to the Death of Goethe* (Cambridge, Mass.: Harvard University Press, 1964), p. 6.
7. Butler, *Tyranny of Greece*, p. 14.
8. Hatfield, *Aesthetic Paganism*, pp. 6–7.
9. Suzanne L. Marchand, *Down from Olympus: Archaeology and Philhellenism in Germany, 1750–1970* (Princeton, N.J.: Princeton University Press, 1996).
10. Parslow, *Rediscovering Antiquity*, p. 27.
11. C. W. Ceram, *Gods, Graves and Scholars: The Story of Archaeology*, trans. E. B. Garside and Sophie Wilkins (London: V. Gollancz, 1971), p. 4. Published originally as *Götter, Gräber und Gelehrte* (Hamburg: Rowohlt, 1949).
12. Parslow, *Rediscovering Antiquity*, p. 104. See also Wolfgang Leppmann, *Winckelmann* (London: Gollancz, 1971), p. 170.
13. Butler, *Tyranny of Greece*, p. 26.
14. Hatfield, *Aesthetic Paganism*, p. 39. For general background see Josef Chytry, *The Aesthetic State: A Quest in Modern German Thought* (Berkeley and London: University of California Press, 1989).

15. J. Eiselein, ed., *Johann Winckelmanns sämtliche Werke* (Donaueschingen, 1825–29), vol. 6, pp. 297–299.
16. For his comments on the Apollo Belvedere, see Hans Zeller, *Winckelmanns Beschreibung des Apollo im Belvedere* (Zurich: Atlantis, 1955).
17. Hatfield, *Aesthetic Paganism*, p. 8.
18. Ibid., pp. 19–20.
19. Ibid., p. 6.
20. Ibid., p. 10.
21. Ibid., p. 6.
22. Ibid., p. 20.
23. Butler, *Tyranny of Greece*, p. 5.
24. Horst Rüdiger, ed., *Winckelmanns Tod: Die Originalberichte* (Wiesbaden: Insel-Verlag, 1959).
25. Hatfield, *Aesthetic Paganism*, p. 1.
26. Ibid., pp. 334–335.
27. It is as if Goethe's successors had taken very seriously his admonition that everyone should be a Greek in his own way: "Jeder sei auf seine Art ein Grieche, aber er sei's." Hatfield, *Aesthetic Paganism*, p. 5.
28. H. B. Garland, *Lessing: The Founder of Modern German Literature* (London: Macmillan, 1963), p. 4.
29. Gustav Sichelschmidt, *Lessing: Der Mann und sein Werk* (Düsseldorf: Droste, 1989); also Peter Pütz, *Die Leistung der Form: Lessings Dramen* (Frankfurt am Main: Suhrkamp, 1986).
30. Victor Lange, *The Classical Age of German Literature, 1740–1815* (London: Edward Arnold, 1982).
31. Gerhard Kaiser, *Klopstock: Religion und Dichtung* (Gütersloh: Gütersloher Verlagshaus Gerd Mohn, 1963), pp. 133–160 (for the theory of genius) and 204ff.
32. For Klopstock's literary ambitions for Germany, see Robert M. Browning, *German Poetry in the Age of Enlightenment: From Brockes to Klopstock* (University Park: Pennsylvania State University Press, 1978), pp. 230–231. See also Adolphe Bossert, *Goethe, ses précurseurs et ses contemporains: Klopstock, Lessing, Herder, Wieland, Lavater; la jeunesse de Goethe* (Paris: Hachette, 1891).
33. Garland, *Lessing*, p. 12.
34. Sichelschmidt, *Lessing*, Chapters 5–8.
35. Garland, *Lessing*, p. 83.
36. Ibid., p. 69.
37. Ibid., p. 32.
38. Ibid., p. 159.
39. Ibid., p. 142.
40. Ibid., pp. 180–181.
41. Pütz, *Leistung der Form*, pp. 242f.
42. Garland, *Lessing*, p. 180.
43. Ibid., p. 57.
44. Ibid., p. 198.
45. Alex Potts, *Flesh and the Ideal: Winckelmann and the Origins of Art History* (New Haven, Conn., and London: Yale University Press, 1994), p. 26.
46. The relatively small—but high-powered—world of the German intellectual community can be seen from the range of Wolf's letters: Siegfried Reiter, ed., *Friedrich August Wolf: Ein Leben in Briefen*. 3 vols. (Stuttgart: Metzler, 1935).
47. Marchand, *Down from Olympus*, p. 19.
48. Ibid., p. 20.
49. Ibid., p. 21.
50. Potts, *Flesh and the Ideal*, p. 25.
51. Ibid., p. 27.
52. Ibid., p. 28.
53. Ibid.
54. Marchand, *Down from Olympus*, p. 31.
55. Potts, *Flesh and the Ideal*, p. 22.

Chapter 4: The Supreme Products of the Age of Paper

1. Peter Hall, *Cities in Civilisation: Culture, Innovation, and Urban Order* (London: Weidenfeld & Nicolson, 1998), p. 69.
2. Ibid., p. 72.
3. Bruford, *Culture and Society*, p. 59.
4. Boyle, *Goethe*, vol. 1, pp. 236–237.
5. Bruford, *Culture and Society*, p. 57.
6. Boyle, *Goethe*, vol. 1, p. 244.
7. Bruford, *Culture and Society*, p. 18.
8. Regine Schindler-Hürlimann, *Wielands Menschenbild: Eine Interpretation des Agathon* (Zurich: Atlantis, 1963).
9. Bruford, *Culture and Society*, p. 42.
10. Jan Cölin, *Philologie und Roman: Zu Wielands erzählerischer Rekonstruktion griechischer Antike im "Aristipp"* (Göttingen: Vandenhoeck & Ruprecht, 1998).
11. Bruford, *Culture and Society*, p. 45.
12. Nicholas Boyle, *Goethe: The Poet and the Age*, vol. 2, *Revolution and Renunciation (1790–1803)* (Oxford: Clarendon Press, 2000).
13. Bruford, *Culture and Society*, pp. 62–63.
14. Ibid., p. 11.
15. Boyle, *Goethe*, vol. 1, p. 170.
16. See Boyle, *Goethe*, vol. 1, pp. 176f. for a comparison of *Werther* and Schiller's *The Robbers* (1781) which, he says, was likewise the object of a Sturm und Drang cult. For Napoleon's criticisms, see Gustav Seibt, *Goethe und Napoleon: Eine historische Begegnung* (Munich: Beck, 2008).
17. Bruford, *Culture and Society*, pp. 13 and 18.
18. Boyle, *Goethe*, vol. 1, p. 267.
19. Dietrich Fischer-Dieskau, *Goethe als Intendant: Theaterleidenschaften im klassichen Weimar* (Munich: Deutscher Taschenbuch Verlag, 2006).
20. Boyle, *Goethe*, vol. 1, p. 104. Habermas, *Structural Transformation*, p. 38.
21. Bruford, *Culture and Society*, p. 97.
22. Ibid., p. 10.
23. Boyle, *Goethe*, vol.1, pp. 592 and 145.
24. Ibid., p. 156.
25. Ibid., p. 164.
26. Ibid., p. 259.
27. Ibid., p. 420.
28. Ibid., p. 443.
29. Ibid., pp. 170–171.
30. Ibid., p. 180.
31. Ibid., p. 515.
32. For Goethe's works in English, see Derek Glass, *Goethe in English: A Bibliography of the Translations in the Twentieth Century*, ed. Matthew Bell and Martin H. Jones (Leeds: Maney Publishing, for the English Goethe Society and the Modern Humanities Research Association, 2005). See also Nicholas Boyle and John Guthrie, eds., *Goethe and the English-speaking World: Essays from the Cambridge Symposium for His 250th Anniversary* (Rochester, N.Y.: Camden House, 2002).
33. Henry Hatfield, *Aesthetic Paganism and German Literature: From Winckelmann to the Death of Goethe* (Cambridge, Mass.: Harvard University Press, 1964), p. x.
34. Bruford, *Culture and Society*, p. 30.
35. Ibid., p. 50.
36. Boyle, *Goethe*, vol.1, 605.
37. Bruford, *Culture and Society*, p. 47.
38. Johann Wolfgang von Goethe, *Faust* (Oxford: Oxford University Press, 1998), Introduction by David Luke, p. ix.
39. David Hawke, *The Faust Myth: Religion and the Rise of Representation* (Basingstoke: Palgrave/ Macmillan, 2007); John Gearey, *Goethe's Faust: The Making of Part 1* (New Haven, Conn., and London: Yale University Press, 1981).
40. Goethe, *Faust*, p. xiv.

41. Ibid., p. xxxv.
42. Ibid., p. vii.
43. Boyle, *Goethe*, vol. 1, p. 346.
44. Max Kommerell, *Der Dichter als Führer in der deutschen Klassik: Klopstock, Herder, Goethe, Schiller, Jean Paul, Hölderlin*. 3rd ed. (Frankfurt am Main: Klostermann, 1982); F. M. Barnard, *Herder's Social and Political Thought: From Enlightenment to Nationalism* (Oxford: Clarendon Press, 1965), p. xix.
45. Barnard, *Herder's Social and Political Thought*, p. 16.
46. Ibid., p. 55.
47. For Herder's view of the difference between apes and men, see H. B. Nisbet, *Herder and the Philosophy and History of Science* (Cambridge: Modern Humanities Research Association, 1970), pp. 250f.
48. Barnard, *Herder's Social and Political Thought*, p. 57.
49. Ibid., p. 59.
50. Ibid., p. 63.
51. Ibid., p. 75.
52. Ibid., p. 93.
53. For Herder's views on evolution, see Nisbet, *Herder and the Philosophy*, pp. 210ff.
54. Barnard, *Herder's Social and Political Thought*, p. 120.
55. Ibid., p. 124.
56. Ibid., p. 147.
57. Johannes von Müller, *Briefwechsel mit Johann Gottfried Herder und Caroline v. Herder, 1782–1808*, ed. K. E. Hoffmann (Schaffhausen: Meier, 1962), 65, III, p. 109.
58. Steven D. Martinson, ed., *A Companion to the Works of Friedrich Schiller* (Rochester, N.Y.: Camden House, 2000), p. 3.
59. Bernt von Heiseler, *Schiller*, trans. and annotated by John Bednall (London: Eyre & Spottiswood, 1962), pp. 52ff. for the first performances. See also Peter-André Alt, *Schiller: Leben—Werk—Zeit*, 2 vols. (Munich: C. H. Beck, 2000), pp. 276ff.
60. Heiseler, p. 126.
61. Ibid., p. 3.
62. Ibid., p. 141.
63. Ibid., p. 9.
64. Rüdiger Safranski, *Friedrich Schiller, oder, Die Erfindung des deutschen Idealismus* (Munich: Hanser, 2004).
65. Martinson, *Companion*, p. 11.
66. Frederick Beiser, *Schiller as Philosopher: A Re-examination* (Oxford: Oxford University Press, 2005), p. 37.
67. Martinson, *Companion*, p. 43.
68. Ibid., p. 54.
69. Ibid., p. 77.
70. Ibid., p. 84.
71. Ibid., p. 207.
72. Ibid., p. 220.

Chapter 5: New Light on the Structure of the Mind

1. Willibald Klinke, *Kant for Everyman* (London: Routledge & Kegan Paul, 1951), p. 43.
2. On Kant's early opposition to Idealism, see *Nova dilucidato*, vol. 1, pp. 411–412. This may be found in *Kant's Gesammelte Schriften*, which was edited by the Royal Prussian (later German) Academy of Sciences, Berlin, and published by George Reiner, subsequently Walter de Gruyter.
3. For Mendelssohn's "philosophical preoccupations" see Alexander Altmann, *Moses Mendelssohn: A Biographical Study* (London and Portland, Oregon: The Littman Library of Jewish Civilisation, 1998), pp. 313ff. Mendelssohn is more fully situated in the history of philosophy in Frederick C. Beiser, *The Fate of Reason: German Philosophy from Kant to Fichte* (Cambridge, Mass.: Harvard University Press, 1987), pp. 92ff.
4. Klinke, *Kant*, p. 254.

5. Ibid., p. 202.
6. Lewis White Beck, *Early German Philosophy: Kant and His Predecessors* (Cambridge, Mass.: Belknap Press of Harvard University Press, 1969), p. 327.
7. Altmann, *Moses Mendelssohn*. See also the same author's *Moses Mendelssohns Frühschriften zur Metaphysik* (Tübingen,: Mohr/Siebeck, 1969); and David Sorkin, *Moses Mendelssohn and the Religious Enlightenment* (London: Peter Halban, 1996), p. xl.
8. Karl Ameriks, ed., *The Cambridge Companion to German Idealism* (Cambridge: Cambridge University Press, 2000), p. 1.
9. Ibid., p. 2.
10. Ibid.
11. *Geschichte der Universität Jena, 1548/58–1958* (Jena: G. Fischer, 1958).
12. Ameriks, *Cambridge Companion*, p. 4.
13. Ibid.
14. Bertrand Russell, *History of Western Philosophy* (London: Routledge, 2005), p. 640.
15. Klinke, *Kant*, p. 78.
16. Ibid., p. 81.
17. Ibid., p. 83. See also Andrew Ward, *Kant: The Three Critiques* (Cambridge, U.K.: Polity Press, 2006).
18. Karl Ameriks, *Kant's Theory of Mind: An Analysis of the Paralogisms of Pure Reason* (Oxford: Clarendon Press, 1982).
19. Klinke, *Kant*, p. 87.
20. For the difficult idea of "ideality," see Ameriks, *Kant's Theory*, pp. 280ff, and Dieter Henrich, *Between Kant and Hegel: Lectures on German Idealism*, ed. David S. Pacini (Cambridge, Mass.: Harvard University Press, 2003).
21. Klinke, *Kant*, p. 82.
22. Paul Guyer, ed., *The Cambridge Companion to Kant* (Cambridge: Cambridge University Press, 1992).
23. Klinke, *Kant*, p. 91.
24. Ibid., p. 97.
25. Ibid., p. 114.
26. Ibid.
27. Falk Wunderlich, *Kant und Bewusstseinstheorien des 18. Jahrhunderts* (Berlin: de Gruyter, 2005).
28. Klinke, *Kant*, p. 128.
29. Ernst Cassirer, *Kant's Life and Thought* (New Haven, Conn., and London: Yale University Press, 1981), pp. 271–273.
30. Cassirer, *Kant's Life*, p. 288.
31. Ibid., p. 303.
32. Ibid., p. 320.
33. Ibid., p. 323.
34. Manfred Frank, *The Philosophical Foundations of Early German Romanticism*, trans. Elizabeth Millán-Zaibert (Albany: State University of New York Press, 2004).
35. Cassirer, *Kant's Life*, p. 333.
36. Pinkard, *German Philosophy*, p. 88.
37. Ibid., p. 89.
38. Henrich, *op. cit.*, pp. 96ff.
39. Pinkard, *German Philosophy*, p. 95.
40. Henrich, *Between Kant and Hegel*, p. 113ff and 127ff.
41. Pinkard, *German Philosophy*, p. 103.
42. Ibid., p. 105.
43. Russell, *History*, pp. 650–651. See also Robert Hanna, *Kant and the Foundations of Analytic Philosophy* (Oxford: Clarendon Press, 2001).
44. For the chronological emergence of Fichte's ideas, see Walter E. Wright's introduction to *The Science of Knowing: J. G. Fichte's 1804 Lectures on the Wissenschaftslehre*, trans. Walter E. Wright (Albany, N.Y.: State University of New York Press, 2005).
45. Pinkard, *German Philosophy*, p. 106.
46. Dieter Henrich, "Die Anfänge der Theorie des Subjekts," in *Zwischenbetrachtungen im Prozess der Aufklärung*, ed. Axel Honneth et al. (Frankfurt: Surhkamp, 1989), pp. 106ff;

also, in English, "Schulz and Post-Kantian Scepticism," Chapter 10 of Henrich, *Between Kant and Hegel.* See also Beiser, *Fate of Reason*, pp. 226ff. (for Reinhold) and 266ff., for Schulze.

47. Fichte's works were collected as *J. G. Fichtes sämmtliche Werke*, edited by J. H. Fichte in 1845–46, and published by Veit of Berlin.
48. Pinkard, *German Philosophy*, p.109.
49. Henrich, *Between Kant and Hegel*, pp. 206ff., also discusses Fichte's theory of imagination, which, he says, is central.
50. Pinkard, *German Philosophy*, p. 123.
51. This has been much examined by specialist historians. For a bibliography, see *Introductions to the Wissenschaftslehre and Other Writings, 1797–1800*, ed. and trans. Daniel Breazeale (Indianapolis, Ind.: Hackett, 1994), pp. xlvff.
52. J. G. Fichte, *Foundations of Transcendental Philosophy*, trans. and ed. Daniel Breazeale (Ithaca, N.Y.: Cornell University Press, 1992).

Chapter 6: The High Renaissance in Music: The Symphony as Philosophy

1. Wolfgang Victor Ruttkowski, *Das literarische Chanson in Deutschland* (Bern/Munich: Francke, 1966).
2. Harold C. Schonberg, *Lives of the Great Composers* (London: Davis-Poynter/Macdonald Futura, 1970/1980), p. 616.
3. Ibid., p. 618.
4. Ibid., p. 620.
5. It was felt that the popularity of the piano in the early 1800s threatened the health of other instruments. See David Gramit, *Cultivating Music: The Aspirations, Interests, and Limits of German Musical Culture, 1770–1848* (Berkeley: University of California Press, 2002), p. 136.
6. Schonberg, *Lives of the Great Composers*, p. 622.
7. Ibid., p. 624.
8. For a musical comparison of Gluck's work just before *Alceste*, see Jack M. Stein, *Poem and Music in the German Lied from Gluck to Hugo Wolf* (Cambridge, Mass.: Harvard University Press, 1971), pp. 29–32. See also Hans Joachim Moser, *Christoph Willibald Gluck: Die Lestung, der Mann, der Vermächtnis* (Stuttgart: Cetta, 1940), p. 323.
9. Schonberg, *Lives of the Great Composers*, p. 624.
10. Ibid., pp. 625–626.
11. For late Haydn, see Hans-Hubert Schönzler, ed., *Of German Music* (London: Oswald Wolff, 1976), p. 92, and Sieghard Brandenburg, ed., *Haydn, Mozart & Beethoven: Studies in the Music of the Classical Period; Essays in Honour of Alan Tyson* (Oxford: Clarendon Press, 1998).
12. Peter Gay, *Mozart* (London: Weidenfeld & Nicolson, 1999), pp. 109f. Also Robert W. Gutman, *Mozart: A Cultural Biography* (London: Secker & Warburg, 2000), pp. 668ff.
13. Schonberg, *Lives of the Great Composers*, p. 628.
14. For a comparison between Gluck and Mozart, see Adolf Goldschmitt, *Mozart: Genius und Mensch* (Hamburg: C. Wegner, 1955), pp. 288ff. And for Gluck's influence on Mozart, see Gutman, *Mozart*, p. 571.
15. Alfons Rosenberg, *Die Zauberflöte: Geschichte und Deutung von Mozarts Oper* (Munich: Prestel, 1964). Hugo Zelzer tells us that between the first night on September 30, 1791, and November 1792, there were more than 100 performances and that the work was performed in Weimar under the direction of Goethe. "German Opera from Mozart to Weber," in Schönzeler, ed., *Of German Music*, p. 127. For a flavor of the reception of Mozart's operas, see Cliff Eissen, *New Mozart Documents: A Supplement to O. E. Deutsch's Documentary Biography* (London: Macmillan, 1991).
16. Schonberg, *Lives of the Great Composers*, p. 630.
17. Ibid., p. 631.
18. David Wyn Jones, *The Symphony in Beethoven's Vienna* (Cambridge: Cambridge University Press, 2006), p. 264.
19. Estéban Buch, *Beethoven's Ninth: A Political History*, trans. Richard Miller (Chicago: Uni-

versity of Chicago Press, 2003). See also Celia Applegate and Pamela Potter, eds., *Music and German National Identity* (Chicago: University of Chicago Press, 2002), p. 8.

20. Schonberg, *Lives of the Great Composers*, p. 632.
21. For Schubert's influence, see Scott Messing, *Schubert in the European Imagination*. 2 vols. (Rochester, N.Y.: University of Rochester Press, 2007). See vol.1, pp. 199f., for the responses to Schubert's death.
22. Charles Fisk, "What Schubert's Last Sonata Might Hold," in Jenefer Robinson, ed., *Music and Meaning* (Ithaca, N.Y.: Cornell University Press, 1997), pp. 179ff. See also Lorraine Byrne, *Schubert's Goethe Settings* (Aldershot: Ashgate, 2003); Hermann Abert, *Goethe und die Musik* (Stuttgart: J. Engelharns Nachfolger, 1922).
23. Some of his views are set out in the *Königliche kaiserliche privilegierte Prager Zeitung*, no. 293, October 20, 1815.
24. Mark Evan Bonds, *Music as Thought: Listening to the Symphony in the Age of Beethoven* (Princeton, N.J.: Princeton University Press, 2006), p. xiii. See also Wyn Jones, *Symphony*, pp. 11–33. For the hierarchy of music, see Gramit, *Cultivating*, pp. 23–24; and Robinson, ed., *Music*.
25. Bonds, *Music as Thought*, p. 1.
26. Ibid., pp. 7 and 17.
27. For Beethoven's earnings from his symphonies, see Ludwig van Beethoven, *Briefwechsel: Gesamtausgabe*, ed. Sieghard Brandenburg, 7 vols. (Munich: G. Henle, 1996), vol. 1, pp. 317ff. For the development of the concert, see Gramit, *Cultivating*, pp. 25 and 138.
28. Bonds, *Music as Thought*, p. 16. For the "depth" of German music, the extent to which its "intellectual" properties set it apart, as a *Sonderweg* in itself, see Applegate and Potter, eds., *Music and German*, pp. 40–42 and 51–55.
29. Applegate and Potter, eds., *Music and German*, pp. 51–52. Bonds, *Music as Thought*, p. 22.
30. For a more materialistic analysis, see Franz Hadamowsky, *Wien, Theatergeschichte von den Anfängen bis zum Ende des Ersten Weltkriegs* (Vienna: Wancura, 1994), pp. 308–310.
31. *E. T. A. Hoffmann's Musical Writings: Kreisleriana, the Poet and the Composer, Music Criticism*, ed. David Charlton, trans. Martyn Clark (Cambridge: Cambridge University Press, 1989).
32. Bonds, *Music as Thought*, pp. 35–40.
33. Again, for a different perspective, concentrating on the music lovers' societies, see Carl Ferdinand Pohl, *Denkschrift aus Anlass des hundertjährigen Bestehens der Tonkünstler-Societät: Im Jahre 1862 reorganisiert als "Haydn," Witwen-und Waisen-Versorgungs-Verein der Tonkünstler in Wien* (Vienna, 1871), pp. 67–69.
34. Applegate and Potter, eds., *Music*, p. 6. For the link to nationalism, see p. 18 and for the Germanness of music, see p. 2. See also Bonds, *Music as Thought*, p. 46.
35. Bonds, *Music as Thought*, p. 51.
36. Abert, *Goethe und die Musik*, and Ruttkowksi, *Literarische Chanson*.
37. For a description of the emergence of the concert hall, see Eduard Hanslick, *Geschichte des Concertwesens in Wien*. 2 vols. (Vienna, 1897), vol. 1, pp. 289f.
38. Applegate and Potter, eds., *Music*, pp. 2 and 9.
39. Bonds, *Music as Thought*, p. 87.
40. Beethoven himself did not share many of these views. See Wyn Jones, *Symphony*, pp. 155ff.
41. Bonds, *Music as Thought*, p. 106.

Chapter 7: Cosmos, Cuneiform, Clausewitz

1. Mott T. Greene, *Geology in the Nineteenth Century: Changing Views of a Changing World* (Ithaca, N.Y.: Cornell University Press, 1982), p. 36.
2. Abraham Gottlob Werner, *Kurze Klassifikation und Beschreibung der verschiedenen Gebirgsarten*. Translation and facsimile of original text, tr. and with an introduction by Alexander Ospovat (New York: Hafner, 1971). Original published 1789.
3. Rachael Laudan, *From Mineralogy to Geology: The Foundations of a Science: 1650–1830* (Chicago: University of Chicago Press, 1987), pp. 48ff. For Werner's theory of color, see Patrick Syme, *Werner's Nomenclature of Colours* (Edinburgh: W. Blackwood, 1821).
4. Laudan, *Mineralogy*, p. 49.
5. Ibid., p. 113ff.

6. Ibid., p. 40.
7. Ospovat, *op. cit.*
8. Laudan, *Mineralogy*, p. 100.
9. Ibid., p. 105.
10. Ibid., p. 111.
11. Marcus du Sautoy, *The Music of the Primes* (London: HarperCollins, 2003/2004), p. 20.
12. See Ludwig Schlesinger, "Über Gauss Jugendarbeiten zum arithmetisch-geometrischen Mittel," *Jahresbericht d. Deutschen Mathematiker-Vereinigung* 20, no. 11–12 (November–December 1911): pp. 396–403.
13. Robert Jordan, "Die verlorene Ceres," *Neueste Nachrichten*, Brunswick, May 1, 1927. "*Determinatio attractionis, quam in punctum quoduis positionis datae exerceret planeta, si ejus massa per totam orbitam, ratione temporis, que singulae partes describuntur, uniformiter esset dispertita.*" *Comment*, Göttingen, IV, 1816–1818, pp. 21–48.
14. Du Sautoy, *Music of the Primes*, p. 109. G. Waldo Dunnington, *Carl Friedrich Gauss: Titan of Science* (New York: Hafner, 1955), pp. 174ff.
15. Morris Kline, *Mathematics for Non-Mathematicians* (New York: Dover, 1967), p. 456.
16. Dunnington, *Carl Friedrich Gauss*, pp. 147–162. See also Catherine Goldstein et al., eds., *The Shaping of Arithmetic: After C. F. Gauss's Disquistiones arithmeticae* (Berlin: Springer, 2007).
17. Dunnington, *Carl Friedrich Gauss*, pp. 139ff.
18. Du Sautoy, *Music of the Primes*, p. 74.
19. For a discussion of the varied forms of homeopathy, see Margery G. Blackie, *The Patient Not the Cure: The Challenge of Homeopathy* (London: Macdonald & Jane's, 1976), pp. 3ff. See also Thomas Lindsay Bradford, *The Life and Letters of Dr. Samuel Hahnemann* (Philadelphia: Boericke & Tafel, 1895).
20. Martin Gumpert, *Hahnemann: The Adventurous Career of a Medical Rebel* (New York: L. B. Fischer, 1945), p. 6.
21. Ibid., p. 22. Bradford, *Life and Letters*, pp. 24–26. For Hahnemann's contemporaries, see Blackie, *Patient*, pp. 25ff.
22. Bradford, *Life and Letters*, p. 35. Blackie, *Patient*, p. 16.
23. Gumpert, *Hahnemann*, p. 68.
24. Ibid., p. 70.
25. Bradford, *Life and Letters*, p. 72.
26. A recent biography describes Humboldt in this way: it is "quite possible that no other European had so great an impact on the intellectual culture of nineteenth-century America." Aaron Sachs, *The Humboldt Current: A European Explorer and His American Disciples* (New York: Oxford University Press, 2007).
27. Herbert Scurla, *Alexander von Humboldt: Eine Biographie* (Düsseldorf: Claasen, 1982), pp. 188–191. See also Hermann Klencke, *Lives of the Brothers Humboldt: Alexander and William*, trans. Juliette Bauer (London: Ingram, Cooke & Co., 1852).
28. Scurla, *Alexander von Humboldt*, pp. 51–57.
29. Ibid., pp. 102ff.
30. Gerard Helferich, *Humboldt's Cosmos* (New York: Gotham, 2004), p. 21. Humboldt was himself called "the second Columbus." Scurla, *Alexander von Humboldt*, p. 415.
31. *Dictionary of Scientific Biography*, VI, p. 550.
32. Scurla, *Alexander von Humboldt*, pp. 138f.
33. Ibid., p. 178. According to Aaron Sachs, he inspired four great American explorers: J. N. Reynolds, Clarence King, George Wallace Melville ,and John Muir. Sachs, *Humboldt Current*, passim.
34. For Humboldt's ideas about *Kosmos* and *Volksbildung*, see Nicolaas A. Rupke, *Alexander von Humboldt: A Metabiography* (Frankfurt and Berlin: Peter Lang, 2005), pp. 38–43.
35. Helferich, *Humboldt's Cosmos*, p. 23.
36. Scurla, *Alexander von Humboldt*, pp. 206–207.
37. For a discussion of the wider importance of Alexander von Humboldt, see: Rupke, *Alexander von Humboldt*, pp. 162–218.
38. C. W. Ceram, *Gods, Graves and Scholars* (London: Book Club Associates, 1967), p. 228.
39. Ibid.
40. Arthur John Booth, *The Discovery and Decipherment of the Trilingual Cuneiform Inscriptions* (London: Longmans, Green and Co., 1902), p. 173. Grotefend's own account has been translated into English in A. H. L. Heere, *Historical Works*, vol. 2 (Oxford, 1833), p. 337. See

also Denise Schmandt-Besserat, *Before Writing*, vol.1, *From Counting to Cuneiform* (Austin: University of Texas Press, 1992).

41. Ceram, *Gods, Graves*, p. 230.
42. Ibid., p. 231.
43. Ibid., p. 233.
44. Hugh Smith, *On Clausewitz: A Study of Military and Political Ideas* (Basingstoke: Palgrave/ Macmillan, 2005), p. viii. Peter Paret says much of it is common sense. Peter Paret, *Understanding War: Essays on Clausewitz and the History of Military Power* (Princeton, N.J.: Princeton University Press, 1992), p. 117.
45. Paret says the advent of nuclear power has made the problems that Clausewitz addressed even more important than in his day. Paret, *Understanding War*, p. 96.
46. Smith, *On Clausewitz*, p. ix.
47. Ibid., p. 3.
48. Wilhelm von Schramm, *Clausewitz: Leben und Werk* (Esslingen am Neckar: Bechtle, 1981), pp. 140ff.
49. Schramm, *Clausewitz*, pp. 363ff.
50. Being director of the war academy concentrated his mind on the general staff. Major von Roder, *Für Euch, meine Kinder!* (Berlin, 1861).
51. Smith, *On Clausewitz*, p. 25. Carl von Clausewitz, "Bemerkungen über die reine und angewandte Strategie des Herrn von Bülow," *Neue Bellona* 9, no. 3 (1805): 271.
52. Schramm, *Clausewitz*, pp. 557ff. Smith, *On Clausewitz*, p. 25.
53. Smith, *On Clausewitz*, p. 27.
54. Ibid.
55. Ibid., p. 44.
56. For his appreciation of history, see Hans Delbrück, "General von Clausewitz," in *Historische und Politische Aufsätze* (Berlin: Walther & Apolant, 1887).
57. Schramm, *Clausewitz*, pp. 135–158 and 220–255. And see Michael Eliot Howard, *Clausewitz* (Oxford and New York: Oxford University Press, 1983).
58. Smith, *On Clausewitz*, pp. 65–66.
59. Schramm, *Clausewitz*, p. 181. Smith, *On Clausewitz*, p. 130.
60. Smith, *On Clausewitz*, p. 134.
61. Ibid., p. 237.
62. Ibid., p. 238.
63. Ibid., p. 239.

Chapter 8: The Mother Tongue, the Inner Voice, and the Romantic Song

1. Kai Hammermeister, *The German Aesthetic Tradition* (Cambridge: Cambridge University Press, 2002), pp. 62–86. See also Friedrich von Schlegel, *The Aesthetic and Miscellaneous Works*, trans. E. J. Millington (London: Bell, 1875). In the nineteenth century, Friedrich Max Müller, a German Orientalist who became the first professor of comparative philology at Oxford, said this: "If I were asked what I consider the most important discovery of the nineteenth century with respect to ancient history of mankind, I should answer by the following short line: Sanskrit *Dyaus Pitar* = Greek Zeùs Πατη = Latin Juppiter = Old Norse Tyr."
2. Manfred Frank, *The Philosophical Foundations of Early German Romanticism*, trans. Elizabeth Millán-Zaibert (Albany: State University of New York Press, 2004). See also Gerald N. Izenberg, *Romanticism, Revolution, and the Origins of Modern Selfhood, 1787–1802* (Princeton, N.J.: Princeton University Press, 1992).
3. Raymond Schwab, *The Oriental Renaissance: Europe's Rediscovery of India and the East 1680–1880* (New York: Columbia University Press, 1984), p. 11.
4. Isaiah Berlin, *The Sense of Reality* (London: Chatto & Windus, 1996), p. 168.
5. Isaiah Berlin, *Freedom and Its Betrayal* (London: Chatto & Windus, 2002), p. 60.
6. Nicholas Halmi, *The Genealogy of the Romantic Symbol* (Oxford: Oxford University Press, 2007), pp. 51–53, 63–65, and 144–147.
7. Berlin, *Sense of Reality*, p. 179.

8. Izenberg, *Romanticism*, especially parts 1 and 2. Izenberg is particularly helpful on the links between politics and psyche and the role of irony. See also Kathleen M. Wheeler, ed., *German Aesthetic and Literary Criticism: The Romantic Ironists and Goethe* (Cambridge: Cambridge University Press, 1984), which contains material from the lesser known Romantics: Novalis, Ludwig Tieck, Karl Solger, and Jean Paul Richter.
9. Butler, *Tyranny of Greece*, p. 6.
10. Berlin, *Freedom and Its Betrayal*, p. 89.
11. Ibid., p. 91.
12. Ibid., p. 96.
13. Manfred Schröter, ed., "Schelling's *Erster Entwurf*," in *Schellings Werke*, 12 vols. (Munich: C. H. Beck, 1927–59), vol. 2, p. 63.
14. See, for example, Karl Jaspers, *Schelling: Grösse und Verhängnis* (Munich: Piper, 1955), p. 154ff.
15. Berlin, *Freedom and Its Betrayal*, p. 98.
16. Ibid.
17. Manfred Frank, *Das Problem "Zeit" in der deutschen Romantik* (Munich: Winkler, 1972), pp. 22–44 and 54–55. Also useful is *The Aesthetic and Miscellaneous Works of Friedrich von Schlegel*, trans. E. J. Millington (London: Henry G. Bohn, 1849). Includes "On the Language and Wisdom of the Indians."
18. Berlin, *Freedom and Its Betrayal*, p. 110.
19. Ibid., p. 111.
20. Ibid., pp. 184–185.
21. Izenberg, *Romanticism*, pp. 18ff.
22. Robert J. Richards, *The Romantic Conception of Life: Science and Philosophy in the Age of Goethe* (Chicago: University of Chicago Press, 2002), p. 18.
23. Ibid., p. 22.
24. Carmen Kahn-Wallerstein, *Schellings Frauen: Caroline und Pauline* (Bern: Francke, 1959).
25. Richards, *Romantic Conception*, p. 102.
26. Ibid., p. 8.
27. Ibid., p. 10.
28. Ibid., p. 12.
29. Friedrich Schelling, *System des transcendentalism Idealismus* (1800), in M. Schröter, *Schellings Werke*, vol. 2, p. 249.
30. Richards, *Romantic Conception*, p. 144.
31. Johann Christian Reil, *Rhapsodien über die Anwendung der psychischen Curmethode auf Geisteszerrüttungen* (Halle: Curtschen Buchhandlung, 1803). See also Henrik Steffens, *Johann Christian Reil: Ein Denkschrift* (Halle: Curtschen Buchhandlung, 1815).
32. Richards, *Romantic Conception*, pp. 267ff.
33. Ibid., pp. 305–306.
34. Karl J. Fink, *Goethe's History of Science* (Cambridge: Cambridge University Press, 1991), p. 9.
35. David Simpson, ed., *German Aesthetic and Literary Criticism: Kant, Fichte, Schelling, Schopenhauer, Hegel* (Cambridge: Cambridge University Press, 1984).
36. Fink, *Goethe's History*, p. 17.
37. Ibid., p. 22.
38. Werner Heisenberg, "Die Goethesche und Newtonsche Farbenlehre im Lichte der Modern Physik," *Geist der Zeit* 19 (1941): 261–275; Jürgen Blasius, "Zur Wissenschaftstheorie Goethes," *Zeitschrift für philsophisches Forschung* 33 (1979): 371–388.
39. Fink, *Goethe's History*, pp. 33–34.
40. Ibid., p. 44.
41. Ibid., p. 45.
42. Rupprecht Matthaei et al., eds. "J. W. Goethe, 'Verhältnis zur Philosophie,'" in *Die Schriften zur Naturwissenschaft.* 11 vols. in 2 parts (Weimar: Böhlau, 1947), part 1, vol. 4, p. 210.

Chapter 9: The Brandenburg Gate, the Iron Cross, and the German Raphaels

1. This is despite the fact that a long overdue (and excellent) catalogue raisonné of Mengs's oeuvre was published in 1999. See Steffi Roettgen, *Anton Raphael Mengs 1728–1779*, 2 vols. (Munich: Hirmir, 1999). The organization of this work brings out the number of religious pictures by Mengs.
2. Thomas Pelzel, *Anton Raphael Mengs and Neoclassicism* (New York: Garland Publishing, 1979), p. 1.
3. Ibid., p. 15.
4. Johann Kirsch, *Die römischen Titelkirchen im Altertum.* Studien zur Geschichte und Kultur des Altertums, IX (Paderborn: F. Schöningh, 1918), pp. 58ff.
5. Pelzel, *Anton Raphael Mengs*, p. 66.
6. Ibid., p. 72.
7. Ibid., p. 86.
8. Carl Justi, *Winckelmann und seine Zeitgenossen.* 3 vols. (Leipzig: F. C. W. Vogel, 1923), vol. 2, p. 382. See also Roettgen, *Anton Raphael Mengs.*
9. Pelzel, *Anton Raphael Mengs*, p. 109.
10. Ibid., p. 111.
11. Ibid., p. 126.
12. G. L. Bianconi, *Elogio storico del Cavaliere Antonio Raffaele Mengs* (Milan, 1780), p. 195. For his time in Spain, see Dieter Honisch, *Anton Raphael Mengs und die Bildform des Frühklassizismus* (Recklinghausen: Aurel Bongers, 1965), pp. 38ff.
13. Pelzel, *Anton Raphael Mengs*, p. 197.
14. Jean Locquin, *La peinture d'histoire en France de 1747 à 1785* (Paris, 1912), p. 104. Quoted in Pelzel, *Anton Raphael Mengs.* See also Hugh Honour, ed., *The Age of Neoclassicism.* Catalog of the Fourteenth Exhibition of the Council of Europe, at the Royal Academy and the Victoria and Albert Museum, London, September 9–November 19, 1972 (London: Arts Council of Great Britain, 1972).
15. Pelzel, *Anton Raphael Mengs*, p. 215.
16. Honour, *Age of Neoclassicism*, p. xxii.
17. Ibid., p. xxiii.
18. Ibid., p. liii.
19. Ibid., p. lxi.
20. Merlies Lammert, *David Gilly, Ein Baumeister des deutschen Klassizismus* (Berlin: Akademie Verlag, 1964), pp. 60ff.
21. Honour, *Age of Neoclassicism*, p. lxii.
22. Ibid.
23. Michael Snodin, ed., *Karl Friedrich Schinkel: A Universal Man.* Exhibition at the Victoria and Albert Museum, July 31–October 27, 1991 (New Haven, Conn.: Yale University Press in association with the Victoria and Albert Museum, 1991).
24. Gottfried Riemann und Christa Hesse, *Karl Friedrich Schinkel: Architekturzeichnungen* (Berlin: Henschel, 1991). See also Helmut Börsch-Supan and Lucius Grisebach, *Karl Friedrich Schinkel: Architektur, Malerei, Kunstgewerbe.* Exhibition, Berlin, 1981 (Berlin: Nicolai, 1981).
25. Louis Schreider, *Das Buch des Eisernen Kreuzes: Die Ordenssammlung* (Berlin, 1971).
26. Gordon Williams, *The Iron Cross: A History, 1813–1957* (Poole: Blandford Press, 1984), p. 12. Williams says that Schinkel's design was preferred to the Kaiser's own.
27. Snodin, ed., *Karl Friedrich Schinkel*, especially the essays by Gottfried Riemann and Alex Potts. See also Riemann und Hesse, *Karl Friedrich Schinkel.* This short book is beautifully illustrated with Schinkel's drawings, illustrated notes, plans, and carefully drawn interiors.
28. Reinhard Wegner, ed., *Karl Friedrich Schinkel, Die Reise nach Frankreich und England in Jahre 1826* (Munich: Deutscher-Kunstverlag, 1990). This contains facsimiles of the original text.
29. Rand Carter, *Karl Friedrich Schinkel: The Last Great Architect* (Chicago: Exedra Books, 1981).
30. Erik Forssman, "Höhere Baukunst," in his *Karl Friedrich Schinkel: Bauwerke und Baugedanken* (Munich: Schnell & Steiner, 1981), pp. 211–233.

31. Oswald Hederer, *Leo von Klenze: Persönlichkeit und Werk* (Munich: Georg D. W. Callwey, 1964). For his work in Bavaria, see pp. 172–180. Klenze was also an accomplished painter; see Norbert Lieb and Florian Hufnagl, *Leo von Klenze: Gemälde und Zeichnungen* (Munich: D. W. Callwey, 1979).
32. For the pictures themselves, see Klaus Gallwitz, ed., *Die Nazarener in Rom: Ein deutscher Künstlerbund der Romantik* (Munich: Prestel, 1981). The exhibition was in Rome and easily confirms the excellent pictorial qualities of these (now) deeply unfashionable painters.
33. Mitchell Benjamin Frank, "Overbeck as the Monk-Artist," in his *German Romantic Painting Redefined: Nazarene Tradition and the Narratives of Romanticism* (Aldershot: Ashgate, 2000), pp. 49ff. For drawings of the artists, see pp. 26–27. See also Fritz Schmalenbach, "Das Overbecksche Familienbild," in *Studien über Malerei und Malereigeschichte* (Berlin: Gebr. Mann, 1972), pp. 77–81.
34. Margaret Howitt, *Friedrich Overbeck: Sein Leben und Schaffen.* 2 vols. (Bern: Herbert Long, 1971), vol. 1, p. 82. Originally published by Herder in Freiburg in 1886.
35. Pelzel, *Anton Raphael Mengs*, p. 21.
36. Ibid., p. 26.
37. Ibid., p. 29.
38. Frank, *German Romantic Painting*, p. 26. See also: Gallwitz, ed., *Nazarener in Rom.*
39. By coincidence, and to their great delight, they discovered an old workman who at one time had helped prepare the plaster for Mengs, and he was able to teach them the rudimentary technique of an almost forgotten craft. Frank, *German Romantic Painting*, p. 26.
40. Frank, *German Romantic Painting*, p. 140.
41. Pelzel, *Anton Raphael Mengs*, p. 40.
42. Frank, *German Romantic Painting*, p. 143.
43. Pelzel, *Anton Raphael Mengs*, p. 56.
44. Ibid., p. 61.
45. Julius Schnorr von Carolsfeld, *Die Bibel in Bildern* (Leipzig: G. Wigand, 1860).
46. Hans Joachim Kluge, *Caspar David Friedrich: Entwürfe für Grabmäler und Denkmäler* (Munich: Deutscher Verlag für Kunstwissenschaft, 1993), pp. 11–14.
47. Ibid., pp. 17ff.
48. See for example, the catalog of the exhibition *The Romantic Vision of Caspar David Friedrich: Paintings and Drawings from the U.S.S.R.*, at the Metropolitan Museum of Art, New York, and the Chicago Art Institute, 1990–1991, distributed by Harry N. Abrams, New York.
49. For a discussion of his symbolism, see Joseph Leo Koerner, *Caspar David Friedrich and the Subject of Landscape* (London: Reaktion Books, 1990), pp. 122f. See also Hubertus Gassner, ed., *Caspar David Friedrich: Die Erfindung der Romantik* (Munich: Hirmer, 2006); and Werner Hofmann, *Caspar David Friedrich: Naturwirklichkeit und Kunstwahrheit* (Munich: C. H. Beck, 2000).
50. Hans-Georg Gadamer thought that Friedrich's emphasis on community confirmed that its disintegration was taking place. Koerner, *Caspar David Friedrich*, p. 130, and Chapter 41, p. 757 of this book.

Chapter 10: Humboldt's Gift: The Invention of Research and the Prussian (Protestant) Concept of Learning

1. R. Steven Turner, "The Prussian Universities and the Research Imperative, 1806–1848" (PhD diss., Princeton University, 1972), p. 1.
2. Ibid., p. 3.
3. Ibid., p. 4.
4. Ibid., p. 8.
5. Johann Friedrich Wilhelm Koch, ed., *Die preussischen Universitäten: Eine Sammlung der Verordnungen, welche die Verfassung und Verwaltung dieser Anstalten betreffen.* 2 vols. (Berlin, 1839–1840), vol. 2, pp. 531–532.
6. Ibid., p. 181.
7. William Clark, *Academic Charisma and the Origins of the Research University* (Chicago: University of Chicago Press, 2006), p. 211.
8. Turner, "Prussian Universities," p. 223.

9. Ibid., p. 229.
10. R. Köpke, "Zum Andenken an Dr John Schulze," *Zeitschrift für das Gymnasialwesen* 23 (1869): 245–256.
11. Turner, "Prussian Universities," pp. 247–248.
12. Clark, *Academic Charisma*, p. 218, for the distinction between erudition and research.
13. Turner, "Prussian Universities," p. 252.
14. Conrad Varrentrapp, *Johannes Schulze und das höhere preussische Unterrichtwesen in seiner Zeit* (Leipzig, 1889), pp. 447–448.
15. Maximilian Lenz, *Die Geschichte der königlichen Friedrich-Wilhelms-Universität zu Berlin.* 4 vols. (Halle, 1910–1919), vol. 2, pp. 470–472.
16. Clark, *Academic Charisma*, p. 237.
17. Turner, "Prussian Universities," p. 270.
18. Ibid., p. 279.
19. Wilhelm von Humboldt, "Ueber die innere und äussere Organisation der höheren wissenschaftlichen Anstalten in Berlin (Unvollendete Denkschrift, geschreiben 1810. Erstmals veröffentlicht 1896)," in Ernst Anrich, ed., *Die Idee der deutschen Universität* (Darmstadt, 1964), pp. 377–378. Quoted in Turner, "Prussian Universities," note 3.
20. Clark, *Academic Charisma*, pp. 178–181. Turner, "Prussian Universities," p. 285.
21. Eduard Fueter, *Die Geschichte der neueren Historiographie* (Munich, 1936), pp. 415ff.
22. Clark, *Academic Charisma*, p. 158.
23. Turner, "Prussian Universities," pp. 293–294.
24. Conrad Bursian, *Geschichte der klassischen Philologie in Deutschland.* 2 vols. (Munich: R. Oldenbourg, 1883), vol. 1, pp. 526–527. Johnson reprint (New York, 1965).
25. Turner, "Prussian Universities," p. 303.
26. Clark, *Academic Charisma*, p. 287.
27. Turner, "Prussian Universities," p. 325.
28. Anrich, *Idee der deutschen Universität*, p. 377.
29. F. W. J. Schelling, *On University Studies*, trans. E. S. Morgan (Athens, Ohio: Ohio University Press, 1966), pp. 26–27.
30. Turner, "Prussian Universities," p. 373.
31. L. Wiese, *Das höhere Schulwesen in Preussen: Historische-statistische Darstellung.* 4 vols. (Berlin, 1864–1902), vol. 1, pp. 420f.
32. According to the historian Max Lenz, the percentage of students at Berlin in the lower faculty studying mathematics and physics grew from 6 percent in 1810 to 16 percent in 1860, and in chemistry from 1 percent to 15 percent during the same period. At the same time, classical philology students grew from 22 percent to 37 percent. Wiese, *Das höhere Schulwesen*, vol. 1, p. 24.
33. Turner, "Prussian Universities," p. 391.
34. Dietrich Gerhard and William Norvin, eds., *Die Briefe Barthold George Niebuhrs.* 2 vols. (Berlin: de Gruyter, 1926), vol. 2, p. 222.
35. Luise Neumann, *Franz Neumann: Erinnerungsblätter von seiner Tochter* (Leipzig: J. C. B. Mohr [P. Siebeck], 1904), p. 360.
36. Turner, "Prussian Universities," p. 403.
37. Ibid., p. 404.
38. See, for example, Justus Liebig, "Der Zustand der Chemie in Preussen," *Annalen der Chemie und Pharmacie* 34 (1884): 123ff. Quoted in Turner, "Prussian Universities," pp. 408 and 419.
39. Helmut Schelsky, *Einsamkeit und Freiheit: Idee und Gestalt der deutschen Universität und ihrer Reformen* (Reinbeck bei Hamburg: Rowohlt, 1963), pp. 131ff.
40. Varrentrapp, *Johannes Schulze*, pp. 350ff.
41. Lenz, *Geschichte*, vol. 3, p. 530.
42. Clark, *Academic Charisma*, pp. 246ff. Turner, "Prussian Universities," p. 453.
43. F. A. W. Diesterweg, *Ueber das Verderben auf den deutschen Universitäten* (Essen, 1836), pp. 1f.

CHAPTER 11: THE EVOLUTION OF ALIENATION

1. Malcolm Pasley, ed., *Germany: A Companion to German Studies* (London: Methuen, 1972), p. 393.
2. See p. 200 of the English translation of Schelling's *The Grounding of Positive Philosophy*, trans. and ed. Bruce Matthews (Albany: State University of New York Press), 2007.
3. Ibid., p. 36. See also Friedrich Schelling, *Ideas for a Philosophy of Nature*, trans. Errol E. Harris and Peter Heath, with an introduction by Robert Stern (Cambridge: Cambridge University Press, 1988).
4. For a good translation, more detailed than there is space for here, see *Georg Wilhelm Friedrich Hegel: Lectures on the Philosophy of Spirit; 1827–1828*, trans. with an introduction by Robert R. Williams (Oxford: Oxford University Press, 2007), pp. 18ff and 165ff.
5. Pasley, ed. *Germany*, pp. 397–398. See also Thomas Sören Hoffmann, *George Wilhelm Friedrich Hegel: Eine Propädeutik* (Wiesbaden: Morix, 2004), pp. 51ff. for his system building, and 278.
6. Pasley, ed. *Germany*, p. 398. For a discussion of Hegel's language, see John McCumber, *The Company of Words: Hegel, Language and Systematic Philosophy* (Evanston, Ill.: Northwestern University Press, 1993), especially part 3, pp. 215ff.; and Klaus Grotsch, ed., *Georg Wilhelm Friedrich Hegel: Gesammelte Werke; in Verbindung mit der Deutschen Forschungsgemeinschaft herausgegeben von der Nordrhein-Westfälischen Akademie der Wissenschaften: Volume 10: Nürnberger Gymnasialkurse und Gymnasialreden (1808–1816)* (Hamburg: Meiner, 2006).
7. Pasley, ed. *Germany*, p. 399.
8. Ibid., p. 401.
9. McCumber, *Company of Words*, p. 328, feels this is a circular, self-referential argument.
10. Hoffmann, *Georg Wilhelm Friedrich Hegel*, pp. 197ff.
11. These different forms, of course, have ethical and economic implications. See Albena Neschen, *Ethik und Ökonomien in Hegels Philosophie und in modernen wirtschaftsethischen Entwürfen* (Hamburg: Meiner, 2008).
12. Pasley, ed. *Germany*, p. 406.
13. David T. McLellan, *The Young Hegelians and Marx* (London: Macmillan, 1969), p. 2.
14. Wilhelm Lang, "Ferdinand Bauer und David Friedrich Strauss," *Preussische Jahrbücher* 160 (1915): pp. 474–504.
15. Heston Harris, *David Friedrich Strauss and His Theology* (Cambridge: Cambridge University Press, 1973), pp. 41ff. For his dismissal from the seminary, see pp. 58ff.
16. C. A. Eschenmayer, *Der Ischariothismus unserer Tage* (Tübingen: Ludwig Friedrich Fues, 1835). See also Jörg F. Sandberger, *David Friedrich Strauss als theologischer Hegelianer: Mit unveröffentlichten Briefen* (Göttingen: Vandenhoeck & Ruprecht, 1972); and David F. Strauss, *The Old Faith and the New: A Confession*, authorized translation from the sixth edition by Mathilde Blind (London: Asher, 1873).
17. McLellan, *Young Hegelians*, p. 88. Ludwig Feuerbach, *Das Wesen des Christentums*. 2 vols. (Berlin: Akademie-Verlag, 1956). Originally published by Otto Wigand in Leipzig, 1841. See chapter 10 for his analysis of mysticism.
18. Pasley, ed. *Germany*, p. 407.
19. Josef Winiger, *Ludwig Feuerbach: Denker der Menschlichkeit; Biographie* (Berlin: Aufbau Taschenbuch, 2004). Winiger describes Feuerbach as the "Luther of Philosophy." See also Marx W. Wartofsky, *Feuerbach* (Cambridge and New York: Cambridge University Press, 1977).
20. McLellan, *Young Hegelians*, pp. 107 and 110.
21. For other influences on Marx, see William Lea McBride, *The Philosophy of Marx* (London: Hutchinson: 1977), pp. 21–48.
22. McBride, *Philosophy of Marx*, p. 38. McLellan, *Young Hegelians*, p. 145.
23. McLellan, *Young Hegelians*, p. 157.
24. Bruce Mazlish, *The Meaning of Karl Marx* (Oxford and New York: Oxford University Press, 1984), p. 13.
25. Ibid., p. 23.
26. Ibid., pp. 37–38.
27. Ibid., p. 45.
28. Ibid., p. 48.

29. Ibid., p. 54.
30. Heinz Frederick Peters, *Red Jenny: A Life with Karl Marx* (London: Allen & Unwin, 1986).
31. Mazlish, *Meaning of Karl Marx*, pp. 59–60. There had been a plan at one stage for the paper to be edited by an equally colorful economist, Friedrich List. List, from Württemberg, had spent time in jail for advocating political reform a little too ardently and had been forced to emigrate to the United States, returning eventually to Leipzig as U.S. Consul. His theories had Keynesian overtones (he advocated *some* government intervention in the economy), but he was chiefly known for his theory of "national economics," that national economies should always be viewed as a whole and that, therefore, the interests of the majority should always come first.
32. Ibid., p. 61.
33. Ibid., p. 63.
34. Bertell Ollmann, *Alienation: Marx's Conception of Man in Capitalist Society* (Cambridge: Cambridge University Press, 1971). See Part 2 for Marx's conception of human nature.
35. Mazlish, *Meaning of Karl Marx*, p. 80.
36. Ibid., p. 84.
37. Ollmann, *Alienation*. See Part 3, pp. 168ff., for the theory of alienation and the labor theory of value.
38. Mazlish, *Meaning of Karl Marx*, p. 90.
39. Ibid., p. 94.
40. Ollmann, *Alienation*, p. 215.
41. Mazlish, *Meaning of Karl Marx*, p. 99.
42. Mark Cowling, ed., *The Communist Manifesto: New Interpretations* (Edinburgh: Edinburgh University Press, 1998).
43. Mazlish, *Meaning of Karl Marx*, p. 104.
44. Ibid., p. 105.
45. For a history of the book, see Francis Wheen, *Marx's "Das Kapital": A Biography* (London: Atlantic Books, 2006).
46. Ollmann, *Alienation*, p. 168.
47. Mazlish, *Meaning of Karl Marx*, p. 111.
48. Ibid., p. 113.
49. Ibid., p. 115.
50. Ibid., p. 150.
51. J. D. Hunley, *The Life and Thought of Friedrich Engels* (New Haven, Conn.: Yale University Press, 1991), p. 1.
52. The attraction of the two men to each other has been explored by Terrell Carver in his *Marx and Engels: The Intellectual Relationship* (London: Wheatsheaf Books, 1983).
53. Hans Peter Bleuel, *Friedrich Engels: Bürger und Revolutionär; Die zeitgerechte Biographie eines grossen Deutschen* (Bern: Scherz, 1981).
54. Hunley, *Life and Thought*, pp. 10 and 14.
55. Engels wasn't alone in his concern. See Michael Levin, *The Condition of England Question: Carlyle, Mill, Engels* (Basingstoke: Macmillan, 1981).
56. Tristram Hunt, *The Frock-Coated Communist: The Revolutionary Life of Friedrich Engels* (London: Allen Lane, 2009), p. 243. Hunley, *Life and Thought*, p. 17.
57. Hunley, *Life and Thought*, p. 24.
58. Carver, *Marx and Engels*, p. 144.
59. Hunley, *Life and Thought*, p. 40.
60. Gérard Bekerman, *Marx and Engels: A Conceptual Concordance*, trans. Terrell Carver (Oxford: Blackwell, 1983).
61. Hunley, *Life and Thought*, p. 108.
62. Ibid., p. 123.
63. Hunt, *The Frock-Coated Communist*, pp. 280–281. Franz Neubauer, *Marx-Engels Bibliographie* (Boppard am Rhein: Boldt, 1979). Carver says their relations were "unruffled" to the end but after Marx's death Engels "established a series of ambiguities that would otherwise have been fairly (though not completely) straightforward issues." For these see the first chapter of his *Marx and Engels*, titled "Second Fiddle?"

Chapter 12: German Historicism: "A Unique Event in the History of Ideas"

1. George G. Iggers, *Leopold von Ranke and the Shaping of the Historical Discipline* (Syracuse, N.Y.: Syracuse University Press, 1990), pp. 38–39.
2. Ibid., p. 40.
3. Ibid., p. 42.
4. Ibid., p. 57.
5. Hermann Klencke, *Lives of the Brothers Humboldt, Alexander and William*, translated and arranged from the German of Klencke by Gustav Schlesier (London: Ingra, Cook & Co., 1852).
6. Iggers, *Leopold von Ranke*, p. 61.
7. Meinecke's views are set out in *Staat und Persönlichkeit* (Berlin: E. S. Mittler & Sohn, 1933), which has chapters on Troeltsch, Stein, Humboldt, and Droysen (see Chapter 21 of this book). See also *Machiavellism: The Doctrine of Raison d'État and Its Place in Modern History*, trans. Douglas Scott (London: Routledge & Kegan Paul, 1957). This book looks at the links between Machiavellism, Idealism, and Historicism in German history, pp. 343ff. In *Cosmopolitanism and the National State*, trans. Robert B. Kimber (Princeton, N.J.: Princeton University Press, 1970), Meinecke looks at Humboldt, Schlegel, and Fichte. In *Historicism: The Rise of a New Historical Outlook* (London: Routledge & Kegan Paul, 1972), he explores how the German movement grew out of the European (English, French, Italian) Enlightenment.
8. See Walther Hofer, *Geschichtsschreibung und Weltanchauung: Betrachtungen zum Werk Friedrich Meineckes* (Munich: R. Oldenbourg, 1950), pp. 232ff., for ideas about causality.
9. G. P. Gooch, *History and Historians in the Nineteenth Century* (London: Longmans, Green & Co., 1913), p. 12.
10. Thorkild Hansen, *Arabia Felix: The Danish Expedition of 1761–1767*, trans. James and Kathleen McFarlane (London: Collins, 1964), p. 34.
11. Gooch, *History and Historians*, p. 23.
12. James M. McGlathery, ed., *The Brothers Grimm and Folklore* (Illinois University Press, 1988), especially pp. 66ff., 91ff., 164ff., and 205ff. For their links to Savigny, see Gabriele Seitz, *Die Brüder Grimm: Leben-Werk-Zeit* (Munich: Winkler, 1984), 37ff. With many amusing drawings.
13. Gooch, *History and Historians*, pp. 55–57.
14. There are many editions of the Grimm folk tales and myths. I have used the "complete edition" with illustrations by Josef Scharl (London: Routledge & Kegan Paul, 1948). The drawings/paintings interspersed in the text maintain the mood of the stories.
15. Gooch, *History and Historians*, pp. 67–68.
16. Ibid., p. 102. See also Wolfgang J. Mommsen, ed., *Leopold von Ranke und die moderne Geschichtswissenschaft* (Stuttgart: Klett-Cotta, 1988), which explores the links between Ranke and Hegel and Ranke and Darwin. With essays by Peter Burke, Rudolf Vierhaus, and Thomas Nipperdey. See also Theodore H. von Laue, *Leopold Ranke: The Formative Years*, Princeton, N.J.: Princeton University Press, 1950.
17. Gooch, *History and Historians*, p. 79.
18. Hanno Helbling, *Leopold von Ranke und der historische Stil* (Zurich: J. Weiss, 1953).
19. Gooch, *History and Historians*, p. 88.
20. Helbling, *Leopold von Ranke*, pp. 70ff.
21. Gooch, *History and Historians*, p. 102.
22. For a discussion of Ranke's view of politics, see Laue, *Leopold Ranke*, pp. 139ff. and pp. 181ff. for his essay on the Great Powers.
23. Iggers, *Leopold von Ranke*, p. 10.
24. For Ranke's legacy, see Hans Heinz Krill, *Die Rankerenaissance: Max Lenz und Erich Marcks: Ein Beitrag zum historisch-politischen Denken in Deutschland, 1880–1935* (Berlin: de Gruyter, 1962). See also Friedrich Meinecke, *Ausgewählter Briefwechsel*, ed. Ludwig Dehio and Peter Classen (Stuttgart: K. F. Koehler, 1962). The many references to Ranke in Meinecke's copious correspondence, more than to any other figure, except Burckhardt and Bismarck, show the influence of the man.
25. Iggers, *Leopold von Ranke*, pp. 18–21.

Chapter 13: The Heroic Age of Biology

1. John Buckingham, *Chasing the Molecule* (Stroud: Sutton, 2004), p. 1.
2. His priority was suspect even at the time of the celebration. See also Susanna Rudofsky and John H. Wotiz, "Psychiatrists and the Dream Accounts of August Kekulé," *Ambix* 25 (1988): 31–38.
3. Buckingham, *Chasing the Molecule*, p. 2.
4. Ibid., p. 29.
5. Jacob Volhard, *Justus von Liebig*. 2 vols. (Leipzig: J. A. Barth, 1909).
6. For Berzelius, see Eran M. Melhado and Tore Frängsmyr, eds., *Enlightenment Science in the Romantic Era: The Chemistry of Berzelius and the Cultural Setting* (Cambridge: Cambridge University Press, 1992). Chapter 5, by Alan J. Rocke, is on Berzelius's role in the development of organic chemistry. Chapter 8, by John Hadley Brooke, looks at dualism and the rise of organic chemistry.
7. Melhado and Frängsmyr, *Enlightenment Science*, pp. 171ff, for Isomorphism.
8. August Wilhelm Hofmann, *The Life-Work of Liebig: Faraday Lecture for 1875* (Madison: University of Wisconsin Press, 1876). Quoted in Buckingham, *Chasing the Molecule*, p. 107.
9. Buckingham, *Chasing the Molecule*, p. 109.
10. Ibid., p. 112.
11. Ibid., p. 115.
12. Ibid., p. 118. Many derivatives of benzene—vanilla and cinnamon among others—had in fact been known since ancient times as pleasant-smelling oils and spices. Von Liebig was to derive the name "benzene" from benzoic acid, obtained from the gum benzoin, a product of the East Indies.
13. E. Mitscherlich, "Über das Benzol und die Säuren der Öl- und Talgarten," *Liebig's Annalen* 9 (1834): 39–56.
14. Buckingham, *Chasing the Molecule*, p. 122.
15. Édouard Grimaux and Charles Gerhardt Jr., *Charles Gerhardt: Sa vie, son oeuvre, sa correspondence* (Paris: Masson, 1900), p. ii.
16. C. A. Russell, *A History of Valency* (Leicester: Leicester University Press, 1971), p. 83.
17. Richard Anschütz, *August Kekulé*. 2 vols. (Berlin: Verlag Chemie, 1929), vol.1, p. 38.
18. Buckingham, *Chasing the Molecule*, p. 185.
19. Ibid., p. 187.
20. Robert Schwarz, *Aus Justus Liebigs und Friedrich Wöhlers Briefwechsel in den Jahren 1829–1873* (Weinheim: Verlag Chemie, 1958), p. 272.
21. For Kolbe, see Alan J. Rocke, *The Quiet Revolution: Hermann Kolbe and the Science of Organic Chemistry* (Los Angeles and Berkeley: University of California Press, 1993). See pp. 258ff. for Kolbe's relations with Kekulé; pp. 353ff. for the "collision" between Kolbe and Hofmann. See also Buckingham, *Chasing the Molecule*, p. 213.
22. See, for example, Hertha von Dechend, *Justus von Liebig: In eigenen Zeugnissen und solchen seiner Zeitgenossen* (Weinheim: Verlag Chemie, 1943), pp. 44ff.
23. *New Dictionary of Scientific Biography*, 4, pp. 310–313.
24. Ibid.
25. Henry Harris, *The Birth of the Cell* (New Haven, Conn.: Yale University Press, 1999).
26. Harris, *Birth of the Cell*, p. 76.
27. Lorenz Oken, *Die Zeugung* (Bamberg and Würzburg: Goebhardt, 1805). Quoted in Harris, *Birth of the Cell*, p. 61.
28. Henry J. John, *Jan Evangelista Purkyně: Czech Scientist and Patriot, 1787–1869* (Philadelphia: The American Philosophical Society, 1959). Chapter 6 is on Goethe and Purkyně, and there is an appendix on Purkyně's contribution to physiology.
29. Harris, *Birth of the Cell*, p. 88.
30. F. Bauer, *Illustrations of Orchidaceous Plants* (London: Ridgeway, 1830–1838).
31. Harris, *Birth of the Cell*, p. 81.
32. J. E. Purkinje, *Bericht über die Versammlung deutscher Naturforscher und Aerzte in Prag im September 1837* (Prague: Opera Selecta, 1948), p. 109.
33. Harris, *Birth of the Cell*, p. 94.
34. *New Dictionary of Scientific Biography*, 6, pp. 356–360.
35. Harris, *Birth of the Cell*, p. 174.

36. Ibid., p. 175.
37. Theodor Schwann, *Mikroskopische Untersuchungen über die Uebereinstimmung in der Struktur und dem Wachstum der Thiere und Pflanzen* (Berlin: Sandersche Buchhandlung, 1839). Quoted in Harris, *Birth of the Cell*, p. 100.
38. Harris, *Birth of the Cell*, p. 4.
39. F. Unger, *Flora* 45 (1832) p. 713.
40. *Dictionary of Scientific Biography*, XIII, pp. 542–543.
41. Ibid., p. 601.
42. Vitezslav, *Gregor Mendel: The First Geneticist* (Oxford: Oxford University Press, 1996).
43. Harris, *Birth of the Cell*, p. 119.
44. *Nachrichten über Leben und Schriften des Herrn Geheimraths Dr. Karl Ernst von Baer, mitgetheilt von ihm selbst* (St. Petersburg: H. Schmitzdorff, 1866), pp. 322ff.
45. Harris, *Birth of the Cell*, pp. 122–127.

Chapter 14: Out from "The Wretchedness of German Backwardness"

1. Hagen Schulze, *The Course of German Nationalism: From Frederick the Great to Bismarck, 1763–1865* (Cambridge: Cambridge University Press, 1991), pp. 43–45.
2. For a discussion of "Staatsnationen" and "Kulturnationen" and the idea of a special path see Hagen Schulze, *Staat und Nation in der europäischen Geschichte* (Munich: C. H. Beck, 1994), pp. 108–125.
3. See, for example, Ernst Cassirer, "Hölderlin und der deutsche Idealismus," in Alfred Kelletat, ed., *Hölderlin: Beiträge zu seinem Verständnis in unserm Jahrhundert* (Tübingen: J. C. B. Mohr [Paul Siebeck], 1961), pp. 79–118.
4. For his relationship to religion, see Wolfgang Schadewelt, "Hölderlins Weg zu den Göttern," in Kelletat, ed., *Hölderlin*, pp. 333–341; Mark Ogden, *The Problem of Christ in the Work of Friedrich Hölderlin* (London: Modern Humanities Research Association. Institute of German Studies, University of London, 1991); and Max Kommerell, *Der Dichter als Führer in der deutschen Klassik: Klopstock, Herder, Goethe, Schiller, Jean Paul, Hölderlin* (Frankfurt am Main: Klostermann, 1982).
5. Heidegger—and others—were attracted to Hölderlin's approach to "Das Volk." See Kommerell, *Dichter als Führer*, pp. 461ff.
6. For a detailed, moving account of the events leading up to the joint suicides, see Joachim Maass, *Kleist: A Biography*, trans. Ralph Manheim (London: Secker & Warburg, 1983), pp. 262–282. See also Lukács, *German Realists*, p. 17.
7. Gerhard Schutz, *Kleist: Eine Biographie* (Munich: C. H. Beck, 2007), pp. 391–395. Heinrich von Kleist, *Five Plays*, trans. with an introduction by Martin Greenberg (New Haven, Conn., and London: Yale University Press, 1988).
8. For Grillparzer's politics, see Bruce Thompson, "Grillparzer's Political Villains," in Robert Pichl, et al., eds., *Grillparzer und die europäische Tradition* (Vienna: Hora, 1987), pp. 101–112.
9. Raoul Auernheimer, *Franz Grillparzer: Der Dichter Österreichs* (Vienna: Ullstein, 1948), pp. 48–61.
10. Franz Grillparzer, *Selbstbiographie*, ed. Arno Dusini (Salzburg and Vienna: Residenz, 1994). An interesting mix of everyday comment and reflections on art and philosophy.
11. Hermann Glaser, ed., *The German Mind of the Nineteenth Century: A Literary and Historical Anthology* (New York: Continuum, 1981), p. 212.
12. Robert Pichl, "Tendenzen der neueren Grillparzer Forschung," in Pichl, et al., eds., *Grillparzer*, pp. 145ff.
13. Marcel Reich-Ranicki has released in Germany his own canon of works, of which *Green Henry* is number 4 (Reclam, Ditzingen, 2003).
14. Even the love element, *especially* the love element, emerges slowly, gently. See Gerhard Neumann, *Archäologie der Passion zum Liebenskonzept in Stifters "Der Nachsommer,"* in Michael Minden, et al., eds., *Stifter and Modernist Symposium* (London: Institute of German Studies, 2006), pp. 60–79; and Lily Hohenstein, *Adalbert Stifter: Lebensgeschichte eines Überwinders* (Bonn: Athenäum, 1952), pp. 226ff.
15. See in particular Michael Minden, "Der grüne Heinrich and the Legacy of Wilhelm Meister,"

in John L. Flood, et al., eds., *Gottfried Keller: 1819–1890* (Stuttgart: Hans-Dieter Heinz Akademischer Verlag, 1991), pp. 29–40; but also Wolfgang Matz, *Adalbert Stifter oder diese fürchterliche Wendung der Dinge: Biographie* (Munich: Deutscher Taschenbuch Verlag, 2005). See also Lukács, *German Realists*, p. 199.

16. Todd Kontje, *The German Bildungsroman: History of a National Genre* (Columbia, S.C.: Camden House, 1993), pp. 26–27.
17. Ritchie Robertson, *Heine* (London: Weidenfeld & Nicolson [Peter Halban], 1988), p. vii. Lukács, *German Realists*, p. 106.
18. Robinson, *Heine*, p. 7. See also Kerstin Decker, *Heinrich Heine: Narr des Glücks; eine Biografie* (Berlin: Propyläen, 2007).
19. Robinson, *Heine*, pp. 10–11.
20. Edda Ziegler, *Heinrich Heine: Leben, Werk, Wirkung* (Zurich: Artemis & Winkler, 1993). See also Robinson, *Heine*, p. 13; and Lukács, *German Realists*, p. 103.
21. For his poems of this period, see S. S. Prawer, *Heine: The Tragic Satirist* (Cambridge: Cambridge University Press, 1961), pp. 141ff.; Marcel Reich-Ranicki, *Der Fall Heine* (Stuttgart: Deutsche Verlags-Anstalt, 1997); and Robinson, *Heine*, p. 20.
22. Robinson, *Heine*, p. 22.
23. Ibid., p. 27.
24. Ibid., p. 81.
25. For a discussion, see Reich-Ranicki, *Der Fall Heine*, pp. 86ff., quoting from Heine's letters.
26. Robinson, *Heine*, p. 87.
27. Ibid., p. 93.
28. For the last poems, see also Prawer, *Heine*, pp. 222f. Lukács, *German Realists*, p. 155.
29. Jan-Christoph Hauschild, *Georg Büchner: Studien und neue Quellen zu Leben, Werk und Wirkung* (Königstein: Athenäeum, 1985), pp. 35ff. and 47f.
30. Raymond Erickson, *Schubert's Vienna* (New Haven, Conn., and London: Yale University Press, 1997), pp. 5f.
31. Ibid., p. 290.
32. Gerbert Frodl, et al., eds., *Wiener Biedermeier: Malerei zwischen Wiener Kongress und Revolution* (Munich: Prestel, 1992), pp. 35–43.
33. Erickson, *Schubert's Vienna*, pp. 40ff. Hans Ottomeyer, et al., eds., *Biedermeier: The Invention of Simplicity; An Exhibition at the Milwaukee Art Museum, the Albertina in Vienna, Deutsche Historische Museum in Berlin*, 2006. Includes a chapter on the rediscovery of Biedermeier and one on the aesthetics of Biedermeier furniture. Probably the definitive work, visually speaking, for the moment. Schonberg, *Lives of the Great Composers*, p. 101. See also George Marek, *Schubert* (London: Robert Hale, 1986), pp. 110–111. Marek says these evenings were, as often as not, "drinking orgies." But they were well attended and, on one occasion, by a princess, two countesses, three baronesses, and a bishop.
34. For *Carnaval*, see Ronald J. Taylor, *Robert Schumann: His Life and Work* (London: Granada, 1982), pp. 113–116 and 127–128. See also John Daviero, *Crossing Paths: Schubert, Schumann, and Brahms* (Oxford: Oxford University Press, 2002); and Alice M. Hanson, *Musical Life in Biedermeier Vienna* (Cambridge: Cambridge University Press, 1985).
35. Taylor, *Robert Schumann*, pp. 320–321; and Schonberg, *Lives of the Great Composers*, p. 148.
36. Clive Brown, *A Portrait of Mendelssohn* (New Haven, Conn., and London: Yale University Press, 2003), pp. 74ff.
37. Ibid., pp. 430–432.
38. Celia Applegate, *Bach in Berlin: Nation and Culture in Mendelssohn's Revival of the St. Matthew Passion* (Ithaca, N.Y.: Cornell University Press, 2005).
39. Eva Kolinsky and Wilfried van der Will, *The Cambridge Companion to Modern German Culture* (Cambridge: Cambridge University Press, 1998), p. 155. See also, for example, Gunter Wiegelmann, et al., *Volkskunde* (Berlin: E. Schmidt, 1977); and Dieter Harmening, et al., eds., *Volkskultur und Geschichte: Festgabe für Josef Dünninger zum 65. Geburtstag* (Berlin: E. Schmidt, 1970).
40. Kolinsky and van der Will, ibid.
41. See, for example, the periodical *Germanistik: Internationales Referatenorgan mit bibliographischen Hinweisen* (Tübingen: Niemeyer). In 1854, a police handbook was producd that blacklisted 6,300 individuals, including the post-Hegelians Arnold Ruge and David Strauss.

Chapter 15: "German Fever" in France, Britain, and the United States

1. Maria Fairweather, *Madame de Staël* (London: Constable, 2005), p. 1.
2. Ibid., p. 4.
3. Ibid., p. 303.
4. Ibid., p. 307.
5. Ibid., p. 375.
6. Ibid., p. 379.
7. Rosemary Ashton, *The German Idea* (Cambridge: Cambridge University Press, 1980), p. 12. See also Hertha Marquardt, *Henry Crabb Robinson und seine deutschen Freunde: Brücke zwischen England und Deutschland im Zeitalter der Romantik.* 2 vols. (Göttingen: Vandenhoeck & Ruprecht, 1964–1967). Jürgen Kedenburg, *Teleologisches Geschichtsbild und theokratische Staatsauffassung im Werke Thomas Carlyles* (Heidelberg: Carl Winter, 1960).
8. Elizabeth M. Vida, *Romantic Affinities: German Authors and Carlyle; A Study in the History of Ideas* (Toronto: University of Toronto Press, 1993). The abridgment of Crabb Robinson's diary, edited and with an introduction by Derek Hudson, was published by Oxford University Press in 1967.
9. Ashton, *German Idea*, p. 4.
10. Ibid., p. 51. Coleridge was at first unconvinced about the merits of *Faust*, and he was not entirely wrong to be worried how it would be received. There was in Britain to begin with a violent rejection of the book; people took against its "immorality," the bargain with God being regarded as especially shocking.
11. See the chapters on *Sartor Resartus* and *Frederick the Great*, "That unutterable horror of a Prussian book," in K. J. Fielding, et al., eds., *Carlyle Past and Present* (London: Vision Books, 1976), pp. 51–60 and 177–197.
12. F. W. Stokoe, in *German Influence in the English Romantic Period, 1788–1818* (Cambridge: Cambridge University Press, 1926), extends the influence to Scott, Shelley, and Byron. He prints a list of books "translated, adapted or initiated from the German" between 1789 and 1805. It includes 167 titles.
13. Fairweather, *Madame de Staël*, p. 176.
14. Ashton, *German Idea*, p. 24.
15. W. H. G. Armytage, *The German Influence on English Education* (London: Routledge & Kegan Paul, 1969), p. 6.
16. Ibid., p. 23.
17. Ibid., p. 32.
18. Ibid., p. 42.
19. Ibid., p. 52.
20. Ibid., p. 54.
21. Ibid., pp. 34 and 45.
22. Hans-Joachim Netzer, *Albert von Sachsen-Coburg und Gotha: Ein deutscher Prinz in England* (Munich: C. H. Beck, 1988), p. 238. Stanley Weintraub, *Albert: Uncrowned King* (London: John Murray, 1997), p. 222. E. J. Feuchtwanger, *Albert and Victoria: The Rise and Fall of the House of Saxe-Coburg-Gotha* (London: Continuum, 2006).
23. Hermione Hobhouse, *Prince Albert: His Life and Work* (London: Hamish Hamilton, 1983), p. viii. The standard work (though now dated) is Theodore Martin, *The Life of HRH the Prince Consort.* 5 vols. (London: Smith, Elder, 1880). See vol. 5, pp. 376ff., for Balmoral and political matters.
24. Hobhouse, *Prince Albert*, p. 64.
25. Franz Bosbach and John R. Davis, eds., *Windsor-Coburg: Geteilter Nachlass—gemeinsames Erbe; eine Dynastie und ihre Sammlungen* (Munich: K. G. Saur, 2007), pp. 49ff., 61ff. and 115ff.
26. John R. Davis, *The Great Exhibition* (Stroud: Sutton, 1999), p. 155.
27. Ibid., p. 114. See also Elisabeth Darby, *The Cult of the Prince Consort* (New Haven, Conn., and London: Yale University Press, 1983).
28. Ulrich von Eyck, *The Prince Consort* (London: Chatto & Windus, 1959), p. 68.
29. Ibid., p. 86.
30. Hobhouse, *Prince Albert*, p. 256.
31. Albert Bernhardt Faust, *The German Element in the United States* (New York: Steuben Society of America, 1927), vol. 1, p. 5.

32. Ibid., p. 33.
33. Ibid., p. 477.
34. Ibid., p. 567.
35. Faust, *German Element*, vol. 2, pp. 202–203.
36. James Morgan Hart, *German Universities: A Narrative of Personal Experience* (New York: Putnam, 1878).
37. Faust, *German Element*, vol. 2, p. 212.
38. Carl Diehl, *Americans and German Scholarship, 1770–1870* (New Haven, Conn., and London: Yale University Press, 1978), pp. 53 and 61. J. Conrad, *Das Universitätsstudium in Deutschland* (Jena, 1884), p. 25. Quoted in Diehl, *Americans*, pp. 63–64.
39. Diehl, *Americans*, p. 116.
40. Ibid., p. 141. And see Jerry Brown, *The Rise of Biblical Criticism* (Middletown, Conn.: Wesleyan University Press, 1969).
41. Hans W. Gatzke, *Germany and the United States: A "Special Relationship?"* (Cambridge, Mass.: Harvard University Press, 1980), p. 30.
42. Faust, *German Element*, vol.1, p. 438.
43. Ibid., vol. 2, p. 261.
44. Ibid., p. 369.
45. Ibid., p. 401.

Chapter 16: Wagner's Other Ring—Feuerbach, Schopenhauer, Nietzsche

1. Bryan Magee, *Wagner and Philosophy* (London: Penguin, 2000/2001), p. 1. This chapter is heavily reliant on Mr. Magee's excellent book.
2. Ibid., p. 3.
3. Joachim Köhler, *Richard Wagner: The Last of the Titans*, trans. Stewart Spencer, (New Haven, Conn., and London: Yale University Press, 2004), p. 140.
4. For Wagner's Hegelianism, see Paul Lawrence Rose, *Wagner: Race and Revolution* (London: Faber & Faber, 1992), pp. 28–31 and 62. Magee, *Wagner and Philosophy*, p. 35.
5. Köhler, *Richard Wagner*, pp. 270–271.
6. Magee, *Wagner and Philosophy*, p. 14.
7. Köhler, *Richard Wagner*, p. 261.
8. Marx W. Wartofsky, *Feuerbach* (Cambridge: Cambridge University Press, 1977), p. 322.
9. Magee, *Wagner and Philosophy*, p. 52.
10. For the way Feuerbach presaged Freud in certain ways, see S. Rawidowicz, *Ludwig Feuerbachs Philosophie: Ursprung und Schicksal* (Berlin: de Gruyter, 1964).
11. Magee, *Wagner and Philosophy*, pp. 72–73.
12. Ibid., p. 76.
13. Ibid., p. 93.
14. Köhler, *Richard Wagner*, pp. 418–419.
15. Magee, *Wagner and Philosophy*, pp. 145–146.
16. Rüdiger Safranski, *Schopenhauer und die wilden Jahre der Philosophie* (Munich: Carl Hanser, 1977), pp. 484ff. For background, I have used Dale Jacquette, ed., *Schopenhauer, Philosophy, and the Arts* (Cambridge: Cambridge University Press, 1996).
17. Magee, *Wagner and Philosophy*, p. 162.
18. Ibid., p. 164.
19. Lawrence Ferrara, "Schopenhauer on Music as the Embodiment of Will," in Jacquette, ed. *Schopenhauer*, pp. 185ff.
20. Magee, *Wagner and Philosophy*, pp. 166–167.
21. Ibid., p. 168.
22. Köhler, *Richard Wagner*, pp. 421–425.
23. Arthur Schopenhauer, *Parerga and Paralipomena: Short Philosophical Essays*, trans. by E. J. F. Payne (Oxford: Clarendon Press, 1974), vol. 2, p. 287. Magee, *Wagner and Philosophy*, p. 171.
24. Magee, *Wagner and Philosophy*, p. 193.

25. Rudolph Sabor, *Richard Wagner: Der Ring des Nibelungen; A Companion Volume* (London: Phaidon, 1997).
26. Ferrara, "Schopenhauer on Music," p. 186.
27. Magee, *Wagner and Philosophy*, p. 209.
28. Köhler, *Richard Wagner*, p. 537. Magee, *Wagner and Philosophy*, p. 209.
29. Magee, *Wagner and Philosophy*, p. 231.
30. Joachim Köhler devotes 35 pp. of his Wagner biography to *Parsifal* and includes many details about a rival to Cosima, pp. 588–623.
31. Magee, *Wagner and Philosophy*, p. 289.
32. He was also a great stylist in the language. See Heinz Schlaffer, *Das entfesselte Wort: Nietzsches Stil und seine Folgen* (Munich: Hanser, 2007).
33. Martin Ruehl, "Politeia 1871: Nietzsche contra Wagner on the Greek State," in Ingo Gildenhard, et al., eds., *Out of Arcadia: Classics and Politics in Germany in the Age of Burckhardt, Nietzsche, and Wilamowitz* (London: Institute of Classical Studies, School of Advanced Study, University of London, 2003), p. 72.
34. Joachim Köhler, *Nietzsche and Wagner: A Lesson in Subjugation*, trans. Ronald Taylor. (New Haven, Conn., and London: Yale University Press, 1998), p. 55. See also George Liébert, *Nietzsche and Music*, trans. David Pellauer and Graham Parkes. (Chicago: University of Chicago Press, 2004). Wagner even asked Nietzsche's help to buy underclothes.
35. Rüdiger Safranski, *Nietzsche: A Philosophical Biography*, trans. Shelley Frisch (London: Granta, 2002), p. 63.
36. Joachim Köhler, *Zarathustra's Secret: The Interior Life of Friedrich Nietzsche*, trans. Ronald Taylor (New Haven, Conn., and London: Yale University Press, 2002), p. 93.
37. Magee, *Wagner and Philosophy*, pp. 299–300.
38. Ibid., p. 306.
39. Ibid., p. 309.
40. Ibid., p. 313.
41. "Schopenhauer as Educator," quoted in Lydia Goehr, "Schopenhauer and the Musicians: An Inquiry into the Sounds of Silence and the Limits of Philosophising about Music," in Jacquette, ed., *Schopenhauer*, p. 216.
42. Magee, *Wagner and Philosophy*, p. 316.
43. Thomas H. Brobjer, *Nietzsche's Philosophical Context: An Intellectual Biography* (Urbana: University of Illinois Press, 2008).
44. Safranski, *Nietzsche*, pp. 184–185.
45. "Your true being does not lie buried deep within you, but rather immeasurably high above you or at least above what you normally take to be your ego." Safranski, *Nietzsche*, p. 260.
46. Franz, Graf zu Solms-Laubach, *Nietzsche and Early German and Austrian Sociology* (Berlin: de Gruyter, 2007). Magee, *Wagner and Philosophy*, p. 319.
47. Magee, *Wagner and Philosophy*, p. 334.
48. Köhler, *Nietzsche and Wagner*, pp. 141ff.
49. Magee, *Wagner and Philosophy*, pp. 336–337.

Chapter 17: Physics Becomes King: Helmholtz, Clausius, Boltzmann, Riemann

The title for this chapter is taken from Iwan Rhys Morus, *When Physics Became King* (London: University of Chicago Press, 2005).

1. *Dictionary of Scientific Biography*, IX, pp. 235–240.
2. Ken Caneva, *Robert Mayer and the Conservation of Energy* (Princeton, N.J.: Princeton University Press, 1993).
3. P. M. Harman, *Energy, Force and Matter: The Conceptual Development of Nineteenth-Century Physics* (Cambridge: Cambridge University Press, 1982), p. 144. J. C. Poggendorff, *Annalen der Physik und Chemie* (Leipzig: J. A. Barth, 1824).
4. Harman, *Energy*, p. 145.
5. Morus, *When Physics Became King*, p. 77.
6. Thomas S. Kuhn, *The Essential Tension: Selected Studies in Scientific Tradition and Change* (Chicago: University of Chicago Press, 1977), pp. 97–98.

7. Harman, *Energy*, p. 1.
8. Morus, *When Physics Became King*, p. 47.
9. Ibid.
10. Ibid., p. 48.
11. Mary Jo Nye, *Before Big Science: The Pursuit of Modern Chemistry and Physics, 1800–1940* (New York: Twayne, 1996), pp. 3, 10–11.
12. Morus, *When Physics Became King*, p. 63.
13. Ibid., p. 55.
14. Marcel Du Sautoy, *The Music of the Primes: Why an Unsolved Problem in Mathematics Matters* (London: Harper Perennial, 2004), p. 95.
15. Christa Jungnickel and Russell McCormmach, *The Intellectual Mastery of Nature* (Chicago: University of Chicago Press, 1986), vol. 1, p. 164. Quoted in Morus, *When Physics*, p. 147. See also Yehuda Elkana, *The Discovery of the Conservation of Energy* (London: Hutchinson, 1974).
16. Morus, *When Physics Became King*, p. 45.
17. Ibid., p. 42.
18. Harman, *Energy*, p. 146.
19. Rudolf Clausius, "Über die Art der Bewegung, welche wir Wärme nennen," *Annalen der Physik und Chemie* 173, no. 3 (1857): 353–380. Quoted in Harman, *Energy*, pp. 147–148.
20. *Dictionary of Scientific Biography*, III, pp. 303–310.
21. Morus, *When Physics Became King*, p. 53. For the Carnot-Clausius link, see George Birtwhistle, *The Principle of Thermodynamics* (Cambridge: Cambridge University Press, 1931), pp. 25–38.
22. Harman, *Energy*, p. 148.
23. Ibid., p. 149.
24. Ibid., p. 150.
25. Lewis Campbell and William Garnett, *The Life of James Clerk Maxwell* (London: Macmillan, 1882), p. 143.
26. Morus, *When Physics Became King*, p. 65.
27. Ibid., p. 68.
28. For more background, see Ted Porter, *The Rise of Statistical Thinking, 1820–1900* (Princeton, N.J.: Princeton University Press, 1983).
29. For accounts in English, see Brian McGuinness, ed., *Ludwig Boltzmann: Theoretical Physics and Philosophical Problems; Selected Writings* (Dordrecht, The Netherlands, and Boston: D. Reidel, 1974), pp. 83–87 and 217–219. See Engelbert Broda, *Ludwig Boltzmann: Mensch, Physiker, Philosoph* (Vienna: Franz Deuticke), 1955, pp. 57–66 and pp. 74ff. for his views on heat death.
30. Carlo Cercignani, *Ludwig Boltzmann: The Man Who Trusted Atoms* (Oxford: Oxford University Press, 1998), especially pp. 120ff., for the statistical interpretation of entropy. This book also contains some amusing cartoons of Boltzmann by Karl Przibram.
31. Carl Boyer, *A History of Mathematics.* 2nd ed., rev. by Uta C. Merzbach (New York: Wiley, 1991), p. 496.
32. Ibid., p. 497.
33. Ibid., p. 507.
34. For the relationship between Klein, Riemann, Dirichlet, and Weierstrass, see the very readable biography by Constance Reid, *Hilbert* (London/Berlin: George Allen and Unwin/ Springer-Verlag, 1970), pp. 65ff.
35. Boyer, *History*, p. 545.
36. Ibid., p. 555.
37. Du Sautoy, *Music of the Primes*, p. 79.
38. For the correspondence between Klein and David Hilbert, with references to Dirichlet, Dedekind, Einstein, Husserl, Nernst, Poincaré and Weierstrass, see Günther Frei, *Der Briefwechsel David Hilbert–Felix Klein (1886–1918)* (Göttingen: Vandenhoeck & Ruprecht, 1985).
39. Reid, *Hilbert*, pp. 45–46.
40. Boyer, *History*, p. 550.

Chapter 18: The Rise of the Laboratory: Siemens, Hofmann, Bayer, Zeiss

1. Werner von Siemens, *Inventor and Entrepreneur: Recollections of Werner von Siemens* (London/Munich: Lund Humphries/Prestel), 1966, p. 23.
2. Ibid., p. 42.
3. For details of Halske, see Georg Siemens, *History of the House of Siemens*, trans. A. F. Rodger (Freiburg/Munich: Karl Alber, 1957), vol. 1, pp. 19f.; and Wilfried Feldenkirchen, *Werner von Siemens: Erfinder und internationaler Unternehmer* (Munich: Piper, 1996).
4. Siemens, *Inventor*, p. 71.
5. Ibid., p. 229.
6. For later developments, see Siemens, *Inventor*, vol. 1, pp. 300ff., and vol. 2, passim.
7. Diarmuid Jeffreys, *Aspirin: The Remarkable Story of a Wonder Drug* (London: Bloomsbury, 2004), pp. 56–57.
8. Ibid., p. 43.
9. Rudolf Benedikt, *The Chemistry of the Coal-Tar Colours*, trans. E. Knecht (London: George Bell, 1886), pp. 1–2.
10. Jeffreys, *Aspirin*, p. 45.
11. John Joseph Beer, *The Emergence of the German Dye Industry* (Urbana: University of Illinois Press, 1959), p. 3.
12. Ibid., p. 10.
13. For the chemical composition/structure of aniline, toluidine, and rosaniline, see Benedikt, *Chemistry*, pp. 76ff.
14. Beer, *Emergence*, pp. 28–29.
15. Ibid., p. 44.
16. Ibid., p. 53.
17. Ibid., p. 57.
18. Ibid., p. 61.
19. Ibid., p. 90.
20. For the links between dyes, colored inks, sweeteners, drugs, and photographic chemicals, see Thomas Beacall, et al., *Dyestuffs and Coal-Tar Products* (London: Crosby Lockwood, 1916).
21. Beer, *Emergence*, p. 97.
22. Ibid., p. 88.
23. Ibid., p. 100.
24. Ibid., p. 115.
25. Ibid., p. 120.
26. See, for example, Josiah E. DuBois, in collaboration with Edward Johnson, *Generals in Grey Suits: The Directors of the International "I.G. Farben" Cartel, Their Conspiracy and Trial at Nuremberg* (London: Bodley Head, 1953).
27. Erik Verg, et al., *Milestones* (Leverkusen: Bayer AG, 1988). Quoted in Jeffreys, *Aspirin*, p. 58.
28. Jeffreys, *Aspirin*, p. 62.
29. Ibid., p. 63.
30. Ibid., p. 64.
31. Ibid., p. 65.
32. Ibid., p. 71.
33. Ibid., p. 72.
34. "Pharmakologisches über Aspirin-Acetylsalicylsäure," *Archiv für die gesammte Physiologie*, 1999. Quoted in Jeffreys, *Aspirin*, p. 73. No author or page references given.
35. Jeffreys, *Aspirin*, p. 73.
36. Diarmuid Jeffreys devotes a chapter of his book about aspirin to what he calls "the aspirin age," about the way Bayer's assets were dispersed in America after World War I. He also traces its role in the IG Farben cartel scandal. In *The Aspirin Age, 1919–1941*, written by Samuel Hopkins Adams but edited by Isabel Leighton (London: Bodley Head, 1950), she distinguishes a time, between the world wars, when—ironically—she seems to feel the world needed the sort of pick-me-up aspirin provides.
37. Edith Hellmuth and Wolfgang Mühlfriedel, *Zeiss 1846–1905*, vol. 1 of *Carl Zeiss: Die Geschichte eines Unternehmens* (Weimar/Cologne/Vienna: Böhlau, 1996), esp. pp. 59–113, "Die wissenschaftliche Grundlegung der modernen Mikroskopfertigung."
38. For an English—but much older—account, see Felix Auerbach, *The Zeiss Works and the Carl*

Zeiss Stiftung in Jena, trans. S. F. Paul and F. J. Cheshire (London: Marshall, Brookes & Chalkley, 1927). This book contains a list of the most important Zeiss inventions.

39. *The Great Age of the Miscroscope* is a catalog produced by the Royal Microscopical Society of the United Kingdom, to mark its 150th anniversary. The society was the first to be formed with a scientific instrument as its focus. The catalog consists of mainly British but also French and German instruments.
40. Just as the microscope is the symbol of the laboratory, so the laboratory is the symbol of science. In *Tales from the Laboratory* (Munich: Iudicium, 2005), editor Rüdiger Görner introduces a series of essays about the influence of science on German literature. See in particular the essay by Dieter Wuttke, "From the Laboratory of a Cultural Historian," about how the spectacular advances of laboratory science in Germany in the nineteenth century opened up and entrenched the division between the sciences and the humanities. As will be seen in later chapters, this had tragic consequences for Germany.

Chapter 19: Masters of Metal: Krupp, Benz, Diesel, Rathenau

1. Peter Batty, *The House of Krupp* (London: Secker & Warburg, 1966), p. 46.
2. Wilhem Berdrow, *Alfred Krupp*. 3 vols. (Berlin: Von Reimar Hobbing, 1927).
3. Ibid., pp. 89ff. Batty, *House of Krupp*, p. 49.
4. Batty, *House of Krupp*, p. 59.
5. Ibid., p. 61.
6. Ibid., p. 64.
7. For the context of this arms race, see Jonathan A. Grant, *Rulers, Guns and Money: The Global Arms Race in the Age of Imperialism* (Cambridge, Mass.: Harvard University Press, 2007). Grant looks systematically at Krupp's dealings with Russia, the Ottoman Empire, Bulgaria, Romania, South America, Japan, Serbia, and Greece.
8. See, for example, Krupp Archive, Essen: WA 7f/886, "Notic Beziehungen zur Türkei," quoted in Grant, *Rulers*, p. 28. Batty, *House of Krupp*, p. 71.
9. Willi A. Boelcke, ed., *Krupp und die Hohenzollern in Dokumenten* (Frankfurt am Main: Akademische Verlagsgesellschaft Athenaion, 1970), for correspondence between Krupp and Bismarck.
10. Batty, *House of Krupp*, p. 72.
11. Ibid., p. 77.
12. Volker R. Berghahn, *Der Tirpitz-Plan: Genesis und Verfall einer innenpolitischen Krisenstrategie unter Wilhelm II* (Düsseldorf: Droste, 1971), pp. 227ff. See also Gary E. Weir, *Building the Kaiser's Navy: The Imperial Navy Office and German Industry in the Von Tirpitz Era* (Shrewsbury: Airlite, 1992), passim.
13. Batty, *House of Krupp*, p. 82.
14. Ibid., p. 83.
15. Peter Gay, *Schnitzler's Century: The Making of Middle Class Culture, 1815–1914* (London: Allen Lane, Penguin Press, 2001), p. 7.
16. Batty, *House of Krupp*, p. 93.
17. Ibid., p. 95.
18. For the appearance of Villa Hügel, see Bernt Engelmann, *Krupp: Legenden und Wirklichkeit* (Munich: Schneckluth, 1969), pp. 208–209. This is a somewhat irreverent book.
19. St. John C. Nixon, *The Antique Automobile* (London: Cassell, 1956), p. 25. David Scott-Moncrieff, with St. John Nixon and Clarence Paget, *Three-Pointed Star: The Story of Mercedes-Benz Cars and Their Racing Successes* (London: Cassell, 1955), pp. 3–19.
20. Nixon, *Antique Automobile,* p. 29.
21. Ibid., p. 33.
22. Scott-Moncrieff, *Three-Pointed Star*, pp. 20–56.
23. Nixon, *Antique Automobile*, p. 35.
24. Scott-Moncrieff , *Three-Pointed Star*, pp. 120–149.
25. Of course, the later history of Daimler-Benz was not without controversy. See Neil Gregor, *Daimler-Benz in the Third Reich* (New Haven, Conn., and London: Yale University Press, 1998). For Maybach, see Scott-Moncrieff, *Three-Pointed Star*, pp. 59ff.
26. For a discussion of German engineers and their social position, see Donald E. Thomas Jr.,

Diesel: Technology and Society in Industrial Germany (Tuscaloosa: University of Alabama Press, 1987), pp. 38ff.

27. Eugen Diesel, *Diesel: Der Mensch, das Werk, das Schicksal* (Hamburg: Hanseatische Verlagsanstalt, 1934), p. 88.
28. Thomas, *Diesel,* pp. 68ff.
29. Hartmut Pogge von Strandmann, ed., *Walther Rathenau, Industrialist, Banker, Intellectual and Politician: Notes and Diaries, 1907–1922* (Oxford: Clarendon Press, 1985), p. 1.
30. Ibid., p. 4.
31. Christian Schölzel, *Walther Rathenau: Eine Biographie* (Paderborn: Ferdinand Schöningh, 2004), p. 28.
32. Pogge von Strandmann, *Walther Rathenau,* p. 14.
33. Schölzel, *Walther Rathenau,* pp. 213ff.
34. Ibid., pp. 81ff.
35. Pogge von Strandmann, *Walther Rathenau,* pp. 16 and 88, and more generally the diary entries for 1911–1914.
36. For details of Rathenau's view of economic policy, see *Walther Rathenau—Gesamtausgabe,* ed. Hans Dieter Hellige and Ernst Schulin. 6 vols. (Munich: G. Müller, 1977–2006).
37. Pogge von Strandmann, *Walther Rathenau,* p. 18. James Joll, in one of three essays on intellectuals in politics, says that the inner contradictions of Germany were mirrored in Rathenau's own nature. James Joll, *Intellectuals in Politics: Three Biographical Essays* (London: Weidenfeld & Nicolson, 1960), p. 70.

Chapter 20: The Dynamics of Disease: Virchow, Koch, Mendel, Freud

1. *New Dictionary of Scientific Biography,* 7, pp. 157–161.
2. Ibid.
3. Ibid.
4. For this side of Virchow, see for example, Rudolf Virchow, *Das Gräberfeld von Koban im Lande der Osseten, Kaukasus: Eine vergleichend-archäologische Studie* (Berlin: A. Asher, 1883).
5. For the relationship between Virchow and Koch, see Frank Ryan, *Tuberculosis: The Greatest Story Never Told* (Bromsgrove: Swift Publishing, 1992), pp. 9f. Bernhard Möllers, *Robert Koch: Persönlichkeit und Lebenswerk, 1843–1910* (Hanover: Schmorl & von Seefeld, 1950), chap. 4, pp. 93–120.
6. *Dictionary of Scientific Biography,* VII, pp. 420–435.
7. For Henle, see Ragnhild Münch, *Robert Koch und sein Nachlass in Berlin* (Berlin: de Gruyter, 2003), p. 7. And Möllers, *Robert Koch,* pp. 23–39.
8. For background on anthrax, see Norbert Gualde, *Resistance: The Human Struggle against Infection,* trans. Steven Randall (Washington, D.C.: Dana, 2006), p. 193, note 4.
9. Möllers, *Robert Koch,* pp. 512–517.
10. Johanna Bleker, "To Benefit the Poor and Advance Medical Science: Hospitals and Hospital Care in Germany, 1820–1870," in Manfred Berg and Geoffrey Cocks, eds., *Medicine and Modernity: Public Health and Medical Care in Nineteenth- and Twentieth-Century Germany* (Washington, D.C.: German Historical Institute/Cambridge University Press, 1997), pp. 17–33. And Möllers, *Robert Koch,* pp. 527–534.
11. *Dictionary of Scientific Biography,* VII, p. 423.
12. Ryan, *Tuberculosis,* pp. 9–13.
13. Münch, *Robert Koch,* pp. 41–46 for the cholera expedition. See also Möllers, *Robert Koch,* pp. 139–147.
14. Thomas Dormandy, *The White Death: A History of Tuberculosis* (London: Hambledon, 1999), p. 132; and for their writings, Münch, *Robert Koch,* pp. 374 and 378.
15. Dormandy, *White Death,* pp. 139–144.
16. *Dictionary of Scientific Biography,* XIV, pp. 183–184. See also the 1978 supplement, pp. 521–524.
17. Möllers, *Robert Koch,* pp. 657–684.
18. See, for example, Vera Pohland, "From Positive-Stigma to Negative-Stigma: A Shift of the Literary and Medical Representation of Consumption in German Culture," in: Rudolf Käser

and Vera Pohland, eds., *Disease and Medicine in Modern German Cultures* (Ithaca, N.Y.: Center for International Studies, Cornell University, 1990).

19. For a note on Schaudinn, see Dormandy, *White Death*, pp. 199n and 265n.
20. Martha Marquardt, *Paul Ehrlich* (London: Heinemann, 1949), p. 160.
21. Robin Morantz Henig, *A Monk and Two Peas: The Story of Gregor Mendel and the Discovery of Genetics* (London: Weidenfeld & Nicolson, 2000), pp. 173ff.
22. For the context of Mendel's discoveries, see Peter J. Bowler, *The Mendelian Revolution: The Emergence of Hereditarian Concepts in Modern Science and Society* (London: Athlone Press, 1989), pp. 93ff. For Klácel, see Henig, *Monk and Two Peas*, pp. 33–36.
23. For the breakdown he suffered in Vienna, see Henig, *Monk and Two Peas*, pp. 46–57.
24. Bowler, *Mendelian Revolution*, p. 100.
25. Ibid., p. 279.
26. Ibid., p. 280.
27. See, for example, Eileen Magnello, "The Reception of Mendelism by the Biometricians and the Early Mendelians (1899–1909)," in Milo Keynes, A. W. F. Edwards, and Robert Peel, eds., *A Century of Mendelism in Human Genetics: Proceedings of a Symposium Organised by the Galton Institute and Held at the Royal Society of Medicine, London 2001* (London/Boca Raton: CRC Press, 2004), pp. 19–32.
28. Bowler, *Mendelian Revolution*, p. 282.
29. Guy Claxton, *The Wayward Mind: An Intimate History of the Unconscious* (London: Little, Brown, 2005), passim.
30. William H. Johnston, *The Austrian Mind: An Intellectual and Social History, 1848–1938* (Berkeley: University of California Press, 1972), p. 236.
31. Giovanni Costigan, *Sigmund Freud: A Short Biography* (London: Robert Hale, 1967), p. 42.
32. Ibid., pp. 68ff.
33. Hugo A. Meynell, *Freud, Marx and Morals* (Totowa, N.J.: Barnes & Noble Books, 1981).

Chapter 21: The Abuses of History

1. Fritz Stern, *The Failure of Illiberalism* (London: Allen & Unwin, 1972), p. xxxvii.
2. Gordon A. Craig, *Germany: 1866–1945* (Oxford and New York: Oxford University Press, 1978/1981), pp. 39ff. See also Friedrich C. Sell, *Die Tragödie des deutschen Liberalismus* (Baden-Baden: Nomos, 1981).
3. See for instance Giles Macdonagh, *The Last Kaiser* (London: Weidenfeld & Nicolson, 2000/Phoenix, 2001), p. 3.
4. Craig, *Germany*, p. 56.
5. Theodor Mommsen, *A History of Rome under the Emperors.* Based on the lecture notes of Sebastian and Paul Hensel, 1882–1886; German edition by Barbara and Alexandre Demandt; English translation by Clare Krojzl, edited and with a new chapter by Thomas Wiedemann. (London: Routledge, 1996).
6. Antoine Guilland, *Modern Germany and Her Historians* (Westport, Conn., Greenwood Press, 1970), p. 156.
7. Ibid., p. 147.
8. Ibid., p. 153.
9. Mommsen, *History of Rome*, p. 297.
10. Guilland, *Modern Germany*, p. 161.
11. Hellmut Seier, *Die Staatsidee Heinrich von Sybels in den Wandlungen der Reichsgründungszeit 1826/71* (Lübeck: Matthiesen, 1961).
12. Guilland, *Modern Germany*, p. 185.
13. Ibid., p. 199.
14. Ibid., p. 219.
15. For another political historian, see Wilfried Nippel, *Johann Gustav Droysen: Ein Leben zwischen Wissenschaft und Politik* (Munich: C. H. Beck, 2008).
16. Seier, *Staatsidee Heinrich von Sybels*, pp. 73ff.
17. Andreas Dorpalen, *Heinrich von Treitschke* (New Haven, Conn.: Yale University Press, 1957), pp. 29–48.
18. Ibid., pp. 226ff.

19. Charles E. McClelland, *The German Historians and England: A Study in Nineteenth-Century Views* (Cambridge: Cambridge Univesity Press, 1971), pp. 168ff. Guilland, *Modern Germany*, p. 272. See also Paul M. Kennedy, *The Rise of the Anglo-German Antagonism* (London: Allen & Unwin, 1980).
20. Walter Bussmann, *Treitschke: Sein Welt- und Geschichtsbild* (Göttingen: Musterschmidt, 1952). Guilland, *Modern Germany*, pp. 273 and 284. Hermann Baumgarten, *Treitschkes deutsche Geschichte* (Strassburg: K. J. Trübner, 1883).
21. Lord Acton, "German Schools of History," *English Historical Review*, 1886.
22. Guilland, *Modern Germany*, p. 309.
23. Ernst Curtius, *Olympia, mit ausgewählten Werken von Pindar, Pausanius, Lukian* (Berlin: Atlantis Verlag, 1935), esp. pp. 67–80, but see also the excellent photographs by Martin Hürlimann.
24. Richard Stoneman, *Land of Lost Gods* (London: Hutchinson, 1987), p. 262.
25. Ibid.
26. Heinrich Schliemann, *Selbstbiographie: Bis zu seinem Tode vervollständigt*, ed. Sophie Schliemann (Wiesbaden: F. A. Brockhaus, 1955), pp. 54ff., 69ff. and 86ff.
27. Stoneman, *Land of Lost Gods*, p. 270.
28. Susan Heuck Allen, *Finding the Walls of Troy: Frank Calvert and Heinrich Schliemann at Hissarlik* (Berkeley: University of California Press, 1999), esp. pp. 72ff. for Calvert's "frauds." See also pp. 85ff.
29. Stoneman, *Land of Lost Gods*, p. 276.
30. For Schliemann and Dörpfeld, see Hermann von Joachim, ed., *Heinrich Schliemann: Grundlagen und Ergebnisse moderner Archäologie 100 Jahre nach Schliemanns Tod* (Berlin: Akademie-Verlag, 1992), pp. 153–160. This book comprises the proceedings of a conference called to consider Schliemann's achievements—and his claims—a century after his death. Ernst Mayr, *Heinrich Schliemann: Kaufmann und Forscher* (Göttingen: Musterschmidt, 1969) explores Schliemann's relations with Max Müller at Oxford, with Rudolf Virchow and Wilhelm Dörpfeld, and a number of philologists.
31. Stoneman, *Land of Lost Gods*, p. 283.
32. Ibid., p. 291.

Chapter 22: The Pathologies of Nationalism

1. Volker R. Berghahn, *Militarism: The History of an International Debate, 1861–1979* (New York: Berg, 1981), p. 9.
2. Paul Kennedy, *The Rise and Fall of the Great Powers* (London: Unwin Hyman, 1988), p. 184.
3. Ibid., p. 184.
4. Nicholas Stargardt, *The German Idea of Militarism: Radical and Socialist Critics 1866–1914* (Cambridge: Cambridge University Press, 1994), pp. 91ff.
5. Kennedy, *Rise and Fall of the Great Powers*, pp. 149–154.
6. Ibid., p. 211.
7. Léon Poliakov, *The Aryan Myth: A History of Racist and Nationalistic Ideas in Europe* (New York: Barnes & Noble Books, 1971/1974), p. 303.
8. Ibid., p. 211.
9. In Stargardt, *The German Idea of Militarism*, the author explains also "the tides of pacificism," 1907–1914. The tide wasn't all one way.
10. Militarism in 1914 is considered later, but see Jeffrey Verhey, *The Spirit of 1914: Militarism, Myth and Mobilisation in Germany* (Cambridge: Cambridge University Press, 2000).
11. Michael B. Gross, *The War against Catholicism* (Ann Arbor: University of Michigan Press, 2004), p. 240. Christoph Weber, *Kirchliche Politik zwischen Rom, Berlin und Trier 1876 bis 1888: Die Beilegung d. preuss. Kulturkampfes* (Mainz: Matthias-Grünewald-Verlag, 1970).
12. Erich Schmidt-Volkmar, *Der Kulturkampf in Deutschland, 1871–1890* (Göttingen: Musterschmidt, 1962), pp. 23–46. Gross, *War against Catholicism*, p. 241.
13. Schmidt-Volkmar, *Kulturkampf*, pp. 106–112.
14. Gross, *War against Catholicism*, p. 41.
15. Ibid., p. 43. Schmidt-Volkmar, *Kulturkampf*, pp. 106ff.

16. Gross, *War against Catholicism*, p. 56.
17. Ibid., p. 133.
18. Ibid., p. 69.
19. Ibid., p. 93.
20. Ibid., p. 109.
21. Ibid., pp. 158–160.
22. Ibid., p. 116.
23. Ibid., p. 213.
24. Schmidt-Volkmar, *Kulturkampf*, pp. 138ff.
25. Gross, *War against Catholicism*, p. 243.
26. Ibid., p. 254.
27. Weber, *Kirchliche Politik*, pp. 76–83.
28. Gross, *War against Catholicism*, p. 255.
29. Alfred Kelly, *The Descent of Darwin: The Popularisation of Darwinism in Germany, 1860–1914* (Chapel Hill: University of North Carolina Press, 1981), p. 5.
30. Ibid., p. 5.
31. Ibid., pp. 21–23.
32. Ibid., p. 40.
33. Ibid., p. 127.
34. Robert J. Evans, "In Search of German Social Darwinism: The History and Historiography of a Concept," in Berg and Cocks, eds., *Medicine and Modernity*, pp. 55–79.
35. Kelly, *Descent of Darwin*, p. 105.
36. Arthur Hermann, *The Idea of Decline in Western History* (New York: The Free Press, 1997), p. 111.
37. For a general survey see Daniel Pick, *Faces of Degeneration: A European Disorder, c. 1848–c. 1918* (Cambridge: Cambridge University Press, 1989), esp. chap. 4, pp. 97–106.
38. Ibid., pp. 176–221.
39. Kelly, *Descent of Darwin*, p. 126.
40. Poliakov, *Aryan Myth*, p. 71.
41. Ibid., pp. 101–105.
42. Kelly, *Descent of Darwin*, p. 191.
43. Ibid., p. 242.
44. Ibid., p. 273.
45. Amos Elon, *The Pity of It All: A Portrait of Jews in Germany, 1743–1933* (London: Allen Lane, Penguin Press, 2003), p. 274.
46. Stern, *The Failure of Illiberalism*, p. 106.
47. Kelly, *Descent of Darwin*, p. 128.
48. Ibid., p. 143.
49. Ibid., pp. 165–167.
50. Pick, *Faces of Degeneration*, p. 135.
51. Ibid.
52. Kelly, *Descent of Darwin*, p. 139.
53. Robert W. Lougee, *Paul de Lagarde, 1827–1891: A Study of Radical Conservatism in Germany* (Cambridge, Mass.: Harvard University Press, 1962), pp. 117ff.
54. Ulrich Sieg, *Deutschlands Prophet: Paul de Lagarde und die Ursprünge des modernen Antisemitismus* (Munich: Hanser, 2007), pp. 203–227.
55. Ibid., pp. 292–325.
56. Lougee, *Paul de Lagarde*, pp. 227–231.
57. Hermann, *Idea of Decline*, p. 54.
58. Stern, *The Failure of Illiberalism*, p. 4. Lougee, *Op. cit.*, pp. 253–254.
59. Geoffrey G. Field, *Evangelist of Race: The Germanic Vision of Houston Stewart Chamberlain* (New York: Columbia University Press, 1981).
60. Hermann, *Idea of Decline*, p. 73.
61. In his letters Chamberlain had corresponded with many leading figures such as Adolf von Harnack, Ludwig Boltzmann, and Christian Ehrenfels. See *Houston Stewart Chamberlain: Briefe und Briefwechsel mit Kaiser Wilhelm II* (Munich: Brudmann, 1928). See also Paul Pretzsch, ed. *Cosima Wagner und Houston Stewart Chamberlain im Briefwechsel 1888–1908* (Leipzig: P. Reclam jun., 1934).

62. Hermann, *Idea of Decline*, p. 75.
63. Stern, *The Politics of Cultural Despair*, pp. xx-xxi.
64. Ibid., p. xxvii.
65. Norbert Elias, *The Germans: Power Struggles and the Development of Habitus in the Nineteenth and Twentieth Centuries*, ed. Michael Schröter, trans. Eric Dunning and Stephen Mennell (Cambridge, U.K.: Polity Press, 1996), pp. ix and 155.

Chapter 23: Money, the Masses, the Metropolis: The "First Coherent School of Sociology"

1. Keith Bullivant, *Realism Today: Aspects of the Contemporary West German Novel* (Leamington Spa, Hamburg, and New York: Berg, 1987), pp. 8–12. Lukács, *German Realists*, p. 323. Hans P. Rickman, *Wilhelm Dilthey: Pioneer of the Human Studies* (London: Paul Elek, 1979), p. 12.
2. Rickman, *Wilhelm Dilthey*, p. 24.
3. Ibid., p. 38.
4. Hellmut Diwald, *Wilhelm Dilthey: Erkenntnistheorie und Philosophie der Geschichte* (Göttingen: Musterschmidt, 1963), pp. 130f.
5. Rickman, *Wilhelm Dilthey*, p. 57.
6. Ilse N. Bulhof, *Wilhelm Dilthey: A Hermeneutic Approach to the Study of History and Culture* (The Hague: Nijhoff, 1980), p. 55 and chapter 3.
7. Rickman, *Wilhelm Dilthey*, p. 70.
8. Diwald, *op. cit.*, pp. 153–169.
9. This progression is discussed in Carlo Antoni, *Vom Historismus zur Soziologie* (Stuttgart: K. F. Koehler, 1950). He begins with Dilthey and includes Weber and Meinecke.
10. Rickman, *Wilhelm Dilthey*, pp. 150–153.
11. Ibid., p. 155.
12. Lewis Coser, *Masters of Sociological Thought: Ideas in Social and Historical Context* (New York: Harcourt, Brace, Jovanovich, 1977).
13. David Frisby, *Georg Simmel* (London: Ellis Horwood Limited and Tavistock Publications, 1984), p. 23.
14. Ibid., pp. 25–26.
15. Ibid., p. 13.
16. Ibid., p. 53.
17. Ibid., p. 71.
18. Margarete Susman, *Die geistige Gestalt George Simmels* (Tübingen: Mohr, 1959), which concentrates on the spiritual side of Simmel. See also Roy Pascal, *From Naturalism to Expressionism: German Literature and Society 1880–1928* (London: Weidenfeld & Nicolson, 1973), p. 157.
19. Hermann von Helmholtz, 1853, "On Goethe's Scientific Researches," lecture delivered before the German Society of Königsberg, trans. E. Atkinson. Reprinted in Hermann von Helmholtz, *Science and Culture: Popular and Philosophical Essays*, ed. David Cahan (Chicago: Chicago University Press, 1995).
20. Frisby, *Georg Simmel*, p. 84.
21. Ibid., p. 93.
22. Ibid., p. 99.
23. Ibid., p. 106.
24. Fritz Ringer, *Max Weber: An Intellectual Biography* (Chicago and London: University of Chicago Press), pp. 36 and 40.
25. Frisby, *Georg Simmel*, pp. 131, 132, and 148.
26. Ferdinand Tönnies, *Community and Civil Society*, ed. Jose Harris, trans. Jose Harris and Margaret Hollis (Cambridge: Cambridge University Press, 2001), p. viii.
27. Tönnies, *Community and Civil Society*, p. xii.
28. Ibid., p. xiv.
29. Ibid., p. xv.
30. Ibid., p. xvii.
31. Ibid., p. xxi.

32. Reiner Grundmann and Nico Stehr, "Why Is Werner Sombart Not Part of the Core of Classical Sociology?" *Journal of Classical Sociology* 1, no. 2 (2001): 257–287.
33. Werner Sombart, *Luxury and Capitalism*, trans. W. R. Dittmar, intro. by Philip Siegelman (Ann Arbor: University of Michigan Press, 1967). In his introduction, Philip Siegelman says Weber and Sombart were the two most gifted descendants of Adam Smith, Ricardo, and Hegel.
34. Friedrich Lenger, *Werner Sombart, 1863–1941: Eine Biographie* (Munich: C.H. Beck, 1995), pp. 115–123. See also Bernhard vom Brocke, ed., *Sombarts "Moderner Kapitalismus": Materialien zur Kritik und Rezeption* (Munich: Deutscher Taschenbuch Verlag, 1987).
35. Grundmann and Stehr, "Why Is Werner Sombart Not Part," p. 261.
36. At one stage he said, "Puritanism *is* Judaism." See the Siegelman introduction in Sombart, *Luxury and Capitalism*, p. xiii.
37. Jeffrey Herf, *Reactionary Modernism: Technology, Culture, and Politics in Weimar and the Third Reich* (Cambridge: Cambridge University Press, 1984).
38. Grundmann and Stehr, "Why Is Werner Sombart Not Part," p. 269. See also Ernst Nolte, *Geschichtsdenken im 20. Jahrhundert: Von Max Weber bis Hans Jonas* (Berlin: Propyläen, 1991), for a different trajectory and a comparison of German thinkers with French, British, and American.
39. Peter Watson, *A Terrible Beauty: The People and Ideas That Shaped the Modern Mind* (London: Weidenfeld & Nicolson, 2000/*The Modern Mind*, New York: HarperCollins, 2000), pp. 45f.
40. M. Rainer Lepsius and Wolfgang J. Mommsen, eds., *Briefe Max Webers* (Tübingen: Mohr [Paul Siebeck], 1990–2008). The letters underline Weber's wide range of correspondents, including Sombart, Tönnies, and Simmel.
41. Harvey Goldmann, *Max Weber and Thomas Mann: Calling and the Shaping of the Self* (Los Angeles and Berkeley: University of California Press, 1988). A useful comparative study.
42. Reinhard Bendix, *Max Weber: An Intellectual Portrait* (London: Heinemann, 1960), examines China, India, and Pakistan, subjecting Weber's concept to a critical appraisal.
43. Hartmut Lehmann and Guenther Roth, eds., *Weber's Protestant Ethic: Origin, Evidence, Context* (Cambridge: Cambridge University Press, 1993). See in particular Thomas Nipperdey's essay, "Max Weber, Protestantism and the Debate around 1900," pp. 73–82.
44. Fritz Ringer, *Max Weber: An Intellectual Biography* (Chicago and London: University of Chicago Press, 2004). See p. 84 for adequate causation, p. 183 for types of authority, and p. 233 for the problem of scientific specialization.
45. Keith Bullivant and Bernhard Spies, "'Die Wiederkehr des immergleich Schlechten?' Cultural Crises in the Work of German Writers in the Twentieth Century," in Ferdinand van Ingen and Gerd Labroisse, eds., *Literaturszene Bundesrepublik—ein Blick von Draussen* (Amsterdam: Rodopi, 1988), pp. 59–78.

Chapter 24: Dissonance and the Most-Discussed Man in Music

1. Jan Swafford, *Johannes Brahms: A Biography* (London: Macmillan, 1998), p. 570.
2. Schonberg, *Lives of the Great Composers*, p. 251.
3. Ibid., p. 252.
4. See the splendid portrait in Swafford, *op. cit.*, p. 49.
5. Schonberg, *Lives of the Great Composers*, p. 254.
6. Christine Jacobsen, ed., *Johannes Brahms: Leben und Werk* (Wiesbaden: Breitkopf & Härtel, 1983), pp. 36ff.
7. Schonberg, *Lives of the Great Composers*, p. 257.
8. Swafford, *Johannes Brahms*, p. 297. See also Daniel Beller-McKenna, *Brahms and the German Spirit* (Cambridge, Mass.: Harvard University Press, 2004), pp. 65ff.
9. Beller-McKenna, *Brahms and the German Spirit*, for Brahms's symphonies and an incipient nationalism in the spirit of Beethoven.
10. Ibid., p. 12.
11. Schonberg, *Lives of the Great Composers*, p. 263.
12. Ibid., p. 264.
13. For his living arrangements, see Frank Walker, *Hugo Wolf: A Biography* (London: Dent, 1968), pp. 55ff.
14. See Walker, *Hugo Wolf*, chapter 10, which explores the work of Mörike and Eichendorff over

many pages. See also Dietrich Fischer-Dieskau, *Hugo Wolf: Leben und Werk* (Berlin: Henschel, 2003), pp. 399 and 445.

15. Susan Youens, *Hugo Wolf: The Vocal Music* (Princeton, N.J., and Oxford: Princeton University Press, 1992), p. 75.
16. For his failure to produce an opera, see Fischer-Dieskau, *Hugo Wolf*, pp. 358–364.
17. Walker, *Hugo Wolf*, p. 443 for the final illness. Schonberg, *Lives of the Great Composers*, p. 269.
18. Schonberg, *Lives of the Great Composers*, p. 274.
19. Hans Fantel, *Johann Strauss, Father and Son, and Their Era* (Newton Abbot: David & Charles, 1971), pp. 32ff.
20. Joseph Wechsberg, *The Waltz Emperors: The Life and Times and Music of the Strauss Family* (London: Weidenfeld & Nicolson), 1973, p. 95.
21. Fantel, *Johann Strauss*, pp. 72ff.
22. Wechsberg, *Waltz Emperors*, p. 166.
23. Schonberg, *Lives of the Great Composers*, pp. 278–279.
24. Ibid., pp. 379–380.
25. Franzpeter Messmer, *Richard Strauss: Biographie eines Klangzauberers* (Zurich: M & T Verlag, 1994), pp. 243ff.
26. Ibid., pp. 171ff.
27. Schonberg, *Lives of the Great Composers*, p. 384.
28. Charles Dowell Youmans, *Richard Strauss's Orchestral Music and the German Intellectual Tradition: The Philosophical Roots of Musical Modernism* (Bloomington: Indiana University Press, 2005). Youmans locates *Guntram* as a turning point in Strauss's thought, the influences here being Max Stirner and Nietzsche. See pp. 86ff.
29. Messmer, *Richard Strauss*, p. 313.
30. George R. Marek, *Richard Strauss: The Life of a Non-Hero* (London: Gollancz, 1967), p. 183.
31. Messmer, *Richard Strauss*, pp. 324ff.
32. Youmans, *Richard Strauss's Orchestral Music*, pp. 136ff.
33. Schonberg, *Lives of the Great Composers*, p. 392.
34. Dika Newlin, *Bruckner, Mahler, Schoenberg* (London: Boyars, 1979), pp. 25ff.
35. Ibid., p. 35.
36. Ibid., p. 119 for the literary influences on Mahler.
37. Ibid., p. 133.
38. Alex Ross, *The Rest Is Noise: Listening to the Twentieth Century* (New York: Farrar, Straus and Giroux, 2007), pp. 19 and 21. Schonberg, *Lives of the Great Composers*, p. 403.
39. William R. Everdell, *The First Moderns* (Chicago and London: University of Chicago Press, 1997), p. 275.
40. James K. Wright, *Schoenberg, Wittgenstein and the Vienna Circle* (Bern: Peter Lang, 2007), pp. 67ff.
41. Michael Cherlin, *Schoenberg's Musical Imagination* (Cambridge: Cambridge University Press, 2007), pp. 44ff.
42. Ross, *The Rest Is Noise*, p. 18.
43. Ethan Haimo, *Schoenberg's Transformation of Musical Language* (Cambridge: Cambridge University Press, 2006), p. 245.
44. See Newlin, *Bruckner, Mahler, Schoenberg*, p. 214, for the "deep background" in Barcelona. And Ross, *The Rest Is Noise*, p. 49.
45. Newlin, *Bruckner, Mahler, Schoenberg*, pp. 234ff.
46. Carl Schorske, *Fin-de-siècle Vienna: Politics and Culture* (London: Weidenfeld & Nicolson, 1980), p. 360.
47. Ross, *The Rest Is Noise*, p. 52.

Chapter 25: The Discovery of Radio, Relativity, and the Quantum

1. Elon, *Pity of It All*, p. 276.
2. Helge Kragh, *Quantum Generations* (Princeton, N.J., and London: Princeton University Press, 1999), p. 13.

3. Ibid., p. 3.
4. Bruce J. Hunt, *The Maxwellians* (Ithaca, N.Y., and London: Cornell University Press, 1991), esp. chap. 8.
5. I have used *New Dictionary of Scientific Biography*, vol. 3, pp. 291–294.
6. Rollo Appleyard, *Pioneers of Electrical Communication* (London: Macmillan, 1930), p. 114. Hunt, *The Maxwellians*, pp. 180–182 and 198–199.
7. See Appleyard, *Pioneers*, p. 119, for a photograph and p. 121 for the gap.
8. *New Dictionary of Scientific Biography*, vol. 3, pp. 291–294.
9. *Physicists' Biographies*, p. 2. http://phisicist.info/
10. Appleyard, *Pioneers*, p. 131.
11. Kragh, *Quantum Generations*, p. 28.
12. Ibid., p. 29.
13. Emilio Segrè, *From X-rays to Quarks: Modern Physicists and Their Discoveries* (San Francisco: W. H. Freeman, 1980), pp. 22–23.
14. *Dictionary of Scientific Biography*, XI, p. 529–521.
15. Kragh, *Quantum Generations*, p. 30.
16. Watson, *Modern Mind/Terrible Beauty*, p. 20.
17. *New Dictionary of Scientific Biography*, vol. 6, pp. 111–115. For the Plancks as a whole, see J. L. Heilbron, *The Dilemmas of an Upright Man: Max Planck as a Spokesman for German Science* (Berkeley and London: University of California Press, 1986).
18. Heilbron, *Dilemmas of an Upright Man*, pp. 6–8.
19. Kragh, *Quantum Generations*, p. 21.
20. Ibid., p. 22.
21. Segrè, *From X-rays to Quarks*, pp. 66–68.
22. For Planck's relationship with Rubens, see Max Planck, *Scientific Autobiography, and Other Papers, with a Memorial Address on Max Planck by Max von Laue*, trans. Frank Gaynor (London: Williams and Norgate, 1950), pp. 39–40.
23. Kragh, *Quantum Generations*, p. 23.
24. Heilbron, *Dilemmas of an Upright Man*, p. 23. Though Planck told his son in 1900 that his work would rank among the great discoveries in physics, p. 55ff.
25. Kragh, *Quantum Generations*, p. 94.
26. Albrecht Fölsing, *Albert Einstein: A Biography*, trans. Ewald Osers (London: Viking, 1997), pp. 32ff.
27. Ibid., pp. 155ff.
28. Kragh, *Quantum Generations*, p. 95.
29. For some of the excitement at that time, see Albert Einstein, *The Collected Papers of Albert Einstein*, vol. 5, *The Swiss Years*, trans. Anna Beck, Don Howard, consultant (Princeton, N.J., and Chichester: Princeton University Press, 1995), which records his letters around 1905. John S. Rigden has devoted *Einstein 1905: The Standard of Greatness* (London: Harvard University Press, 2005) to just that year.
30. Fölsing, *Albert Einstein*, p. 165. See also Albert Einstein, *A Stubbornly Persistent Illusion: The Essential Scientific Works of Albert Einstein*, ed. with commentary by Stephen Hawking (Philadelphia and London: Running Press, 2007).
31. Everdell, *First Moderns*, p. 30.
32. Kragh, *Quantum Generations*, p. 32.
33. *Dictionary of Scientific Biography*, IV, pp. 123–127.
34. Kragh, *Quantum Generations*, p. 38.
35. Joseph W. Dauben, *Georg Cantor: His Mathematics and the Philosophy of the Infinite* (Cambridge, Mass.: Harvard University Press, 1979), chap. 6, p. 125.
36. Kragh, *Op. cit.*, p. 39.
37. *New Dictionary of Scientific Biography*, vol. 2, pp. 29–36.
38. Kragh, *Quantum Generations*, p. 41.
39. Michael Dummet, *Frege: Philosophy of Mathematics* (London: Duckworth, 1991), pp. 141f.
40. Kragh, *Quantum Generations*, p. 46.
41. Klein played an extraordinary role in German—and world—mathematics, which is explored in Lewis Pyenson, *Neohumanism and the Persistence of Pure Mathematics in Wilhelmine Germany* (Philadelphia: American Philosophical Society, 1983), which relates mathematics to Bildung. Constance Reid, in *Hilbert*, p. 19, describes the correspondence between Klein and

Hilbert as "nervous." See pp. 48ff for the Göttingen of the time. See also Günther Frei, ed., *Der Briefwechsel David Hilbert–Felix Klein (1886–1918)* (Göttingen: Vandenhoeck & Ruprecht, 1985).

42. Reid, *Hilbert*, pp. 74ff. Jeremy Gray, *The Hilbert Challenge* (Oxford: Oxford University Press, 2000) is devoted to this event and the reaction.

Chapter 26: Sensibility and Sensuality in Vienna

1. Peter Gay, *Schnitzler's Century*, pp. 64–65. Clive James, *Cultural Amnesia: Notes in the Margin of My Time* (London: Picador, 2007), p. 699.
2. Christian Brandstätter, ed., *Vienna 1900 and the Heroes of Modernism* (London: Thames & Hudson, 2006), pp. 335–342.
3. E. E. Yates, *Schnitzler, Hofmannsthal, and the Austrian Theatre* (New Haven, Conn., and London: Yale University Press, 1992), pp. 1–5.
4. Friedrich Rothe, *Karl Kraus: Die Biographie* (Munich: Piper, 2003), pp. 171–216. Edward Timms, *Karl Kraus, Apocalyptic Satirist: The Post-War Crisis and the Rise of the Swastika* (New Haven, Conn., and London: Yale University Press, 2005). See the early pages for the "dream of German domination."
5. Arthur Schnitzler, *The Road into the Open = Der Weg ins Freie*, trans. Roger Byers (Berkeley and Oxford: University of California Press, 1992). James, *Cultural Amnesia*, pp. 702 and 764–76.
6. Watson, *Modern Mind/Terrible Beauty*, p. 29.
7. Ulrich Weinzierl, *Hofmannsthal: Skizzen zu seinem Bild* (Vienna: Zsolnay, 2005), pp. 147ff.
8. Benjamin Bennet, *Hugo von Hofmannsthal: The Theatre of Consciousness* (Cambridge: Cambridge University Press, 1988), pp. 272ff.
9. Franz Clemens Brentano, *The Origin of Our Knowledge of Right and Wrong*, ed. Oskar Kraus, trans. Roderick M. Chisholm and Elizabeth H. Schneewind, English edition ed. Roderick M. Chisholm (London: Routledge & Kegan Paul, 1969), p. 75.
10. *Stanford Encyclopaedia of Philosophy*, Center for the Study of Language and Information, Stanford University, Calif., 94305, http://plato.stanford.edu/, entry on Wilhelm Wundt, p. 15 of 17.
11. Dermot Moran, *Edmund Husserl: Founder of Phenomenology* (Cambridge: Polity Press 2005), pp. 94–129.
12. Archives Husserl à Louvain, *Geschichte des Husserl-Archivs/Husserl-Archive Leuven = History of the Husserl-Archives* (Dordrecht: Springer, 2007).
13. David S. Luft, *Eros and Inwardness in Vienna: Weininger, Musil, Dorderer* (Chicago and London: University of Chicago Press, 2003), p. 49.
14. Ross, *The Rest Is Noise*, p. 38.
15. Harry Oosterhuis, *Stepchildren of Nature: Krafft-Ebing, Psychiatry and the Making of Sexual Identity* (Chicago and London: University of Chicago Press, 2000), pp. 25–36.
16. Watson, *Modern Mind/Terrible Beauty*, p. 34.
17. Hans Gross (1847–1915) was the creator of modern crime detection, another discipline that arose with the growth of major metropolises. It was Gross who started the systematic examination of footprints, the trajectory of blood stains, the study of underworld argot, and the relevance of x-rays for detection. See Ronald Martin Howe, *Criminal Investigation: A Practical Textbook for Magistrates, Police Officers and Lawyers*, adapted from the *System der Kriminalistic of Dr. Hans Gross* by John Adam and J. Collyer Adam (London: Sweet & Maxwell, 1949), p. 84 for bloodstains, p. 125 for fingerprints, p. 207 for footprints.
18. Brandstätter, ed., *Vienna 1900*, pp. 239–260.
19. Werner Oechslin, *Otto Wagner, Adolf Loos, and the Road to Modern Architecture*, trans. Lynette Widder (Cambridge: Cambridge University Press, 2002), p. 112.
20. Brandstätter, ed., *Vienna 1900*, pp. 293–407.
21. Burckhardt Rukschcio, *Adolf Loos: Leben und Werk* (Salzburg: Residenz, 1987).
22. Brandstätter, ed., *Vienna 1900*, pp. 93–109, 111–119, for the Klimt group. Serge Lemoine and Marie-Amélie zu Salm-Salm, ed., *Vienna 1900: Klimt, Schiele, Moser, Kokoschka* (Aldershot: Lund Humphries, 2005), p. 37 for an excellent view of the Ringstrasse.
23. Lemoine and Salm-Salm, ed., *Vienna 1900*, p. 41. See also Tobias G. Natter and Gerbert

Frodl, eds., *Klimt's Women* (New Haven, Conn., and London: Yale University Press [Cologne: DuMont], 2000), pp. 25–31.
24. Johnston, *Austrian Mind*, p. 357.
25. Watson, *Modern Mind/Terrible Beauty*, p. 36.
26. John T. Blackmore, *Ernst Mach: His Work, Life, and Influence* (Berkeley: University of California Press, 1972).
27. Brigitte Hamann, *Hitler's Vienna* (Oxford: Oxford University Press, 1999), p. 25.
28. Ibid., p. 80.
29. Ibid., p. 237.
30. Schorske, *Fin-de-siècle Vienna*, pp. 184–246.
31. Heinrich Schnee, *Karl Lueger: Leben und Wirken eines grossen Sozial- und Kommunalpolitikers: Umrisse einer politischen Biographie* (Berlin: Duncker & Humblot, 1960), pp. 91ff.
32. Hamann, *Hitler's Vienna*, p. 326.

Chapter 27: Munich/Schwabing: Germany's "Montmartre"

1. Ronald Hayman, *Thomas Mann: A Biography* (London: Bloomsbury, 1996), p. 163.
2. Maria Makela, *The Munich Secession: Art and Artists in Turn-of-the-Century Munich* (Princeton, N.J., and London: Princeton University Press, 1990), p. 3.
3. Paul Raabe, *The Era of German Expressionism*, trans. J. M. Ritchie (Woodstock, N.Y.: Overlook Press, 1965/1974), p. 79.
4. Christian Lenz, *The Neue Pinakothek Munich* (Munich: Beck [London: Scala], 2003), pp. 8–11.
5. Makela, *Munich Secession*, p. 13. See also Rainer Metzger, *Munich: Its Golden Age of Art and Culture, 1890–1920*, picture ed., Christian Branstätter (London: Thames & Hudson, 2009).
6. Ibid., p. 15.
7. Ibid., p. 74.
8. Ibid., p. 81.
9. Ibid.
10. Barbara C. Gilbert, ed., *Max Liebermann: From Realism to Impressionism* (Los Angeles: Skirball Cultural Center, and Seattle: University of Washington Press, 2005), pp. 167ff., for the disillusionment in Liebermann's art.
11. For his time in Berlin, see Sigrid Achenbach and Matthis Eberle, *Max Liebermann in seiner Zeit.* Exhibition catalog (Munich: Prestel, 1979), pp. 72ff.
12. Wolfgang Venzmer, *Adolf Hölzel: Leben und Werk; Monographie mit Verzeichnis der Ölbilder, Glasfenster und ausgewählter Pastelle* (Stuttgart: Deutsche Verlags-Anstalt, 1982). See pp. 16–19 for Hölzel in Dachau.
13. Makela, *Munich Secession*, p. 105.
14. Heinrich Voss, *Franz von Stuck 1863–1928: Werkkatalog d. Gemälde: Mit e. Einf. in seinen Symbolismus* (Munich: Prestel, 1973). See pp. 20–30 for a discussion of the theme of "sin."
15. Makela, *Munich Secession*, p. 112.
16. Winfried Nerdinger, ed., *Richard Riemerschmid: Vom Jugendstil zum Werkbund; Werke und Dokumentation* (Munich: Prestel, 1982), pp. 13ff.
17. Ibid., pp. 34–38.
18. Makela, *Munich Secession*, p. 125.
19. Alan Windsor, *Peter Behrens: Architect and Designer* (London: Architectural Press, 1981), pp. 77ff.
20. Frederic J. Schwartz, *The Werkbund: Design Theory and Mass Culture before the First World War* (New Haven, Conn., and London: Yale University Press, 1996). See pp. 44–60 for a discussion of art, craft, and alienation.
21. Tilmann Buddensieg, *Industriekultur: Peter Behrens und die AEG, 1907–1914* (Berlin: Mann, 1981).
22. He complained all his life that, with his "moustachioed personality," he always looked more like a commercial traveler than a writer. Klaus Harpprecht, *Thomas Mann: Eine Biographie* (Reinbeck: Rowohlt, 1995), pp. 58ff.

23. Nigel Hamilton, *The Brothers Mann: The Lives of Heinrich and Thomas Mann, 1871–1950 and 1875–1955* (London: Secker & Warburg, 1978), p. 49.
24. Willi Jasper, *Der Bruder: Heinrich Mann; Eine Biographie* (Munich: Hanser, 1992), pp. 51–60.
25. Hayman, *Thomas Mann*, p. 73.
26. Hans Wysling, ed., *Letters of Heinrich and Thomas Mann*, trans. Don Reneau with additional translations by Richard and Carla Winston (Berkeley and London: University of California Press, 1998). James, *Cultural Amnesia*, p. 429.
27. Hayman, *Thomas Mann*, p. 62.
28. Karin Verena Gunnemann, *Heinrich Mann's Novels and Essays: The Artist as Political Educator* (Rochester, N.Y., and Woodbridge: Camden House, 2002), pp. 51ff.
29. Robert Eben Sackett, *Popular Entertainment, Class, and Politics in Munich, 1900–1923* (Cambridge, Mass.: Harvard University Press, 1982), p. 11. See also Peg Weiss, *Kandinsky in Munich: The Formative Jugendstil Years* (Princeton, N.J.: Princeton University Press, 1979), pp. 19ff.
30. Friedrich Rothe, *Frank Wedekinds Dramen: Jugendstil und Lebensphilosophie* (Stuttgart: Metzler, 1968). See pp. 68–92 for Schopenhauer and Nietzsche.
31. Peter Jelavich, *Munich and Theatrical Modernism: Politics, Playwriting, and Performance, 1890–1914* (Cambridge, Mass., and London: Harvard University Press, 1985), pp. 167–185. See p. 170 for a photograph of the dancers dressed as executioners.
32. Eugen Roth, *Simplicissimus: Ein Rückblick auf die satirische Zeitschrift* (Hanover: Fackelträger-Verlag, 1954).
33. Jelavich, *Munich and Theatrical Modernism*, pp. 74ff. and 101ff.
34. Johannes Eichner, *Kandinsky und Gabriele Münter: Von Ursprüngen moderner Kunst* (Munich: F. Bruckmann, 1957). See pp. 26–35 for Münter.
35. Hartwig Fischer and Sean Rainbird, eds., *The Path to Abstraction* (London: Tate Publishing, 2006), p. 209.
36. Vivian Endicott Barnett and Armin Zweite, eds., *Kandinsky: Watercolours and Drawings* (Munich: Prestel, 1992), pp. 9ff. See also Reinhard Zimmermann, *Die Kunsttheorie von Wassily Kandinsky* (Berlin: Gebr. Mann, 2002).
37. Mark Roskill, *Klee, Kandinsky and the Thought of Their Time: A Critical Perspective* (Urbana: University of Illinois Press, 1992), pp. 54ff.
38. For Marc, Jawlensky, and others, see Armin Zweite, ed., *The Blue Rider in the Lenbachhaus München: Masterpieces by Franz Marc, Vassily Kandinsky, Gabriele Münter, Alexei Jawlensky, August Macke, Paul Klee* (Munich: Prestel, 1989), pp. 29 and 194.
39. W. Kandinsky, *Über das Geistige in der Kunst*, Bern: Benteli, 1952.
40. See Esther da Costa Meyer and Fred Wasserman, eds., *Schoenberg, Kandinsky, and the Blue Rider* (New York: Jewish Museum; London: Scala, 2003), pp. 79–94 for a direct linking of abstraction and emancipated dissonance. See also Gerald N. Izenberg, *Modernism and Masculinity: Mann, Wedekind, Kandinsky through World War I* (Chicago and London: University of Chicago Press, 2000), chaps. 2 and 3, for a direct linking between Wedekind and abstraction, Thomas Mann and sexuality.

Chapter 28: Berlin Busybody

1. David Clay Large, *Berlin* (New York: Basic Books, 2000), p. 1.
2. Ibid., p. 2.
3. Gordon A. Craig, *Theodor Fontane: Literature and History in the Bismarck Reich* (Oxford: Oxford University Press, 1999). See pp. 96ff. for Fontane's own view of Bismarck.
4. Ibid., p. 109. Large, *Berlin*, p. 7.
5. Large, *Berlin*, p. 9.
6. Ulrike Laufer and Hans Ottmeyer, *Gründerzeit 1848–1871: Industrie & Lebensträume zwischen Vormärz und Kaiserreich* (Dresden: Sandstein, 2008), pp. 95ff. for the banks.
7. Large, *Berlin*, p. 14.
8. Fritz Stern, *Gold and Iron: Bismarck, Bleichröder, and the Building of the German Empire* (New York: Knopf, 1977), pp. 106ff.
9. Large, *Berlin*, pp. 18–19.

10. Ibid., p. 20.
11. Godela Weiss-Sussex and Ulrike Zitzlsperger, eds., *Berlin: Kultur und Metropole in den zwanziger und seit den neunziger Jahren* (Munich: Iudicium, 2007). See pp. 183–194 for an examination of the "myth" of Berlin and pp. 155–167 for the way Berlin is remembered visually.
12. Heinz Ohff, *Theodor Fontane: Leben und Werk* (Munich: Piper, 1995), pp. 363–368.
13. Large, *Berlin*, pp. 24–26.
14. Ohff, *Theodor Fontane*, p. 368.
15. Large, *Berlin*, pp. 49–50.
16. Christian von Krockow, *Kaiser Wilhelm II und seine Zeit: Biographie einer Epoche* (Berlin: Siedler, 1999), pp. 92–114 and 163–184. See also Christopher Clark, *Kaiser Wilhelm* (Harlow: Longman, 2000).
17. Large, *Berlin*, pp. 59–60.
18. John C. G. Röhl, *Wilhelm II: Der Aufbau der persönlichen Monarchie 1888–1900* (Munich: C. H. Beck, 2001), pp. 221–231.
19. Annika Mombauer and Wilhem Deist, eds., *The Kaiser: New Research on Wilhelm II's Role in Imperial Germany* (Cambridge: Cambridge University Press, 2003).
20. Large, *Berlin*, p. 63.
21. Hans Daiber, *Gerhart Hauptmann oder der letze Klassiker* (Vienna-Munich-Zurich: Fritz Molden, 1971), pp. 47–59.
22. Margaret Sinden, *Gerhart Hauptmann: The Prose Plays* (Toronto: University of Toronto Press, 1957), pp. 149ff. for plays about "common people."
23. Eberhard Hilscher, *Gerhart Hauptmann* (Berlin: Verlag der Nation, 1969), pp. 131–154.
24. For Hauptmann's letters to Brahms, see Martin Machatzke, ed., *Gerhart Hauptmann: Tagebücher, 1897 bis 1905* (Frankfurt am Main: Propyläen, 1987), pp. 545f. and 594f.
25. Röhl, *Wilhelm II*, pp. 1008–1016. Large, *Berlin*, p. 64.
26. Helene Thimig-Reinhardt, *Wie Max Reinhardt lebte* (Percha am Stamberger See: R. S. Schulz, 1973), pp. 77–87.
27. Oliver M. Sayler, *Max Reinhardt and His Theatre* (New York: Brentano's, 1924), p. 92.
28. Franz Herre, *Kaiser Wilhelm II: Monarch zwischen den Zeiten* (Cologne: Kiepenheuer & Witsch, 1993).
29. Large, *Berlin*, p. 65.
30. For Strauss's relationship with Bülow, see Willi Schuh and Franz Trenner, *Correspondence: Hans von Bülow and Richard Strauss*, trans. Anthony Gishford (London: Boosey & Hawkes, 1955), p. 68, including a plan (not realized) to turn an Ibsen play into an opera.
31. Gisold Lammel, *Adolph Menzel und seine Kreise* (Dresden: Verlag der Kunst, 1993), especially the pictures on pp. 152–153.
32. Large, *Berlin*, p. 69.
33. Ibid., p. 73.
34. Ibid., p. 71.
35. Georg Brühl, *Die Cassirers: Streiter für den Impressionismus* (Leipzig: Edition Leipzig, 1991). See pp. 105ff. for Paul.
36. Peter Paret, *The Berlin Secession: Modernism and Its Enemies in Imperial Germany* (Cambridge, Mass.: Belknap Press of Harvard University Press, 1980), p. 39.
37. Nell Roslund Walden, *Herwarth Walden: Ein Lebensbild* (Berlin: F. Kupferberg, 1963), pp. 45f.
38. Magdalena M. Moeller, *Die "Brücke": Meisterwerke aus dem Brücke-Museum Berlin* (Munich: Hirmer, 2000), pp. 1–40. See also Carol S. Eliel, *The Apocalyptic Landscapes of Ludwig Meidner* (Munich: Prestel, 1989).
39. Wilhelm von Bode, *Mein Leben* (Berlin: H. Reckendorf, 1930).
40. See Wilhelm von Bode, *Rembrandt und seine Zeitgenossen: Charakterbilder der grossen Meister der holländischen und flämischen Malerschule im siebzehnten Jahrhundert* (Leipzig: E. A. Seemann, 1923), as an example of his scholarship.
41. Bernhard Maaz, ed., *Nationalgalerie Berlin: Das 19 Jahrhundert; Bestandskatalog der Skulpturen* (Leipzig: Seemann, 2006), pp. 20f.
42. Large, *Berlin*, p. 77.
43. Ibid., p. 81.

Chapter 29: The Great War between Heroes and Traders

1. Maureen Healy, *Vienna and the Fall of the Habsburg Empire* (Cambridge: Cambridge University Press, 2004), p. 2.
2. Matthew Stibbe, *German Anglophobia and the Great War, 1914–1918* (Cambridge: Cambridge University Press, 2001), p. 49. Elias, *The Germans*, p. 181.
3. Roger Chickering, *Imperial Germany and the Great War, 1914–1918* (Cambridge: Cambridge University Press, 1998), p. 134.
4. Stibbe, *German Anglophobia*, p. 50.
5. Ibid., p. 51.
6. Ibid., p. 52.
7. Ibid., p. 54.
8. Hans Heinz Krill, *Die Rankerenaissance: Max Lenz und Erich Marcks; Ein Beitrag zum historisch-politischen Denken in Deutschland, 1880–1935* (Berlin: de Gruyter, 1962), pp. 6–12 and 67–69 for Lenz and 174–187 for his ideas of nationality; pp. 42ff. for Marcks; pp. 211ff. for the role of propaganda in World War I.
9. For Meinecke, see Stefan Meinecke, *Friedrich Meinecke: Persönlichkeit und politisches Denken bis zum Ende des ersten Weltkrieges* (Berlin: de Gruyter, 1995); also Stibbe, *German Anglophobia*, p. 63.
10. Arden Bucholz, ed. and trans., *Delbrück's Modern Military History* (Lincoln, Neb., and London: University of Nebraska Press, 1997).
11. Anton Mirko Koktanek, *Oswald Spengler in seiner Zeit* (Munich: Beck, 1968), p. 183. See also H. Stuart Hughes, *Oswald Spengler: A Critical Estimate* (New York: Scribners, 1952), p. 57.
12. Detlef Felken, *Oswald Spengler: Konservativer Denker zwischen Kaiserreich und Diktatur* (Munich: Beck, 1988), pp. 68–76.
13. Marchand, *Down from Olympus*, p. 240.
14. Stibbe, *German Anglophobia*, p. 74.
15. Ibid., p. 75.
16. Ibid., p. 78.
17. Volker Berghahn, *Perspectives on History* (the newsmagazine of the American Historical Association, September 10, 2007): http://www.historians.org/perspectives/issues/2000/0003/0003mem/cfm.
18. Ibid.
19. Fritz Stern, *Failure of Illiberalism*, p. 152. See Fritz Fischer, *World Power or Decline: The Controversy over Germany's Aims in the First World War*, trans. Lancelot L. Farrar, Robert Kimber, and Rita Kimber (New York: W. W. Norton, 1974), for a reconsideration of the issue a decade later. "Treason" is discussed on p. viii.
20. Martha Hanna, *The Mobilization of Intellect: French Scholars and Writers during the Great War* (Cambridge, Mass.: Harvard University Press, 1996), p. 8.
21. Hanna, *Mobilization of Intellect*, p. 12.
22. Stuart Wallace, *War and the Image of Germany: British Academics, 1914–1918* (Edinburgh: John Donald, 1988), p. 7.
23. Ibid., p. 38. Ariel Roshwald and Richard Stites, eds., *European Culture and the Great War: The Arts, Entertainment, and Propaganda* (Cambridge: Cambridge University Press, 2002), p. 44.
24. Hanna, *Mobilization of Intellect*, p. 22.
25. Marchand, *Down from Olympus*, pp. 245–246.
26. Ibid., p. 258.
27. John Dewey, *German Philosophy and Politics* (New York: Henry Holt, 1915), p. 35.
28. Ibid., p. 14.
29. Ibid., p. 17.
30. Ibid., pp. 30–31.
31. Ibid., p. 45.
32. Ibid., p. 37.
33. Ibid., pp. 62–63.
34. Ibid., p. 73.
35. Ibid., p. 100.
36. George Santayana, *Egotism in German Philosophy* (London: J. M. Dent, 1916), p. xiii.

37. Ibid., p. xviii.
38. Ibid., p. 170.
39. Ibid., p. 62.
40. Ibid., p. 89.
41. Ibid., p. 103.
42. Ibid., p. 130.
43. Ibid., p. 168.
44. Trevor Dupuy, *A Genius for War* (London: Macdonald and Jane's, 1977), p. 5. David Stone, in *Fighting for the Fatherland: The Story of the German Soldier from 1648 to the Present Day* (London: Conway, 2006), says, however, that toward the end of the war the Germans could renew their units with fresh men, much more than—in this case—the French. "In mid-March 1918 almost 200 German divisions stood ready to inflict that final crushing blow that would at least enable Germans to achieve their historic destiny" (p. 284). Then came the "stab-in-the-back."
45. Dupuy, *Genius for War*, p. 177.
46. Alexander Watson, *Enduring the Great War: Combat, Morale and Collapse in the German and British Armies, 1914–1918* (Cambridge: Cambridge University Press, 2008), p. 240.
47. Dupuy, *Genius for War*, p. 7.
48. Ibid., pp. 9–10.
49. David Charles, *Between Genius and Genocide: The Tragedy of Fritz Haber, Father of Chemical Warfare* (London: Jonathan Cape, 2005), pp. 156–157.

Chapter 30: Prayers for a Fatherless Child: The Culture of the Defeated

1. Watson, *Modern Mind/Terrible Beauty*, p. 152. (The title for this chapter is taken from Wolfgang Schivelbusch, *The Culture of Defeat: On National Trauma, Mourning and Recovery* (London: Granta, 2003).
2. Patrick Bridgwater, *The German Poets of the First World War* (London: Croom, Helm, 1985), Foreword. Roshwald and Stites, eds., *European Culture*, p. 32.
3. Karl Ludwig Schneider, *Der bildhafte Ausdruck in den Dichtungen Georg Heyms, Georg Trakls und Ernst Stadlers: Studien zum lyrischen Sprachstil des deutschen Expressionismus* (Heidelberg: C. Winter, 1961); Eduard Lachmann, *Kreuz und Abend: Eine Interpretation der Dichtungen Georg Trakls* (Salzburg: O. Müller, 1954).
4. Bridgwater, *German Poets*, p. 16.
5. Ibid., p. 191.
6. Ibid., p. 169.
7. Jeremy Adler, ed., *August Stramm: Alles ist Gedicht; Briefe, Gedichte, Bilder, Dokumente* (Zurich: Arche, 1990), pp. 95ff. for the poems, 9ff. for his letters from the war. Photo frontispiece.
8. Francis Sharp says few poets "of any nationality or language have come to such serenely poetic terms with the holocaust of twentieth-century warfare." Francis Sharp, *The Poet's Madness: A Reading of Georg Trakl* (Ithaca, N.Y.: Cornell University Press, 1981), pp. 188.
9. Bridgwater, *German Poets*, p. 171.
10. Ibid., p. 44.
11. Ibid., p. 172.
12. Ibid.
13. Roshwald and Stites, eds., *European Culture*, pp. 38–39 for film, p. 50 for Toller, pp. 150–151 for Kraus.
14. Deirdre Bair, *Jung: A Biography* (London: Little, Brown, 2004), pp. 207, 257.
15. Ibid., pp. 316–321. William McGuire, *The Freud/Jung Letters: The Correspondence between Sigmund Freud and C. G. Jung*, trans. Ralph Manheim and R. F. C. Hull (London: Hogarth Press: Routledge & Kegan Paul, 1974).
16. Paul Lerner, "Rationalising the Therapeutic Arsenal: German Neuropsychiatry in World War I," in Berg and Cocks, eds., *Medicine and Modernity*, pp. 121–128.
17. David R. Oldroyd, *Thinking about the Earth* (London: Athlone Press, 1996), p. 250.
18. Roger M. McCoy, *Ending in Ice: The Revolutionary Idea and Tragic Expedition of Alfred Wegener* (Oxford: Oxford University Press, 2006), p. 31.

19. Joachim Schulte, et al., eds., *Ludwig Wittgenstein: Philosophische Untersuchungen: Kritisch-genetische Edition* (Frankfurt am Main: Suhrkamp, 2001). A good introduction.
20. Gordon Baker, *Wittgenstein, Frege and the Vienna Circle* (Oxford: Basil Blackwell, 1988), pp. 51ff. and 101ff.
21. Pasquale Frascella, *Understanding Wittgenstein's Tractatus* (London: Routledge, 2007), chapters 2, 4, and 6.
22. Baker, *Wittgenstein*, pp. 101ff.
23. Simon Glendinning, *The Idea of Continental Philosophy* (Edinburgh: Edinburgh University Press, 2006), pp. 282ff.
24. Herman, *The Idea of Decline*, p. 228.
25. Watson, *Modern Mind/Terrible Beauty*, pp. 171–172.
26. Detlef Felken, *Oswald Spengler: Konservativer Denker zwischen Kaiserreich und Diktatur* (Munich: Beck, 1988), pp. 58ff.
27. Watson, *Modern Mind/Terrible Beauty*, p. 173.
28. Ibid.
29. Walter Lacquer, *New York Times Book Review*, May 15, 1983, p. 1.
30. Hayman, *Thomas Mann*, p. 289.
31. Lacquer, *New York Times Book Review*, p. 2.
32. Ibid.
33. Ibid.
34. Eugen Egger, *Hugo Ball: Ein Weg aus dem Chaos* (Otten: Otto Walter, 1951), pp. 41ff.
35. Tom Sandqvist, *Dada East: The Romanians of Cabaret Voltaire* (Cambridge, Mass.: MIT Press, 2006), pp. 90ff. See also Dominique Noguez, *Lénine dada: Essai* (Paris: Robert Laffont, 1989).
36. Peter Schifferli, ed., *Dada: Die Geburt des Dada; Dichtung und Chronik der Gründer/Mit Photos und Dokumenten. [In Zusammenarbeit mit] Hans Arp, Richard Huelsenbeck [und] Tristan Zara* (Zurich: Im Verlag der Arche, 1957). Leah Dickerman, with essays by Brigid Doherty, et al., *Dada: Zurich, Berlin, Hanover, Cologne, New York, Paris* (Washington, D.C.: National Gallery of Art in Association with Distributed Art Publishers, 2005).
37. Eric Robertson, *Arp: Painter, Poet, Sculptor* (London: Yale University Press, 2006), pp. 36ff.
38. Dorothea Dietrich, *The Collages of Kurt Schwitters: Tradition and Innovation* (Cambridge: Cambridge University Press, 1993), pp. 37ff. See also Kate Traumann Steinitz, *Kurt Schwitters: Erinnerungen aus den Jahen, 1918–30* (Zurich: Arche, 1963).
39. Walter Mehring, *Berlin Dada: Eine Chronik mit Photos und Dokumenten* (Zurich: Arche, 1959).
40. Uwe M. Schneede, *George Grosz: His Life and Work*, trans. Susanne Flatauer (London: Gordon Fraser, 1979), pp. 14–15.
41. Matthias Eberle, *World War I and the Weimar Artists: Dix, Grosz, Beckmann, Schlemmer* (New Haven, Conn., and London: Yale University Press, 1985), pp. 1–21.
42. Wolfgang Schivelbusch, *The Culture of Defeat: On National Trauma, Mourning and Recovery*, trans. Jefferson Chase (London: Granta, 2003), p. 247.
43. Norbert Elias, *The Germans: Power Struggles and the Development of Habitus in the Nineteenth and Twentieth Centuries*, ed. Michael Schröter, trans. Eric Dunning and Stephen Mennell (Cambridge: Polity Press, 1996), p. 7.

Chapter 31: Weimar: "Unprecedented Mental Alertness"

1. Otto Freundlich, *Before the Deluge: A Portrait of Berlin in the 1920s* (London: Harper, 1995), p. 175. For excellent photographs of Berlin in the 1920s, see Rainer Metzger, *Berlin in the Twenties: Art and Culture 1918–1933* (London: Thames & Hudson, 2007), passim. Elias, *The Germans*, pp. 214ff.
2. See, for example, Hans-Jürgen Buderer, *Neue Sachlichkeit: Bilder auf der Suche nach der Wirklichkeit; Figurative Malerei der zwanziger Jahre* (Munich: Prestel, 1994); and Bärbel Schrader, *The Golden Twenties: Art and Literature in the Weimar Republic*, trans. Katherine Vanovitch (New Haven, Conn., and London: Yale University Press, 1988).
3. Ian Roberts, *German Expressionist Cinema: The World of Light and Shadow* (London: Wallflower, 2008), p. 25, gives new archival research on Wiene.

4. S. S. Prawer, *Caligari's Children: The Film as Tale of Terror* (Oxford and New York: Oxford University Press, 1980), pp. 8ff.
5. For Gropius, see Kathleen James-Chakraborty, eds., *Bauhaus Culture: From Weimar to the Cold War* (Minneapolis, Minn., and London: University of Minnesota Press, 2006), pp. 26ff.
6. Lutz Schöbe and Wolfgang Thöner, *Stiftung Bauhaus Dessau: Die Sammlung* (Ostfildern-Ruit: Hatje, 1995), pp. 29f.
7. Ibid., pp. 32–33.
8. Lee Congden, *Exile and Social Thought: Hungarian Exiles in Germany and Austria, 1919–1933* (Princeton, N.J.: Princeton University Press, 1991), p. 181.
9. Dick Geary, *Karl Krautsky* (Manchester: Manchester University Press, 1987), p. 58.
10. Rolf Wiggershaus, *Die Frankfurter Schule: Geschichte, theoretische Entwicklung, politische Bedeutung* (Munich: Hanser, 1986), pp. 36ff.
11. Wiggershaus, *Die Frankfurter Schule*, contains portraits of the main characters between pp. 55 and 123.
12. Robert E. Norton, *Secret Germany: Stefan George and His Circle* (Ithaca, N.Y., and London: Cornell University Press, 2002), p. 688. See also Thomas Karlauf, *Stefan George: die Entdeckung des Charisma; Biographie* (Munich: Karl Blessing, 2007).
13. Judith Ryan, *Rilke, Modernism and Poetic Tradition* (Cambridge: Cambridge University Press, 1999), p. 111.
14. For Rilke's debt to Freud, see Adrian Stevens and Fred Wagner, eds., *Rilke und die Moderne: Londoner Symposion* (Munich: Iudicium, 2000), pp. 49ff.
15. Joseph Mileck, *Hermann Hesse: Biography and Bibliography* (Berkeley and London: University of California Press, 1977), vol. 1, p. 4.
16. Karl Corino, *Robert Musil: Eine Biographie* (Reinbeck bei Hamburg: Rowohlt, 2003), pp. 993ff. See David S. Luft, *Eros and Inwardness in Vienna: Weininger, Musil, Doderer* (Chicago and London: University of Chicago Press, 2003), pp. 115–125 for general background.
17. For the links between Kafka and Musil, see Reiner Stach, *Kafka: The Decisive Years*, trans. Shelley Frisch (New York: Harcourt, 2005), pp. 401–412.
18. Wolfgang Jeske, *Lion Feuchtwanger oder der arge Weg der Erkenntnis: Eine Biographie* (Stuttgart: Metzler, 1984), pp. 238ff.
19. Christine Barker and R. W. Last, *Erich Maria Remarque* (London: Oswald Wolff, 1979), p. 13.
20. Hilton Tims, *Erich Maria Remarque: The Last Romantic* (London: Constable, 2003), p. 53.
21. Barker and Last, *Erich Maria Remarque*, p. 60.
22. John Willett, *The New Sobriety: 1917–1933; Art and Politics in the Weimar Period* (London: Thames & Hudson, 1978), p. 193.
23. Barker and Last, *Erich Maria Remarque*, pp. 151–152. James, *Cultural Amnesia*, pp. 55 and 400.
24. Andreas Jacob, *Grundbegriffe der Musiktheorie Arnold Schönbergs* (Hildesheim: Olms, 2005), vol. 1, p. 374. For Kessler, see Harry Kessler, *Berlin in Lights: The Diaries of Harry Kessler* (New York: Grove, 2000).
25. Freundlich, *Before the Deluge*, p. 180.
26. Ross, *The Rest Is Noise*, pp. 206–207.
27. John Jarman, *The Music of Alban Berg* (London and Boston: Faber, 1979), pp. 15ff. and 80ff. See Kathryn Bailey, *The Life of Webern* (Cambridge: Cambridge University Press, 1998), pp. 116ff., for more details about the musical culture of the time in Germany.
28. Ross, *The Rest Is Noise*, p. 207.
29. For other American influences at that time, see Elizabeth Harvey, "Culture and Society in Weimar Gremany: The Impact of Modernism and Mass Culture," in Mary Fulbrook, ed., *Twentieth-Century Germany: Politics, Culture and Society 1918–1990* (London: Arnold, 2001), p. 62.
30. Hans Mayer, *Brecht* (Frankfurt am Main: Suhrkamp, 1996), pp. 323ff.
31. Walter Lacquer, *Weimar, A Cultural History, 1918–1933* (London: Weidenfeld & Nicolson, 1974), p. 153.
32. John Fuegi, *Brecht & Co.: Biographie* (Hamburg: Europäische Verlagsanstalt, 1997), pp. 271ff.
33. Ross, *The Rest Is Noise*, p. 192.
34. Foster Hirsch, *Kurt Weill on Stage: From Berlin to Broadway* (New York: Knopf, 2002), p. 12.

See also Jürgen Schebera, *Kurt Weill 1900–1950: Eine Biographie in Texten, Bildern und Dokumenten* (Mainz: Schott, 1990), pp. 77ff.

35. Keith Bullivant, ed., *Culture and Society in the Weimar Republic* (Manchester: Manchester University Press, 1997), pp. 50ff., for the differences between Brecht and Alfred Döblin in *Berlin Alexanderplatz.*
36. See Thomas J. Saunders, *Hollywood in Berlin: American Cinema and Weimar Germany* (Berkeley and London: University of California Press, 1994), where the author argues that, in the 1920s, Germany was poised to overtake America in film production.
37. Dietrich Scheunemann, ed., *Expressionist Film: New Perspectives* (Rochester, N.Y.: Camden House, 2003), p. 25.
38. Ibid., p. 38.
39. Harvey, "Culture and Society," pp. 68ff.
40. Patrick McGilligan, *Fritz Lang: The Nature of the Beast* (London: Faber, 1997), p. 148.
41. Charlotte Chandler, *Nobody's Perfect: Billy Wilder; A Personal Biography* (New York and London: Simon & Schuster, 2002), p. 60.
42. Luzi Korngold, *Erich Wolfgang Korngold: ein Lebensbild* (Vienna: Elisabeth Lafite, 1967), pp. 62ff.
43. Carl Zuckmayer, *A Part of Myself*, trans. Richard and Clara Winston (London: Secker & Warburg, 1970), p. 32.
44. Marlene Dietrich, *ABC meines Lebens* (Berlin: Blanvalet, 1963). See in particular the entries under Hollywood and Billy Wilder,.
45. Guido Knopp, *Hitler's Women—and Marlene*, trans. Angus McGeoch (Stroud: Sutton, 2003), p. 266.

Chapter 32. Weimar. The Golden Age of Twentieth-Century Physics, Philosophy, and History

1. John Cornwell, *Hitler's Scientists* (London: Viking, 2003; Penguin, 2004), p. 111.
2. Ibid., p. 114.
3. Charles P. Enz, *No Time to Be Brief: A Scientific Biography of Wolfgang Pauli* (Oxford: Oxford University Press, 2002), pp. 84ff.
4. David C. Cassidy, *Uncertainty: The Life and Science of Werner Heisenberg* (New York: W. H. Freeman, 1992), pp. 127f.
5. Leo Corry, *David Hilbert and the Axiomatisation of Physics (1898–1918): From Grundlagen der Geometrie to Grundlagen der Physik* (Dordrecht: Kluwer, 2004).
6. Nancy Thorndike Greenspan, *The End of the Certain World* (New York: Wiley, 2005).
7. Einstein thought there were two kinds of scientific theory—"principled" theories, like the equivalence between gravity and acceleration, where reality "unfurls" from basic principles, and "constructive" theories, like quantum theory, where the underlying principle has yet to be found. Interest in Einstein's more philosophical views has been growing recently. See Amanda Gefter, "Power of the Mind," *New Scientist* 2529 (December 10, 2005): 54–55.
8. Cornwell, *Hitler's Scientists*, p. 104.
9. Walter Isaacson, *Einstein: His Life and Universe* (New York and London: Simon & Schuster, 2007).
10. Cornwell, *Hitler's Scientists*, p. 110.
11. Glyn Jones, *The Jet Pioneers: The Birth of Jet-Powered Flight* (London: Methuen, 1989), pp. 41–49.
12. Ibid., pp. 142ff.
13. Corry, *David Hilbert.*
14. John W. Dawson Jr., *Logical Dilemmas: The Life and Work of Kurt Gödel* (Wellesley, Mass.: A.K. Peters, 1997), p. 55.
15. Watson, *Modern Mind/Terrible Beauty*, p. 271. Furthermore, as Roger Penrose has pointed out, Gödel's "open-ended mathematical intuition is fundamentally incompatible with the existing structure of physics."
16. Michael Stöltzner and Thomas Uebel, eds., *Wiener Kreis: Texte zur wissenschaftlichen Weltauffassung von Rudolf Carnap, Otto Neurath, Moritz Schlick, Philipp Frank, Hans Hahn, Karl*

Menger, Edgar Zilsel und Gustav Bergmann (Hamburg: Meiner, 2006). See the introduction, pp. ix–civ, 315ff. and 362ff. for Carnap, 503f. for Gödel.

17. Ben Rogers, *A. J. Ayer* (London: Chatto & Windus, 1999), p. 86.
18. Ibid.
19. Ibid., p. 87.
20. Paul Arthur Schilpp, *The Philosophy of Rudolf Carnap* (La Salle, Ill.: Open Court; London: Cambridge University Press, 1963), especially pp. 183f., 385ff., and 545f. See also A. W. Carus, *Carnap and Twentieth-Century Thought: Explication and Enlightenment* (Cambridge: Cambridge University Press, 2007), pp. 91–108 and 185–207.
21. I have used Rüdiger Safranski, *Ein Meister aus Deutschland: Heidegger und seine Zeit* (Munich: Hanser, 1994), pp. 145ff.
22. Michael Grossheim, *Von Georg Simmel zu Martin Heidegger: Philosophie zwischen Leben und Existenz* (Bonn and Berlin: Bouvier, 1991), pp. 14–18.
23. Michael E. Zimmerman, *Heidegger's Confrontation with Modernity: Technology, Politics, and Art* (Bloomington: Indiana University Press, 1990).
24. Max Scheler, *The Nature of Sympathy*, trans. Peter Heath, intro. by W. Stark (London: Routledge & Kegan Paul, 1954), pp. 96–102.
25. Ibid.
26. To situate Cassirer, see Michael Friedman, *A Parting of the Ways: Carnap, Cassirer, and Heidegger* (Chicago: Open Court, 2000), pp. 1–10 and 129–144.
27. That year he took part in a famous debate with Martin Heidegger at Davos in Switzerland, where they locked horns over Kant, a year or so after Heidegger had published *Being and Time*. Safranski, *Meister aus Deutschland*, pp. 183–188.
28. Ernst Cassirer, *Philosophie der symbolischen Formen, Text und Anmerkungen bearbeitet von Claus Ronsenkarnz* (Hamburg: Felix Meiner, 2001), pp. 43ff. and 193.
29. Silvia Ferretti, *Cassirer, Panofsky and Warburg: Symbol, Art, and History*, trans. Richard Pierce (New Haven, Conn., and London: Yale University Press, 1989), pp. 122ff.
30. Percy Ernst Schramm, *Hitler, the Man and the Military Leader*, trans., ed., and with an intro. by Donald S. Detwiler (London: Allen Lane, Penguin Press, 1972), p. 9.
31. Alain Boureau, *Kantorowicz: Stories of a Historian*, trans. Stephen G. Nichols and Gabrielle M. Spiegel, foreword by Martin Jay (Baltimore: Johns Hopkins University Press, 2001), p. 2.
32. Norbert Elias, *The Germans*.
33. There has been an explosion of Elias scholarship recently. See Richard Kilminster, *Norbert Elias: Post-Philosophical Sociology* (New York and Abingdon: Routledge, 2007), which has a section on Elias and Weimar culture (pp. 10–14), a chapter on Elias and Mannheim, and one devoted to *The Civilising Process*. See also Stephen Menell, *Norbert Elias: Civilisation and the Human Self-Image* (Oxford: Basil Blackwell, 1989), with chapters on "Sports and Violence," "Civilisation and De-civilisation," "Involvement and Detachment."
34. Peter Schäfer and Gary Smith, *Gershom Scholem: Zwischen den Disziplinen* (Frankfurt am Main: Suhrkamp, 1995).
35. Susan A. Handelman, *Fragments of Redemption: Jewish Thought and Literary Theory in Benjamin, Scholem, and Levinas* (Bloomington: Indiana University Press, 1991), pp. 109ff.

Chapter 33: Weimar: "A Problem in Need of a Solution"

1. Peter Gay, *Freud: A Life for Our Times* (London: MAX, 2006), p. 546.
2. Watson, *Modern Mind/Terrible Beauty*, p. 273.
3. Paul-Laurent Assoun, *Freud and Nietzsche*, trans. Richard L. Collier Jr.(London: Athlone Press, 2000), pp. 70–82 and 137–156.
4. Renos K. Papadopoulos, et al., *Jung in Modern Perspective* (Hounslow, Middlesex: Wildwood, 1984), p. 203.
5. Bernard J. Paris, *Karen Horney: A Psychoanalyst's Search for Self-understanding* (New Haven, Conn., and London: Yale University Press, 1994), pp. 92ff.
6. George L. Mosse, *The Crisis of German Ideology: Intellectual Origins of the Third Reich* (New York: H. Fertig, 1998).
7. For Nietzsche and Darwin, see John Richardson, *Nietzsche's New Darwinism* (Oxford: Oxford

University Press, 2004), pp. 78ff., 81f., 95f., and 146f. Gregory Moore, *Nietzsche, Biology, and Metaphor* (Cambridge: Cambridge University Press, 2002), pp. 115ff., for Nietzsche and "the nervous age."

8. For Hitler and Nietzsche, see Jacob Golomb and Robert S. Wistrich, eds., *Nietzsche, Godfather of Fascism? On the Uses and Abuses of Philosophy* (Princeton, N.J., and Oxford: Princeton University Press, 2002), pp. 90–106.
9. See Moore, *Nietzsche, Biology*, for a chapter on "The Physiology of Power."
10. See Charles R. Bambach, *Heidegger's Roots: Nietzsche, National Socialism, and the Greeks* (Ithaca, N.Y., and London: Cornell University Press, 2003), pp. 12ff. for the myths of the homeland, and 112ff. for Heidegger's concept of Mitteleuropa.
11. Frank-Lothar Kroll, *Utopie als Ideologie: Geschichtsdenken und politisches Handeln im Dritten Reich* (Paderborn: Schöningh, 1998), pp. 72–77.
12. Roger Griffin, *Modernism and Fascism: The Sense of a Beginning under Mussolini and Hitler* (Basingstoke: Palgrave, 2007).
13. Fritz Stern, *The Politics of Cultural Despair* (Berkeley and London: University of California Press, 1961/1974), p. 184.
14. Ibid., p. 189.
15. Ibid., pp. 191–192.
16. Ibid., pp. 194–196.
17. Ibid., p. 220.
18. Ibid., pp. 257–259.
19. Anthony Phelan, ed., *The Weimar Dilemma: Intellectuals in the Weimar Republic* (Manchester: University of Manchester Press, 1985), especially Keith Bullivant, "The Conservative Revolution," pp. 47–70. See also Jeffrey Herf, *Reactionary Modernism: Technology, Culture and Politics in Weimar and the Third Reich* (Cambridge: Cambridge University Press, 1984), p. 109. Elias, *The Germans*, p. 212.
20. Watson, *Modern Mind/Terrible Beauty*, p. 300.
21. Bernd Widdig, *Culture and Inflation in West Germany* (Berkeley and London: University of California Press, 2001), p. 140.
22. Ernst Reinhard Piper, *Alfred Rosenberg: Hitlers Chefideologe* (Munich: Karl Blessing, 2005), pp. 179ff. for the *Mythus*.
23. Ibid., pp. 212–231. See also Cecil Robert, *The Myth of the Master Race: Alfred Rosenberg and Nazi Ideology* (London: Batsford, 1972).
24. Julien Benda, *The Treason of the Learned*, trans. Richard Aldington (Boston: Beacon, 1955), p. xxi.
25. Ibid., pp. 13–14.
26. Ibid., p. 18.
27. Ibid., pp. 30–32.
28. Ibid., p. 41.
29. Ibid., p. 42.
30. Ibid., pp. 55–59.
31. Ibid., p. 86.
32. Ibid., p. 104.
33. Ibid., p. 116.
34. Ibid., pp. 117–120.
35. Ibid., p. 141,
36. Ibid., pp. 145–147.
37. For contemporary reactions, see Robert J. Niess, *Julien Benda* (Ann Arbor: University of Michigan Press, 1956), pp. 168ff.

Chapter 34: Nazi Aesthetics: The "Brown Shift"

1. Frederic Spotts, *Hitler and the Power of Aesthetics* (London: Hutchinson, 2002), pp. 11–15.
2. Ibid., pp. 152 and 156.
3. For Ernst Barlach's prolonged fight, see Peter Paret, *An Artist against the Third Reich: Ernst Barlach, 1933–1938* (Cambridge: Cambridge University Press, 2003), pp. 77–108 and 110ff.
4. For his ideas about transforming Linz by art and architecture, see Hanns Christian Löhr, *Das*

braune Haus der Kunst: Hitler und der "Sonderauftrag Linz"; Visionen, Verbrechen, Verluste (Berlin: Akademie-Verlag, 2005), pp. 1–18.

5. Peter Adam, *The Arts of the Third Reich* (London: Thames & Hudson, 1992), pp. 129ff., "The Visualisation of National Socialist Ideology." See also Berthold Hinz, *Art in the Third Reich*, trans. by Robert and Rita Kimber (Oxford: Basil Blackwell, 1980).
6. Paret, *An Artist against the Third Reich*, pp. 109–138, specifically discusses "un-German art."
7. After it closed in Munich, the degenerate art show traveled to Berlin and a number of other German cities. That exhibition was a one-off but the House of German Art show became an annual affair—until 1945. Rudolf Herz, *Hoffmann & Hitler: Fotographie als Medium des Führer-Mythos* (Munich: Klinkhardt & Biermann, 1994), pp. 170ff. and 260ff. An excellent study of how the Führer was presented visually.
8. Victor Klemperer, *The Language of the Third Reich: A Philologist's Notebook*, trans. Martin Brady (New York: Continuum, 2000/2006), p. 63.
9. Ibid., p. 72.
10. In architecture, the Greek ideal still ruled. See Alex Scobie, *Hitler's State Architecture: The Impact of Classical Antiquity* (University Park and London: Pennsylvania State University Press for the College Art Association, 1990), pp. 1ff. for Hitler and classical antiquity, 93ff. for Speer's theory of "ruin value."
11. Jay Baird, *To Die for Germany: Heroes in the Nazi Pantheon* (Bloomington and Indianapolis: University of Indiana Press, 1992), p. 161.
12. Oron J. Hale, *The Captive Press in the Third Reich* (Princeton, N.J.: Princeton University Press, 1964), pp. 67ff. and 76–93 for control of the press.
13. Baird, *To Die for Germany*, p. 132.
14. Ibid., p. 133.
15. Ibid., p. 137.
16. See also Jay W. Baird, *Hitler's War Poets: Literature and Politics in the Third Reich* (Cambridge: Cambridge University Press, 2008).
17. Baird, *To Die for Germany*, p. 154.
18. Ibid., p. 157.
19. Ibid., p. 167.
20. Mary-Elizabeth O'Brien, *Nazi Cinema as Enchantment: The Politics of Entertainment in the Third Reich* (Rochester, N.Y.: Camden House, 2005), pp. 118ff. and 160ff. In regard to radio, in the winter of 1936 the German broadcasting authority announced that the main feature of future programming was to "create joy and solidify the community." Part of this plan included broadcasts on German peasantry "along with agricultural news," under the general title "Peasantry and Landscape."
21. Antje Ascheid, *Hitler's Heroines: Stardom and Womanhood in Nazi Cinema* (Philadelphia: Temple University Press, 2003), chapters 2, 3, and 4.
22. Baird, *To Die for Germany*, p. 200.
23. Ibid., pp. 186–192.
24. Karl-Heinz Schoeps, *Literature and Film in the Third Reich* (Rochester, N.Y.: Camden House, 2004). Very useful references.
25. Baird, *To Die for Germany*, p. 197.
26. For Nazi influence on film outside Germany, see Roel Vande Winkel and David Welch, eds., *Cinema and the Swastika: The International Expansion of Third Reich Cinema* (Basingstoke: Palgrave Macmillan, 2007), which traces Nazi influence as far afield as Brazil, Croatia, Greece, Norway, and the United States, pp. 306ff. For more discussion of German-American film relations in the 1933–1940 period, see Sabina Hake, *Popular Cinema of the Third Reich* (Austin: University of Texas Press, 2001), pp. 128–148.
27. Erik Levi, *Music in the Third Reich* (London: Macmillan, 1994), p. 71.
28. Michael H. Kater, *The Twisted Muse: Musicians and Their Music in the Third Reich* (New York and Oxford: Oxford University Press, 1997), pp. 14–21.
29. Levi, *Music in the Third Reich*, p. 40.
30. There were in addition several works of "scholarship" aiming to show that German music had a racial element, that it was Nordic in origin, the Nordic races alone being capable of the heroic virtues as represented, for instance, in the music of Beethoven. One monograph, published by Karl Blessinger in 1939, claimed that German music had declined in three stages—

via Mendelssohn, Meyerbeer, and Mahler. No prizes for guessing what they had in common. Levi, *Music in the Third Reich*, pp. 53–56.

31. Levi, *Music in the Third Reich*, p. 70.
32. Michael H. Kater, *Composers of the Nazi Era: Eight Portraits* (New York and Oxford: Oxford University Press, 2000), pp. 197–198.
33. For Hindemith, see chapter 2 of Kater, *Composers of the Nazi Era*. See p. 111 for Carl Orff and p. 144 for Hans Pfitzner.
34. Kater, *The Twisted Muse*, pp. 22–39. Levi, *Music in the Third Reich*, p. 118.
35. Levi, *Music in the Third Reich*, p. 179.
36. Ibid., p. 181.
37. Kater, *The Twisted Muse*, pp. 41–42.
38. Brigitte Hamann, *Winifred Wagner: A Life at the Heart of Hitler's Bayreuth*, trans. Alan Bance (London: Granta, 2005).
39. Kater, *The Twisted Muse*, pp. 188ff. And for Furtwängler's fight, see Fred K. Prieberg, *Trial of Strength: Wilhelm Furtwängler and the Third Reich*, trans. Christopher Dolan (London: Quartet, 1991), chapters 2 and 3.
40. Levi, *Music in the Third Reich*, p. 203. See also Misha Aster, *Das "Reichsorchester": Die Berliner Philharmoniker und der Nationalsozialismus* (Munich: Siedler, 2007).
41. Glen W. Gadberry, ed., *Theatre in the Third Reich: The Pre-war Years* (New York: Greenwood, 1995), p. 2.
42. Ibid., pp. 6–9.
43. Ibid., p. 124.
44. Ibid., p. 103.
45. Ibid., p. 115.
46. Ibid., p. 81.
47. The Nazis also had something to say about women's fashion. See Irene Guenther, *Nazi Chic?: Fashioning Women in the Third Reich* (Oxford: Berg, 2004), esp. chap. 5 on "purifying" the German clothes industry.
48. Armin Strohmeyr, *Verlorene Generation: Dreissig vergessene Dichterinnen und Dichter des "anderen Deutschland"* (Zurich: Atrium, 2008), for the "lost generation" of satirists, songwriters, socialists, and historians.

Chapter 35: Scholarship in the Third Reich: "No Such Thing as Objectivity"

1. Steven P. Remy, *The Heidelberg Myth: The Nazification and De-Nazification of a German University* (Cambridge, Mass.: Harvard University Press, 2002), p. 16.
2. Ibid., p. 22.
3. Ibid., p. 24.
4. Ibid., p. 43.
5. Ibid., p. 26.
6. Claudia Koonz, *The Nazi Conscience* (Cambridge, Mass.: Belknap Press of Harvard University Press, 2003), p. 196.
7. Remy, *Heidelberg Myth*, p. 33. Elias, *The Germans*, p. 383.
8. Remy, *Heidelberg Myth*, p. 50.
9. Philipp Lenard, "The Limits of Science," in George L. Mosse, ed., *Nazi Culture: Intellectual, Cultural and Social Life in the Third Reich*, trans. by Salvator Attanasia, et al. (Madison: University of Wisconsin Press, 1966), pp. 201–205.
10. Remy, *Heidelberg Myth*, p. 56.
11. Ibid., p. 60.
12. Koonz, *Nazi Conscience*, p. 205.
13. Remy, *Heidelberg Myth*, p. 84.
14. James R. Dow and Hannjost Lixfeld, eds., *The Nazification of an Academic Discipline: Folklore in the Third Reich* (Bloomington: Indiana University Press, 1994).
15. George S. Williams, *The Longing for Myth in Germany: Religious and Aesthetic Culture from Romanticism to Nietzsche* (Chicago: University of Chicago Press, 2004), p. 1.
16. Dow and Lixfeld, eds., *Nazification*, p. 21.

17. Fritz Joachim Raddatz, *Gottfried Benn, Leben, niederer Wahn: Eine Biographie* (Berlin: Propyläen, 2001), pp. 48ff.
18. Dow and Lixfeld, eds., *Nazification*, pp. 42–46.
19. Ibid., pp. 57–59.
20. Ibid., p. 80.
21. Ibid., pp. 189ff and 198.
22. Peter Padfield, *Himmler: Reichsführer-SS* (London: Macmillan, 1990), pp. 166ff.
23. Christopher Hale, *Himmler's Crusade: The True Story of the 1938 Nazi Expedition into Tibet* (London: Bantam Press, 2003), pp. 207ff.
24. Ibid., p. 211.
25. Ibid., p. 233.
26. Cornwell, *Hitler's Scientists*, p. 25.
27. Ibid., pp. 30–32.
28. Ibid., pp. 130–131.
29. Albert Einstein, *The Born-Einstein Letters: Friendship, Politics, and Physics in Uncertain Times; Correspondence between Albert Einstein and Max and Hedwig Born from 1916 to 1955 with Commentaries by Max Born*, trans. Irene Born (Basingstoke: Macmillan, 2005), pp. 113ff.
30. Roger Highfield and Paul Carter, *The Private Lives of Albert Einstein* (London: Faber & Faber, 1993), pp. 240–241, for lesser-known details of Einstein's more personal difficulties.
31. Cornwell, *Hitler's Scientists*, p. 130. For an unusual view of Einstein, see Dennis P. Ryan, ed., *Einstein and the Humanities* (New York and London: Greenwood Press, 1987), with chapters on the moral implications of relativity, poetic responses to relativity, and relativity and psychology.
32. Cornwell, *Hitler's Scientists*, p. 140.
33. Michael H. Kater, *Doctors under Hitler* (Chapel Hill: Universty of North Carolina Press, 1989), pp. 19ff. and 63ff.
34. Ibid., pp. 177ff. See also Charles McClelland, "Modern German Doctors: A Failure of Professionalisation?" in Berg and Cocks, eds., *Medicine and Modernity*, pp. 81–97.
35. Geoffrey Cocks, *Psychotherapy in the Third Reich: The Göring Institute* (New York and Oxford: Oxford University Press, 1985), pp. 53–60. See also Laurence A. Rickels, *Nazi Psychoanalysis* (Minneapolis and London: University of Minnesota Press, 2002).
36. Cocks, *Psychotherapy*, p. 87.
37. Jarrell C. Jackman and Carla M. Borden, *The Muses Flee Hitler: Cultural Transfer and Adaptation, 1930–1945* (Washington, D.C.: Smithsonian Institution Press, 1983), pp. 205ff.
38. Ibid., p. 25.
39. Ibid.
40. David Simms, "The Führer Factor in German Equations," review of Sanford L. Segal, *Mathematicians under the Nazis* (Princeton, N.J.: Princeton University Press, 2003), *Times Higher Education Supplement*, September 17, 2004, p. 28.
41. Sanford L. Segal reports in *Mathematicians under the Nazis* that no mathematician played a part in the Resistance, that the National Socialists in fact had little interest in mathematics, that Otto Blumenthal remained as editor of *Mathematische Annalen* up to 1939 but died in Theresienstadt in 1944. Heinrich Behnke, who had a Jewish son to protect, managed to found a "school of several complex variables," which was to become "the spur for a postwar revival of German mathematics under Friedrich Hirzenbruch." See also Jackman and Borden, eds., *Muses Flee Hitler*, pp. 221ff.
42. Cornwell, *Hitler's Scientists*, p. 168.
43. Ibid., p. 170.
44. Werner Heisenberg, *Physics and Beyond: Encounters and Conversations*, trans. Arnold J. Pomeranz (New York: Harper & Row, 1971), p. 166.
45. Otto Hahn, *A Scientific Autobiography*, trans. and ed. Willy Ley (London: McGibbon & Kee, 1967), p. 85.
46. Otto Hahn, *My Life*, trans. Ernst Kaiser and Eithne Wilkins (London: Macdonald, 1970), p. 149.
47. Otto Frisch, *What Little I Remember* (Cambridge: Cambridge University Press, 1979), pp. 120ff.
48. Cornwell, *Hitler's Scientists*, pp. 208–210.

49. Watson, *Modern Mind/Terrible Beauty*, pp. 392–393.
50. Ibid.
51. Rudolf Peierls, *Atomic Histories* (Woodbury, N.Y.: American Institute of Physics, 1997), pp. 187–194.
52. Koonz, *Nazi Conscience*, p. 58.
53. Ibid. See also Clemens Kauffmann, *Leo Strauss zur Einführung* (Hamburg: Junius, 1997).
54. Götz Aly and Susanne Heim, *Architects of Annihilation: Auschwitz and the Logic of Destruction* (London: Weidenfeld & Nicolson, 1991), p. 58.
55. Ibid., p. 54.
56. Ibid., p. 61.
57. Ibid., p. 86.
58. Ibid., p. 95.
59. Ibid., p. 179.
60. Ibid., p. 166.
61. Ulf Schmidt, *Karl Brandt: The Nazi Doctor; Medicine and Power in the Third Reich* (London: Hambledon Continuum, 2007), pp. 125ff.

Chapter 36: The Twilight of the Theologians

1. I have used Brian Moynahan, *The Faith* (London: Aurum, 2002), p. 675.
2. Ibid.
3. F. X. J. Homer, "The Führer's Faith: Hitler's Sacred Cosmos," in F. X. J. Homer and Larry D. Wilcox, eds., *Germany and Europe in the Era of Two World Wars: Essays in Honour of Oron James Hale* (Charlottesville: University Press of Virginia, 1986), pp. 61–78.
4. Alistair McGrath, *The Making of Modern German Christology: From the Enlightenment to Pannenberg* (Oxford: Blackwell, 1986), p. 5
5. Wilhelm Pauck, *Harnack and Troeltsch: Two Historical Theologians* (New York: Oxford University Press, 1968), p. 117, for Troeltsch's address at Harnack's funeral.
6. McGrath, *Making of Modern German Christology*, p. 61.
7. Franz L. Neumann, et al., *The Cultural Migration: The European Scholar in America* (Philadelphia: University of Pennsylvania Press, 1953), p. 140.
8. Johannes Hemleben, *Rudolf Steiner und Ernst Haeckel* (Stuttgart: Verlag Freies Geistesleben, 1965), pp. 38ff. Geoffrey Ahern, *Sun at Midnight: The Rudolf Steiner Movement and the Western Esoteric Tradition* (Wellingborough: Aquarian Press, 1984), p. 87ff.
9. Ahern, *Sun at Midnight*, p. 64.
10. See the account of a Vienna congress in 1922 in Guenther Wachsmuth, *The Life and Works of Rudolf Steiner: From the Turn of the Century to His Death* (New York: Whittier, 1955), p. 445.
11. Bruce L. McCormack, *Karl Barth's Critically Dialectical Theology: Its Genesis and Development, 1909–1936* (Oxford: Clarendon Press, 1995), pp. 38ff., for a description of "The theological situation at the turn of the century."
12. Martin Rumscheidt, *Revelation and Theology: An Analysis of the Barth-Harnack Correspondence of 1923* (Cambridge: Cambridge University Press, 1972), pp. 31–34 and 75–78.
13. Eberhard Busch, *Karl Barth: His Life from Letters and Autobiographical Texts*, trans. by John Bowden (London: SCM Press, 1976), pp. 38ff.
14. McGrath, *Making of Modern German Christology*, p. 94.
15. Busch, *Karl Barth*, pp. 92f. and 117f.
16. Zdravko Kujundzija, *Boston Collaborative Encyclopaedia of Western Theology*, entry on Barth, p. 16. http://people.bu.edu/wwildman/bce/
17. Busch, *Karl Barth*, pp. 120f. McCormack, *Karl Barth's Critically Dialectical Theology*, pp. 209ff.
18. McCormack, *Karl Barth's Critically Dialectical Theology*, p. 371. Kimlyn J. Bender, *Karl Barth's Christological Ecclesiology* (Aldershot: Ashgate, 2005), pp. 95f.
19. Busch, *Karl Barth*, p. 245.
20. Kujundzija, *Boston Collaborative Encyclopaedia*, p. 17.
21. McCormack, *Karl Barth's Critically Dialectical Theology*, p. 449.
22. Martin Evang, *Rudolf Bultmann in seiner Frühzeit* (Tübingen: Mohr [Paul Siebeck], 1988),

pp. 211f. Bernd Jaspert, ed., *Karl Barth–Rudolf Bultmann Letters, 1922–1966*, trans. Geoffrey W. Bromley (Edinburgh: T&T Clark, 1982).

23. John MacQuarrie, *The Scope of Demythologising: Bultmann and His Critics* (London: SCM Press, 1960), pp. 65ff. and 151ff.
24. David L. Edwards, "Rudolf Bultmann: Scholar of Faith," *Christian Century*, September 1–8, 1976, pp. 728–730. McGrath, *Making of Modern German Christology*, p. 135.
25. MacQuarrie, *Scope of Demythologising*, pp. 186ff.
26. Busch, *Karl Barth*, p. 141.
27. For his links to Erich Fromm, Sidney Hook and others, and his comparison of psychology and sociology to the "spiritual vacuum," see Raymond F. Bulman, *A Blueprint for Humanity: Paul Tillich's Theology of Culture* (Lewisburg: Bucknell University Press, 1981), in particular pp. 128ff.
28. Sabine Leibholz-Bonhoeffer, *The Bonhoeffers: Portrait of a Family* (London: Sidgwick & Jackson, 1971), p. 17.
29. For the debt to Barth, see Ronald Gregor Smith, *World Comes of Age: A Symposium on Dietrich Bonhoeffer* (London: Collins, 1967), pp. 93ff.
30. Eberhard Bethge, *Dietrich Bonhoeffer: Theologe. Christ. Zeitgenosse* (Munich: Kaiser, 1967), pp. 183f.
31. Ibid., p. 1036.
32. Ibid., pp. 803–811. See also Eberhard Bethge, *Bonhoeffer: Exile and Martyr* (London: Collins, 1975).
33. James Brabazon, *Albert Schweitzer: A Biography* (Syracuse, N.Y.: Syracuse University Press, 2000), pp. 64ff.
34. Ibid., pp. 110ff.
35. Ibid., pp. 443ff.
36. Maurice Friedman, *Encounter on the Narrow Ridge: A Life of Martin Buber* (New York: Paragon House, 1991).
37. Moynahan, *Faith*, p. 678.
38. Ernst Christian Helmreich, *The German Churches under Hitler: Background, Struggle, and Epilogue* (Detroit: Wayne State University Press, 1979), p. 123. J. S. Conway, *The Nazi Persecution of the Churches, 1933–1945* (London: Weidenfeld & Nicolson, 1968), p. 2.
39. Richard Steigmann-Gall, *The Holy Reich* (Cambridge: Cambridge University Press, 2003), p. 1.
40. Ibid., p. 6.
41. Ibid., p. 37.
42. Ibid., p. 42.
43. Ibid., p. 234.
44. Ibid., p. 111.
45. Robert P. Ericksen, *Theologians under Hitler* (New Haven, Conn., and London: Yale University Press, 1985), p. 52.
46. Ibid., page 56.
47. Paul Althaus, *Die Ethik Martin Luthers* (Gütersloh: Gütersloher Verlagshaus G. Mohn, 1965).
48. Ericksen, *Theologians under Hitler*, p. 103.
49. Emanuel Hirsch, *Das Wesen des reformatorischen Christentums* (Berlin: de Gruyter, 1963), pp. 105ff. Ericksen, *Theologians*, pp. 155–165.
50. Moynahan, *Faith*, p. 680.
51. James Bentley, *Martin Niemöller* (Oxford: Oxford University Press, 1984), pp. 81f and 143ff.

Chapter 37: The Fruits, Failures, and Infamy of German Wartime Science

1. Remy, *Heidelberg Myth*, pp. 85–86.
2. Ibid., pp. 95–96.
3. Alfred Weber, for example, published a work of 423 pages in 1935, *Kulturgeschichte als Kultursoziologie* (Munich: Piper).
4. Fritz Ernst wrote an article about Karl the Bold of Burgundy (1433–77) containing these

lines: "What he lacked was balance . . . He allowed a glut of hate and ambition to consume him, without drawing lasting strength from it . . . So never did he have an inner freedom . . . despite elementary military mistakes he held himself to be a great field commander." Remy, *Heidelberg Myth*, p. 113.

5. Remy, *Heidelberg Myth*, pp. 222–223.
6. Ibid., p. 231.
7. Cassidy, *Uncertainty*, p. 420.
8. Ibid., p. 435. See also Rainer Karlsch, *Hitlers Bombe: Die geheime Geschichte der deutschen Kernwaffenversuche* (Munich: Deutsche Verlags-Anstalt, 2005), p. 72 for Houtermans.
9. Paul Lawrence Rose, *Heisenberg and the Nazi Atomic Bomb Project: A Study in German Culture* (Berkeley and London: University of California Press, 1998).
10. Karlsch, *Hitlers Bombe*, pp. 54f. and 107f.
11. Eduard Schönleben, *Fritz Todt, der Mensch, der Ingenieur, der Nationalsozialist: Ein Bericht über Leben und Werk* (Oldenburg: G. Stalling, 1943), pp. 108ff.
12. Cornwell, *Hitler's Scientists*, p. 317.
13. Cassidy, *Uncertainty*, pp. 397ff., for a chapter on "German Physics." See also Karlsch, *Hitlers Bombe*, pp. 266–270. Karlsch claims the Germans actually built an atomic reactor at Gottow, a village outside Berlin and tested a device on the island of Rügens in March 1945.
14. Although Speer had canceled the German atomic bomb project, the Allies didn't know that, and, at the time of the D-Day landings in 1944, General Leslie Groves, commanding officer of the Manhattan Project, worried that the Germans "would prepare an impenetrable radioactive defence against our landing troops." That didn't materialize and the special group Alsos under the émigré Dutch scientist Samuel Goudsmit, which had been set up to follow the advance troops and investigate German scientific achievements, soon discovered that their bomb research was well behind that of the Allies, even though the institute in Berlin had transferred to a safer location, at Haigerloch in the Swabian Alps. Alsos also discovered that three of the German atom physicists, Walter Gerlach, Kurt Diebner, and Karl Wirtz, had transferred some of their uranium and heavy water to Haigerloch but had left the remainder in the German capital where, to the Allies's dismay and Stalin's pleasure, it was discovered by the NKVD (Peoples' Commissariat for Internal Affairs) on April 24, 1945. Cornwell, *Hitler's Scientists*, p. 334.
15. Cornwell, *Hitler's Scientists*, p. 253.
16. Ibid., p. 289.
17. Erik Bergaust, *Satellite* (London: Lutterworth Press, 1957), p. 28, for more of Oberth's ideas.
18. Cornwell, *Hitler's Scientists*, p. 256.
19. Steven Rose, ed., *C. B. W. Chemical and Biological Warfare: London Conference on C. B. W.* (London and Toronto: Harrap, 1968), passim.
20. I have used Diarmuid Jeffreys, *Hell's Cartel: I.G. Farben and the Making of Hitler's War Machine* (London: Bloomsbury, 2008), chaps. 10 and 12. See also Stephan H. Lindner, *Inside IG Farben: Hoechst during the Third Reich*, trans. Helen Schoop (Cambridge: Cambridge University Press, 2008), chap. 4.4, pp. 307ff. for Farben's drug experiments on human subjects.
21. Jeffreys, *Hell's Cartel*, pp. 321ff.
22. Many other inmates had their eyes injected with dyes or were shot with poisoned bullets to see how quickly the poisons worked. Ute Deichman, *Biologists under Hitler*, trans. Thomas Dunlap (Cambridge, Mass.: Harvard University Press, 1996).
23. Franz M. Wuketits, *Konrad Lorenz: Leben und Werk eines grossen Naturforschers* (Munich: Piper, 1990), pp. 108ff.
24. Alex Nisbett, in his 1976 biography of Lorenz, gives a version of this episode that is much kinder to his subject. Alex Nisbett, *Konrad Lorenz* (London: Dent & Sons, 1976), pp. 78–79. In 1988 Lorenz published *The Waning of Humaneness*, trans. Robert Warren Kickert (London: Unwin Hyman), without any apparent trace of irony. It had originally appeared in German in 1983.
25. In a textbook, *Human Heredity*, originally published in Germany in 1927, Fischer wrote an entire section on "Racial Differences in Mankind," and Fritz Lenz wrote on "Psychological Differences between the Leading Races of Mankind."
26. For an excellent overview of the medical, legal, and moral issues swirling around sterlization and euthanasia in the Third Reich, and many useful references, see Gisela Bock, "Sterilisa-

tion and 'Medical' Massacres in National Socialist Germany: Ethics, Politics and the Law," in Berg and Cocks, eds., *Medicine and Modernity*, pp. 149–172.

27. Dupuy, *Genius for War*, p. 253.

Chapter 38: Exile, and the Road into the Open

1. On figures, see, for example, Donald Peterson Kent, *The Refugee Intellectual* (New York: Columbia University Press, 1953), pp. 11–16; Jean-Michel Palmier, *Weimar in Exile: The Anti-Fascist Emigration in Europe and America*, trans. David Fernbach (London: Verso, 2006), pp. 11–15. Volkmar von Zühlsdorff, *Hitler's Exiles: The German Cultural Resistance in America and Europe*, trans. Martin H. Bott (London: Continuum, 2004).
2. There is now a sizable literature on German refugees, with a good deal of overlap between titles. Chapters 39 and 40 of this book are devoted to their longer-term cultural and intellectual impact in the United States and the United Kingdom, and my chief sources are given there. Also recommended are Steffen Pross, *In London treffen wir uns wieder* (Berlin: Eichborn, 2000), the best account in German, useful for that reason alone. Charmian Brinson, et al., eds., *"England? Aber wo liegt es?": Deutsche und österreichische Emigranten in Grossbritannien* (Munich: Iudicium, 1996). Reinhold Brinkmann and Christoph Wolff, eds., *Driven into Paradise: The Musical Migration from Nazi Germany to the United States* (Berkeley and London: University of California Press, 1999), contains a good chapter on Korngold, pp. 223f., and on Black Mountain College, p. 279. Tom Ambrose, *Hitler's Loss: What Britain and America Gained from Europe's Cultural Exiles* (London: Peter Owen in association with the European Jewish Publication Society, 2001). The Council for Assisting Refugee Academics has recently been commemorated in Jeremy Seabrook's *The Refugee and the Fortress: Britain and the Flight from Tyranny* (London: Palgrave Macmillan, 2008).
3. Varian Fry, *Surrender on Demand* (Boulder, Colo.: Johnson Books, 1997); and Andy Marino, *American Pimpernel: The Man Who Saved the Artists on Hitler's Death List* (London: Hutchinson, 1999).
4. Rosemary Sullivan, *Villa Air-Bel: The Second World War, Escape, and a House in France* (London: John Murray, 2006), pp. 83ff. and 251ff.
5. Watson, *Modern Mind/Terrible Beauty*, p. 356.
6. Ibid., p. 357. The historian Lawrence Wechsler has gone so far as to prepare an "alternative" Hollywood map, displaying the addresses of intellectuals and scholars, as opposed to the more conventional map showing the homes of movie stars.
7. Colin Loader, *The Intellectual Development of Karl Mannheim: Culture, Politics, and Planning* (Cambridge: Cambridge University Press, 1985), p. 19. Mannheim also became the editor of the International Library of Sociology and Social Reconstruction, a large series of books published by George Routledge whose authors included Harold Lasswell, professor of political science at Chicago, E. F. Schumacher, Raymond Firth, Erich Fromm, and Edward Shils.
8. Loader, *Intellectual Development*, p. 162.
9. Thomas K. McCraw, *Prophet of Innovation: Joseph Schumpeter and Creative Destruction* (Cambridge, Mass.: Belknap Press of Harvard University Press, 2007), p. 248.
10. Yuichi Shionoya, *Schumpeter and the Idea of Social Science: A Metatheoretical Study* (Cambridge: Cambridge University Press, 1997), p. 124.
11. McCraw, *Prophet of Innovation*, p. 255.
12. Stephen F. Frowen, ed., *Hayek: Economist and Social Philosopher; A Critical Retrospect*, Basingstoke: Macmillan, 1997, pp. 63ff and 237ff.
13. Andrew Gamble, *Hayek: The Iron Cage of Liberty* (Cambridge: Polity Press, 1996), pp. 59ff.
14. Malachi Haim Hacohen, *Karl Popper, the Formative Years, 1902–1945: Politics and Philosophy in Interwar Vienna* (Cambridge: Cambridge University Press, 2000), pp. 186ff.
15. Anthony O'Hear, ed., *Karl Popper: Philosophy and Problems* (Cambridge: Cambridge University Press, 1995), pp. 45ff and 75ff.
16. Hacohen, *Karl Popper*, pp. 383ff.
17. O'Hear, ed., *Karl Popper*, p. 225.
18. Watson, *Modern Mind/Terrible Beauty*, pp. 374–375.
19. James, *Cultural Amnesia*, pp. 48ff.
20. Martin Mauthner, *German Writers in French Exile* (London and Portland, Ore.: Valentine

Mitchell, in association with the European Jewish Publication Society, 2007), p. 58. For Clive James's stylish dismissal of Walter Benjamin, see his *Cultural Amnesia*, pp. 47–55.

21. Mauthner, *German Writers*, p. 60.

Chapter 39: The "Fourth Reich": The Effect of German Thought on America

1. Allan Bloom, *The Closing of the American Mind* (London: Penguin, 1987), pp. 148–149.
2. Ibid., p. 152.
3. Franz L. Neumann, et al., *The Cultural Migration: The European Scholar in America* (Philadelphia: Pennsylvania University Press, 1953), pp. 34–35.
4. Some of the general references on prominent German refugees/exiles in the cultural/scientific spheres were given in Chapter 38, note 2. To them may be added Jean Michel Palmier, *Weimar in Exile: The Antifascist Emigration in Europe and America*, trans. David Fernbach (London: Verso, 2006), a very solid, systematic study of more than 600 pages, with sections on the press, publishing, and literature, on the theater, academics, and on Hollywood at war. Erhard Bahr and Carolyn See, *Literary Exiles & Refugees in Los Angeles: Papers Presented at a Clark Library Seminar, 14 April 1984* (Los Angeles: William Andrews Clark Memorial Library, 1988), has two sections, one on Weimar exiles and one on English expatriates. Hartmut Lehmann and James J. Sheehan, eds., *An Interrupted Past: German-speaking Refugee Historians in the United States after 1933* (Washington, D.C., and Cambridge: German Historical Institute and Cambridge University Press, 1991), includes a chapter on German historians in the OSS (Office of Strategic Services), and chapters on Hajo Holborn, Ernst Kantorowicz, and Theodor Mommsen. Mitchell G. Ash and Alfons Söllner, eds., *Forced Migration and Scientific Change: Émigré German-speaking Scientists and Scholars after 1933* (Washington, D.C., and Cambridge: German Historical Institute and Cambridge University Press, 1996), is admirably detailed. Joachim Radkau, *Die deutsche Emigration in den USA* (Düsseldorf: Bertelsmann Universitätsverlag, 1971). Helge Pross, *Die deutsche akademische Emigration nach den Vereinigten Staaten 1933–1941* (Berlin: Dunker & Humblot, 1955).
5. Anthony Heilbut, *Exiled in Paradise: German Refugee Artists and Intellectuals in America, from the 1930s to the Present* (Berkeley: University of California Press, 1983 and 1997, with a new postscript), pp. 44, 46, 51, and 65.
6. Ibid., p. 77.
7. Ibid., p. 130.
8. Ibid.
9. Lewis A. Coser, *Refugee Scholars in America: Their Impact and Their Experiences* (New Haven, Conn., and London: Yale Univesity Press, 1984), p. 35.
10. Ibid., p. 47.
11. Lawrence J. Friedman, *Identity's Architect: A Biography of Erik H. Erikson* (London: Free Association Books, 1999), p. 157.
12. Ibid., pp. 149ff.
13. Ibid., p. 156.
14. Nina Sutton, *The Other Side of Madness*, trans. David Sharp and the author (London: Duckworh, 1995), pp. 120ff. See also Bruno Bettelheim, *Recollections and Reflections* (London: Thames & Hudson, 1990).
15. Sutton, *Other Side of Madness*, p. 269.
16. Heilbut, *Exiled in Paradise*, p. 209.
17. Sutton, *Other Side of Madness*, pp. 268f.
18. Coser, *Refugee Scholars*, p. 70.
19. Lawrence Wilde, *Erich Fromm and the Quest for Solidarity* (New York: Palgrave Macmillan, 2004), pp. 19–36.
20. Coser, *Refugee Scholars*, p. 72.
21. Daniel Burston, *The Legacy of Erich Fromm* (Cambridge, Mass., and London: Harvard University Press, 1991), pp. 133ff. for the "pathology of normalcy."
22. Paul Robinson, *The Freudian Left: Wilhelm Reich, Geza Roheim, Herbert Marcuse* (Ithaca, N.Y., and London: Cornell University Press, 1990), chap. 1, on "Freudian radicalism."

23. David Seelow, *Radical Modernism and Sexuality: Freud, Reich, D. H. Lawrence and Beyond* (New York and Basingstoke: Palgrave Macmillan, 2005), pp. 47ff.
24. Ash and Söllner, eds., *Forced Migration*, p. 269.
25. For Hans Jonas, see Hans Jonas, *Technik, Medizin und Ethik: zur Praxis des Prinzips Verantwortung* (Frankfurt am Main: Insel, 1987), pp. 90f. for the role of research in modern society; and David J. Levy, *Hans Jonas: The Integrity of Thinking* (Columbia, Mo., and London: University of Missouri Press, 2002), p. 77 for responsibility in a technological age. For Löwith, see Karl Löwith, *My Life in Germany before and after 1933: A Report* (London: Athlone Press, 1994), pp. 111–119 for his time in Japan.
26. Andrew Jamieson and Ron Eyerman, *Seeds of the Sixties* (Berkeley and London: University of California Press, 1994), p. 47.
27. Ibid., p. 50.
28. Heilbut, *Exiled in Paradise*, pp. 403–404.
29. Ibid., p. 412.
30. Elisabeth Young-Bruehl, *Why Arendt Matters* (New Haven, Conn., and London: Yale University Press, 2006), p. 73 for the relevance of Arendt after 9/11.
31. Richard Wolin and John Abromeit, *Heideggerian Marxism/Herbert Marcuse* (Lincoln: University of Nebraska Press, 2005), pp. 176ff.
32. Timothy J. Lukes, *The Flight into Inwardness: An Exposition and Critique of Herbert Marcuse's Theory of Liberative Aesthetics* (Selinsgrove, Pa.: Susquehanna University Press, 1985), p. 46.
33. Jamieson and Eyerman, *Seeds of the Sixties*, pp. 124–125. See also Robert Pippin, et al., eds., *Marcuse: Critical Theory & the Promise of Utopia* (London: Macmillan Education, 1988), pp. 143ff. and 169ff.
34. Anne Norton, *Leo Strauss and the Politics of American Empire* (New Haven, Conn., and London: Yale University Press, 2004).
35. Heinrich Meier, *Carl Schmitt & Leo Strauss: The Hidden Dialogue*, trans. J. Harvey Lomax (Chicago and London: University of Chicago Press, 1995).
36. Daniel Tanguay, *Leo Strauss: An Intellectual Biography* (New Haven, Conn., and London: Yale University Press, 2007), pp. 99ff. See also Mark Blitz, *Leo Strauss, the Straussians and the American Regime* (New York: Rowman & Littlefield, 1999).
37. Mark Lilla, "The Closing of the Straussian Mind," *New York Review of Books*, November 4, 2004, pp. 55–59.
38. Coser, *Refugee Scholars*, p. 205.
39. Jan-Werner Müller, *A Dangerous Mind: Carl Schmitt in Post-War European Thought* (New Haven, Conn., and London: Yale University Press, 2003), pp. 194–206.
40. His later career was, inevitably perhaps, less remarkable, a raft of memoirs and consultancies, not all of them successful: he was, for instance, a director of Hollinger International, the chief executive of which, Conrad Black, was jailed for six years for fraud in 2007. It was widely perceived that the board on which Kissinger served did not exert sufficient oversight of the company, enabling Black to commit the crimes of which he was convicted.
41. Stefan Müller-Doohm, *Adorno: A Biography*, trans. Rodney Livingstone (Cambridge: Polity Press, 2005), pp. 267–277. Adorno himself "disdained" biography as an intellectual form. See also Coser, *Refugee Scholars*, p. 160.
42. Detlev Clausen, *Theodor W. Adorno: One Last Genius*, trans. Rodney Livingstone (Cambridge, Mass., and London: Belknap Press of Harvard University Press, 2008), p. 222 for discussions with Horkheimer.
43. Ibid., pp. 135–144.
44. Müller-Doohm, *Adorno*, pp. 336ff and 374ff.
45. Coser, *Refugee Scholars*, p. 107.
46. Ibid., p. 114.
47. Paul Lazarsfeld, William H. Sewell, and Harold L. Wilensky, *The Uses of Sociology* (London: Weidenfeld & Nicolson, 1968).
48. Coser, *Refugee Scholars*, p. 164.
49. Ibid., p. 166.
50. Peter Drucker, *Post-capitalist Society* (Oxford: Butterworth-Heinemann, 1993), pp. 17ff.
51. Coser, *Refugee Scholars*, p. 298.
52. Ibid., p. 299.

53. Michael Friedman and Richard Creath, eds., *The Cambridge Companion to Carnap* (Cambridge: Cambridge University Press, 2007), pp. 65–80.
54. A. W. Carus, *Carnap and Twentieth-century Thought: Explication and Enlightenment* (Cambridge: Cambridge University Press, 2007), pp. 139ff.
55. Friedman and Creath, eds., *Cambridge Companion to Carnap*, pp. 176–199.
56. Carus, *Carnap*, pp. 209ff.
57. Coser, *Refugee Scholars*, p. 304.
58. Ash and Söllner, eds., *Forced Migration*, p. 285.
59. Raymond Bulman, *A Blueprint for Humanity: Paul Tillich's Theology of Culture* (Lewisburg, Pa.: Bucknell University Press, 1981), pp. 112ff. for ontological *versus* technological reason.
60. Coser, *Refugee Scholars*, p. 318. See Jürgen Haffer, *Ornithology, Evolution, and Philosophy: The Life and Science of Ernst Mayr 1904–2005* (Berlin: Springer, 2008), for the life and work of Ernst Mayr. Born in Kempten and educated at Greifswald and Berlin, Mayr became an influential biological philosopher, especially on the implications of evolution, and a professor at Harvard. Among his students was Jared Diamond.
61. Ash and Söllner, eds., *Forced Migration*, p. 155.
62. Lehmann and Sheehan, eds., *Interrupted Past*, for all these figures; see also Ash and Söllner, eds., *Forced Migration*, pp. 75 and 87.
63. Hann Schissler, "Explaining History: Hans Rosenberg," pp. 180ff., and Robert E. Lerner, "Ernst Kantorowicz and Theodor E. Mommsen," pp. 188ff., in Ash and Söllner, eds., *Forced Migration*.
64. Coser, *Refugee Scholars*, p. 279.
65. Lehman and Sheehan, eds., *Interrupted Past*, p. 176.
66. For the German historians who worked in the OSS, see Barry M. Katz, "German Historians in the Office of Strategic Services," in Ash and Söllner, eds., *Forced Migration*, pp. 136ff.
67. His students included Leonard Krieger. Gerhard A. Ritter has looked at "German Émigré Historians between Two Worlds: Hajo Holborn, Dietrich Gerhard, Hans Rosenberg," *German Historical Institute Bulletin* 39 (Fall 2006): 23ff. See also Otto P. Pflanze, "The Americanisation of Hajo Holborn," in Ash and Söllern, eds., *Forced Migration*, pp. 170ff.
68. Coser, *Refugee Scholars*, p. 279.
69. Fritz Stern, *Dreams and Delusions* (New Haven, Conn., and London: Yale University Press, 1987), p. 327.
70. Coser, *Refugee Scholars*, p. 269.
71. Ibid., p. 255.
72. Erwin Panofsky, *Meaning in the Visual Arts* (New York: Overlook Press, 1974), p. 332.
73. Michael Ann Holly, *Panofsky and the Foundations of Art History* (Ithaca, N.Y., and London: Cornell University Press, 1984), pp. 21ff.
74. Ibid., pp. 158ff.
75. Hans Hofmann (introduction by Sam Hunter), *Hans Hofmann*, New York: Harry N. Abrams, 1979, p. 10*fn*.
76. Heilbut, *Exiled in Paradise*, p. 137.
77. Ibid., p. 141.
78. Ibid., p. 222.
79. Joseph Horowitz, *Wagner Nights: An American History* (Berkeley: University of California Press, 1994), quoted in Joseph Horowitz, *Artists in Exile: How Refugees from Twentieth-century War and Revolution Transformed the American Performing Arts* (New York: HarperCollins, 2008), p. xvi.
80. C. D. Innes, *Erwin Piscator's Political Theatre: The Development of Modern German Drama* (Cambridge: Cambridge University Press, 1972), p. 69. And see George Buehler, *Berthold Brecht, Erwin Piscator: Ein Vergleich ihrer theoretischen Schriften* (Bonn: Bouvier, 1978), pp. 126–131 for a comparison of Brecht and Piscator.
81. John Fuegi, *Brecht & Co.: Biographie* (Hamburg: Europäische Verlagsanstalt, 1997), pp. 636–643.
82. Heilbut, *Exiled in Paradise*, p. 176. See also Ronald Speirs, ed., *Brecht's Poetry of Political Exile* (Cambridge: Cambridge University Press, 2000).
83. Fuegi, *Brecht & Co.*, pp. 610–611.
84. Heilbut, *Exiled in Paradise*, p. 299.
85. All the quotes are from Heilbut, *Exiled in Paradise*, p. 321.

86. Peter Goodchild, *Edward Teller: The Real Dr. Strangelove* (London: Weidenfeld & Nicolson, 2004), p. 26. The Ian Jacobs quote about the relative merits of German scientists is given in Andrew Roberts, *The Storm of War* (London: Allen Lane, 2009), p. 573.
87. See Edward Teller, *A Twentieth-century Journey in Science and Politics*, with Judith L. Shoolery (Cambridge, Mass.: Perseus Publishing, 2001), pp. 177–178, for the initial falling-out. And see Edward Teller, *Better a Shield than a Sword: Perspectives on Defense and Technology* (New York: Free Press; London: Collier Macmillan, 1987), pp. 115f., for the beginnings of secrecy in physics.
88. Silvan S. Schweber, *In the Shadow of the Bomb: Bethe, Oppenheimer, and the Moral Responsibility of the Scientist* (Princeton, N.J., and Chichester: Princeton University Press, 2000), pp. 107–114. See also Kati Marton, *The Great Escape: Nine Jews Who Fled Hitler and Changed the World* (New York and London: Simon & Schuster, 2007), pp. 184–187.
89. Goodchild, *Edward Teller*. Neumann and Oskar Morgenstern were experts on game theory, considered important for a grounding in strategy. See John von Neumann and Oskar Morgenstern, *Theory of Games and Economic Behaviours* (Princeton, N.J.: Princeton University Press, 1953), pp. 46ff. and 587ff. Neumann also helped develop computers.
90. John W. Dawson, *Logical Dilemmas: The Life and Work of Kurt Gödel* (Wellesley, Mass.: A. K. Peters, 1997), pp. 176–178.
91. Palle Yourgrau, *A World without Time: The Forgotten Legacy of Gödel and Einstein* (New York: Basic Books, 2005), pp. 94–95.
92. Ibid., p. 6.
93. Ibid., p. 115.
94. David Ketter and Herbert Lauer, eds., *Exile, Science, and Bildung* (New York: Palgrave, 2005), pp. 2–3.

Chapter 40: "His Majesty's Most Loyal Enemy Aliens"

1. General references on German exiles of note are given in Chapter 38, note 2, and Chapter 39, note 4. In addition, sources for Great Britain include Gerhard Hirschfeld, ed., *Exile in Great Britain: Refugees from Hitler's Germany* (Leamington Spa: Bergs for the German Historical Institute, London, 1984), which is mainly concerned with the political and industrial effects. William Abbey, et al., eds., *Between Two Languages: German-speaking Exiles in Great Britain, 1933–1945* (Stuttgart: Hans-Dieter Heinz, 1995), includes an interesting section on Germans who tried to warn Britain about the Nazis, and a section on German writers and dramatists writing in English. Panikos Panayi, ed., *Germans in Britain since 1500* (London: Hambledon Press, 1996), gives a longer-term perspective.
2. Daniel Snowman, *The Hitler Emigrés* (London: Chatto & Windus, 2002), pp. 12–13.
3. Rudolf Bing, *5,000 Nights at the Opera* (London: Hamish Hamilton, 1972), p. 86.
4. For Dartington, see William Glock, *Notes in Advance* (Oxford: Oxford University Press, 1991), pp. 57–77. See also John Hodgson, *Mastering Movement: The Life and Work of Rudolf Laban* (London: Methuen, 2001).
5. Rudolf Ernst Peierls, *Atomic Histories* (Woodbury, N.Y.: AIP Press, 1997), pp. 187ff, for his own role.
6. Snowman, *Hitler Emigrés*, p. 104.
7. Georgina Ferry, *Max Perutz and the Secret of Life* (London: Chatto & Windus, 2007), pp. 63–65.
8. Helen Fry, *The King's Most Loyal Enemy Aliens: Germans Who Fought for Britain in the Second World War* (Stroud: Sutton, 2007).
9. Snowman, *Hitler Emigrés*, p. 135.
10. Charles Drazin, *Korda: Britain's Only Movie Mogul* (London: Sidgwick & Jackson, 2002), pp. 221–229.
11. Snowman, *Hitler Emigrés*, p. 135.
12. Ibid., pp. 169–170.
13. Richard Woodfield, ed., *Reflections on the History of Art: Views and Reviews* (Oxford: Phaidon, 1987), p. 231. Gombrich somewhat overshadowed Norbert Lynton (born Loewenstein in Berlin in 1927), who became director of exhibitions at the Arts Council, and in his many books did so much to promote contemporatry British art, with studies of Kenneth Armitage, Victor Pasmore, and William Scott.
14. Snowman, *Hitler Emigrés*, p. 276. For the Neuraths, see *Times* (London), April 18, 2009, pp.

44–45.

15. E. J. Hobsbawm, *Interesting Times: A Twentieth-century Life* (London: Allen Lane, 2002), p. 335.
16. Ben Barkow, *Alfred Wiener and the Making of the Holocaust Library* (London: Valentine Mitchell, 1997). See pp. 51 and 104 for the conception of the idea.
17. Christhard Hoffmann, ed., *Preserving the Legacy of German Jewry: A History of the Leo Baeck Institute, 1955–2005* (Tübingen: Mohr Siebeck, 2005).
18. Glock, *Notes in Advance*, pp. 78–86.
19. Alison Garnham, *Hans Keller and the BBC: The Musical Conscience of British Broadcasting, 1959–1979* (Aldershot: Ashgate, 2003), pp. 63ff.
20. See, for example, Martin Esslin, *The Field of Drama: How the Signs of Drama Create Meaning on Stage and Screen* (London: Methuen, 1987), which concentrates on Shakespeare, Ibsen, Goethe, Schiller, and Beckett.
21. Snowman, *Hitler Emigrés*, pp. 404, 408.
22. Muriel Nissel, *Married to the Amadeus: Life with a String Quartet* (London: Giles de la Mare, 1998), p. 7.
23. Ronald Grierson, *A Truant Disposition* (Faversham, Kent: Westgate, 1992).
24. Elisabeth Young-Bruehl, *Anna Freud: A Biography* (London: Macmillan, 1989), pp. 246–257. Uwe Henrik Peters, *Anna Freud: Ein Leben für das Kind* (Munich: Verlegt bei Kindler, 1979), pp. 238–251.
25. Young-Bruehl, *Anna Freud*, pp. 24–29.
26. Ibid., pp. 163–184.
27. Phyllis Grosskurth, *Melanie Klein: Her World and Her Work* (London: Hodder & Stoughton, 1986). Pearl King and Riccardo Steiner, *The Freud-Klein Controversies, 1941–1945* (London: Tavistock/Routledge, 1991), contains long transcripts of many meetings where the disagreements were thoroughly aired, with other well-known psychiatrists taking part—Michael Balint, Edward Glover, Susan Isaacs, and John Bowlby.
28. Julia Kristeva, *Melanie Klein*, trans. Ross Guberman (New York: Columbia University Press, 2001), p. 73.
29. Richard Kilminster, *Norbert Elias: Post-philosophical Sociology* (Abingdon and New York: Routledge, 2007), pp. 72ff.
30. Norbert Elias, *The Germans*.
31. Siniša Malesević and Mark Haugaard, *Ernest Gellner and Contemporary Social Thought* (Cambridge: Cambridge University Press, 2007), pp. 125–139 and 168–186.
32. He came on the *Kindertransport*—8,000–10,000 children were shipped out before September 1939. See Mark Harris and Deborah Oppenheimer, *Into the Arms of Strangers: Stories of the Kindertransport* (London: Bloomsbury, 2000) for firsthand accounts and vivid photographs.
33. Catherine Lampert, Norman Rosenthal, and Isabel Carlisle, *Frank Auerbach: Paintings and Drawings, 1954–2001* (London: Royal Academy of Arts, 2001), p. 111. See also Robert Hughes, *Frank Auerbach* (London: Thames & Hudson, 1990).

Chapter 41: "Divided Heaven": From Heidegger to Habermas to Ratzinger

1. Steve Crawshaw, *An Easier Fatherland: Germany in the Twenty-first Century* (London: Continuum, 2004), p. 25.
2. Ibid., p. 15.
3. Wolfgang Schivelbusch, *In a Cold Crater* (Berkeley and London: University of California Press, 1998), p. 2.
4. Eva Kolinsky and Wilfried van der Will, eds., *The Cambridge Companion to Modern German Culture* (Cambridge: Cambridge University Press, 1998), p. 297.
5. J. Christoph Bürkle, *Hans Scharoun und die Moderne: Ideen, Projekte, Theaterbau* (Frankfurt am Main: Campus, 1986), pp. 141ff.
6. Kolinsky and van der Will, eds., *The Cambrdige Companion to Modern German Culture*, p. 299.
7. Schivelbusch, *In a Cold Crater*, p. 2.
8. Ibid., p. 86.
9. Ibid., pp. 143–144.
10. Gina Thomas, ed., *The Unresolved Past: A Debate in German History; A Conference Sponsored*

by the Wheatland Foundation, Chaired and Introduced by Ralf Dahrendorf (London: Weidenfeld & Nicolson in association with the Wheatland Foundation, 1990), p. 49.

11. Siegfried Kracauer, *From Caligari to Hitler: A Psychological History of the German Film* (Princeton, N.J.: Princeton University Press, 1947).
12. See also Siegfried Kracauer, *The Mass Ornament: Weimar Essays*, trans. and introduced by Thomas Y. Levin (Cambridge, Mass.: Harvard University Press, 1995). Dedicated to Adorno. Besides the title essay, others include "Bestsellers and Their Audience"; "The Biography as an Art Form of the New Bourgeoisie"; "The Group as a Bearer of Ideas"; "The Hotel Lobby." All very prescient. For Kracauer on modernity, see David Frisby, *Fragments of Modernity: Theories of Modernity in the Work of Simmel, Kracauer, and Benjamin* (Cambridge: Polity Press, 1985).
13. Blackbourn and Eley, *The Peculiarities of German History*, pp. 29–30.
14. Jan-Werner Müller, *Another Country: German Intellectuals, Unification and National Identity* (New Haven, Conn., and London: Yale University Press, 2000), p. 8.
15. Ibid., p. 33.
16. Georg Lukács, *The Destruction of Reason*, trans. Peter Palmer (London: Merlin Press, 1980), pp. 403ff. for the vitalism argument, pp. 755ff. for the "alternative path." Ralf Dahrendorf, *Science and Democracy in Germany* (London: Weidenfeld & Nicolson, 1967/1968).
17. Dahrendorf, *Science and Democracy*, p. 46.
18. Ibid., p. 64.
19. Ibid., p. 131.
20. Ibid., p. 147.
21. Ibid., pp. 157–158.
22. Ibid., p. 202.
23. Dahrendorf also found interesting the differences in psychology between Germans and Americans. In one study Americans and Germans were asked whether they were lonely and, if so, what that meant to them. Americans associated loneliness with "weak," "sick," "sad," "shallow," and "cowardly," whereas Germans associated it with "big," "strong," "courageous," "healthy," and "deep." Dahrendorf felt that Germans valued the private virtues for the strength of the inner life, whereas in America, and to a lesser extent in Britain, the opposite was true, that people were more involved with the public virtues, with public argument and social conflict, which is why loneliness was seen in a bad/sad light. Dahrendorf, *Science and Democracy*, pp. 287–288.
24. Dahrendorf, *Science and Democracy*, pp. 342–343.
25. Fritz K. Ringer, *The Decline of the German Mandarins: The German Academic Community, 1890–1933* (Cambridge, Mass.: Harvard University Press, 1969).
26. Frederic Lilge, *The Abuse of Learning: The Failure of the German University* (New York: Macmillan, 1948), p. 69.
27. Ringer, *Decline of the German Mandarins*, p. 114.
28. Ibid., p. 20.
29. Ibid., pp. 34–35.
30. Ibid., p. 60.
31. Ibid., pp. 104–105.
32. Ibid., pp. 126 and 140.
33. Ibid., p. 146.
34. Ibid., p. 212.
35. Ibid., p. 224.
36. Ibid., p. 247.
37. Ibid., p. 254.
38. Ibid., p. 423.
39. Elias even went so far as to compose his own "Churchillian" "blood, toil, tears, and sweat" speech for Germans, in which he repeated Dahrendorf's point that their conflicts needed to be settled in a democratic way, not the way of the *satisfaktionsfähige Gesellschaft*. See Elias, *The Germans*, p. 409 for the speech. See also Ringer, *Decline of the German Mandarins*, p. 444.
40. Alfons Söllner, "Normative Westernisation? The Impact of Rémigrés on the Formation of Political Thought in Germany," in Jan-Werner Müller, ed., *German Ideologies since 1945: Studies in the Political Thought and Culture of the Bonn Republic* (New York and Basingstoke: Palgrave Macmillan, 2003), pp. 40ff.
41. For the reopening of the universities, Heidelberg in particular, see James A. Mumper, "The

Re-opening of Heidelberg University, 1945–46: Major Earl L. Crum and the Ambiguities of American Post-war Policy," in Homer & Wilcox, eds., *Germany and Europe*, pp. 238–239.

42. Edward N. Zalta, principal editor, *Stanford Encyclopaedia of Philosophy*, entry on Karl Jaspers, p. 5 of 18. http://plato.stanford.edu/
43. Karl Jaspers, *Nachlass zur philosophischen Logik*, ed. Hans Saner and Marc Hänggi (Munich: Piper, 1991).
44. Suzanne Kirkbright, *Karl Jaspers: A Biography; Navigations in Truth* (New Haven, Conn., and London: Yale University Press, 2004), p. 209.
45. Ibid., pp. 203ff.
46. Charles B. Guignon, *The Cambridge Companion to Heidegger* (Cambridge: Cambridge University Press, 2006), pp. 70–96.
47. Rüdiger Safranski, *Ein Meister aus Deutschland: Heidegger und seine Zeit* (Munich: Hanser, 1994), pp. 332ff.
48. Tom Rockmore, *On Heidegger's Nazism and Philosophy* (London: Harvester Wheatsheaf, 1992), p. 282 for the effect of Nazism on Heidegger's philosophy.
49. Richard Wolin, *Heidegger's Children* (Princeton, N.J.: Princeton University Press, 2001), p. xii. Rockmore, *On Heidegger's Nazism*, pp. 244ff. But see also John Macquarrie, *An Existential Theology: A Comparison of Heidegger and Bultmann* (London: SCM Press, 1955), pp. 16, 18, and 84.
50. Rockmore, *On Heidegger's Nazism*, p. 204, for Nazism and technology. Guignon, ed., *Cambridge Companion to Heidegger*, pp. 345–372.
51. James W. Ceaser, *Reconstructing America: The Symbol of America in Modern Thought* (New Haven, Conn., and London: Yale University Press, 1997), chap. 7.
52. Ibid., p. 187–192.
53. Ibid., p. 195.
54. Martin Jay, "Taking on the Stigma of Authenticity: Adorno's Critique of Genuineness," *New German Critique* 97, vol. 33, no. 1 (Winter 2006): 15–30.
55. Safranski, *Meister aus Deutschland*, p. 24; and p. 407ff. for Heidegger's debate at Davos with Cassirer. See also Michael Friedman, *Parting of the Ways*, pp. 129–144.
56. Wolin, *Heidegger's Children*, p. 72.
57. Ibid., p. 81.
58. Ibid., p. 95.
59. Jean Grondin, *Hans-Georg Gadamer: A Biography*, trans. Joel Weinsheimer (New Haven, Conn., and London: Yale University Press, 2003).
60. Ibid., pp. 283ff.
61. Jess Malpas, *Stanford Encyclopaedia of Philosophy*, entry on Gadamer, p. 7/16. http://plato.stanford.edu/
62. Safranski, *Meister aus Deutschland*, p. 289.
63. Robert Bernasconi, ed., *Hans-Georg Gadamer: The Relevance of the Beautiful*, trans. Nicholas Walker (Cambridge: Cambridge University Press, 1986), pp. 123–130.
64. Timothy Clark, *The Poetics of Singularity: The Counter-Culturalist Turn in Heidegger, Derrida, Blanchot, and the Later Gadamer* (Edinburgh: Edinburgh University Press, 2005), pp. 61ff.
65. Malpas, *Stanford Encyclopaedia*, p. 12/16.
66. Robert Wuthnow, *Cultural Analysis: The Work of Peter L. Berger, Mary Douglas, Michel Foucault, and Jürgen Habermas* (Boston, Mass., and London: Routledge, 1984), p. 16.
67. Deborah Cook, *Adorno, Habermas, and the Search for a Rational Society* (London: Routledge, 2004), pp. 112–123.
68. Wuthnow, *Cultural Analysis*, p. 181.
69. Ibid., pp. 197–198.
70. Ibid., p. 190.
71. Cook, *Adorno, Habermas*, pp. 66ff.
72. Wuthnow, *Cultural Analysis*, p. 195.
73. Ibid., pp. 224–225.
74. Nick Crossley and John Michael Roberts, *After Habermas: New Perspectives on the Public Sphere* (Oxford: Blackwell: *Sociological Review*, 2004), pp. 131–155 considers the Internet as a public space and its prospects for a "transnational democracy."
75. Konrad H. Jarausch, *After Hitler: Recivilising Germans, 1945–1995* (Oxford: Oxford University Press, 2006).

76. Ibid., p. 16.
77. Ibid., p. 48.
78. Ibid., p. 63.
79. Ibid., p. 16.
80. Müller, ed., *German Ideologies*, p. 122. Jarausch, *After Hitler*, p. 100.
81. Müller, *German Ideologies*, p. 147.
82. Jarausch, *After Hitler*, p. 167.
83. Rolf Hochhuth, *Täter und Denker: Profile und Probleme von Cäsar bis Jünger* (Stuttgart: Deutsche Verlags-Anstalt, 1987), pp. 41f.
84. Crawshaw, *Easier Fatherland*, p. 42.
85. Ibid., p. 49.
86. Jarausch, *After Hitler*, p. 186.
87. Ibid., p. 197.
88. Rainer Taëni, *Rolf Hochhuth* (Munich: Beck, 1977). Jarausch, *After Hitler*, p. 225.
89. Ruth A. Starkman, ed., *Transformations of the New Germany* (New York and Basingstoke: Palgrave Macmillan, 2006), p. 133.
90. Müller, *German Ideologies*, p. 192.
91. Ibid., p. 194.
92. Starkman, ed., *Transformations*, p. 37.
93. Hans Jürgen Syberberg, *Die freudlose Gesellschaft: Notizen aus dem letzten Jahr* (Munich and Vienna: Hanser, 1981).
94. Starkman, ed. *Transformations*, pp. 40–45.
95. Denis Calandra, *New German Dramatists: A Study of Peter Handke, Franz Xaver Kroetz, Rainer Werner Fassbinder, Heiner Müller, Thomas Brasch, Thomas Bernhard, and Botho Strauss* (London: Macmillan, 1983), pp. 150–161.
96. Starkman, ed. *Transformations*, p. 48.
97. Ibid., p. 234.
98. Although Martin Walser, *Leben und Schreiben: Tagebücher* (Reinbek bei Hamburg, Rowohlt, 2005–2007), ends in 1973, it contains poems and drawings by Walser as part of his diary. He seems to have been a Picasso *manqué.*
99. Müller, *German Ideologies*, p. 58.
100. Starkman, ed. *Transformations*, p. 60.
101. Ibid., p. 61.
102. Friedman, *Parting of the Ways*, pp. 21ff.
103. Ibid., pp. 41ff.
104. I have used Tracey Rowland, *Ratzinger's Theology* (Oxford: Oxford University Press, 2008), p. 5.
105. Joseph Ratzinger (Pope Benedict XVI), *Christianity and the Crisis of Cultures*, trans. Brian McNeill (San Fransisco: Ignatius Press, 2006), pp. 25ff.
106. Ibid., pp. 47–53.
107. Rowland, *Ratzinger's Theology*, p. 69. Ratzinger, *Christianity*, pp. 61–64.
108. Rowland, *Ratzinger's Theology*, p. 69.
109. Ibid., p. 72.

Chapter 42: Café Deutschland: "A Germany Not Seen Before"

1. A. Dirk Moses, *German Intellectuals and the Nazi Past* (Cambridge: Cambridge University Press, 2007), pp. 58–61.
2. Alexander and Margarete Mitscherlich, *Die Unfähigkeit zu trauern: Grundlagen kollektiven Verhaltens* (Munich: Piper, 1967).
3. Keith Bullivant, *The Future of German Literature* (Oxford/Providence, R.I.: Berg, 1994), p. 37.
4. Nicholas Boyle, *German Literature: A Very Short Introduction* (Oxford: Oxford University Press, 2008), p. 143.
5. Lothar Huber and Robert C. Conrad (editors), *Heinrich Böll on Page and Screen: The London Symposium* (London: Institute of German Studies, University of London, 1997), pp. 17ff.
6. Ibid., pp. 65ff.

7. Michael Jürgs, *Bürger Grass: Biographie eines deutschen Dichters* (Munich: C. Bertelsmann, 2002), pp. 138ff., for early thoughts on Oskar.
8. Crawshaw, *Easier Fatherland*, p. 28.
9. Ibid., p. 87.
10. Jürgs, *Bürger Grass*, pp. 144f.
11. In English: Martin Walser, *Runaway Horse*, trans. Leila Vennewitz (London: Secker and Warburg, 1980). See also Martin Walser, *The Inner Man*, trans. Leila Vennewitz (London: Deutsch, 1986).
12. Bullivant, *Future of German Literature*, p. 30.
13. Hans Höller, *Ingeborg Bachmann: Briefe einer Freundschaft* (Munich: Piper, 2004).
14. Ingeborg Bachmann, *Darkness Stolen: The Collected Poems*, trans. and intro. Peter Filkins (Brookline, Mass.: Zephyr Press, 2006), p. xx. Elfriede Jelinek, *Oh Wildnis, oh Schutz vor ihr* (Hamburg: Rowohlt, 1985).
15. Bullivant, *Realism Today*, p. 222.
16. Fulbrook, ed., *German National Identity*, p. 258.
17. Christa Wolf, *Das dicht besetzte Leben: Briefe, Gespräche und Essays*, ed. Angela Drescher (Berlin: Aufbau Taschenbuch Verlag, 2003).
18. Sara Kirsch, *Sämtliche Gedichte* (Munich: Deutsche Verlagsanstalt, 2005).
19. Magenau, *Christa Wolf*, pp. 192ff. Rob Burns, ed., *German Cultural Studies* (Oxford: Oxford University Press, 1995), p. 177. Compare with Jörg Magenau, *Christa Wolf: Eine Biographie* (Berlin: Kindler, 2002), pp. 328ff.
20. Burns, ed., *German Cultural Studies*, p. 189.
21. Bernd Neumann, *Uwe Johnson, mit zwölf Porträts von Diether Ritzert* (Hamburg: Europäische Verlagsanstalt, 1994), pp. 269ff. for Johnson's 1960s.
22. Crawshaw, *Easier Fatherland*, p. 199.
23. Bullivant, *Future*, p. 98.
24. Ibid., p. 91.
25. Crawshaw, *Easier Fatherland*, p. 100.
26. Gitta Honegger, *Thomas Bernhard: The Making of an Austrian* (New Haven, Conn., and London: Yale University Press, 2001), pp. 128ff.
27. Tom Bower, "My Clash with Death-Camp Hanna," *Sunday Times* (London), February 15, 2009.
28. Crawshaw, *Easier Fatherland*, p. 200.
29. Günther Grass, *Peeling the Onion*, trans. Michael Henry Heim (London: Harvill Secker, 2007).
30. Crawshaw, *Easier Fatherland*, p. 205.
31. Volker Weidermann, *Das Buch der verbrannten Bücher* (Cologne: Kiepenheuer & Witsch, 2008), pp. 300ff.
32. Burns, ed., *German Cultural Studies*, p. 192.
33. Boyle, *German Literature*, p. 145.
34. Mark William Roche, *Gottfried Benn's Static Poetry: Aesthetic and Intellectual-Historical Interpretations* (Chapel Hill and London: University of North Carolina Press, 1991).
35. Ingeborg Bachmann, *Briefe einer Freundschaft*, ed. Hans Höller. Foreword by Hans Werner Henze. (Munich: Piper, 2004). Some letters in English. Henze found the English "ugly" but liked the drinks in London, p. 335. Nice solecism (or is it?) about musicians "cueing" at bus-stops.
36. Paul Celan, *Poems*, trans. and intro. Michael Hamburger (Manchester: Carcanet New Press, 1980). Bilingual edition; see p. 51 for "Death Fugue."
37. Hans Magnus Enzensberger, *Selected Poems*, trans. author and Michael Hamburger (Newcastle upon Tyne: Bloodaxe Books, 1994).
38. Johannes R. Becher, *Macht der Poesie: Poetische Konfession* (Berlin: Aufbau, 1951), for his views on poetry in the depths of the Cold War.
39. Günter Kunert, *Erwachsenenspiele: Erinnerungen* (Munich: Hanser, 1997).
40. Volker Braun, *Lustgarten, Preussen: Ausgewählte Gedichte* (Frankfurt am Main: Suhrkamp, 1996). Wolf Biermann, *Preussischer Ikarus: Lieder, Balladen, Gedichte, Prosa* (Cologne: Kiepenheuer & Witsch, 1978). Sarah Kirsch, *Sämtliche Gedichte* (Munich: Deutsche Verlagsanstalt, 2005), p. 249 for "Katzenleben," p. 405 for "Bodenlos."
41. Bullivant, *Future*, p. 167.

42. J. Kaiser, "Die deutsche Literature war nicht zerrissen," *Süddeutsche Zeitung*, October 2–3, 1990, quoted in Bullivant, *Future of German Literature*, p. 172.
43. Volker Weidermann, *Lichtjahre: Eine kurze Geschichte der deutschen Literatur von 1945 bis heute* (Cologne: Kiepenheuer & Witsch, 2006), p. 245.
44. For the Fassbinder "legends" following his death at 37, see David Barnett, *Rainer Werner Fassbinder and the German Theatre* (Cambridge: Cambridge University Press, 2005), pp. 1ff.
45. Bullivant, *Future*, p. 218.
46. Kolinsky and van der Will, eds., *Cambridge Companion to Modern German Culture*, p. 223.
47. Susan Manning, *Ecstasy and the Demon: Feminism and Nationalism in the Dances of Mary Wigman* (Berkeley and London: University of California Press, 1993).
48. *Theater heute*, December 1989, p. 6.
49. Kolinsky and van der Will, eds., *Cambridge Companion to Modern German Culture*, p. 311.
50. Ibid., p. 312.
51. Burns, ed., *German Cultural Studies*, p. 317.
52. Ian Buruma, "Herzog and His Heroes," *New York Review of Books*, July 19, 2007, pp. 24–26.
53. Alexander Graf, *The Cinema of Wim Wenders: The Celluloid Highway* (London: Wallflower Press, 2002), pp. 48–54.
54. Kolinsky and van der Will, eds., *Cambridge Companion to Modern German Culture*, pp. 311–312.
55. Ibid., p. 233.
56. Ross, *The Rest Is Noise*, p. 10.
57. Werner Oehlmann, *Das Berliner Philharmonische Orchester* (Kassel/Basel/Tours/London: Bärenreiter-Verlag, 1974), pp. 117ff. and 127ff.
58. Roger Vaughan, *Herbert von Karajan: A Biographical Portrait* (London: Weidenfeld & Nicolson, 1986), p. 116.
59. Kolinsky and van der Will, eds., *Cambridge Companion to Modern German Culture*, p. 252.
60. Hans Werner Henze, *Music and Politics: Collected Writings, 1953–1981*, trans. Peter Labany (London: Faber, 1982), p. 196 for "The task of revolutionary music."
61. Michael Kurtz, *Stockhausen: A Biography*, trans. Richard Toop (London: Faber, 1992), pp. 110ff.
62. Ibid., pp. 210ff.
63. Werner Haftmann, *Wols Aufzeichungen: Aquarelle, Aphorismen, Zeichnungen* (Cologne: M. Du Mont Schauberg, 1963), pp. 10 and 32, for a note by Jean-Paul Sartre.
64. Nicolas Charlet, *Yves Klein*, trans. Michael Taylor (Paris: Adam Biro/Vilo International, 2000), pp. 18ff.
65. Alain Borer, *The Essential Joseph Beuys*, ed. Lothar Schirmer (London: Thames & Hudson, 1996).
66. For a discussion of Beuys's materials, see Richard Demarco, "Three Pots for the Poorhouse," in *Joseph Beuys: The Revolution Is Us*, catalog for an exhibition in Liverpool, 1993–1994, published by the trustees of the collection of the Tate Gallery, Liverpool, 1993.
67. Götz Adriani, Winfried Konnertz, and Karin Thomas, *Joseph Beuys* (Cologne: Dumont Buchverlag, 1994). Definitive photographs.
68. Günther Gereken, "Holz-(schnitt)-wege," in *Anselm Kiefer*, exhibition at the Groningen Museum, Groningen, 1980–1981.
69. A. R. Penck, "Auf Penck zurückblickend (1978)," in *A. R. Penck—Y. Zeichnungen bis 1975*, exhibition at the Kunstmuseum, Basel, 1978.
70. Rainer Fetting, *Holzbilder* (wood paintings), exhibition at the Marlborough Gallery, New York, 1984.
71. Kolinsky and van der Will, eds., *Cambridge Companion to Modern German Culture*, p. 280.
72. Crawshaw, *Easier Fatherland*, p. 92.
73. Personal interview, Wolf-Hagen Krauth, Prussian Academy of Sciences, Berlin, April 9, 2008.
74. History of the Kaiser Wilhelm Society in the Third Reich. http://www.mpiwg-berlin.mpg.de/KWG/projects_e.htm
75. Peter Blundell-Jones, *Hans Scharoun* (London: Phaidon, 1995), pp. 94–102.
76. Christine Hoh-Slodczyk, et al., *Hans Scharoun: Architekt in Deutschland, 1893–1972* (Munich: Beck, 1992), pp. 98–101.
77. J. Christoph Bürkle, *Hans Scharoun und die Moderne: Ideen, Projekte, Theaterbau* (Frankfurt am Main: Campus, 1986), for photographs and plans.

78. Crawshaw, *Easier Fatherland*, p. 115.
79. Dietrich Schwanitz, *Bildung: Alles, was man wissen muss* (Munich: Wilhelm Goldmann Verlag, 2002).

Conclusion: German Genius: The Dazzle, Deification, and Dangers of Inwardness

The quotation at the head of this chapter is taken from Thomas Mann, *Betrachtungen eines Unpolitischen* (Berlin: G. Fischer, 1918/1922). Translated by W. D. Morris as *Reflections of a Non-Political Man* (New York: F. Ungar, 1983). The quotation appears in W. H. Bruford, *The German Tradition of Self-Cultivation: Bildung from Humboldt to Thomas Mann* (Cambridge: Cambridge University Press, 1975), p. vii.

1. W. H. Auden, *Collected Shorter Poems, 1930–1944* (London: Faber, 1950), pp. 171–175.
2. R. P. T. Davenport-Hines, *Auden* (London: Heinemann, 1995), p. 157.
3. Benjamin Nelson, ed., *Freud and the 20th Century* (London: George Allen & Unwin, 1958), p. 13.
4. Ibid., p. 14.
5. Ibid., p. 17.
6. Frank Furedi, *Therapy Culture: Cultivating Vulnerability in an Uncertain Age* (London: Routledge, 2004). And see, for example, Dennis Hayes, "Happiness Drives Education from the Classroom," *Times Higher Education Supplement*, September 14, 2007, p. 22.
7. Richard Lapierre, *The Freudian Ethic* (London: George Allen & Unwin, 1960), p. 60.
8. Watson, *Modern Mind/Terrible Beauty*, p. 601.
9. See, for example, Alexandra Blair, "Expulsion of Under-Fives Triples in a Year," *Times* (London), April 20, 2007, p. 17; and Alexandra Frean, "Emphasis on Emotions Creates 'Can't Do' Students," *Times* (London), June 12, 2008, p. 13.
10. I have used Bernd Magnus and Kathleen M. Higgins, eds., *The Cambridge Companion to Nietzsche* (Cambridge: Cambridge University Press, 1996), p. 282.
11. Ibid., p. 309–310.
12. Ibid., p. 314.
13. Ibid., p. 2.
14. Ibid., p. 4.
15. Ibid., p. 225.
16. Lawrence Scaff, *Fleeing the Iron Cage* (Berkeley and London: University of California Press, 1989), p. 226.
17. Ibid., p. 230.
18. Magnus and Higgins, eds., *Cambridge Companion to Nietzsche*, p. 80.
19. Ibid., p. 82.
20. Ibid., p. 172.
21. Barbara Tuchman, *The Proud Tower* (London: Folio Society, 1995), p. 284.
22. Hew Strachan, *The Outbreak of the First World War* (Oxford: Oxford University Press, 2004), p. 183.
23. James Bowen, *A History of Western Education*. 3 vols. (London: Methuen, 1981), vol. 1, pp. 321 and 345.
24. *Times* (London), March 23, 2006, p. 9.
25. Gertrude Himmelfarb, *The Roads to Modernity: The British, French, and American Enlightenments* (New York: Vintage, 2005), p. 51.
26. See the note at the head of this chapter.
27. Erica Carter, "Culture, History and National Identity in the Two Germanies, 1945–1999," in Fulbrooke, *Twentieth-Century Germany*, p. 266.
28. Müller, *Another Country*, pp. 172–173.
29. Ibid., p. 71.
30. Ibid., p. 161.
31. Ibid., p. 189.
32. Müller, *German Ideologies since 1945*, p. 131.
33. Müller, *Another Country*, pp. 179 and 196.
34. Ringer, *Decline of the German Mandarins*, p. 29.

35. Scaff, *Fleeing the Iron Cage*, p. 96.
36. Müller, *Another Country*, *op. cit.*, p. 23.
37. Ibid., p. 155.
38. Ibid., p. 175.
39. Wuthnow, *Cultural Analysis*, p. 189.
40. Schivelbusch, *Culture of Defeat*, p. 231.
41. Hannah Arendt, *Burden of Our Time*, p. 320–321.
42. Ibid., p. 324.
43. Ibid., p. 326.
44. Niall Ferguson, *The War of the World: History's Age of Hatred* (London: Allen Lane, 2006), p. 243.
45. Jarausch, *After Hitler*, p. 100.
46. Ibid., pp. 139f.
47. Müller, *Another Country*, p. 131.
48. Burns, *German Cultural Studies*, pp. 253 and 257.
49. *Daily Telegraph* (London), June 11, 2006, p. 24.
50. *Times Higher Education Supplement*, March 27, 2008, p. 19.
51. *Times Higher Education Supplement*, May 22, 2008, p. 19.
52. Jürgen Habermas, *The Future of Human Nature* (Cambridge: Polity Press, 2003), p. 38.
53. Ibid., p. 39.
54. Ibid., p. 48.
55. Ibid., p. 56.
56. Ibid., pp. 79–81.
57. Ibid., p. 87.
58. The memories of these policies were reinforced in May 2008 when Dr. Hans-Joachim Sewering, who had been linked with the murder of 900 children at Eglfing-Haar extermination camp, received the German Federation of Internal Medicine's highest honor, the Günter-Budelmann medal. Johannes Rau, "Der Mensch ist jetzt Mitspieler der Evolution geworden," *Frankfurter Allgemeine Zeitung*, May 19, 2001.
59. Crawshaw, *Easier Fatherland*, p. 219.

Index

A

Abbe, Ernst, 365, 366, 387
Abbt, Thomas, 78
Abitur (examination), 109, 234
Abraham, Karl, 613, 753
Abur of Learning, The: The Failure of the German University (Lilge), 765–766
Abu Telfan (Raabe), 440
academic institutions, 359. *See also Gymnasien;* Kaiser Wilhelm Institutes; universities
  Ritterakademien, 50
  technical institutes, 168
  during Weimar Republic, 570
academies of sciences, 226
Academy of Antiquities (Kassel), 100
Ace in the Hole (film), 591
Achenwall, Gottfried, 73
Actions for Dancers (Aktionen für Tänzer) (Wigman), 804
Acton, John Emerich Edward Dalberg, 317, 409, 538
Addresses to the German Nation (Reden an die deutsche Nation) (Fichte), 262
Adenauer, Konrad, 753, 759, 772, 791
Adorno, Theodor, 572, 631, 702, 716, 724–726, 741, 768–769, 771–772, 799
AEG, 379–380, 509
Aenesidemus (Schulze), 148
Aesthetic Paganism in German Literature (Hatfield), 101
aesthetics
  Nazi, 629–648
  perfectibility as concept of, 74–75
  in revival of eighteenth century, 94
After Hitler (Umkehr, Die) (Jarausch), 778–781
Age of Capitalism, The (Hobsbawm), 749
Age of Constantine the Great, The (Zeit Konstantins des Grossen) (Burckhardt), 91
Age of Extremes, The (Hobsbawm), 749
Age of Industry, The (Hobsbawm), 749
Age of Revolution, The (Hobsbawm), 749
Akhmatova, Anna, 801
Aktion T4, 671–672
Albers, Joseph, 702, 734
Albert, Prince (England), 318–321, 357
Alceste (Gluck), 155
Alder Wright, C. R., 364
Alexander (Baumann), 637
Alexander, Franz, 570, 613
alienation, concept of, 239–259, 251–252, 572, 612, 675, 803
Allegory of History (Mengs), 210
Allen, William, 20
Allgemeine Literatur-Zeitung (periodical), 146
Allgemeine musikalische Zeitung, 162, 164
All Quiet on the Western Front (Im Westen Nichts Neues) (Remarque), 579–581
Also Sprach Zarathustra (Nietzche), 336
Altenstein, Karl, 108, 236–237, 244
Altertumswissenschaft, 52, 107
Althaus, Paul, 685–686
Aly, Götz, 669, 671, 672
Amadeus String Quartet, 752
American Friends of German Freedom, 700
American Journal of Sociology, 445
American Judaism (Glazer), 8
American Philosophical Society, 326
American Psychoanalytic Association, 716
Americks, Karl, 138
Ancestress, The (Die Ahnfrau) (Grillparzer), 294

"Andenken" (Remembrance) (Hölderlin), 291–292
Anders, Günther, 771, 808
Anderson, Benedict, 58, 114, 837
"An die Kulturwelt," 33–34
Angestellten, Die (The Salaried Classes) (Kracauer), 588–589, 760
"Angststurm" (Stramm), 550
Anheisser, Siegfried, 643
Annalen der Chemie und Pharmacie, 342
Annalen der Physik, 482
Annalen der Physik und Chemie, 342, 343, 345
anthropology, 385
anti-Catholicism, 421–425
anti-Semitism
  Benda on, 622
  coining of term, 522
  in European countries, 858n45
  investigation by Adorno, 725
  of Jung, 661
  of Kittel, 685
  of Moeller van den Bruck, 617–618
  of Pan-German League, 421
  political parties and, 434
  pre-World War II, 521–522, 767
  reactions to, 491, 492, 499
  of Treitschke, 408
  in Vienna, 501
  of Weininger, 495
Anzeiger des Westens, Der (newspaper), 326
Apartment, The (film), 591
Apitz, Bruno, 795
Apologia for the Reasonable Worshippers of God (Apologie oder Schutzschrift für die vernünftigen Verehrer Gottes) (Reimarus), 103–104
Appeal to the Cultural World. *See* Manifesto of the 93
Appleyard, Rollo, 478
archaeology, 95, 100, 385, 404, 410–415, 539
Architectural Review, 748
architecture, 211–215, 305, 490, 496, 510, 523, 570–571, 814–815
  Bauhaus, 571, 629
Archiv der reinen und angewandten mathematik (journal), 226
archives, 263, 267, 407, 409
Archives Interdites (Combe), 16
Archiv für die Physiologie (journal), 83
Ardenne, Manfred von, 692
Arendt, Hannah, 553, 651, 701, 719, 720–721, 770, 818, 836, 841–842
armaments, 370, 371–373, 404, 417
  atomic bomb research, 691, 913n14
armies. *See* military
Arndt, Ernst Moritz, 309, 801
Arnheim, Rudolf, 734
Arnim, Achim von, 230
Arnold, Thomas, 317
Arp, Hans, 563, 564–565
art and artists, 73–76. *See also* painting
  Albert and, 319
  contemporary, 844
  as creators, 194–195
  emigration, 630, 734–735
  function of, 197
  Innerlichkeit and, 403
  Kant on, 144
  Kitsch, 568
  in Munich, 504–505, 515, 516, 517
  Schopenhauer on, 333
  in Third Reich, 621, 629–631, 632–633
  Wagner on, 328–329
  Winckelmann on antiquity, 98–99
  World War II and, 13, 15
Art and Illusion (Gombrich), 747
art history, 95–100, 218, 732–733
ARTnews (magazine), 15
Aryan mystique, 429–432, 435
Aschheim, Stephen, 823
Ashton, Rosemary, 314
aspirin, 365, 887n36
Assistance Fund for German Science, 595
Astel, Karl, 671
Astor, David, 750
atheism, 67, 332
Athenäum (periodical), 119
athletic clubs, 429
Auch eine Philosophie der Geschichte zur Bildung der Menschheit (Herder), 123
Auden, W. H., 296, 578, 810, 817–819
Auer, Fritz, 814
Auerbach, Frank, 755–756
Aufklärung, 69, 261, 832
Aurora Press, 735
Ausländer, Rose, 800
Austerlitz (Sebald), 797–798
Austria, 304, 567. *See also* Vienna
Austro-Prussian War (1866), 372, 408, 418
Authoritarian Personality, The (Adorno), 725
automobiles, 375–379
Aveling, Edward, 426
Ayer, A. J., 602, 603, 631
Ažbe, Anton, 515

B

Baader, Andreas, 780
Baader-Meinhof terrorism, 22, 793
Baal (Brecht), 586
Bach, Johann Sebastian, 41–43, 60, 153–154
Bacharach, Burt, 593
Bachmann, Inbegorg, 794, 800
Badische Beobachter (periodical), 425
Baeck, Leo, 750
Baer, Karl Ernst von, 37, 286–287, 853
Bagge, Harald, 287

Bahr, Hermann, 490
Baird, Jay, 634
Ball, Hugo, 564, 565
Baran, Paul, 727
Barbie, Klaus, 17
Barker, Christine, 579
Barlach, Ernst, 617, 633
Barmen Declaration (1934), 677
Barnack, Oscar, 367
Baron, Hans, 731
Barr, Alfred, 701
Barren, William, 604
Barry, Martin, 287
Barth, Karl, 676–677, 769, 855
Bartholdy, Salomon, 218
Bartók, Béla, 702, 703
Baselitz, Georg, 812
BASF, 361, 695
Bassarids, The (Die Bassariden) (opera, Henze and Auden), 810, 819
Battéoli Commission, 17
Battle of Teutoburg Forest, The (Die Hermannsschlacht) (Kleist), 293
Batty, Peter, 370, 371
Bauer, Bruno, 247, 248, 249
Bauer, F. C., 245
Bauer, Franz, 282
Bauer, Peter, 746
Bauhaus, 571, 629
Baumann, Hans, 618, 634, 636–638
Baumeister, Willi, 700
Baumgarten, Alexander, 74
Baumgarten, Hermann, 421
Baumgarten, Sigmund, 70
Bäumler, Alfred, 650
Bausch, Pina, 804
Bayer, 360–361, 362, 695, 887n36
Bayer, Friedrich, 362
Bayreuth festivals, 329, 644
Becher, J. R., 801
Becker, Thorsten, 796
Beckmann, Max, 629, 630, 633, 699
Beer, John, 358–359, 362
Beethoven, Ludwig van, 59, 158–159, 162–163, 239
Before Dawn (Vor Sonnenaufgang) (Hauptmann), 524
Beggar's Opera, The (Gay), 586
Begriffsschrift (Frege), 486
Behler, Ernst, 823
Behltle, Charles, 366–367
Behnisch, Günter, 814
Behrens, Peter, 215, 509–510, 516, 571
Being and Time (Sein und Zeit) (Heidegger), 603–604, 651
"Beiträge zur Phytogenesis" (Schleiden), 284
Bell, Daniel, 296
Benda, Julien, 622–625, 766, 841
Benedict, Ruth, 716, 717
Benjamin, Walter, 449, 609, 619, 669, 709
Benn, Gottfried, 454, 655, 710, 790, 799–800
Benz, Karl, 375–376
Berek, Max, 367
Berenhorst, Georg Heinrich von, 187
Berg, Alban, 302, 515, 568, 584, 585, 643
Berger, Peter, 730
Bergmann, Carl, 287
Bergstraesser, Arnold, 650, 768–769
Berlin, 519–545. *See also* University of Berlin
  Academy for the Science of Judaism, 655
  Academy of Arts and Sciences, 61
  Academy of Fine Arts, 527
  Academy of Sciences, 356
  after World War II, 758
  architecture, 211–215
    contemporary, 814–815
    rebuilding of Cathedral, 219
  art and artists (*See also* Berlin Sezession)
    Dada, 565
    performers, 584
  art burning of March 1938, 633
  Association of Berlin Artists, 527
  Berliner-Kritische Association, 244
  book-burning in 1933, 580, 798
  Brandenburg Gate, 211–212, 213
  Doktor Club, 245, 248
  economy after Franco-Prussian war, 520
  education
    Artillery and Engineering School, 355
    Friedrich-Wilhelms Institute, 383
    Friedrich-Wilhelm-Universität, 529
    War College, 185
  Film Festival (1961), 806
  Germaine de Staël in, 312
  Holocaust Memorial, 784, 785–786, 815
  Kreuzberg war memorial, 213–214
  medical institutions, 384, 388
  museums, 528–529
    Altes Museum, 214, 528
    Ethnological Museum, 414
    Kaiser-Friedrich-Museum, 528
    National Gallery, 528, 630
    Neues Museum, 528
    Royal Museum, 415
  music
    innovation, 59
    opera, 467, 643, 805
    Philharmonic Orchestra, 525–526, 642, 645, 809
  Royal Academy, 527
  Sound and Smoke cabaret, 525
  Soviet Union and, 759
  theater
    Deutsches Theater, 524, 525, 589, 592
    Freie Bühne (Free Stage) movement, 524
    Grosses Schauspielhaus, 647

Berlin (*cont.*)
Schauspielhaus, 214
Theater des Volkes, 647
Wall, 759
Berlin, Isaiah, 193–194, 196
Berlin Alexanderplatz (Döblin), 619
Berliner Illustrierte Zeitung (newspaper), 427
Berliner Abendblätter (newspaper), 293
Berlin Physics Society, 482
Berlin Psychoanalytic Institute, 570, 613
Berlin Psychoanalytic Society, 661
Berlin Psychological Institute, 661
Berlin Sezession, 507, 516, 527–528
Berne, Eric, 720
Bernhardi, Friedrich von, 409, 534, 539
Berning, Heinrich, 696
Bernstein, J. G., 228
Bertalanffy, Ludwig von, 567
Berzelius, Jöns Jakob von, 273, 276
Bethe, Hans, 659, 738–739
Bethmann-Hollweg, Theobad von, 533
Bettelheim, Bruno, 662, 717–718, 818
Bettinelli, Saverio, 92
Bettmann, Otto, 735
Beuys, Joseph, 812, 849
Beyer, Frank, 806
Bianconi, Giovanni Lodovico, 207
Bible, 66, 70–71, 105, 245, 679
"Biedermann's Evening Socialising" (Scheffel), 305
Biedermeier culture, 304–307
Biermann, Wolf, 796, 801
Bierwisch, Manfred, 813
Big Heat, The (film, Lang), 590
Bild, Das (newspaper), 6
Bildung, 53–54, 833–834, 838
Americanization of, 740–742
Berlin University and, 87
education as, 830
Goethe's pursuit of, 118
Herder and, 124
Humboldt, Wilhlem von, on, 109–110, 262–263
inwardness and, 832–833
music listening and, 163
neohumanism and, 109
Pietism and, 87
Schiller on, 130
secularization of, 87–88
Wieland and, 113
Bildung (Schwanitz), 815
Bildungsstaat, concept of, 77
Bildungstrieb, 82
Bildunsbürgertum, 270
Billiards at Half Past Nine (Billard um Halb Zehn) (Böll), 791
Billinger, Richard, 648
Bilse, Benjamin, 525
Bing, Rudolf, 744, 746, 751
bioethics, 384–386
biology, 68, 78–84, 201, 271–287, 615, 660
Birkeland, Kristian, 361
Birth of Tragedy out of the Spirit of Music, The (Die Geburt der Tragödie aus der Geist der Musik) (Nietzsche), 336–337
Bismarck, Otto von, 371–372, 373, 385, 404, 407, 519
Bitomsky, Hartmut, 808
Blackbourn, David, 27
Black Lines (Kandinsky), 517
Black Mountain College, 702, 734
Blanning, Tim, 43, 54, 56, 59, 60
Bleichröder, Gerson, 520
Bleker, Johanna, 387
Bloch, Felix, 659, 738
Bloom, Allan, 713–714, 740, 833–834
Bloom, Harold, 296
Blue Angel, The (film), 513, 591–592
Blum, Martin, 781
Blumenbach, Johann Friedrich, 81–82, 86
Bock, Hieronymus, 79, 80
Bode, Wilhelm von, 433, 528–529
Bodmer, Johann Jakob, 57, 74, 124
Boeckh, August, 235, 345
Boehm, Anthony, 316
Boehm, Max Hildebert, 655
Böhm, Karl, 645
Böhmer, Caroline. *See* Schlegel, Caroline
Böhmer, Johann Friedrich, 266
Bohr, Niels, 596, 597–598, 691
Bohrer, Karl Heinz, 782, 832
Böll, Heinrich, 28, 780, 791–792
Bollingen Press, 735
Bölsche, Wilhelm, 426, 427
Boltzmann, Ludwig, 348, 483, 854
Bonds, Mark Evans, 161, 162
Bonhoeffer, Dietrich, 680–681, 818, 855
Book of Burnt Books, The (Das Buch der verbrannten Bücher) (Weidermann), 29, 798
Book of Songs (Buch der Lieder) (Heine), 298
Bopp, Franz, 191, 192
Bormann, Martin, 620
Born, Max, 598, 659, 745, 753
Börne, Ludwig, 303
Borsig, August, 375
Bosch, Carl, 361, 666
botany, 79–80, 85, 203
Bothe, Walter, 653
Böttger, J. F., 168
Bötticher, Paul Anton. *See* Lagarde, Paul de
Boumann, Johann, 213
Boureau, Alain, 607
Bourgeois, Der (Sombart), 453–454
Bousquet, René, 17
Bowen, James, 829
Bower, Tom, 797

Boyer, Carl, 349
Boyle, Nicholas, 29, 115, 122, 800
Bracher, K. D., 22
Brahm, Otto, 524
Brahms, Johannes, 133, 173, 292, 459–462, 465, 525, 558
Brainin, Norbert, 752
Brandt, Max, 587
Brandt, Willy, 733
Braun, Volker, 796, 801, 802
Braun, Wernher von, 693, 695
Brecht, Bertolt, 584, 585–587, 588, 590, 651
  on Adorno, 724
  anti–World War I poems, 552
  Berliner Ensemble and, 803
  collaboration with Auden, 818
  friendship with Benjamin, 619
  on life in East Germany, 801
  in Novembergruppe, 568
  refugee in the United States, 702, 715, 736–737
  return to Europe, 738
Breitinger, Johann Jakob, 57
Breker, Arno, 632
Brentano, Clemens, 212, 230
Brentano, Franz Clemens, 492–493
Brentano, Lujo, 533
Breuer, Josef, 395, 396, 492–493
Breuer, Stefan, 454–455
Bridgwater, Patrick, 547
Bright Picture (Kandinsky), 517
Brinkmann, Carl, 650
Britain
  academics studying in Germany, 538
  Anthropological Society of London, 430
  dye industry in, 359
  German influence on University of London, 317
  on German scholarship, 538
  Great Exhibition of 1851 (London), 320
  interest in German culture, 314–318
  Kindertransport to, 744–745
  London International Exhibition (1862), 356
  London Philharmonia Orchestra, 809
  London Philharmonic Orchestra, 747
  refugees, 699, 743–756
    Academic Assistance Council and, 663, 745
    Artists' Refugee Committee, 699
    BBC and, 750
    in Cambridge, 746
    Free German League of Culture, 699, 746
    internment on Isle of Man, 745, 754
    in Oxford, 746
  visited by Tönnies, 450
  World War I propaganda of two Germanies, 534
British Arts and Crafts movement, 510
British Association for the Advancement of Science, 752
British Psychoanalytic Society, 754
Britten, Benjamin, 292
Brod, Max, 578
Brodie, Bernard, 184
Broken Jug, The (Der Zerbrochene Krug) (Kleist), 293
Brotherhood of St. Luke, 215–217. *See also* Düreristen
Brotstudium, concept of, 229
Brown, Robert, 282, 284
Browning, Christopher, 20
Brücke, Die, 516, 528
Brücke, Ernst, 234
Bruckmann, Peter, 510
Bruckner, Anton, 459, 469
Bruford, William, 119, 833
Brügmann, Walter, 646
Brunfels, Otto, 79
Brussig, Thomas, 799
Buber, Martin, 609, 682–683, 685
Bubis, Ignatz, 784
Büchner, Georg, 290, 301–303, 585
Büchner, Ludwig, 301, 426
Buckingham, John, 272–273, 279
Buddenbrooks (T. Mann), 511–512
Bühler, Emil, 375
Bühler, Karl, 717
Bullitt, William, 662
Bullivant, Keith, 36, 452, 618, 790, 793
Bülow, Adam Heinrich Dietrich von, 187
Bülow, Hans von, 459, 460, 462, 466, 525–526
Bultmann, Rudolf, 677, 678–679, 769, 855
"Bummelmeir's Complaint" (Scheffel), 305
Bunsen, Robert, 348
Burckhardt, Jacob, 22, 91, 250–251, 410, 441
bureaucracy
  educated middle class and, 830
  Prussian, 45, 48–49, 108
  Weber on, 457
Burgh, John, 746
Burke, Peter, 92
Burnouf, Émile, 184
Burns, Mary, 257
Burte, Hermann, 634
Butler, E. M., 97, 100
Buxtehude, Dietrich, 153

C

Cabinet of Dr. Caligari, The (film), 568
Café Deutschland (Immendorf), 813
Café Müller (Wigman), 804
cafés, 470, 490, 504, 521, 563–564
Cahn, Arnold, 362
Calleo, David, 421
Calvert, Frank, 412

cameras, 367
Camera Work (journal), 517
Campbell, Thomas, 317
Campendonck, Heinrich, 699
Canetti, Elias, 746
Cantor, Georg, 37, 174, 485–486
Cantor, Norman, 35
Cantril, Hadley, 727
Capa, Robert, 735
Capitalism, Socialism and Democracy (Schumpeter), 704
Carlyle, Thomas, 314, 315
Carnap, Rudolf, 601, 602, 631, 729–730
Carnaval (Schumann), 307
Carow, Heiner, 806
cartels, 361–362
Carter, Erica, 832
Case of J. Robert Oppenheimer, The (In der Sache J. Robert Oppenheimer) (Kipphardt), 736
Cassirer, Bruno, 527, 606
Cassirer, Ernst, 136, 143, 564, 605–607, 723, 741, 746
Cassirer, Paul, 527, 606
Castle, The (Das Schloss) (Kafka), 577, 578
Cat and Mouse (Katz und Maus) (Grass), 792
Catholic Church
  anti-Catholicism, 421–425
  ban of *Myth of the Twentieth Century*, 621
  Hitler and, 683
  monastic orders, 423, 425, 687
  Rosenberg on, 620
Caucasian Chalk Circle, The (Der kaukasische Kreidekreis) (Brecht), 736
Celan, Paul, 800
Cellar, The: A Withdrawal (Der Keller: eine Entziehung) (Bernhardt), 797
Central Council of Jews in Germany, 784
Central European University, 755
Centre for the Study of Nationalism (Prague), 755
ceremonies of the whole, 647, 836
Chamberlain, Houston Stewart, 421, 434, 435–436, 449, 452, 892n61
Chamisso, Adelbert von, 305
Chemisches Journal, 226
chemistry, 168, 271, 272, 279, 358
Chickering, Roger, 428, 536
Childhood and Society (Erikson), 717
choral societies, 164, 325
Christianity, 673, 677–678, 683–685, 686, 788, 853. *See also* Catholic Church; Protestantism
  Positive Christianity, 684
Cicerone, The (Der Cicerone) (Burckhardt), 91
Cieszkowski, August von, 245
Cincinnati Volksblatt (newspaper), 326
Civilisation of the Renaissance in Italy, The (Der Kultur der Renaissance in Italien) (Burckhardt), 91, 92
Civilising Process, The (Elias), 609, 754
Civilization and Its Discontents (Das Unbehagen in der Kultur) (Freud), 611–612
Civil Service law (1933), 660
Class, Heinrich, 421
Classical Scholarship: A Summary (Darstellung der Altertumswissenschaft) (F. A. Wolf), 106, 107
Clausewitz, Carl Philipp Gottlieb von, 184–188
Clausius, Rudolf, 234, 345–348, 479, 819, 854
Clement, Ricardo. *See* Eichmann, Adolf
climatology, 180
Closing of the American Mind, The (A. Bloom), 833–834
coffee houses. *See* cafés
Coffin, Henry Sloane, 680
Cold Amnesty (Die kalte Amnestie) (Friedrich), 798
Coldness, The: An Isolation (Die Kälte: eine Isolation) (Bernhard), 797
Coleridge, Samuel Taylor, 314–315, 883n10
Collin (Heym), 796
Colm, Gerhard, 726
Cologne, 517, 757
Combes, Sonia, 16
Commentary (magazine), 720
Communio (journal), 787
Communist League, 253
Communist Manifesto, The (Manifest des kommunistischen Partei) (Marx and Engels), 253–254, 257
community, redemptive. *See* redemptive community
Community and Civil Society (Gemeinschaft und Gesellschaft) (Tönnies), 450–452
Concept of the Corporation, The (Drucker), 728
Concept of the Political, The (Der Begriff des Politischen) (Schmitt), 668–669, 723
Concerning German Literature (Friedrich the Great), 62
Concerning the Spiritual in Art (Kandinsky), 517
Condition of the Working Class in England (Die Lage der arbeitenden Klasse in England) (Engels), 251, 257
Confessing Church (Bekennende Kirche), 678, 680
Congress of German Naturalists and Physicians, 365
Conquest, Robert, 759
consciousness, 125, 144, 147. *See also* self-consciousness
Conservative Revolution, 452, 618–619
Constitution of Liberty, The (Hayek), 749
Contact (magazine), 747

Contemporary Crisis in Geman Physics, The (Die gegenwärtige Krise der deutschen Physik) (Stark), 599
Continuity and Irrational Numbers (Dedekind), 485
Conze, Alexander, 415
Conze, Werner, 670
Copenhagen (Frayn), 691
Corinth, Lovis, 515, 528
Cornelius, Peter von, 217–218, 218–219, 239
Cornwell, John, 660, 691
Correns, Karl, 286, 391
Coser, Lewis, 716–717, 726, 728, 730
Courage to Be, The (Tillich), 730
Courant, Richard, 487, 659, 664
Crabb Robinson, Henry, 314, 315
Crabwalk (Im Krebsgang) (Grass), 798
Craig, Gordon, 33, 402, 403–404
Cranko, John, 804
Crawshaw, Steve, 22, 23, 792, 798
Creation, Redemption, and the Last Judgment (Cornelius), 239
Crell, Lorenz, 226
Crelle, August Leopold, 349
Crelle's Journal, 349–350
crime detection, beginning of modern, 897n17
Cristofori, Bartolomeo, 154
critical method
  adoption by Marx, 251
  transformation of, 231
critical theory
  Frankfurt method of, 573
  Habermas and, 775–776
Critique of Judgment (Kritik der Urteilskraft) (Kant), 143–145, 161, 201
Critique of Practical Reason (Kritik der praktischen Vernunft) (Kant), 142–143
Critique of Pure Reason (Kritik der reinen Vernunft) (Kant), 139–142
Crookes, William, 479
Cross on the Mountains (Friedrich), 221, 239
Cullen, William, 175–176
Cultural Alliance for the Democratic Revival of Germany, 758
cultural pessimism, tradition of, 616–618, 675, 837–842
culture
  German concept of (*See Kultur*)
  Nietzsche's critique of contemporary, 339
Culture of Defeat, The (Die Kultur der Niederlage) (Schivelbusch), 566
Cuno, Theodor, 256
Curtius, Ernst, 410–411, 415
Czechowksi, Heinz, 801

D

Dada, 563–565
Dahlmann, Friedrich Christoph, 408, 410
Dahrendorf, Ralf, 452, 732, 748, 761, 762–765, 842, 920n23
Daimler, Gottlieb, 376–377
Daimler, Paul, 377, 378
Dali, Salvador, 221
Damm, Christian Tobias, 95
Dannhauser, Josef, 305–306
Danton's Death (Dantons Tod) (Büchner), 302
Darnton, Robert, 56
Darwin, Charles, 180, 430, 615
Darwinism, 426–428, 427, 431
Dawson, John, Jr., 601
Daxelmüller, Christoph, 655
Dead City, The (Die tote Stadt) (opera, Korngold), 591
Death in Venice (Der Tod in Venedig) (T. Mann), 513–314
Debit and Credit (Soll und Haben) (Freytag), 439
Debreyne, P. J. C., 395
Decision (magazine), 817
"Declaration of Notables" (Berlin), 522
Decline of the German Mandarins, The (Ringer), 765–768
Decline of the West, The (Der Untergang des Abendlandes) (Spengler), 560, 620, 839
Dedekind, Richard, 174, 485
defense of the wolves (Verteidigung der Wölfe) (Enzensberger), 800
degeneracy/degeneration, 428, 429
Degenerate Art exhibition, 632–633, 908n7
degenerate art law (May 1938), 633
Degeneration (Entartung) (Norday), 429
Dehmel, Richard, 510, 532
Deichman, Ute, 696
De l'Allemagne (Staël), 311, 313, 323
Delbrück, Hans, 409, 533–534
Delius, Frederick, 336
Demnig, Gunter, 815, 849
Democracy in Germany (Dahrendorf), 748
Democratic and the Authoritarian State, The (Newmann), 762
"Denial of the Vagina, The" (Horney), 613
Denying the Holocaust (Lipstadt), 7
Deputy, The (Der Stellvertreter) (Hochhuth), 736, 802
design
  Biedermeier furniture, 305–306
  Jugendstil and, 508–509
  in Vienna, 496–499
Dessoir, Max, 395
Destruction of Reason, The (Die Zerstörung der Venunft) (Lukács), 762
Deutsche Chronik (Schubart), 55
Deutsche Demokratische Republik (DDR). *See* East Germany
Deutsche Forschungsgemeinschaft, 690
Deutsche Frage, Die (Röpke), 762

Deutsche Gelehrtenrepublik, Die (Klopstock), 102
Deutsche Grammatik (J. Grimm), 230
Deutsche Hochschule für Politik (German Institute for Politics), 570
Deutsche Merkur, Der (monthly), 113
Deutsche Mythologie (J. Grimm), 266
Deutschen, Die (Moeller van den Bruck), 617
Deutsche Problem, Das (Ritter), 762
Deutscher, Isaac, 750
Deutscher Sozialismus (Sombart), 454
Deutscher Werkbund, 510
Deutsches Wörterbuch (Grimm brothers), 230
Deutsch-Französische Jahrbücher (German-French Annuals), 249
Deutschland, Deutschland über alles (Tucholsky), 582
Deutschlands Schicksal (Hirsch), 686
Deutschland und der Weltkrieg (Hintze, Meinecke, Oncken, and Schumacher), 534
Dewey, John, 539–542
Diagnosis of Our Time (Mannheim), 703–704
Dialectic of Enlightenment (Dialektik der Aufklärung) (Adorno and Horkheimer), 725, 775
Dialectics of Secularisation, The (Die Dialektik der Säkularisierung) (Habermas and Ratzinger), 787
Diary of a Ne'er-do-well (Aus dem Leben eines Taugenichts) (Eichendorff), 290
Dibelius, Martin, 650
Dichterliebe (Schumann), 307
"Dichtung und Wahrheit" (Auden), 818
Dido (Charlotte von Stein), 118
Diederichs, Eugen, 614
Diehl, Carl, 323, 324
Diesel, Rudolf, 378
Dieterle, William, 736
Dietl, Helmut, 808
Dietrich, Marlene, 513, 581, 592, 593
Dill, Ludwig, 507
Dilthey, Wilhelm, 441–444, 447, 455, 605, 675, 682, 750, 854
Dinter, Artur, 674
Dirichlet, Peter Gustav Lejeune, 174, 234, 350, 485
Discourse der Mahlern (periodical), 57
Discovery of the Unconscious, The (Ellenberg), 393
Dismissal of Bismarck, The (film), 593
Divided Heaven, The (film), 806
Diwald, Helmut, 11
Dix, Otto, 553, 565, 629, 700
Döblin, Alfred, 619, 738
Doctor Faustus (T. Mann), 737, 836
Documenta (Kassel), 811
Dog Years (Hundejahre) (Grass), 792
Dohnanyi, Christoph von, 680
Dohnanyi, Hans von, 680
Dohnanyi, Klaus von, 680
Dokupil, Jiri Georg, 813
Doll, Richard, 665
Döllinger, Johann, 317
Don Carlos (Schiller), 128
Don Juan (R. Strauss), 465, 466
Don't Mention the War (Ramsden), 5
Doppler, Christian Johann, 391
Dorn, Walter, 49
Dornberger, Walter, 658
Dörpfeld, Wilhelm, 410, 413–414
Double Ego, The (Das Doppel-Ich) (Dessoir), 395
Double Indemnity (film), 591
Double Life (Doppelleben) (Benn), 710
Dow, James R., 654
"Dread of Women, The" (Horney), 613
Dream, a Life, The (Der Traum, ein Leben) (Grillparzer), 294
Dreser, Heinrich, 364, 365, 856
Droste-Hülshoff, Annette von, 305
Droysen, Johann Gustav, 410, 424, 445, 839
Drucker, Peter, 728
Du Bois-Reymond, Emil Heinrich, 234
Dudow, Slatan, 588, 805
Duino Elegies (Duineser Elegien) (Rilke), 574–575
Duisberg, Carl, 362, 363–364
Dumont, Louis, 845
Dupuy, Trevor, 543–544, 698
Dürer, Albrecht, 43
Düreristen, 217–220
Dürrenmatt, Friedrich, 803
du Sautoy, Marcus, 174
dye industry, 358–359, 387

E

Earth Spirit (Erdgeist) (Wedekind), 515
East Germany, 781, 782, 813
  films, 805–806
  literature, 794–796
Ebert, Carl, 744
Eckhart, Dietrich, 436, 562
Eckstein, Ferdinand, 193
economics and refugees in the United States, 727–728
Eddington, Arthur, 172, 658–659
education. *See also* academic institutions
  budget during Third Reich, 689
  difference between school and university, 229
  enforced school attendance in Prussia, 829
  Friedrich the Great and, 63
  influence in Britain, 316–317, 318
  influence in the United States, 323
  influence of universities of Halle and Göttingen, 54

Kant on, 143
Lessing on, 124
Mannhein on, 704
military church and, 48
physics institutes, 476
Pietism and, 47
reforms, 227, 233–234
state and schools relationship, 108–109
technological, 234
Effi Briest (Fontane), 440–441
Efficiency of Our Race and the Protection of the Weak, The (Die Tüchtigkeit unsrer Rasse und der Schutz der Schwachen) (Ploetz), 433–434
Effusions of an Art-Loving Monk (Herzensergiessungen eines kunstliebenden Klosterbruders) (Wackenroder and Tieck), 216
Egotism in German Philosophy (Santayana), 542
Ehrenfels, Christian von, 493, 494–495
Ehrlich, Paul, 278, 388, 390–391, 853
Eich, Günter, 799
Eich, Hermann, 762
Eichendorff, Joseph von, 200, 290, 463
Eichengrün, Arthur, 364, 856, 865
Eichhorn, Karl Friedrich, 264–266
Eichmann, Adolf, 695, 721–722
Eichmann in Jerusalem (Arendt), 721–722
Eichrodt, Ludwig, 305
"Eighteenth Brumaire of Louis Bonaparte, The" (Marx), 248
Einstein, Albert, 482–484, 598, 599, 738
contemporary world of ideas and, 819
Gödel and, 739–740
Hilbert's collaboration with, 487
persecution of, 552, 595, 658–659
Popper student of, 706
on scientific theory, 905n7
Einstein's German World (Stern), 35
Eisenmann, Peter, 815
Eisenstaedt, Alfred, 735
Eisler, Hanns, 641
Eisner, Lotte, 760
Elegy for Young Lovers (Elegie für junge Liebende) (opera, Henze), 810
Elektra (opera, R. Strauss and Hofmannsthal), 465, 466, 467–468
Eley, Geoffrey, 27
Elgio storica del Cabaliere Antonio Raffaelle Mengs (Bianconi), 207
Elias, Norbert, 31, 437, 531, 566, 568, 607, 608–609, 698, 754, 768, 848
Eliot, George, 314, 316, 331
Eliot, T. S., 34, 840
Elkana, Yehuda, 9–10
Ellenberger, Henri, 393
Elman, Mischa, 702
Elmhirst, Leonard and Dorothy, 744
Elon, Amos, 431, 475
Emergency Rescue Committee (ERC), 700
Emergency Society of German Scholars Abroad, 663
Emil and the Detectives (Emil und die Detektive) (Kästner), 583
Emmerich, Wolfgang, 655, 801
Emperor, Rome and Renovatio (Kaiser, Rom und Renovatio) (Schramm), 607–608
Empty Fortress, The (Bettelheim), 717
Endell, August, 508
End of Economic Man, The (Drucker), 728
Engels, Friedrich, 188, 245, 249, 251, 256–259, 853–854
engineering, 356, 844
engines, 375, 378, 600, 692, 693
"Englishness of English Art, The" (lectures by Pevsner), 748
Enigma, 693
Enlightenment. *See Aufklärung*
Entarte Musik exhibition (Düsseldorf, 1938), 643
Entartete Kunst exhibition (Degenerate Art), 632–633, 908n7
Enzensberger, Hans Magnus, 761, 794, 796, 800, 802, 808
Epistle to the Romans (Römerbrief) (Barth), 677
Erb, Rainer, 801
Erdmannsdorff, Friedrich, 211
Ericksen, Robert, 685, 686
Erickson, Raymond, 304–305
Erikson, Erik, 717
Ernst, Max, 222, 699, 701, 735
Ernst August Konstantin (Weimar), 112
Ernst Mach Society, 602. *See also* Vienna Circle
Eros and Civilization (Marcuse), 722
Erwartung (Schoenberg), 472
Esalen Institute, 719–720
Esau, Abram, 691
Escape from Freedom (Fromm), 718–719
Essay Concerning Human Understanding (Locke), 68
"Essay on a New Principle for Ascertaining the Curative Power of Drugs, with a Few Glances at Those Hitherto Employed" (Hahnemann), 176
Essence of Christianity, The (Das Wesen des Christentums) (Feuerbach), 245–246, 329, 674
Essence of Judaism, The (Baeck), 750
Essinger, Anna, 755
Esslin, Martin, 745, 751
Ethics (Ethik) (Bonhoeffer), 680–681
Ettinghausen, Richard, 734
Ettingshausen, Andreas von, 285, 293
eugenics, 433–434
Euler, Leonhard, 171, 351
Europäische Triarchie, Die (Hess), 247

Evans, Richard, 12, 20, 21, 28, 428
events of 1968, 22, 719, 779, 785, 832, 842
Everett, Edward, 323
Every Tom, Dick, and Harry Novel (Hinze-Kunze-Roman) (Braun), 796
evolution, pre-Darwinian idea of, 202–203
evolutionism, 84–88
Exhortation to the Germans to Exercise Their Reason and Their Language Better (Ermahnung an die Deutschen, ihren Verstand und ihre Sprache besser zu üben) (Leibniz), 57
"Exile, Science and Bildung" (conference, 2002), 741
Extinction: A Degeneration (Auslöschung: ein Zerfall) (Bernhard), 797
Eyde, Sam, 361
Eyes of the Mummy, The (film), 589

F
Fabian (Kästner), 583
Fackel, Die (periodical), 490
Failure of Illiberalism, The (Stern), 401
"Farewell" (Benn), 800
Fassbinder, Rainer Werner, 803, 807
Faust (Goethe), 116, 120–121, 217–218
Faust, Albert, 323
Dr. Faust legend, 105, 121
Fear Eats the Soul (film), 807
Fechner, Gustav, 394
Fehling, Jürgen, 648
Feigl, Herbert, 631
Feininger, Andreas, 735
Feininger, Lyonel, 630, 702
Fenichel, Otto, 570, 613, 718–719
Ferguson, Niall, 841
fertilizers, 271–272, 279–281
Fetting, Rainer, 812
Feuchtwanger, Lion, 578–579, 586, 701, 702, 855–856
Feuerbach, Ludwig, 245–246, 296, 329–330, 853
Fichte, Johann Gottlieb, 148–151, 262
  Benda on, 623
  de Staël and, 312–313
  on German Nation, 262
  Idealism of, 138
  influenced by Kant, 138, 148
  influence on Engels, 256
  on role of professors, 226, 228, 229, 233
  Schinkel and, 212
  on self, 195–196
  understanding of the will, 197
  at University of Berlin, 227, 228
  Wissenschaftslehre, 119
Fifth International Congress for Genetics (1927), 697
filmmaking
  after fall of Berlin Wall, 808
  after World War II, 758, 805–809, 806–808
  contemporary, 844
  directors in Hollywood, 587, 589, 590
  documentaries, 808
  in East Germany, 805–806
  Expressionism in, 568–569
  Gegenwartsfilm, 805
  Heimatfilme, 806
  under Hitler, 638–639
  impact of Germans in Britain, 744
  Kuratorium junger deutscher Film, 806–807
  music by Korngold, 591
  New German Cinema (Das neue Kino), 806
  "rubble" films, 806
  Verbotsfilme, 806
  during Weimar Republic, 587–593, 760
  during World War I, 552
Fink, Eugene, 823
Fink, Karl, 203, 204
Finkelstein, Norman G., 10
Fire, The (Der Brand) (Friedrich), 798
Fischer, Eugen, 697
Fischer, Fritz, 536–537, 761
Fischer, Joschka, 2, 733, 780, 783, 784
Fischer, Karl, 429
Fischer, Samuel, 575
Fitzgerald, F. Scott, 581
Five Cantos/August 1914 (Fünf Gesänge) (Rilke), 574
Fliegende Blätter (magazine), 305
Flute Concert, The (Menzel), 526
Focus (magazine), 29
folklore, Third Reich and, 654–657
Fontane, Theodor, 440, 520, 522
Fontes Rerum Germanicarum (Böhmer), 266
Forster, Georg, 178
Förster-Nietzsche, Elisabeth, 561, 676
Fortunatus and His Sons (Hofmannsthal), 492
Forty-Eighters, 324
Forty-Fivers, 785, 789
49th Parallel (film), 746
Foster, Norman, 815
Foundations of Arithmetic (Die Grundlagen der Arithmetik) (Frege), 486
Foundations of the Nineteenth Century, The (Die Grundlagen des Neunzehnten Jahrhunderts) (Chamberlain), 436, 620
Foundations of the Whole Doctrine of Science, The (Die Grundlage der gesamten Wissenschaftslehre) (Fichte), 119, 148–149, 150
Founding of the German Empire, The (Die Begründung des deutschen Reichs) (Sybel), 407
"Frage nach der Technik, Die" (lecture, Heidegger), 770
France. *See also* Franco-Prussian War (1870–71)

anti-French position of Mommsen, 405
anti-Semitism in, 20, 499
dye industry in, 359
Free League of Artists, 699
impact of French Revolution, 129, 197, 261
intelligentsia in, 54–55
interest in German culture, 311–314
Liebermann in, 506
Paris, discussion on shelling of, 373
refugees in, 699
Franck, James, 545
Francke, August Hermann, 47, 52, 125, 316
Franco-Prussian War (1870–71), 385, 386, 418, 421, 519
Frank, Hans, 652
Frank, Philipp, 602
Frankfurt am Main, 59, 757, 814
Frankfurt Institute, 572, 630–631, 679–680, 702, 718–719, 722, 725
freedom, concept of, 76
Fichte and, 149–150
in German philosophy, 26
Hayek on, 706
influence of Ranke on, 270
Kant and, 150
moral, 446
Tönnies on, 451
Frege, Gottlob, 486, 487, 559, 609, 729–730
Freiligrath, Ferdinand, 247, 290, 299, 304
Freischütz, Der (Weber), 160
Frenkel-Brunswik, Else, 662
Freud, Anna, 717, 753, 754
Freud, Sigmund, 37, 395–397, 611–612
break with Jung, 553–555
in Britain, 753
contemporary world of ideas and, 819, 820–822, 827
emigration, 662
German Academy of the Arts and Sciences in Exile and, 700
influence of, 468, 706
in the United States, 713
Mahler as patient of, 469–470
student of Brentano, 492–493
on the United States, 716
in Vienna after World War I, 567
Freud, Sophie, 567
Freudianism and Marxism, 572–573
Freundlich, Otto, 699
Freyer, Hans, 454, 650
Freytag, Gustav, 430, 439
Frick, Wilhelm, 611
"Friday's Child" (Auden), 818
Friedlander, Max, 734
Friedländer, Saul, 784
Friedländer, Walter, 734
Friedrich II the Great (Prussia), 45, 49, 60–63
Bach, Johann Sebastian, and, 41
biographies, 315, 608
on contemporary culture, 62
study by Ranke, 268
Friedrich August I (Saxony), 168
Friedrich August II (Saxony), 208
Friedrich Wilhelm I (Prussia), 45, 48
Friedrich Wilhelm II (Prussia), 211
Friedrich Wilhelm III (Prussia), 213–214, 227
Friedrich Wilhelm IV (Prussia), 219, 299, 304, 320–321
Friedrich, Caspar David, 220–222, 472, 508, 807, 852
Friedrich, Jörg, 798
Friedrich, Walter, 480
Friend of Health, The (Freund der Gesundheit) (Hahnemann), 175
Friendship (Pforr), 217
Frings, Josef, 787
Frisch, Karl von, 660
Frisch, Max, 793
Frisch, Otto, 666
Fritsch, Theodor, 673
Froberger, Johann Jacob, 153
From Caligari to Hitler (Kracauer), 588–589, 760
From Hegel to Nietzsche (Löwith), 772
Fromm, Erich, 572, 608, 631, 701, 718–719
Front Page, The (film), 591
Fry, Helen, 756
Fry, Varian, 700–701, 728
Fuchs, Klaus, 745
Fuchs, Leonhart, 79, 80
Fuchsel, J. C., 169
Füger, Fridrich Heinrich, 215
Führerprinzip (film), 593
Fundamental Law of Arithmetic (Grundgesetze der Arithmetik) (Frege), 559
Funk, Leo, 377
Furedi, Frank, 821
Fürsten und Völker von Süd-Europa (Ranke), 267
Furtwängler, Wilhelm, 641, 644, 645, 809
Fuseli, Henry (Johann Heinrich Füssli), 215
Fustel de Coulanges, Numa Denis, 624
Future of an Illusion, The (Die Zukunft einer Illusion) (Freud), 611
Future of Human Nature, The (Die Zukunft der menschlichen Natur) (Habermas), 845–847
Future of Industrial Man, the (Drucker), 728

G

Gadamer, Hans-Georg, 291, 720, 772, 773–775, 837
Gaffky, Georg, 387, 388
Gagen, Friedrich von, 230
Gaines, James, 41
"Galius Dei" (T. Mann), 503

Gang vor die Hunde, Der (Kästner), 583
Gärtner, Friedrich von, 219
Gas I (Kaiser), 552
Gatterer, Johann Christoph, 71, 862n4
Gauss, Carl Friedrich, 170–174, 177, 350, 353, 485, 852
Gawthrop, Richard, 46, 69
Gay, John, 586
Gay, Peter, 573, 604, 607
Gay-Lussac, Joseph-Louis, 273
Geiger, Hans, 691
Gellert, Christian, 57–58
Gellner, Ernest, 746, 754–755
Gemeinschaft, concept of, 451, 763
Genesis (Werth), 434
genius, concept of, 75–76, 145
Genius for War, A (Dupuy), 543–544
Gentz, Heinrich, 212, 213
geology, 66, 167, 168–169, 203, 557
George, Stefan, 292, 471, 547, 574, 608, 618, 682
Georgekreis (around Stefan George), 574, 608
Gerhardt, Charles Frédéric, 276
German Academy of the Arts and Sciences in Exile, 699–700
German Central Work Station for Folk and Cultural Landscape Research, 654
German Chainsaw Massacre, The (film), 808
German Democratic Republic (GDR). *See* East Germany
German Economy in the Nineteenth Century, The (Die Deutsche Volkswirtschaft im neunzehnten Jahrhundert) (Sombart), 453
German Historical Institute (Washington, D.C.), 36
German History (Lamprecht), 33
German History in the Time of the Reformation (Deutsche Geschichte im Zeitalter der Reformation) (Ranke), 268
Germania anno zero (film), 805
German Idea of Freedom, The (Krieger), 762
German Ideologie, The (Die deutsche Ideologie) (Marx and Engels), 252, 257
Germanien (magazine), 657
German language, 3, 56, 57
German Navy League (Deutscher Flottenverein), 420
Germanness, 45, 669–672
German Orchestra (New York), 325
German Philosophy and Politics (Dewey), 539–542
Germans, The (Elias), 31
"German Schools of History, The" (Acton), 317
Germantown (Pennsylvania), 323
Germantown Zeitung (periodical), 325
German unification. *See also* reunification
  demand for, 94, 289
  Disraeli on, 859n56
German Wartime Society 1939–1945 (Potsdam Institute of Military History), 786
Germany. *See also* East Germany; German unification; reunification
  change of meaning in Beethoven's time, 164
  effect of decentralization on periodicals, 56
  Friedrich and liberation movements, 221–222
  image in Britain, 2
  self-image, 401
  theory of two Germanies, 537–538
"Germany, You Eternal Flame" (Schumann), 635
Germany and the Next War (Deutschland und der nächste Krieg) (Bernhardi), 539
Germany in Autumn (film), 807
Germany's Aims in the First World War (Griff nach der Weltmacht) (Fischer), 536–537
Gershenkron, Alexander, 727
Geschichte des Agathon (Wieland), 113
Gesellschaft, Die (monthly), 510
Gesellschaft as opposed to *Gemeinschaft,* 451, 763
Gesellschaft und Demokratie in Deutschland (Dahrendorf), 452
Gesner, Johann Matthias, 52
Gewissen (journal), 617
Gierke, Otto von, 533
Gildersleeve, Basil, 324
Gilly, David, 212
Gilly, Friedrich, 212–213
Giving of the Keys (Mengs), 210
Glazer, Nathan, 8
Globke, Hans, 759
Glock, William, 750–751
Gluck, Christoph Willibald, 155–156
Gneist, Rudolf von, 326
"Goat Song, Swelling Up" ("Anschwellender Bockgesang") (B. Strauss), 783
Gobineau, Joseph-Arthur de, 434, 615
Gödel, Kurt, 601–602, 631, 739–740, 841
Goebbels, Joseph
  Berlin Philharmonic Orchestra and, 642
  campaign against *All Quiet on the Western Front,* 580
  on meeting of Hitler and Chamberlain, 436
  minister for Popular Enlightenment and Propaganda, 629
  Moeller van den Bruck and, 618
  power of radio and, 638
  propaganda, 639
  theater and, 646, 647
Goehr, Walter, 747
Goethe, Johann Wolfgang von, 58, 114–122
  biology and, 201
  on community, 836
  Germaine de Staël and, 312

Herder and, 123
Humboldt, Alexander von, and, 177
on infiltration of French culture into Prussia, 62
influences on, 99, 191
Karl August of Weimar and, 114, 116
Niebuhr and, 264
Persian and, 192
poems set to music, 160, 463
portraits of, 215
on purpose of life, 833
Schiller and, 128–129
urphenomena, 203–205
at Weimar court, 114
Goethe's Elective Affinities (Goethes Wahlverwandschaften) (Benjamin), 619
Goldene Spiegel, Der (The Golden Mirror) (Wieland), 112–113
Goldhagen, Daniel, 18, 21. *See also Hitler's Willing Executioners*
Goldscheid, Rudolf, 447
Goldsmid, Isaac Lyon, 317
Goldstein, Eugen, 479
Gombrich, Ernst, 567, 733, 744, 745, 747
Gooch, G. P., 263, 266–267
Good-Bye, Lenin! (film), 808
Goodman, Nelson, 729
Goosepluckers, The (Liebermann), 506
Göring, Hermann, 435, 646, 668, 695
Göring, M. H., 661
Gorky, Maxim, 823
Gospel of John, The (Das Evangelium des Johannes) (Bultmann), 679
Gothic Revival, 216
Gottsched, Christoph, 57, 124
Gottsched, Johann, 102
Götz von Berlichingen with the Iron Hand (Götz von Berlichingen mit der eisernen Hand) (Goethe), 58, 114
Gould, Stephen Jay, 177, 852
Grabbe, Christian Dietrich, 303
Grass, Günter, 28, 761, 780, 781, 792–793, 798, 832
Graupner, Gottlieb, 325
Grauzone morgens (Grayzone in the Morning) (Grünbein), 802
Great Society of Mannheim, 703–704
Green Henry (Der grüne Heinrich) (Keller), 296, 297
Greenspan, Nancy Thorndike, 598
Green Table, The (grüne Tisch, Der) (Jooss), 804
Grierson, Ronald (Rolf Hans Griessmann), 752–753
Grillparzer, Franz, 290, 293–294, 558
Grimm, Jacob, 230, 265, 309, 310
Grimm, Wilhelm, 230, 309, 310
Gropius, Walter, 510, 568, 571, 744, 757
Gross, Hans, 897n17
Gross, Jan, 18
Gross, Michael, 422, 423
Gross, Walter, 651
Grosz, George, 565, 629, 734–735
Grotefend, Georg Friedrich, 181–184
Grube, Nikolai, 844
Grünbein, Durs, 802
Gründerzeit, 374
Gründgens, Gustaf, 648
Grundmann, Reiner, 453
Grundzüge der wissenschaftlichen Botanik (Schleiden), 284
Gruner, Ludwig, 319
Gruppe 47 (Group 47), 790
Gumpert, Martin, 174, 176
Günther, H. F. K., 621
Gutzkow, Karl, 303
Gymnasien, 109, 233–234
Gypsy Blood (Carmen) (film), 589

H

Haber, Fritz, 361, 533, 545, 660
Haberler, Gottfried von, 727
Habermas, Jürgen, 11, 494, 761, 775–777, 849
debate with Gadamer, 775
on events of 1968, 779
on Heidegger, 775, 845–847
on parallels between religious fundamentalism and Nazism, 784–785
on reading societies, 56
redemptive community and, 837
on television, 809
on Weimar court, 116
Haeckel, Ernst, 421, 426, 427, 428, 533
Haffner, Sebastian, 745, 750
Hagen, Paul (Karl Frank), 700
Hahn, Otto, 545, 666, 691
Hahnemann, Samuel Christian Friedrich, 174–177
Halftime (Halbzeit) (Walser), 793
Haller, Albrecht von, 80–81
Halske, Johann Georg, 356
Hamann, Brigitte, 501
Hamann, Johann Georg, 77
Hamburg, 59, 757
Hamburgische Dramaturgie, Die (periodical), 103, 131
Hamlet-Maschine (Müller), 803
Hammer and Anvil (Hammer und Amboss) (Spielhagen), 439–440
Handel, Georg Friedrich, 60, 153–154
Handel and Haydn Society, 325
Handke, Peter, 796, 804, 807–808
Hangmen Also Die (film), 590, 737
Hanna, Martha, 538
Hanslick, Eduard, 469
Harbou, Thea von, 590

Hardenberg, Friedrich von. *See* Novalis
Hardenberg, Karl August von, 108, 227, 236–237
Harfouch, Corinna, 797
Harnack, Adolf von, 533, 674, 675, 676
Harris, Henry, 282
Harteck, Paul, 690–691
Hartley, Marsden, 525
Hartmann, Eduard von, 394
Hartmann, Heinz, 602
Hartmann, Nicolai, 773
Hartnack, Edmund, 386
Harvard University, 323, 324
Harvey, Elizabeth, 588–589
Harvey, Lilian, 639
Harz Journey, The (Die Harzreise) (Heine), 298
Haskins, Charles Homer, 92
Hastings, Max, 25, 786
Hata, Sachahiro, 390
Hatfield, Henry, 100, 101
Hauff, Reinhard, 807
Haunted Screen, The (Eisner), 760
Hauptmann, Elisabeth, 586
Hauptmann, George, 784
Hauptmann, Gerhart, 404, 506, 524, 532, 586
Hausegger, Siegmund von, 644
Hauser, Arnold, 703
Hausmann, Raoul, 565
Haydn, Franz Joseph, 59, 156–157, 159
Hayek, Friedrich von, 567, 705–706, 727, 746, 749
Hayman, Ronald, 586
Hearst, Stephen, 745, 751
Heartfield, John, 565, 582, 699
Heaven Has No Favorites (Remarque), 581
Hebbel, Friedrich, 296
Heckel, Erich, 700
Hecker, Waldemar, 516
Hegel, Georg Wilhelm Friedrich, 241–244
  contemporary world of ideas and, 819
  criticism
    by Feuerbach, 246
    by Popper, 707–708
  as early Romantic, 200
  friendship with Hölderlin, 291
  Heine student of, 298
  Hitler and, 615
  Idealism of, 138
  influence of, 248, 251, 256
  influences on, 99, 148
  Lewes and, 316
  Santayana on, 542–543
  sculpture by Wickmann, 215
  speculative philosophy of, 240
  on Winckelmann, 100
Heibut, Anthony, 716
Heidegger, Martin, 291, 292, 454, 494, 603–604, 650, 651–652, 770–772, 845, 863n21
  existentialism and, 603
  Habermas on, 775, 845–847
  influence on Bultmann, 678
  National Socialism and, 672
  on Nietzsche, 823
  relevance of, 847–848
  students of, 722, 723, 773
Heidegger's Children (Wolin), 720, 772
Heiden, Konrad, 701
Heilbut, Anthony, 724–725, 736
Heilige Geschichte (Hess), 247
Heim, Susanne, 669, 671, 672
Heimat (film), 807
Heine, Heinrich, 160, 192, 247, 290, 298–301, 303, 771
Heine, Thomas, 511
Heinkel, Ernst, 600
Heinroth, Johann Christian August, 394
Heisenberg, Werner, 597–599, 665, 691, 692, 841
Heissenbüttel, Helmut, 803
Helm, Georg, 477
Helmer, Olaf, 631
Helmholtz, Hermann von, 234, 343, 344–345, 348, 445, 477, 486, 854
Hempel, Carl, 631
Henckel, Florian, 808
Henle, Jacob, 386
Hennings, Emmy, 564
Henze, Hans Werner, 810, 819
Hepp, Paul, 362–363
Herbart, Johann Friedrich, 394
Herder, Caroline, 120
Herder, Johann Gottfried von, 77, 79, 99, 122–127, 138, 192, 195, 309, 310
Herf, Jeffrey, 454
Heroes Like Us (Helden wie wir) (Brussig), 799
Heroes versus Traders (Sombart), 839
Hertz, Gustav, 659
Hertz, Heinrich Rudolf, 477–478, 599, 854
Herwegh, Georg, 247, 290, 304, 329329
Herzl, Theodor, 470, 490, 499, 682
Herzog, Werner, 807
Hess, Moses, 245, 247
Hesse, Hermann, 575–576
Heydrich, Reinhard, 652
Heym, Georg, 547
Heym, Stefan, 796
Heyne, Christian Gottlob, 52, 100, 106, 181, 228, 230
Hieronymus Holzschuher (Dürer), 528–529
Higgins, Kathleen, 823
Higher Schools and Universities in Germany (Pattison), 318
Hilbert, David, 487, 597, 600–601, 662
Hillgruber, Andreas, 11

Hilpert, Heinz, 646, 648
Himmelfarb, Gertrude, 830, 832–833
Himmelweit, Hilde, 746
Himmler, Heinrich, 20, 435, 620, 656–657, 670, 684
Hindemith, Paul, 292, 568–569, 586, 587, 643–644, 663
Hindenburg, Karl Friedrich, 226
Hinrichs, Carl, 45
Hinsberg, Oskar, 363
Hinschius, Paul, 422
Hintze, Otto, 409, 533, 534
Hirsch, Emanuel, 686
Hirschman, Albert, 727–728
Hirt, August, 696
Hirth, George, 509
Historical Evaluation and Research Organization (HERO), 544
historicism, 69–73, 163, 261–170
  influence in the United States, 713, 731
  law and, 264–266
  Winckelmann and, 100
Histories of the Romanic and Germanic Peoples from 1491 to 1514 (Geschichte der romanischen und germanischen Völker von 1494 bis 1514) (Ranke), 267
Historikerstreit, 10–15, 733
Historische Zeitschrift (journal), 406, 408
history, 30, 70–72, 263
  Benda on, 623–624
  critical tradition, 231
  influence of refugees in the United States, 730–733
  as man's spiritual struggle to overcome nature, 76
  nationalism and, 404–410
  National Socialism and, 615, 653–654
  standard history curriculum, 436
  at University of Göttingen, 51
  in Weimar Germany, 607–609
History and Class Consciousness (Geschichte und Klassenbewusstsein) (Lukács), 775
History of European Thought in the Nineteenth Century (Merz), 170
History of German Law and Institutions (Deutsche Staats- und Rechtsgeschichte) (Eichhorn), 264–265
History of Germany in the Nineteenth Century (Deutsche Geschichte im 19. Jahrhundert) (Treitschke), 321, 408–309
History of Modern Germany, A (Holborn), 732
History of Prussia (Geschichte Preussens) (Voigt), 266
History of Roman Law in the Middle Ages (Geschichte des römischen Rechts im Mittelalter) (Savigny), 265
History of Rome (Römische Geschichte) (Mommsen), 405
History of Rome (Römische Geschichte) (Niebuhr), 231
History of the Art of Antiquity, The (Geschichte der Kunst des Altertums) (Winckelmann), 30, 97–98, 207
History of the First Crusade, The (Geschichte des ersten Kreuzzuges) (Sybel), 405–406
History of the French Revolution (Geschichte der Revolutionszeit 1789–95) (Sybel), 406
History of the German People (Geschichte des teutschen Volkes) (Luden), 266
History of the Popes (Die römischen Päpste) (Ranke), 268
History of the Popes from the Close of the Middle Ages (Geschichte der Päpste seit dem Ausgang des Mittelalters) (Pastor), 607
History of the Synoptic Tradition, The (Die Geschichte der synoptischen Tradition) (Bultmann), 678–679
History of the Thirty Years War (Geschichte des Dreissigjährigen Krieges) (Schiller), 130–131
"Hitler" (Schumann), 635
Hitler, Adolf, 614–616
  on art, 630
  caricature by Beckmann, 630
  education, 559
  in German Workers Party, 562
  meeting with Chamberlain, 436
  at opening night of *Salome,* 467
  polycratic dictatorship, 697–698
  religion and, 673–687
  Schramm and, 608
  in Vienna, 501
  will, 631–632
Hitler, a Film from Germany (Hitler, ein Film aus Deutschland) (Syberberg), 782, 807
"Hitlergeist und Wissenschaft" (The Hitler Spirit and Science) (Stark and Lenard), 599
Hitler's Professors (Weinreich), 759–760
Hitler's Willing Executioners (Goldhagen), 18–22, 733
Hitler: The Man and the Military Leader (Hitler als militärische Führer) (Schramm), 608
Hitler Youth Quex (film), 640
Hobhouse, Hermione, 319
Hobsbawm, Eric, 748–749
Höch, Hanna, 565
Hochhuth, Rolf, 736, 780, 781, 802
Höchst, 360, 695
Hoffmann, Eric Achille, 390
Hoffmann, E. T. A., 162–163, 200, 831
Hoffmann, Felix, 364, 856
Hoffmann von Fallersleben, August Heinrich, 299, 304
Hofmann, August Wilhelm von, 318, 357, 358, 491–492

Hofmann, Hans, 734
Hofmannsthal, Hugo von, 467–468, 490, 492, 567, 612, 619
Holborn, Hajo, 570, 730, 731–732
Holbrooke, Richard, 733
Hölderlin, Friedrich, 138, 200, 241, 291–292, 548
Hollein, Hans, 814
Holocaust, denial of, 6–10
Holocaust in American Life, The (The Holocaust and Collective Memory) (Novick), 7–8
Holocaust Industry, The (Finkelstein), 10
Holocaust Library (London), 749
Holst, Imogen, 752
Holy Family, The (Die heilige Familie) (Engels and Marx), 257
Hölzel, Adolf, 507
Homeopathy (journal), 174
Honecker, Erich, 801, 806
Hörbiger, Hanns, 658
Horen, Die (periodical), 129
Horkheimer, Max, 572, 631, 702, 725, 741
Horney, Karen, 570, 612–613, 662
Hottinger, Johann Konrad, 216
Hour of True Feeling, The (Die Stunde der Wahren Empfindung) (Handke), 796
Hour We Knew Nothing of Each Other, The (Die Stunde da wir nichts von einander wußten) (Handke), 804
House of Jewish Learning, 682
Houtermans, Fritz, 691–692
Howard, Thomas, 54
Hoyer, Hermann, 632
Huber, Kurt, 655
Huch, Ricarda, 193
Hufbauer, Karl, 226
Hugenberg, Alfred, 420–421
Hughes, Everett C., 453
Hulse, Michael, 115, 116
Hülsenbeck, Richard, 564, 565
Human Condition, The (Arendt), 721
Humann, Carl, 414–415
Humann, Franz, 414
Humboldt, Alexander von, 81, 138, 177–181, 212, 427, 852, 871n26, 871n33
Humboldt, Wilhelm von, 86, 87, 108–110, 124, 138, 227, 851–852
  on *Bildung,* 832
  de Staël and, 312
  educational reforms, 226, 227–228, 233–234
  on French nation, 262
  on purpose to world history, 262
  Schinkel and, 212
  student of Heyne, 228
Humperdinck, Engelbert, 532
Hunley, J. D., 259
Hunsrückenstrasse (Uecker), 811
Hunt, The (Stuck), 508
Hunt, Tristram, 257, 259
Husserl, Edmund, 37, 486–487, 493–494, 605, 652
  students of, 603, 722, 723, 773
Hutton, James, 169
Huxley, Michael, 803, 804
Hyperion (Hölderlin), 291
hypnotherapy, 556

I
I and Thou (Ich und Du) (Buber), 682–683
"Iconoclasm in German Philosophy" (Oxenford), 331
Idealism, 138, 139, 732
  emergence of, 136
  influence on *Wissenschaftsideologie,* 228
  instrumental music and, 162
  non-Germans and, 845
Ideas for the Philosophy of History of Humanity (Ideen zur Philosophie der Geschichte der Menschheit) (Herder), 261
IG Farben, 361, 695
Iggers, George, 263, 269, 270
Illiger, Johannes, 81
"Illustrations of the Dynamical Theory of Gases" (Maxwell), 347
imagined communities, 58, 114
"Im Granatloch" (Schnack), 551
Immendorf, Jörg, 812, 813
Inability to Mourn, The (Die Unfähigkeit zu Trauern) (Mitscherlich and Mitscherlich), 24, 785, 789–790
Indian Summer (Der Nachsommer) (Stifter), 295
individualism, types of, 449–452
individuality, change in meaning of, 193–196
Informed Heart, The (Bettelheim), 717
Innerlichkeit. See inwardness
insecticides, 360–361
Institute for Advanced Study (Princeton), 659
Institute for Research on the Jewish Question (Frankfurt), 656
Institute of East European Economic Studies (Königsberg), 670
Institute of Fine Arts (New York University), 702, 733–734
Institute of Theoretical Physics (Copenhagen), 596. *See also* Bohr, Niels
instruments, musical, 154, 869n5
Intellectual History Association, 656
International Congress of Mathematicians (Paris, 1900), 487
International General Medical Society for Psychotherapy, 661
International Research Council, 595
Interpretation of Dreams, The (Die Traumdeutung) (Freud), 395, 397, 827

In the Beginning Was the Word (Hoyer), 632
In the Land of Cockaigne (Im Schlaraffenland) (H. Mann), 511, 512
Intrigue and Love (Kabale und Liebe) (Schiller), 128, 132
"Inventur" (Eich), 799
inwardness, 87, 110, 194, 403, 437, 713, 790, 830–833, 838
Iphigenie auf Tauris (Goethe), 117
Irving, David, 6–7
Iselin, Isaak, 76
Isherwood, Christopher, 743, 818
Israel in Egypt (Handel), 154
Italia and Germania (Overbeck), 217
Italienische Forschungen (Rumohr), 218

J

Jachmann, Reinhold, 135
Jacobi, Carl Gustav Jacob, 234–235, 350
Jacobi, Friedrich Heinrich, 138, 147, 149
Jacobi, Lotte, 735
Jacobs, Ian, 738
Jacob the Liar (film), 806
Jaehne, Friedrich, 695
Jahn, Friedrich Ludwig, 309
Jahoda, Marie, 753
Jahrbuch der Psychoanalyse, 555
Jahrbücher für wissenschaftliche Kritik (periodical), 244
Jakobson, Roman, 700
James, Clive, 489, 583, 709, 859n49
James, William, 662
Janco, Marcel, 564
Janet, Pierre, 394–395
Jannings, Emil, 592, 648
Janowitz, Hans, 568
Jarausch, Konrad H., 778–781, 842
Jargon of Authenticity, The (Jargon der Eigentlichkeit) (Adorno), 771–772
Jaspers, Karl, 608, 652, 767, 769–770, 823, 832, 856
Jawlensky, Alexei, 515, 516, 564
Jay, Martin, 630–631
Jeffreys, Diarmuid, 357, 363
Jelinek, Elfriede, 794
Jellinek, Emil, 378
Jellinek, Walter, 650
Jena, Battle of (1806), 108, 138, 185
Jenninger, Philip, 12
Jens, Walter, 794
Jewish Colonial Trust, 499
Jewish Question, The (lecture, Kittel), 685
Jewish-Theological Seminar (Breslau), 655
Jewish Volkskunde, 655–656
Jews. *See also* anti-Semitism
  in Berlin, 520
  Chamberlain on, 436
  downplaying of, in Holocaust, 8–9
  Heine, Heinrich, as, 298, 300
  held responsible for 1873 crash, 521
  identification as "degenerates," 434
  limited form of citizenship, 108
  as physicians, 661
  refugees in London, 745
  resurgence of Jewish culture under Weimar Republic, 619
Jews and Modern Capitalism, The (Die Juden und das Wirtschaftsleben) (Sombart), 452
Joachim, Joseph, 460, 462
Jodl, Alfred, 608
Joël, Karl, 447
Johnson, Alvin, 701, 726
Johnson, Uwe, 796
Johnst, Hans, 646
Johnston, William, 567
Jomini, Antoine Henri de, 187
Jones, Ernest, 394, 662, 754
Jones, Mumford, 159
Jones, William, 190
Jonny spielt auf (Krenek), 587
Jooss, Kurt, 744, 745, 747, 804
Joseph in Prison (Mengs), 209
Joule, James Prescott, 344
Journal für die reine und angewandte Mathematik, 349–350
journalism, 325, 750
Journal of Practical Medicine, 176
journals, professional, 52, 226, 234, 476
J. R. Becher Institute for Literature (Leipzig), 801
Judgment of Paris, The (Mengs), 209
Jüdin von Toledo, Die (Grillparzer), 294
Jugend (journal), 509
Jugendstil, 498, 508–510, 516
Jung, Carl, 553–555, 612, 661
Jung, Edgar, 618
Jünger, Ernst, 454, 618, 669, 709–710, 771, 790
Jünger, Friedrich Georg, 770–771
Junges Deutschland (Young Germany), 303–304
Jung Wien (Young Vienna), 490
Jurisprudence (Klimt), 498

K

Kaczynski, Jaroslaw, 18
Kaczynski, Lech, 18
Kafka, Franz, 577–578, 818
Kaiser, Georg, 552
Kaiser, Joachim, 802
Kaiser Proclamation in Versailles (Werner), 526–527
Kaiser Wilhelm Institutes, 529, 689, 696, 697, 814, 838
Kalnein, Wend von, 211
Kandinsky, Wassily, 472, 508, 515–518, 832

Kansteiner, Wulf, 23
Kant, Hermann, 795
Kant, Immanuel, 59, 136, 139–146
  Bildungstrieb and, 82–83
  biology and, 201
  on Blumenbach, 81
  on community, 836
  contemporary world of ideas and, 819
  Dewey on, 540–541
  on difference between animals and man, 833
  French on, 537
  Herder student of, 123
  Höderlin and, 291
  on ideal reality, 201
  influence of, 149, 171, 315, 331–332, 443, 682
  observation of self, 150
  Santayana on, 542
  on self, 195
Kantorowicz, Ernst, 35, 574, 607, 608, 731, 836
Kapital, Das (Marx and Engels), 254–255, 258–259, 819–820
Kaprow, Alan, 734
Karajan, Herbert von, 809
Karl August, Duke of Sachs-Weimar-Eisenach, 112, 114, 116
Karpinska, Luise von, 394
Kassandra (C. Wolf), 795
Kästner, Erich, 581, 582–584
Kastner, Wilhelm, 273
Katona, George, 728
Kaufmann, Fritz, 556
Kautsky, Karl, 572
Kazin, Alfred, 820–821
Kekulé, August, 271, 277–278, 853
Keller, Gottfried, 37, 296–297, 463
Keller, Hans, 745, 750–751
Keller, Heinrich, 215
Kellner, Carl, 366
Kelly, Alfred, 416, 427, 428
Kelly, Michael, 464
Kempowski, Walter, 799
Kennedy, Paul, 418, 544
Kershaw, Ian, 22, 23
Kersting, Georg Friedrich, 215
Kessler, Henry, Count, 584
Keylor, William, 537
Keynes, John Maynard, 559, 704, 746
Kiefer, Anselm, 812
Kielinger, Thomas, 843
Kielmeyer, Carl Friedrich, 81, 83–84, 86
Kindertransport (Children's Transport) to Britain, 744–745
Kinder- und Hausmärchen (Grimm brothers), 230, 265–266
King Candaules (Hofmannsthal), 492
King's Most Loyal Enemy Aliens, The (Fry), 756
King's Two Bodies, The (Kantorowicz), 608
Kipphardt, Heinar, 736
Kirchner, Ernst Ludwig, 528, 630, 633, 700
Kirchoff, Gustav, 477
Kirsch, Sarah, 801
Kissinger, Henry, 724, 733, 916n40
Kittel, Gerhard, 685
Klácel, Matthew, 391
Klages, Ludwig, 618
Klarsfeld, Serge, 16
Klauer, Martin, 215
Klee, Paul, 515, 629, 630
Kleiber, Erich, 585, 645
Klein, Felix Christian, 350, 352–353
Klein, Fritz, 533
Klein, Melanie, 570, 613, 753–754
Klein, Yves, 811
Kleist, Heinrich von, 200, 290, 293
Klemperer, Otto, 641, 645, 701, 702
Klemperer, Victor, 633–634
Klenze, Leo von, 212, 215, 219
Klimt, Ernst, 497
Klimt, Gustav, 470, 497–498, 507–508, 558
Klinger, Max, 507–508, 532
Klopstock, Friedrich Gottlieb, 102
Kluge, Alexander, 807
Knappertsbusch, Hans, 645
Knife in the Head (film), 807
Knipping, Paul, 480
Knopp, Guido, 23
Knorr, Ludwig, 360
Knowledge and Human Interests (Erkenntnis und Interesse) (Habermas), 776
Koch, Joseph Anton, 215
Koch, Robert, 386–389, 853
Koestler, Arthur, 278, 750
Koffka, Kurt, 716
Kogon, Eugen, 779
Kohl, Helmut, 733, 780
Köhler, Barbara, 802
Köhler, Joachim, 328
Köhler, Johann David, 70
Köhler, Wolfgang, 661, 714, 716, 817
Kojève, Alexandre, 669
Kokoschka, Oskar, 552, 699, 744, 746
Kolbe, Hermann, 279
Kollwitz, Käthe, 527, 552, 629
Kootz, Samuel, 735
Korda, Alexander, 744, 745–746
Korngold, Erich, 591, 736
Kosmos (A. von Humboldt), 180
Kracauer, Siegfried, 588–589, 741, 760–761
Krafft, Adam, 43
Krafft-Ebing, Richard von, 496
Kragh, Helge, 475
Kraus, Karl, 470, 490, 515, 552, 553, 567, 586
Krauss, Werner, 648
Krautheimer, Richard, 734

Krebs, Hans, 745, 746
Kreisler, Fritz, 702
Krenek, Ernst, 587
Krieck, Ernst, 650–651, 652, 653, 672
Krieger, Leonard, 732, 762
Kris, Ernst, 734
Kristeller, Paul Oskar, 731, 741
Kritik. See critical method
Kritische Journal der Philosophie, 241
Kronecker, Leopold, 486
Krupp, Alfred, 369–375
Krupp, Hermann, 370
Kubizek, August, 501
Kuhle Wampe (film), 588
Kuhn, Helmut, 773
Kuhn, Thomas, 342, 344
Kultur, 30–31, 34
  as collective consciousness of a *Volk,* 125
  factor in World War I, 531–532
  as opposed to *Zivilisation,* 31, 532, 535–536, 541, 838
  Spengler on, 561
Kulturbund. *See* Cultural Alliance for the Democratic Revival of Germany
Kulturbund deutscher Juden, 641
Kulturkampf, 422
Kulturstaat, theory of, 236
Kunert, Günter, 796, 801
Küng, Hans, 786
"Kunstwerk im Zeitalter seiner technischen Reproduzierbarkeit, Das" (Benjamin), 709
Kuntze, Reiner, 801

L

Laak, Dirk van, 779
Laban, Rudolf, 744, 804
laboratories, 355–367, 476
Labouring Men (Hobsbawm), 749
Lachmann, Karl Konrad, 232
Lacquer, Walter, 563
Lagarde, Paul de, 434–435, 614, 839
Lagrange, Joseph-Louis, 353
Lam, Wilfredo, 701
Lamentations of Germany, The (Vincent), 44
Lamprecht, Karl, 33, 533
Landes, David, 732
Lane, Allen, 748
Lane, George, 324
Lang, Fritz, 587, 589–590, 702
Langbehn, Julius, 432–433, 452, 522, 614, 839
Lange, Bernd Lutz, 781
Langhans, Carl Gotthard, 211–212, 213
language. *See also* German language
  Herder on, 124
  as identifier of *Volk* or nationality, 124–125
  importance at University of Göttingen, 51
  as main focus of education, 110
Language, Truth and Logic (Ayer), 631
Lanner, Josef, 464
Laplace, Pierre-Simon, 139, 171
Large, David Clay, 520
Lasch, Christopher, 821–822
Lasker, Edward, 521
Lassen, Christian, 184
Lasson, Adolf, 535
Last, R. W., 579, 583
Last Command, The (film), 593
Last Days of Mankind, The (Die letzten Tage der Menschheit) (Kraus), 490, 553
Laterndl (cabaret), 746
Laube, Heinrich, 303
Laudan, Rachel, 170
Laue, Max von, 480, 659, 665, 692
Lavery, John, 505
law
  codification of Prussian law, 63
  as defining achievement of civilization, 265
  historicism and, 264–266
  legal justification for the Nazi laws, 650
  natural law, 72, 194
Lawyers, The (Die Juristen) (Hochhuth), 780, 802
Lazarsfeld, Paul, 716, 726–727, 741
Lazarus, Moritz, 445
League of German Societies for Folklore, 655
League of the Just, 250, 253
Leander, Zarah, 639
Le Corbusier (Charles-Édouard Jeanneret), 510, 571
Lecture Hall, The (Die Aula) (H. Kant), 795
Lectures on the Philosophy of History (Vorlesungen über die Philosophie der Geschichte) (Hegel), 243
Leeuwenhoek, Anton van, 281
Legend of Paul and Paula, The (film), 806
Lehar, Franz, 584
Lehmann, Karl, 734
Lehrhaus Movement, 619
Leibniz, Gottfried, 57, 69, 79, 85, 88, 123–125
Leistikow, Walter, 528
Leitz, Ernst, 365, 366, 367
Lenard, Philipp, 533, 599, 649, 652, 672
Lenau, Nikolaus, 771
Lenoir, Timothy, 37, 83
Lenz (Büchner), 302
Lenz (Schneider), 793–794
Lenz, Johann Michael Reinhold, 77
Lenz, Max, 409, 533, 876n32
Leo Baeck Institute for the Study of the History and Culture of German-speaking Jewry, 750
Leonce und Lena (Büchner), 302
Lepenies, Wolf, 30–33
Lessing, Gotthold Ephraim, 78, 101–105, 124, 131, 215, 623

Letters from Berlin (Briefe aus Berlin) (Heine), 298
Letters on the Kantian Philosophy (Briefe über die kantische Philosophie) (Reinhold), 147
Levassor, Émile, 377
Levi, Erik, 642, 644, 645, 810
Levi, Hermann, 643
Levi-Civita, Tullio, 663
Lewes, George Henry, 314, 315–316
Lewin, Kurt, 716, 717
Libermann, Max, 33, 505–507, 515, 527, 528, 532, 629
Licht (Light) (operas, Stockhausen), 810
Liebenfels, Jörg Lanz von, 434
Liebig, Justus von, 272, 275, 279–281, 318, 341, 358, 366, 853
Liebknecht, Karl, 565
"Lied der Deutschen, Das" (Hoffmann von Fallersleben), 299
Lieder (Schubert), 160, 305
"Lieutenant Gustl" (Schnitzler), 491
Life of Jesus, The (Das Leben Jesu) (Strauss), 245, 256, 316
Light Years (Lichtjahre) (Weidermann), 798–799
Lilge, Frederic, 765–766
Linde, Carl von, 378
Lindner, Richard, 735
Link, Heinrich Friedrich, 81
Lion Has Wings, The (film), 746
Lipstadt, Deborah, 7
List, Friedrich, 325, 878n31
Liszt, Franz, 133, 460, 466
literacy rates, 56, 426, 829
literature. *See also* Nobel Prize for Literature; poetry
  Asphaltliteraten, 635
  in East Germany, 794–796
  German-Americans and, 325
  novella, 296–297
  novels
    Bildung in, 833
    Bildungsroman, 119, 295, 297
    contemporary, 844
    Entwicklungsroman, 440
    sociocritical, 304
    Vaterromane (father novels), 791, 794
  professional, 232
  Trümmerliteratur, 790
  tuberculosis in, 389
Lives of Others, The (film), 808
Lixfeld, Hannjost, 654
Locarno Pact (1925), 595
Locquin, Jean, 210
Loeffler, Friedrich, 387, 388
Logical Foundation of Probability (Carnap), 729
Logical Investigations (Logische Untersuchungen) (Husserl), 494
Logical Structure of the World, The (Der logische Aufbau der Welt) (Carnap), 729
Logic of Scientific Discovery, The (Logik der Forschung) (Popper), 706, 749
Logic of the Cultural Sciences, The (Zur Logik der Kulturwissenschaften) (E. Cassirer), 606
Loos, Adolf, 212, 215, 497
Lorenz, Konrad, 696–697
Lorre, Peter, 586, 590, 702
Lost Honour of Katharina Blum (Die verlorene Ehre der Katharina Blum) (Böll), 792
Lost Weekend, The (film), 591
Love Is Colder than Death (film), 807
Love-Life in Nature (Das Liebesleben in der Natur) (Bölsche), 427
Love Parade, The (film), 589
Lovett, Martin, 752
Lowenthal, Leo, 572, 608
Löwenthal, Richard, 750
Löwith, Karl, 720, 768, 772–773
Loyalty Law (1952), 772
Lübbe, Hermann, 23, 784
Lubitsch, Ernst, 552, 587, 589
Luden, Heinrich, 266
Ludendorff, Erich, 560, 566
Ludwig I (Bavaria), 218
Ludwig I (Hesse), 273
Ludwig II (Bavaria), 329
Ludwig Feuerbach and the End of Classical German Philosophy (Ludwig Feuerbach und der Ausgang der klassischen deutschen Philosophie) (Engels), 258
Lueger, Karl, 434, 501–502
Luisa Miller (Verdi), 128
Lukács, Georg, 37, 298, 449, 703, 762
Lulu (opera, Berg), 515, 585, 643
Lulu (Wedekind), 515, 585
Lunasharski, Anatoly, 823
Lüpertz, Markus, 812
Lust (Jelinek), 794
Luther, Martin, 26, 87
Luxemburg, Rosa, 565, 820

M

M (film), 590
Macaulay, Thomas Babington, 264
Mach, Ernst, 495, 500, 576
Machinist Hopkins (Brandt), 587
Machlup, Fritz, 727
Macke, August, 552
Magee, Bryan, 327, 331, 334
Magic Flute, The (Die Zauberflöte) (Mozart), 158, 747
Magic Mountain, The (Der Zauberberg) (T. Mann), 562, 575
Magnus, Bernd, 823
Mahler, Gustav, 336, 459, 467, 469–470, 558, 591

Mahler-Werfel, Alma, 465, 701, 702
Maier, Charles, 11, 28
Major, Traugott, 215
Makela, Maria, 505
Malkin, Joseph, 702
Management: Tasks, Responsibilities, Practices (Drucker), 728
Man for Himself (Fromm), 719
Manifesto of the 93, 33–34, 532, 676–677, 840
  reaction abroad, 537, 539
Mann, Erika, 648, 817
Mann, Heinrich, 510–514, 562–563, 701, 702, 735
Mann, Klaus, 648, 710, 737–738, 817
Mann, Thomas, 21, 510–514, 575–576, 618, 654, 741
  on *Decline of the West,* 561
  "Gedanken im Krieg," 34
  influenced by early Romantics, 200
  on *Kultur,* 532, 535–536
  on Lagarde, 435
  meaning of German for, 37
  on Munich, 503
  on redemptive community, 836
  refugee in the United States, 700, 702, 737–738
  on Stifter, 296
  on Wagner and Schopenhauer, 327
  on World War I, 562
Mannheim, Karl, 37, 605, 608, 631, 703–704, 741, 745, 754, 914n7
Mannheim Gas Engine Company, 375
Man without Qualities, The (Der Mann ohne Eigenschaften) (Musil), 576–577, 855
Marat/Sade (Weiss), 803
Marc, Franz, 517, 552, 832
Marchand, Suzanne, 106, 107, 832
Marcks, Erich, 533
Marcuse, Herbert, 720, 722, 779
  attempt to merge Freudianism and Marxism, 572–573
  Frankfurt school and, 572
  students of the 1960s and, 719
Maria Stuart (Schiller), 131, 132
Marr, Wilhelm, 522
Marrus, Michael R., 16
Marx, Eleanor, 249, 256
Marx, Karl, 246–247, 250–256
  contemporary world of ideas and, 819–820
  Hess and, 247
  impact of Strauss' *Life of Jesus* on, 245
  influence of, 571, 713
  opposition to, 704, 707–708
Marxism, 240
  Darwinism and, 428
  Freudianism and, 572–573
  Habermas on, 776
Masaryk, Jan, 701
Masaryk, Tomáš, 492–493
Maser, Werner, 616
Maslich, Bruce, 247
masochism, coining of term, 496
Masson, André, 701
Mass Psychology of Fascism, The (Die Massenpsychologie des Faschismus) (Reich), 719
Mathematical Reviews (journal), 663–664
mathematics, 349–353, 484–487
  Gauss and, 171–173, 852
  intellectural overlap with physics, 600–601
  matrix math, 597
  under Nazis, 910n41
  versus poetry, 76–77
  at University of Göttingen, 51
Mathis der Maler (Hindemith), 644
Matussek, Matthias, 1–2, 5
Matussek, Thomas, 1, 3–4, 6
Maude, F. N., 188
Maupertuis, Pierre de, 61
Max Planck Institute, 775, 814
Max Weber and Karl Marx (Löwith), 772
Max Weber und die deutsche Politik (Mommsen), 762
Maxwell, James Clark, 347, 476
May, Joe, 716
Maybach, Wilhelm, 376, 377, 378
Mayer, Julius Robert von, 341–342, 344
May Laws, 425
Mayr, Ernst, 80, 84, 917n60
Mazlish, Bruce, 252, 253, 254
McDougall, William, 538, 615
McGrath, Alistair, 674
McLellan, David, 247, 257
McLuhan, Marshall, 619
Meaning in History (Löwith), 772
mechanists. *See* atheism
Meckel, Christoph, 796
Meckel, Johann Friedrich, 81
medicine, 383–397
  antibiotics, 390
  cancer, German scholarship on, 664–665
  homeopathy, 174–177
  physicians and Nazi party, 661
Medicine (Klimt), 498
Medizinische Reform, Die (weekly), 384
Meidner, Ludwig, 528, 699, 746
Meier, Christopher, 12
Meinecke, Friedrich, 263, 409, 532, 533, 534, 535, 731, 836
Meinecke, Thomas, 799
Meiner, Ludwig, 630
Meiners, Christoph, 430
Meinhof, Ulrike, 780
Mein Kampf (Hitler), 614, 632, 661
Meistersinger von Nürnberg, Die (Wagner), 329, 334

Meitner, Lise, 666
Memoirs of an Infantry Officer (Sassoon), 579
Mendel, Gregor, 37, 86, 286, 391–393, 819
Mendelssohn, Felix, 133, 239, 298, 307, 308–309
Mendelssohn, Moses, 74–75, 78, 102–103, 136, 137
Mengele, Josef, 697
Mengs, Anton Raphael, 207–208, 215
Mengs, Ismael, 207
Menschen am Sonntag (film), 588, 590
Menzel, Adolf von, 526
Merck, Johann, Heinrich, 77
Merkur (magazine), 782
Merry Vineyard, The (Der fröhliche Weinberg) (Zuckmayer), 592
Merz, John Theodore, 170
Messiah, The (Der Messias) (Klopstock), 102
Metamorphosis (Die Verwandlung) (Kafka), 577, 578
Metropolis (film), 590
Metropolis and Mental Life, The (Die Großstadt und das Geistesleben) (Simmel), 448–449
Metternich, Klemens von, 304
Meyer, Carl, 568
Meyer, Eduard, 533
Meyer, Karl, 374
Meyer, Rudolf, 296
Michaelis, Johann D., 71, 72–73, 200, 227
Michael Kohlhaas (Kleist), 293
Michelet, Jules, 92, 395
Micrographia (Hooke), 281
microscope, 283–284, 365–367, 386, 387, 390
Middendorf, Helmut, 812–813
middle class
  Biedermeier culture and, 306
  educated, 403, 837
    cultural pessimism and, 840
    versus the rest, 828–830
  political exclusion, 403
  use of Darwinism as ideological weapon, 426–428
Midsummer Night's Dream, A (film), 736
Mies van der Rohe, Ludwig, 215, 510, 571, 758
military
  aerial warfare, 545
  chemical warfare, 544–545
  deference to in Berlin, 521
  influence of Napoleon on, 186, 417
  institutionalising of, 543–544
  military and naval personnel of the powers, 420
  Pietist influence on, 48
  Prussian, 417–418
  religion and, 683
  warship tonnage, 420
  during World War I, 902n44
  during World War II, 698
Military Society, 185
Miller, Martin, 746
Milliband, Ralph, 746
Mind (journal), 755
Minima Moralia (Adorno), 726
mining, 117, 167–168
Mirat, Crescence-Eugénie (Mathilde), 300–301
Mises, Ludwig von, 727
Mitscherlich, Alexander and Margarete, 24, 785, 789–790
Mitscherlich, Eilhardt, 234, 275
Mitterrand, François, 15, 17
Möbius, August Ferdinand, 174
Modern Capitalism (Der moderne Kapitalismus) (Sombart), 452, 453
modernism
  antimodernism of Lagarde, 435
  as Jewish modernism in Vienna, 501
  reactionary modernism, 454
modernity
  anti-modernity in German universities in the 1920s, 767
  hatred of, 432–437
  Philosophy of Money as first sociological analysis of, 448
  skepticism about, 848
  Weber, Max, on, 825–826
Modern Man in Search of a Soul (Jung), 612
Moeller van den Bruck, Arthur, 616–618, 771, 840, 859n45
Moholy-Nagy, László, 571, 701, 741
Moleschott, Jacob, 427
Moltke, Gebhardt von, 2
Moltke, Helmuth von, the elder, 188, 418, 519
Mommsen, Hans, 404, 445
Mommsen, Theodor, 231, 404–405, 408, 433
Mommsen, Wolfgang, 404, 762
Monge, Gaspard, 353
Monte Carlo (film), 589
Monthly Register, 314
Monumenti antichi inediti (Unpublished Relics of Antiquity) (Winckelmann), 97
Moore, Marianne, 296
morality, 68, 142, 442
Morell, Theo, 672
Morgenstern, Oskar, 727, 918n89
Morgue (Benn), 710
Mörike, Eduard, 305
Morocco (film), 593
Morris, Charles, 729
Morris, William, 571
Moser, Claus, 745, 746, 752
Moses, A. Dirk, 23–24, 36, 785, 843
Mosse, George L., 614, 615
Mötke, Eduard, 463
Moynahan, Bryan, 686
Mozart, Wolfgang Amadeus, 59, 155, 157–158

Mühlfeld, Richard, 462
Mühsam, Erich, 504
Müller, Heiner, 803
Müller, Jan-Werner, 779, 832, 842
Müller, Johannes, 234, 283, 383
Müller, Max, 318
Müller, Wilhelm, 305
Müller's Archiv (journal), 284
Munch, Edvard, 527, 812
Münchhausen, Gerlach Adolf von, 50–51
Mundt, Theodor, 303
Munich, 503–518
  art and artists
    Art Academy, 515
    Artists Association, 504–505
    Bavarian Academy of Fine Arts, 770
    House of German Art (Haus der deutschen Kunst), 632
    Neue Künstlervereinigung München (New Artists' Association) (NKVM), 516, 517
    Royal Academy of Fine Arts, 504
  German industrial exhibition of 1854, 504
  Ludwigskirche, 219
  Municipal Archaeological Institute, 632
  museums
    Glyptothek, 219
    Neue Pinakothek, 504
  Olympic Games Complex, 814
  reconstruction after World War II, 757
  Schwabing, 504
Munich Putsch of 1923, 614
Munich Sezession, 503–504, 505, 509
Münter, Gabriele, 516, 517
Murders Are Among Us, The (film), 805
Murnau, F. W., 587
Murrow, Edward R., 664
Museum of Modern Art (New York), 733, 760
music, 58–59, 153–165, 459–473. *See also* opera
  after World War II, 809–810
  atonality, 471, 584
  baroque, 43
  composers inspired by Schiller, 133
  conductor as dominant force, 160, 308
  contemporary, 844
  development of full orchestral scores, 155
  German influence in the United States, 325
  innovation in, 59
  inwardness and, 831
  Italian Renaissance and, 93–94
  Jewish musicians, 641
  Mannheim school of composers, 155
  modern repertoire, 308
  nineteenth century achievement, 30
  performance of sacred texts in the vernacular, 59–60
  poetry and, 831
  public concerts, 58–59, 161
  race and, 908n30
  refugees in Britain, 751–752
  refugees in the United States, 735–736
  Schopenhauer on, 334
  serial, 584
  symphony music, 59
    Brahms and, 459, 461–462
    emergence of, 161
    Mozart, 158
  during Third Reich, 641–645, 642, 643
  Tristan und Isolde as beginning of modern music, 335
  waltz, 463–464
  during Weimar Republic, 583–587
music festivals, 164, 329, 644
Musikalische Patriot, Der (periodical), 60
Musikvereins, 325
Musil, Robert, 576–577, 735, 823, 855
Mutter, Anne-Sophie, 809
Mylius, Christlob, 102
Mysticism of Paul the Apostle, The (Mystik des Apostels Paulus), 682
Myth of the State, The (E. Cassirer), 606–607
Myth of the Twentieth Century, The (Der Mythus des 20. Jahrhunderts) (Rosenberg), 620

N

"Nachruf" (Obituary) (Schneider), 802
Nachtsheim, Hans, 696
Nägeli, Carl, 285–286
Naked amidst Wolves (Nackt unter Wölfen) (Apitz), 795
Napoleon, 185–186, 212, 227, 261–262
Napoleon III, 372–373
Nathan der Weise (Nathan the Wise) (Lessing), 104, 314
nation, concept of, 262
National Broadcasting Company (Germany), 638
nationalism, 403, 417–437
  cultural, 290, 437
  of Fichte, 196
  of Moeller van den Bruck, 617
  versus rationalism, 619
  scholarly, 266
  Wagner and, 330
National Peace League and the Society for Racial Hygiene, 428
National Socialism, 562. *See also* Third Reich
  coherence of systems of thought, 634
  community and, 836
  cultural/intellectual bases, 619–621
Nationalsozialistische Deutsche Arbeiterpartei (NSDAP), 620
Nations and Nationalism (Gellner), 755
Natorp, Paul, 773

Nature (journal), 653, 667, 690
nature, meaning of, 72, 201
Naturphilosophie, 201, 228, 342, 426–428
navy, 420, 421
Nazarenes. *See* Düreristen
Nazism. *See* National Socialism; Third Reich
Negri, Pola, 589
Nehamas, Alexander, 823
Neighbours (Gross), 18
neoclassicism, 209, 210, 211, 215
Neo-Conservatives, 617
neohumanism, 108, 109, 228
Neologists, 70–71
Nernst, Walter, 665
Netherlands, refugees in, 506, 507, 699
Neue Sachlichkeit (New Objectivity) movement, 568, 583
Neues vom Tage (Daily News) (Hindemith), 587
Neue Zeitschrift für Musik (journal), 461
Neumann, Franz, 234, 236, 570, 572, 762, 768
Neumann, John von, 487, 739
Neumann, Sigmund, 570
Neumannm, Franz, 714
Neurath, Otto, 567, 602, 603
Neurotic Personality of Our Time, The (Horney), 613
Nevelson, Louise, 734, 735
New American Library, 735
New English Arts Club, 699
Newes, Tilly, 515
Newman, Ernest, 466
New Republic (magazine), 720
New School for Social Research (New York), 701, 720, 726, 736
newspapers, 829
  after World War II, 758
  in German in the United States, 325–326
New Sufferings of Young W., The (Die neuen Leiden des jungen W.) (Plenzdorf), 795
New Vienna Daily (newspaper), 472
New World, scientific discovery of, 177–181
New Yorker Staats-Zeitung, Die (newspaper), 326
Nicolai, Friedrich, 102
Niebuhr, Barthold, 231, 264
Niebuhr, Reinhold, 679
Niemöller, Martin, 686–687
Nierendorf, Karl, 735
Nietzsche, Friedrich, 296
  anticipating Freud's ideas, 394
  contemporary world of ideas and, 819, 822–823
  criticism of Cantor, 486
  Fischer, Fritz, on, 536
  influence of, 468, 517, 682
  overlap with Weber, 826
  use of term Aryan, 430
  Wagner and, 335–338
Nietzsche contra Wagner (Nietzsche), 337
Nightingale, Florence, 317–318
nihilism, 147, 823–825
Nine Books of Prussian History (Neun Bücher preussische Geschichte) (Ranke), 268
Ninotchka (film), 589, 590–591
Nipperdey, Thomas, 30, 46, 58, 197, 227, 228, 232, 279, 305, 427, 446, 456
Nissel, Siegmund, 752
Nissen, Hans J., 844
Nixon, St. John, 376
Nobel Peace Prize, 681, 682, 814
Nobel Prize for Economics, 749
Nobel Prize for Literature, 404, 792, 814
Nobel Prize for Physics, 596, 598, 599
Nobel Prize for Physiology or Medicine, 389, 391, 814
Noddack, Ida and Walter, 666
Nolde, Emil, 568, 630, 633
Nolte, Ernst, 11, 748
No Man's Land (Dorle und Wolf) (Walser), 793
No More Parades (Ford), 579
Nonne, Max, 556
Non: Pour le Mérite (film), 640
Nordau, Max, 428, 431, 839
North German Confederation, 424
Norton, Anne, 723
Notes Towards the Definition of Culture (Eliot), 34, 840
Novalis (Friedrich von Hardenberg), 138, 191, 200
Novembergruppe, 568–569, 585
Novick, Peter, 7–8, 9, 10, 732
Nye, Mary Jo, 343

O

Oberhausen Manifesto, 806
Oberländer, Theodor, 670
Oberth, Hermann, 693
Obrist, Hermann, 508, 509, 515, 516
Observe and Question, 13
Obsoleteness of Man, The (Die Antiquiertheit des Menschen) (Anders), 771
Oesterle, Anna, 656
Offenbach, Jacques, 463
Offending the Audience (Publikumsbeschimpfung) (Handke), 804
Office of Ancestral Inheritance, 656
Ohain, Han von, 600
Ohm, Martin, 355
Oken, Lorenz, 85, 228, 282, 430
Oliphant, Mark, 667
Öllinger, Hans, 14
Olympia (film), 639
Oncken, Hermann, 534
One Dimensional Man (Marcuse), 722, 779

On Humanism (Über den Humanismus) (Heidegger), 770
On Nationalism and the Jewish Question (Über Nationalismus und die Judenfrage) (Jünger), 709
On Social Differentiation (Über soziale Differenzierung) (Simmel), 446–447
On the Aesthetic Education of Man (Über die ästhetische Erziehung des Menschen) (Schiller), 129–130
On the Concept of Number (Über den Begriff der Zahl) (Husserl), 486
On the Connection between the Animal and Spiritual Nature of Man (Schiller), 127
"On the Consequences of a Theorem in the Theory of Trigonometric Series" (Cantor), 486
On the Conservation of Force (Helmholtz), 344
"On the Hindus" (Jones), 190
"On the Hypotheses which Lie at the Foundation of Geometry" (Riemann), 350
"On the Jewish Question" (Marx), 249
"On the Kind of Motion that We Call Heat" (Clausius), 346–347
On the Marble Cliffs (Auf den Marmorklippen) (Jünger), 709
On the Misfortune and Fortune of Art in Germany after the Last War (Vom Unglück und Glück der Kunst in Deutschland nach dem letzten Krieg) (Syberberg), 782
"On the Moving Force of Heat, and the Laws Regarding the Nature of Heat That Are Deductible Therefrom" (Clausius), 345
On the Natural History of Destruction (Luftkrieg und Literatur) (Sebald), 798
On the Origin of Species (Darwin), 259, 393, 426
On the Vocation of the Scholar (Fichte), 228
On the World Soul (Von der Weltseele) (F. Schelling), 202
On War (Vom Kriege) (Clausewitz), 184, 185–188
Opel, Fritz von, 693
Open Society and Its Enemies, The (Popper), 706, 707
opera, 155–156
  Aryan myths and, 435
  forms of, 328
  under Hitler, 644
  Italian, 155
  Mozart, 157, 158, 643
  Weber, 160
  Zeitoper, 587
Operation Paperclip, 695
Oper und Drama (Wagner), 328
Oppenheim, Bram, 746
optics, 203, 204–205, 366
Orators, The (Auden), 818
Ordinary Men (Browning), 20
Orfeo ed Euridice (Gluck), 155
Organic Chemistry in Its Applications to Agriculture and Physiology (Liebig), 280
Organon of Homeopathic Medicine, The (Die Organon der rationellen Heilkunde) (Hahnemann), 176
Oriental Renaissance, The (Schwab), 181, 189
Origin of Continents and Oceans, The (Die Entstehung der Kontinente und Ozeane) (Wegener), 557
Origin of German Tragic Drama, The (Ursprung des deutschen Trauerspiels) (Benjamin), 619
Origin of Kingship in Germany, The (Die Entstehung des deutschen Königstums) (Sybel), 406
Origin of the Family, Private Property and the State, The (Der Ursprung der Familie, des Privateigentums und des Staats) (Engels), 258
Origins of Totalitarianism, The (Arendt), 720–721
Origo, Iris, 755
Ossietsky, Carl von, 582
Ostini, Abbate Pietro, 217
Ostwald, Ludwig, 477
Otto, Frei, 814
Ottomans and the Spanish Monarchy of the Sixteenth and Seventeenth Centuries (Ranke), 267
Our Daily Bread (film), 805
Our Sons (Michelet), 395
"Outlines of a Critique of Political Economy" (Engels), 250
Overbeck, Johann Friedrich, 215, 217, 218
Owen, Richard, 85
Owen, Wilfred, 551
Oxenford, John, 331
Oxpovat, Alex, 169
Oyster Princess, The (film), 589
Oz, Amos, 9
Ozawa, Seiji, 809
Ozment, Steven, 43

P

Pacelli, Eugenio (Pius XII), 668, 683
Pachelbel, Johann, 60, 153
painting
  Abstract Expressionists, 734
  after World War II, 810–813
  Biedermeier, 306
  fresco paintings, 218, 220, 239, 875n39
  German-Americans and, 325
  landscape, 221
    in Dachau, 507
Painting with Skiff (Kandinsky), 516–517
paleontology, 170

Pallas, Peter Simon, 169
Pan (periodical), 510
Pandora (ballet), 747
Pandora's Box (Die Büchse der Pandora) (Wedekind), 515
Pan-German League (Alldeutscher Verband), 420–421, 501
Pan-German Thule Society, 562
Pankok, Bernhard, 508, 509
Panofsky, Erwin, 570, 607, 702, 714, 733, 734, 735, 741
Papon, Maurice, 17
Papst, Eugen, 636
Park, Robert, 453
Parsifal (film), 782
Parsifal (Wagner), 327, 329, 334, 339
Parsley, Malcolm, 157
Partisan Review (magazine), 720
Passavant, Johann David, 215
Past and Present (journal), 749
Pasteur, Louis, 388
Pastor, Ludwig von, 433, 607
Pastorius, Franz Daniel, 323
Pastors' Emergency League, 680
pastors' sons, role of, 834, 837
Pater, Walter, 315
patriotism, 94, 230, 622
Patterson, Michael, 803, 804
Pattison, Mark, 318
Pauli, Wolfgang, 596–597, 738, 841
Paulsen, Friedrich, 226, 450, 542
Paxton, Robert, 16
Péan, Pierre, 17
Peano, Giuseppe, 486
Peasants at Midday (Kirchner), 633
Pechstein, Max, 630
Peculiarities of German History, The (Blackbourn and Eley), 27
Peierls, Rudolf, 667, 745, 746
Pelzel, Thomas, 209
Penck, A. R., 812
Penn, William, 321–322
People of Seldwyla, The (Die Leute von Seldwyla) (Keller), 297
Perfection of Technology (Die Perfektion der Technik) (Jüger), 770–771
Perkin, William, 357
Perls, Fritz, 719–720
Perls, Hugo, 735
perpetual peace, Kant's notion of, 145–146
Pertz, Georg Heinrich, 266
Perutz, Max, 745, 746, 753
Peter, Friedrich, 14
Pevsner, Nicholas, 745, 748
Pfaff, Johann Friedrich, 226
Pfannkuche, A. H. T., 428
Pfeil, L. H., 316
Pfister's Mill (Pfisters Mühle) (Raabe), 440
Pfitzner, Hans, 644, 645
Pforr, Franz, 215
Phaidon Press, 747
Phalanx group, 516
pharmaceutical industry, 360, 364, 387
PhD degree, 53, 835
Phenomenology of the Spirit (Phänomenologie des Geistes) (Hegel), 241, 243
philhellenism, 100, 108
Philological Museum, The (periodical), 317
philology, 230, 317
Philosophia botanica (Linnaeus), 117
Philosophical Faith (Der philosophische Glaube angesichts der Offenbarung) (Jaspers), 769
Philosophical Investigations (Wittgenstein), 753
Philosophical Magazine, 347
Philosophie der Mythologie (Schelling), 191
Philosophische Gespräche (M. Mendelssohn), 137
Philosophische Rundschau (journal), 773
philosophy. *See also* freedom, concept of
  Hegelianism in universities, 237
  influence of refugees in the United States, 729–730
  logical positivism, 602–603, 631
  music and, 161–163
  speculative, 239–240, 831, 838
  theory of logic, 559
  in Weimar Germany, 603–607
Philosophy (Klimt), 498
Philosophy of Arithmetic (Philosophie der Arithmetik) (Husserl), 486–487
Philosophy of Money, The (Die Philosophie des Geldes) (Simmel), 447, 449
Philosophy of Symbolic Forms (Philosophie der Symbolischen Formen) (E. Cassirer), 606
photography, 387, 735
physics, 341–353, 475–484, 666–667
  anti-relativists, 599
  Aryan, 653
  contemporary, 844
  emigration of physicists, 659–660
  intellectual overlap with mathematics, 600–601
  theoretical, 854
  in Weimar Germay, 506–603
Physikalisch-Technische Reichsanstalt, 343
Physiology Society (Berlin), 388
Pia Desideria (Spener), 47
Piano Player, The (Die Klavierspielerin) (Jelinek), 794
Picture Bible (Schnorr), 220
Picture Book for Good Children (Bilderbuch für artige Kinder) (H. and T. Mann), 511
Picture Puzzle (Suchbild) (Meckel), 796
Piene, Otto, 811
Pierrot lunaire (Schoenberg), 472–473
Pietas Hallensis (Woodward), 316

Pietism, 45–46, 67–68, 87, 321, 830
Pinax (Bauhin), 80
Pinkard, Terry, 146
Piscator, Erwin, 584, 586, 701, 736, 803
Planck, Max, 345–346, 352, 476, 480–482, 533, 552, 598, 659, 660, 665, 819
Playfair, Nigel, 586
"Plea for Forgetting, A" (Elkana), 9–10
Pledge, The (Die Bürgschaft) (Becker), 796
Plenzdorf, Ulrich, 795
Plessner, Helmuth, 762
Ploetz, Alfred, 431, 433–434
Plough, Sword and Book (Gellner), 755
Plücker, Julius, 234
poetry
  Biedermeier, 305
  in East Germany, 801–802
  as link between language and nature, 205
  versus mathematics, 76–77
  as the mother tongue, 192
  music and, 831
  political, 304
  superiority of, 291–292
  Tendenzdichter, 299
  in West Germany, 799–800
  during World War I, 547–550
Poggendorff, Johann Christian, 342, 343, 345
Pogge von Strandmann, Hartmut, 381
Poland, 18, 669–670
Polanyi, Karl, 727
Polanyi, Michael, 567
Poliakov, Leon, 431
Political Collapse of Europe, The (Holborn), 731–732
political science and refugees in the United States, 720–724
"Politicisation of the Intelligentsia, The" (Benjamin), 619
Politics of Cultural Despair, The (F. Stern), 762
Polke, Sigmar, 756, 811
Pollock, Friedrich, 572, 631
Pommer, Erich, 568, 589, 758
Poncelet, Jean-Victor, 353
Popper, Karl, 567, 602, 706–708, 749
population, Nazi classification system, 670–671
porcelain manufacturing, 168, 306
Porter, Noah, 324
Porzellan: Poem vom Untergang meiner Stadt (Grübein), 802
Potonié, Henry, 85
Pötsch, Leopold, 559
Potsdam Institute for Military History, 24, 843
Poverty of Historicism, The (Popper), 707
Powell, Michael, 744
Power (Jud Süss) (Feuchtwanger), 579
Preminger, Otto, 587
Pressburger, Emeric, 744, 746
Preussentum und Pietismus (Hinrichs), 46
Preussische Jahrbücher, 408, 521–522
Preyer, Wihelm, 427
Primitive Rebels (Hobsbawm), 749
Prince Frederick of Homburg (Prinz Friedrich von Homburg) (Kleist), 200, 293
Princeton Institute for Advanced Study, 734
Princeton University, German scholars at, 631
"Principles of Communism" (Engels), 259
Private and Public Happiness (Hirschman), 728
Problems of Poetry, The (Probleme der Lyrik) (Benn), 800
Professor Unrat (H. Mann), 512–513, 592
Prolegomena to Homer (Prolegomena ad Homerum) (F. A. Wolf), 106, 232
Protestant Ethic and the Spirit of Capitalism, The (Die Protestantische Ethik und der Geist des Kapitalismus) (Weber), 456
Protestantism, 198, 300, 424, 456, 830
Protogaea (Leibniz), 85
Prüfer, Kurt, 695
Prussia. *See also* Friedrich II the Great
  music, 58–60
  Pietism in, 45–49
  reading revolution, 55–58
  rise of the university, 49–55
Prussian Academy, 480–482
Prussian Academy of Art, 527
Prussian Experience, The (Rosenberg), 49
"Prussian Universities and the Research Imperative, 1806 to 1848, the" (Turner), 225
psychiatry, World War I and, 553
psychoanalysis, 553, 556
  in Britain, 753–754
  under Third Reich, 661–662
  in the United States, 716–720
psychology
  community and Gestalt psychology, 836
  influence of refugees in the United States, 716–720
  Nazification of, 661
  theory of *Gestalt,* 493
"Psychology of Money, The" (Simmel), 447
Psychology of the Unconscious, The (Jung), 555–556
Psychopathia Sexualis (Krafft-Ebing), 496
Psychopathology of Everday Life, The (Freud), 556–557
publication, freedom of, 263
public health, 385, 386
public hygiene policy, 175
publishing, 829
  after Thirty Years' War, 56
  periodicals, 103, 829
  refugees in Britain, 747–748
  refugees in the United States, 735

Pufendorf, Samuel, 44
Pulzer, Peter, 21, 753
"Purity of the Reich, The" (Schumann), 635
Purkyně (Purkinje), Jan Evangelista, 37, 234, 282, 853
Putnam, Hilary, 729
Pütter, Johann Stephan, 73

Q

Qualification and Curriculum Authority (Britain), 4
quantum theory, 480, 483, 597–598
Quest of the Historical Jesus (Von Reimarus zu Wrede) (Schweitzer), 681–682
Quine, Willard Van Orman, 603, 729, 730
Quinet, Edgar, 189

R

Raabe, Wilhelm, 440
Rabbi von Bacharah (Heine), 300
racial theory, 430, 434–436
radar, 693
radio, 478, 600, 758, 908n20
Radio Times (magazine), 2
Rahner, Karl, 787
railways, 370–371, 417
Ramsden, John, 5
Ranke, Leopold von, 26, 91, 231, 266–269, 405
Rankl, Karl, 751
Rapaport, David, 662
Rathenau, Emil, 379, 509
Rathenau, Walther, 379–380, 509, 659
Ratzinger, Joseph Alois (Benedict XVI), 5, 786–788, 849
Rau, Johannes, 847
Rauch, Christian, 215
Rauschenberg, Robert, 734
Read, Herbert, 622
Reader, The (Der Vorleser) (Schlink), 797
reading revolution, 55–56, 831, 837
Realism, 440, 524
redemptive community, 836–837, 839
"Rede vor Arbeitern in Wien" (T. Mann), 200
Redlich, Fritz, 727, 728
Reflections of a Nonpolitical Man (Betrachtungen eines Unpolitischen) (T. Mann), 563
Reflections on Christa T. (Nachdenken über Christa T.) (C. Wolf), 795, 833
Reflections on the Philosophy of the History of Mankind (Ideen zur Philosophie der Geschichte der Menschheit) (Herder), 123
refugees
  in Britain, 699, 743–756
  Emergency Society of German Scholars Abroad, 663
  impression of American people, 716
  in the Netherlands, 506, 507, 699
  remigrants, 769
  in the United States, 720–724, 727–728, 729–730, 730–733
    Committee on Displaced Foreign Psychologists, 662
    Emergency Rescue Committee, 700
Reger, Max, 292, 459, 469
Reich, Wilhelm, 570, 613, 701, 718–719
Reich Board for Industrial Rationalization, 669
Reich Commissioner for the Strengthening of German Nationhood, 670
Reichenbach, Hans, 631, 729
Reichmann, Frieda, 718–719
Reich-Ranicki, Marcel, 803
Reichsdramaturg, 646
Reichskommissar für Festigung deutschen Volkstums, 657
Reich Union for Folk-Nation and Homeland, 656
Reil, Johann Christian, 83, 86, 202–203, 228
Reill, Peter Hanns, 69, 72–73, 74
Reimarus, Hermann, 71, 103–104
Rein, Adolf, 650
Reinhardt, Max, 467, 514, 524–525, 533, 647, 700, 736
Reinhold, Karl Leonhard, 138, 147, 149
Reitz, Edgar, 807
relativity, theories of, 483–484, 487
religion. *See also* Catholic Church; Christianity; Protestantism
  advent of doubt, 65, 837
  call for a new, 435
  capitalism and, 457
  Feuerbach on, 246
  Mannhein on, 704
  Nazis and, 621, 683
Remak, Robert, 287
"Remarks on the Forces of Inanimate Nature" ("Bemerkungen über die Kräfte der unbelebten Natur") (Mayer), 342
Remarque, Erich Maria, 579–581, 587, 738
Rembrandt as Teacher (Rembrandt als Erzieher) (Langbehn), 432–433
Remy, Steven, 649, 650, 654
Renan, Ernst, 431
Renner, Karl, 14
research
  concept of, 53, 146, 227
  institutionalization of, 834–836, 838, 851
  under Third Reich
    funding, 689
    unethical, 696
Resewitz, Friedrich Gabriel, 75
reunification
  architecture and, 815
  comedy films, 808
  Unification Day (October 3, 1990), 780

Reunion, The (Barlach), 633
Reuter, Ernst, 663
Revolt of the Weavers, The (Kollwitz), 527
Revue des deux mondes, 538
Rhapsodien über die Anwendung der psychischen Curmethode auf Geisteszerrüttungen (Reil), 202
Rheinische Zeitung (newspaper), 249
Richard, Jules, 601
Richard, Robert J., 199–200
Richter, Gerhard, 811
Richter, Hans, 564, 588, 735
Richter, Jean Paul, 138
Rie, Lucie, 745
Riefenstahl, Leni, 638–639
Rieff, Philip, 821
Riehl, Alois, 533
Riehl, Wilhelm Heinrich, 444–445, 452, 655
Riemann, Bernhard, 174, 350–352, 353
Riemerschmid, Richard, 508, 509, 510
Rienzi (Wagner), 328
Rilke, Rainer Maria, 511, 547, 574–575, 604
Ring der Norda, 434
Ring des Nibelungen, Der (Wagner), 327, 329, 330
Ringer, Fritz, 765–768, 833, 834
Rise and Fall of the City of Mahagonny, The (Aufstieg und Fall der Stadt Mahagonny) (Brecht and Weill), 587
Rise of Political Anti-Semitism in Germany, The (Pulzer), 21
Ritter, Alexander, 466
Ritter, George, 762
Ritter, Gerhard, 761
Ritter, Karl, 634, 639–641
Ritterbusch, Paul, 690
Rivers, Larry, 734
Road Back, The (Remarque), 581
Road into the Open, The (Der Weg ins Freie) (Schnitzler), 491
Road to Serfdom, The (Hayek), 706
Robbers, The (Die Räuber) (Schiller), 127–128, 315
Robertson, Ritchie, 301
Rochlitz, Friedrich, 164
rocket science, 693–694
Rogall, Georg Friedrich, 48
Rogoff, Irit, 813
Rolland, Romain, 563, 576
Romanticism, 77, 192, 194, 199, 298, 393
  inwardness and, 831
  Kant's theory of genius and, 145
  racial theory and, 430
  term coined by de Staël, 311, 313
Romantic School, The (Die Romantischer Schule) (Heine), 299
Röntgen, Wilhelm, 479–480, 854
Röpke, Wilhelm, 762
Rorty, Richard, 823
Rose, Hajo, 699
Rose, Max, 375
Rosenberg, Alfred, 435, 436, 562, 620, 630, 656
Rosenberg, Hans, 49, 731
Rosenberg, Isaac, 551
Rosenkavalier, Der (R. Strauss and Hofmannsthal), 465–466, 468
Rosenstock, Sami (Tristan Tzara), 564
Rosenzweig, Franz, 682
Ross, Alex, 471
Ross, Ludwig, 410
Rossellini, Roberto, 805
Rostal, Max, 752
Rottenblatt, Joseph, 745
Round Table, The (Menzel), 526
Rousso, Henry, 16
Rubens, Heinrich, 481
Rubinstein, Artur, 702
Rudolph, Hermann, 11–12
Rudolphi, Karl Asmund, 228, 283
Ruge, Arnold, 245, 249
Rumbledump, The (Schüdderump, Der) (Raabe), 440
Rumohr, Friedrich von, 218
Rumpff, Carl, 362
Runaway Horse (Ein fliehendes Pferd) (Walser), 793
Russell, Bertrand, 148, 350, 494, 552, 559
Rust, Bernhard, 653, 662, 691
Ryle, Gilbert, 755

S

Sacher-Masoch, Leopold von, 496
Sachs, Hans, 43, 717
Salome (R. Strauss), 465, 467
Salzburg, 59, 809
Sane Society, The (Fromm), 719
Santayana, George, 99, 539, 542–543
Sartor Resartus (Carlyle), 315
Sassoon, Siegfried, 579
Sauer, Christopher, 325
Savigny, Friedrich Carl von, 212, 228, 265, 405
Scaff, Lawrence, 825
Sceptical Generation, The (Die skeptische Generation) (Schelsky), 789
Schadow, Friedrich, 215, 218
Schadow, Johann Gottfried, 211, 215, 217
Schadow, Rudolf, 217
Schadow, Wilhelm, 217
Schaffer, Yvonne, 647
Scharnhorst, Gerhard von, 185, 187
Scharoun, Hans, 814
Schaudinn, Fritz, 390
Scheffel, Josef Victor von, 305
Scheler, Max, 532, 533, 604–605, 855

Schelling, Friedrich, 85
  on art, 162
  biology and, 201, 202
  on Blumenbach, 81
  on common mythology, 191
  on Fichte, 149
  friendship with Hölderlin, 291
  Idealism of, 138
  influences on, 148, 191
  on music, 831
  Naturphilosophie and, 228
  on role of professors, 226, 228
  speculative philosophy of, 240–241
  spiritual self-development, 196–197
  at Tübingen, 241
  on the unconscious, 197
  on *Wissenschaft,* 229
"Schelling und die Offenbarung" (Engels), 256–257
Schelsky, Helmut, 761, 789
Schemann, Ludwig, 434, 435
Scherchen, Hermann, 641
Scherpe, Klaus, 784
Schick, Gottlieb, 215
Schidlof, Hans, 745, 752
Schiele, Egon, 567
Schillebeeckx, Edward, 787
Schiller, (Johann Christoph) Friedrich, 108, 112, 127–133, 138, 160, 177, 191
Schilling, Klaus, 697
Schillings, Max von, 641, 644–645
Schindewolf, Otto Heinrich, 85
Schinkel, Karl Friedrich, 212–215, 221, 414, 852–853
Schivelbusch, Wolfgang, 566, 757, 841
"Schlachtfeld" (Battlefield) (Stramm), 549–550
Schlegel, August Wilhelm, 138
  de Staël on, 312
  early Romantic, 200
  influenced by Winckelmann, 99
  on music, 162
  Orientalism and, 190–193
Schlegel, Caroline (Böhmer), 138, 200
Schlegel, Dorothea (Veit), 138
Schlegel, Friedrich, 138, 217
  Aryan myth and, 430
  influenced by Winckelmann, 99
  on marriage of biology and poetry, 202
  Orientalism and, 191
  Romanticism and, 197, 200
  on *Wilhelm Meister,* 119, 197–198
Schleicher, Rüdiger, 680
Schleiden, Matthias Jakob, 283–284, 366, 853
Schleiermacher, Friedrich, 138
  biography by Dilthey, 441
  as early Romantic, 200
  influence on Engels, 256
  neohumanism and, 228
  Orientalism and, 191
  on role of professors, 226, 228
  on source of all religion, 193
  on universities, 229
  at University of Berlin, 227, 228
Schlemmer, Oskar, 629, 700
"Schlesischen Weber, Die" (Heine), 300
Schleter, Andreas, 213
Schlick, Moritz, 559, 601, 602, 707, 729
Schliemann, Heinrich, 385, 410, 411–414
Schlingensief, Christoph, 808
Schlink, Bernhard, 28
Schlöndorff, Volker, 807
Schlosser, Raine, 646
Schlözer, August Ludwig con, 73
Schmidt, Helmut, 733, 808–809
Schmidt-Rottluff, Karl, 700
Schmitt, Carl, 454, 668–669, 723
Schmitthenner, Paul, 652, 653, 672, 689
Schmoller, Gustav, 533
Schnack, Anton, 548, 550–552
Schneider, Peter, 793, 796, 802
Schnitger, Arp, 154
Schnitzler, Arthur, 489, 490–491, 859n49
Schnorr von Carolsfeld, Julius, 218, 219–220
Schocken Books, 735
Schoenberg, Arnold, 336, 459, 470–473, 584–585, 587, 641, 702
scholarship
  originality and, 232–233
  origins of modern, 105–110
Scholem, Gershom, 607, 608, 609, 718–719
Schonberg, Harold, 37, 43, 156, 461–462, 463
Schonbühne, Die (magazine), 581
Schönerer, George, 501
Schönhoven, Klaus, 780
Schönlein, Johann L., 383
Schopenhauer, Arthur, 191, 327, 330–333, 336, 338–339, 394, 430
Schott, Otto, 366
Schott Glass Manufacturing Company, 390
Schrader, Gerhard, 695
Schramm, Percy Ernst, 35, 607–608
Schröder, Gerhard, 784
Schröder, Richard, 781
Schrödinger, Erwin, 30, 598, 600, 659, 708–709
Schtonk! (film), 808
Schubart, Christian, 55
Schubert, Franz, 133, 159–160, 239, 298, 305, 306, 464
Schubert's Vienna (Erickson), 304–305
Schuckmann, Kaspar Friedrich von, 228
Schüfften, Eugen, 590
Schulze, Alfred Otto Wolfgang (Wols), 810–811
Schulze, G. E. L., 148

Schulze, Hagen, 289
Schulze, Ingo, 799
Schulze, Johannes, 237
Schumacher, Ernst (Fritz), 510, 750
Schumacher, Hermann, 534
Schumann, Gerhard, 634–636
Schumann, Robert, 133, 298, 306–307, 461
Schumann-Heink, Ernestine, 468
Schumpeter, Joseph, 704–705
Schütz, Heinrich, 153
Schwab, Raymond, 181, 189
Schwanitz, Dietrich, 815
Schwann, Theodor, 283, 284–285, 853
Schwarzenegger, Arnold, 723–724
Schweigsame Frau, Die (opera, Strauss and Zweig), 641
Schweitzer, Albert, 681–682
Schwitters, Kurt, 564, 565, 630, 699, 744, 745, 746
Schygulla, Hanna, 807
science. *See also* scientists; *Wissenschaft*
  cultural *versus* natural, 77
  division in pure and applied, 234
  educated middle class and, 830
  French view during World War I, 538
  Goethe's contributions to, 203–205
  humanities *versus*, 838
  Mach on, 500
  military and, 544
  Nazi concept of, 657–665
  nineteenth century achievement, 30
  popularizers, 426–428, 457
  Romantic interest in, 201–203
  Wilhelm I and, 529
  during World War II, 689–698
Science (journal), 663
scientific method, 205, 707
scientific societies, 829
scientists
  Nobel Prizes since 1945, 814 (*See also entries starting with Nobel Prize*)
  during Third Reich, 658, 738–739
Sebald, W. G., 28, 296, 797–798
Sebaldt von Werth, Max, 434
Seduction of Culture in German History, The (Lepenies), 31
Seghers, Anna, 795
Self-Accusation (Selbstbezichtigung) (Handke), 804
self-consciousness, 58
  Fichte on, 149
  language and, 125
self-realization, notion of, 741–742
Sell, Ernest, 358
Selle, Götz von, 51
seminars, 52–53, 233–236
Semler, Johann Salomo, 71
Semmelweiss, Ignaz, 386
Sendschreiben (Open Letters) (Winckelmann), 97
Sennett, Richard, 754
Sense of Order, The (Gombrich), 747
Serkin, Rudolf, 584
Seven Deadly Sins, The (Dix), 630
Seven Year Itch, The (film), 591
Sewering, Jans-Joachim, 926n58
sex, 395, 427, 495, 496
Sex and Character (Geschlecht und Charakter) (Weininger), 495
Sezession. *See* Berlin Sezession; Munich Sezession
Shadows in Paradise (Remarque), 581
Sidgwick, Henry, 120, 124, 538, 790
Sieff, Israel, 747–748
Siegfried (opera, Wagner), 329, 334
Siemens, Werner von, 355, 362
Siemens & Halske Telegraph Construction Company, 356
Signs of Life (film), 807
Silberman, Andreas, 154
Silberman, Gottfried, 154
Simmel, Georg, 37, 445–449, 452, 855
  Breuer, Stefan, on, 454
  influence of, 455, 682
    in the United States, 713
  review of *Rembrandt as Teacher*, 433
  Scheler student of, 605
  student of Treitschke, 409
  Troeltsch and, 675
Simplicissimus (weekly), 511, 514
Sin (Stuck), 508
Siodmak, Robert, 587, 588, 590
Sisenstein, Sergei, 638
Sites of Memory, Sites of Mourning (Winter), 732
Smiling Lieutenant, The (film), 589
Smith, Hugh, 185
Snowman, Daniel, 743, 744, 746–747
Social Darwinism, 427, 428, 614
Socialist Decision, The (Die sozialistische Entscheidung) (Tillich), 679
Social Research (journal), 726
Social Teaching of the Christian Churches, The (Die Soziallehren der christlichen Kirchen und Gruppen) (Troeltsch), 674–675
Society and Democracy in Germany (Gesellschaft und Demokratie in Deutschland) (Dahrendorf), 762–765
Society for Jewish Folklore, 655
Society for Racial Hygiene, 434
Society for the Abolition of Slavery, 326
Society for the Study of Early German History (Berlin), 266
Society of Arts (Britain), 320
Society of German Naturalists and Doctors, 283

Sociologie (Soziologie) (Simmel), 448–449
sociology, 163–165, 439–458, 726, 741
Söderbaum, Kristina, 639
Soldaten, Die (opera, Zimmermann), 810
"Soldier's Prayer" (Schumann), 636
Solms, Bernhard, 646
Solzhenitsyn, Alexander, 792
Sombart, Werner, 452–454, 532, 653, 836, 839
Some Like It Hot (film), 591
Sommerfeld, Arnold, 665
Sonderweg, 25–27, 32, 828
Songs of the Reich (Die Lieder vom Reich) (G. Schumann), 635
Sonnefels, Joseph von, 55
Sonnemann, Emmy, 648
Sons and Daughters of the Jewish Deportees of France, 16
Sontheimer, Kurt, 761
Soros, George, 755
Sorrows of Young Werther, The (Die Leiden des jungen Werthers) (Goethe), 115–116
Sound of Music, The (film), 14
Speer, Albert, 658, 692, 757, 913n14
Speier, Hans, 726
Spemann, Adolf, 634
Spencer, Charles, 745
Spencer, Herbert, 431
Spener, Philipp Jacob, 47
Spengler, Oswald, 449, 452, 454, 560–562, 618, 620
  nationalist cultural pessimism and, 839, 840
  on the United States, 771
  on World War I, 534–535
Sperr, Martin, 803
Spiegel, Der (weekly), 1–2
Spielhagen, Friedrich von, 439
Spinoza, Baruch, 442
Spiro, Eugen, 699
Sport im Bild (periodical), 580
Spranger, Eduard, 535
Spring Awakening (Frühlings Erwachen) (Wedekind), 514
SS-Staat, Der (Kogon), 779
Stadler, Anton, 157
Staël, Germaine de, 311–313
Stapel, William, 655
Stargardt, Nicholas, 418
Stark, Johannes, 599
state, 54–55, 150, 269
State Pietism, 48, 49
Static Poems (Statische Gedichte) (Benn), 799–800
Staudinger, Hans, 726
Staudte, Wolfgang, 805
Stead, Wickham, 845
Steffens, Henrik, 226, 228
Stehr, Nico, 453
Steiger, Peter, 296
Steiger, Robert, 122
Steigmann-Gall, Richard, 684
Stein, Charlotte von, 117
Stein, Gertrude, 516
Stein, Heinrich Friedrich Karl vom und zum, 266
Stein, Lorenz von, 245
Stein, Peter, 804
Steiner, George, 753
Steiner, Rudolf, 676
Steinheil, Carl August von, 174
Stekel, Wilhelm, 553
Stephenson, George, 376
Steppenwolf (Hesse), 576
Stern (magazine), 657
Stern, Ernst, 516
Stern, Fritz, 20, 21, 22, 35, 401, 433, 436, 537, 616–617, 732–733, 762
Stern, Moritz, 343
Sternberg, Josef von, 513, 592
Sterne (film), 805
Sterne, Carus, 426
Stibbe, Matthew, 531
Stieglitz, Alfred, 517
Stifter, Adalbert, 290, 295–296
Stiller (Frisch), 793–794
Stinnes, Hugo, 375
Stockhausen, Karlheinz, 784, 810
stock market crash (1873), 374, 521
stock market crash (1929), 611
Storm of Steel (In Stahlgewittern) (Jünger), 618
Story of Art, The (Gombrich), 747
Stoss, Veit, 43
Stramm, August, 548, 549–550
Strategy of Economic Development (Hirschman), 728
Strauss, Botho, 782–783, 803
Strauss, David, 245
Strauss, Eduard, 465
Strauss, Johann, I, 173, 459
Strauss, Johann, II, 459, 463, 464–465
Strauss, Leo, 608, 669, 690, 720, 722–723, 740
Strauss, Pauline, 465
Strauss, Richard, 459, 465–469, 492, 809
  under Hitler, 641
  Hölderlin's poetry set to music by, 292
  influences on, 133, 336
  on Johann Strauss II, 465
  on Korngold, 591
  on Schoenberg, 470
  waltz and, 173
  Wilhelm I and, 526
Streicher, Julius, 661
Stresemann, Gustav, 421
Strobach, Hermann, 655

Structural Transformation of the Public Sphere, The (Der Strukturwandel der Öffentlichkeit) (Habermas), 779
Stuart, Gisela, 5
Stuck, Franz von, 507–508, 515
Studies in Hysteria (Studien über Hysterie) (Freud and Breuer), 395
Stuhlinger, Ernst, 695
"Stukalied" (song), 640
Stukas (film), 640–641
Sturm, Der (periodical), 549
Stürmer, Michael, 12
Sturm und Drang (Storm and Stress) movement, 77, 114
Submissive Belgium (report), 13
Success (Erfolg) (Feuchtwanger), 579, 856
Sullivan, Arthur, 463
Sulzer, Johann, 78
Summer in the City (film), 807
Sunday Circle, 703
Sunset Boulevard (film), 591
Sutter, Joseph, 216
Swales, Martin, 795
swastika, 562
Sybel, Heinrich von, 405–407, 421–422, 445
Syberberg, Hans Jürgen, 782, 807
Symbols of Transformation (Wandlungen und Symbole der Libido) (Jung), 555
Szilard, Leo, 660, 666, 690, 738, 745

T

Taeuber, Sophie, 564
Tanztheater, 804
Tauber, Richard, 747
Teaching Institute for Judaic Studies (Breslau), 655
telegraph, 172, 356
Telemann, Georg Philipp, 60
telescopes, 366
television, 600
television culture, 808–809
Teller, Edward, 738–739
Temkin, Oswei, 85
Tendenzdichter (committed poets), 299
ter Meer, Fritz, 695
Testament of Dr. Mabuse, The (film), 590
theater, 802–805
  amateur theatricals at Weimar Court, 116
  in Berlin, 523–524
  Biedermeier, 306
  under Hitler, 645–648
  Lessing and, 102
  magazines on, 581
  musical, 735–736
  of "ordinary people," 803
  subsidies in Germany, 804–805
  Unified Theater Law, 647
  in Weimar Berlin, 584
  Wieland on, 113–114
  World War I and, 552–553
theologians, 673–687
  early twentieth century renaissance, 674–680, 855
  under Hitler, 683–687
Theory and Practice (Theorie und Praxis) (Habermas), 776
Theory of Chronic Diseases (Hahnemann), 176
"Theory of Knowledge in the Exact Sciences" (Königsberg, 1931), 601
Theozoology (Liebenfels), 434
Third or Fourth Man, The (Der dritte oder der vierte Mensch) (A. Weber), 770
Third Reich, 629–648
  art, 621, 629–631, 632–633
  Chamber for Arts and Culture, 630
  Confessing Church (Bekennende Kirche), 678, 680
  demographic economics, 669–672
  education funding, 689
  Evangelical Church of the German Nation, 677–678
  filmmaking, 593, 638–641
  First National Socialist Theater Festival (Dresden, 1934), 646
  folklore, 654–657
  German Faith Movement, 687
  Heidigger and, 604
  Hitler on miscegenation, 615
  Lorenz's research and, 697
  mass murder, science of, 695
  mathematics, 653
  mental patients, murder of, 672
  music, 641–645
  paganism and, 687
  philosophy, 631
  physicians, 661
  physics, 652–653, 658–660
  poetry, 634–637
  psychoanalysis, 661–662
  Reich Commissioner for Solidifying the German Folk-Nation, 657
  science, 657
  theater, 645–648
  theology, 683–687
  unethical research, 696
  universities and scholars, 649–652
Third Reich, The (Das dritte Reich) (Moeller van den Bruck), 617–618, 839
33 Moments of Happiness (33 Augenblicke des Glücks) (Schulze), 799
This People Israel (Baeck), 750
Thoma, Hans, 532
Thoma, Ludwig, 511
Thomasius, Christian, 57, 136, 137
Thomas Paine (Johnst), 646
Thorak, Josef, 632

"Thoughts in Wartime" ("Gedanken im Krieg") (T. Mann), 34, 535–536, 562
Three Comrades (Remarque), 581
Three Faces of Fascism (Nolte), 748
Threepenny Opera, The (Die Dreigroschenoper) (Brecht and Weill), 586–587
Thyssen, August, 375
Ticknor, George, 323
Tieck, Christian Friedrich, 215
Tieck, Ludwig, 138, 216
Tiedemann, Friedrich, 86
Tier rang gewaltig mit Tier (Schnack), 551
Tillich, Paul, 604, 631, 675, 679–680, 686, 714, 720, 730
Time to Love and a Time to Die, A (Remarque), 581
Tin Drum, The (Die Blechtrommel) (Grass), 792
Tirpitz, Alfred von, 421
Tischbein, Heinrich Wilhelm, 118, 210, 215
Toch, Ernst, 641
Tod eines Kritikers (Death of a Critic) (Walser), 28
"Todesfuge" (Celan), 800
Todt, Fritz, 692
Toller, Ernst, 618, 552552
"Tomorrow Belongs to Me" ("Morgen Gehört mir") (Baumann), 636–638
Tonio Kröger (T. Mann), 512
Tönnies, Ferdinand, 445, 449–452, 453, 454, 455, 682, 713, 836
Torquato Tasso (Goethe), 117
Totem and Taboo (Totem und Tabu) (Freud), 611
Touch the Water, Touch the Wind (Oz), 9
Touvier, Paul, 17
Tractatus Logico-Philosophicus (Wittgenstein), 552, 558, 559
Trakl, Georg, 547, 548–549
Tram Stop (Strassenbahnhaltestelle) (Beuys), 812
Traveller, if You Come to Spa (Wanderer, kommst du nach Spa . . .) (Böll), 791
Treason of the Learned, The (Benda), 622–625, 766
Treatise on the Materia Medica (Cullen), 175–176
Treitschke, Heinrich von, 321, 405, 408–410, 424, 445, 477, 521–522, 839
Treviranus, Georg Reinhold, 81
Trial, The (Der Prozess) (Kafka), 577, 578
Tristan und Isolde (Wagner), 327, 329, 334, 335
Tristia ex Ponto (Grillparzer), 294
Triumph of the Therapeutic, The (Rieff), 821
Triumph of the Will (film), 638–639
Troeltsch, Ernst, 263, 532, 674–675
Trotha, Margarethe von, 807, 808
Truth about Genius (Versuch über das Genie) (Resewitz), 75
Truth and Method (Wahrheit und Methode) (Gadamer), 773–774
Tschermak, Erich, 391
Tschudi, Hugo von, 529
tuberculosis, 386, 387–388, 389
Tucholsky, Kurt, 581–582, 619
Turner, R. Steven, 225, 226, 228, 232, 233, 236, 237
Twain, Mark, 522
Twilight of the Idols (Götzendämmerung) (Nietzsche), 337
Two Fundamental Problems of the Theory of Knowledge (Die beiden Grundprobleme der Erkenntnistheorie) (Popper), 706
Two Views (Zwei Ansichten) (U. Johnson), 796
Tyranny of Greece over Germany, The (Butler), 100

U

Über den Bildungstrieb und das Zeugungsgeschäfte (Blumenbach), 81
Über die Geschichte der Menscheit (Iselin), 76
Über die Sprache und Weisheit der Indier (F. Schlegel), 191
"Über formal unentscheidbare Sätze der *Principia Mathematica* und verwandter Systeme" (Gödel), 601
Übermensch, 339
Uecker, Günther, 811
Uhde, Fritz von, 507
Uhlman, Fred, 746
Ulmer, Edgar G., 590
unconscious, the, 196–198, 393–397
Unger, Franz, 86, 285, 392
Unicorn, The (Das Einhorn) (Walser), 793
United States
- German influence on music, 325
- German rocket scientists in, 695
- German views on, 771
- Germany and, 321–326
- refugees and immigrants, 321, 324, 699, 700, 713–742

Unity in Science, 631
universities, 829
- Association of German Universities, 649
- compared to British universities, 49–50
- Dahrendorf on, 763–764
- Nazism and, 649, 689–690
- as research institutes, 225–226
- Ringer on, 767
- rise of, 49–55, 831

Universities—American, English, Geman (Flexner), 741
University in Exile, 720, 723
University of Berlin, 529
- American students at, 323
- founding, 109, 227

humanism at, 94
professors at, 107, 150, 241, 264, 384, 388, 404, 408, 441, 445, 480
students, 876n32
University of Bonn, 228
American students at, 323
science seminar at, 234
Sybel teaching at, 405, 407
University of Breslau, 50, 282
University of Erlangen, 50
University of Göttingen, 47, 50, 226, 487
American students at, 323
Hainbund circle, 102
impact of, 50–51
mathematics at, 352, 484
physics research at, 597
University of Halle, 47, 50, 106
American students at, 323
mathematics at, 484
seminar at, 52
University of Heidelberg, 653
American students at, 323
550th anniversary under Hitler, 652
mathematics at, 484
professors at, 241, 775
reopening after World War II, 769–770
University of Jena, 146–147
mathematics at, 484
professors at, 148–150, 241
University of Königsberg, 48, 147
British influence in, 136
Jacobi, C. G. J., at, 236
reforms at, 227
University of Marburg, 406, 678–679
University of Munich, 406
University of Vienna, 348, 391, 469
University of Würzburg, 384, 479
Unloved Germans, The (Eich), 762
Unseen Revolution, The (Drucker), 728
"Untergang der Titanic, Der" (Enzensberger), 802
Unwin, Stanley, 747
uprisings of 1848, 384, 402
"Uranium Club," 691
Urlaub auf Ehrenwort (film), 640
"Ursprung der Kunstwerks, Der" (lectures, Heidegger), 773–774
Uses of Enchantment, The (Bettelheim), 717, 718

V

Valentin, Curt, 735
values, 194, 605
Veit, Dorothea, 217
Veit, Johann, 217
Veit, Philipp, 217, 218
Venus im Pelz (Sacher-Masoch), 496
Venus of Willendorf (Piene), 811
Vereinigte Werkstätten für Kunst und Handwerk, 509
Vergangenheitsbewältigung, 759
Verspätete Nation, Die (Plessner), 762
"Versuch einer Kritik aller Offenbarung, Ein" (Fichte), 148
Versuche über Pflanzenhybriden (Mendel), 286, 393
Vichy France and the Jews (Marrus and Paxton), 16
Vichy Syndrome, The (Rousso), 16
Vienna, 489–502
after World War I, 567–568
café society, 490
music
Academy of Singing, 461
Brahms in, 461
Conservatory, 469
innovation, 59
State Opera, 809
Wehrmann im Eisen statue, 531, 545
Vienna Circle, 500, 602
Gödel and, 601
Popper and, 706
refugees to the United States, 631, 729–730
Viertel, Berthold, 746
Views of a Clown (Ansichten eines Clowns) (Böll), 791
Vincent, Edgar, 584
Vincent, Philip, 44
Virchow, Rudolf, 366, 383–386, 428, 431, 853
Viret, Pierre, 67
Vischer, Peter, 43
Visit, The (Der Besuch der alten Dame) (Dürrenmatt), 803
vital force, concept of, 273–274
Vogel, Ludwig, 216
Vogt, Carl, 426, 427
Voigt, August, 504
Voigt, Georg, 92
Voigt, J. C. W., 117
Voigt, Johannes, 266
Volk
culture, 309, 310, 654–657
elevation to mystical entity under Hitler, 684
Heidegger and, 604
Herder and, 125
term, 309–310
Völkerpsychologie, 405
Völkischer Beobachter (The People's Observer) (newspaper), 620
Volksstück, 803
Volkstum und Heimat (journal), 656
Volk und Rasse (journal), 654
Voltaire, 61, 67, 92
"Von der Lebenskraft" (Reil), 83
Von Heute auf Morgen (Schoenberg), 587
Vormärz, 304

Voss, Heinrich, 508
Vossische Zeitung (newspaper), 102, 331, 580, 582
Vossler, Karl, 533
Vranitsky, Franz, 15
Vries, Hugo de, 391
Vulpius, Christiane, 118
VW-Complex (film), 808

W

Waagen, Gustav Friedrich, 212, 214
Wackenroder, Wilhelm Heinrich, 216
Wagner, Cosima, 467
Wagner, Franz Josef, 6, 164–165
Wagner, Otto, 496–497
Wagner, Richard, 328
  community and, 836
  influence of, 471
  influences on, 200, 327, 329–330
  Nietsche and, 335–338
  operas as re-creation of original Aryan myths, 435
  racial ideas and, 434
  Strauss, Richard, and, 466
  Wilhlem I and, 526
Wagner, Rudolph, 81, 283, 287
Wagner, Siegfried, 532, 645
Wagner Case, The (Nietzsche), 337
Walden, Herwarth, 528
Waldheim, Kurt, 14
Waldseemüller, Martin, 321
Wallace, Stuart, 538
Wallenstein trilogy (Schiller), 131–132
Wall Jumper, The (Der Mauerspringer) (Schneider), 796
Wallot, Paul, 523
Walser, Martin, 28, 761, 782, 783–784, 793, 832, 836
Walter, Bruno, 558, 645, 702
Walther, Helmut, 692–693
Wanderer above the Sea of Fog, The (Friedrich), 221
Wandlung, Die (The Transformation) (Toller), 552–553
Warburg, Aby, 570
Warburg Institute of Art History (Hamburg), 570, 630, 747
Ward, Adolphus William, 317
Ward, Lester, 446
War of the World, The (Ferguson), 841
Warttemberg, 44
Wasserman, August von, 388, 390
Waves of Sea and Love, The (Des Meeres und der Liebe Wellen) (Grillparzer), 294
Way of All Flesh, The (film), 592
Weavers, The (Die Weber) (Hauptmann), 524
Weber, Alfred, 532, 608, 649, 690, 770
Weber, Carl Maria von, 160, 173, 328
Weber, Karl, 95
Weber, Marianne, 690
Weber, Max, 455–458, 649
  Breuer, Stefan, on, 455
  on community, 836
  contemporary world of ideas and, 819, 822–823, 825–826
  Dahrendorf on, 764
  influence of, 713
  Pan-German League and, 421
  on *The Philosophy of Money*, 447
  Schumpeter's opposition to, 704
  on Sombart, 453
  on World War I, 535
Weber, Wilhelm, 173, 350
Webern, Anton von, 585, 643
Wedekind, Frank, 511, 514–515, 563, 585
Weerth, Georg, 304
Wegener, Alfred, 557–558
Wehler, Hans-Ulrich, 25–28
Weidenfeld, George, 745, 747–748
Weidermann, Volker, 29, 798–799
Weierstrass, Karl, 486
Weill, Kurt, 568, 586, 590, 641
Weimar (city), 111, 114, 128, 312
Weimar Republic, 567–593
  as betrayal of German political ideals, 34
  cinema, 587–593
  doctors in, 661
  music, 583–587
Weingartner, Felix von, 533
Weininger, Otto, 495
Weinrich, Max, 759–760
Weiss, Peter, 780, 793, 803
Weisse Rose (White Rose) resistance group (Munich), 690
Weisskopf, Victor, 739
Weizsäcker, Carl Friedrich von, 691
Welles, Orson, 593
Welt, Die (periodical), 843
Weltbühne, Die (magazine), 581, 582, 583
Wenders, Wim, 807–808
Werefkin, Marianne von, 515, 516
Werfel, Franz, 564, 701, 702, 735
Werner, Abraham Gottlob, 117, 167–170, 178
Werner, Anton von, 526–527
Wertheimer, Max, 716, 732
Weskott, Johann Friedrich, 362
Western Cannon, The (Bloom), 296
Westminster Review, 331
Westöstlicher Divan (Goethe), 192
Westphalia, Treaty of (1648), 44, 289
Weyl, Hermann, 487, 659
Weyrauch, Wolfgang, 799
What Does the German Worker Read? (Was liest der deutsche Arbeiter?) (Pfannkuche), 428
What I Have Heard about Adolf Hitler (survey, 1977), 780

"What Is Life" (lectures, Schrödinger), 708
What Remains (Was bleibt) (C. Wolf), 795
Wheen, Francis, 819
white generation *(weiße Jahrgänge)*, 761
Whitney, William Dwight, 324
Whittle, Frank, 600
Why Is There No Socialism in the United States? (Warum gibt es in den Vereinigten Staaten keinen Sozialismus?) (Sombart), 452
Wickmann, Karl, 215
Wiederbelebung des classischen Altertums, Die (Voigt), 92
Wiedermann, Gustav, 343
Wieland, Christoph, 112–113
Wienberg, Rudolf, 303
Wiene, Robert, 568
Wiener, Alfred, 749–750
Wiesenthal, Simon, 14
Wigman, Mary, 804
Wilamowitz-Moellendorff, Ulrich von, 105, 533
Wilder, Billy, 587, 588, 589, 590–591, 593
Wilhelm I (emperor), 418, 519
  Albert and, 321
  Krupp and, 371
  modern music and, 525–526
  Reinhardt's production and, 525
Wilhelm II (emperor), 321
  dislike of Berlin, 522–523
  Werner tutor of, 527
  World War I and, 536
Wilhelm Meister (Goethe), 119–120, 197–198
Wilhelm Tell (Schiller), 131
Wilkins, Maurice, 709
Williams, Raymond, 619
Willis, Thomas, 68
Willkom, Ernst, 304
Willstätter, Richard, 533, 595–596
"Wille zum Glück, Der" (T. Mann), 511
Winckelmann, Johann Joachim, 30, 95–100, 207, 208–209, 210, 211
Wind, Edgar, 753
Wings of Desire (film), 807–808
Winkler, Heinrich, 843
Winter, Jay, 732
Wintergerst, Joseph, 216
Winterreise, Die (Schubert), 307
Wirth, Herman, 655, 656
Wissenschaft, 53. *See also* science
  Berlin University and, 87
  Bildung versus, 838
  Britain and, 538
  Schelling on, 229
"Wissenschaft als Beruf" (Science as a Vocation) (Weber), 457
Wissenschaft des Judentums, 619
Wissenschaftlichkeit as part of *Bildung*, 110
Wissenschaftsideologie, 228, 229, 230, 237
Wissenschaftslehre (Fichte), 119, 148–149, 150
Wistar, Caspar, 326
Wittgenstein, Ludwig, 494, 552, 558–560, 561, 601, 753
Wittkower, Rudolf, 734, 753
Wöhler, Friedrich, 272, 273, 274, 275, 853
Wojtyla, Karol (John Paul II), 604
Wolf, Christa, 795, 796, 833
Wolf, Friedrich August, 105–108, 124, 832
  classical philology and, 230
  critical approach of, 232
  on role of professors, 226, 228
  student of Heyne, 228
  at University of Berlin, 227, 228
Wolf, Hugo, 37, 305, 459, 462–463, 854
Wolf, Konrad, 806
Wolff, Abraham, 48
Wolff, Albert, 215
Wolff, Christian, 88, 136, 137
Wolff, Kaspar Friedrich, 281
Wolff, Kurt and Helen, 735
Wolin, Richard, 720, 772
Wollheim, Gert, 699
Wollheim, Richard, 753
Wols (Alfred Otto Wolfgang Schulze), 810–811
Woltmann, Ludwig, 431, 614
women's movement, 425, 794
women writers in East Germany, 795
Woodward, Josiah, 316
Woolsey, Thomas Dwight, 324
"Word for the Germans, A" (Eliot), 316
Words and Things (Gellner), 755
Worker, The (Der Arbeiter) (Jünger), 709
Works of Art of the Future, The (Das Kunstwerk der Zukunft) (Wagner), 328
World as Will and Representation, The (Die Welt als Wille und Vorstellung) (Schopenhauer), 327, 331, 394
World War I, 531–545
  artists in Zurich during, 563–564
  novels about, 579
  poetry of, 547, 548–549
World War II
  birthdate and, 761
  Dietrich, Marlene, during, 593
  music during, 809
Woyzeck (Büchner), 301–303
Woyzeck (opera, Berg), 302, 585
Wundt, Wilhelm, 33, 493, 533, 554, 682
Wyler, William, 736

X

x-rays, 480

Y

Yale University, German-trained professors at, 324

YIVO Institute for Jewish Research (Vilna), 759
You and Me (film), 590
Youth Movement, 429

Z

Zadkine, Ossip, 702
Zander, Michael, 746
Zeiss, Carl Friedrich, 365–366
Zeit, Die (newspaper), 490
Zeitgeist, 125, 497
Zeitgenossen, Die (Moeller van den Bruck), 617
Zeitschrift für Sozialwissenschaft (journal), 726
Zemlinsky, Alexander von, 591
Zentralblatt für Mathematik und ihre Grenzgebiete (journal), 663
Zero (art group), 811
Ziegler, Adolf, 632, 643
Ziegler, Wilhelm, 656
Zimbalist, Efrem, 702
Zimmer, Ernst, 292
Zimmermann, Bernd Alois, 810
Zinnemann, Fred, 587, 588, 590
Zionism, 499, 608, 609, 682
Zivilisation versus Kultur, 31, 532, 535–536, 838
Zmarzlik, Hans-Günter, 428
Zollverein, 370, 424
Zsigmondy, Richard, 390
Zuckmayer, Carl, 513, 586, 592, 738
"Zum ewigen Frieden" (Of Eternal Peace) (Kant), 199
Zur Geschichte und Literatur (periodical), 103
"Zur Kritik neuerer Geschichtsschreiber" (Ranke), 267
Zuse, Konrad, 693
Zwanzigste Jahrhundert, Das (magazine), 510–511
Zweig, Stefan, 49, 490, 641, 700, 746
Zwischen den Zeiten (journal), 677

ALSO BY PETER WATSON

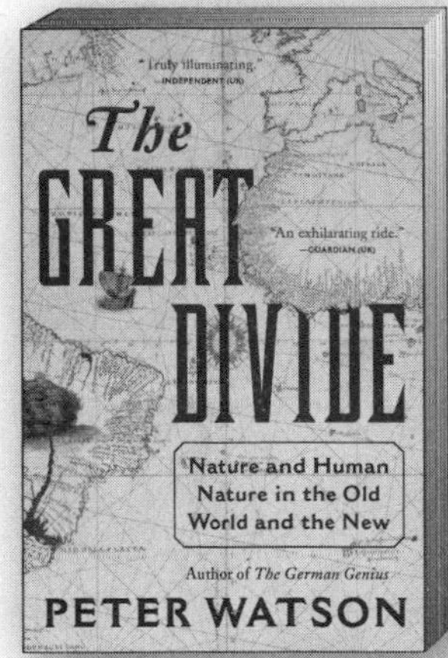

THE GREAT DIVIDE

Nature and Human Nature in the Old World and the New

Available in Paperback and eBook

Exploring the development of humankind between the Old World and the New—from 15,000 BC to AD 1500—Watson offers a groundbreaking new understanding of human history.

THE GERMAN GENIUS

Europe's Third Renaissance, the Second Scientific Revolution, and the Twentieth Century

Available in Paperback and eBook

Peter Watson goes back through time to explore the origins of the German genius, explaining how and why it flourished, and how it shaped our lives.

IDEAS

A History of Thought and Invention, From Fire to Freud

Available in Paperback and eBook

Peter Watson's hugely ambitious and stimulating history of ideas from deep antiquity to the present day, offers an illuminated path to a greater understanding of our world and ourselves.

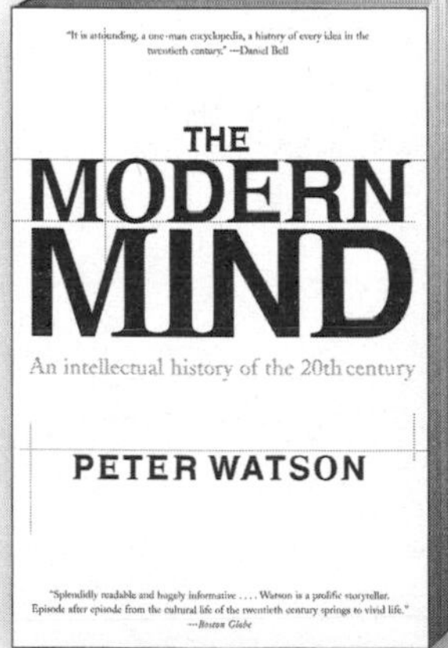

THE MODERN MIND

An Intellectual History of the 20th Century

Available in Paperback and eBook

"Splendidly readable and hugely informative. . . . Watson is a prolific storyteller. Episode after episode from the cultural life of the twentieth century springs to vivid life." —*Boston Globe*

Available wherever books are sold.